LOVELOCK and RUND—Tensors, Differential Forms, and Variational Principles

MARTIN—Nonlinear Operators and Differential Equations in Banach Spaces

MELZAK—Companion to Concrete Mathematics

MELZAK—Invitation to Geometry

NAYFEH—Perturbation Methods

NAYFEH and MOOK—Nonlinear Oscillations

ODEN and REDDY—An Introduction to the Mathematical Theory of Finite Elements

PASSMAN—The Algebraic Structure of Group Rings

PRENTER—Splines and Variational Methods

RIBENBOIM—Algebraic Numbers

RICHTMYER and MORTON—Difference Methods for Initial-Value Problems, 2nd Edition

RIVLIN—The Chebyshev Polynomials

RUDIN—Fourier Analysis on Groups

SAMELSON—An Introduction to Linear Algebra

SCHUMAKER—Spline Functions: Basic Theory

SHAPIRO—Introduction to the Theory of Numbers

SIEGEL—Topics in Complex Function Theory

        Volume 1—Elliptic Functions and Uniformization Theory

        Volume 2—Automorphic Functions and Abelian Integrals

        Volume 3—Abelian Functions and Modular Functions of Several Variables

STAKGOLD—Green's Functions and Boundary Value Problems

STOKER—Differential Geometry

STOKER—Nonlinear Vibrations in Mechanical and Electrical Systems

STOKER—Water Waves

WHITHAM—Linear and Nonlinear Waves

WOUK—A Course of Applied Functional Analysis

ZAUDERER—Partial Differential Equations of Applied Mathematics

PARTIAL DIFFERENTIAL EQUATIONS OF APPLIED MATHEMATICS

PARTIAL DIFFERENTIAL EQUATIONS OF APPLIED MATHEMATICS

ERICH ZAUDERER
Polytechnic Institute of New York

A Wiley-Interscience Publication

JOHN WILEY & SONS

New York · Chichester · Brisbane · Toronto · Singapore

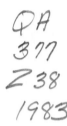

Library of Congress Cataloging in Publication Data:

Zauderer, Erich.
  Partial differential equations of applied mathematics.
  (Pure and applied mathematics, ISSN 0079-8185)
  "A Wiley-Interscience publication."
  Bibliography: p.
  Includes index.
  1. Differential equations, Partial. I. Title.
II. Series: Pure and applied mathematics (John Wiley &
Sons)
QA377.Z38 1983   515.3'53   82-21855
ISBN 0-471-87517-1

Printed in the United States of America

10 9 8 7 6 5 4

**To my wife, Naomi
and my sons, Joshua and David**

PREFACE

The study of partial differential equations of applied mathematics involves the formulation of problems that lead to partial differential equations, the classification and characterization of equations and problems of different types, and the examination of exact and approximate methods for the solution of these problems. Each of these aspects is considered in this book.

The first chapter is concerned with the formulation of problems that give rise to equations representative of the three basic types (parabolic, hyperbolic, and elliptic) to be considered in this book. These equations are all obtained as limits of difference equations that serve as models for discrete random walk problems. These problems are of interest in the theory of Brownian motion, and this relationship is examined. Although some elementary concepts from probability theory are used in this chapter, the remaining chapters proceed independently of this one.

Chapter 2 deals with first order partial differential equations and presents the method of characteristics for the solution of initial value problems for these equations. Problems that arise or can be interpreted in a wave propagation context are emphasized. First order equations also play an important role in the methods presented in Chapter 9.

In Chapter 3 partial differential equations are classified into different types and simplified "canonical" forms are obtained for second order linear equations in two independent variables. The concept of characteristics is introduced for higher order equations and systems of equations, and its significance for equations of different types is examined. In addition, the question of what types of auxiliary conditions are to be placed on solutions of partial differential equations so that the resulting problems are reasonably formulated is considered. Finally, some physical concepts such as energy conservation and dispersion which serve to distinguish equations of different types are discussed.

Chapter 4 presents the method of separation of variables for the solution of problems given in bounded spatial regions. This leads to a discussion of eigenvalue problems for partial differential equations and the one-dimensional version thereof known as the Sturm–Liouville problem. Eigenfunction expansions, in general, and Fourier series, in particular, are considered and applied to the solution of homogeneous and inhomogeneous problems for linear partial differential equations of second order. It is also shown that eigenfunction expansions can be used for the solution of nonlinear problems by considering a nonlinear heat conduction problem.

In Chapter 5 the Fourier, Fourier sine, Fourier cosine, Hankel, and Laplace transforms are introduced and used to solve various problems for partial differential equations given over unbounded regions in space or time. As the solutions of these problems are generally obtained in an integral form that is not easy to evaluate, approximation methods for the evaluation of Fourier and Laplace integrals are presented.

Not all problems encountered in applied mathematics lead to equations with smooth coefficients or have solutions that have as many derivatives as required by the order of the partial differential equations. Consequently, Chapter 6 discusses methods whereby the concept of solution is weakened by replacing the differential equations by integral relations that reduce the number of derivatives required of solutions. Also, methods are presented for dealing with problems given over composite media that can result in singular coefficients. Finally, the method of energy integrals is discussed and shown to yield information regarding the uniqueness and dependence on the data of solutions of partial differential equations.

Green's functions, which are discussed in Chapter 7, depend on the theory of generalized functions for their definition and construction. Therefore, a brief but self-contained discussion of generalized functions is presented in this chapter. Various methods for determining Green's functions are considered and it is shown how initial and boundary value problems for partial differential equations can be solved in terms of these functions.

Chapter 8 contains a number of topics. It begins with a variational characterization of the eigenvalue problems considered in Chapter 4, and this is used to verify and prove some of the properties of eigenvalues and eigenfunctions stated in Chapter 4. Furthermore, the Rayleigh–Ritz method which is based on the variational approach is presented and it yields an approximate determination of eigenvalues and eigenfunctions in cases where exact results are unavailable. The classical Riemann method for solving initial value problems for second order hyperbolic equations is briefly discussed, as are maximum and minimum principles for equations of elliptic and parabolic types. Finally, a number of partial differential equations of mathematical physics are studied, among which the equations of fluid dynamics and Maxwell's equations of electromagnetic theory are discussed at length.

Chapter 9, which concludes the book, is the longest chapter. It deals with perturbation and asymptotic methods for solving both linear and nonlinear partial differential equations. In recent years these methods have become an important tool for the applied mathematician in simplifying and solving complicated problems for linear and nonlinear equations. Regular and singular perturbation methods and boundary layer theory are discussed. Linear and nonlinear wave propagation problems associated with the reduced wave equation that contains a large parameter are examined. These include the scattering and diffraction of waves from various obstacles and the problem of beam propagation in nonlinear optics. It is also shown how singularities that can arise for solutions of hyperbolic equations can be analyzed without having to solve the full problem given for these equations. Finally, an asymptotic simplification procedure is presented that permits the replacement of linear and nonlinear equations and systems by simpler equations that retain certain essential features of the solutions of the original equations.

The Bibliography contains a list of references as well as additional reading. The entries are arranged according to the chapters of the book and they provide a collection of texts and papers that discuss some or all of the material covered in each chapter, possibly at a more elementary or advanced level than that of the text.

The material in this book has been developed and expanded from a set of lectures I have given over a number of years in a course on partial differential equations. It is intended for advanced undergraduate and beginning graduate students in applied mathematics, the sciences, and engineering. The student is assumed to have completed a standard calculus sequence including elementary ordinary differential equations, and to be familiar with some elementary concepts from advanced calculus, vector analysis, and matrix theory. (For instance, the concept of uniform convergence, the divergence theorem, and the determination of eigenvalues and eigenvectors of a matrix are assumed to be familiar to the student.) Although a number of equations and problems considered are physically motivated, a knowledge of the physics involved is not essential for the understanding of the mathematical aspects of the solution of these problems.

In writing this book I have not assumed that the student has been previously exposed to the theory of partial differential equations at some elementary level and that this book represents the next step. Thus I have included such standard solution techniques as the separation of variables and eigenfunction expansions together with the more advanced methods described earlier. However, in contrast to the more elementary presentations of this subject, this book does not dwell at great length on the method of separation of variables, the theory of Fourier series or integrals, the Laplace transform, or the theory of Bessel or Legendre functions. Rather, the standard results and methods are presented briefly but from a more general and advanced point of view. Consequently, it has been possible to present a variety of approaches and methods for solving problems for linear and nonlinear equations and systems without having the length of the book become excessive.

There is more than enough material in the book to be covered in a year-long course. For a shorter course it is possible to use the first part of Chapter 3 and Chapters 4 and 5 as a core, and to select additional material from the other chapters as time permits.

The book contains many examples. Very often new approaches or methods are brought out in the form of an example. Thus the examples should be accorded the same attention as the remainder of the text. There are more than 500 exercises in the book, many of which come with answers. They are placed at the end of each chapter and are listed according to the sections in the chapter. With a few exceptions, no substantially new theories or concepts are introduced in the exercises. For the most part, the exercises are based on material developed in the text, and the student should attempt to solve as many of them as possible to test his or her mastery of the subject.

ERICH ZAUDERER

New York City
March 1983

CONTENTS

PARTIAL DIFFERENTIAL EQUATIONS OF APPLIED MATHEMATICS

1

RANDOM WALKS
AND PARTIAL
DIFFERENTIAL
EQUATIONS

It is traditional to begin a course on partial differential equations of applied mathematics with derivations of the basic types of equations to be studied based on physical principles. Conventionally, the problem of the vibrating string or the process of heat conduction is considered and the corresponding wave or heat equation is derived. (Such derivations and relationships will be discussed later in the text.) We have chosen instead to use some elementary random walk problems as a means for deriving and introducing prototypes of the equations to be studied in the text: (1) *the diffusion equation*, (2) *the wave equation*, or more specifically, *the telegrapher's equation*, and (3) *Laplace's equation*. The equations describing the random walks are difference equations whose continuum limits are the aforementioned partial differential equations. Only elementary and basic concepts from probability theory will be required for our discussion of the random walk problems and these concepts will not be used in the sequel.

Our discussion of the limiting processes will be somewhat heuristic and formal but they can be rigorously justified. We shall not be discussing numerical methods for solving partial differential equations in this text, but the discrete formulation of the differential equations based on the random walk problems yields a useful insight into the differences in the numerical approaches required for each of the differential equations considered. Furthermore, some basic properties of each of the three types of equations to be derived are elementary consequences of the discrete random walk formulations. These properties will be verified directly for each of the limiting partial differential equations.

1.1. THE DIFFUSION EQUATION AND BROWNIAN MOTION

We begin by considering the *unrestricted one-dimensional random walk* problem. A particle starts at the origin of the x-axis and executes random steps or jumps each of length δ to the right or to the left. Let x_i be a random variable that assumes the value δ if the particle moves to the right at the ith step and the value $-\delta$ if it moves to the left. We assume that each step is independent of the others, so that the x_i are (identically distributed) independent random variables.

Let the probability that the particle moves to the right or left equal p or q, respectively. Since these probabilities are identical for each step, we have Prob$(x_i = \delta) \equiv P(x_i = \delta) = p$ and $P(x_i = -\delta) = q$. The particle must move either right or left so that $p + q = 1$. The position of the particle at the nth step is given by

$$(1.1) \qquad X_n = x_1 + x_2 + \cdots + x_n.$$

Using the binomial distribution, the probability that the particle is located at a fixed point after a given number of steps can be determined explicitly. Since we are more interested in the continuum limit of the random walk problem as the step length $\delta \to 0$ and the number of steps $n \to \infty$, we shall not consider the exact solution of the above problem. We do, however, determine the mathematical expectation and the variance of the random variable X_n. They shall be related below to certain physically significant constants when the above random walk model is connected with the theory of Brownian motion of a particle. The expectation of X_n yields the expected or mean location of the particle at the nth step whereas the variance of X_n is a measure of how much the actual location of the particle varies around the expected location.

EXAMPLE 1.1.   EXPECTATION AND VARIANCE: THE DISCRETE CASE

For a discrete valued random variable x that assumes the values a_m with probability p_m, the expected value or the *mathematical expectation* is defined as

$$(1.2) \qquad E(x) = \sum_m a_m p_m.$$

Thus for the random variable x_i we have

$$(1.3) \qquad E(x_i) = (+\delta)P(x_i = \delta) + (-\delta)P(x_i = -\delta) = (p - q)\delta.$$

Since $E(x)$ is a linear function of x we have

$$(1.4) \qquad E(X_n) = E\left(\sum_{i=1}^{n} x_i\right) = \sum_{i=1}^{n} E(x_i) = (p - q)\delta n$$

in view of (1.3). In the physical literature the mathematical expectation is expressed as

$$(1.5) \qquad\qquad E(x) = \langle x \rangle$$

and this notation will be used below. Thus (1.4) yields

$$(1.6) \qquad\qquad \langle X_n \rangle = (p - q)\delta n.$$

Note that if $p = q = \frac{1}{2}$, so that the particle is equally likely to move to the right or to the left, we have $\langle X_n \rangle = 0$. Then the expected location of the particle after n steps is the origin (i.e., the starting point). If $p > q$, the expected location is to the right of the origin, whereas the converse is true if $p < q$.

The *variance* of a random variable x is defined in terms of the expectation as

$$(1.7) \qquad V(x) = \langle (x - \langle x \rangle)^2 \rangle = \langle x^2 \rangle - \langle x \rangle^2,$$

where the last equation is an easy consequence of the properties of the expectation. It can also be shown that the variance of a sum of independent random variables equals the sum of the variances of the random variables. Thus since $X_n = \sum_{i=1}^n x_i$ and the x_i are independent random variables,

$$(1.8) \qquad V(X_n) = V\left(\sum_{i=1}^n x_i \right) = \sum_{i=1}^n V(x_i) = \sum_{i=1}^n \left[\langle x_i^2 \rangle - \langle x_i \rangle^2 \right].$$

But

$$(1.9) \quad \langle x_i^2 \rangle = (+\delta)^2 P(x_i = \delta) + (-\delta)^2 P(x_i = -\delta) = \delta^2(p + q) = \delta^2$$

since $p + q = 1$. Recalling (1.3) we easily obtain

$$(1.10) \qquad V(X_n) = \sum_{i=1}^n \left[\delta^2 - (p - q)^2 \delta^2 \right] = 4pq\delta^2 n,$$

on using the identity $1 = (p + q)^2$.

The foregoing results will now be applied to yield a mathematical description of one-dimensional *Brownian motion*. This motion refers to the ceaseless, irregular, and (apparently) random motion of small particles immersed in a liquid or gas. A given particle is assumed to undergo one-dimensional motion due to random collisions with smaller particles in the fluid or gas. In a given unit of time many collisions occur. Each collision is assumed to be independent of the others and to impart a displacement of the particle of length δ to the right or to the left. The apparent randomness of the motion is characterized by assuming that each collision independently moves the particle to the right or to the left with probability p or q, respectively. Clearly, the random walk problem described can be used to simulate the observed motion of the particle.

Experimentally, it is found for a particle immersed in a fluid or gas (undergoing one-dimensional motion) that the average or mean displacement of the particle per unit time equals c, whereas the variance of the observed displacement around the average equals $D > 0$. We assume that there are r collisions per unit time (note that r is not known a priori but is assumed to be large). Then for our random walk model of the observed motion we must have (approximately) after r steps

$$(1.11) \qquad (p - q)\delta r \approx c,$$

$$(1.12) \qquad 4pq\delta^2 r \approx D,$$

in view of the expressions (1.6) and (1.10) for the mean and the variance.

Since the motion of the particle appears to be continuous, we must examine the limit as the step length $\delta \to 0$ and the number of steps $r \to \infty$. This must occur in such a way that c and D [as given in (1.11) and (1.12)] remain fixed in the limit. Now if $p \neq q$ and $p - q$ does not tend to zero as $\delta \to 0$ and $r \to \infty$, we have from (1.11)

$$(1.13) \qquad \delta r \to \frac{c}{p - q},$$

which implies

$$(1.14) \qquad 4pq\delta^2 r \to \left(\frac{4cpq}{p - q} \right)\delta \to 0.$$

However, $4pq\delta^2 r$ must tend to $D \neq 0$ in the limit as $\delta \to 0$ and $r \to \infty$, in view of (1.12). Therefore, we must have $p - q \to 0$ in the limit. Combined with $p + q = 1$ this implies that both p and q tend to $\frac{1}{2}$ in the limit. If $p = q = \frac{1}{2}$ in the discrete model, we have $c = 0$. But if $p - q \neq 0$ so that $c \neq 0$, the particle exhibits a *drift* to the right or to the left depending on the sign of c. Additionally, if $D = 0$ there is no variation around the average displacement c and the motion of the particle must appear to be deterministic and not random or irregular.

For r steps to occur in unit time, each individual step of length δ must occur in $1/r = \tau$ units of time whereas n steps occur in $n/r = n\tau$ time units. To describe the motion of the particle that starts at the point $x = 0$ at the time $t = 0$, we use the random walk model to obtain the probability that the particle is at the position x at the time t. That is, we have approximately after n steps

$$(1.15) \qquad X_n = x \qquad \text{and} \qquad n\tau = t.$$

If $x > 0$, for example, we must have

$$(1.16) \qquad X_n = k\delta = x \qquad \text{and} \qquad n\tau = t$$

where k equals the excess of the number of steps taken to the right over those taken to the left. A similar expression with $k < 0$ is valid if $x < 0$.

We define

$$(1.17) \qquad v(x, t) = P(X_n = x) \qquad \text{at the time} \qquad t = n\tau$$

to be the probability that at the (approximate) time t, the particle is located (approximately) at the point x. As indicated, we could determine an explicit expression for $v(x, t)$ on using the binomial distribution. However, since we are interested in a continuum limit of the random walk problem as δ and τ tend to zero, we construct a difference equation satisfied by $v(x, t)$. Again, we do not solve the difference equation but show that in the continuum limit it tends to a partial differential equation that serves as a model for the Brownian motion of a particle.

The probability distribution $v(x, t)$ satisfies the difference equation

$$(1.18) \qquad v(x, t + \tau) = pv(x - \delta, t) + qv(x + \delta, t).$$

This states that *the probability that the particle is at x at the time $t + \tau$ equals the probability that it was at the point $x - \delta$ at the time t multiplied by the probability p that it moved to the right in the following step plus the probability that the particle was at the point $x + \delta$ at the time t multiplied by the probability q that it moved to the left in the following step*. The plausibility of the above equation is apparent.

Expanding in a Taylor series with remainder we have

$$(1.19) \qquad \begin{cases} v(x, t + \tau) = v(x, t) + \tau v_t(x, t) + O(\tau^2) \\ v(x \pm \delta, t) = v(x, t) \pm \delta v_x(x, t) + \tfrac{1}{2}\delta^2 v_{xx}(x, t) + O(\delta^3) \end{cases}$$

where $O(y^k)$ means that $\lim_{y \to 0} y^{-k} O(y^k)$ is finite. Substituting (1.19) into (1.18) and using $p + q = 1$ we readily obtain

$$(1.20)$$

$$v_t(x, t) = \left[(q - p)\frac{\delta}{\tau} \right] v_x(x, t) + \frac{1}{2}\left(\frac{\delta^2}{\tau} \right) v_{xx}(x, t) + O(\tau) + O\left(\frac{\delta^4}{\tau} \right).$$

As $\delta \to 0$ and $\tau \to 0$, (1.20) tends to the limiting partial differential equation

$$(1.21) \qquad \frac{\partial v}{\partial t} = -c\frac{\partial v}{\partial x} + \frac{1}{2}D\frac{\partial^2 v}{\partial x^2},$$

on using (1.11) and (1.12) with $r = 1/\tau$. In view of (1.16) which requires x and t to remain fixed in the limit, we must also have $|k| \to \infty$ and $n \to \infty$ such that $k\delta \to x$ and $n\tau \to t$. Further, $v(x, t)$ must now be interpreted as a *probability density* associated with the continuous random variable x at the time t, rather than as a probability distribution in the discrete random walk model. This means that the probability that x lies in a given interval $[a, b]$ is given by

$$(1.22) \qquad P(a \leqslant x \leqslant b) = \int_a^b v(x, t)\, dx \qquad \text{at the time } t.$$

The equation (1.21) is known as a *diffusion equation* and D is called the *diffusion coefficient*. We examine the significance of the coefficients c and D in (1.21) in Example 1.3, with calculations based directly on (1.21) rather than on the results obtained from the random walk model.

To completely determine the solution $v(x, t)$ of (1.21) we need to specify initial conditions for the density function. Since the particle was initially (i.e., at the time $t = 0$) assumed to be located at the point $x = 0$, it has unit probability of being at $x = 0$ at the time $t = 0$ and zero probability of being elsewhere at that time. This serves as initial data for the difference equation (1.18). In terms of the density function $v(x, t)$ that satisfies (1.21), we still must have $v(x, 0) = 0$ for $x \neq 0$, and the density must be concentrated at $x = 0$. Since the total probability at $t = 0$ satisfies

$$(1.23) \qquad P(-\infty < x < \infty) = \int_{-\infty}^{\infty} v(x, 0)\, dx = 1,$$

we see that $v(x, 0)$ behaves like the *Dirac delta function*; that is,

$$(1.24) \qquad v(x, 0) = \delta(x)$$

[the delta function $\delta(x)$ is not to be confused with the step length δ introduced in the random walk problem.]

EXAMPLE 1.2.   THE DIRAC DELTA FUNCTION

The Dirac delta function is discussed more fully in Chapter 7. For the purposes of our present discussion, we may characterize it as the limit of a sequence of discontinuous functions $\delta_\epsilon(x)$ defined as

$$(1.25) \qquad \delta_\epsilon(x) = \begin{cases} \dfrac{1}{2\epsilon}, & |x| < \epsilon \\ 0, & |x| > \epsilon \end{cases}$$

Each $\delta_\epsilon(x)$ has unit area under the curve and in the limit as $\epsilon \to 0$, $\delta_\epsilon(x) \to 0$

for all $x \neq 0$. Thus

$$(1.26) \qquad \int_{-\infty}^{\infty} \delta(x)\, dx = \lim_{\epsilon \to 0} \int_{-\infty}^{\infty} \delta_\epsilon(x)\, dx = 1.$$

It also follows for continuous functions $f(x)$ that

$$(1.27) \qquad \int_{-\infty}^{\infty} f(x)\delta(x)\, dx = f(0)$$

since $f(x)\delta(x) = f(0)\delta(x)$, as $\delta(x)$ vanishes for all $x \neq 0$. Further, $\delta(x - \zeta)$ vanishes for all $x \neq \zeta$ so that $f(x)\delta(x - \zeta) = f(\zeta)\delta(x - \zeta)$ and if $\delta(x)$ is replaced by $\delta(x - \zeta)$ in (1.27), we obtain $f(\zeta)$ instead of $f(0)$.

The combined equations (1.21) and (1.24) constitute an initial value problem for the partial differential equation (1.21). It is derived later and can be shown directly by substitution that the function

$$(1.28) \qquad v(x, t) = \frac{1}{\sqrt{2\pi Dt}} \exp\left[-\frac{(x - ct)^2}{2\,Dt} \right],$$

is a solution of the initial value problem (1.21) and (1.24). For fixed t, the probability density function (1.28) is the density function of a normal or *Gaussian distribution* with *mean* $\langle x \rangle = ct$ and *variance* $\langle (x - \langle x \rangle)^2 \rangle = Dt$. In view of the exponential decay of $v(x, t)$, we find that the density is concentrated around the curve $x - ct = 0$, so that the particle appears to move with the (drift) velocity $dx/dt = c$. This also follows from the equation $\langle x \rangle / t = c$. The variance of the particle location around the path $x = ct$ [in the (x, t)-plane] increases linearly in t and is given by Dt.

Although the foregoing results for the mean and the variance are well-known for the normal distribution, they may be verified directly by considering the limit of the discrete random walk problem. In Example 1.3 these results for the mean and the variance are obtained directly from the partial differential equation (1.21).

The random walk model considered above yields (1.6) and (1.10) as the mean and the variance. As $n \to \infty$, $\delta \to 0$, and $1/r \to 0$, we obtain from (1.6)

$$(1.29) \qquad \langle x \rangle = \lim \langle X_n \rangle = \lim(p - q)\delta n = \lim(p - q)\delta r \frac{n}{r}$$

$$= \lim[(p - q)\delta r] n\tau = ct$$

on using (1.11) and (1.16). Similarly, (1.10) yields

$$(1.30) \qquad \langle (x - \langle x \rangle)^2 \rangle = \lim \langle (X_n - \langle X_n \rangle)^2 \rangle = \lim 4pq\delta^2 n$$

$$= \lim 4pq\delta^2 r\, n\tau = Dt.$$

Using (1.29) and (1.30) and applying the central limit theorem of probability theory to X_n (which is a sum of n independent random variables) as $n \to \infty$, we can conclude directly that the limiting distribution for x is Gaussian with mean ct and variance Dt.

In the following example we show how the mean and the variance of the random variable x with density $v(x, t)$ can be obtained directly from differential equation (1.21) and the initial conditions (1.24) without actually solving for $v(x, t)$.

EXAMPLE 1.3.   EXPECTATION AND VARIANCE: THE CONTINUOUS CASE

If x is a continuous random variable, the kth moment of x is defined as

$$(1.31) \qquad \langle x^k \rangle = \int_{-\infty}^{\infty} x^k v(x, t)\, dx$$

where $v(x, t)$ is the probability density function at the time t. We assume that $v(x, t)$ and its x-derivatives vanish sufficiently rapidly at infinity, so that all terms evaluated at infinity in the following integrations vanish. [In fact, (1.28) shows that $v(x, t)$ vanishes exponentially at infinity.]

First we show $\langle x^0 \rangle = 1$ for all t so that $v(x, t)$ is, in fact, a probability density for all $t > 0$. Integrating in (1.21) we have

$$(1.32) \quad \int_{-\infty}^{\infty} v_t\, dx = \frac{d}{dt} \langle x^0 \rangle = -c \int_{-\infty}^{\infty} v_x\, dx + \frac{1}{2} D \int_{-\infty}^{\infty} v_{xx}\, dx = 0$$

since v and v_x are assumed to vanish at infinity. Thus $\langle x^0 \rangle$ is constant in time. But (1.23) shows that $\langle x^0 \rangle = 1$ at $t = 0$ so that $\langle x^0 \rangle = 1$ for all time $t > 0$.

Next, we consider the expected value $\langle x \rangle$ of the continuous random variable x. We multiply (1.21) by x and integrate from $-\infty$ to $+\infty$. This gives

$$(1.33) \qquad \int_{-\infty}^{\infty} x v_t\, dx = \frac{d}{dt} \int_{-\infty}^{\infty} x v\, dx = \frac{d}{dt} \langle x \rangle = -c \int_{-\infty}^{\infty} x v_x\, dx$$

$$+ \frac{1}{2} D \int_{-\infty}^{\infty} x v_{xx}\, dx = c \int_{-\infty}^{\infty} v\, dx = c \langle x^0 \rangle = c$$

on integrating by parts, using (1.32) and $\langle x^0 \rangle = 1$. At the time $t = 0$ since $v(x, 0) = \delta(x)$, the property (1.27) of the delta function implies that $\langle x \rangle = 0$ at $t = 0$. Therefore, we obtain an initial value problem for $\langle x \rangle$,

$$(1.34) \qquad \frac{d}{dt} \langle x \rangle = c, \qquad \langle x \rangle \Big|_{t=0} = 0$$

with the solution

$$(1.35) \qquad \langle x \rangle = ct$$

which agrees with the result (1.29).

The variance of x is given by

$$(1.36) \qquad V(x) = \langle x^2 \rangle - \langle x \rangle^2.$$

The second moment $\langle x^2 \rangle$ clearly vanishes at $t = 0$ in view of (1.24) and (1.27). Multiplying (1.21) by x^2 and integrating, we have

$$(1.37) \quad \int_{-\infty}^{\infty} x^2 v_t \, dx = \frac{d}{dt} \langle x^2 \rangle = -c \int_{-\infty}^{\infty} x^2 v_x \, dx + \frac{1}{2} D \int_{-\infty}^{\infty} x^2 v_{xx} \, dx$$

$$= 2c \int_{-\infty}^{\infty} xv \, dx + D \int_{-\infty}^{\infty} v \, dx = 2c \langle x \rangle + D \langle x^0 \rangle,$$

on integrating by parts. Using (1.35) and $\langle x^0 \rangle = 1$ we obtain the initial value problem for $\langle x^2 \rangle$

$$(1.38) \qquad \frac{d}{dt} \langle x^2 \rangle = 2c^2 t + D, \qquad \langle x^2 \rangle \Big|_{t=0} = 0$$

which yields

$$(1.39) \qquad \langle x^2 \rangle = c^2 t^2 + Dt.$$

Thus the variance is given by

$$(1.40) \qquad V(x) = \langle x^2 \rangle - \langle x \rangle^2 = c^2 t^2 + Dt - c^2 t^2 = Dt,$$

which agrees with (1.30).

The foregoing results, that is, the random walk problem and the limiting initial value problem, provide a valid description of the motion of a Brownian particle in an unbounded region. If the region is bounded on one or both sides, certain boundary conditions must be added. We now consider two random walk problems in which the particle is restricted to move in the region $x < l$, and at $x = l > 0$ there is either an absorbing or reflecting boundary.

At an *absorbing boundary*, once the particle reaches the boundary at $x = l$ it is absorbed and can no longer move into the region $x < l$. With $v(x, t)$ as the probability that the particle is (approximately) at the point x at the time t we have at $x = l$,

$$(1.41) \qquad v(l, t + \tau) = pv(l - \delta, t),$$

since the particle cannot reach $x = l$ from the right. Using the Taylor series (1.19) gives

$$(1.42) \qquad (1 - p)v(l, t) = O(\tau) + O(\delta).$$

Since $p \to \frac{1}{2}$ as δ and τ tend to zero, we have in the limit

$$(1.43) \qquad\qquad v(l, t) = 0,$$

as the boundary condition for the density function at an absorbing boundary.

For the case of a *reflecting boundary*, the boundary is assumed to have the properties of an elastic barrier. As the particle reaches the barrier it moves beyond $x = l$ and in the next step returns to $x = l$ with probability q. Thus the boundary condition for the probability $v(x, t)$ is

$$(1.44) \qquad\qquad v(l, t + \tau) = pv(l - \delta, t) + qv(l, t).$$

Again using (1.19) we have

(1.45)
$$v(l, t) + \tau v_t(l, t) + O(\tau^2) = (p + q)v(l, t) - p\delta v_x(l, t) + O(\delta^2).$$

As $\delta \to 0$ and $\tau \to 0$ we must have $\delta^2/\tau \to D \neq 0$ which implies that $\tau/\delta \to 0$. Thus dividing by δ and going to the limit in (1.45) gives

$$(1.46) \quad \lim\left(\frac{\tau}{\delta}\right)v_t + \lim O\left(\frac{\tau^2}{\delta}\right) + \lim pv_x(l, t) + \lim O(\delta) = \frac{1}{2}v_x(l, t) = 0,$$

and the boundary condition for the density function at a reflecting boundary is

$$(1.47) \qquad\qquad v_x(l, t) = 0.$$

Within the interval $-\infty < x < l$, the density function $v(x, t)$ satisfies the diffusion equation (1.21) with the initial condition (1.24) at $t = 0$. At $x = l$, $v(x, t)$ satisfies either the boundary condition (1.43) or (1.47). The solutions of these initial and boundary value problems are considered in Chapter 5.

We conclude this section by noting a number of consequences for the solutions of initial and initial boundary value problems for the diffusion equation (1.21) that follow from the difference equation formulation of the associated random walk problems. To begin with, we observe that the difference equations (1.18), (1.41), and (1.44) imply that the solutions of initial and initial boundary value problems evolve in time. That is, the solution at any point at a given time t depends only on initial and boundary data given at earlier times. This characterizes the *causality property* that only past events influence future events.

Secondly, we consider the initial value problem or the initial boundary value problem with boundary condition (1.43). The difference equation (1.18) for $v(x, t + \tau)$ represents it as a weighted average of v evaluated at two points at an earlier time t. (We have a weighted average since $p > 0$, $q > 0$, and $p + q = 1$). Since v vanishes initially everywhere except at one point and v vanishes on the boundary, we conclude that $0 \leqslant v \leqslant 1$ everywhere and the maximum value of v (i.e., $v = 1$), as well as the minimum value of v (i.e., $v = 0$), is attained either on the boundary or on the initial line. This maximum (and minimum) principle carries over to the diffusion equation for which it is shown to be valid in Chapter 8. The effect of the diffusion process is to distribute the densities in the interior in a fairly uniform fashion so that the maximum and minimum values occur on the initial or boundary line.

Finally, we note that the requirement that δ^2/τ tends to a finite nonzero limit as δ and τ tend to zero implies that $\delta/\tau \to \infty$ in the same limit. This means that the speed of the particle in Brownian motion (which is given by the limit of δ/τ) is infinite. This fact is also implied by the solution (1.28) of the initial value problem (1.21) and (1.24). It shows that $v(x, t)$ is instantaneously nonzero for all x when $t > 0$ even though $v(x, t)$ vanishes for all $x \neq 0$ at $t = 0$. Thus there is a nonzero probability, however small it may be, that the particle is located in the neighborhood of any point as soon as t increases from zero.

The foregoing limitation of the theory of Brownian motion based on solution of the diffusion equation was already noticed by Einstein who was the first to derive a diffusion equation to describe Brownian motion. He recognized that the diffusion equation yields a valid model only as t gets large. Since (with $c = 0$) (1.21) also represents the equation of heat conduction, with v equal to the temperature and D interpreted appropriately, the same difficulty with regard to the interpretation of the physical processes involved occurs for the heat equation.

A different discussion of the theory of Brownian motion was given by Ornstein and Uhlenbeck in an effort to overcome this shortcoming. They base their theory on the so-called *Langevin equation* for the velocity of the particle in Brownian motion. Their results are valid for small times and reduce to the foregoing results for the density function as $t \to \infty$.

In the following section we show, on the basis of a different random walk model, how to construct a limiting differential equation that yields a finite speed for the particle undergoing Brownian motion for small t and reduces to the preceding results as $t \to \infty$.

1.2. THE TELEGRAPHER'S EQUATION AND DIFFUSION

The random walk problem considered in the previous section was shown to imply an infinite speed for a particle in Brownian motion in the continuum limit modeled by the diffusion equation (1.21). It may be reasonably argued that this is a consequence of the basic assumption that each step in the random walk is independent of the previous steps. Thus in the limit as the step length, as well as the time lapse between steps, tends to zero, the probability that the particle moves right or left tends to $\frac{1}{2}$. The particle is, therefore, equally likely to move to the right or to the left regardless of the direction in which it was moving previously. The limiting path is, consequently, totally irregular and the particle cannot be said to have a fixed finite velocity.

An assumption in the random walk problem that might be expected to yield a smoother path of motion for the particle in the limit (at least in the initial stages) is that there exists a positive correlation between two adjacent steps. This correlation is expected to increase to a maximum value of unity as the step length and the time between steps tend to zero. The correlation implies a tendency for the particle to continue moving in a given direction once it begins to move in that direction. If, initially, probabilities for motion to the right or left are established, the particle will maintain its tendency to move in a fixed direction for a certain time at a finite speed. After a while, the inherent randomness of the process reduces the motion to that obtained in the previous section. The foregoing assumption was introduced into the random walk problem by R. Fürth in a study of Brownian motion. He showed that it implies a finite velocity for the particle at small times and yields the results of Section 1.1 for large times.

The *correlated random walk* was independently considered by G. I. Taylor in a discussion of diffusion processes. It was reexamined by S. Goldstein who formulated a difference equation characterizing the random walk and constructed its limiting partial differential equation. We shall use the results of Taylor and Goldstein in our discussion but retain the notation of Section 1.1. The correlated random walk is described in the following example. (Although this random walk can be characterized as a stationary Markov process, we prefer not to do so to avoid having to introduce more advanced concepts from probability theory.)

EXAMPLE 1.4.   THE CORRELATED RANDOM WALK

The particle is assumed to start at $x = 0$ and move to the right or to the left with probability equal to $\frac{1}{2}$ initially. Subsequently, the particle has a constant probability of persistence in or reversal of direction at each step. At the ith step the random variable x_i assumes the values $+\delta$ or $-\delta$ and is readily seen to do so with probability equal to $\frac{1}{2}$. However, the steps are no longer assumed to be independent, as was the case in the preceding section. With X_n again giving the position of the particle after n steps [as in (1.1)] we now have for the mean value of X_n,

$$(1.48) \qquad \langle X_n \rangle = \left\langle \sum_{i=1}^{n} x_i \right\rangle = \sum_{i=1}^{n} \langle x_i \rangle = 0,$$

since the equal likelihood of a step to the right or left implies $\langle x_i \rangle = 0$.
For the variance of X_n we have

$$(1.49) \qquad V(X_n) = \langle (X_n - \langle X_n \rangle)^2 \rangle = \langle X_n^2 \rangle$$

on using (1.48), so that the variance equals the second moment of X_n. Also, we easily obtain

$$(1.50) \qquad V(x_i) = \langle (x_i - \langle x_i \rangle)^2 \rangle = \langle x_i^2 \rangle = \delta^2.$$

The second moment $\langle X_n^2 \rangle$ is given as

$$(1.51) \qquad \langle X_n^2 \rangle = \left\langle \left(\sum_{i=1}^{n} x_i \right)^2 \right\rangle = \left\langle \sum_{i=1}^{n} x_i^2 + 2 \sum_{\substack{i,j=1 \\ i<j}}^{n} x_i x_j \right\rangle$$

$$= \sum_{i=1}^{n} \langle x_i^2 \rangle + 2 \sum_{\substack{i,j=1 \\ i<j}}^{n} \langle x_i x_j \rangle = n\delta^2 + 2 \sum_{\substack{i,j=1 \\ i<j}}^{n} \langle x_i x_j \rangle,$$

and we evaluate it in terms of the correlation coefficient between x_i and x_j.

The *correlation coefficient* between two random variables x_i and x_{i+k} is defined as

$$(1.52) \qquad \rho(x_i, x_{i+k}) = \frac{\langle x_i x_{i+k} \rangle - \langle x_i \rangle \langle x_{i+k} \rangle}{\sqrt{V(x_i)V(x_{i+k})}},$$

and it is seen to vanish if the x_i are independent random variables. We shall assume that partial correlations between two nonadjacent random variables x_i and x_j (i.e., $|j - i| > 1$) equal zero. Since the x_i are identically distributed random variables, the correlation coefficient between any two adjacent random variables is equal, so that

$$(1.53) \qquad \rho(x_i, x_{i+1}) \equiv \rho.$$

Further, the correlation between x_i and x_{i+k} (with $k > 1$) occurs only through the intermediate random variables $x_{i+1}, x_{i+2}, \ldots, x_{i+k-1}$ since partial correlations are assumed to vanish. Thus

$$(1.54) \qquad \rho(x_i, x_{i+k}) = \rho^k.$$

Since $V(x_i) = \delta^2$ for all i and $\langle x_i \rangle = 0$, we obtain from (1.52)

$$(1.55) \qquad \langle x_i x_{i+k} \rangle = \delta^2 \rho^k.$$

Introducing (1.55) into (1.51) gives

$$(1.56) \qquad \langle X_n^2 \rangle = \delta^2 \left[n + 2(n - 1)\rho + 2(n - 2)\rho^2 + \cdots + 2\rho^{n-1} \right].$$

This series is easily summed in terms of the finite geometric series, and we have

$$(1.57) \qquad \langle X_n^2 \rangle = \delta^2 \left[n + \frac{2n\rho}{1-\rho} - \frac{2\rho(1-\rho^n)}{(1-\rho)^2} \right].$$

Proceeding as in the Section 1.1, we set $n = t/\tau$ and consider the limit as $\delta \to 0$, $\tau \to 0$, and $n \to \infty$ with t fixed and $X_n \to x$. Equation (1.57) takes the form

$$(1.58) \qquad \langle X_n^2 \rangle = \left(\frac{\delta}{\tau} \right)^2 \left[\frac{(1+\rho)}{(1-\rho)} \tau t - \frac{2\rho(1-\rho^{t/\tau})\tau^2}{(1-\rho)^2} \right].$$

We assume that $\lim \delta/\tau = \gamma$, the finite velocity of the particle. For (1.58) to have a finite nonzero limit as $\tau \to 0$ we must have

$$(1.59) \qquad \lim_{\tau \to 0} \frac{\tau}{1-\rho} = \frac{1}{2\lambda}$$

where λ is a nonzero positive constant. [The factor 2 in (1.59) is introduced for convenience.] This is consistent with the fact that as $\tau \to 0$ the correlation coefficient ρ must tend to unity. It then follows that the limit of $\rho^{t/\tau}$ as $\tau \to 0$ is

$$(1.60) \qquad \lim_{\tau \to 0} \rho^{t/\tau} = e^{-2\lambda t},$$

and (1.58) tends to

$$(1.61) \qquad \langle x^2 \rangle = \gamma^2 \left[\frac{t}{\lambda} - \frac{1}{2\lambda^2}(1 - e^{-2\lambda t}) \right].$$

An expression of the form (1.61) for the variance of x was obtained by Ornstein and Uhlenbeck and Fürth in their (improved) theories of Brownian motion. It was also assumed in their theories (as we have done) that the mean displacement $\langle x \rangle$ of the particle equals zero. For large values of t, (1.61) reduces to

$$(1.62) \qquad \langle x^2 \rangle \approx \left(\frac{\gamma^2}{\lambda} \right) t$$

which agrees with the result (1.30) for the mean square displacement of the particle in Brownian motion (in the case where $\langle x \rangle = 0$) if we set

$$(1.63) \qquad D = \frac{\gamma^2}{\lambda},$$

where D is the diffusion coefficient. For small values of t, on expanding $e^{-2\lambda t}$ in a Taylor series we obtain

$$(1.64) \qquad \langle x^2 \rangle \approx \gamma^2 t^2.$$

This shows that in the initial stages of the motion

$$(1.65) \qquad \frac{\sqrt{\langle x^2 \rangle}}{t} \approx \gamma$$

which means that the motion of the particle is essentially uniform with speed γ. Ornstein and Uhlenbeck using Langevin's equation obtained a density function for the random variable x that gives the displacement of the particle in Brownian motion. It reduces to the density function (1.28) for the normal distribution as $t \to \infty$, has zero mean, and its variance is given by (1.61). They also constructed a differential equation that their density function satisfies but were not able to show that their equation reduces to the diffusion equation (1.21) (with $c = 0$) as $t \to \infty$.

Following the approach of S. Goldstein, we now show how to construct a difference equation and its limiting partial differential equation that characterize the correlated random walk described.

Let $\alpha(x, t)$ be the probability that a particle is at the point x at the time t and arrived there from the left, whereas $\beta(x, t)$ is the probability that a particle is at x at the time t and arrived there from the right. Thus $\alpha(x, t)$ and $\beta(x, t)$ characterize right and left moving particles, respectively. Also let p be the probability that the particle persists in its direction after completing a step, whereas q is the probability that it reverses its direction after completing a step. The probabilities p and q are assumed not to vary from step to step, and we have $p + q = 1$. Thus if a particle arrives at x from the left, p is the probability that it continues to the right in the next step, whereas q is the probability that it reverses itself and goes to the left in the next step. (Note that p and q were defined differently in the previous section.)

With steps of length δ occurring in time intervals of length τ, we immediately obtain the following coupled system of difference equations for $\alpha(x, t)$ and $\beta(x, t)$,

$$(1.66) \qquad \alpha(x, t + \tau) = p\alpha(x - \delta, t) + q\beta(x - \delta, t)$$
$$(1.67) \qquad \beta(x, t + \tau) = p\beta(x + \delta, t) + q\alpha(x + \delta, t)$$

on using the preceding definitions of α, β, p, and q.

The assumption introduced above that as $\tau \to 0$ the correlation coefficient $\rho \to 1$ [see (1.59)] implies that as $\tau \to 0$, the probability p of persistence in direction should tend to unity, whereas the probability q of reversal should tend to zero. This means that we should have for small τ,

$$(1.68) \qquad\qquad p = 1 - \hat{\lambda}\tau + O(\tau^2)$$

$$(1.69) \qquad\qquad q = \hat{\lambda}\tau + O(\tau^2)$$

where $\hat{\lambda}$ is the rate of reversal of direction. Now, it can be shown that the correlation coefficient ρ is related to p and q as

$$(1.70) \qquad\qquad \rho = p - q.$$

Thus since $p + q = 1$, (1.68)–(1.69) yield

$$(1.71) \qquad \rho = p - q = 1 - 2q = 1 - 2\hat{\lambda}\tau + O(\tau^2).$$

Recalling (1.59), we have

$$(1.72) \qquad \lim_{\tau \to 0} \frac{\tau}{1 - \rho} = \frac{1}{2\lambda} = \lim_{\tau \to 0} \frac{\tau}{2\hat{\lambda}\tau + O(\tau^2)} = \frac{1}{2\hat{\lambda}}$$

so that $\hat{\lambda} = \lambda$.

It was observed by Kac that in view of the equations (1.68) and (1.69) and the independence of the probabilities p and q for each step, the probability of reversal of direction in a given time span is determined by what is known as a *Poisson process*. This fact will be exploited later when an elementary property of the Poisson process is used.

Introducing Taylor expansions for α and β in (1.66) and (1.67), using (1.68) and (1.69), and taking the limit as δ and τ tend to zero, with $\delta/\tau \to \gamma$, we easily obtain the following coupled system of partial differential equations:

$$(1.73) \qquad\qquad \frac{\partial \alpha}{\partial t} + \gamma \frac{\partial \alpha}{\partial x} = -\lambda\alpha + \lambda\beta$$

$$(1.74) \qquad\qquad \frac{\partial \beta}{\partial t} - \gamma \frac{\partial \beta}{\partial x} = \lambda\alpha - \lambda\beta.$$

In the limit, $\alpha(x, t)$ and $\beta(x, t)$ are to be interpreted as probability density functions for right and left moving particles, respectively.

Since at time $t = 0$ the particle is located at $x = 0$ and is equally likely to move to the right or to the left, the probabilities α and β vanish for $x \neq 0$ and equal $\frac{1}{2}$ at $x = 0$. In terms of the density functions $\alpha(x, t)$ and $\beta(x, t)$ this yields the initial conditions

$$(1.75) \qquad\qquad \alpha(x, 0) = \beta(x, 0) = \tfrac{1}{2}\delta(x),$$

(1.79) becomes

$$(1.86) \qquad \epsilon^2 \frac{\partial^2 v}{\partial \sigma^2} + 2\lambda \epsilon \frac{\partial v}{\partial \sigma} - \gamma^2 \frac{\partial^2 v}{\partial x^2} = 0.$$

If we assume that v does not vary rapidly with respect to σ, we may discard the term $\epsilon^2(\partial^2 v/\partial \sigma^2)$ compared to $2\lambda \epsilon(\partial v/\partial \sigma)$ since $\epsilon \ll 1$ implies that $\epsilon^2 \ll \epsilon$. Thus (1.79) can be approximated for large t by the equation

$$(1.87) \qquad \frac{\partial v}{\partial t} = \left(\frac{\gamma^2}{2\lambda} \right) \frac{\partial^2 v}{\partial x^2},$$

where σ was replaced by t in (1.86) after discarding the second time derivative term. This equation is to be compared with the diffusion equation (1.21) where we must put $c = 0$. It has already been shown in (1.62)–(1.63) that with $D = \gamma^2/\lambda$, the mean square displacements obtained from the random walk models in this and the previous section agree for large t. Consequently, if the parameters are identified in this manner, (1.87) is identical with (1.21). The validity of the diffusion equation is limited to large values of t in its use as a model for Brownian motion as we have seen. However, (1.79) may be taken to represent a valid model for Brownian motion for all $t \geq 0$. For small values of t it yields the required finite particle velocities, and for large t it reduces to the diffusion equation.

To examine the behavior of the solution of the initial value problem (1.79)–(1.81) for small t, we note that the speed of the particle is γ. Thus since the particle is at $x = 0$ at the time $t = 0$ and it can travel to the right or the left, it can never reach the set of points x for which $|x| > \gamma t$. The density function $v(x, t)$ should, therefore, vanish for $|x| > \gamma t$ and this will be verified when the full solution of (1.79)–(1.81) is given in a later chapter. The location of the particle on one of the lines $x = \pm \gamma t$ can only be the result of the particle never having reversed its direction from the time $t = 0$ on, when it started on one of the paths $x = \pm \gamma t$ in the (x, t)-plane. The probability of reversal of direction is characterized by a Poisson process as described by (1.68)–(1.69) with rate of reversal λ. It is well-known for this process with rate λ, that the probability of nonreversal (i.e., persistence) in direction is given by $e^{-\lambda t}$ at time t. As the particle is equally likely to be on the line $x = \gamma t$ or $x = -\gamma t$, we conclude that the probability that the particle is on either line is

$$(1.88) \qquad v(x, t)|_{x = \pm \gamma t} = \tfrac{1}{2} e^{-\lambda t}.$$

For small values of t, $e^{-\lambda t} \approx 1$ so that most of the probability is concentrated on the lines $x = \pm \gamma t$. Therefore, we have essentially deterministic motion with speed γ along the lines $x = \pm \gamma t$. As t increases, $e^{-\lambda t}$ decays rapidly and the density $v(x, t)$ begins to become more concentrated in the interior region $|x| < \gamma t$. Eventually, the motion appears to become totally random and can be described by the random walk model and the limiting diffusion equation given in the previous section.

In a further analysis of the effect of the randomness assumptions that led to the partial differential equations (1.21) and (1.79), it is of interest to consider the reduced equations obtained from (1.21) and (1.79) when D and λ, respectively, are equated to zero. Since D is a measure of the variance around the mean particle path, we expect that the reduced equation with $D = 0$ should yield deterministic motion with the mean speed c. Also, λ is the rate of direction reversal of the particle, so that if $\lambda = 0$, the particle should move with speed γ either along the path $x = \gamma t$ or $x = -\gamma t$.

Putting $D = 0$ in (1.21) gives

$$(1.89) \qquad \frac{\partial v}{\partial t} + c \frac{\partial v}{\partial x} = 0.$$

With the initial condition (1.24), the formal solution of (1.89) is

$$(1.90) \qquad v(x, t) = \delta(x - ct),$$

where the Dirac delta function $\delta(x - ct)$ vanishes for $x - ct \neq 0$. If we formally differentiate (1.90), we find that it satisfies (1.89) and (1.90) equals $\delta(x)$ at $t = 0$. The density function $v(x, t)$ is concentrated on the path $x - ct = 0$ so that the particle moves (deterministically) along that path. Equivalently, the particle moves with the fixed velocity c. The case where $D \neq 0$ in (1.21) (i.e., the diffusion equation) may, therefore, be characterized as representing a random motion around a deterministic path for the particle given by $x = ct$ in the (x, t)-plane. The equation (1.89) is called a *wave equation* representing unidirectional wave motion with velocity c, for reasons that will become apparent when such equations are discussed in Chapter 2.

Putting $\lambda = 0$ in (1.79) yields the *wave equation*

$$(1.91) \qquad \frac{\partial^2 v}{\partial t^2} - \gamma^2 \frac{\partial^2 v}{\partial x^2} = 0.$$

With the initial conditions (1.80)–(1.81), the formal solution of (1.91) is

$$(1.92) \qquad v(x, t) = \tfrac{1}{2}\delta(x - \gamma t) + \tfrac{1}{2}\delta(x + \gamma t).$$

This result may be obtained from (1.89)–(1.90) by noting that with $\lambda = 0$, the system (1.73)–(1.74) reduces to two unidirectional wave equations

$$(1.93) \qquad \begin{cases} \dfrac{\partial \alpha}{\partial t} + \gamma \dfrac{\partial \alpha}{\partial x} = 0 \\[2mm] \dfrac{\partial \beta}{\partial t} - \gamma \dfrac{\partial \beta}{\partial x} = 0 \end{cases}$$

with initial data $\alpha(x, 0) = \beta(x, 0) = \frac{1}{2}\delta(x)$. Recalling that $v = \alpha + \beta$ and the result (1.90) immediately yields (1.92). The definition of the Dirac delta function implies, by way of (1.92), that the particle is restricted to move along the deterministic path $x = \gamma t$ or $x = -\gamma t$ with speed γ. The factor $\frac{1}{2}$ before each delta function is a consequence of the random choice of direction at the time $t = 0$ for the particle (i.e., either to the right or to the left). Once the particle chooses a direction it must continue to move in that direction, since $\lambda = 0$ implies that the probability of reversal of direction is zero. If $\lambda \neq 0$, we have seen in the foregoing that there is some initial directionality to the particle motion, but this rapidly disappears and the motion becomes completely random. The wave equation (1.91) permits two directions of motion (to the right and to the left) with speed γ.

We shall not introduce or discuss boundary value problems based on the random walk model formulated in this section. A full discussion of initial and initial boundary value problems for the telegrapher's and related equations will be given in later chapters. We conclude by noting that the difference equation formulation of the random walk problem in this section again indicates, as was the case in the preceding section, that the solution at time t depends only on data for the problem given at earlier times. The fact that the particle speed γ is finite implies, as was indicated, that the density function $v(x, t)$ with the data (1.80)–(1.81) concentrated at the origin vanishes when $|x| > \gamma t$. This is true for the solutions of the telegrapher's and wave equations. The existence of a finite (maximal) speed of propagation of disturbances is a fundamental property of hyperbolic partial differential equations of which the wave and telegrapher's equations are prototypes. This property is not shared by diffusion equations (which are equations of parabolic type), as shown in the preceding section. (The classification of equations into various types will be given in Chapter 3.)

1.3. LAPLACE'S EQUATION AND GREEN'S FUNCTION

In the random walk problems of the preceding sections we were concerned with finding the probability that a particle located initially at the point $x = 0$ and moving in a random manner is located at a point x at a later time t. Consequently, we were dealing with time dependent problems. In this section we consider two time independent random walk problems in plane regions. The one and three-dimensional versions of these problems will be considered in the exercises.

Each of the problems we study involves a random walk in a bounded plane region with an absorbing boundary. The first problem examines the probability that a particle starting at some point in the region reaches a certain point on the boundary and is absorbed before it reaches and is absorbed by the remaining portion of the boundary. The second problem is essentially concerned with the probability that the particle reaches a fixed interior point before it is absorbed by the boundary. The number of steps required for the particle to reach the fixed boundary or interior point is not relevant. That is, the time it takes for the particle to reach that point is not important so that these problems are time independent. They may be thought to represent stationary or steady-state versions of appropriate modifications of the problems considered in Sections 1.1 and 1.2. In fact, the time independent form of the diffusion and telegrapher's equations (in two space dimensions) is Laplace's equation which we now derive.

Let A represent the bounded region under consideration and ∂A its boundary curve. We enclose A and its boundary by a rectangle with sides $x = a$, $x = b$, $y = c$, and $y = d$ where $a < b$ and $c < d$. With δ as the step length in the random walk, we assume the intervals $[a, b]$ and $[c, d]$ can be subdivided into the set of points $x_k = a + \delta k$ and $y_l = c + \delta l$, respectively, with $0 \leqslant k \leqslant n$, $0 \leqslant l \leqslant m$, $x_n = b$, and $y_m = d$. Each point that lies within A (i.e., does not lie on ∂A) is called an interior point. Each interior point (x_k, y_l) has four neighboring points in four perpendicular directions. If one of the neighboring points lies on ∂A or is exterior to A, we call it a boundary point. The points in the rectangle that are neither interior nor boundary points will not be considered.

In the first random walk problem we ask for the probability that a particle starting at an interior point of A reaches the fixed boundary point (x_i, y_j) before it reaches any other boundary point. Let $v(x, y)$ be the probability that the particle starts at the interior point (x, y) and reaches the boundary point (x_i, y_j). We assume that the particle is equally likely to move to any of its four neighboring points from the interior point (x, y). Thus the probability that it moves to any of its four neighbors equals $1/4$. Since the probability that the particle reaches the boundary point (x_i, y_j) from the point (x, y) can be expressed in terms of the probability that it moves to any of its four neighboring points and reaches (x_i, y_j) from one of these points, we obtain the difference equation

(1.94)
$$v(x, y) = \tfrac{1}{4}\big[v(x + \delta, y) + v(x - \delta, y) + v(x, y + \delta) + v(x, y - \delta)\big].$$

If (x, y) is a boundary point, we have

(1.95)
$$v(x, y) = \begin{cases} 1, & (x, y) = (x_i, y_j) \\ 0, & (x, y) \neq (x_i, y_j) \end{cases}$$

since when the particle is at a boundary point $(x, y) \neq (x_i, y_j)$ it is absorbed and cannot reach the point (x_i, y_j). If one of the neighboring points of (x, y) is a boundary point, (1.95) is to be used in (1.94).

In the limit as the step length $\delta \to 0$, the number of points in the subdivisions of $[a, b]$ and $[c, d]$ tend to infinity, and the boundary points defined actually lie on the boundary ∂A. Using Taylor's formula gives

$$(1.96) \quad v(x \pm \delta, y) = v(x, y) \pm \delta v_x(x, y) + \tfrac{1}{2}\delta^2 v_{xx}(x, y) + O(\delta^3)$$

$$(1.97) \quad v(x, y \pm \delta) = v(x, y) \pm \delta v_y(x, y) + \tfrac{1}{2}\delta^2 v_{yy}(x, y) + O(\delta^3).$$

Inserting (1.96)–(1.97) into (1.94) dividing by δ^2 and letting $\delta \to 0$ yields

$$(1.98) \qquad \frac{\partial^2 v}{\partial x^2} + \frac{\partial^2 v}{\partial y^2} = 0,$$

which is known as *Laplace's equation*. The function $v(x, y)$ is now interpreted as a probability density.

We further assume that arc length s is defined on the boundary ∂A and that as $\delta \to 0$, the point (x_i, y_j) tends to the (boundary) point $(\hat{x}, \hat{y})$ on ∂A. The boundary conditions (1.95) are easily found to take the form

$$(1.99) \qquad v(x, y) = 0, \qquad (x, y) \in \partial A, \qquad (x, y) \neq (\hat{x}, \hat{y})$$

$$(1.100) \qquad \int_{\partial A} v(x, y)\, ds = 1.$$

For example, if $(\hat{x}, \hat{y})$ is a point in an open interval on the x-axis that comprises a portion of the boundary ∂A, we can set

$$(1.101) \qquad v(x, y) = \delta(x - \hat{x})$$

in that interval [where $\delta(x - \hat{x})$ is the Dirac delta function] and $v(x, y) = 0$ elsewhere on ∂A.

The second random walk problem we consider asks for the probability that a particle starting at an interior point (x, y) in the region A reaches a fixed interior point (ξ, η) before it reaches a boundary point and is absorbed. The region A is subdivided as in the first problem and interior and boundary points are defined as before with the step length again equal to δ. Since the problem is time independent, and the particle does not stop its motion once it reaches (ξ, η) for the first time, it is possible for the particle to pass through the point (ξ, η) more than once before it reaches and is absorbed at the boundary. Consequently, if the particle begins its motion at (ξ, η), it has unit probability of reaching (ξ, η) since it is there already. However, it can also move to one of its four neighboring points and reach (ξ, η) from there if the neighbor is not a boundary point. Therefore, if we introduce a function $w(x, y)$ that characterizes the prospects of a particle reaching (ξ, η) from the starting point (x, y), we cannot consider $w(x, y)$ to be a probability distribution since it may assume values exceeding unity. In particular, $w(\xi, \eta) \geq 1$, as we have seen.

The preceding random walk problem (as well as the first one of this section) was considered by Courant, Friedrichs, and Lewy in an early paper on difference methods for the equations of applied mathematics. They introduced the probabilities that a particle starting at (x, y) reaches (ξ, η) in $0, 1, 2, \ldots,$ $n, \ldots$ steps and defined a function of (x, y) that equals the sum of all these probabilities. This function was defined to give the "mathematical expectation" that the particle starting at (x, y) reaches (ξ, η) before being absorbed at the boundary. We shall take the function $w(x, y)$ introduced above to be defined in the manner given by these authors. (We note that the term "mathematical expectation" was used in a somewhat different context in the previous sections.) The function $w(x, y)$ satisfies a difference equation as we now show.

If the point (x, y) is a boundary point we must have

$$(1.102) \qquad w(x, y) = 0, \qquad (x, y) \text{ a boundary point,}$$

since the particle is absorbed at a boundary point and cannot reach (ξ, η) from there. If (x, y) is an interior point not equal to (ξ, η), we obtain the difference equation

$$(1.103)$$
$$w(x, y) = \tfrac{1}{4}\left[w(x - \delta, y) + w(x + \delta, y) + w(x, y - \delta) + w(x, y + \delta)\right],$$

since the particle is equally likely (and, therefore, has probability $\tfrac{1}{4}$) to move to any one of its four neighboring points and reach (ξ, η) from there. If $(x, y) = (\xi, \eta)$, we have

$$(1.104)$$
$$w(\xi, \eta) = 1 + \tfrac{1}{4}\left[w(\xi - \delta, \eta) + w(\xi + \delta, \eta) + w(\xi, \eta - \delta) + w(\xi, \eta + \delta)\right],$$

since the particle has unit probability of reaching (ξ, η) considering that it is there to begin with and has the additional possibility of reaching (ξ, η) by moving to one of its four neighboring points and returning to (ξ, η) from there.

For small δ, (1.103)–(1.104) take the form

$$(1.105) \qquad \frac{\partial^2 w}{\partial x^2} + \frac{\partial^2 w}{\partial y^2} = \begin{cases} O(\delta^2), & (x, y) \neq (\xi, \eta) \\ -\dfrac{4}{\delta^2} + O(\delta^2), & (x, y) = (\xi, \eta) \end{cases}$$

on using (1.96) and (1.97). This shows that $w(x, y)$ satisfies Laplace's equation at interior points $(x, y) \neq (\xi, \eta)$ as $\delta \to 0$. At the point (ξ, η), the right side of (1.105) blows up as $\delta \to 0$. Now as $\delta \to 0$, $w(x, y)$ is to be understood as a density function, so that the integral of $w(x, y)$ over some small neighborhood of (x, y) characterizes the property in which we are interested. If we consider a square with center at (ξ, η) and with side proportional to the step length δ, the area of the square multiplied by $4/\delta^2$ tends to a finite nonzero limit as $\delta \to 0$.

Since the right side of (1.105) vanishes for $(x, y) \neq (\xi, \eta)$ and its integral over the aforementioned square has a finite nonzero limit as $\delta \to 0$, we conclude that it must be proportional in the limit to the two-dimensional Dirac delta function.

The two-dimensional Dirac delta function with singular point (ξ, η) is defined as (see Section 7.2):

(1.106) $$\delta(x - \xi)\delta(y - \eta) = 0, \qquad (x, y) \neq (\xi, \eta)$$

(1.107) $$\iint_R \delta(x - \xi)\delta(y - \eta)\, dx\, dy = 1$$

where R is any open region containing the point (ξ, η). Since (1.106) implies that for continuous $f(x, y)$, $f(x, y)\delta(x - \xi)\delta(y - \eta) = f(\xi, \eta)\delta(x - \xi)\delta(y - \eta)$, we have the further property

(1.108) $$\iint_R f(x, y)\delta(x - \xi)\delta(y - \eta)\, dx\, dy = f(\xi, \eta).$$

As $\delta \to 0$, the boundary points of the discrete problem tend to points on ∂A and (1.102) now states that $w(x, y)$ vanishes on ∂A. To analyze the properties of the solution of this boundary value problem, it is convenient to replace $w(x, y)$ by the so-called *Green's function* $K(x, y; \xi, \eta)$ for this problem. The Green's function for Laplace's equation with homogeneous boundary conditions is defined to be a solution of

(1.109) $$\frac{\partial^2 K}{\partial x^2} + \frac{\partial^2 K}{\partial y^2} = -\delta(x - \xi)\delta(y - \eta), \qquad (x, y) \in A$$

which satisfies the boundary condition

(1.110) $$K(x, y; \xi, \eta) = 0, \qquad (x, y) \in \partial A.$$

As was shown above, the density function $w(x, y)$ differs from the Green's function only by a constant factor. Thus both w and K are solutions of (special) inhomogeneous forms of Laplace's equation. (Green's functions are a useful tool for solving boundary value problems for partial differential equations and are discussed in Chapter 7.)

The two random walk problems considered in this section are not completely unrelated, since one problem asks for the probability that an interior point is reached and the other seeks the probability that a boundary point is reached. In the following example we show that the two problems are indeed connected, by establishing a relationship between the density function $v(x, y)$ for the first random walk problem and the Green's function $K(x, y; \xi, \eta)$ for the second random walk problem.

EXAMPLE 1.6.   GREEN'S THEOREM

We formally apply *Green's second theorem* to the functions $v(x, y)$ and $K(x, y; \xi, \eta)$. Integrating over the region A and its boundary ∂A we have

$$(1.111) \qquad \iint_A \left(v \nabla^2 K - K \nabla^2 v \right) dx \, dy = \int_{\partial A} \left(v \frac{\partial K}{\partial n} - K \frac{\partial v}{\partial n} \right) ds$$

where $\nabla^2 = \partial^2/\partial x^2 + \partial^2/\partial y^2$ is the Laplacian operator, $\partial/\partial n$ is a derivative in the direction of the exterior normal to the boundary ∂A, and s is arc length on ∂A. We have $\nabla^2 v = 0$ and $\nabla^2 K = -\delta(x - \xi)\delta(y - \eta)$, in view of (1.98) and (1.109). Further, (1.99)–(1.100) imply that $v(x, y) \, \partial K(x, y; \xi, \eta)/\partial n$ evaluated on the boundary equals $v(x, y) \, \partial K(\hat{x}, \hat{y}; \xi, \eta)/\partial n$. Combined with the fact that K vanishes on ∂A we obtain from (1.111)

$$(1.112) \qquad v(\xi, \eta) = -\int_{\partial A} v \frac{\partial K}{\partial n} ds = -\frac{\partial K(\hat{x}, \hat{y}; \xi, \eta)}{\partial n},$$

where (1.100) was used. This shows that if the Green's function K can be determined, the boundary value problem (1.98)–(1.100) can be solved since (ξ, η) is an arbitrary interior point in A.

Equation (1.112) has the following interpretation. The probability density $v(\xi, \eta)$ characterizes the probability that particles starting in a neighborhood of (ξ, η) reach the boundary point $(\hat{x}, \hat{y})$. The normal derivative $\partial K(\hat{x}, \hat{y}; \xi, \eta)/\partial n$ is a measure of the flux of particle density associated with particles originating near the boundary point $(\hat{x}, \hat{y})$ and reaching the interior point (ξ, η). Therefore, $-\partial K/\partial n$ is a measure of the flux in the reverse direction and this should essentially equal $v(\xi, \eta)$.

We conclude this section with a number of comments regarding the properties of boundary value problems for Laplace's equation that are suggested by the difference equation treatment considered above.

The difference equation (1.94) together with the boundary condition (1.95), when applied at all interior points in the subdivision of A, yields a simultaneous system of equations for the functions $v(x_k, y_l)$ where (x_k, y_l) is a typical interior point. All the boundary data play a role in the solution for v at the interior points. This contrasts with the situation encountered in the preceding sections. Though the solution was required for all $t > 0$, the solution for fixed t did not depend on data for later values of t.

Further, the difference equation (1.94) characterizes the value of $v(x, y)$ at the center of a square as the mean or average of its values at the four vertices. With (x, y) as its center, the square has the points $(x \pm \delta, y)$ and $(x, y \pm \delta)$ as its vertices. This mean value property carries over to functions $v(x, y)$ satisfying Laplace's equation. It is shown later that any solution $v(x, y)$ of

(1.98) equals the average of its values on a circle with center at (x, y) as long as the circle is interior to the region A.

Finally, the above mean value property for the difference equation implies that the maximum and minimum values of $v(x, y)$ must be assumed at boundary points of A. This follows since $v(x, y)$ is the average of its neighboring values and, therefore, cannot be greater than or less than those values. This maximum and minimum principle will be shown to be valid for certain boundary value problems for Laplace's equation.

EXERCISES FOR CHAPTER 1

Section 1.1

1.1.1. Verify, by direct substitution, that the function $v(x, t)$ given in (1.28) is a solution of the diffusion equation (1.21).

1.1.2. Show that the function $v(x, t)$ given in (1.28) satisfies the following conditions when $t > 0$,

(a) $\int_{-\infty}^{\infty} v(x, t)\, dx = 1$.

(b) $\int_{-\infty}^{\infty} x v(x, t)\, dx = ct$

(c) $\int_{-\infty}^{\infty} x^2 v(x, t)\, dx = c^2 t^2 + Dt$.

Conclude thereby that $v(x, t)$ is a probability density function with mean ct and variance Dt.

Hint:

$$\int_{-\infty}^{\infty} e^{-x^2}\, dx = \sqrt{\pi}\,.$$

1.1.3. Again considering the function $v(x, t)$ of (1.28), show that the limit as $t \downarrow 0$ of $v(x, t)$ is zero when $x \neq 0$. Conclude from this and the result of Exercise 1.1.2(a) that $\lim_{t \downarrow 0} v(x, t) = \delta(x)$, so that the initial condition (1.24) is satisfied.

1.1.4. Show that if $l > 0$, the functions

$$v_{\pm}(x, t) = \frac{1}{\sqrt{2\pi Dt}} \left\{ \exp\left[-\frac{(x)^2}{2 Dt} \right] \mp \exp\left[-\frac{(x - 2l)^2}{2 Dt} \right] \right\}$$

satisfy the diffusion equation (1.21) (with $c = 0$), the initial condition (1.24), and, respectively, satisfy the boundary conditions (1.43) and (1.47). (The construction of these solutions may be based on the method of images which is discussed in Section 7.5).

1.1.5. Derive the diffusion equation in two dimensions

$$v_t = \tfrac{1}{2} D(v_{xx} + v_{yy}),$$

by constructing an appropriate difference equation in the manner leading to (1.18), but assuming the particle is equally likely to move to the points $(x \pm \delta, y)$ and $(x, y \pm \delta)$ from the point (x, y). Use the Taylor series and define $D = \lim_{\substack{\tau \to 0 \\ \delta \to 0}}(\delta^2/2\tau)$ to obtain the diffusion equation from the difference equation.

1.1.6. Derive the diffusion equation in three dimensions

$$v_t = \tfrac{1}{2}D(v_{xx} + v_{yy} + v_{zz})$$

with $D = \lim_{\substack{\tau \to 0 \\ \delta \to 0}}(\delta^2/3\tau)$, by using assumptions similar to those given in Exercise 1.1.5.

1.1.7. With $x = k\delta$ and $t = n\tau$, where $k = 0, \pm 1, \pm 2, \ldots$ and $n = 0, 1, 2, \ldots$, $v(x, t) = v(k\delta, n\tau)$ gives the probability that the particle is at the point $k\delta$ at the time $n\tau$.

(a) Express the difference equation (1.18) in terms of the function $v(k\delta, n\tau)$.

(b) Given the initial conditions $v(0, 0) = 1$ and $v(k\delta, 0) = 0$ for $k \neq 0$, solve the difference equation recursively for all k with $n = 1, 2, 3,$ and 4.

(c) Show that for each of the foregoing values of n we have $\sum_{k=-\infty}^{\infty} v(k\delta, n\tau) = 1$.

(d) Show that the solution of the initial value problem of part (b) is

$$v(k\delta, n\tau) = \begin{cases} \dfrac{n!}{\left(\dfrac{n+k}{2}\right)!\left(\dfrac{n-k}{2}\right)!}p^{(n+k)/2}q^{(n-k)/2}; & \dfrac{n \pm k}{2} = 0, 1, 2, \ldots \\ \\ 0; & \text{otherwise} \end{cases}.$$

Using this solution, verify the results of part (b).

1.1.8. Extending the results in the text, assume that the probability that a particle located at the point x moves to the right at the ith step is

$$P(x_i = \delta) = p(x) = \tfrac{1}{2}[1 + b(x)\delta],$$

and the probability that it moves to the left is

$$P(x_i = -\delta) = q(x) = \tfrac{1}{2}[1 - b(x)\delta],$$

with $b(x)$ given such that $0 \leq p, q \leq 1$ for all x under consideration. The probabilities p and q are now functions of position.

(a) Assuming the other conditions are unchanged, obtain the difference equation for the probability distribution $v(x, t)$ in the form

$$v(x, t + \tau) = p(x - \delta)v(x - \delta, \tau) + q(x + \delta)v(x + \delta, \tau).$$

(b) Using the Taylor series and the result $\lim_{\substack{\tau \to 0 \\ \delta \to 0}}(\delta^2/\tau) = D$, obtain the limiting diffusion equation

$$\frac{\partial v}{\partial t} = D\left\{ -\frac{\partial}{\partial x}[b(x)v] + \frac{1}{2}\frac{\partial^2 v}{\partial x^2} \right\}.$$

[Note that for $b =$ constant this reduces to (1.21) with c defined as $c = bD$, and it yields the Brownian motion described in the text.]

1.1.9. With $b(x) = -x$, the model in Exercise 1.1.8 describes Brownian motion with the particle subjected to an elastic "restoring" force. As $|x|$ increases, the probability that the particle moves further away from the origin decreases.

(a) Show that this "restoring" property follows from the definition of $p(x)$ and $q(x)$.

(b) Using the methods of Example 1.3 and retaining the initial condition (1.24), determine the first three moments of the random variable x for this case.

(c) Obtain the expectation and the variance using the results of part (b) and compare them with the expectation and variance given in the text for the Brownian motion of a "free" particle.

Section 1.2

1.2.1. Verify that (1.56) sums to (1.57).

1.2.2. Show that the limit in (1.60) is correct.

1.2.3. Obtain the results of Example 1.5 for the indicated moments by solving the given initial value problems for the ordinary differential equations.

1.2.4. Consider the solution (1.28) of the initial value problem (1.24) for the diffusion equation (1.21). Using the approach of Exercise 1.1.3, show that the limit of this solution as $D \to 0$ is the delta function (1.90).

1.2.5. (a) Let $v = Ve^{-\lambda t}$ in the telegrapher's equation (1.79) and show that V satisfies the equation

$$V_{tt} - \gamma^2 V_{xx} - \lambda^2 V = 0.$$

(b) Put $z = \sqrt{\gamma^2 t^2 - x^2}$ and show that $V(x, t) = W(z)$ satisfies the ordinary differential equation

$$W''(z) + \frac{1}{z}W'(z) - \left(\frac{\lambda^2}{\gamma^2}\right)W(z) = 0.$$

[Note that the region $z \geqslant 0$ is where the solution $v(x, t)$ was shown to be nonzero in the text.]

(c) Let $W(z) = (1/\sqrt{z})w(z)$ and show that $w(z)$ satisfies the equation

$$w''(z) + \left[1/4z^2 - (\lambda^2/\gamma^2)\right]w(z) = 0.$$

(d) Observe that for $|x| < \gamma t$, as $\gamma t \to \infty$ we have

$$z = \sqrt{\gamma^2 t^2 - x^2} \approx \gamma t - \frac{x^2}{2\gamma t},$$

and, consequently, $z \to \infty$ as $\gamma t \to \infty$. Thus we can approximate the equation for $w(z)$ when γt is large, by

$$w''(z) - \left(\frac{\lambda^2}{\gamma^2}\right)w(z) \approx 0.$$

Using (1.63), show that as $\gamma t \to \infty$ we have approximately $v(x, t) \approx (a/\sqrt{\gamma t})e^{-(x^2/2 Dt)}$ where a is an arbitrary constant. This has the form of the solution (1.28) of the diffusion equation with $c = 0$.

1.2.6. Put $x = k\delta$ and $t = n\tau$, as in Exercise 1.1.7 and set $\alpha(x, t) = \alpha(k\delta, n\tau)$ and $\beta(x, t) = \beta(k\delta, n\tau)$. Show that the difference equations (1.66)–(1.67) take the form

$$\begin{cases} \alpha[k\delta, (n + 1)\tau] = p\alpha[(k - 1)\delta, n\tau)] + q\beta[(k - 1)\delta, n\tau] \\ \beta[k\delta, (n + 1)\tau] = p\beta[(k + 1)\delta, n\tau)] + q\alpha[(k + 1)\delta, n\tau] \end{cases}$$

with the initial conditions $\alpha(0, 0) = \beta(0, 0) = \frac{1}{2}$, and $\alpha(k\delta, 0) = \beta(k\delta, 0) = 0$, for $k \neq 0$.

1.2.7. (a) Solve the difference equations of Exercise 1.2.6 recursively for all k and for $n = 1, 2, 3,$ and 4.

(b) Show that $\sum_{k=-\infty}^{\infty}\alpha(k\delta, n\tau) = \sum_{k=-\infty}^{\infty}\beta(k\delta, n\tau) = \frac{1}{2}$ for the above values of n.

1.2.8. (a) Solve the difference equations of Exercise 1.2.6 for $\alpha(\pm n\delta, n\tau)$ and $\beta(\pm n\delta, n\tau)$. Show that $\alpha(n\delta, n\tau) = \beta(-n\delta, n\tau) = \frac{1}{2}p^{n-1}$ and $\alpha(-n\delta, n\tau) = \beta(n\delta, n\tau) = 0$ for $n \geq 1$.

(b) With $x = \pm n\delta$ and $t = n\tau$ so that $x = \pm(\delta/\tau)t$, show that as $n \to \infty$, $\delta \to 0$, $\tau \to 0$, and $\delta/\tau \to \gamma$, we have $\alpha(\gamma t, t) \to \frac{1}{2}e^{-\lambda t}$, $\beta(-\gamma t, t) \to \frac{1}{2}e^{-\lambda t}$. These results are in agreement with (1.88) since $v = \alpha + \beta$.

1.2.9. With $v(k\delta, n\tau) = \alpha(k\delta, n\tau) + \beta(k\delta, n\tau)$ show that v satisfies the difference equation

$$v[k\delta, (n + 1)\tau] = pv[(k - 1)\delta, n\tau] + pv[(k + 1)\delta, n\tau]$$
$$- (p - q)v[k\delta, (n - 1)\tau],$$

with the initial conditions $v(0,0) = 1$, $v(k\delta, 0) = 0$, $k \neq 0$, and $v(-\delta, \tau) = v(\delta, \tau) = \frac{1}{2}$, $v(k\delta, \tau) = 0$, $k \neq \pm 1$. [We observe that with $p = q = \frac{1}{2}$, the correlation coefficient $\rho = p - q$ vanishes and the difference equation reduces to (1.18). If $\rho = p - q \neq 0$, the probability $v(k\delta, n\tau)$ depends not only on the location of the particle at the previous step but the previous two steps.]

1.2.10. (a) Solve the difference equation of Exercise 1.2.9 for $v(k\delta, n\tau)$ with $n = 2, 3$, and 4 and for all k.

(b) Solve the equation for $v(\pm n\delta, n\tau)$, for $n \geq 2$.

(c) Show that $v(k\delta, n\tau) = 0$ for $|k| > n$.

1.2.11. Generalize the model of Brownian motion presented in this section by assuming the probabilities p and q of persistence and reversal depend on the direction of motion of the particle. Let p^+ and q^+ represent the probabilities of persistence and reversal in direction if the particle is moving to the right and let p^- and q^- represent the same probabilities for leftward motion. (Note that $p^+ + q^+ = p^- + q^- = 1$ and that in the text we assumed $p^+ = p^-$ and $q^+ = q^-$.) Let $\alpha(x, t)$ and $\beta(x, t)$ be defined as in the text and show that they now satisfy the difference equations

$$\alpha(x, t + \tau) = p^+ \alpha(x - \delta, t) + q^- \beta(x - \delta, t)$$

$$\beta(x, t + \tau) = p^- \beta(x + \delta, t) + q^+ \alpha(x + \delta, t).$$

1.2.12. (a) With $p^\pm = 1 - \lambda^\pm \tau + O(\tau^2)$ and $q^\pm = \lambda^\pm \tau + O(\tau^2)$, show that the limiting partial differential equations for the system of difference equations of Exercise 1.2.11 as $\delta \to 0$ and $\tau \to 0$ are (with $\delta/\tau \to \gamma$)

$$\alpha_t + \gamma \alpha_x = -\lambda^+ \alpha + \lambda^- \beta$$

$$\beta_t - \gamma \beta_x = \lambda^+ \alpha - \lambda^- \beta.$$

(b) Putting $v = \alpha + \beta$, show that $v(x, t)$ satisfies the telegrapher's equation

$$v_{tt} - \gamma^2 v_{xx} + (\lambda^+ + \lambda^-) v_t - \gamma(\lambda^+ - \lambda^-) v_x = 0.$$

1.2.13. (a) Apply the method of Example 1.5 to show that as $t \to \infty$, the expectation $\langle x \rangle$, and variance $V(x)$ of the continuous random variable characterized by the equation in Exercise 1.2.12(b) are given as

$$\langle x \rangle \approx \left[\frac{\gamma(\lambda^+ - \lambda^-)}{\lambda^+ + \lambda^-} \right] t$$

$$V(x) = \langle x^2 \rangle - \langle x \rangle^2 \approx 2\gamma^2 \left[\frac{1}{\lambda^+ + \lambda^-} - \frac{(\lambda^+ - \lambda^-)^2}{(\lambda^+ + \lambda^-)^3} \right] t.$$

(b) Compare the results of part (a) with those for the Brownian motion model of Section 1.1. Explain why the coefficient of t in the expression for $\langle x \rangle$ may be characterized as a "drift" velocity.

(c) Show that $V(x) \approx \gamma^2 t^2$ for small t.

1.2.14. Generate the results of Exercise 1.2.13 for the expectation and variance by proceeding as follows. Assuming x and t are large and of the same order of magnitude (say, $x = \hat{x}/\epsilon$ and $t = \hat{t}/\epsilon$ where $0 < \epsilon \ll 1$), we can neglect the second derivative terms in the equation for $v(x, t)$ given in Exercise 1.2.12, and obtain in a first approximation,

$$v_t - \left[\frac{\gamma(\lambda^+ - \lambda^-)}{\lambda^+ + \lambda^-} \right] v_x = 0.$$

Using this equation to obtain an expression for v_{tt}, show that in the next approximation we obtain the following diffusion equation for $v(x, t)$:

$$v_t = \left[\frac{\gamma(\lambda^+ - \lambda^-)}{\lambda^+ + \lambda^-} \right] v_x + \gamma^2 \left[\frac{1}{\lambda^+ + \lambda^-} - \frac{(\lambda^+ - \lambda^-)^2}{(\lambda^+ + \lambda^-)^3} \right] v_{xx}.$$

Demonstrate that this equation implies the results of Exercise 1.2.13(a).

Section 1.3

1.3.1. (a) Show that in the one-dimensional case, if the particle is equally likely to move to the right or to the left, (1.94) is replaced by

$$v(x) = \tfrac{1}{2} [v(x - \delta) + v(x + \delta)].$$

(b) Show that in the three-dimensional case, when the particle is equally likely to move to any of its six neighboring points, we obtain

$$v(x, y, z) = \tfrac{1}{6} [v(x + \delta, y, z) + v(x - \delta, y, z) + v(x, y - \delta, z)$$
$$+ v(x, y + \delta, z) + v(x, y, z - \delta) + v(x, y, z + \delta)].$$

1.3.2. Show that the limiting differential equations for Exercise 1.3.1(a) and (b) are

(a) $v_{xx} = 0$;

(b) $v_{xx} + v_{yy} + v_{zz} = 0$,
respectively.

1.3.3. Consider the interval $0 \leqslant x \leqslant 1$. Let $v(x)$ be the probability that the particle reaches the boundary point $x = 1$ and is absorbed before it reaches and is absorbed at the boundary $x = 0$.

(a) Show that the appropriate boundary conditions in this case are $v(0) = 0$ and $v(1) = 1$.

(b) Show that $v(x) = x$, which satisfies the boundary conditions of part (a), is a solution of both the difference equation of Exercise 1.3.1(a) and the (one-dimensional) differential equation of Exercise 1.3.2(a).

1.3.4. We define the Green's function $K(x; \xi)$ in the one-dimensional case, for the interval $0 \leqslant x \leqslant 1$, to be the solution of the boundary value problem

$$\frac{\partial^2 K(x; \xi)}{\partial x^2} = -\delta(x - \xi); \qquad K(0; \xi) = K(1; \xi) = 0; \qquad 0 < \xi < 1.$$

(a) By integrating the equation for $K(x; \xi)$ over a small neighborhood of $x = \xi$ show that

$$\left. \frac{\partial K}{\partial x} \right|_{x \uparrow \xi}^{x \downarrow \xi} = -1$$

(i.e., the jump in $\partial K/\partial x$ at $x = \xi$ is -1). Assuming that $K(x; \xi)$ is continuous at $x = \xi$, define $K(x, \xi)$ as $K(x; \xi) = K_1(x; \xi),\ 0 \leqslant x \leqslant \xi$, and $K(x; \xi) = K_2(x; \xi),\ \xi \leqslant x \leqslant 1$. Show that we then obtain the equations

$$\frac{\partial^2 K_1}{\partial x^2} = 0,\, 0 < x < \xi; \qquad \frac{\partial^2 K_2}{\partial x^2} = 0, \qquad \xi < x < 1$$

and the supplementary conditions

$$K_1(0; \xi) = K_2(1; \xi) = 0; \qquad K_1(\xi; \xi) = K_2(\xi; \xi);$$

$$\frac{\partial K_2}{\partial x}(\xi; \xi) - \frac{\partial K_1}{\partial x}(\xi; \xi) = -1.$$

(b) Show that the solution of part (a) is

$$K_1(x; \xi) = (1 - \xi)x; \qquad K_2(x; \xi) = (1 - x)\xi.$$

1.3.5. (a) Apply a one-dimensional form of Green's theorem (1.111) to show that with $v(x)$ and $K(x; \xi)$ defined as in the two preceding exercises we have

$$v(\xi) = -\frac{\partial K(1; \xi)}{\partial x}.$$

(b) Demonstrate that the result of part (a) is correct by using the explicit forms for the functions $v(x)$ and $K(x; \xi)$.

1.3.6. Solve the difference equation (1.94) in the square $0 \leqslant x \leqslant 3$ and $0 \leqslant y \leqslant 3$ at the points with coordinates $x = 1, 2$ and $y = 1, 2$, assuming $\delta = 1$ and $(x_i, y_j) = (0, 1)$ in (1.95).

1.3.7. Show that $v(x, y) = 1$ is a solution of the boundary value problem for Laplace's equation in the square $0 \leqslant x \leqslant 1$, $0 \leqslant y \leqslant 1$, with the boundary conditions $v(x, 0) = 1$ and $\partial v / \partial n = 0$ on the other three sides of the square. Noting that $\partial v / \partial n = 0$ corresponds to an "elastic" boundary condition, interpret the solution.

1.3.8. Show that $v(x, y) = (x + iy)^n$, where $i = \sqrt{-1}$, is a solution of Laplace's equation (1.98) for all integers $n > 0$.

1.3.9. Consider $v = (x + iy)^2$ and write it in the form $v(x, y) = f(x, y) + ig(x, y)$, where f and g are the real and imaginary parts of the function $v(x, y)$. Show that $f(x, y)$ and $g(x, y)$ have no relative maxima or minima in the (x, y)-plane and that both satisfy Laplace's equation.

1.3.10. Show that $v(x, y, z) = (x + iy \cos \theta + iz \sin \theta)^n$ is a solution of Laplace's equation in three dimensions (see Exercise 1.3.2) for all integers $n > 0$ and for $0 \leqslant \theta < 2\pi$.

1.3.11. Write $v = (x + iy \cos \theta + iz \sin \theta)^2$ in the form $v = F(x, y, z) + iG(x, y, z)$, where F and G are real valued functions. Show that F and G are solutions of Laplace's equation and that they have no relative maxima and minima in (x, y, z)-space.

2

FIRST ORDER PARTIAL DIFFERENTIAL EQUATIONS

2.1. INTRODUCTION

This chapter initiates our study of methods for solving partial differential equations. We begin with single first order equations, that is, equations in which the highest derivative of the dependent variable is of first order. We restrict our discussion to problems involving two independent variables and deal almost exclusively with the method of characteristics for solving these problems. The presentation is thereby simplified, but as shown in the exercises, this method is easily extended to handle equations with more independent variables. Other solution methods for first order equations exist, but apart from the method of the complete integral, they are not discussed here.

First order equations can occur directly as models for physical processes or can arise as approximations to higher order equations or systems of equations. Thus the unidirectional wave equation (1.89), which is studied in Example 2.2, is an approximation to the diffusion equation of Section 1.1, in case the diffusion coefficient is negligibly small. The quasilinear wave equation (2.53), or more generally (2.82), serves as an approximation to Euler's equations of fluid dynamics (see Example 8.10) or represents a simple model for traffic flow. The nonlinear eiconal equation (2.125) plays an important role in the theory of geometrical optics and represents an approximation to wave optics as shown in Section 9.4. In fact, linear, quasilinear, and nonlinear first order equations play significant roles in the approximation methods for higher order equations which are discussed in Chapter 9.

The formulation of mathematical models can lead naturally to systems of first order equations, as in the case of the correlated random walk of Section

1.2 and the system (1.73)–(1.74) or as in Euler's hydrodynamic equations of Example 8.10. Alternatively, as shown in Example 2.1, systems can be constructed through the reduction of higher order equations. Conversely, linear systems of equations with constant coefficients can always be reduced to single higher order equations, as was shown in Section 1.2 for the special system (1.73)–(1.74) and as can be demonstrated in general. The methods of this section cannot be used for the solution of first order systems except in special cases, such as in the derivation of d'Alembert's solution of the wave equation in Example 2.3. Although we do not present any general methods for solving first order systems in this text, specific problems involving linear and nonlinear systems will be studied.

EXAMPLE 2.1.   REDUCTION OF HIGHER ORDER EQUATIONS TO SYSTEMS

The *wave equation* (1.91) can be written in the factored form

$$(2.1) \quad \frac{\partial^2 v}{\partial t^2} - \gamma^2 \frac{\partial^2 v}{\partial x^2} = \left(\partial_t^2 - \gamma^2 \partial_x^2 \right) v = \left(\partial_t + \gamma \partial_x \right)\left(\partial_t - \gamma \partial_x \right) v = 0,$$

where the differential operators ∂_x, ∂_t, and $\partial_t \pm \gamma \partial_x$ are defined in an obvious way. If we set $(\partial_t - \gamma \partial_x)v = u$, (2.1) implies that $(\partial_t + \gamma \partial_x)u = 0$ and we obtain the system

$$(2.2) \quad \begin{cases} \dfrac{\partial v}{\partial t} - \gamma \dfrac{\partial v}{\partial x} = u \\[2mm] \dfrac{\partial u}{\partial t} + \gamma \dfrac{\partial u}{\partial x} = 0 \end{cases}.$$

Though this system is coupled, it is possible to solve the second equation independently of the first. The solution $u(x, t)$ can then be introduced into the first equation and $v(x, t)$ can be determined. A solution of the system (2.2) is found in Example 2.3 using the method of characteristics.

The *damped wave equation* (1.79) after a transformation of the dependent variable of the form $w = e^{\lambda t}v$ becomes

$$(2.3) \quad \frac{\partial^2 w}{\partial t^2} - \gamma^2 \frac{\partial^2 w}{\partial x^2} - \lambda^2 w = 0.$$

This factorization method can be applied to yield

$$(2.4) \quad \begin{cases} \dfrac{\partial w}{\partial t} - \gamma \dfrac{\partial w}{\partial x} = u \\[2mm] \dfrac{\partial u}{\partial t} + \gamma \dfrac{\partial u}{\partial x} = \lambda^2 w \end{cases}.$$

In this case both first order equations must be solved simultaneously, and the methods developed in this chapter do not lead to a straightforward solution of (2.4).

If we formally apply the factorization technique to *Laplace's equation* (1.98), we obtain

$$(2.5) \qquad \frac{\partial^2 v}{\partial x^2} + \frac{\partial^2 v}{\partial y^2} = \left(\partial_x^2 + \partial_y^2 \right) v = \left(\partial_x + i\partial_y \right)\left(\partial_x - i\partial_y \right) v = 0$$

where $i = \sqrt{-1}$. Proceeding as for the wave equation leads to a first order coupled system with real and imaginary coefficients. Since we generally require real solutions of Laplace's equation, this reduction procedure is unsatisfactory. A system of equations more customarily associated with Laplace's equation can be obtained by putting $\partial v/\partial x = u$ and $\partial v/\partial y = w$. Assuming the mixed partial derivatives of v are equal, we easily obtain on using (2.5),

$$(2.6) \qquad \begin{cases} \dfrac{\partial w}{\partial x} = \dfrac{\partial u}{\partial y} \\[2mm] \dfrac{\partial w}{\partial y} = -\dfrac{\partial u}{\partial x} \end{cases}.$$

These are the *Cauchy–Riemann equations* which are well known in the theory of complex variables. Clearly, they must be solved simultaneously.

For the *diffusion equation* (1.21), no simple factorization is possible. However, if we put $u = \partial v/\partial x$, we obtain the system

$$(2.7) \qquad \begin{cases} \dfrac{\partial v}{\partial x} = u \\[2mm] \dfrac{\partial v}{\partial t} - \dfrac{D}{2}\dfrac{\partial u}{\partial x} = -cu \end{cases}$$

Again, this system must be solved simultaneously, and no obvious benefit appears from writing the diffusion equation in system form.

2.2. LINEAR FIRST ORDER PARTIAL DIFFERENTIAL EQUATIONS

The most general first order linear partial differential equation has the form

$$(2.8) \qquad a(x, t)\frac{\partial v}{\partial x} + b(x, t)\frac{\partial v}{\partial t} = c(x, t)v + d(x, t),$$

where a, b, c, and d are given functions of x and t. At each point (x, t) where the vector $[a, b]$ is defined and nonzero, the left side of (2.8) is (essentially) a directional derivative of $v(x, t)$ in the direction of $[a, b]$. The equations

$$(2.9) \qquad \frac{dx}{ds} = a(x, t); \qquad \frac{dt}{ds} = b(x, t),$$

determine a family of curves $x = x(s)$, $t = t(s)$ whose tangent vector $[x'(s), t'(s)]$ coincides with the direction of the vector $[a, b]$ at each point where $[a, b]$ is defined and nonzero. Therefore, the derivative of $v(x, t)$ along these curves becomes

$$(2.10) \qquad \frac{dv}{ds} = \frac{dv[x(s), t(s)]}{ds} = \frac{\partial v}{\partial x}\frac{dx}{ds} + \frac{\partial v}{\partial t}\frac{dt}{ds}$$

$$= a\frac{\partial v}{\partial x} + b\frac{\partial v}{\partial t} = cv + d,$$

on using the chain rule and (2.8)–(2.9).

The family of curves $x = x(s)$, $t = t(s)$, $v = v(s)$ determined by the solution of the system of ordinary differential equations (2.9)–(2.10) are called the *characteristic curves* of the partial differential equation (2.8). Since the equations (2.9) can be solved independently of (2.10), the curves in the (x, t)-plane determined from (2.8) are occasionally also referred to as characteristic curves or *characteristic base curves*. The approach we develop to solve (2.8) by using the solutions of (2.9)–(2.10) is called the *method of characteristics*. It is based on the geometric interpretation of the partial differential equation (2.8).

The existence and uniqueness theory for ordinary differential equations, assuming certain smoothness conditions on the functions a, b, c, and d, guarantees that exactly one solution curve $(x(s), t(s), v(s))$ of (2.9)–(2.10) (i.e., a characteristic curve) passes through a given point (x_0, t_0, v_0) in (x, t, v)-space. As a rule, we are not interested in determining a general solution of the partial differential equation (2.8) but rather a specific solution $v = v(x, t)$ [i.e., a surface in (x, t, v)-space] that passes through or contains a given curve C. This problem is known as the *initial value problem* for (2.8).

The method of characteristics for solving the initial value problem for (2.8) proceeds as follows. We assume the initial curve C is given parametrically as

$$(2.11) \qquad x = x(\tau), \qquad t = t(\tau), \qquad v = v(\tau)$$

for a given range of values of the parameter τ. Every value of τ fixes a point on C through which a unique characteristic curve passes. The family of characteristic curves determined by the points of C may be parametrized as

$$(2.12) \qquad x = x(s, \tau), \qquad t = t(s, \tau), \qquad v = v(s, \tau),$$

with $s = 0$ corresponding to the initial curve C. That is, we have $x(0, \tau) = x(\tau)$, $t(0, \tau) = t(\tau)$, and $v(0, \tau) = v(\tau)$.

The equations (2.12), in general, yield a parametric representation of a surface in (x, t, v)-space that contains the initial curve C. Assuming the equations $x = x(s, \tau)$ and $t = t(s, \tau)$ can be inverted to give s and τ as (smooth) functions of x and t, these functions can be introduced into the equation $v = v(s, \tau)$. The resulting function $v = V(x, t)$ satisfies (2.8) in a neighborhood of the curve C, the initial condition (2.11) {i.e., $V[x(\tau), t(\tau)] = v(\tau)$}, and is the unique solution of the given initial value problem. More precise conditions under which the method of characteristics yields unique solutions are given in Section 2.3. Figure 2.1 shows how the *integral surface* is constructed in terms of the initial and characteristic curves.

If the foregoing method does not lead to a solution, the initial value problem may not have a solution at all or it may have infinitely many solutions. The latter situation arises if the initial curve C is itself a characteristic curve, in what is known as a *characteristic initial value problem*. The following examples illustrate the use of the method of characteristics in solving initial value problems for linear first order equations. The role of characteristics in relation to the nonexistence or nonuniqueness of solutions of initial value problems is reexamined from a different point of view in Section 3.2. Finally, we observe that although the method of characteristics reduces the problem of solving first order partial differential equations to that of solving a system of ordinary differential equations, the solution of this (generally) nonlinear system is most often not an elementary task.

EXAMPLE 2.2.   UNIDIRECTIONAL WAVE MOTION

It was stated in Chapter 1 that the Equation (1.89), that is,

(2.13)
$$\frac{\partial v}{\partial t} + c\frac{\partial v}{\partial x} = 0,$$

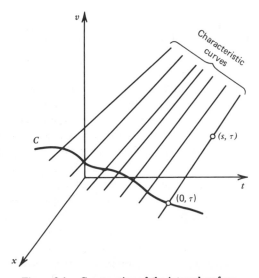

Figure 2.1.   Construction of the integral surface.

represents *unidirectional wave motion*. (The coefficient c is assumed to be constant.) We examine the initial value problem for (2.13) with the initial condition at $t = 0$

$$(2.14) \qquad\qquad v(x, 0) = F(x),$$

where $F(x)$ is a given function.

To apply the method of characteristics, we parametrize the initial curve C as follows,

$$(2.15) \qquad\qquad x = \tau, \qquad t = 0, \qquad v = F(\tau).$$

The characteristic equations [i.e., (2.9)–(2.10)] become

$$(2.16) \qquad\qquad \frac{dx}{ds} = c, \qquad \frac{dt}{ds} = 1, \qquad \frac{dv}{ds} = 0.$$

Solving (2.16) subject to (2.15) with $s = 0$ corresponding to the initial curve, gives

$$(2.17) \qquad x(s, \tau) = cs + \tau, \qquad t(s, \tau) = s, \qquad v(s, \tau) = F(\tau).$$

Using the first two equations to solve for s and τ as functions of x and t yields

$$(2.18) \qquad\qquad s = t, \qquad \tau = x - ct.$$

Substituting this result in the equation for v in (2.17), we obtain

$$(2.19) \qquad\qquad v(s, \tau) = F[\tau(x, t)] = F[x - ct].$$

Clearly, if $F(x)$ is differentiable, the solution $v(x, t) = F(x - ct)$ satisfies (2.13) as well as the initial condition (2.14). The formal solution $v(x, t) = \delta(x - ct)$ of (1.89) given in Chapter 1 follows from (2.19) since $v(x, 0) = \delta(x)$ in that problem.

If we consider the initial function $v(x, 0) = F(x)$ to represent a *wave form*, the solution $v(x, t) = F(x - ct)$ shows that a point x for which $x - ct = $ constant, will always occupy the same position on the wave form. If $c > 0$, that point x moves to the right with the speed $dx/dt = c$. Since x is a typical point, we see that the entire initial wave form $F(x)$ moves to the right without changing its shape with speed c (if $c < 0$, the direction of motion is reversed). Consequently, (2.13) characterizes unidirectional wave motion with velocity c.

The characteristic base curves for (2.13) are given by $x - ct = \tau$ in view of (2.18). For each value of the parameter τ, (2.17) shows that v is constant on the characteristic base curves. This implies that the initial data are transmitted along the characteristic base curves, in the sense that whatever the value of v is at $t = 0$ at some point x, it is retained at all points x that lie on the characteristic curve through that initial point. As a result, the initial data are transmitted at the characteristic velocity $dx/dt = c$ (to the right or to the left) as t increases from zero.

EXAMPLE 2.3.   D'ALEMBERT'S SOLUTION OF THE WAVE EQUATION

In this example we derive the solution of the initial value problem for the wave equation (2.1) which was first obtained by d'Alembert. We encountered the wave equation in Section 1.2 as a special case of the telegrapher's equation and it describes many physical phenomena as we shall see, but it is often first introduced as an equation describing the transverse displacement of a tightly stretched string (it is discussed from this point of view in Example 4.9). The constant γ^2 then represents certain physical properties of the string and $v(x, t)$ is the vertical displacement of a point x on the string at the time t. Appropriate initial conditions for the wave equation are

$$(2.20) \qquad v(x,0) = f(x); \qquad \frac{\partial v(x,0)}{\partial t} = g(x).$$

For the case of a vibrating string, these represent its initial displacement and velocity. It is then assumed that the string is sufficiently long that disturbances arising at the ends of the string do not affect its motion within the time span in which it is observed.

We solve the initial value problem (2.1) and (2.20) by using the system (2.2). The initial value of $v(x, t)$ in the system is given as in (2.20), but the initial condition for $u(x, t)$ is obtained from the first equation of (2.2) as

$$(2.21) \qquad u(x,0) = g(x) - \gamma f'(x).$$

Using (2.19) we immediately obtain

$$(2.22) \qquad u(x, t) = g(x - \gamma t) - \gamma f'(x - \gamma t)$$

since $F(x) = g(x) - \gamma f'(x)$ in this case. Substituting into the first equation in (2.2) gives

$$(2.23) \qquad \frac{\partial v}{\partial t} - \gamma \frac{\partial v}{\partial x} = g(x - \gamma t) - \gamma f'(x - \gamma t)$$

with the initial condition $v(x,0) = f(x)$.

From (2.9)–(2.10) we obtain the characteristic equations for (2.23) as

$$(2.24) \qquad \frac{dx}{ds} = -\gamma, \qquad \frac{dt}{ds} = 1, \qquad \frac{dv}{ds} = g - \gamma f',$$

with the initial curve parametrized as

$$(2.25) \qquad\qquad x = \tau, \qquad t = 0, \qquad v = f(\tau).$$

The solutions of (2.24)–(2.25) yield the system of curves

$$(2.26) \qquad x = -\gamma s + \tau, \qquad t = s, \qquad v = \int_0^s (g - \gamma f')\, d\sigma + f(\tau).$$

Now, both g and f' are functions of $x - \gamma t$, and the first two equations in (2.26) imply that $x - \gamma t = -2\gamma s + \tau$. Thus $v(s, \tau)$ can be expressed as

$$(2.27) \quad v(s, \tau) = \int_0^s [g(-2\gamma\sigma + \tau) - \gamma f'(-2\gamma\sigma + \tau)]\, d\sigma + f(\tau).$$

The change of variables $\lambda = -2\gamma\sigma + \tau$ in (2.27) yields

$$(2.28) \quad v(s, \tau) = -\frac{1}{2\gamma} \int_\tau^{-2\gamma s + \tau} [g(\lambda) - \gamma f'(\lambda)]\, d\lambda + f(\tau)$$

$$= -\frac{1}{2\gamma} \int_\tau^{-2\gamma s + \tau} g(\lambda)\, d\lambda + \frac{1}{2} f(-2\gamma s + \tau) + \frac{1}{2} f(\tau).$$

Again the first two equations in (2.26) imply that $\tau = x + \gamma t$ and $-2\gamma s + \tau = x - \gamma t$, so that in terms of the variables x and t, the solution (2.28) is

$$(2.29) \quad v = V(x, t) = \frac{1}{2}[f(x + \gamma t) + f(x - \gamma t)] + \frac{1}{2\gamma} \int_{x - \gamma t}^{x + \gamma t} g(\lambda)\, d\lambda.$$

This is *d'Alembert's solution* of the initial value problem for the wave equation. It can be verified by direct substitution that it satisfies (2.1) and the initial conditions (2.20).

Let $G(z) = \int^z g(\lambda)\, d\lambda$, that is, G is the antiderivative of g. Then (2.29) can be expressed as

$$(2.30) \qquad v(x, t) = \left[\frac{1}{2} f(x + \gamma t) + \frac{1}{2\gamma} G(x + \gamma t) \right]$$

$$+ \left[\frac{1}{2} f(x - \gamma t) - \frac{1}{2\gamma} G(x - \gamma t) \right].$$

Since $\gamma > 0$, the first bracketed term represents a wave traveling to the left with speed γ and the second bracketed term, a wave traveling to the right with speed γ. Each wave travels without change of shape. However, the presence of these wave forms in the solution of the wave equation is not always apparent because of the interference caused by the interaction of these traveling waves when they are superposed or summed as in (2.30).

EXAMPLE 2.4. AN EQUATION WITH SINGULAR COEFFICIENTS

The partial differential equation

(2.31)
$$x\frac{\partial u}{\partial x} + t\frac{\partial u}{\partial t} = cu,$$

where c is a constant, may be said to have singular coefficients since the coefficient vector $[x, t]$ of the derivative terms vanishes at $(x, t) = (0, 0)$. As the initial condition for (2.31) we take

(2.32)
$$u(x, 1) = f(x),$$

and examine the effect of the singular coefficients on the solution of the initial value problem.

The initial curve C can be parametrized as

(2.33)
$$x = \tau, \qquad t = 1, \qquad u = f(\tau)$$

and the characteristic equations are

(2.34)
$$\frac{dx}{ds} = x, \qquad \frac{dt}{ds} = t, \qquad \frac{du}{ds} = cu.$$

The solutions of (2.34) satisfying the initial conditions (2.33) at $s = 0$ are

(2.35)
$$x(s, \tau) = \tau e^s, \qquad t(s, \tau) = e^s, \qquad u(s, \tau) = f(\tau)e^{cs}.$$

Solving for s and τ in terms of x and t, gives

(2.36)
$$s = \log t; \qquad \tau = \frac{x}{t},$$

which is valid for $t > 0$. Inserting this into the equation for u, we obtain as the solution of the initial value problem (2.31)–(2.32)

(2.37)
$$u(x, t) = f\left(\frac{x}{t}\right)t^c.$$

In contrast to the two examples considered previously, this solution is not valid for all x and t but generally has a singularity at $t = 0$. For example, if f is identically constant and c is a positive integer, the solution is well behaved for all x and t, whereas if c is negative, it blows up at $t = 0$.

EXAMPLE 2.5. A CHARACTERISTIC INITIAL VALUE PROBLEM

We continue our discussion of (2.31) of the preceding example. To simplify the analysis somewhat, we let $c = 1$ in (2.31).

The first two equations in (2.34) determine the characteristic base curves in the (x, t)-plane. They may be combined to yield the single equation

$$(2.38) \qquad \frac{dx}{dt} = \frac{x}{t}.$$

This separable ordinary differential equation has the solution

$$(2.39) \qquad x = \alpha t,$$

where α is an arbitrary constant. The characteristic curves $x = \alpha t$ form a pencil of straight lines passing through the origin.

The characteristic equation for u takes the form

$$(2.40) \qquad \frac{du}{dt} = \frac{u}{t}$$

and has the solution

$$(2.41) \qquad u = \beta t$$

where β is an arbitrary constant.

Thus a characteristic initial value problem is obtained by specifying that $u = \beta t$ on the line $x = \alpha t$ with fixed α and β. If we parametrize the initial curve C as

$$(2.42) \qquad x = \tau, \qquad t = \frac{\tau}{\alpha}, \qquad u = \left(\frac{\beta}{\alpha}\right)\tau,$$

the characteristic equations (2.34) (with $c = 1$) and the initial conditions $x(0, \tau) = \tau$, $t(0, \tau) = \tau/\alpha$, $u(0, \tau) = (\beta/\alpha)\tau$ yield

$$(2.43) \qquad x(s, \tau) = \tau e^s, \qquad t(s, \tau) = \left(\frac{\tau}{\alpha}\right)e^s, \qquad u(s, \tau) = \left(\frac{\beta}{\alpha}\right)\tau e^s.$$

The first two equations in (2.43) cannot be uniquely inverted to give s and τ as functions of x and t. We do note, however, that $u(x, t) = (\beta/\alpha)x$ and $u = \beta t$ are both solutions of the initial value problem so that the problem does not have a unique solution. Furthermore, it follows from (2.41), since the constant β can vary from one characteristic to the next, that

$$(2.44) \qquad u(x, t) = g\left(\frac{x}{t}\right)t$$

is a general solution of (2.31) (with $c = 1$) for arbitrary differentiable $g(z)$. As long as $g(z)|_{z=\alpha} = \beta$, (2.44) is a solution of the foregoing characteristic initial value problem. Consequently, this problem has infinitely many solutions.

However, if initial conditions for u are given on the characteristic base curve $x = \alpha t$ such that u cannot be expressed as $u = \beta t$, the problem does not have a solution, for (2.31) (with $c = 1$) imposes the form $u = \beta t$ on all solutions evaluated on $x = \alpha t$. Technically, this does not constitute a characteristic initial value problem since the initial curve in (x, t, u)-space is not characteristic.

2.3   QUASILINEAR FIRST ORDER PARTIAL DIFFERENTIAL EQUATIONS

A first order partial differential equation of the form

$$(2.45) \qquad a(x, t, u)\frac{\partial u}{\partial x} + b(x, t, u)\frac{\partial u}{\partial t} = c(x, t, u),$$

is said to be quasilinear since it is linear in the derivative terms but may contain nonlinear expressions of the form $u\partial u/\partial t$ or u^2. Such equations occur in a variety of nonlinear wave propagation problems and in other contexts but we will emphasize the wave propagation aspect in the examples. Proceeding as in Section 2.2, we interpret (2.45) geometrically in order to construct a solution.

We assume that a solution $u = u(x, t)$ of (2.45) can be found and examine the properties of the solution as implied by (2.45). The solution $u = u(x, t)$ is called an *integral surface*, and we express it in implicit form as $F(x, t, u) = u(x, t) - u = 0$. The gradient vector $\nabla F = [\partial u/\partial x, \partial u/\partial t, -1]$ is normal to the integral surface $F(x, t, u) = 0$. By transposing the term c in (2.45) we can express the resulting equation as a scalar or dot product,

$$(2.46) \qquad a\frac{\partial u}{\partial x} + b\frac{\partial u}{\partial t} - c = [a, b, c]\cdot\left[\frac{\partial u}{\partial x}, \frac{\partial u}{\partial t}, -1\right] = 0.$$

The vanishing of the dot product of the vector $[a, b, c]$ and the gradient vector ∇F implies that these vectors are orthogonal. Accordingly, the vector $[a, b, c]$ lies in the tangent plane of the integral surface $u = u(x, t)$ at each point in the (x, t, u)-space where ∇F is defined and nonzero.

At each point (x, t, u), the vector $[a, b, c]$ determines a direction that is called the *characteristic direction*. As a result, the vector $[a, b, c]$ determines a characteristic direction field in (x, t, u)-space, and we can construct a family of curves that have the characteristic direction at each point. If the parametric form of these curves is $x = x(s)$, $t = t(s)$, and $u = u(s)$, we must have

$$(2.47) \qquad \frac{dx}{ds} = a(x, t, u), \qquad \frac{dt}{ds} = b(x, t, u), \qquad \frac{du}{ds} = c(x, t, u),$$

since $[dx/ds, dt/ds, du/ds]$ is the tangent vector along the curves. The *characteristic equations* (2.47) differ from those in the linear case, since the equations for x and t are not, in general, uncoupled from the equation for u. The solutions of (2.47) are called the *characteristic curves* of the quasilinear equation (2.45).

Assuming a, b, and c are sufficiently smooth and do not all vanish at the same point, the theory of ordinary differential equations guarantees that a unique characteristic curve passes through each point (x_0, t_0, u_0). The *initial value problem* for (2.45) requires that $u(x, t)$ be specified on a given curve in (x, t)-space. This determines a curve C in (x, t, u)-space referred to as the initial curve. To solve this initial value problem we pass a characteristic curve through each point of the initial curve C. If these curves generate a surface, this integral surface is the solution of the initial value problem. We now state and outline the proof of a theorem that gives conditions under which a unique solution of the initial value problem for (2.45) can be obtained.

Let, a, b, and c in (2.45) *have continuous partial derivatives in all three variables. Suppose the initial curve C given parametrically as* $x = x(\tau)$, $t = t(\tau)$, *and* $u = u(\tau)$ *has a continuous tangent vector and that*

$$(2.48) \quad \Delta(\tau) \equiv \frac{dt}{d\tau} a[x(\tau), t(\tau), u(\tau)] - \frac{dx}{d\tau} b[x(\tau), t(\tau), u(\tau)] \neq 0,$$

on C. Then, there exists one and only one solution $u = u(x, t)$, *defined in some neighborhood of the initial curve C, that satisfies* (2.45) *and the initial condition* $u[x(\tau), t(\tau)] = u(\tau)$.

The proof of this theorem proceeds along the following lines. The characteristic system (2.47) with initial conditions at $s = 0$ given as $x = x(\tau)$, $t = t(\tau)$, and $u = u(\tau)$ has a unique solution of the form

$$(2.49) \qquad x = x(s, \tau), \qquad t = t(s, \tau), \qquad u = u(s, \tau),$$

with continuous derivatives in s and τ, and with

$$(2.50) \quad x(0, \tau) = x(\tau), \qquad t(0, \tau) = t(\tau), \qquad u(0, \tau) = u(\tau).$$

This follows from the existence and uniqueness theory for ordinary differential equations. The Jacobian of the transformation $x = x(s, \tau)$, $t = t(s, \tau)$ at $s = 0$ is

$$(2.51) \quad \frac{\partial(x, t)}{\partial(s, \tau)}\bigg|_{s=0} = \begin{vmatrix} \dfrac{\partial x}{\partial s} & \dfrac{\partial x}{\partial \tau} \\[2mm] \dfrac{\partial t}{\partial s} & \dfrac{\partial t}{\partial \tau} \end{vmatrix}_{s=0} = \frac{dt}{d\tau} a - \frac{dx}{d\tau} b\bigg|_{s=0} = \Delta(\tau)$$

and it is nonzero in view of the assumption (2.48). By the continuity assumption, the Jacobian does not vanish in a neighborhood of the initial curve.

Therefore, the implicit function theorem guarantees that we can solve for s and τ as functions of x and t near the initial curve.

Then

$$(2.52) \qquad u(s, \tau) = u[s(x, t), \tau(x, t)] = U(x, t),$$

is a solution of (2.45). This is readily seen on substituting $u[s(x, t), \tau(x, t)]$ into (2.45), using the chain rule and the characteristic equations (2.47). The uniqueness of the solution follows since any two integral surfaces that contain the same initial curve must coincide along all the characteristic curves passing through the initial curve. This is a consequence of the uniqueness theorem for the initial value problem for (2.47). This completes our proof.

The condition (2.48) essentially means that the initial curve C is nonchararacteristic or is not the envelope of characteristic curves. If the initial curve is characteristic, solutions exist but are not unique. If the initial curve is the envelope of characteristic curves (see the appendix), the integral surface may not be differentiable along the initial curve. If the initial curve is noncharacteristic everywhere except at a discrete set of points, special problems with the solution arise near these points. We will now consider a single example of a quasilinear equation, the simplest case of unidirectional nonlinear wave motion. The initial value problem for this equation is discussed in great detail and various features of nonlinear wave motion, which distinguish it from linear wave motion, are emphasized.

EXAMPLE 2.6.   UNIDIRECTIONAL NONLINEAR WAVE MOTION

The simplest quasilinear equation characterizing one-directional nonlinear wave motion is

$$(2.53) \qquad \frac{\partial u}{\partial t} + u \frac{\partial u}{\partial x} = 0.$$

In this example we consider the initial value problem for (2.53) with

$$(2.54) \qquad u(x, 0) = f(x).$$

where $f(x)$ is a given smooth function. First a general discussion of the solution of (2.53)–(2.54) is given and then three specific choices for $f(x)$ are considered.

To solve the initial value problem we parametrize the initial curve as

$$(2.55) \qquad x = \tau, \qquad t = 0, \qquad u = f(\tau).$$

Applying the condition (2.48) we obtain

$$(2.56) \qquad \Delta(\tau) = -1,$$

so that $\Delta(\tau)$ is nonzero along the entire initial curve. The characteristic equations are

$$(2.57) \qquad \frac{dx}{ds} = u, \qquad \frac{dt}{ds} = 1, \qquad \frac{du}{ds} = 0,$$

with initial conditions at $s = 0$ given by (2.55). Denoting the solutions by $x(s, \tau)$, $t(s, \tau)$, and $u(s, \tau)$, we find that $du/ds = 0$ implies $u(s, \tau)$ is constant along the characteristic curves. Therefore,

$$(2.58) \qquad u(s, \tau) = u(0, \tau) = f(\tau),$$

on using (2.55). Inserting the expression for u in the equation $dx/ds = u$ we can immediately solve for $x(s, \tau)$ and $t(s, \tau)$ to obtain

$$(2.59) \qquad x(s, \tau) = \tau + sf(\tau), \qquad t(s, \tau) = s.$$

If the conditions of the theorem are satisfied, this system can be inverted near $s = 0$ to give $s = t$ and $\tau = \tau(x, t)$. Then (2.58) yields the solution

$$(2.60) \qquad u = f[\tau(x, t)] = U(x, t).$$

Since $s = t$ and $\tau = x - sf(\tau) = x - tu$ in view of (2.58), the solution of (2.53) can also be given in implicit form as

$$(2.61) \qquad u = f[x - tu].$$

We consider the solution $u = u(x, t)$ of (2.53) to represent a wave. The *wave form* at the time t is given by the curve $u = u(x, t)$ in the (x, u)-plane with t as a parameter and $u = f(x)$ as the initial wave form. The (numerical) value of $u(x, t)$ gives the height of the wave at the point x and the time t. To determine the motion of the wave, we must find the velocity dx/dt of each point on the wave. The first two characteristic equations in (2.57) imply that

$$(2.62) \qquad \frac{dx}{dt} = u.$$

The greater the amplitude $|u(x, t)|$ of the wave, the greater the speed of the corresponding point x on the wave. [This contrasts with the situation for linear wave motion determined by equation (2.13) where $dx/dt = c = $ constant, so that each point and, consequently, the entire wave form moves with a single speed.] If $u(x, t) > 0$, the point x moves to the right, if $u(x, t) = 0$, it remains fixed, whereas if $u(x, t) < 0$, it moves to the left. Thus if $u(x, t)$ takes on both positive and negative values, individual points maintain a fixed direction of motion but different portions of the wave form can move either to the right or to the left. Therefore, the wave motion need not be totally unidirectional as is the case for linear wave motion. Assuming $u > 0$, points x where $u(x, t)$ has larger values and the wave is higher move more rapidly to the right than points x where $u(x, t)$ has smaller values and the wave is lower. If, initially, there are higher portions of the wave form located to the left or the rear of lower portions, the higher points may eventually catch up with and pass the lower points, at which time the wave is said to *break*. At the time t of breaking, the function $u(x, t)$ becomes multivalued and is no longer a valid solution of (2.53). Similar difficulties can occur if $u(x, t)$ is not restricted to be positive. The geometric aspects of the breaking process are demonstrated in Example 2.7.

In many physical processes described by the quasilinear wave equation (2.53), the function $u(x, t)$ represents a physical quantity such as density, which is intrinsically expected to be single valued. Thus when the wave $u(x, t)$ breaks and becomes multivalued, the equation (2.53) describing the physical process is no longer an acceptable model for the physical process. In general, this means that certain higher derivative terms that were neglected in the derivation of the quasilinear equation become significant and must be retained. It will be shown in Section 9.6 for a model equation, known as *Burgers' equation*, how the addition of a second derivative term to (2.53) can yield solutions valid for all time. A similar situation has already been observed in our discussion of the diffusion equation (1.21) in Chapter 1. With delta function initial data, the solution of the diffusion equation given in (1.28) is a smooth function for all x and for $t > 0$. However, if the diffusion coefficient D is equated to zero and the first order unidirectional wave equation (1.89) results, the solution (1.90) remains sharply singular for all $t > 0$. This indicates that the inclusion of higher derivative terms tends to smooth out the solution.

An alternative method for dealing with the breaking phenomenon is to introduce a (so-called) *shock wave* discontinuity that extends the validity of the solution beyond the breaking time. Since the choice of the correct shock wave depends heavily on the physics of the problem, this method is not discussed here (see, however, Section 9.6).

It must be emphasized that as far as the first order equation (2.53) is concerned, the breaking or nonbreaking of the wave is purely a function of the initial data and cannot be avoided. It is, therefore, of interest to determine the time when the wave $u(x, t)$ first begins to break. We consider two methods for doing so.

First, we use implicit differentiation to determine the slope of the wave form $u = u(x, t) = f(x - tu)$ at the time t. The slope $\partial u/\partial x$ is readily found to be

(2.63)
$$\frac{\partial u}{\partial x} = \frac{f'(x - tu)}{1 + tf'(x - tu)}.$$

From (2.59) we see that the characteristic base curves for (2.53) are the family of straight lines $x - tf(\tau) = \tau$, where τ is a parameter. When the denominator on the right side of (2.63) [i.e., $1 + tf'(x - tu) = 1 + tf'(\tau)$] first vanishes, the slope $\partial u / \partial x$ becomes infinite and the wave begins to break. Since the denominator vanishes when

$$(2.64) \qquad\qquad t = -\frac{1}{f'(\tau)},$$

the *breaking time* is determined from the value of τ at which t has its smallest nonnegative value.

The second method for determining the breaking time considers the characteristic curves (2.58)–(2.59) of the equation (2.53). We have shown that $u = f(\tau)$ on the characteristic base curves

$$(2.65) \qquad\qquad x - tf(\tau) = \tau,$$

where τ is a parameter. If for two or more values of τ the straight lines (2.65) intersect, u will in general be multivalued at the intersection point, for $u(x, t)$ must equal $f(\tau)$ at that point, and $f(\tau)$ may have different values on each of the lines (2.65) that intersect at the point. To find the possible intersection points we must determine the envelope of the family of straight lines (2.65). As shown in the appendix, if $F(x, t, \tau) = 0$ is a one-parameter family of curves, the envelope of this family is obtained by eliminating the parameter τ from the system $F(x, t, \tau) = 0$ and $F_\tau(x, t, \tau) = 0$. In our case, $F = x - tf(\tau) - \tau = 0$ and $\partial F / \partial \tau = -tf'(\tau) - 1 = 0$. The equation $\partial F / \partial \tau = 0$ shows that the times at which the characteristic curves touch the envelope are given by $t = -1/f'(\tau)$, so that the initial breaking time is in agreement with that determined from the first method.

We note that if $f'(\tau) > 0$ for all τ, the right side of (2.64) must always be negative, so that the wave never breaks. Geometrically, this means that the initial wave form $u = f(x)$ is a monotonically increasing function of x. Since the larger the values of u, the faster the corresponding points on the wave move, no point x on the wave can overtake any other point and no breaking occurs. However, if $f'(\tau) < 0$ for all τ, the initial wave form is a monotonically decreasing function. Thus points in the rear portion of the wave move faster than points in the front portion and eventually overtake them. When this happens, the wave breaks. If $f'(\tau)$ takes on both positive and negative values, wave breaking generally occurs.

The foregoing results concerning the breaking of the wave $u = f(x - tu)$ can be interpreted from a different point of view. From (2.64) and (2.65) we conclude that the breaking time is determined from the point on the curve $x = \tau - f(\tau)/f'(\tau)$, $t = -1/f'(\tau)$, $u = f(\tau)$ where t has its smallest value. As is easily verified, this curve has the property that $\Delta(\tau)$ [as defined in (2.48)] vanishes on it. Thus if this curve were chosen as an initial curve for (2.53), the theorem stated for the general case would not guarantee that this initial value problem has a unique (smooth) solution. In fact, (2.63) implies that the solution would not be differentiable along the initial curve. We infer from the above that even when the initial data guarantee a unique solution for the given quasilinear equation, this solution breaks down along a curve where the condition (2.48) (i.e., $\Delta(\tau) \neq 0$) is violated.

The breaking curve determined is not a characteristic curve for (2.53) since u is not constant on it. For the sake of completeness, we show that if the initial curve for (2.53) is characteristic, the initial value problem has infinitely many solutions. Let $u = a = $ constant on the line $x = ct$. [This curve is a characteristic for (2.53) since with $t = s$, $x = cs$ the characteristic equations (2.57) are satisfied.] Then, $u = f[x - tu]$ is a solution of the characteristic initial value problem as long as $f(0) = a$ but f is an otherwise arbitrary but smooth function.

We continue our discussion of (2.53) in the following example and select three specific initial wave forms and examine the resulting wave motions.

EXAMPLE 2.7.   UNIDIRECTIONAL WAVE MOTION: SPECIFIC INITIAL DATA

In this example three special choices of initial values for (2.53) (i.e., $\partial u / \partial t + u \partial u / \partial x = 0$) are given and the corresponding solutions analyzed.

   1.   The initial value is

$$(2.66) \qquad u(x,0) = f(x) = -x.$$

Since $f'(x) = -1$, we expect (in view of the foregoing discussion) that the wave will break and the solution $u(x, t)$ will become multivalued. In fact (since $f(\tau) = -\tau$), (2.64) gives $t = 1$ as the breaking time.

Now, (2.61) gives the implicit form the solution as

$$(2.67) \qquad u = f(x - tu) = -[x - tu]$$

from which it follows that

$$(2.68) \qquad u(x, t) = \frac{x}{t - 1}.$$

We easily verify that (2.68) satisfies (2.53) and the initial condition (2.66), as well as the fact that it blows up at the time $t = 1$.

The motion of the wave $u(x, t) = x/(t - 1)$ is indicated in Figure 2.2. As t increases, the wave form $u(x, 0) = -x$ executes a clockwise rotation around the origin in the (x, u)-plane. Since $u = 0$ at $x = 0$, that point (i.e., $x = 0$) is stationary. Also, $|u|$ increases linearly with $|x|$ and points x further away from the origin have a linearly increasing velocity, yielding the effect indicated in the figure. At $t = 1$ the wave form $u(x, t)$ coincides with the u-axis and becomes infinitely multivalued.

The breakdown of the solution and its multivaluedness at $t = 1$ may also be determined by considering the characteristic base curves, which for this initial

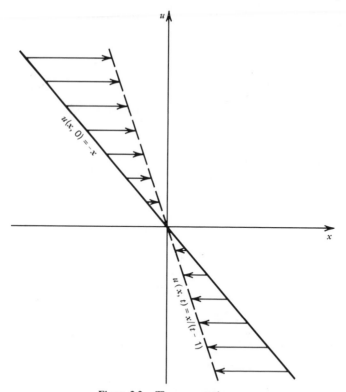

Figure 2.2.   The wave motion.

value problem are given by

$$(2.69) \qquad\qquad x = (1 - t)\tau,$$

with τ as a parameter. When $t = 1$, these characteristic lines all intersect at the point $(x, t) = (0, 1)$. Since each of the lines carries a different value of τ, and $u(x, t)$ is constant on each of these lines, it must become infinitely multivalued at $t = 1$. This is indicated in Figure 2.3.

If the initial condition (2.66) is replaced by $f(x) = x$, we have $f(\tau) = \tau$ and $f'(\tau) = 1$. Then, according to (2.64), the wave never breaks. This fact is borne out by the solution $u(x, t) = x/(t + 1)$ of this problem, which is defined for all $t > 0$.

2.   The initial condition is

$$(2.70) \qquad\qquad u(x,0) = f(x) = 1 - x^2.$$

Using (2.61) gives the implicit form of the solution as

$$(2.71) \qquad\qquad u = f(x - tu) = 1 - (x - tu)^2.$$

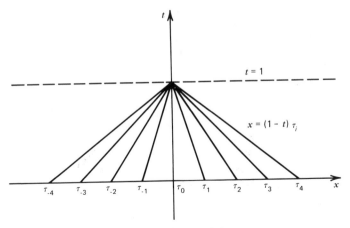

Figure 2.3. The characteristic curves.

Solving this quadratic equation for u gives two branches

(2.72)
$$u(x, t) = \frac{x}{t} - \frac{1 \pm \sqrt{1 + 4t(t - x)}}{2t^2}.$$

To satisfy the initial condition (2.70) we must choose the minus sign in (2.72) and obtain

(2.73)
$$u(x, t) = \frac{x}{t} - \frac{1 - \sqrt{1 + 4t(t - x)}}{2t^2}.$$

Differentiating (2.73) with respect to x, we find that the slope $\partial u / \partial x$ becomes infinite when the radical in (2.73) vanishes; that is, when

(2.74)
$$x = t + \frac{1}{4t}.$$

As long as $x < t + 1/4t$, the radical in (2.73) is real valued and the solution $u(x, t)$ is well defined. An easy calculation shows that $x < t + 1/4t$ for all $x < 1$ since $t > 0$. At $x = 1$, the radical vanishes when $t = \frac{1}{2}$ and as x increases towards $+\infty$, the time t at which it vanishes decreases to zero. Thus if the initial condition is given over the infinite interval $-\infty < x < \infty$, the wave begins to break immediately at $x = +\infty$.

The characteristic base curves for this problem are [see (2.65)]

(2.75)
$$x - t(1 - \tau^2) = \tau,$$

where τ is a parameter. The solution $u(x, t)$ is constant along the characteristics. The envelope of this family of characteristic curves is readily found to be

$$(2.76) \qquad x - t - \frac{1}{4t} = 0$$

which is identical with (2.74). Since two neighboring curves of the characteristic family intersect on the envelope corresponding to different values of τ, the solution $u(x, t)$ becomes double valued at the envelope. In fact, the radical term in the solution (2.72) vanishes on the curve (2.76) in the (x, t)-plane, so that $u(x, t)$ splits into the two branches given in (2.72) along the envelope (2.76). An indication of how the wave $u(x, t)$ propagates is given in Figure 2.4.

If the initial data (2.70) are restricted to the interval $-\infty < x \leqslant a$ where $a > 1$, the foregoing discussion implies that the resulting wave will not break immediately. However, some care must be exercised in determining the breaking time. On applying the formula (2.64) to this problem, we obtain $t = 1/2\tau$, so that the wave breaks when $t = 1/2a$. Yet the solution (2.73) with $x \leqslant a$ and $t > 0$ predicts the earlier breaking time $t = \frac{1}{2}(a - \sqrt{a^2 - 1})$. This discrepancy may be resolved by observing that for this problem, (2.73) is valid only in the region $x \leqslant a + (1 - a^2)t$, $t > 0$. That is, the region is bounded by the characteristic base curve $x = a + (1 - a^2)t$ that passes through the point $(x, t) = (a, 0)$. In that region the breaking time is easily found to be $t = 1/2a$.

3.   The initial condition is

$$(2.77) \qquad u(x, 0) = f(x) = \sin x.$$

The implicit form of the solution is given as

$$(2.78) \qquad u = f(x - tu) = \sin(x - tu),$$

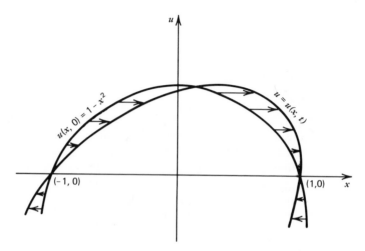

Figure 2.4.   The wave motion.

and the characteristic curves are

$$(2.79) \qquad\qquad x - t\sin\tau - \tau = 0.$$

Using (2.64) we have

$$(2.80) \qquad\qquad t = -\frac{1}{\cos\tau},$$

so that the (first) breaking time occurs when $t = 1$, since when

$$(2.81) \qquad\qquad \tau = (2n+1)\pi, \qquad n = 0, \pm1, \pm2, \dots$$

we have $\cos\tau = -1$. At all other values of τ, t as given by (2.80) either exceeds unity or is negative.

A qualitative picture of the wave motion is indicated in Figure 2.5 where the interval $0 \leqslant x \leqslant 2\pi$ is considered. The critical nature of the point $(x, u) = (\pi, 0)$ is apparent since the wave form rotates in a clockwise motion near that point.

This concludes our discussion of first order quasilinear partial differential equations. The equation

$$(2.82) \qquad\qquad \frac{\partial u}{\partial t} + c(u)\frac{\partial u}{\partial x} = 0,$$

where $c(u)$ is an arbitrary function of u is the quasilinear equivalent of the linear equation (2.13). Certain properties of the solutions of this equation will be brought out in the exercises. If we multiply across by $c'(u)$ in (2.82), an

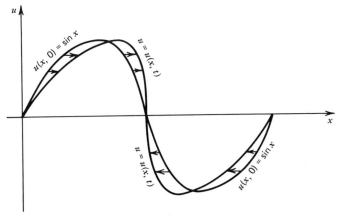

Figure 2.5. The wave motion.

equation for $c(u)$ is obtained that has the form (2.53) but with u replaced by $c(u)$. Thus we do not expect the behavior of the solutions of (2.82) to be qualitatively much different from those of (2.53).

2.4. NONLINEAR FIRST ORDER PARTIAL DIFFERENTIAL EQUATIONS

In this section the method of characteristics is extended to deal with nonlinear first order partial differential equations in two independent variables. Although solutions of linear and quasilinear equations can be obtained as special cases of the general method developed below, these equations are of sufficient special interest to merit the separate discussion accorded them. Following a general discussion of the method of characteristics, we consider several examples among which the classical eiconal equation of geometrical optics is analyzed in some detail.

In its most general form the first order partial differential equation can be written as

$$(2.83) \qquad\qquad F(x, t, u, u_x, u_t) = 0,$$

where $u_x = \partial u/\partial x$ and $u_t = \partial u/\partial t$. Let $p = u_x$ and $q = u_t$ and consider an integral surface $u = u(x, t)$ satisfying (2.83). Its normal vector has the form $[u_x, u_t, -1] = [p, q, -1]$, and (2.83) requires that at the point (x, t, u) the components p and q of the normal vector satisfy the equation

$$(2.84) \qquad\qquad F(x, t, u, p, q) = 0.$$

Since an integral surface $u = u(x, t)$ of (2.83) is not known a priori, we think of (2.84) as characterizing a collection of admissible solutions or integral surfaces $u = u(x, t)$. That is, their normal vectors at the points (x, t, u) must satisfy (2.84). Each normal vector determines a tangent plane to the surface, and (2.84) is seen to generate a one-parameter family of tangent planes (to possible integral surfaces) at each point in (x, t, u)-space.

For example, if (2.83) has the form $u_x u_t - 1 = 0$, then $F = pq - 1 = 0$ and $q = 1/p$. As p ranges through all real values, $q = 1/p$ determines a one-parameter family of normal vectors $[p, q, -1] = [p, 1/p, -1]$ at each point (x, t, u). Similarly, if $F = xu + tu_t^2 - (\sin x)u_x + 1 = 0$ then at (x, t, u) we have $F(x, t, u, p, q) = xu + tq^2 - (\sin x)p + 1 = 0$ which yields a set of values $q = q(p)$. Note that q cannot always be specified as a single valued function of p. However, we shall always assume that one branch of the possible set of solutions $q = q(p)$ has been chosen.

In general, the (tangent) planes determined by p and q envelop a cone known as the *Monge cone*. Then, if $u = u(x, t)$ is a solution of (2.83), it must be tangent to a Monge cone at each point (x, t, u) on the surface. The intersection of the Monge cone with the surface determines a field of directions on the surface known as the *characteristic directions*, as shown in Figure 2.6.

To find the characteristic directions we proceed as follows. The planes determined by the set of normal directions $[p, q, -1]$ at a point (x_0, t_0, u_0) satisfy the equations

$$(2.85) \quad H[x, t, u, p, q] = u - u_0 - p(x - x_0) - q(p)(t - t_0) = 0.$$

It is assumed in (2.85) that $F[x_0, t_0, u_0, p, q] = 0$ is solved for q as a function of p and one of the set of possible solutions expressed as $q = q(p)$ is selected. The envelope of the planes (2.85) (where p is a parameter) is determined by eliminating p from (2.85) and the equation

$$(2.86) \qquad \frac{\partial H}{\partial p} = -(x - x_0) - \frac{dq}{dp}(t - t_0) = 0.$$

If we solve for $p = p(x, t)$ from (2.86) and substitute the result in (2.85) we obtain the equation of the Monge cone.

For example, if $F = pq - 1 = 0$, then $q = 1/p$ and $dq/dp = -1/p^2$. From (2.86) we obtain $q^2 = 1/p^2 = (x - x_0)/(t - t_0)$. Substituting in (2.85) gives the equation of the Monge cone as $(u - u_0)^2 = 4(x - x_0)(t - t_0)$.

Continuing our discussion, we observe that $F[x_0, t_0, u_0, p, q] = 0$ implies

$$(2.87) \qquad \frac{dF}{dp} = \frac{\partial F}{\partial p} + \frac{dq}{dp}\frac{\partial F}{\partial q} = F_p + q'(p)F_q = 0.$$

Since $q'(p) = -F_p/F_q$, substitution in (2.86) yields

$$(2.88) \qquad \frac{x - x_0}{F_p} = \frac{t - t_0}{F_q}.$$

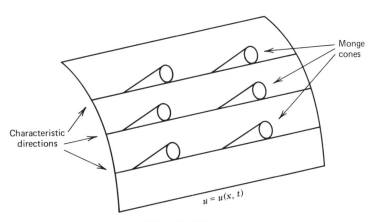

Figure 2.6. The Monge cones.

Using (2.88) in (2.85) gives

(2.89) $$\frac{u - u_0}{t - t_0} = p\frac{x - x_0}{t - t_0} + q = p\frac{F_p}{F_q} + q = \frac{pF_p + qF_q}{F_q}.$$

Combining (2.88) and (2.89) yields

(2.90) $$\frac{u - u_0}{pF_p + qF_p} = \frac{x - x_0}{F_p} = \frac{t - t_0}{F_q}.$$

Recalling that (x, t, u) are the running variables in the tangent planes (2.85) and that the denominators in (2.90) are constants evaluated at (x_0, t_0, u_0) on the integral surface, we see that a fixed direction on each tangent plane (for particular values of p and q) is determined by (2.90). The direction is determined by the vector $[F_p, F_q, pF_p + qF_q]$. As p and q range through all their values, these directions, known as the *characteristic directions*, determine the family of lines (2.90) that generate the Monge cone at (x_0, t_0, u_0).

We now construct equations for curves $x = x(s)$, $t = t(s)$, and $u = u(s)$ that have a characteristic direction at each point, (i.e., the direction of the vector $[F_p, F_q, pF_p + qF_q]$). The ordinary differential equations for these curves are given as

(2.91) $$\frac{dx}{ds} = F_p$$

(2.92) $$\frac{dt}{ds} = F_q$$

(2.93) $$\frac{du}{ds} = pF_p + qF_q.$$

Since p and q can vary from point to point, the curves (2.91)–(2.93) must be chosen such that they lie on a single surface $u = u(x, t)$. To achieve this result, we assume that a surface $u = u(x, t)$ is given so that p and q (i.e., u_x and u_t) are known. Then we determine the values that p and q must have along curves $x = x(s)$, $t = t(s)$, $u = u[x(s), t(s)]$ that lie on that surface. Since we must also have $p = p[x(s), t(s)]$ and $q = q[x(s), t(s)]$ on that surface, we obtain

(2.94) $$\frac{dp}{ds} = p_x x'(s) + p_t t'(s) = p_x F_p + p_t F_q$$

(2.95) $$\frac{dq}{ds} = q_x x'(s) + q_t t'(s) = q_x F_p + q_t F_q$$

on using (2.91)–(2.92).

Further, if $u = u(x, t)$ is a solution of (2.83), we have $F[x, t, u, p, q] = 0$, which implies

$$(2.96) \qquad \frac{dF}{dx} = F_x + F_u p + F_p p_x + F_q q_x = 0$$

$$(2.97) \qquad \frac{dF}{dt} = F_t + F_u q + F_p p_t + F_q q_t = 0.$$

But $u_x = p$ and $u_t = q$ imply that $u_{xt} = p_t = u_{tx} = q_x$. Substituting $p_t = q_x$ into (2.96)–(2.97) and inserting the result in (2.94)–(2.95) yields

$$(2.98) \qquad \frac{dp}{ds} = -F_x - F_u p$$

$$(2.99) \qquad \frac{dq}{ds} = -F_t - F_u q.$$

The five equations (2.91)–(2.93) and (2.98)–(2.99) now constitute a completely self-contained system of equations for the functions $x(s)$, $t(s)$, $u(s)$, $p(s)$, and $q(s)$. [The integral surface $u = u(x, t)$ no longer needs to be given a priori in order to specify the values of p and q.] They are known as the *characteristic equations* for the differential equation (2.83).

The *initial value problem* for (2.83) requires that the integral surface $u = u(x, t)$ contain a curve C, the initial curve, which we give parametrically as

$$(2.100) \qquad x = x(\tau), \qquad t = t(\tau), \qquad u = u(\tau).$$

In terms of the functions $x(s)$, $t(s)$, and $u(s)$, these are taken to correspond to initial values given at $s = 0$, as was done in the previous sections. However, in contrast to the situation for linear and quasilinear equations, we must now also require initial values for $p(s)$ and $q(s)$ if we hope to obtain a unique solution of the characteristic equations and, consequently, of the initial value problem for (2.83). The initial values $p(\tau)$ and $q(\tau)$ on the curve C cannot be arbitrary, since p and q must be components of the normal vector to the integral surface $u = u(x, t)$.

If we let $p = p(\tau)$ and $q = q(\tau)$ be the initial values of p and q on the curve C, these values must be determined from the equation

$$(2.101) \qquad F[x(\tau), t(\tau), u(\tau), p(\tau), q(\tau)] = 0,$$

and the so-called *strip condition*

$$(2.102) \qquad \frac{du(\tau)}{d\tau} = p(\tau) \frac{dx(\tau)}{d\tau} + q(\tau) \frac{dt(\tau)}{d\tau}.$$

These equations follow from the fact that $p(\tau)$ and $q(\tau)$ must be components of the normal vectors to the surface $u = u(x, t)$ evaluated along the initial curve C. The data $p(\tau)$ and $q(\tau)$ are determined by solving (2.101) and (2.102) simultaneously.

Given the characteristic equations and initial conditions for x, t, u, p, and q, together with the requirement (see the appendix)

$$(2.103) \qquad\qquad \frac{dx}{d\tau}F_q - \frac{dt}{d\tau}F_p \neq 0,$$

on the initial curve C, we can obtain a unique solution of the initial value problem for (2.83) in a neighborhood of the initial curve C. We note that the assignment of x, t, and u together with p and q on the curve C means that we are specifying an *initial strip*, that is, a space curve $x(\tau)$, $t(\tau)$, $u(\tau)$ together with a family of tangent planes at each point having a normal vector with components $[p(\tau), q(\tau), -1]$, as shown in Figure 2.7. This explains why (2.102) is called the strip condition.

The solutions $x(s, \tau)$, $t(s, \tau)$, and $u(s, \tau)$ of the characteristic equations for fixed τ determine a space curve. The functions $p(s, \tau)$ and $q(s, \tau)$ determine a tangent plane with normal vector $[p, q - 1]$ at each point of the space curve. The combination of a curve and its tangent planes is called a *characteristic strip*. The integral surface $u = u(x, t)$ is constructed by piecing together these characteristic strips to form a smooth surface. Its equation is found by solving for s and τ as functions of x and t from $x = x(s, \tau)$ and $t = t(s, \tau)$ and inserting the results in $u = u(s, \tau)$.

In the following examples we demonstrate the use of the method of characteristics for solving nonlinear equations by applying it to several specific initial value problems.

EXAMPLE 2.8. A NONLINEAR EQUATION

We consider the equation

$$(2.104) \qquad\qquad F(x, t, u, u_x, u_t) = u_x u_t - 1 = 0$$

with the initial condition

$$(2.105) \qquad\qquad u(x, 0) = x.$$

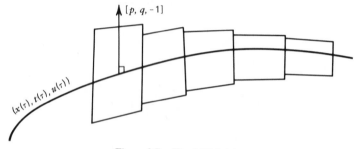

Figure 2.7. The initial strip.

In parametric form the initial conditions are given as

$$(2.106) \qquad x = \tau, \qquad t = 0, \qquad u = \tau.$$

The equations (2.101)–(2.102) for determining $p(\tau)$ and $q(\tau)$ are

$$(2.107) \qquad \begin{cases} p(\tau)q(\tau) - 1 = 0 \\ \qquad\qquad 1 = p(\tau), \end{cases}$$

and, therefore,

$$(2.108) \qquad p(\tau) = 1, \qquad q(\tau) = 1.$$

Also, on the initial curve we have $x'(\tau)F_p - t'(\tau)F_q = p = 1$ so that (2.103) is satisfied.

The characteristic equations are found to be

$$(2.109) \qquad \frac{dx}{ds} = q; \qquad \frac{dt}{ds} = p; \qquad \frac{du}{ds} = 2pq; \qquad \frac{dp}{ds} = 0; \qquad \frac{dq}{ds} = 0.$$

The initial curve (2.106) corresponds to $s = 0$. Since p and q are constant on the characteristics we have

$$(2.110) \qquad p(s, \tau) = p(0, \tau) = 1; \qquad q(s, \tau) = q(0, \tau) = 1.$$

Solving for x, t, and u and using (2.106) and (2.110) gives

$$(2.111) \qquad x(s, \tau) = s + \tau, \qquad t(s, \tau) = s, \qquad u(s, \tau) = 2s + \tau.$$

We invert the first two equations to determine s and τ as functions of x and t and obtain

$$(2.112) \qquad s = t, \qquad \tau = x - t.$$

Inserting this result into the third equation in (2.111) gives

$$(2.113) \qquad u(x, t) = x + t,$$

which clearly satisfies (2.104) and the initial condition (2.105).

EXAMPLE 2.9. A NONLINEAR WAVE EQUATION

We consider the equation

$$(2.114) \qquad F(x, t, u, u_x, u_t) = u_t + u_x^2 = 0,$$

with the initial condition

$$(2.115) \qquad\qquad u(x,0) = ax,$$

where a is a constant. The equation (2.114) is of the general form $u_t + c(u, u_x)u_x = 0$, of which two special cases $c = $ constant and $c = u$ were considered in the previous sections. We might expect that (2.114) has some features in common with the linear and quasilinear wave equations considered earlier and refer to (2.114) as a wave equation. The nonlinearity of the x-derivative term in (2.114) implies, as we demonstrate, that the solution $u = u(x, t)$ is not constant along the characteristic curves. However, the velocity of specific points x on the wave is related to the nature of the wave form at those points as was the case for the quasilinear equation (2.53).

To solve this initial value problem we parametrize the initial curve as

$$(2.116) \qquad\qquad x = \tau, \qquad t = 0, \qquad u = a\tau.$$

From (2.101)–(2.102) we obtain

$$(2.117) \qquad\qquad q(\tau) + p(\tau)^2 = 0; \qquad a = p(\tau)$$

which yields as the initial data for p and q,

$$(2.118) \qquad\qquad p(\tau) = a, \qquad q(\tau) = -a^2.$$

Since $x'(\tau)F_q - t'(\tau)F_p = 1$, condition (2.103) is also satisfied.

The characteristic equations for this problem are found to be

$$(2.119) \qquad \frac{dx}{ds} = 2p, \qquad \frac{dt}{ds} = 1, \qquad \frac{du}{ds} = q + 2p^2,$$

$$\frac{dp}{ds} = 0, \qquad \frac{dq}{ds} = 0.$$

Using the data (2.117)–(2.118), we obtain

$$(2.120) \quad x(s, \tau) = 2as + \tau; \qquad t(s, \tau) = s; \qquad u(s, \tau) = a^2s + a\tau$$

and

$$(2.121) \qquad\qquad p(s, \tau) = a, \qquad q(s, \tau) = -a^2.$$

Solving for $s = s(x, t)$, $\tau = \tau(x, t)$ from the first two equations in (2.120) gives

$$(2.122) \qquad\qquad s = t, \qquad \tau = x - 2at,$$

and the solution of (2.114)–(2.115) becomes

$$(2.123) \qquad\qquad u(x, t) = a[x - at].$$

The solution $u(x, t) = a[x - at]$ represents a plane wave moving with velocity *a.* If $a > 0$, it moves to the right, whereas if $a < 0$, it moves to the left as shown in Figure 2.8.

Although the solution $u = a[x - at]$ behaves like the solution of a linear unidirectional wave equation with velocity a, the wave $u = a[x - at]$ is not constant on the characteristic base curves, $x - 2at = \tau$. Points on these curves travel with velocity $2a$, whereas points on the wave $u = a[x - at]$ have velocity a.

In fact, if we replace the parameter s by t in the equations (2.119) we obtain

$$(2.124) \qquad \frac{dx}{dt} = 2p; \qquad \frac{du}{dt} = q + 2p^2; \qquad \frac{dp}{dt} = 0; \qquad \frac{dq}{dt} = 0.$$

Since $p = $ constant and $p = u_x$, we conclude that points x move with a velocity equal to twice the slope of the wave form $u = u(x, t)$. However, the fact that $u \neq $ constant on the characteristics implies that this velocity is not that of points on the wave moving parallel to the x-axis. This is consistent with the foregoing results. It does indicate that if the initial wave form $u(x, 0) = f(x)$ has variable slope, points x where $f(x)$ has positive slope move to the right, whereas points where $f(x)$ has negative slope move to the left. Also, the greater the magnitude of the slope, the greater the velocity. This process continues throughout the wave motion and can lead to the breakdown of the solution even for smooth initial wave forms. This difficulty did not arise for the problem (2.114)–(2.115) since the initial slope is constant.

In the exercises another initial value problem for (2.114) is considered where the problem of the breaking of the wave $u = u(x, t)$ does occur.

EXAMPLE 2.10.   THE EICONAL EQUATION

The *eiconal equation of geometrical optics* in two space dimensions has the form

$$(2.125) \qquad u_x^2 + u_y^2 = n^2,$$

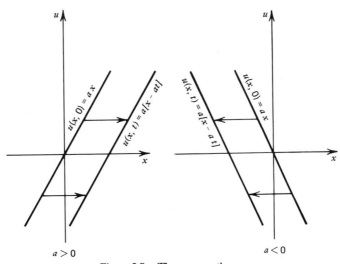

Figure 2.8.   The wave motion.

and we shall assume that $n = $ constant. We present a derivation of the eiconal equation in the exercises that exhibits its connection with wave propagation problems. In this example we replace the variable t of (2.83) by the variable y so that $q = u_y$ and y takes the place of t in (2.83)–(2.103).

As the initial condition for (2.125) we take

$$(2.126) \qquad u(x, y)|_{x=y} = u(y, y) = ay.$$

where $a = $ constant. Parametrically, this can be expressed as

$$(2.127) \qquad x = \tau, \qquad y = \tau, \qquad u = a\tau.$$

The initial values $p(\tau)$ and $q(\tau)$ on the curve (2.127) are determined from (2.101)–(2.102) as

$$(2.128) \qquad p^2 + q^2 - n^2 = 0,$$
$$(2.129) \qquad a = p(\tau) + q(\tau).$$

Equation (2.128) states that the vector $[p/n, q/n]$ is a unit vector. Let $\theta = $ constant and set

$$(2.130) \qquad p(\tau) = n\cos\theta, \qquad q(\tau) = n\sin\theta,$$

so that (2.128) is satisfied and then (2.129) implies

$$(2.131) \qquad \cos\theta + \sin\theta = \sqrt{2}\,\sin\!\left(\theta + \frac{\pi}{4}\right) = \frac{a}{n}.$$

[We select one of the solutions of (2.131). If, for example, $a/n = 1$, we can have $\theta = 0$ or $\theta = \pi/2$.] The condition (2.103) requires that

$$(2.132) \qquad 2(\sin\theta - \cos\theta) \neq 0.$$

Since this expression vanishes only at $\theta = \pi/4$ or $\theta = 5\pi/4$, we must exclude those values of θ and the related values of a; that is, $a = \pm n\sqrt{2}$.

The characteristic equations are

$$(2.133) \qquad \frac{dx}{ds} = 2p; \qquad \frac{dy}{ds} = 2q; \qquad \frac{du}{ds} = 2p^2 + 2q^2; \qquad \frac{dp}{ds} = 0; \qquad \frac{dq}{ds} = 0.$$

Thus p and q are constant on the characteristics so that

$$(2.134) \qquad p(s, \tau) = p(0, \tau) = n\cos\theta; \qquad q(s, \tau) = q(0, \tau) = n\sin\theta,$$

and we obtain the straight line characteristics:

$$(2.135) \qquad x(s, \tau) = 2ns\cos\theta + \tau; \qquad y(s, \tau) = 2ns\sin\theta + \tau$$
$$(2.136) \qquad u(s, \tau) = 2n^2 s + (\sin\theta + \cos\theta)n\tau.$$

where (2.131) was used. Solving for s and τ as functions of x and y from (2.135) yields

$$(2.137) \qquad s = \frac{y - x}{2n(\sin\theta - \cos\theta)}; \qquad \tau = \frac{x\sin\theta - y\cos\theta}{\sin\theta - \cos\theta}.$$

Inserting this into (2.136) gives the *plane wave* solution

$$(2.138) \qquad u(x, y) = n(x\cos\theta + y\sin\theta).$$

It can be verified directly that (2.138) satisfies the eiconal equation and the initial condition (2.126), on using (2.131). A more general plane wave solution of the eiconal equation can immediately be constructed from (2.138) in the form

$$(2.139) \qquad u(x, y) = u_0 + n[(x - x_0)\cos\theta + (y - y_0)\sin\theta],$$

where x_0, y_0, and u_0 are constants.

A second initial value problem for the eiconal equation, where the values are given at a point rather than on a curve, leads to a *singular solution* of (2.125) which is called a *cylindrical wave*. Let the initial point be $(x, y, u) = (x_0, y_0, u_0)$ so that the initial values are

$$(2.140) \qquad x(\tau) = x_0, \qquad y(\tau) = y_0, \qquad u(\tau) = u_0.$$

The "strip condition" (2.102) is satisfied automatically for all p and q. Thus $p(\tau)$ and $q(\tau)$ are determined from the single equation

$$(2.141) \qquad F[x_0, y_0, u_0, p(\tau), q(\tau)] = p(\tau)^2 + q(\tau)^2 - n^2 = 0.$$

Let

$$(2.142) \qquad p(\tau) = n\cos\tau, \qquad q(\tau) = n\sin\tau,$$

with $\tau = $ constant and (2.141) is satisfied. Proceeding as in the preceding initial value problem we again find that $p(s, \tau)$ and $q(s, \tau)$ are constant on the characteristics and

$$(2.143) \qquad x(s, \tau) = 2ns\cos\tau + x_0; \qquad y(s, \tau) = 2ns\sin\tau + y_0$$

$$(2.144) \qquad u(s, \tau) = 2n^2 s + u_0.$$

Squaring and summing the equations for x and y in (2.143) and solving for s readily yields

$$(2.145) \qquad u(x, y) = u_0 + n\sqrt{(x - x_0)^2 - (y - y_0)^2}$$

as the *singular solution* of the eiconal equation. It clearly satisfies (2.125) everywhere except at the initial point (x_0, y_0) where the derivatives u_x and u_y are singular. In fact, (2.145) determines a cone with vertex at (x_0, y_0, u_0), which is identical with the Monge cone through that point.

In the context of geometrical optics, if $u = u(x, y)$ is a solution of the eiconal equation, the level curves $u(x, y) = $ constant represent *wave fronts*, whereas the characteristic base curves $x = x(s, \tau)$ and $y = y(s, \tau)$, for fixed τ represent *light rays*. Thus the level curves of the solutions (2.138) or (2.139) are straight lines. In terms of the physical three-dimensional problem for which we are considering two-dimensional cross sections, the straight lines correspond to planes. Hence they are called *plane waves*. Similarly, the level curves of the solution (2.145) are circles that in the three-dimensional problem correspond to cylinders and result, therefore, in *cylindrical waves*. The tangent vectors to the characteristic base curves (i.e., the light rays) are found from (2.133) to be $[dx/ds, dy/ds] = 2[p, q]$. These curves are straight lines since p and q are constant for fixed τ. Further, the normal vectors to the level curves $u(x, y) = $ constant are $\nabla u = [u_x, u_y] = [p, q]$. This shows that the light rays are normal to the wave fronts.

The intensity of the light in geometrical optics is characterized by the convergence or divergence of the light rays. As the rays converge the intensity increases, whereas the intensity decreases as the rays diverge. The singular point (x_0, y_0) in the above initial value problem is known as a *focal point* or a focus, since all the rays are seen to converge at that point or diverge from it. It represents a point of high intensity of light. Another region of high intensity of light is given by curves that are envelopes of the characteristic base curves or the light rays. Such curves are known as *caustic curves*. Although the solutions of the eiconal equations break down at foci or caustic curves, other methods not based solely on the eiconal equation lead to valid results in such regions. They are discussed in Chapter 9.

To complete our discussion of the eiconal equation we show how the singular solution (2.145) may be used to solve a general initial value problem for the eiconal equation. The method we use is related to the *complete integral* method for solving first order partial differential equations. A complete integral of the equation (2.83) is a solution of the form

$$(2.146) \qquad\qquad M[x, t, u, a, b] = 0,$$

where a and b are arbitrary constants. In our problem the solution (2.145), in fact, contains three arbitrary constants, x_0, y_0, and u_0. We now indicate how (2.145) can be used to solve the initial value problem (2.127). For simplicity, we assume that the constants a and n are chosen such that $a = n$. This implies that we either have $\theta = 0$ or $\theta = \pi/2$ in (2.138), so that $u(x, y) = nx$ or $u(x, y) = ny$.

We set $x_0 = \tau$, $y_0 = \tau$, and $u_0 = n\tau$ in (2.145) so that $(x_0, y_0, u_0) = (\tau, \tau, n\tau)$ represents the initial curve (2.127). This yields the family of solutions

$$(2.147) \qquad\qquad u(x, y) = n\tau + n\sqrt{(x - \tau)^2 + (y - \tau)^2}.$$

Expressing the solution in implicit form as

$$(2.148) \qquad H(x, y, u, \tau) = u - n\tau - n\sqrt{(x - \tau)^2 + (y - \tau)^2} = 0$$

and differentiating with respect to τ gives

$$(2.149) \qquad \frac{\partial H}{\partial \tau} = -n + n\left[\frac{(x - \tau) + (y - \tau)}{\sqrt{(x - \tau)^2 + (y - \tau)^2}}\right] = 0$$

where (2.148)–(2.149) are the equations that determine the envelope of the family of solutions (2.147). We easily conclude from (2.149) that

$$(2.150) \qquad (x - \tau)(y - \tau) = 0,$$

so that either $\tau = x$ or $\tau = y$.

With $\tau = x$ in (2.148) we have

$$(2.151) \qquad H = u - nx - n\sqrt{(y - x)^2} = u - nx - n|y - x| = 0,$$

and with $\tau = y$

$$(2.152) \qquad H = u - ny - n\sqrt{(y - x)^2} = u - ny - n|y - x| = 0.$$

Two continuously differentiable solutions can be obtained from the above in the form

$$(2.153) \qquad u(x, y) = nx,$$

and

$$(2.154) \qquad u(x, y) = ny,$$

which correspond to the values $\theta = 0$ and $\theta = \pi/2$ in (2.138), respectively. Both solutions clearly satisfy the initial condition $u = nx = ny$ on the line $x = y$.

The fact that there are two solutions to the initial value problem (2.125)–(2.126) merits discussion. It signifies that merely specifying u on a curve without also specifying p and q (i.e., an initial strip) does not yield a unique solution, in general. The eiconal equation, $F = p^2 + q^2 - n^2 = 0$ is a quadratic expression for which $p = \pm\sqrt{1 - q^2}$, so that for every choice of q there are generally two values of p. This means that there are, in general, two differentiable solutions passing through a given initial curve. The importance of the existence of these two solutions in applications of the eiconal equation is indicated in the exercises, and is emphasized in our discussion of the eiconal equation in Chapter 9.

APPENDIX: ENVELOPES OF CURVES AND SURFACES

Let $F(x, t, \tau) = 0$ be a one-parameter family of curves in the (x, t)-plane with parameter τ. The *envelope* of the family of curves is determined from the equations

$$F(x, t, \tau) = 0; \qquad \frac{\partial F}{\partial \tau}(x, t, \tau) = 0.$$

Solving for $\tau = \tau(x, t)$ from the second equation and substituting in the first equation gives the envelope

$$F[x, t, \tau(x, t)] = 0.$$

If F is a function of more than two variables and depends on a parameter τ, the foregoing procedure for constructing the envelope also applies.

We now consider an example. Let a curve be defined as $t = f(x)$. If $[\tau, f(\tau)]$ is a fixed point on that curve, the tangent line to the curve at that point has the equation

$$F(x, t, \tau) = t - f(\tau) - (x - \tau)f'(\tau) = 0.$$

As the parameter τ varies, we have a one-parameter family of tangent lines to the curve. Now,

$$\frac{\partial F}{\partial \tau} = -(x - \tau)f''(\tau) = 0$$

implies if $f''(\tau) \neq 0$ for all τ, that $\tau = x$. Inserting this in the equation $F(x, t, \tau) = 0$ yields the envelope

$$F(x, t, \tau)|_{\tau = x} = t - f(x) = 0.$$

Thus the given curve $t = f(x)$ is the envelope of its family of tangent lines. The family of curves is tangent to the envelope at, at least, one point.

As an example of a three-dimensional problem we consider the family of spheres

$$F(x, y, \tau) = (x - \tau)^2 + y^2 + z^2 - a^2 = 0,$$

whose radii equal a and whose centers are the points $(\tau, 0, 0)$, where τ is a parameter. Since

$$\frac{\partial F}{\partial \tau} = 2(x - \tau) = 0,$$

we find that $\tau = x$. The envelope of the family of spheres is, therefore, given by

$$F[x, y, z, \tau(x)] = y^2 + z^2 - a^2 = 0.$$

This is the equation of a circular cylinder of radius a, whose axis coincides with the x-axis.

If the family of curves is given in parametric form

$$x = x(s, \tau), \qquad t = t(s, \tau),$$

we let s be the running parameter along a single curve and τ be the parameter specifying a member of the family. Geometrically, the curves of the family are tangent to the envelope at their points of intersection, so we may characterize τ as a parameter along the envelope curve (each value of τ specifies a point of intersection on that curve). Now, $[\partial x/\partial s, \partial t/\partial s]$ is a tangent vector of a member of the family for each fixed τ, whereas $[\partial t/\partial \tau, -\partial x/\partial \tau]$ is a normal vector of the envelope curve when s is evaluated at the envelope. Since the tangent vector of a member of the family of curves is orthogonal to the normal vector of the envelope at the point of intersection, we have for their dot product

$$\Delta(s, \tau) = \left[\frac{\partial x}{\partial s}, \frac{\partial t}{\partial s}\right] \cdot \left[\frac{\partial t}{\partial \tau}, -\frac{\partial x}{\partial \tau}\right] = x_s t_\tau - t_s x_\tau = 0.$$

This equation describes the envelope of the family.

The envelope equation is to be compared with the conditions (2.48) and (2.103). Noting (2.51), (2.91), and (2.92) we see that both of the conditions (2.48) and (2.103) are equivalent to

$$\Delta(0, \tau) = x_s t_\tau - t_s x_\tau|_{s=0} \neq 0.$$

In terms of the initial value problem for the equations (2.45) and (2.83) the condition $\Delta(0, \tau) \neq 0$ requires that the initial curve C be neither a characteristic curve nor an envelope of characteristic curves. Otherwise the solution may not exist or, if it does exist, need not be unique. Additionally, even if the initial curve satisfies the condition (2.48) or (2.103), when the family of characteristics forms an envelope, the solution breaks down, as is seen for the quasilinear equation (2.53) and is shown for the eiconal equation in Chapter 9.

For example, the circle $x^2 + t^2 = a^2$ in parametric form is

$$x = a \cos \tau, \qquad t = a \sin \tau,$$

and the tangent lines are

$$x(s, \tau) = a \cos \tau - as \sin \tau; \qquad t(s, \tau) = a \sin \tau + as \cos \tau,$$

as is readily seen. Here s is the running parameter along a tangent line and τ characterizes a particular tangent line. We have

$$\Delta(s, \tau) = x_s t_\tau - t_s x_\tau = a^2 s,$$

so that $\Delta(s, \tau) = 0$ implies that $s = 0$ is the envelope. Then, $x(0, \tau) = a \cos \tau$, $t(0, \tau) = a \sin \tau$ yields the circle $x^2 + t^2 = a^2$, as was expected.

EXERCISES FOR CHAPTER 2

Section 2.1

2.1.1. Show that if we set $u = v_t$ and $w = v_x$ in the wave equation (2.1), we obtain the system

$$u_x = w_t; \qquad u_t = \gamma^2 w_x.$$

Note that this system is coupled, but in contrast to (2.2), neither of the equations can be solved independently of the other. Thus the manner in which an equation of higher order is represented as a system can play a significant role.

2.1.2. Show that the functions u and w in the Cauchy–Riemann equations (2.6) both satisfy Laplace's equation.

Section 2.2

2.2.1. Solve the initial value problem for the damped unidirectional wave equation,

$$v_t + c v_x + \lambda v = 0, \qquad v(x, 0) = F(x),$$

where $\lambda > 0$ and $F(x)$ is given.

2.2.2. (a) Solve the initial value problem for the inhomogeneous equation

$$v_t + c v_x = f(x, t), \qquad v(x, 0) = F(x)$$

where $f(x, t)$ and $F(x)$ are specified functions.

(b) Solve this problem in the special case where $f(x, t) = xt$ and $F(x) = \sin x$.

2.2.3. Discuss the solution of the wave equation (2.1) in the following cases:

(a) $v(x, 0) = f(x) = x$; $v_t(x, 0) = g(x) = 0$.

(b) $f(x) = 0$; $g(x) = x$.

(c) $f(x) = \sin x$; $g(x) = -\gamma \cos x$.

(d) $f(x) = \sin x$; $g(x) = \gamma \cos x$.

Based on the results obtained, observe that solutions of the wave equation may or may not have the form of traveling waves.

2.2.4. Consider the inhomogeneous wave equation

$$v_{tt} - \gamma^2 v_{xx} = F(x, t).$$

(a) Apply the method of Example 2.1 to reduce it to the system

$$v_t - \gamma v_x = u$$
$$u_t + \gamma u_x = F.$$

(b) Solve the initial value problem for the inhomogeneous wave equation with the initial data $v(x, 0) = f(x)$, $v_t(x, 0) = g(x)$.

2.2.5. Solve the "signaling" problem for (2.13) in the region $x > 0$ with the boundary condition $v(0, t) = G(t)$ for $-\infty < t < +\infty$.

2.2.6. Show that the initial and boundary value problem for (2.13) can be solved if $c > 0$ in the quarter plane $x > 0$, $t > 0$ with the data $v(x, 0) = F(x)$ and $v(0, t) = G(t)$ [where $F(0) = G(0)$], if F and G are arbitrary functions. However, if $c < 0$, the problem can be solved only if F and G are related in a special way.

2.2.7. Solve the initial value problem for the equation

$$tu_x + xu_t = cu; \qquad u(x, x) = f(x),$$

where $c =$ constant and $f(x)$ is specified.

2.2.8. Show that the initial value problem

$$u_t + u_x = x; \qquad u(x, x) = 1$$

has no solution. Observe that the initial curve $t = x$ is a characteristic base curve and explain why this is not a characteristic initial value problem.

2.2.9. (a) Show that the problem

$$u_t + cu_x = F(x, t); \qquad u\left(x, \frac{1}{c}x\right) = f(x)$$

is a characteristic initial value problem if $f'(x) = (1/c)F[x, (1/c)x]$.

(b) Verify that

$$u(x, t) = \frac{1}{c} \int_{x-ct}^{x} F\left[r, \frac{r - x + ct}{c}\right] dr + g(x - ct)$$

is a solution of the characteristic initial value problem of part (a) if $g(0) = f(0)$ but $g(z)$ is an otherwise arbitrary function.

(c) Let $F(x, t) = \cos(x + t)$ in part (a). Determine the appropriate choice of $f(x)$ and obtain the general solution of the problem.

2.2.10. Generalize the method of characteristics to problems in three dimensions. Given the linear equation

$$av_x + bv_y + cv_z = dv + e,$$

where a, b, c, d, e, and v are functions of (x, y, z), obtain the characteristic equations

$$\frac{dx}{ds} = a; \qquad \frac{dy}{ds} = b; \qquad \frac{dz}{ds} = c; \qquad \frac{dv}{ds} = dv + e.$$

The initial value problem for this case specifies an initial *hypersurface* given parametrically in (x, y, z, v)-space, as

$$x = x(\lambda, \tau), \qquad y = y(\lambda, \tau), \qquad z = z(\lambda, \tau), \qquad v = v(\lambda, \tau);$$

with λ and τ as the parameters. The family of characteristic curves is now given as

$$x = x(s, \lambda, \tau), \qquad y = y(s, \lambda, \tau), \qquad z = z(s, \lambda, \tau), \qquad v = v(s, \lambda, \tau),$$

with s as the running parameter along a curve and (λ, τ) as a two parameter family specifying the individual curves. If the equations for (x, y, z) can be inverted to yield (s, λ, τ) as smooth functions of (x, y, z), the function $v = v(x, y, z)$ obtained thereby is a solution of the initial value problem. Discuss situations in which the method of characteristics may not give a solution or yields a nonunique solution.

2.2.11. Using the method of characteristics solve the initial value problem

$$v_x + v_y + v_z = 0; \qquad v(x, y, 0) = f(x, y),$$

where $f(x, y)$ is given.

 (a) Parametrize the initial data and let $x = \lambda$, $y = \tau$, $z = 0$, and $v = f(\lambda, \tau)$ and set up the characteristic equations.

 (b) Show that the solution of these equations is

$$x = s + \lambda, \qquad y = s + \tau, \qquad z = s, \qquad v = f(\lambda, \tau).$$

 (c) Conclude that the solution of the initial value problem is

$$v(x, y, z) = f(x - z, y - z).$$

2.2.12. Extend the method of characteristics to linear equations in n-dimensions. Given the equation

$$\sum_{i=1}^{n} a_i(x_1, \ldots, x_n) \frac{\partial v}{\partial x_i} = b(x_1, \ldots, x_n)v + c(x_1, \ldots, x_n)$$

where $v = v(x_1, \ldots, x_n)$, obtain the characteristic system

$$\frac{dx_i}{ds} = a_i, \quad (i = 1 \ldots, n); \qquad \frac{dv}{ds} = bv + c.$$

The initial values for this problem are given as,

$$x_i = x_i(\tau_1, \ldots, \tau_{n-1}), \qquad v = v(\tau_1, \ldots, \tau_{n-1})$$

at $s = 0$, with $\tau_1, \ldots, \tau_{n-1}$ as parameters. The characteristic curves are

$$x_i = x_i(s, \tau_1, \ldots, \tau_{n-1}), \qquad v = v(s, \tau_1, \ldots, \tau_{n-1})$$

where $i = 1, 2, \ldots, n$. If $s, \tau_1, \ldots, \tau_{n-1}$ can be solved for in terms of $x_1, \ldots, x_n$, then $v = v(x_1, \ldots, x_n)$ is a solution of the initial value problem.

2.2.13. Use the method of characteristics to solve the initial value problem

$$\sum_{i=1}^{n} v_{x_i} = 0, \qquad v(x_1, x_2, \ldots, x_{n-1}, 0) = f(x_1, x_2, \ldots, x_{n-1}),$$

where f is given.

(a) Show that the solution of the characteristic equations can be given as

$$x_i = s + \tau_i, \quad (i = 1, \ldots, n-1);$$
$$x_n = s, \quad v = f(\tau_1, \ldots, \tau_{n-1}).$$

(b) Conclude that the solution of the problem is

$$v(x_1, \ldots, x_n) = f(x_1 - x_n, x_2 - x_n, \ldots, x_{n-1} - x_n).$$

Section 2.3

2.3.1. Using implicit differentiation, verify that $u = f(x - tu)$ is a solution of the wave equation (2.53).

2.3.2. Consider the damped quasilinear wave equation

$$u_t + uu_x + cu = 0,$$

where c is a positive constant.

(a) Using the method of characteristics, construct a solution of the initial value problem with $u(x, 0) = f(x)$ leaving the results in parametric form. Discuss the wave motion and the effect of the damping.

(b) Determine the breaking time of the solution by finding the envelope of the characteristic curves. With τ as the parameter on the initial line show that unless $f'(\tau) < -c$, no breaking occurs. Contrast this result with that for the undamped case discussed in the text.

2.3.3. Solve the initial value problem for the equation

$$u_t + cu_x + u^2 = 0; \qquad u(x,0) = x,$$

where c is a constant.

2.3.4. Obtain the solution of the initial value problem

$$u_t + c(u)u_x = 0, \qquad u(x,0) = f(x),$$

where c is a function of u, in the implicit form $u = f[x - tc(u)]$. Discuss the solution in the cases where $c'(u) > 0$ and $c'(u) < 0$.

2.3.5. Using the implicit form of the solution obtained in Exercise 2.3.4, determine the breaking time of the wave.

2.3.6. Solve the initial value problem

$$u_t + u^2u_x = 0, \qquad u(x,0) = x.$$

Determine the breaking time of the solution and compare it with that obtained through the result of Exercise 2.3.5.

2.3.7. Solve the initial value problem

$$u_t + uu_x = x; \qquad u(x,0) = f(x),$$

using the method of characteristics.

(a) Using the parameters s and τ as defined in the text, show that the solution can be expressed as $x(s, \tau) = \frac{1}{2}[f(\tau) + \tau]e^s - \frac{1}{2}[f(\tau) - \tau]e^{-s}$, $t = s$, and $u(s, \tau) = \frac{1}{2}[f(\tau) + \tau]e^s + \frac{1}{2}[f(\tau) - \tau]e^{-s}$.

(b) Obtain the solution in the form $u = u(x, t)$ when $f(x) = 1$ and when $f(x) = x$.

2.3.8. Consider the initial value problem

$$u_t + uu_x = 0; \qquad u(x,0) = u_0 + \epsilon F(x),$$

where $u_0 = $ constant, $0 < \epsilon \ll 1$, and $F(x)$ is uniformly bounded for all x. With $\epsilon = 0$, $u = u_0$ represents a constant solution of the wave equation. The initial condition represents a small perturbation around the constant u_0, so we look for the solution in the form of a perturbation series around the constant state $u = u_0$.

(a) Let

$$u = u_0 + \epsilon u_1(x, t) + \epsilon^2 u_2(x, t) + \cdots$$

and substitute this series formally into the equation for u. Collecting like powers of ϵ and equating their coefficients to zero, obtain a recursive system of equations for the functions u_i $(i \geqslant 1)$. [Determine only the equations for u_1 and u_2.] Show that the appropriate initial conditions for the u_i are $u_1(x, 0) = F(x)$, $u_i(x, 0) = 0$, $i \geqslant 2$.

(b) Show that the solutions for u_1 and u_2 are

$$u_1(x, 0) = F(x - u_0 t); \qquad u_2(x, t) = -tF'(x - u_0 t) F(x - u_0 t).$$

(c) Since $u = u_0 + \epsilon u_1 + \epsilon^2 u_2 + \ldots$, show that when $-\epsilon t F' \approx 1$, the term $\epsilon^2 u_2$ is of the same order of magnitude as ϵu_1 and the terms in the series do not get smaller as was assumed. Compare the time t at which the series breaks down with the breaking time given in the text, as applied to the initial value problem.

(d) Obtain the foregoing results for u_1 and u_2 by inserting the series for u directly into the implicit form of the solution given in (2.61), expanding the function f in a series in ϵ and comparing like powers of ϵ.

2.3.9. Generalize the method of characteristics to deal with the three-dimensional quasilinear equation

$$au_x + bu_y + cu_z = d,$$

where a, b, c, and d are functions of x, y, z, and u. Show that the characteristic equations are

$$\frac{dx}{ds} = a; \qquad \frac{dy}{ds} = b; \qquad \frac{dz}{ds} = c; \qquad \frac{du}{ds} = d;$$

discuss appropriate initial conditions.

2.3.10. Solve the initial value problem

$$u_x + u_y + u_z = u^2; \qquad u(x, y, 0) = x + y.$$

Section 2.4

2.4.1. Solve the initial value problem

$$u_x^2 u_t - 1 = 0; \qquad u(x, 0) = x.$$

2.4.2. Use the method of characteristics to solve

$$u_t + u_x^2 = t; \qquad u(x, 0) = 0.$$

2.4.3. Solve the initial value problem

$$u_t + u_x^2 + u = 0; \qquad u(x,0) = x,$$

using the method of characteristics. (The solution may be left in parametric form.)

2.4.4. Consider the wave equation in two dimensions

$$v_{xx} + v_{yy} = \frac{1}{c^2} v_{tt},$$

where c may be a function of (x, y).

(a) Let $v(x, y, t) = V(x, y)e^{-i\omega t}$ ($i = \sqrt{-1}$) and show that $V(x, y)$ satisfies the reduced wave equation.

$$V_{xx} + V_{yy} + \left(\frac{\omega^2}{c^2}\right)V = 0.$$

(The constant ω is the "angular" frequency of the solution.) Since c represents the speed of wave propagation (as we shall see later in the text), we introduce a constant reference speed c_0, say, the speed of light, and define the *index of refraction n* as $n(x, y) = c_0/c(x, y)$ and the *wave number k* as $k = \omega/c_0$. Then the reduced wave equation takes the form

$$V_{xx} + V_{yy} + k^2 n^2 V = 0.$$

(b) Assume that $k \gg 1$ and look for a solution of the reduced wave equation in the form $V = Ae^{iku}$, where A and u are functions of (x, y). Show that on substituting this form in the reduced wave equation, collecting like powers of k, and equating the coefficients of k^2 and k to zero, we obtain the eiconal equation (2.125) for $u(x, y)$ and the following equation for $A(x, y)$,

$$2A_x u_x + 2A_y u_y + (u_{xx} + u_{yy})A = 0.$$

This is known as a transport equation for A (which represents an amplitude term).

(c) Show that if n, A = constant and u is the plane wave solution (2.138), $V(x, y)$ is an exact solution of the reduced wave equation. Otherwise, the function $V = Ae^{iku}$ represents what is known as the *geometrical optics* approximation to the solution of the reduced wave equation. (A further discussion of this approximation is given in Section 9.4.)

2.4.5. Let $u = \sqrt{x^2 + y^2}$. Solve the transport equation for $A(x, y)$ given in Exercise 2.4.4, subject to the condition $A(x, y) = 1$ on the circle $x^2 + y^2 = 1$.

Show that $A(x, y)$ is a singular at the origin $(0, 0)$, which represents a focal point for $u(x, y)$.

2.4.6. Consider the eiconal equation

$$u_x^2 + u_y^2 = n^2(x, y); \qquad n(x, y) = \begin{cases} n_1, & x < 0 \\ n_2, & x > 0 \end{cases},$$

with $n_2 > n_1$ and both equal to a constant.

(a) Given the boundary condition $u(0, y) = n_1 y \cos \theta$ where $\theta =$ constant, solve for all possible $u(x, y)$ in the regions $x < 0$ and $x > 0$. (Hint: They will be plane wave solutions.) These solutions play a role in the problem of a plane interface between two media with differing indices of refraction.

(b) Obtain expressions for the directions of the rays found in part (a), as given by ∇u and evaluated on the interface $x = 0$. These formulas represent what are known a *Snell's laws* of reflection and refraction.

2.4.7. Extend the results of the text for nonlinear equations to n dimensions. With the equation given as

$$F(x_1, x_2, \ldots, x_n, u, u_{x_1}, \ldots, u_{x_n}) = 0,$$

obtain the characteristic equations with $p_i = u_{x_i}$,

$$\frac{dx_i}{ds} = F_{p_i}; \qquad \frac{du}{ds} = \sum_{i=1}^{n} p_i F_{p_i}; \qquad \frac{dp_i}{ds} = -(F_u p_i + F_{x_i}),$$

where $i = 1, 2, \ldots, n$. Also, obtain conditions equivalent to (2.101)–(2.102).

2.4.8. Obtain plane wave solutions and singular solutions equivalent to (2.139) and (2.145), respectively, for the eiconal equation in three dimensions,

$$u_x^2 + u_y^2 + u_z^2 = n^2, \qquad n = \text{constant}.$$

2.4.9. Solve the initial value problem

$$u_t + u_x^2 = 0; \qquad u(x, 0) = -x^2.$$

Show that the solution breaks down when $t = \frac{1}{4}$.

2.4.10. Differentiate (2.114) with respect to x and let $v = u_x$. Show that v satisfies an equation of the form (2.82) and interpret the results of Example 2.9 accordingly.

2.4.11. Implicitly differentiate $F(x, t, u, p, q)$ with respect to s, and use (2.91)–(2.93) and (2.98)–(2.99) to conclude that F is constant along the characteristics.

3

CLASSIFICATION OF
EQUATIONS AND
CHARACTERISTICS

The telegrapher's and wave equations, the diffusion equation, and Laplace's equation, derived in Chapter 1, are prototypes of the three basic types of partial differential equations encountered most often in applications. They are equations of hyperbolic, parabolic, and elliptic type, respectively. The classification of equations and systems of equations, in general, into these three (or possibly other) types is considered in this chapter.

Furthermore, formulations of initial and boundary value problems appropriate to equations of different types are given. Additionally, characteristics that were shown to play an important role in the theory of first order partial differential equations are reintroduced from a somewhat different point of view. Their significance in the theory of hyperbolic partial differential equations is demonstrated. Finally, certain basic concepts relating to second order linear equations with constant coefficients are introduced and discussed.

3.1. LINEAR SECOND ORDER PARTIAL DIFFERENTIAL EQUATIONS

We begin by considering the general second order linear partial differential equation in two independent variables

$$(3.1) \qquad Au_{xx} + 2Bu_{xy} + Cu_{yy} + Du_x + Eu_y + Fu = G,$$

where the terms $A, B \cdots G$ are given real valued functions of x and y. Special cases of (3.1) were derived in Chapter 1 (with y replaced by t in some instances). It will now be shown that the principal part of (3.1), that is, the terms containing second derivatives of u, can be transformed by a change of independent variables into a form similar to that of the wave, diffusion, and Laplace's equation considered in Chapter 1. This indicates that these equations are, in fact, prototypes of second order linear equations of the general form (3.1). The transformed versions of these equations are referred to as *canonical forms*.

Without loss of generality, it may be assumed that $A(x, y) \neq 0$ in some region R and we divide by A in (3.1). Then, with $\partial_x = \partial/\partial x$ and $\partial_y = \partial/\partial y$, we express the principal part of the differential operator in (3.1) as follows,

$$(3.2) \qquad \partial_x^2 + \left(\frac{2B}{A}\right)\partial_x\partial_y + \left(\frac{C}{A}\right)\partial_y^2 = \left(\partial_x - \omega^+\partial_y\right)\left(\partial_x - \omega^-\partial_y\right)$$

$$+ \left(\frac{\partial\omega^-}{\partial x} - \omega^+\frac{\partial\omega^-}{\partial y}\right)\partial_y,$$

where ω^+ and ω^- are defined as [on comparing both sides of (3.2)]

$$(3.3) \qquad\qquad \omega^+ + \omega^- = -\frac{2B}{A}; \qquad \omega^+\omega^- = \frac{C}{A}.$$

Solving the system (3.3) for ω^+ and ω^- gives

$$(3.4) \qquad\qquad \omega^\pm(x, y) = \frac{-B \pm \sqrt{B^2 - AC}}{A}$$

In obtaining (3.2) we have followed the procedure used in Section 2.1, where the differential operators for the wave and Laplace's equation were factored. Since the coefficients A, B, and C may depend on x and y, we have the additional term involving ∂_y in (3.2). Instead of using (3.2) to express (3.1) as a system of two first order equations (as was done in Section 2.1), we use (3.2) to simplify the form of the given equation (3.1).

Since we are generally interested in obtaining real valued solutions of (3.1), the possibility of using the factorization (3.2) to simplify (3.1) depends on whether $\omega^\pm(x, y)$ are real or complex valued functions of x and y. This is determined by the sign of the discriminant $B^2 - AC$ in (3.4). We shall assume that $B^2 - AC$ is either of one sign or vanishes identically throughout the given region R. On that basis we classify (3.1) as belonging to one of the following three types in the region R:

(3.5)	$B^2 - AC > 0$,	*hyperbolic type*;
(3.6)	$B^2 - AC = 0$,	*parabolic type*;
(3.7)	$B^2 - AC < 0$,	*elliptic type*.

Applying these criteria to the equations of Chapter 1, we find that the wave (and telegrapher's), diffusion, and Laplace's equation are of hyperbolic, parabolic, and elliptic type, respectively. We now determine canonical forms for (3.1) of each of the types (3.5)–(3.7).

3.1.a   Equations of Hyperbolic type

When $B^2 - AC > 0$ in the region R, the functions $\omega^\pm(x, y)$ are real valued and distinct in that region. We may then express the operators $\partial_x - \omega^\pm \partial_y$ as directional derivatives. Along the family of curves

$$(3.8) \qquad \frac{dy}{dx} = -\omega^\pm(x, y),$$

we have, for any differentiable function $v = v(x, y)$,

$$(3.9) \qquad \frac{dv}{dx} = \frac{\partial v}{\partial x} + \frac{dy}{dx}\frac{\partial v}{\partial y} = \left(\partial_x - \omega^\pm \partial_y\right)v,$$

on using (3.8).

The (one-parameter) families of curves determined by the solutions of (3.8) are called the *characteristic curves* of (3.1). They form two independent families of curves in the (x, y) – plane since $dy/dx = -\omega^\pm$ with $\omega^+ \neq \omega^-$ implies that they intersect nontangentially. We express these curves as

$$(3.10) \qquad \xi = \xi(x, y); \qquad \eta = \eta(x, y),$$

where $\xi =$ constant corresponds to the curves $y' + \omega^+ = 0$, whereas $\eta =$ constant corresponds to the curves $y' + \omega^- = 0$.

We introduce ξ and η as new coordinates in (3.1), and this is referred to as the *characteristic coordinate system*. On the characteristic curves $\xi =$ constant and $\eta =$ constant,

$$(3.11) \qquad 0 = \xi_x + y'(x)\xi_y = \xi_x - \omega^+\xi_y,$$

$$(3.12) \qquad 0 = \eta_x + y'(x)\eta_y = \eta_x - \omega^-\eta_y.$$

Also, if $u = u(\xi, \eta) = u[\xi(x, y), \eta(x, y)]$, we have

$$(3.13) \qquad \frac{\partial u}{\partial x} = u_\xi\xi_x + u_\eta\eta_x; \qquad \frac{\partial u}{\partial y} = u_\xi\xi_y + u_\eta\eta_y,$$

so that

(3.14)
$$\frac{\partial u}{\partial x} - \omega^+ \frac{\partial u}{\partial y} = \left(\partial_x - \omega^+ \partial_y\right)u = \left(u_\xi \xi_x + u_\eta \eta_x - \omega^+ u_\xi \xi_y - \omega^+ u_\eta \eta_y\right)$$
$$= \left(\xi_x - \omega^+ \xi_y\right)u_\xi + \left(\eta_x - \omega^+ \eta_y\right)u_\eta$$
$$= \left[\left(\eta_x - \omega^+ \eta_y\right)\partial_\eta\right]u,$$

in view of (3.11). Similarly, (3.12) implies that

(3.15) $$\frac{\partial u}{\partial x} - \omega^- \frac{\partial u}{\partial y} = \left(\partial_x - \omega^- \partial_y\right)u = \left[\left(\xi_x - \omega^- \xi_y\right)\partial_\xi\right]u.$$

It may be assumed that $\xi_y \neq 0$ and $\eta_y \neq 0$. Using (3.11)–(3.12) to express ξ_x and η_x in terms of ξ_y and η_y, we obtain in view of (3.14)–(3.15),

(3.16) $$\partial_x - \omega^+ \partial_y = -\eta_y(\omega^+ - \omega^-)\partial_\eta$$
(3.17) $$\partial_x - \omega^- \partial_y = \xi_y(\omega^+ - \omega^-)\partial_\xi,$$

which are both nonzero operators since $\omega^+ \neq \omega^-$. Thus

(3.18)
$$\left(\partial_x - \omega^+ \partial_y\right)\left(\partial_x - \omega^- \partial_y\right) = \left[-\eta_y(\omega^+ - \omega^-)\partial_\eta\right]\left[\xi_y(\omega^+ - \omega^-)\partial_\xi\right]$$
$$= -\xi_y \eta_y(\omega^+ - \omega^-)^2 \partial^2_{\xi\eta} - \eta_y(\omega^+ - \omega^-)\partial_\eta\left[\xi_y(\omega^+ - \omega^-)\right]\partial_\xi.$$

Substituting this result in (3.1), as modified in (3.2), and expressing everything in terms of ξ and η, we obtain (on dividing by the nonzero coefficient of $u_{\xi\eta}$)

(3.19) $$u_{\xi\eta} + au_\xi + bu_\eta + cu = d,$$

where a, b, c, and d are specified functions of ξ and η. This is one of the canonical forms for (3.1) when it is of hyperbolic type.
The transformation

(3.20) $$\xi = \alpha + \beta; \qquad \eta = \alpha - \beta$$

in (3.19) yields an alternate canonical form for the hyperbolic case,

(3.21) $$u_{\alpha\alpha} - u_{\beta\beta} + \tilde{a}u_\alpha + \tilde{b}u_\beta + \tilde{c}u = \tilde{d},$$

with specified functions $\tilde{a}$, $\tilde{b}$, $\tilde{c}$, and $\tilde{d}$ of α and β. We see that in either the form (3.19) or (3.21) the principal parts of the canonical forms have constant coefficients. The telegrapher's and wave equations of Section 1.2 are of the form (3.21).

3.1.b Equations of Parabolic Type

When $B^2 - AC = 0$ in the region R, we have $\omega^+ = \omega^- = \omega$ and (3.2) becomes

$$(3.22) \quad \partial_x^2 + \left(\frac{2B}{A}\right)\partial_x\partial_y + \left(\frac{B^2}{A^2}\right)\partial_y^2 = \left(\partial_x - \omega\partial_y\right)^2 + \left(\frac{\partial\omega}{\partial x} - \omega\frac{\partial\omega}{\partial y}\right)\partial_y$$

with $\omega = -B/A$ in view of (3.4). Since there is only one value of ω, we obtain only one family of characteristic curves defined by

$$(3.23) \qquad\qquad \frac{dy}{dx} = -\omega(x, y) = \frac{B}{A}.$$

Let

$$(3.24) \qquad\qquad \xi = \xi(x, y); \qquad \eta = \eta(x, y),$$

where $\xi = $ constant is the family of characteristic curves determined from (3.23) and $\eta = $ constant is an arbitrarily chosen independent family of curves, such that the Jacobian determinant of the transformation (3.24) (i.e, $\xi_x\eta_y - \xi_y\eta_x$) is nonzero in the region R.

Proceeding as for the hyperbolic case and putting $u = u(\xi, \eta)$, we obtain (after some simplification) the canonical form for (3.1) in the parabolic case

$$(3.25) \qquad\qquad u_{\eta\eta} + au_\xi + bu_\eta + cu = d,$$

where a, b, c, and d are specified functions of ξ and η. Again, the coefficients in the principal part of (3.25) are constant. The diffusion equation of Section 1.1 has the form (3.25).

3.1.c Equations of Elliptic Type

When $B^2 - AC < 0$, ω^+ and ω^- as defined in (3.4) are complex valued in the region R. Hence the characteristics defined by (3.8) are complex curves. Assuming the terms $A, B, \ldots, G$ in (3.1) can be defined for complex arguments, the change of variables (3.10) may still be applied to bring (3.1) into the form (3.19). However, the variables ξ and η are complex, so that the usefulness of the canonical form (3.19) is questionable.

The further transformation

$$(3.26) \qquad\qquad \xi = \alpha + i\beta; \qquad \eta = \alpha - i\beta$$

in (3.19) (with $i = \sqrt{-1}$) introduces real valued variables α and β. (This follows since ω^+ and ω^- are complex conjugates in the elliptic case, when the terms A, B, and C are real as assumed.) We then easily obtain the canonical form

$$(3.27) \qquad u_{\alpha\alpha} + u_{\beta\beta} + au_\alpha + bu_\beta + cu = d$$

for the elliptic case, where a, b, c, and d are specified real valued functions of α and β. [The canonical form (3.27) can also be obtained by another method (which we do not consider) that does not require the introduction of complex variables.] Laplace's equation of Section 1.3 has the form (3.27).

If the terms A, B,..., F in (3.1) are identically constant, the above canonical forms for the hyperbolic, parabolic, and elliptic cases are valid over the entire domain of definition of the inhomogeneous term G, and they all contain constant coefficients. In that case a further simplification of the equation can be achieved, as shown in the exercises.

EXAMPLE 3.1.   AN EQUATION OF MIXED TYPE

The equation

$$(3.28) \qquad u_{xx} + yu_{yy} = 0,$$

is said to be of *mixed type*, since it is hyperbolic for $y < 0$, parabolic for $y = 0$, and elliptic for $y > 0$. This follows immediately from the fact that $B^2 - AC = -y$ for this equation. We now obtain canonical forms for (3.28) in all three regions of the (x, y) − plane given above.

The parabolic region is the simplest to deal with since at $y = 0$ we immediately obtain the canonical form

$$(3.29) \qquad u_{xx} = 0.$$

In this case $\omega = 0$, so that the characteristic curve determined from $dy/dx = 0$ is $y = 0$. That is, the x-axis is the characteristic curve, and it represents a curve across which a transition from hyperbolic to elliptic type takes place. For this reason, the solution of initial and boundary value problems for (3.28) in a neighborhood of the x-axis is rather difficult since (3.28) is of three different types in such a neighborhood.

In the hyperbolic region $y < 0$ we have

$$(3.30) \qquad \omega^\pm(x, y) = \pm\sqrt{-y},$$

and the characteristic equations

$$(3.31) \qquad y'(x) = -\omega^\pm = \mp\sqrt{-y}$$

yield the two characteristic families

$$(3.32) \qquad \xi = x + 2\sqrt{-y} = \text{constant},$$

$$(3.33) \qquad \eta = x - 2\sqrt{-y} = \text{constant}.$$

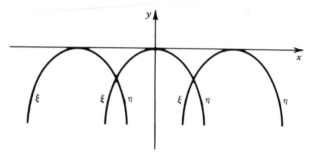

Figure 3.1. The characteristic curves.

From (3.2) we obtain for the differential operator in (3.28)

$$(3.34) \qquad \partial_x^2 + y\partial_y^2 = \left(\partial_x - \sqrt{-y}\,\partial_y\right)\left(\partial_x + \sqrt{-y}\,\partial_y\right) - \tfrac{1}{2}\partial_y,$$

and from (3.18)

$$(3.35) \qquad \left(\partial_x - \sqrt{-y}\,\partial_y\right)\left(\partial_x + \sqrt{-y}\,\partial_y\right) = 4\partial_{\xi\eta}^2$$

Since $\partial_y = \xi_y\partial_\xi + \eta_y\partial_\eta = -(1/\sqrt{-y})[\partial_\xi - \partial_\eta]$ and $\xi - \eta = 4\sqrt{-y}$, we obtain the canonical form for (3.28) in the hyperbolic region as

$$(3.36)$$

$$u_{xx} + yu_{yy} = \left(\partial_x - \sqrt{-y}\,\partial_y\right)\left(\partial_x + \sqrt{-y}\,\partial_y\right)u - \frac{1}{2}\partial_y u$$

$$= \left[4\partial_{\xi\eta}^2 + \frac{2}{\xi - \eta}\left(\partial_\xi - \partial_\eta\right)\right]u = 4\left[u_{\xi\eta} + \frac{1}{2(\xi - \eta)}\left(u_\xi - u_\eta\right)\right] = 0.$$

The characteristic curves for the hyperbolic case are the two branches of the parabolas $y = -\tfrac{1}{4}(x - c)^2$, where c is a constant, as shown in Figure 3.1. The branches with the positive slopes give the curves $\xi = $ constant, whereas $\eta = $ constant gives the branches with the negative slopes. Note that both branches are tangent to the x-axis, which is the single characteristic curve in the parabolic region. In fact, the x-axis is the envelope of the characteristic curves for the hyperbolic region $y < 0$.

In the elliptic region $y > 0$, we have $\omega^{\pm}(x, y) = \pm \sqrt{-y} = \pm i\sqrt{y}$, so that the characteristic curves are complex. The equations (3.32)–(3.33) remain valid but take the form

$$(3.37) \qquad \xi = x + 2i\sqrt{y}\,; \qquad \eta = x - 2i\sqrt{y}\,.$$

Using (3.26) gives

$$(3.38) \qquad \alpha = x; \qquad \beta = 2\sqrt{y}\,.$$

With α and β as new variables, we have

$$(3.39) \qquad u_{xx} = u_{\alpha\alpha}; \qquad u_y = \frac{1}{\sqrt{y}} u_\beta; \qquad u_{yy} = \frac{1}{y} u_{\beta\beta} - \frac{1}{2y^{3/2}} u_\beta$$

so that the canonical form is

$$(3.40) \qquad u_{xx} + y u_{yy} = u_{\alpha\alpha} + u_{\beta\beta} - \frac{1}{\beta} u_\beta = 0.$$

We conclude our discussion of this example with the observation that the canonical forms (3.36) and (3.40) in the hyperbolic and elliptic cases have coefficients that are singular when $\xi = \eta$ and $\beta = 0$, respectively. Both singular regions correspond to the x-axis, across which the equation (3.28) undergoes a transition from hyperbolic to parabolic to elliptic type. The lack of validity of the hyperbolic and elliptic canonical forms at the x-axis is signalled by the singularity of the coefficients there. Although the study of equations of mixed type is of interest in a number of applications, we shall deal (almost) exclusively with equations of a single type in the region under consideration.

3.2. CHARACTERISTIC CURVES

To introduce and motivate our discussion of characteristic curves for second order partial differential equations, we begin by reconsidering linear first order equations and their characteristics. However, we now discuss them from a different point of view than that of Section 2.2.

The characteristic (base) curves of the linear first order equation

$$(3.41) \qquad a(x, y)u_x + b(x, y)u_y = c(x, y)u + d(x, y)$$

can be obtained as solutions of

$$(3.42) \qquad \frac{dy}{dx} = \frac{b}{a}.$$

[The characteristic equations (2.9), where t is replaced by y, can be combined and expressed in the form (3.42).] We suppose that $y = h(x)$ is a solution of (3.42) and look for a solution u of (3.41) with the initial value $u[x, h(x)] = f(x)$. Using the chain rule and (3.41)–(3.42), we conclude that $adu/dx = af'(x) = cf + d$ on the characteristic base curve $y = h(x)$. Unless $f(x)$ satisfies this compatibility condition, the initial value problem has no solution. If $f(x)$ does satisfy the compatibility equation, the curve $y = h(x)$, $u[x, h(x)] = f(x)$ in (x, y, u)-space is a characteristic curve for (3.41) and there are infinitely many solutions or integral surfaces $u = u(x, y)$ of that equation that contain the characteristic curve. This is a consequence of our discussion in Section 2.1, where solutions of first order equations were generated by characteristic curves.

If we ask for a solution of (3.41) that is continuous along the curve $y = h(x)$ but has a discontinuous derivative across it, then this curve must be a solution of (3.42) (i.e., it is a characteristic base curve), as demonstrated below. This is to be expected since the only way two different integral surfaces of (3.41) can be pieced together continuously, but with jumps in their (limiting) normal vectors across a curve, is when that curve is a characteristic curve, as follows from Chapter 2 where integral surfaces were generated from characteristic curves.

It is possible, however, to construct a (formal) solution of (3.41) that together with its first derivatives has a jump discontinuity across a curve $y = g(x)$, without having the curve be a characteristic base curve. Yet, as the example discussed indicates, solutions for which the curve of discontinuity is not characteristic must be constructed by artificial means and do not represent the solution of an initial value problem for (3.41). Solutions of (3.41) that are discontinuous across characteristics are called weak solutions. Weak solutions of second order equations are discussed later in the text and much of that discussion is readily adapted to deal with first order equations.

A simple example that brings out the main points of the foregoing discussion is furnished by the equation $u_x + u_y = 0$. Its characteristic base curves are the solutions of $y' = 1$ and they are given as $y - x = $ constant. With $y = h(x) = x$ and $u(x, x) = f(x)$, we must choose $f(x) = $ constant to satisfy the compatibility condition. If we let $f(x) = 0$, we find that $u = \alpha(y - x)$, where α is an arbitrary constant, is a family of solutions of the initial value problem with $u(x, x) = 0$ (that is, there are infinitely many solutions). The function $u = \alpha_1(y - x)$, $y \leq x$ and $u = \alpha_2(y - x)$, $y > x$, where $\alpha_1 \neq \alpha_2$, represents a solution that is continuous at the characteristic base curve $y = x$, but whose first derivatives have jump discontinuities across it. The function $u = \beta_1(y - x)$, $y < 0$, and $u = \beta_2(y - x)$, $y > 0$, is a solution of the given equation for $y \neq 0$, but it is not continuous across $y = 0$ when $\beta_1 \neq \beta_2$. Also, $u_x = -\beta_1$ for $y < 0$ and $u_x = -\beta_2$ for $y > 0$. However, the curve $y = 0$ is not a characteristic base curve. Yet if we select $x = 0$, $u(0, y) = \beta_1 y$ with $y \leq 0$, and $x = 0$, $u(0, y) = \beta_2 y$ with $y > 0$ as an initial curve for $u_x + u_y = 0$, the (formal) solution is given as $u(x, y) = \beta_1(y - x)$, $y \leq x$ and $u(x, y) = \beta_2(y - x)$, $y > x$. [This follows from the method of characteristics or by using the general solution $u = F(x - y)$ of the equation.] The solution is continuous everywhere and its derivative has a jump discontinuity across the characteristic base curve $y = x$ and not across $y = 0$. Thus even though the prescribed initial curve lies on the originally given discontinuous integral surface, the integral surface obtained with that curve as an initial condition has a derivative discontinuity across a characteristic and does not coincide with the given discontinuous integral surface.

We now show that curves across which first derivatives of continuous solutions of (3.41) have discontinuities must be characteristic base curves. Let $y = h(x)$ be the curve of discontinuity and represent it implicitly as $\varphi(x, y) = y - h(x) = 0$. We consider the family of curves $\varphi(x, y) = $ constant and the family of orthogonal trajectories $\psi(x, y) = $ constant [the orthogonal trajectories are solutions of the equation $y' = -1/h'(x)$]. Introducing the (ξ, η)-coordinate system defined by

$$(3.43) \qquad\qquad \xi = \varphi(x, y); \qquad \eta = \psi(x, y)$$

we obtain for $u = u(\xi, \eta) = u[\xi(x, y), \eta(x, y)]$,

$$(3.44) \qquad\qquad u_x = u_\xi \xi_x + u_\eta \eta_x = \varphi_x u_\xi + \psi_x u_\eta,$$

$$(3.45) \qquad\qquad u_y = u_\xi \xi_y + u_\eta \eta_y = \varphi_y u_\xi + \psi_y u_\eta.$$

Thus

$$(3.46) \quad au_x + bu_y = \left(a\varphi_x + b\varphi_y\right)u_\xi + \left(a\psi_x + b\psi_y\right)u_\eta = cu + d.$$

Now let $u_1(x, y)$ and $u_2(x, y)$ represent solutions of (3.46) defined in the regions $\xi \geqslant 0$ and $\xi \leqslant 0$, respectively, with $u_1 = u_2$ along the curve $\xi = y - h(x) = 0$. With $u = u_1$ for $\xi \geqslant 0$ and $u = u_2$ for $\xi < 0$, the solution u and the derivative u_η are continuous on the curve $\xi = 0$. However, u_ξ represents a normal derivative across the curve $\xi = 0$ and u_ξ has a jump discontinuity there by assumption (note that u_η is an interior derivative along the curve $\xi = $ constant). Since u_1 and u_2 are solutions of (3.41) for $\xi \neq 0$ we have

$$(3.47) \quad \left[\left(a\varphi_x + b\varphi_y\right)u_\xi + \left(a\psi_x + b\psi_y\right)u_\eta - cu - d\right]\Bigg|_{u=u_1, \xi<0}^{u=u_2, \xi>0} = 0,$$

where we consider the difference of the bracketed expression across the curve $\xi = 0$ along the orthogonal curves $\eta = $ constant. In the limit as $\xi \to 0$ we obtain from (3.47)

$$(3.48) \qquad\qquad \left(a\varphi_x + b\varphi_y\right)[u_\xi]|_{\xi=0} = 0$$

where $[u_\xi]|_{\xi=0}$ is the jump in u_ξ across the curve $\xi = 0$. The other terms vanish since they are continuous at $\xi = 0$. By assumption $[u_\xi]|_{\xi=0} \neq 0$, so that (3.48) implies

$$(3.49) \qquad\qquad a\varphi_x + b\varphi_y|_{\xi=0} = -ah'(x) + b = 0,$$

since $\xi = 0$ corresponds to the curve $\varphi(x, y) = y - h(x) = 0$. As a result, $h'(x) = y'(x) = b/a$ so that $y = h(x)$ is a solution of the characteristic equation (3.42) as was to be shown. That is, discontinuities in derivatives of continuous solutions $u = u(x, y)$ of (3.41) can occur only across characteristic (base) curves.

We now demonstrate that if initial data $u(x, y)$ are prescribed for the equation (3.41) along a characteristic (base) curve $y = h(x)$, it is impossible to solve for u_x and u_y uniquely along that curve, if a solution exists at all. Thus if there is a solution, no unique tangent plane can be defined along the initial curve and it is not possible to specify a unique integral surface containing the initial curve that satisfies the initial value problem. (This result has been obtained previously, based on the method of characteristics, but we now obtain it in a different way, one that can be easily generalized to deal with higher order equations.)

Given the initial value $u[x, h(x)] = f(x)$ on the curve $y = h(x)$ that satisfies (3.42), we have

$$(3.50) \qquad \frac{du[x, h(x)]}{dx} = u_x + h'(x)u_y = f'(x),$$

Combining (3.50) with (3.41) evaluated on $y = h(x)$ yields the simultaneous system for u_x and u_y,

$$(3.51) \qquad \begin{cases} u_x + h'(x)u_y = f'(x) \\ au_x + bu_y = cf(x) + d. \end{cases}$$

The determinant of coefficients of this system vanishes; that is,

$$(3.52) \qquad D = \begin{vmatrix} 1 & h'(x) \\ a & b \end{vmatrix} = -ah'(x) + b = 0$$

since $y'(x) = h'(x) = b/a$ in view of (3.42). Consequently, either u_x and u_y cannot be determined at all on $y = h(x)$ or, at best, they can be solved for nonuniquely if $f(x)$ satisfies a consistency condition. This condition is readily found to be $af'(x) = cf(x) + d$, which is identical with the compatibility condition given above. Conversely, the above discussion shows that a curve $y = h(x)$ on which u is prescribed and for which u_x and u_y either cannot be determined or ·an only be determined nonuniquely must be characteristic since we must then have $D = 0$ in (3.52). The determinant D vanishes only if $y = h(x)$ is a solution of (3.42).

To determine the characteristic curves for the second order equation (3.1), we apply the first of the two techniques developed for first order equations. (The second approach is presented in the exercises.) We shall specify the possible curves $y = h(x)$ across which the second derivatives of a solution

$u = u(x, y)$ of (3.1) can have discontinuities. The solution $u(x, y)$ and its first derivatives are assumed to be continuous across these curves.

Let $\varphi(x, y) = y - h(x) = 0$ and define the orthogonal family of curves $\xi = \text{constant}$ and $\eta = \text{constant}$ as in (3.43). In terms of the variables ξ and η, (3.1) takes the form

(3.53)
$$\left(A\varphi_x^2 + 2B\varphi_x\varphi_y + C\varphi_y^2 \right) u_{\xi\xi}$$
$$+ 2\left[A\varphi_x\psi_x + B(\varphi_x\psi_y + \varphi_y\psi_x) + C\varphi_y\psi_y \right] u_{\xi\eta}$$
$$+ \left(A\psi_x^2 + 2B\psi_x\psi_y + C\psi_y^2 \right) u_{\eta\eta} + \ldots = G,$$

where the dots represent first derivative and undifferentiated terms in u. Since u and u_ξ are assumed to be continuous across $\xi = y - h(x) = 0$, we conclude that u_η, $u_{\xi\eta}$, and $u_{\eta\eta}$ are also continuous since $\partial/\partial\eta$ is an interior derivative operator along $\xi = 0$.

Again we consider solutions $u_1(x, y)$ and $u_2(x, y)$ of (3.1) defined in the regions $\xi \geq 0$ and $\xi \leq 0$, respectively. Evaluating (3.53) in each of the above regions, taking the difference of these equations, and going to the limit as $\xi \to 0$, we obtain

(3.54)
$$\left(A\varphi_x^2 + 2B\varphi_x\varphi_y + C\varphi_y^2 \right)[u_{\xi\xi}]|_{\xi=0} = 0$$

where $[u_{\xi\xi}]|_{\xi=0}$ is the jump in the second derivative of u across the curve $\xi = 0$. All other terms vanish in view of their assumed continuity across that curve. Since $[u_{\xi\xi}]|_{\xi=0}$ is taken to be nonzero, we must have

(3.55)
$$A\varphi_x^2 + 2B\varphi_x\varphi_y + C\varphi_y^2 = 0.$$

This is known as the *characteristic equation* for (3.1).

Assuming (as was done in Section 3.1) that the coefficient A in (3.1) is nonzero, we can factor (3.55) as

(3.56)
$$A\left(\varphi_x - \omega^+\varphi_y \right)\left(\varphi_x - \omega^-\varphi_y \right) = 0,$$

with ω^+ and ω^- defined as in (3.4). If ω^+ and ω^- are real valued, we see that a curve $\xi = \varphi(x, y) = 0$ or, equivalently, $y = h(x)$ across which $u_{\xi\xi}$ is discontinuous must be one of the characteristic curves (3.8). This follows since $\varphi = y - h(x)$ implies that $\varphi_x - \omega^\pm\varphi_y = -h'(x) - \omega^\pm$. Thus one of the two equations $y' = h' = -\omega^\pm$ must be satisfied in view of (3.56).

Using similar arguments, it can be shown that discontinuities in higher derivatives of a solution must also occur only across characteristic curves.

EXAMPLE 3.2.   THE WAVE, DIFFUSION, AND LAPLACE'S EQUATIONS

In this example we discuss the characteristic equations (3.55) for the wave, diffusion, and Laplace's equations derived in Chapter 1.

For the wave equation (1.91), the characteristic equation has the form

$$(3.57) \qquad \varphi_x^2 - \gamma^2\varphi_t^2 = (\varphi_x + \gamma\varphi_t)(\varphi_x - \gamma\varphi_t) = 0.$$

Equating each factor to zero separately gives as the characteristic curves

$$(3.58) \qquad \varphi(x, t) = x \pm \gamma t = \text{constant},$$

which are straight lines in the (x, t)-plane with slopes $dt/dx = \pm 1/\gamma$. Since the telegrapher's equation (1.79) has the same principal part as the wave equation (1.91), it has the same characteristic curves (3.58).

For the diffusion equation (1.21) we have the characteristic equation

$$(3.59) \qquad \left(\frac{D}{2}\right)\varphi_x^2 = 0,$$

which implies that $\varphi = \varphi(t) = $ constant, so that the straight lines $t = $ constant are the characteristics.

Laplace's equation (1.98) yields the characteristic equation

$$(3.60) \qquad \varphi_x^2 + \varphi_y^2 = 0.$$

For real valued $\varphi(x, y)$ we must have $\varphi_x = \varphi_y = 0$ so that no real solutions other than $\varphi \equiv$ constant exist. Thus Laplace's equation has no real characteristic curves.

The results of the above example are typical for equations of hyperbolic, parabolic, and elliptic type. Elliptic equations have no real characteristic curves so that solutions cannot have discontinuous derivatives. Consequently, solutions $u = u(x, y)$ are extremely smooth functions and this is consistent with the nature of elliptic equations that describe equilibrium processes where everything has already smoothed itself out.

Parabolic equations of the type we generally consider, have the lines $t = $ constant as characteristics, and discontinuities in derivatives must occur across these lines. Since we are concerned with the evolution of solutions as t (i.e., the time variable) increases, if these discontinuities occur initially when $t = 0$, they cannot be spread into the region $t > 0$ and again solutions are smooth functions.

It is in the theory of equations of hyperbolic type that characteristics play the most significant role. The preceding example shows that for the wave and telegrapher's equations the characteristics extend from the (initial) line $t = 0$ into the region $t > 0$. Thus in their role as curves across which discontinuities or singularities in the solution occur, they act as carriers of singular initial data for the solution and the effects of these singularities are felt for all time. The importance of characteristics in the solution of various problems for hyperbolic equations is demonstrated throughout the text.

3.3. CLASSIFICATION OF EQUATIONS IN GENERAL AND THEIR CHARACTERISTICS

We begin by considering the second order linear partial differential equation in n variables

$$(3.61) \qquad \sum_{i=1}^{n} \sum_{j=1}^{n} a_{ij} \frac{\partial^2 u}{\partial x_i \partial x_j} + \sum_{i=1}^{n} b_i \frac{\partial u}{\partial x_i} + cu + d = 0,$$

where u, a_{ij}, b_i, c, and d are functions of $x_1, \ldots, x_n$. This equation cannot, in general, be reduced to a simple canonical form over a full region, as was done in the case of two independent variables in Section 3.1. For the purpose of classifying (3.61) into different types, we shall generalize the factorization procedure of Section 3.1. It will then be seen that if the coefficients a_{ij} are identically constant or if we restrict ourselves to a single point in $(x_1, \ldots, x_n)$-space, it is possible to bring the principal part of (3.61) (i.e., the highest derivative terms) into a canonical form.

With $\partial_{x_i} = \partial/\partial x_i$, $i = 1, \ldots, n$, and $\partial_x^T = [\partial_{x_1}, \ldots, \partial_{x_n}]$, as a row (gradient) vector (which is the transpose of the column vector ∂_x), it is easily seen that we may express the principal part of the differential operator in (3.61) as

$$(3.62) \qquad \sum_{i=1}^{n} \sum_{j=1}^{n} a_{ij} \partial_{x_i} \partial_{x_j} = \partial_x^T A \partial_x + \cdots,$$

where the dots stand for first derivative terms and the n by n matrix A has the coefficients a_{ij} as its elements. Note that if the a_{ij} are constants, the first derivative terms in (3.62) are absent. Since we assume that mixed partial derivatives of u are equal (i.e., $u_{x_i x_j} = u_{x_j x_i}$), (3.61) may be arranged so that its coefficients have the property that $a_{ij} = a_{ji}$ and we assume that this has been done. Thus the matrix A is symmetric and it is assumed to be real valued in the present discussion.

It is well known from matrix theory that a real valued symmetric matrix A has only real eigenvalues $\lambda_1, \lambda_2, \ldots, \lambda_n$ (counted with their multiplicities) and that there exists a corresponding orthonormal set of n eigenvectors $r_1, \ldots, r_n$. Forming the matrix R with the eigenvectors r_i as its n columns, we find that R is an orthogonal matrix with the property

$$(3.63) \qquad R^T A R = D = \begin{bmatrix} \lambda_1 & & 0 \\ & \ddots & \\ 0 & & \lambda_n \end{bmatrix}.$$

That is, using R we can diagonalize the matrix A.

We now introduce the directional derivative operators

(3.64) $\partial_{\xi_i} = \mathbf{r}_i^T \partial_x, \qquad i = 1, \ldots, n,$

and form the vector operator

(3.65) $\partial_\xi = R^T \partial_x.$

[The operators ∂_{ξ_i} may be compared with the operators ∂_ξ and ∂_η in
(3.16)–(3.17).] Since R is an orthogonal matrix so that $R^T = R^{-1}$, (i.e., its
transpose equals its inverse), we have from (3.65) $\partial_x = R\partial_\xi$. Introducing this
expression into (3.62) gives

(3.66) $\partial_x^T A \partial_x = \left(R\partial_\xi \right)^T A \left(R\partial_\xi \right) = \partial_\xi^T \left(R^T A R \right) \partial_\xi + \cdots$

$$= \partial_\xi^T D \partial_\xi + \cdots = \sum_{i=1}^{n} \lambda_i \partial_{\xi_i}^2 + \cdots,$$

where the dots represent first derivative operators (since the elements in the
preceding matrices need not be constant) and where (3.63) has been used.

Recalling the classification method for equations in two independent variables given in Section 3.1, we find that in terms of the result (3.66), the elliptic
case corresponds to the situation when λ_1 and λ_2 have the same sign. The
hyperbolic case requires that λ_1 and λ_2 have opposite signs whereas the
parabolic case occurs when λ_1 or λ_2 equals zero. For the n dimensional case in
(3.61), we base our classification of the equation at a point P in $(x_1, \ldots, x_n)$-
space on the result obtained in (3.66). Thus it is characterized by the properties
of the eigenvalues of A.

With $\lambda_1, \lambda_2, \ldots, \lambda_n$ as the eigenvalues of A, we introduce the following
classification for the equation (3.61):

(3.67) $\begin{cases} \lambda_i > 0, & \text{all } i, \\ \text{or} \\ \lambda_i < 0, & \text{all } i, \end{cases}$ *elliptic type.*

(3.68) $\begin{cases} \text{One of the } \lambda_i < 0 \text{ or } \lambda_i > 0; \\ \text{All other } \lambda_i \text{ have opposite sign;} \end{cases}$ *hyperbolic type.*

(3.69)   One or more of the $\lambda_i = 0$;   *parabolic type.*

For the two and three-dimensional cases the above classification exhausts
all possibilities. However, if $n \geqslant 4$ in (3.61), it may happen that two or more of
the λ_i have one sign, whereas two or more of the remaining λ_i have the
opposite sign. Such equations are said to be of ultrahyperbolic type. Since they
do not occur often in applications they are not studied in this text.

If (3.61) is of fixed type at every point in a region, the equation is said to be of that type in the region. If the type changes within a region, the equation is said to be of mixed type. When the coefficients a_{ij} of the principal part of (3.61) are identically constant, the equation is clearly of one type everywhere in $(x_1, \ldots, x_n)$-space.

As basic examples of second order partial differential equations we have

$$(3.70) \qquad \nabla^2 u = u_{xx} + u_{yy} + u_{zz} = 0; \qquad \textit{elliptic type.}$$

$$(3.71) \qquad u_{tt} - \nabla^2 u = u_{tt} - u_{xx} - u_{yy} - u_{zz} = 0; \textit{ hyperbolic type.}$$

$$(3.72) \qquad u_t - \nabla^2 u = u_t - u_{xx} - u_{yy} - u_{zz} = 0; \textit{ parabolic type.}$$

These are the higher dimensional Laplace's, wave, and diffusion equations. The operator $\nabla^2 = \partial^2/\partial x^2 + \partial^2/\partial y^2 + \partial^2/\partial z^2$ is the *Laplacian operator* in three dimensions. A further example is given by *Schrödinger's equation*,

$$i\hbar u_t = -\left(\frac{\hbar^2}{2m}\right)\nabla^2 u + Vu,$$

where $i = \sqrt{-1}$, $\hbar$ is Planck's constant, m is the mass and V is a given potential function. This equation is clearly of parabolic type.

Once we have exhibited how second order equations in n variables are to be classified, it is appropriate to ask whether there exist transformations of the independent variables (similar to those considered in Section 3.1) that bring (3.61) into a canonical form, say, of the type given in (3.66) but where the principal part has constant coefficients. It is shown in the exercises that this is not possible, in general, if $n \geq 3$ in (3.61). However, if the coefficients a_{ij} of the principal part are identically constant, simple canonical forms do exist. In fact (noting (3.64)), the linear transformation

$$(3.73) \qquad \xi_i = \mathbf{r}_i^T \mathbf{x}, \qquad i = 1, \ldots, n,$$

where $\mathbf{x}^T = [x_1, \ldots, x_n]$, transforms the principal part of (3.61) into the form

$$(3.74) \qquad \sum_{i=1}^{n}\sum_{j=i}^{n} a_{ij} u_{x_i x_j} = \sum_{i=1}^{n} \lambda_i u_{\xi_i \xi_i},$$

where the λ_i are the (constant) eigenvalues of A. Clearly, an elementary further transformation of the variables $\xi_1, \ldots, \xi_n$ can reduce the principal part to a form in which the coefficients are either 0, $+1$, or -1.

To determine the characteristics for (3.61) we look for surfaces $\varphi(x_1, \ldots, x_n) = $ constant across which second derivatives of solutions of (3.61) can have discontinuities. This can be done by introducing n (independent) families of surfaces $\varphi^{(i)}(x_1, \ldots, x_n) = $ constant (with $\varphi^{(1)} \equiv \varphi$) and the corresponding coordinate system $\eta_i = \varphi^{(i)}(x_1, \ldots, x_n)$, $i = 1, \ldots, n$. Transforming from the x_i coordinates to the η_i coordinates in (3.61) we easily obtain, with $\eta_1 = \varphi$,

$$(3.75) \qquad \left[\sum_{i=1}^{n} \sum_{j=1}^{n} a_{ij} \varphi_{x_i} \varphi_{x_j} \right] u_{\varphi\varphi} + \ldots = 0,$$

where the dots represent second derivative terms in the variables $\eta_2, \ldots, \eta_n$ as well as first derivative terms and undifferentiated terms in all variables. Now $u_{\varphi\varphi}$ is a second derivative of u across the surface $\varphi = $ constant. Thus if $\varphi = $ constant satisfies the equation

$$(3.76) \qquad \sum_{i=1}^{n} \sum_{j=1}^{n} a_{ij} \varphi_{x_i} \varphi_{x_j} = 0,$$

$u_{\varphi\varphi}$ cannot be uniquely determined in terms of u_φ and u, as well as interior derivatives of u, given on the surface $\varphi = $ constant. Consequently, the only surfaces across which u may have discontinuities in its second derivatives are solutions of (3.76). These surfaces are the *characteristic surfaces* for (3.61), and (3.76) is the *characteristic equation*.

For example, Laplace's equation (3.70) has the characteristic equation

$$(3.77) \qquad \varphi_x^2 + \varphi_y^2 + \varphi_z^2 = 0,$$

and this equation has no real solutions. Thus Laplace's equation has no real characteristics. This indicates that its solutions are smooth functions. The three-dimensional wave equation (3.71) yields the characteristic equation

$$(3.78) \qquad \varphi_t^2 - \varphi_x^2 - \varphi_y^2 - \varphi_z^2 = 0.$$

Among the important solutions of (3.78) are the plane waves,

$$(3.79) \qquad \varphi = \omega t - \kappa_1 x - \kappa_2 y - \kappa_3 z = \text{constant},$$

where $\omega^2 = \kappa_1^2 + \kappa_2^2 + \kappa_3^2$, and a singular solution, the characteristic cone

$$(3.80) \qquad \varphi = (t - t_0)^2 - (x - x_0)^2 - (y - y_0)^2 - (z - z_0)^2 = 0,$$

where $P_0 = (x_0, y_0, z_0, t_0)$ is a given point. The three-dimensional diffusion equation (3.72) has the characteristic equation

$$(3.81) \qquad \varphi_x^2 + \varphi_y^2 + \varphi_z^2 = 0,$$

which yields the characteristic surfaces

(3.82) $$\varphi(x, y, z, t) = t = \text{constant.}$$

EXAMPLE 3.3.   CLASSIFICATION OF AN EQUATION OF MIXED TYPE

We consider the equation

(3.83)
$$u_{x_1 x_1} + 2(1 + cx_2) u_{x_2 x_3} = u_{x_1 x_1} + (1 + cx_2) u_{x_2 x_3} + (1 + cx_2) u_{x_3 x_2} = 0$$

where c is a constant and apply the above classification procedure. From (3.62) we have

(3.84) $$\partial_{x_1}^2 + (1 + cx_2) \partial_{x_2} \partial_{x_3} + (1 + cx_2) \partial_{x_3} \partial_{x_2} = \partial_x^T A \partial_x - c \partial_{x_3}$$

where

(3.85) $$A = \begin{bmatrix} 1 & 0 & 0 \\ 0 & 0 & (1 + cx_2) \\ 0 & (1 + cx_2) & 0 \end{bmatrix}; \qquad \partial_x = \begin{bmatrix} \partial_{x_1} \\ \partial_{x_2} \\ \partial_{x_3} \end{bmatrix}.$$

The eigenvalues and the corresponding orthonormalized eigenvectors of A are

(3.86) $$\lambda_1 = 1, \mathbf{r}_1 = \begin{bmatrix} 1 \\ 0 \\ 0 \end{bmatrix}; \qquad \lambda_2 = 1 + cx_2, \mathbf{r}_2 = \begin{bmatrix} 0 \\ \dfrac{1}{\sqrt{2}} \\ \dfrac{1}{\sqrt{2}} \end{bmatrix};$$

$$\lambda_3 = -(1 + cx_2), \mathbf{r}_3 = \begin{bmatrix} 0 \\ \dfrac{1}{\sqrt{2}} \\ \dfrac{-1}{\sqrt{2}} \end{bmatrix}.$$

The orthogonal matrix R is

(3.87) $$R = \begin{bmatrix} 1 & 0 & 0 \\ 0 & \dfrac{1}{\sqrt{2}} & \dfrac{1}{\sqrt{2}} \\ 0 & \dfrac{1}{\sqrt{2}} & \dfrac{-1}{\sqrt{2}} \end{bmatrix},$$

and it may be verified that $R = R^T = R^{-1}$ and that

$$(3.88) \qquad R^T A R = \begin{bmatrix} 1 & 0 & 0 \\ 0 & (1 + cx_2) & 0 \\ 0 & 0 & -(1 + cx_2) \end{bmatrix} = D.$$

Further,

$$(3.89) \qquad \partial_\xi = R\partial_x = \begin{bmatrix} \partial_{x_1} \\ \dfrac{1}{\sqrt{2}}\left(\partial_{x_2} + \partial_{x_3}\right) \\ \dfrac{1}{\sqrt{2}}\left(\partial_{x_2} - \partial_{x_3}\right) \end{bmatrix} = \begin{bmatrix} \partial_{\xi_1} \\ \partial_{\xi_2} \\ \partial_{\xi_3} \end{bmatrix},$$

so that, since R is a constant matrix,

(3.90)

$$\partial_x^T A \partial_x = \left(R\partial_\xi\right)^T A\left(R\partial_\xi\right) = \partial_\xi^T D \partial_\xi = \left[\partial_{x_1}, \frac{1}{\sqrt{2}}\left(\partial_{x_2} + \partial_{x_3}\right), \frac{1}{\sqrt{2}}\left(\partial_{x_2} - \partial_{x_3}\right)\right]$$

$$\cdot \begin{bmatrix} 1 & 0 & 0 \\ 0 & (1 + cx_2) & 0 \\ 0 & 0 & -(1 + cx_2) \end{bmatrix} \begin{bmatrix} \partial_{x_1} \\ \dfrac{1}{\sqrt{2}}\left(\partial_{x_2} + \partial_{x_3}\right) \\ \dfrac{1}{\sqrt{2}}\left(\partial_{x_2} - \partial_{x_3}\right) \end{bmatrix}$$

$$= \partial_{x_1}^2 + \left(\frac{1 + cx_2}{2}\right)\left(\partial_{x_2} + \partial_{x_3}\right)^2 - \left(\frac{1 + cx_2}{2}\right)\left(\partial_{x_2} - \partial_{x_3}\right)^2$$

$$+ \frac{c}{2}\left(\partial_{x_2} + \partial_{x_3}\right) - \frac{c}{2}\left(\partial_{x_2} - \partial_{x_3}\right)$$

$$= \lambda_1 \partial_{\xi_1}^2 + \lambda_2 \partial_{\xi_2}^2 + \lambda_3 \partial_{\xi_3}^2 + \frac{c}{\sqrt{2}} \partial_{\xi_2} - \frac{c}{\sqrt{2}} \partial_{\xi_3},$$

where we have used (3.86) and (3.89).

In view of (3.68)–(3.69), we conclude that (3.83) is parabolic when $x_2 = -1/c$ (if $c \neq 0$) and is hyperbolic in the half-spaces $x_2 > -1/c$ and $x_2 < -1/c$. If $c = 0$, all the eigenvalues are constant with $\lambda_1 = \lambda_2 = 1$ and $\lambda_3 = -1$, so that (3.83) is hyperbolic everywhere. Applying the transformation (3.73), we have

$$(3.91) \qquad \xi_1 = x_1; \qquad \xi_2 = \frac{1}{\sqrt{2}}\left(x_2 + x_3\right); \qquad \xi_3 = \frac{1}{\sqrt{2}}\left(x_2 - x_3\right),$$

and (3.83) takes on the canonical form

$$(3.92) \qquad u_{\xi_1\xi_1} + u_{\xi_2\xi_2} - u_{\xi_3\xi_3} = 0.$$

This has the form of the wave equation in two dimensions with ξ_3 as the time variable.

We continue our discussion by considering the classification of first order systems of linear partial differential equations in two independent variables. The equations may be written as

$$(3.93) \qquad A\frac{\partial \mathbf{u}}{\partial x} + B\frac{\partial \mathbf{u}}{\partial y} = C\mathbf{u} + \mathbf{d},$$

where A, B, and C are n by n matrix functions of x and y and $\mathbf{u}(x, y)$ and $\mathbf{d}(x, y)$ are n-component vectors. Our classification procedure is based on the properties of the characteristic curves that can occur for (3.93).

Now (3.93) represents a natural generalization of the linear first order equation (3.41) with the scalar dependent variable replaced by a vector variable. Consequently, it is appropriate to formulate an initial value problem for (3.93) that assigns the value $\mathbf{u} = \mathbf{f}(x)$ on the curve $y = h(x)$. Proceeding as in Section 3.2, the characteristic curves of (3.93) are defined to be those curves on which $\mathbf{u}_x$ and $\mathbf{u}_y$ cannot be uniquely specified (if indeed they can be determined at all) in terms of the initial data given.

On the initial curve $y = h(x)$ we have $\mathbf{u}[x, h(x)] = \mathbf{f}(x)$. Thus

$$(3.94) \qquad \frac{d\mathbf{u}[x, h(x)]}{dx} = \mathbf{u}_x + h'(x)\mathbf{u}_y = \mathbf{f}'(x).$$

Solving for $\mathbf{u}_x$ in terms of $\mathbf{u}_y$ and substituting in (3.93), gives

$$(3.95) \qquad [B - h'(x)A]\mathbf{u}_y = C\mathbf{f}(x) - A\mathbf{f}'(x) + \mathbf{d},$$

where A, B, C, and $\mathbf{d}$ are evaluated on the curve $y = h(x)$. Now (3.95) represents a system of equations for $\mathbf{u}_y$ [on the curve $y = h(x)$], and it has a unique solution only if the determinant of the coefficient matrix of $\mathbf{u}_y$ is nonzero, that is,

$$(3.96) \qquad \det[B - h'(x)A] \equiv |B - h'(x)A| \neq 0.$$

A curve $y = h(x)$ for which this determinant vanishes identically is called a *characteristic curve* for (3.93). If the initial conditions $\mathbf{u}[x, h(x)] = \mathbf{f}(x)$ are such that the system (3.95) has a (nonunique) solution, we have a *characteristic initial value problem*. In any case, $y = h(x)$ is called a characteristic curve.

To determine the full set of (possible) characteristic curves, we express the characteristic determinant (3.96) in the form

$$(3.97) \qquad\qquad |B[x, y(x)] - y'(x)A[x, y(x)]| = 0,$$

with $y = y(x)$ as the characteristic curve. This determinant yields an nth degree algebraic equation for $y'(x)$ and this is the *characteristic equation* for (3.93). [Note that if we set $\varphi(x, y) = y - y(x) = 0$, then (3.97) can be written as $|\varphi_x A + \varphi_y B| = 0$. The characteristic equation for (3.93) is often written in this form.]

If all the roots $y'(x)$ of (3.97) are real and distinct in some region, the system (3.93) is said to be *totally hyperbolic* in that region. Denoting these roots by the functions $\omega_i(x)$ $(i = 1,\dots, n)$ we obtain n families of characteristic curves as solutions of the equations.

$$(3.98) \qquad\qquad y'(x) = \omega_i(x); \qquad i = 1, 2,\dots, n.$$

If all the roots $\omega_i(x)$ of (3.97) are complex valued, the system (3.93) is said to be of *elliptic type*. Then there are no real characteristic curves as carriers of possible discontinuities in derivatives of solutions and the solutions are expected to be smooth functions. The case where some roots are real and others are complex will not be considered. When all the roots are real but one or more is a multiple root, additional conditions which we now consider must be given to determine if (3.93) is of hyperbolic or parabolic type.

We shall now assume that the coefficient matrix B in (3.97) is nonsingular throughout the region under consideration so that it has an inverse. Multiplying through by the inverse matrix B^{-1} in (3.93), we obtain an equation of the form

$$(3.99) \qquad\qquad \mathbf{u}_y + A\mathbf{u}_x = C\mathbf{u} + \mathbf{d},$$

that is, we have effectively put $B = I$, the identity matrix. Then, with $B = I$ in (3.97) and λ_i $(i = 1,\dots, n)$ as the eigenvalues of A counted with their multiplicities, we have $\lambda_i = 1/\omega_i$, $i = 1,\dots, n$. (We note that under the present assumptions we cannot have $\omega_i = 0$.) If all the eigenvalues are real and there exist n linearly independent eigenvectors $\mathbf{r}_i$ $(i = 1,\dots, n)$ for the matrix A, the system (3.99) is of *hyperbolic type*. The multiplicity of the eigenvalues plays no role, but it is true that when all the eigenvalues are distinct there exist n linearly independent eigenvectors. In particular, if A is a (real) symmetric matrix, there are always n linear independent eigenvectors. However, if there are multiple (real) eigenvalues and fewer than n linearly independent eigenvectors for A, the system (3.99) is of *parabolic type*.

When (3.99) is of hyperbolic type it can be transformed into the following *canonical form*. We form the matrix R whose column vectors $\mathbf{r}_1,\dots, \mathbf{r}_n$ are the eigenvectors of A. Then let

(3.100) $\mathbf{u} = R\mathbf{v}$

in (3.99) and since R is nonsingular, we have for $\mathbf{v}$,

(3.101) $\mathbf{v}_y + (R^{-1}AR)\mathbf{v}_x = \hat{C}\mathbf{v} + \hat{\mathbf{d}}$,

where $R^{-1}AR$ is a diagonal, not necessarily constant, matrix whose diagonal elements are the eigenvalues of A. (This follows since $R^{-1}AR$ is a similarity transformation which diagonalizes the matrix A.) The matrix $\hat{C}$ equals $R^{-1}C - R^{-1}R_y - R^{-1}AR_x$, where R_x and R_y are derivatives of the matrix R, and $\hat{\mathbf{d}} = R^{-1}\mathbf{d}$.

EXAMPLE 3.4.   FIRST ORDER SYSTEMS IN TWO VARIABLES

In Section 2.1 the wave, diffusion, and Laplace's equations were reduced to first order systems. We now reconsider their classification in system form based on the above discussion.

The system (2.2) for the wave equation may be written as

(3.102) $\mathbf{u}_t + A\mathbf{u}_x = C\mathbf{u}$

where

$$\mathbf{u} = \begin{bmatrix} v \\ u \end{bmatrix}, \qquad A = \begin{bmatrix} -\gamma & 0 \\ 0 & \gamma \end{bmatrix}, \qquad C = \begin{bmatrix} 0 & 1 \\ 0 & 0 \end{bmatrix}.$$

Putting $y = t$ in (3.97) we have

(3.103) $|I - t'(x)A| = \begin{vmatrix} 1 + \gamma t'(x) & 0 \\ 0 & 1 - \gamma t'(x) \end{vmatrix} = 1 - \gamma^2 [t'(x)]^2 = 0,$

as the characteristic equation because $B = I$ in this case. Since $t'(x) = \pm 1/\gamma$, we obtain, as expected, the characteristic curves $x \pm \gamma t = $ constant. The roots of the characteristic equation are $\omega_1 = 1/\gamma$ and $\omega_2 = -1/\gamma$, and these are real and distinct. Thus the system (3.102) is totally hyperbolic.

The Cauchy–Riemann equations (2.6) have the form

(3.104) $\mathbf{u}_x + B\mathbf{u}_y = 0$,

where

$$B = \begin{bmatrix} 0 & -1 \\ 1 & 0 \end{bmatrix}, \qquad A = I, \qquad \mathbf{u} = \begin{bmatrix} w \\ u \end{bmatrix}.$$

Thus (3.97) becomes

$$(3.105) \quad |B - y'(x)A| = \begin{vmatrix} -y'(x) & -1 \\ 1 & -y'(x) \end{vmatrix} = 1 + [y'(x)]^2 = 0.$$

Since $y'(x) = \pm i$ $(i = \sqrt{-1})$, there are no real characteristics and the system (3.104) is elliptic.

The system (2.7) representing the diffusion equation has the form

$$(3.106) \qquad\qquad \hat{A}\mathbf{u}_x + \hat{B}\mathbf{u}_t = \hat{C}\mathbf{u}$$

where

$$\hat{A} = \begin{bmatrix} -\dfrac{D}{2} & 0 \\ 0 & 1 \end{bmatrix}; \quad \hat{B} = \begin{bmatrix} 0 & 1 \\ 0 & 0 \end{bmatrix}; \quad \hat{C} = \begin{bmatrix} -c & 0 \\ 1 & 0 \end{bmatrix} \quad \mathbf{u} = \begin{bmatrix} u \\ v \end{bmatrix}.$$

We multiply across in (3.106) by

$$\hat{A}^{-1} = \begin{bmatrix} -\dfrac{2}{D} & 0 \\ 0 & 1 \end{bmatrix}$$

and obtain

$$(3.107) \qquad\qquad \mathbf{u}_x + B\mathbf{u}_t = C\mathbf{u},$$

where

$$B = \begin{bmatrix} 0 & -\dfrac{2}{D} \\ 0 & 0 \end{bmatrix}, \quad C = \begin{bmatrix} \dfrac{2c}{D} & 0 \\ 1 & 0 \end{bmatrix}.$$

This has the basic form of (3.99) and has the characteristic equation

$$(3.108) \qquad |B - t'(x)I| = \begin{vmatrix} -t'(x) & -\dfrac{2}{D} \\ 0 & -t'(x) \end{vmatrix} = [t'(x)]^2 = 0.$$

Thus $t'(x) = 0$ is a double root and the characteristic curves are $t = $ constant. Since the matrix B has only one linearly independent eigenvector $\mathbf{r} = \begin{bmatrix} 1 \\ 0 \end{bmatrix}$ (as is easily shown), the system (3.107) is of parabolic type.

The system

$$(3.109) \qquad\qquad \mathbf{u}_t + \mathbf{u}_x = C\mathbf{u},$$

where $C = \begin{bmatrix} 0 & 1 \\ 1 & 0 \end{bmatrix}$ and $\mathbf{u} = \begin{bmatrix} u \\ v \end{bmatrix}$ is equivalent to the single equation

$$(3.110) \qquad u_{xx} + 2u_{xt} + u_{tt} - u = 0.$$

Comparing (3.109) with (3.99), we set $t = y$ and $A = I$. The identity matrix I has the double real eigenvalues $\lambda_1 = \lambda_2 = 1$ and the linearly independent eigenvectors $\mathbf{r}_1 = \mathbf{i}$ and $\mathbf{r}_2 = \mathbf{j}$. Thus (3.109) is a hyperbolic system. Yet according to the classification procedure of Section 3.1, [with $y = t$ in (3.1)], we have $A = B = C = 1$ in (3.110) so that $B^2 - AC = 0$ and (3.110) is of parabolic type. This example shows that the definition of hyperbolicity given above for systems of equations is somewhat more general than given for single equations.

Turning to the consideration of a single linear partial differential equation of order m in n independent variables, we have

$$(3.111) \qquad \sum_{i_1=1}^{n} \sum_{i_2=1}^{n} \cdots \sum_{i_m=1}^{n} a_{i_1, i_2, \ldots, i_m} \frac{\partial^m u}{\partial x_{i_1} \cdots \partial x_{i_m}} + \cdots = 0,$$

where we write only the principal part of the differential equation. The dots represent lower derivative and undifferentiated terms in $u(x_1, \ldots, x_n)$. Adapting the procedure given above for the second order equation (3.61), it is readily shown that the characteristic surfaces $\varphi(x_1, \ldots, x_n) = $ constant are solutions of the characteristic equation

$$(3.112) \qquad \sum_{i_1=1}^{n} \cdots \sum_{i_m=1}^{n} a_{i_1, \ldots, i_m} \varphi_{x_{i_1}} \cdots \varphi_{x_{i_m}} = 0.$$

These are surfaces in n-space across which mth order derivatives of solutions of (3.111) can have discontinuities. If there are no real surfaces that satisfy (3.112), the equation (3.111) is of elliptic type. Classification into other types is somewhat more complicated and is not considered. However, specific equations of higher order are discussed in the text.

As an example we consider the *biharmonic equation* in two variables,

$$(3.113) \qquad \nabla^2 \nabla^2 u = u_{xxxx} + 2u_{xxyy} + u_{yyyy} = 0.$$

Its characteristic equation is

$$(3.114) \qquad \varphi_x^4 + 2\varphi_x^2 \varphi_y^2 + \varphi_y^4 = \left(\varphi_x^2 + \varphi_y^2 \right)^2 = 0.$$

Clearly, it has no real solutions so that (3.113) is elliptic.

Next, we consider a system of equations in n variables

$$(3.115) \qquad \sum_{i=1}^{n} A^i \frac{\partial \mathbf{u}}{\partial x_i} + B\mathbf{u} + \mathbf{c} = \mathbf{0},$$

where A^i and B are k by k matrix functions of $x_1, \ldots, x_n$ and $\mathbf{u}$ and $\mathbf{c}$ are k-component column vector functions of $x_1, \ldots, x_n$. The characteristic surfaces for (3.115) are surfaces $\varphi(x_1, \ldots, x_n) = $ constant on which the derivatives $\mathbf{u}_{x_i}$ $(i = 1, \ldots, n)$ cannot be specified uniquely (if at all) for given initial data $\mathbf{u} = \mathbf{f}$ on those surfaces. It can be shown that such surfaces must be solutions of the characteristic equations

$$(3.116) \qquad \det\left[\sum_{i=1}^{n} A^i \frac{\partial \varphi}{\partial x_i} \right] \equiv \left| \sum_{i=1}^{n} A^i \varphi_{x_i} \right| = 0.$$

Again, if no real surfaces satisfying (3.116) exist, (3.115) is of elliptic type. We shall have occasion to characterize systems as being of hyperbolic or parabolic type in specific examples and exercises but do not classify them in general. We also do not discuss the classification of systems of higher order equations. Technically, they can always be reduced to first order systems.

Finally, it is necessary to discuss nonlinear equations and systems. If the principal parts of these equations are linear and the nonlinearities are confined to the lower order terms, the classification of these equations proceeds as above. For example, the equation

$$(3.117) \qquad \nabla^2 u = e^u$$

is of elliptic type, whereas the system

$$(3.118) \qquad \begin{cases} u_t - \gamma u_x = u^2 + v \\ v_t + \gamma v_x = uv \end{cases}$$

where $\gamma = $ constant is of hyperbolic type.

As to equations where nonlinearities occur in the principal parts, we deal only with quasilinear equations. That is, the principal part is linear in the highest derivative terms, but it may have coefficients that contain lower derivative or undifferentiated terms. The classification proceeds as for the linear case, but it depends on the specific solution under consideration. The characteristics also depend on the specific solution. (This fact was already observed in our discussion of first order quasilinear equations in Chapter 2.)

As an example we consider the system

$$(3.119) \qquad \begin{cases} u_t - v_x = 0 \\ v_t - c'(u)u_x = 0 \end{cases}$$

where $c(u)$ is a given differentiable function. (This system occurs in a number of branches of continuum mechanics in one form or another.) Identifying y with t, (3.119) has the form (3.99) with

$$A = \begin{bmatrix} 0 & -1 \\ -c'(u) & 0 \end{bmatrix}, \qquad \mathbf{u} = \begin{bmatrix} u \\ v \end{bmatrix}, \qquad C = 0, \qquad \mathbf{d} = \mathbf{0}.$$

The eigenvalues of the matrix A are $\lambda_1 = \sqrt{c'(u)}$ and $\lambda_2 = -\sqrt{c'(u)}$. If the solution u is such that $c'(u) > 0$, we find that (3.119) is hyperbolic. If $c'(u) = 0$, the system is clearly parabolic, whereas it is elliptic if $c'(u) < 0$. In the hyperbolic case, the characteristic curves are determined from the equations $dx/dt = \pm \sqrt{c'(u)}$. They are obviously dependent on the particular solution u under consideration.

Putting $v = w_t$ and $u = w_x$ in (3.119), we obtain the second order quasilinear equation

$$(3.120) \qquad\qquad w_{tt} - c'(w_x)w_{xx} = 0.$$

Considering a specific solution w in (3.120) and recalling the results of Section 3.1 we have

$$(3.121) \qquad\qquad B^2 - AC = c'(w_x).$$

Thus (3.120) is hyperbolic if $c'(w_x) > 0$, parabolic if $c'(w_x) = 0$, and elliptic if $c'(w_x) < 0$. This is consistent with the above results for the system (3.119) since $u = w_x$.

We have occasion throughout the text to consider specific quasilinear partial differential equations and systems apart from those already studied in Chapter 2.

3.4. FORMULATION OF INITIAL AND BOUNDARY VALUE PROBLEMS

As we have seen in our discussion of the model equations derived in Chapter 1 and the first order partial differential equations of Chapter 2, we are not merely interested in finding arbitrary functions that satisfy the given differential equations. Rather, we ask for specific solutions that satisfy certain auxiliary conditions associated with the given problem.

The diffusion and telegrapher's or wave equations that are of parabolic and hyperbolic types, respectively, both contain a time dependence. We have found it appropriate in our discussion in Chapter 1, to prescribe values of the solutions of these equations at an initial time $t = 0$. For the diffusion equation, initial values for the density function $v(x, t)$ were assigned, whereas for the telegrapher's and the wave equation, both $v(x, t)$ and $\partial v(x, t)/\partial t$ were prescribed at $t = 0$. In case the values of x are unrestricted, this constitutes an *initial value problem* for each of these equations.

When the values of x are restricted to lie in a bounded or semi-infinite interval, $v(x, t)$, $\partial v(x, t)/\partial x$, or possibly a linear combination of both must be prescribed on the boundaries for all time $t \geqslant 0$, as was seen in our discussion of the diffusion equation in Chapter 1. Similar conditions are appropriate for the telegrapher's or wave equation when x is similarly restricted. The combined prescription of $v(x, t)$ and/or its derivatives on the initial line $t = 0$ and on the boundary line(s), constitutes an *initial and boundary value problem* for each of these equations for $v(x, t)$.

Laplace's equation was shown in Chapter 1 to characterize an equilibrium or steady-state situation where time dependence plays no role. The unknown function $v(x, y)$ was specified on the boundary of the region under consideration. This constitutes a *boundary value problem* for Laplace's equation.

As we have seen, the equations discussed in Chapter 1 are representative of second order equations of parabolic, hyperbolic, and elliptic type. As a general rule (in terms of the equations we shall consider), parabolic and hyperbolic equations are characteristic of problems that contain a time dependence. Initial value or initial and boundary value problems are appropriate for such equations depending on whether spatial boundaries occur for such problems. Elliptic equations represent equilibrium or steady-state situations in regions with boundaries (in which time dependent effects play no role), and boundary value problems are appropriate for such equations.

The number of initial and/or boundary conditions that should be assigned for a given differential equation depends on several factors. It may be stated that if the initial line or plane $t = 0$ is noncharacteristic and the equation contains k time derivatives, then the function and its first $k - 1$ time derivatives must be prescribed at $t = 0$. This is valid if we deal with a vector or scalar function. The dependence of the number of boundary conditions on the order of the differential equations is more complicated and requires separate discussions for different classes of equations.

More generally, if the data for the problem are not given at $t = 0$ or on some spatial boundary for all time, it becomes necessary to examine if they have the character of initial or boundary data. This determination again depends on the given partial differential equation. When data are given in a region that has the character of an initial curve or surface, the problem is known as a *Cauchy problem*. Also, for certain equations, boundary data are referred to as *Dirichlet* or *Neumann data*, and these are discussed at the appropriate place in the text. Further, we consider data given on characteristic curves and surfaces and it is necessary to determine under what circumstances such problems can be solved. (We have already encountered certain problems of this type in Chapter 2 and in the preceding sections of this chapter.)

The problem of deciding what form of initial and/or boundary data are appropriate for given partial differential equations is fairly complicated. A set of guidelines was proposed by Hadamard, who listed three requirements that must be met when formulating an initial and/or boundary value problem. A problem for which the differential equation and the data lead to a solution satisfying these requirements is said to be *well posed* or *correctly posed*. If it does not meet these requirements, it is *incorrectly posed*.

Hadamard's conditions for a well posed problem are:

1. *The solution must exist.*
2. *The solution must be uniquely determined.*
3. *The solution must depend continuously on the initial and/or boundary data.*

The first two conditions require that the equation plus the data for the problem must be such that one and only one solution exists. The third condition states that a slight variation of the data for the problem should cause the solution to vary only slightly. Thus since data are generally obtained experimentally and may be subject to numerical approximations, we require that the solution be stable under small variations in initial and/or boundary values. That is, we cannot permit wild variations to occur in the solution if the data are altered slightly. These conditions represent reasonable requirements for a problem arising in a physical context.

Thus for any given differential equation defined over a certain region, one must check whether the data assigned for the problem meet the Hadamard criteria. It was seen in Chapter 1 that the equations we derived were naturally associated with a set of initial and/or boundary conditions. These conditions are, in fact, appropriate for the differential equations that were considered and the problems given in Chapter 1 are well posed. It may well be argued that any problem arising in a physical context comes with built-in data, so that there is no need to decide which data are relevant. However, since any partial differential equation representing a physical process is a mathematical model obtained, in general, under various simplifying assumptions, it is not a priori obvious that the formulation of the mathematical problem is reasonable or well posed.

The problems we consider in this text are sufficiently standard that their appropriate formulations are well understood. Nevertheless, we comment occasionally on questions of uniqueness or continuous dependence on data.

It should be noted that certain problems representing physical processes are, in fact, incorrectly posed in the sense of Hadamard, and this does not appear to be due to a weakness in the mathematical model. Such problems have been studied and methods for dealing with such problems are continuing to be developed. We do not consider such problems in this text.

We now present two examples of incorrectly posed problems and discuss further questions relating to this matter in Section 3.5.

EXAMPLE 3.5. INCORRECTLY POSED PROBLEMS

Boundary value problems are, as a rule, not well posed for hyperbolic and parabolic equations. This follows because these are, in general, equations whose solutions evolve in time and their behavior at later times is predicted by

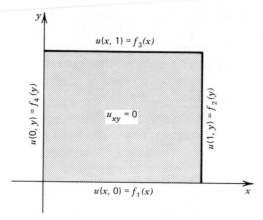

Figure 3.2. A boundary value problem.

their previous states. Thus a boundary value problem that arbitrarily prescribes the solution at two or more separate times is not reasonable.

As a simple example, we consider the hyperbolic equation $u_{xy} = 0$ in the unit square $0 < x < 1$ and $0 < y < 1$ with boundary values assigned on the sides of the square. We show that this problem has no solution if the data are prescribed arbitrarily. Since $u_{xy} = 0$ implies that $u_x(x, y) = $ constant, we have $u_x(x, 0) = u_x(x, 1)$. Taking note of the boundary conditions given in Figure 3.2, we have $u_x(x, 0) = f_1'(x)$ and $u_x(x, 1) = f_3'(x)$. Thus unless $f_1(x)$ and $f_3(x)$ are prescribed such that $f_1'(x) = f_3'(x)$, the boundary value problem cannot be solved. Therefore, it is incorrectly posed.

For Laplace's equation, the Cauchy problem is, in general, not well posed. This is shown by the following example credited to Hadamard. We consider the equation $u_{xx} + u_{yy} = 0$ in the region $y > 0$ with the Cauchy data $u(x, 0) = 0$ and $u_y(x, 0) = (\sin nx)/n$. The solution is easily obtained as $u(x, y) = [\sinh(ny)\sin(nx)]/n^2$. Now, as $n \to \infty$, $u_y(x, 0) \to 0$ so that for large n the Cauchy data $u(x, 0)$ and $u_y(x, 0)$ can be made arbitrarily small in magnitude. However, the solution $u(x, y)$ oscillates with an amplitude that grows exponentially like e^{ny} as $n \to \infty$. Thus arbitrarily small data can lead to arbitrarily large solutions and the solution is unstable. This violates the third condition of Hadamard requiring continuous dependence of the solution on the data.

3.5. STABILITY THEORY, ENERGY CONSERVATION, AND DISPERSION

In this section we consider certain general properties of partial differential equations that distinguish equations of different types beyond or apart from the classification process given in the previous sections. We recall that classification depends only on the form of the *principal part* of the equation. Here we examine the role played by the *lower order terms* in the equation in determining the behavior of the solutions.

For simplicity we restrict our discussion to second order linear partial differential equations with constant coefficients. The general ideas carry over to higher order linear equations and systems of equations with constant coefficients. Also certain aspects can be generalized to apply to equations with variable coefficients and to nonlinear equations as indicated later in the text.

Given the linear second order homogeneous equation in two variables

$$(3.122) \qquad Au_{xx} + 2Bu_{xt} + Cu_{tt} + Du_x + Eu_t + Fu = 0,$$

where $A, B, \ldots, F$ are real constants and $u = u(x, t)$, we look for exponential solutions of the form

$$(3.123) \qquad u(x, t) = a(k)\exp[ikx + \lambda(k)t],$$

where $a(k)$ is a constant and $i = \sqrt{-1}$. The parameter k is assumed to be real and $\lambda(k)$ must be chosen such that (3.123) satisfies (3.122). The solutions $u(x, t)$ for each k are called *normal modes* of the equation (3.122). Since the coefficients in (3.122) are assumed to be real, the real and imaginary parts of the normal modes (3.123) are also solutions of (3.122). (If α is a complex value quantity and is written as $\alpha = \alpha_1 + i\alpha_2$, where α_1 and α_2 are real valued, then $\alpha_1 = \text{Re}[\alpha]$ and $\alpha_2 = \text{Im}[\alpha]$ are the real and imaginary parts of α, respectively.)

We assume that t represents time and discuss the behavior of the normal modes in the region $t \geq 0$. As shown in later chapters, the solution of an initial value problem for (3.122) with initial data at $t = 0$ can be represented as a superposition of normal modes either by the use of Fourier series or integrals. Thus the behavior of the solution of the initial value problem is largely determined by that of the normal modes.

Inserting (3.123) into (3.122) yields the quadratic equation

$$(3.124) \qquad C\lambda^2 + (2iBk + E)\lambda + [-Ak^2 + iDk + F] = 0.$$

Solving for $\lambda = \lambda(k)$, we let $\lambda(k)$ represent either of the two possible solutions of (3.124). Then the normal mode can be written as

$$(3.125) \quad u(x, t) = a(k)\exp[i(kx + \text{Im}\,\lambda(k)t)]\exp[\text{Re}\,\lambda(k)t].$$

The magnitude of $u(x, t)$ is given as

$$(3.126) \qquad |u(x, t)| = |a(k)|\exp[\text{Re}\,\lambda(k)t].$$

Assuming $|a(k)|$ is bounded for all k, we see that the growth of $|u(x, t)|$ as t increases from zero is determined by the expression $\text{Re}\,\lambda(k)$.

We consider two possibilities. Either Re $\lambda(k)$ is bounded above for all real k or it is unbounded. We define the constant Ω to be the least upper bound (denoted as *lub*) of Re $\lambda(k)$ as k ranges through all its values; that is,

$$(3.127) \qquad \Omega = lub \, \text{Re}\, \lambda(k), \qquad -\infty < k < \infty.$$

If Re $\lambda(k)$ is not bounded above, we set $\Omega = +\infty$ and this is the case we now study. The case when $\Omega < +\infty$ is considered later.

Assuming $\Omega = +\infty$ we put $a(k) = 1/\lambda^2(k)$ in the normal mode (3.123) and consider the initial value problem for the normal mode with data at $t = 0$. We have, on evaluating (3.123) and its derivative at $t = 0$,

$$(3.128) \qquad u(x,0) = \left[\frac{1}{\lambda^2(k)}\right]e^{ikx}; \qquad u_t(x,0) = \left[\frac{1}{\lambda(k)}\right]e^{ikx}.$$

Since Re $\lambda(k)$ is unbounded, there exist values of k for which $|u(x,0)| = 1/|\lambda(k)|^2$ and $|u_t(x,0)| = 1/|\lambda(k)|$ are arbitrarily small and yet $|u(x,t)|$ [as given in (3.126)] is arbitrarily large for any $t > 0$. This follows since $1/|\lambda(k)|$ decays algebraically whereas $\exp[\text{Re}\,\lambda(k)t]$ grows exponentially. Thus Hadamard's stability criterion for the initial value problem for (3.122) is violated if $\Omega = +\infty$, and the initial value problem is not well posed. If $\Omega < +\infty$, it can be shown that the initial value problem for (3.122) is well posed.

Before considering the case where $\Omega < +\infty$, we consider an example in which the foregoing discussion is applied to specific equations of elliptic, parabolic, and hyperbolic types.

EXAMPLE 3.6. WELL POSEDNESS FOR ELLIPTIC, PARABOLIC, AND HYPERBOLIC EQUATIONS

Given the elliptic equation,

$$(3.129) \qquad u_{xx} + u_{tt} + \rho u = 0,$$

with $\rho = $ constant, the $\lambda(k)$ in the normal mode solution (3.123) has the form

$$(3.130) \qquad \lambda(k) = \pm\sqrt{k^2 - \rho}.$$

Selecting the positive root, we see that $\lambda(k)$ is real for $k^2 \geqslant \rho$ and that Re $\lambda(k) \to +\infty$ as $|k| \to +\infty$. Consequently, $\Omega = +\infty$ and the Cauchy problem for (3.129) with data at $t = 0$ is not well posed.

The parabolic equation

$$(3.131) \qquad \rho u_{xx} + u_t = 0,$$

where $\rho = $ constant, has the form of the diffusion or heat equation if $\rho < 0$

and is known as the backward diffusion or heat equation if $\rho > 0$. One form can be obtained from the other if the time direction is reversed; that is, if t is replaced by $-t$. Normal mode solutions for (3.131) yield

$$(3.132) \qquad\qquad \lambda(k) = \rho k^2.$$

Now if $\rho < 0$, we have $\Omega = 0$, whereas if $\rho > 0$, we have $\Omega = +\infty$ so that the Cauchy problem is not well posed if $\rho > 0$. The appropriate data (in view of our discussion) would be $u(x,0) = (1/\rho k^2)e^{ikx}$ with $a(k) = 1/\lambda(k)$, since only one time derivative occurs in (3.131). The normal mode solution has the form

$$(3.133) \qquad\qquad u(x,t) = \left[\frac{1}{\rho k^2}\right]\exp[ikx + \rho k^2 t],$$

and it is apparent that for $\rho > 0$ and $k \to \infty$, $|u(x,0)|$ is small and $|u(x,t)|$ is large.

The hyperbolic equation

$$(3.134) \qquad\qquad u_{tt} - u_{xx} + \rho u = 0,$$

where $\rho = $ constant, has normal mode solutions (3.123) with $\lambda(k)$ given as

$$(3.135) \qquad\qquad \lambda(k) = \pm i\sqrt{k^2 + \rho}\,.$$

For sufficiently large $|k|$, if $\rho < 0$, we have $\operatorname{Re}\lambda(k) = 0$. With $\rho \geqslant 0$, $\operatorname{Re}\lambda(k)$ vanishes for all k. Thus $\Omega < +\infty$ for (3.134) and the Cauchy problem is well posed.

In all the foregoing cases we have $|\lambda(k)| \to \infty$ as $|k| \to \infty$, so that if data of the form (3.128) are chosen and $\Omega < +\infty$, $|u(x,0)|$ and $|u_t(x,0)|$ can be made arbitrarily small by choosing $|k| \gg 1$. But $|u(x,t)|$ is also small since $\exp[\operatorname{Re}\lambda(k)t]$ does not grown unboundedly with k for fixed values of t.

The constant Ω defined in (3.127) is known as the *stability index* for the differential equation. We have already seen that when $\Omega = +\infty$, solutions of the initial value problem for (1.322) are unstable. That is, solutions whose magnitude is initially arbitrarily small can grow arbitrarily large even at finite times. In the following discussion we assume that $\Omega < +\infty$ and ask whether solutions of initially bounded magnitude can grow unboundedly as $t \to +\infty$. Equations for which this can happen are said to be unstable. Otherwise they are said to be stable. Thus even if the Cauchy problem is well posed, the equation may be unstable.

Physically, instability indicates that, even in the absence of external effects due to forcing terms in (3.122), internal mechanisms generate a growth in the solution as time increases. If the Cauchy problem is not well posed (i.e., $\Omega = +\infty$), the validity of the mathematical model must be reexamined. If the Cauchy problem is well posed, instability plays an important role if the equation was derived by a linearization procedure under which it was assumed that solutions remain small in magnitude for all time. Examples of stability analyses where the present results play a role are given later in the text.

Given the normal mode solution (3.123) and its magnitude (3.126), we note the following results. If $\Omega < 0$, then $|u(x, t)| \to 0$ as $t \to \infty$ for all k. If $\Omega > 0$ there are normal modes with k near Ω, for which $|u(x, t)| \to \infty$ as $t \to \infty$. If $\Omega = 0$, there may be values of k for which $|u(x, t)|$ is bounded but does not tend to zero as $t \to \infty$, although $|u(x, t)| \to 0$ for the remaining values of k. [We remark that since Ω is a least upper bound it may happen that Re $\lambda(k)$ never equals zero even if $\Omega = 0$.]

Noting the preceding discussion, we introduce the following classification for (3.122). If $\Omega < 0$, the equation is said to *strictly stable*. If $\Omega > 0$, the equation is *unstable*. If $\Omega = 0$, the equation is said to be *neutrally stable*, but it may be unstable in this case.

A simple example indicating the problems in the neutrally stable case is given by the equation

$$(3.136) \qquad\qquad u_{xx} + 2u_{xt} + u_{tt} = 0,$$

for which $\lambda(k) = -ik$ so that Re $\lambda(k) = 0 = \Omega$. The equation has solutions

$$(3.137) \qquad\qquad u(x, t) = ae^{ik(x-t)} + bte^{ik(x-t)},$$

where a and b are arbitrary constants. In particular, the solution of the Cauchy problem $u(x, 0) = 0$ and $u_t(x, 0) = 1$ is given by $u(x, t) = t$. Thus even though the data are uniformly bounded in magnitude, the solution grows unboundedly as $t \to \infty$. It is generally true that for a neutrally stable case, if there is instability, the growth of the solution will be algebraic rather than exponential. [We remark that the second term on the right of (3.137) is not a normal mode solution of (3.136) as defined in (3.123).]

It must again be emphasized, that for the data at $t = 0$ of the normal mode solution (3.123), we have $|u(x, 0)| = |a|$ and $|u_t(x, 0)| = |a\lambda|$, which are uniformly bounded for all x and fixed k. However, $|u(x, t)| = |a|\exp[\text{Re } \lambda(k)t]$ and if $\Omega < \infty$, $|u(x, t)|$ can tend to infinity only as $t \to \infty$. This is in contrast to the case where $\Omega = +\infty$ and $|u(x, t)|$ can become arbitrarily large for fixed t.

If Re $\lambda(k) = 0$ for all k, (3.122) is said to be an equation of *conservative type*. Then we have

$$(3.138) \qquad\qquad |u(x, t)| = |a| = |u(x, 0)|,$$

in view of (3.126), so that the amplitude $|u|$ of the normal mode solution is constant in time. Since $|u|^2$ is generally a measure of the energy of the normal mode solution, we find that the energy is conserved in time and say that the equation (3.122) is conservative.

In the conservative case, $\lambda(k)$ may be expressed as $\lambda(k) = -i\omega(k)$, where $\omega(k)$ is real valued for all k. The normal mode solution then take the form

$$(3.139) \qquad u(x, t) = a(k)\exp\{i[kx - \omega(k)t]\}.$$

Inserting (3.139) into (3.122) yields

$$(3.140) \qquad \omega = \omega(k).$$

which is known as the *dispersion relation* for the differential equation. If $\omega(k)$ is real valued for all real k as we have assumed and, additionally, $\omega''(k) \neq 0$ (i.e., ω is not a linear function of k), the equation (3.122) is said to be of *dispersive type*. [We note that regardless of the assumptions made above on $\omega(k)$, (3.140) is known as the dispersion relation for (3.122).]

Recalling the discussion in Example 2.2, we find that (3.139) for fixed k has the form of a wave that travels with velocity $dx/dt = \omega(k)/k$. The term

$$(3.141) \qquad \theta = kx - \omega(k)t$$

in (3.139) is called the phase of the normal mode and $dx/dt = \omega(k)/k$ is the *phase velocity*. If $\omega''(k) \equiv 0$, we assume $\omega(0) = 0$ and find that $\omega(k) = ck$ where $c = $ constant. Thus the phase velocity $dx/dt = ck/k = c$, has a constant value c for all normal modes. Since the general solution, as we have indicated, can be constructed as a superposition of normal modes it will also represent a wave traveling with velocity c. [In general, $\omega(k)$ has two values so that the general solution is a sum of two waves.]

However, if $\omega''(k) \neq 0$, the phase velocity $dx/dt = \omega(k)/k$ will be different for different values of k so that different normal modes have different velocities. The general solution obtained as a superposition of the normal modes yields a *wave* that disperses since its components or modes all travel at different velocities. We shall see in Section 5.7 that the relevant concept for *dispersive wave motion* is not the phase velocity $\omega(k)/k$, but the *group velocity* defined as $d\omega(k)/dk$.

Finally, if the stability index $\Omega \leqslant 0$ and Re $\lambda(k)$ is negative for all except a finite number of values of k, the equation (3.122) is said to be of *dissipative type*. In that case (3.126) shows that $|u(x, t)|$ decays to zero as $t \to \infty$ for all but a finite number of values of k and the energy $|u(x, t)|^2$ is dissipated as $t \to \infty$. When $\Omega < 0$ all solutions tend to zero as $t \to \infty$. However, if $\Omega = 0$, those modes for which Re $\lambda(k) = 0$ do not decay as $t \to \infty$ and they are expected to constitute the major contribution to the solution for large values of t. This fact is demonstrated when dissipative equations are studied in Section 5.7.

We now consider two examples of equations of dissipative and dispersive type. The telegrapher's equation derived in Chapter 1 is an equation of dissipative type. The Klein–Gordon equation which occurs in relativistic physics is an equation of dispersive type. Both these equations are commonly used as model equations of dissipative and dispersive type.

EXAMPLE 3.7.   THE TELEGRAPHER'S AND KLEIN–GORDON EQUATIONS

The *telegrapher's equation* (1.79)

$$(3.142) \qquad u_{tt} - \gamma^2 u_{xx} + 2\hat{\lambda} u_t = 0,$$

[where λ has been replaced by $\hat{\lambda}$ to avoid confusion with $\lambda = \lambda(k)$], yields for the normal mode solutions (3.123)

$$(3.143) \qquad \lambda(k) = -\hat{\lambda} \pm \sqrt{\hat{\lambda}^2 - \gamma^2 k^2}.$$

Since $\hat{\lambda} > 0$ by assumption, the real part of $\lambda(k)$ is easily seen to be negative for all $k \neq 0$. At $k = 0$ either $\lambda(0) = 0$ or $\lambda(0) = -2\hat{\lambda}$ so that $\Omega = 0$ for (3.142). Consequently, the telegrapher's equation is neutrally stable and is of dissipative type. Note that when $\hat{\lambda} = 0$ (3.142) reduces to the wave equation and $\lambda(k) = \pm i\gamma k$ in that case. Then $\mathrm{Re}\,\lambda(k) = 0$ for all k so that the wave equation is also neutrally stable and is of conservative type. For this reason when $\hat{\lambda} > 0$, (3.142) is often called the *damped wave equation*.

All normal mode solutions of (3.142) are damped as $t \to \infty$ except for the solution corresponding to $\lambda(0) = 0$. It is shown in Section 5.7 that the main contribution to the solution of the Cauchy problem for (3.142) as $t \to \infty$ comes from the normal modes with $k \approx 0$.

The *Klein–Gordon equation* has the form

$$(3.144) \qquad u_{tt} - \gamma^2 u_{xx} + c^2 u = 0,$$

where γ and c are constants. For its normal modes (3.123) we have

$$(3.145) \qquad \lambda(k) = \pm i\sqrt{\gamma^2 k^2 + c^2} = \pm i\omega(k),$$

so that the dispersion relation is

$$(3.146) \qquad \omega = \omega(k) = \sqrt{\gamma^2 k^2 + c^2}.$$

Since $\omega(k)$ is real valued for all k and $\omega''(k) \not\equiv 0$, we conclude that, in addition to being neutrally stable, the Klein–Gordon equation is of conservative and dispersive type. Again, it may be noted that if $c = 0$ in (3.144), it reduces to the wave equation for which $\omega(k) = k$. Since $\omega''(k) = 0$ in this case, the wave equation is not of dispersive type.

To conclude our discussion of stability theory, it must be emphasized that the foregoing definitions are relevant, in general, only for the initial value problem for (3.122). When initial and boundary value problems are considered, the parameter k may be restricted to a discrete set of values. The solution of the problem is then given as a superposition of a discrete set of normal modes that correspond to the aforementioned values of k. Stability is then defined in terms of the growth or decay of these normal modes. If one or more of the normal modes is unbounded as $t \to \infty$, the problem is unstable. If all the normal modes decay as $t \to \infty$, the problem is stable. Examples of stability analyses for initial and boundary value problems are given later in the text.

EXERCISES FOR CHAPTER 3

Section 3.1

3.1.1. Show that the equation

$$u_{xx} + 4u_{xy} + 3u_{yy} + 3u_x - u_y + 2u = 0,$$

is of hyperbolic type. Determine its characteristic curves and bring it into the canonical form (3.19). Introduce the further transformation

$$u(\xi, \eta) = \exp(\alpha\xi + \beta\eta)v(\xi, \eta),$$

and choose the constants α and β to eliminate the first derivative terms in the resulting canonical form.

3.1.2. Show that the equation

$$u_{xx} + 2u_{xy} + u_{yy} + 5u_x + 3u_y + u = 0,$$

is of parabolic type and bring it into the canonical form (3.25). Using the transformation of the dependent variable given in Exercise 3.1.1, show that the terms involving u_η and u in the canonical form can be eliminated.

3.1.3. Show that the equation

$$u_{xx} - 6u_{xy} + 12u_{yy} + 4u_x - u = \sin(xy),$$

is of elliptic type and obtain its canonical form (3.27). Proceeding as in the above exercises, show that the first derivative terms in the canonical form can be eliminated.

3.1.4. Determine the regions where *Tricomi's equation*

$$u_{xx} + xu_{yy} = 0,$$

is of elliptic, parabolic, and hyperbolic types. Obtain its characteristics and its canonical form in the hyperbolic region.

3.1.5. Show that the equation

$$u_{xx} + yu_{yy} + \tfrac{1}{2}u_y = 0$$

has the simple canonical form $u_{\xi\eta} = 0$ in the region where it is of hyperbolic type. Use this result to show that it has the general solution

$$u = f\left(x + 2\sqrt{-y}\right) + g\left(x - 2\sqrt{-y}\right),$$

where f and g are arbitrary functions, in the hyperbolic region.

3.1.6. Classify the following equations into hyperbolic, elliptic, or parabolic type.

    (a) $5u_{xx} - 3u_{yy} + (\sin x)u_x + e^{xy^2}u_y + u = 0.$

    (b) $u_{yy} - 10u_x + 4u_y + (\cosh x)u = 0.$

    (c) $10u_{xx} + u_{yy} - u_x + [\log(1 + x^2)]u = 0.$

3.1.7. Classify the following equations into hyperbolic, elliptic, or parabolic type.

    (a) $e^{xy}u_{xx} + (\cosh x)u_{yy} + u_x - u = 0.$

    (b) $[\log(1 + x^2 + y^2)]u_{xx} - [2 + \cos x]u_{yy} = 0.$

    (c) $u_{yy} + [1 + x^2]u_x - u_y + u = 0.$

3.1.8. Show that if $B = 0$ in (3.1), the equation is of fixed type if $AC < 0$, $AC = 0$, or $AC > 0$ everywhere in the region where (3.1) is defined.

Section 3.2

3.2.1. If $\varphi(x, y) = $ constant is a family of characteristics for (3.41), show that (3.46) reduces to

$$\left(a\psi_x + b\psi_y\right)u_\eta = cu + d.$$

Assuming u is continuous across $\xi = \varphi(x, y) = 0$ while u_ξ has a jump discontinuity there, demonstrate that the jump $[u_\xi]$ at $\xi = 0$ satisfies the ordinary differential equation

$$\left(a\psi_x + b\psi_y\right)[u_\xi]_\eta = c[u_\xi],$$

by differentiating (3.46) with respect to ξ. This result determines the variation of the jump $[u_\xi]$ along the characteristic $\xi = 0$.

3.2.2. Apply the results of Exercise 3.2.1 to the equation $u_x + u_y = 0$ considered in the text. Show that the jump in the solution obtained in the text is given as $[u_\xi] = \alpha_1 - \alpha_2$ across $\xi = 0$ and that this satisfies the condition determined in Exercise 3.2.1 for the variation of the jump.

3.2.3. Consider the equation

$$u_x + u_y - cu = 0,$$

where c = constant. Show that the appropriate (ξ, η) coordinates in (3.43) are

$$\xi = y - x, \qquad \eta = y + x,$$

where $\xi = y - x$ = constant are the characteristics. Obtain the equation

$$2u_\eta - cu = 0$$

and the equation for the jump $[u_\xi]$,

$$2[u_\xi]_\eta = c[u_\xi].$$

Verify that $u(x, y) = \beta_1(y - x)\exp[\frac{1}{2}c(y + x)]$, $y \leqslant x$ and $u(x, y) = \beta_2(y - x)\exp[\frac{1}{2}c(y + x)]$, $y > x$ are solutions of the given equation for $y \neq x$ and that the jump in u_ξ across $\xi = y - x = 0$ satisfies the equation for the jump (β_1 and β_2 are constants).

3.2.4. If $\varphi(x, y)$ = constant is a family of characteristics for (3.1) show that (3.53) reduces to the form

$$2\beta u_{\xi\eta} + \gamma u_{\eta\eta} + \delta u_\eta + \rho u_\xi + \lambda u + \sigma = 0.$$

If $u_{\xi\xi}$ has a jump across $\xi = \varphi(x, y) = 0$ and the terms in the above equation are continuous across $\xi = 0$, conclude on differentiating the equation with respect to ξ that the jump $[u_{\xi\xi}]$ satisfies

$$2\beta[u_{\xi\xi}]_\eta + \rho[u_{\xi\xi}] = 0; \qquad \xi = 0,$$

assuming smooth coefficients.

3.2.5. Given the hyperbolic equation

$$u_{xx} - u_{yy} = 0,$$

show that $u(x, y) = \beta_1(y - x)^2$; $y \leqslant x$ and $u(x, y) = \beta_2(y - x)^2$, $y > x$ are solutions with discontinuities second derivatives across the characteristic $y = x$ if the constants β_1 and β_2 are unequal. If $\xi = y - x$, show that $[u_{\xi\xi}]$ satisfies the appropriate form of the equation for the jump across $\xi = 0$ given in the preceding exercise.

3.2.6. By differentiating (3.53) as often as necessary with respect to ξ, show that jumps in $\partial^n u / \partial \xi^n$, $n \geqslant 3$ must also occur across characteristics, assuming the appropriate lower order derivatives are continuous.

3.2.7. If u and the normal derivative $\partial u / \partial n$ are specified on a curve $y = h(x)$, show how to specify u_x and u_y on that curve. Then, assuming $u_x[x, h(x)] = \alpha(x)$ and $u_y[x, h(x)] = \beta(x)$, where α and β are known functions, obtain the equations

$$u_{xx} + u_{xy}h' = \alpha'; \qquad u_{xy} + u_{yy}h' = \beta'$$
$$Au_{xx} + 2Bu_{xy} + Cu_{yy} = -D\alpha - E\beta - Ff + G,$$

for the specification of u_{xx}, u_{xy}, and u_{yy} along the curve $y = h(x)$. {We assume $u[x, h(x)] = f(x)$.} Show that if

$$Ah'^2 - 2Bh' + C = 0,$$

the second derivatives of u cannot be specified uniquely on $y = h(x)$, if they can be determined at all. If we set $\varphi = y - h(x)$, show that the above equation becomes the characteristic equation (3.55).

3.2.8. Consider the hyperbolic equation

$$u_{xx} - u_{yy} = 0.$$

Apply the method of the preceding exercise and show that u_{xx}, u_{xy}, and u_{yy} can be determined nonuniquely on a characteristic if the compatibility condition $\alpha'h' - \beta' = 0$ is satisfied. Construct a set of initial conditions for u and $\partial u / \partial n$ on the characteristic $y - x = 0$ such that the compatibility conditions are satisfied and show that this problem for the given equation has infinitely many solutions.

3.2.9. Show that the compatibility condition that guarantees that the problem in Exercise 3.2.7 has a nonunique solution along the characteristic $y = h(x)$ is

$$C\beta' + A\alpha'h' = (G - D\alpha - E\beta - Ff)h',$$

if A and C are not both zero. Determine the appropriate condition if $A = C = 0$.

Section 3.3

3.3.1. Show that the equation

$$3u_{x_1x_1} - 2u_{x_1x_2} + 2u_{x_2x_2} - 2u_{x_2x_3} + 3u_{x_3x_3} + 5u_{x_2} - u_{x_3} + 10u = 0$$

is of elliptic type by determining that the matrix A [see (3.62)] has the eigenvalues $\lambda_1 = 1$, $\lambda_2 = 3$, and $\lambda_3 = 4$. Introduce the transformation (3.73) and obtain the equation

$$u_{\xi_1\xi_1} + 3u_{\xi_2\xi_2} + 4u_{\xi_3\xi_3} + \frac{9}{\sqrt{6}}u_{\xi_1} + \frac{1}{\sqrt{2}}u_{\xi_2} - \frac{4}{\sqrt{3}}u_{\xi_3} + 10u = 0.$$

Construct transformations of the independent and dependent variables that replace the coefficients of the second derivative terms by unity and eliminate the first derivative terms.

3.3.2. Classify the following equations into elliptic, parabolic, or hyperbolic type.

(a) $u_{xx} + 2u_{yz} + (\cos x)u_z - e^{y^2}u = \cosh z$.

(b) $u_{xx} + 2u_{xy} + u_{yy} + 2u_{zz} - (1 + xy)u = 0$.

(c) $7u_{xx} - 10u_{xy} - 22u_{yz} + 7u_{yy} - 16u_{xz} - 5u_{zz} = 0$.

(d) $e^z u_{xy} - u_{xx} = \log[x^2 + y^2 + z^2 + 1]$.

3.3.3. Show that all linear second order equations of elliptic type with constant coefficients can be brought into the form

$$\sum_{i=1}^{n} u_{x_i x_i} + cu = F(x_1, \dots, x_n).$$

3.3.4. Show that all linear second order equations of hyperbolic type (in $n + 1$ variables) with constant coefficients can be transformed into

$$\sum_{i=1}^{n} u_{x_i x_i} - u_{x_0 x_0} + cu = F(x_0, x_1, \dots, x_n).$$

3.3.5. Determine the regions where

$$u_{xx} - 2x^2 u_{xz} + u_{yy} + u_{zz} = 0$$

is of hyperbolic, elliptic, or parabolic type.

3.3.6. Demonstrate that the equations

(a) $\nabla \cdot (p\nabla u) + qu = F$,

(b) $\rho u_t - \nabla \cdot (p\nabla u) + qu = F$,

(c) $\rho u_{tt} - \nabla \cdot (p\nabla u) + \lambda u_t + qu = F$,

are of elliptic, parabolic, and hyperbolic types, respectively, in a region R where ρ and p are positive. [Assume that $u = u(x, y, t)$ or $u = u(x, y, z, t)$ in the hyperbolic and parabolic cases, and $u = u(x, y)$ or $u = u(x, y, z)$ in the elliptic case.]

3.3.7. Considering that if we disregard the order of differentiation, the function $u(\xi_1, \dots, \xi_n)$ has exactly $\frac{1}{2}n(n - 1)$ mixed partial derivatives of order n, show that if $n > 3$ there are, in general, an insufficient number of equations of transformation leading from (3.62) to (3.66) to eliminate the mixed partial derivative terms in $\xi_1, \dots, \xi_n$ throughout a given region. If $n = 3$, the mixed partial derivative terms can be eliminated but the coefficients of the $u_{\xi_i \xi_i}$ terms cannot, in general, be made to equal plus or minus one, as was done in the two-dimensional case.

3.3.8. Use the transformation (3.91) to bring the equation (3.83) of example 3.3 into the form

$$u_{\xi_1\xi_1} + \left[1 + \frac{c}{\sqrt{2}}(\xi_2 + \xi_3)\right]u_{\xi_2\xi_2} - \left[1 + \frac{c}{\sqrt{2}}(\xi_2 - \xi_3)\right]u_{\xi_3\xi_3} = 0.$$

Attempt a transformation of this equation to bring it into the form (3.92).

3.3.9. Show that the system $A\mathbf{u}_x + B\mathbf{u}_y = C\mathbf{u} + \mathbf{d}$ where

$$A = \begin{bmatrix} 1 & 0 \\ -3 & -1 \end{bmatrix}; \quad B = \begin{bmatrix} 2 & 1 \\ -1 & -1 \end{bmatrix},$$

is of elliptic type.

3.3.10. Determine the characteristic curves of the (totally) hyperbolic system $\mathbf{u}_y + A\mathbf{u}_x = C\mathbf{u}$, where

$$A = \begin{bmatrix} 1 & 0 & 1 \\ 0 & 2 & 3 \\ 0 & 0 & -1 \end{bmatrix}, \quad C = \begin{bmatrix} -2 & 1 & 5 \\ 0 & 3 & 7 \\ 1 & -3 & -10 \end{bmatrix}.$$

Reduce this system to the canonical form (3.101).

3.3.11. Show that $\mathbf{u}_y + A\mathbf{u}_x = \mathbf{0}$ where

$$A = \begin{bmatrix} 1 & 1 & 0 \\ 1 & 1 & 0 \\ 0 & 0 & 2 \end{bmatrix}$$

is a hyperbolic system and reduce it to the (diagonal) canonical form (3.101).

3.3.12. Use the canonical form obtained in Exercise 3.3.11 to show that the general solution of the equation $\mathbf{u}_y + A\mathbf{u}_x = \mathbf{0}$ given in that exercise is

$$\mathbf{u}(x, y) = \begin{bmatrix} \frac{1}{\sqrt{2}}[f_1(x) + f_2(x - 2y)] \\ \frac{1}{\sqrt{2}}[-f_1(x) + f_2(x - 2y)] \\ f_3(x - 2y) \end{bmatrix}.$$

Use this result to solve the initial value problem for $\mathbf{u}_y + A\mathbf{u}_x = \mathbf{0}$ with the initial condition

$$\mathbf{u}(x, 0) = \begin{bmatrix} \sin x \\ 1 \\ e^x \end{bmatrix}.$$

3.3.13. Show that the system $\mathbf{u}_y + A\mathbf{u}_x = 0$, where

$$A = \begin{bmatrix} 1 & 2e^x \\ 2e^{-x} & 1 \end{bmatrix},$$

is totally hyperbolic. Determine its characteristic curves and reduce it to the canonical form (3.101).

3.3.14. If the constant–coefficient system $\mathbf{u}_y + A\mathbf{u}_x = 0$ is totally hyperbolic (with $\mathbf{u}$ as an n-component vector) show how to obtain a general solution by using the canonical form (3.101).

3.3.15. Show that the equation

$$u_{ttt} - c^2 u_{xxt} + \alpha\left(u_{tt} - a^2 u_{xx} \right) = 0$$

is of hyperbolic type and determine its characteristic curves. (It is of hyperbolic type if the characteristic curves are real and distinct.)

Section 3.4

3.4.1. Obtain the solution of the Cauchy problem for Laplace's equation given in Example 3.5 by expanding $u(x, y)$ as a power series

$$u(x, y) = \sum_{k=0}^{\infty} \frac{1}{k!} \frac{\partial^k u(x,0)}{\partial y^k} y^k.$$

Determine the derivatives $\partial_y^k u(x,0)$ by differentiating Laplace's equation $u_{yy} = -u_{xx}$ along the initial line $y = 0$ and using the Cauchy data given there.

3.4.2. Construct the solution of the Cauchy problem for the heat equation $u_t = c^2 u_{xx}$ with the initial condition $u(x,0) = f(x)$, as a power series

$$u(x, t) = \sum_{n=0}^{\infty} \frac{1}{n!} \frac{\partial^n u(x,0)}{\partial t^n} t^n,$$

assuming that $f(x)$ is sufficiently differentiable. [Determine the derivatives $\partial_t^n u(x,0)$ from the heat equation and the initial condition.]

3.4.3. Use the method of Exercise 3.4.2 to show that the solution of

$$u_t = c^2 u_{xx}; \qquad u(x,0) = \cos x$$

is given as $u(x, t) = \cos(x)\exp(-c^2 t)$.

3.4.4. Obtain a solution of the telegrapher's equation

$$u_{tt} - \gamma^2 u_{xx} + 2\lambda u_t = 0,$$

where γ and λ are positive constants, in the form of a power series in t, if the Cauchy data are given as $u(x, 0) = \cos x$, $u_t(x, 0) = 0$.

3.4.5. Show that the exterior boundary value problem for Laplace's equation

$$u_{xx} + u_{yy} + u_{zz} = 0, \qquad \rho > a,$$

where $\rho^2 = x^2 + y^2 + z^2$ and $u(x, y, z)$ is specified on the sphere $\rho = a$ as $u|_{\rho=a} = A$ (A = constant) has (at least) two solutions:

$$u = A; \qquad u = \frac{Aa}{\rho}.$$

(The solution can be shown to be unique if we require that $u \to 0$ as $\rho \to \infty$.)

3.4.6. Apply the divergence theorem to the expression $\nabla \cdot \nabla u$ to show that the boundary value problem for $\nabla^2 u = 0$ in the region A with $\partial u / \partial n = f$ on the boundary ∂A of A has no solution unless

$$\int_{\partial A} f d\sigma = 0.$$

Also, observe that if a solution does exist for this problem, it is not unique since the solution $u = a$, (a = constant) can be added to the given result.

Section 3.5

3.5.1. Show that the stability index Ω for the hyperbolic equation

$$u_{tt} - \gamma^2 u_{xx} - c^2 u = 0$$

is given as $\Omega = c$. (The constant c is assumed to be positive.) Thus although the Cauchy problem for this equation is well posed, the equation is unstable.

3.5.2. Consider the hyperbolic equation

$$u_{tt} - c^2 u_{xx} + u_t - \alpha u_x = 0.$$

By examining the solutions $\lambda = \lambda(k)$ of (3.124) as specialized to the above equation, show that the equation is unstable if $c^2 < \alpha^2$. (Hint: Look at small values of k.)

3.5.3. Demonstrate that the following equations are all of dispersive type by considering normal mode solutions of the forms (3.139). Also, determine the relevant dispersion relations.

    (a) $u_{tt} - \gamma^2 u_{xx} - \alpha^2 u_{xxtt} = 0.$
    (b) $u_{tt} - \gamma^2 u_{xxxx} = 0.$
    (c) $u_t + \alpha u_x + \beta u_{xxx} = 0.$

3.5.4. Consider the hyperbolic equation

$$u_{ttt} - \gamma^2 u_{xxt} + u_{tt} - c^2 u_{xx} = 0.$$

Obtain the relationship $\lambda = \lambda(k)$ for the normal mode solutions (3.123) of the above equation. Using the known criterion that the polynomial $P(\lambda) = \lambda^3 + a_1 \lambda^2 + a_2 \lambda + a_3$ has roots with negative real parts if $a_3 > 0$, $a_1 > 0$, and $a_1 a_2 > a_3$, conclude that the given hyperbolic equation is neutrally stable and is of dissipative type if $\gamma^2 > c^2$.

3.5.5. Given the system of equations

$$A\mathbf{u}_x + B\mathbf{u}_t + C\mathbf{u} = 0,$$

where A, B, and C are constant n by n matrices and $\mathbf{u}$ is an n-component vector, normal mode solutions are given as

$$\mathbf{u}(x, t) = \mathbf{a}(k)\exp[ikx + \lambda(k)t],$$

where $\mathbf{a}(k)$ is a constant vector. Show that $\mathbf{u}$ is a solution of the given equation if we have

$$|ikA + \lambda B + C| = 0.$$

Explain how concepts of well-posedness and stability may be introduced for the given system on the basis of the above result.

3.5.6. Apply the method of Exercise 3.5.5 to the system (1.73)–(1.74) and to the systems given in Example 3.4 [replacing y by t in the Cauchy–Riemann system (3.104)]. Discuss stability and well-posedness questions for these systems and compare the results with those obtained for the scalar equivalents of these systems.

3.5.7. With $\mathbf{k} = [k_1, k_2, k_3]$ and $\mathbf{x} = [x, y, z]$, obtain the appropriate equation $\lambda = \lambda(\mathbf{k})$ for the normal mode solutions

$$u(\mathbf{x}) = a(\mathbf{k})\exp[i\mathbf{k} \cdot \mathbf{x} + \lambda(\mathbf{k})t]$$

of the following equations:

(a) $u_{tt} - \gamma^2 \nabla^2 u + \hat{\lambda} u_t = 0,$

(b) $u_{tt} - \gamma^2 \nabla^2 u + c^2 u = 0,$

(c) $u_{tt} + \nabla^2 u = 0,$

(d) $u_t - \gamma^2 \nabla^2 u = 0,$

where $\hat{\lambda} > 0$, all the coefficients are constants, and ∇^2 is the Laplacian operator in three dimensions. Discuss well-posedness and stability for these equations in the manner of Section 3.5.

4

INITIAL AND BOUNDARY VALUE PROBLEMS IN BOUNDED REGIONS

4.1. INTRODUCTION

This chapter deals with initial and boundary value problems for elliptic, parabolic, and hyperbolic equations given over bounded spatial regions. The basic method for solving these problems is the technique of separation of variables. This method requires that eigenvalue problems for differential equations be studied and expresses solutions as series of eigenfunctions. Only scalar problems in one, two, or three space dimensions will be treated in this chapter. The one-dimensional eigenvalue problem, known as the *Sturm–Liouville problem*, will be studied in detail. Higher dimensional eigenvalue problems will be considered in Chapter 8.

The separation of variables method is applicable to the study of linear homogeneous equations with homogeneous boundary conditions. Techniques for solving inhomogeneous problems, such as Duhamel's Principle and eigenfunction expansions, will also be presented in this chapter. Further, it will be demonstrated by means of a detailed example in Section 4.7 how eigenfunction expansions may be used in the solution of nonlinear problems.

To achieve a unity of presentation, we consider only second order equations of each of the three basic types given in a special form. This form, however, is representative of a large class of partial differential equations of physical interest. All equations derived in Chapter 1 are either of this form or can be reduced to this form. To indicate how these equations which may have variable coefficients can arise in applications, we now present a brief derivation of a parabolic equation of the form to be considered. This derivation, in contrast to those given in Chapter 1, is more representative of the approaches generally used in deriving partial differential equations as models for physical processes.

Throughout this chapter, G will represent a bounded interval in one space dimension or a bounded region in two or three dimensions and ∂G will represent its boundary. (In one dimension ∂G is just the end points of the interval G.)

In the following derivation we restrict our discussion to three space dimensions and let $\hat{G}$ be an arbitrary closed region within G, with $\partial \hat{G}$ as the boundary surface. Let ds be an element of surface area of $\partial \hat{G}$ and dv be a volume element. The function $u(x, t)$ where x is a point in space and t is time is assumed to represent a scalar physical quantity such as temperature, for example.

With $\mathbf{n}$ as the exterior unit normal vector on the boundary $\partial \hat{G}$, $p(x)\nabla u \cdot \mathbf{n}\,ds$ represents the flux or rate of flow of the quantity $u(x, t)$ though ds at the time t. The positive function $p(x)$ is given and assumed to be time independent. The operator ∇ is the gradient operator in three-dimensional space. The time rate of change of $u(x, t)$ in an element dv at the time t is given by $\rho(x)(\partial u / \partial t)\,dv$, where the positive quantity $\rho(x)$ is given and time independent. Additional effects occurring in the element dv at the time t are given as $H(x, t)\,dv$. The function $H(x, t)$ is assumed to be of the form $H(x, t) = -q(x)u(x, t) + \hat{F}(x, t)$, where for notational convenience in our discussion we set $\hat{F}(x, t) = \rho(x)F(x, t)$. Again, $q(x)$ is a given nonnegative time independent function and qu represents (internal) effects due to changes proportional to $u(x, t)$, whereas $\hat{F}(x, t)$ represents external influences on the medium under consideration. (See Section 8.2 for the consequences of permitting q to be negative.)

The basic physical balance or conservation law for the arbitrary region $\hat{G}$ requires that

$$(4.1) \qquad \iint_{\hat{G}} \rho \frac{\partial u}{\partial t}\,dv = \int_{\partial\hat{G}} p\nabla u \cdot \mathbf{n}\,ds + \iint_{\hat{G}} H\,dv = \int_{\partial\hat{G}} p\nabla u \cdot \mathbf{n}\,ds$$

$$- \iint_{\hat{G}} qu\,dv + \iint_{\hat{G}} \rho F\,dv.$$

This states that *the time rate of change of* $u(x, t)$ *in the region* $\hat{G}$ equals *the flux of* $u(x, t)$ *through the boundary of* $\hat{G}$ and *the changes due to internal effects proportional to* $u(x, t)$, as well as *external effects acting throughout* $\hat{G}$ (i.e., the effects must balance each other out).

Applying the divergence theorem to the surface integral in (4.1) yields

$$(4.2) \qquad \int_{\partial\hat{G}} p\nabla u \cdot \mathbf{n}\,ds = \iint_{\hat{G}} \nabla \cdot (p\nabla u)\,dv.$$

Inserting (4.2) into (4.1) and combining terms gives

$$(4.3) \qquad \iint\limits_{\hat{G}} \left[\rho \frac{\partial u}{\partial t} - \nabla \cdot (p\nabla u) + qu - \rho F \right] dv = 0.$$

Assuming a continuous integrand, the arbitrariness of the region $\hat{G}$ implies the vanishing of the integrand in (4.3) (see Exercise 8.1.9). Consequently, we obtain the inhomogeneous parabolic equation

$$(4.4) \qquad \rho \frac{\partial u}{\partial t} - \nabla \cdot (p\nabla u) + qu = \rho F,$$

which is valid at an arbitrary point in the region G. A similar argument in two space dimensions leads to the same equation with ∇ interpreted as a two-dimensional gradient operator. In one space dimension, the resulting equation has the form

$$(4.5) \qquad \rho \frac{\partial u}{\partial t} - \frac{\partial}{\partial x}\left(p \frac{\partial u}{\partial x} \right) + qu = \rho F.$$

In view of our continual use of the above and related equations, we define the differential operator L in two or three dimensions as

$$(4.6) \qquad Lu = -\nabla \cdot (p\nabla u) + qu,$$

and in one space dimension as

$$(4.7) \qquad Lu = -\frac{\partial}{\partial x}\left(p \frac{\partial u}{\partial x} \right) + qu.$$

It is assumed that ρ and p are positive functions and that q is a nonnegative function. Then the *parabolic equations* (4.4) and (4.5) take the form

$$(4.8) \qquad \rho \frac{\partial u}{\partial t} + Lu = \rho F.$$

If the problem leading to (4.8) is stationary or time independent, that is, F and all conditions given on the problem depend on x only, we may look for a solution in the form $u = u(x)$ and (4.8) reduces to

$$(4.9) \qquad Lu = \rho F.$$

This is an equation of *elliptic type*.

Further, it can be shown that a large class of *hyperbolic equations* of interest in applications can be expressed as

$$(4.10) \qquad \rho \frac{\partial^2 u}{\partial t^2} + Lu = \rho F.$$

For example, in one space dimension the equation governing the longitudinal vibration of a rod has the form (4.10). However, a derivation of this and other such equations will not be given here. The telegrapher's equation of Chapter 1 is not of the form (4.10) since it contains a $\partial u/\partial t$ term. However, a simple change of the dependent variable reduces it to the above form. In fact, any equation of the form (4.10) that contains an additional term $\hat\rho\partial u/\partial t$ with $\hat\rho$ independent of t, can be transformed into the form (4.10) by a change of variable if $\hat\rho/\rho = $ constant.

In the following sections we shall consider initial and/or boundary value problems for the equations (4.8)–(4.10). Appropriate initial values for the parabolic and hyperbolic equations (4.8) and (4.10) will be introduced as required. However, the boundary values on ∂G for all three equations (4.8)–(4.10) are assumed to be of a standard form common in applications.

In the case of two or three space dimensions we consider boundary conditions of the form

$$(4.11) \qquad \alpha(x)u + \beta(x)\left.\frac{\partial u}{\partial n}\right|_{\partial G} = B(x,t),$$

where $\alpha(x)$, $\beta(x)$, and $B(x,t)$ are given functions evaluated on the boundary region ∂G. [Note that in the parabolic and hyperbolic cases, the region G and its boundary ∂G are extended parallel to themselves in (x,t)-space. In the elliptic case $B = B(x)$ is independent of t.] The expression $\partial u/\partial n$ denotes the exterior normal derivative on ∂G. The boundary condition (4.11) relates the values of u on ∂G and the flux of u through ∂G. We require that $\alpha(x) \geqslant 0$, $\beta(x) \geqslant 0$, and $\alpha + \beta > 0$ on ∂G. If (1) $\alpha \neq 0$, $\beta = 0$, (2) $\alpha = 0$, $\beta \neq 0$, (3) $\alpha \neq 0$, $\beta \neq 0$, (4.11) are known as boundary conditions of the *first*, *second*, and *third kind*, respectively. Boundary conditions of the first and second kind are often referred to as *Dirichlet* and *Neumann conditions*.

For the one-dimensional case, G is assumed to represent the interval $0 < x < l$ and ∂G comprises the points $x = 0$ and $x = l$. The boundary conditions (4.11) in the hyperbolic and parabolic cases take the form

$$(4.12) \qquad \begin{cases} \alpha_1 u(0,t) - \beta_1 u_x(0,t) = g_1(t) \\ \alpha_2 u(l,t) + \beta_2 u_x(l,t) = g_2(t) \end{cases},$$

where the conditions on (α, β) given above carry over in a natural way to the pairs (α_1, β_1) and (α_2, β_2) both of which are constant. The minus sign in the condition at $x = 0$ in (4.12) follows since the exterior normal derivative at that point is $\partial u/\partial n = -\partial u/\partial x$.

When F and B (or g_1 and g_2) in (4.8)–(4.12) are nonzero, the equations and boundary conditions are of inhomogeneous type. The separation of variables method introduced in the following section applies only to the homogeneous cases when F and B are both zero.

4.2. SEPARATION OF VARIABLES

In this section we present the *method of separation of variables* and apply it to the solution of initial and/or boundary value problems for homogeneous versions of the equations and boundary conditions introduced in the preceding section.

In the *hyperbolic case* we consider the homogeneous equation for $u = u(x, t)$,

$$(4.13) \qquad \rho(x)\frac{\partial^2 u}{\partial t^2} + Lu = 0, \qquad x \in G, t > 0,$$

where the bounded region G, the coefficient $\rho(x)$, and the operator L are defined as in Section 4.1. The homogeneous boundary conditions are, in two or three dimensions,

$$(4.14) \qquad \alpha(x)u + \beta(x)\left.\frac{\partial u}{\partial n}\right|_{\partial G} = 0, \qquad t > 0,$$

and in one dimension

$$(4.15) \qquad \begin{cases} \alpha_1 u(0, t) - \beta_1 u_x(0, t) = 0 \\ \alpha_2 u(l, t) + \beta_2 u_x(l, t) = 0 \end{cases}, \qquad t > 0$$

On the boundary ∂G the coefficients in (4.14)–(4.15) must satisfy the conditions given Section 4.1. The initial conditions for (4.13) are

$$(4.16) \qquad u(x, 0) = f(x); \qquad u_t(x, 0) = g(x); \qquad x \in G.$$

In the *parabolic case*, the equation for $u(x, t)$ is given as

$$(4.17) \qquad \rho(x)\frac{\partial u}{\partial t} + Lu = 0, \qquad x \in G, t > 0,$$

with the boundary conditions (4.14) or (4.15) and the initial condition

$$(4.18) \qquad u(x, 0) = f(x), \qquad x \in G.$$

In the *elliptic case*, we introduce a function $u = u(x, y)$, where x is a point in the region G (which is now one or two dimensional) and y is a scalar variable given over the interval $0 < y < \hat{l}$. The equation for $u(x, y)$ has the form

$$(4.19) \qquad \rho(x)\frac{\partial^2 u}{\partial y^2} - Lu = 0, \qquad x \in G, 0 < y < \hat{l}.$$

The boundary conditions in x are of the form (4.14) or (4.15). These are combined with the boundary conditions at $y = 0$ and $y = \hat{l}$,

(4.20) $u(x, 0) = f(x); \qquad u(x, \hat{l}) = g(x); \qquad x \in G.$

EXAMPLE 4.1.   EQUATIONS WITH CONSTANT COEFFICIENTS

We assume that x is a one-dimensional variable and that $\rho(x) = p(x) = 1$ and $q(x) = 0$ in (4.13), (4.17), and (4.19), and formulate the appropriate problems for each of these equations. The regions wherein the solutions are to be determined and the initial and/or boundary conditions for these problems are presented in Figures 4.1–4.3.

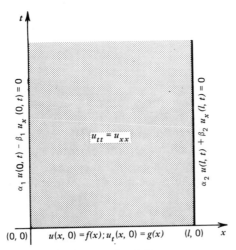

Figure 4.1.   Hyperbolic case.

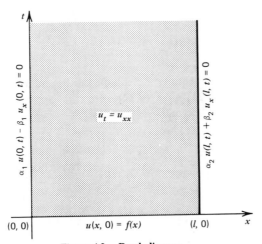

Figure 4.2.   Parabolic case.

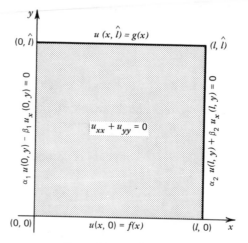

Figure 4.3.   Elliptic case.

It should be noted that for the hyperbolic and parabolic cases we are considering initial and boundary value problems, whereas in the elliptic case we have a strict boundary value problem. The equation (4.19) for the elliptic problem is written in the indicated form to permit a unified presentation of the separation of variables technique for all three cases.

The separation of variables method asks for a solution of (4.13) and (4.17) in the form

$$(4.21) \qquad\qquad u(x, t) = M(x)N(t),$$

and of (4.19) in the form

$$(4.22) \qquad\qquad u(x, y) = M(x)N(y),$$

with the function $M(x)$ required to satisfy the boundary conditions (4.14) or (4.15). Substituting (4.21) and (4.22) into the appropriate equations and dividing through by ρMN yields

$$(4.23) \qquad\qquad \frac{N''(t)}{N(t)} = -\frac{LM(x)}{\rho(x)M(x)}, \text{ hyperbolic case.}$$

$$(4.24) \qquad\qquad \frac{N'(t)}{N(t)} = -\frac{LM(x)}{\rho(x)M(x)}, \text{ parabolic case.}$$

and

$$(4.25) \qquad -\frac{N''(y)}{N(y)} = -\frac{LM(x)}{\rho(x)M(x)}, \qquad \text{elliptic case.}$$

Since the left and right sides of these equations depend on different variables, they cannot be nonconstant functions of their respective variables. Thus each side of the equations must equal a constant. We denote this (separation) constant by $-\lambda$. As a result, we obtain the following equations for M and N,

$$(4.26) \qquad LM(x) = \lambda\rho(x)M(x)$$

and

$$(4.27) \qquad \begin{cases} N''(t) + \lambda N(t) = 0, & \text{hyperbolic case} \\ N'(t) + \lambda N(t) = 0, & \text{parabolic case} \\ N''(y) - \lambda N(y) = 0, & \text{elliptic case} \end{cases}$$

In addition to being a solution of (4.26), $M(x)$ is required to satisfy the boundary conditions (4.14) or (4.15). [Consequently, $u(x, t) = M(x)N(t)$ or $u(x, y) = M(x)N(y)$ will also satisfy these boundary conditions.] Since both the equation and the boundary conditions for $M(x)$ are homogeneous, $M(x) = 0$ is a solution of the problem. This solution must be rejected since it is of no value in solving the given problem for u. Thus we require nonzero solutions of the boundary value problem for $M(x)$, and such solutions exist only for certain values of the parameter λ in (4.26).

The problem of determining the nonzero $M(x)$ is known as an *eigenvalue problem*. The values of λ for which nonzero $M(x)$ exist are known as *eigenvalues*. The corresponding solutions of the eigenvalue problem for $M(x)$ are called *eigenfunctions*. In the one-dimensional case, the eigenvalue problem is known as the *Sturm-Liouville problem* in the mathematical literature and this problem is discussed in the following section.

The differential operator L [or more precisely, $(1/\rho L)$], which occurs in (4.26), when applied to functions that satisfy the boundary conditions (4.14) or (4.15) has the property of being a *self-adjoint* as well as a *positive* operator. These properties are defined and discussed in the following example, after which important consequences of these properties for the eigenvalue problem associated with the operator L are examined. Although a large part of our discussion is presented in notation appropriate for the two and three-dimensional cases, the general results are valid for the one-dimensional case as well.

EXAMPLE 4.2.   SELF - ADJOINT AND POSITIVE OPERATORS

We consider the (smooth) functions u and w defined over the two or three-dimensional region G and satisfying the boundary condition (4.14) on ∂G. It will be shown that for these functions

$$(4.28) \qquad \iint\limits_{G} [wLu - uLw] \, dv = \int_{\partial G} p\left[u\frac{\partial w}{\partial n} - w\frac{\partial u}{\partial n}\right] ds = 0,$$

where $\partial/\partial n$ is the exterior normal derivative.

To prove (4.28) we note that

$$(4.29) \qquad w\nabla \cdot (p\nabla u) = \nabla \cdot (pw\nabla u) - p\nabla w \cdot \nabla u.$$

Interchanging w and u in (4.29) and subtracting one expression from the other gives

$$(4.30) \quad w\nabla \cdot (p\nabla u) - u\nabla \cdot (p\nabla w) = \nabla \cdot (pw\nabla u) - \nabla \cdot (pu\nabla w).$$

In view of (4.6) (i.e., the expression for L), we obtain

$$(4.31)$$

$$\iint\limits_{G} [wLu - uLw] \, dv = -\iint\limits_{G} \nabla \cdot [pw\nabla u - pu\nabla w] \, dv$$

$$= -\int_{\partial G} p[w\nabla u - u\nabla w] \cdot \mathbf{n} \, ds = \int_{\partial G} p\left[u\frac{\partial w}{\partial n} - w\frac{\partial u}{\partial n}\right] ds,$$

on applying the divergence theorem. Both u and w satisfy the boundary condition (4.14) so that

$$(4.32) \qquad \begin{cases} \alpha u + \beta \dfrac{\partial u}{\partial n}\Big|_{\partial G} = 0 \\[2mm] \alpha w + \beta \dfrac{\partial w}{\partial n}\Big|_{\partial G} = 0 \end{cases}.$$

We may think of (4.32) as a simultaneous system of homogeneous equations for α and β at each point of ∂G. Since $\alpha + \beta > 0$ on ∂G by assumption, the above system must have a nonzero solution. This can occur only if the determinant of the coefficients of the system (i.e., $u \, \partial w/\partial n - w \, \partial u/\partial n$) vanishes on ∂G. Consequently, the surface integral in (4.31) vanishes as was to be shown.

Given two functions f and g defined and integrable over the region G, the *inner product* of these two functions with *weight* $\rho(x) > 0$ is defined to be

$$(4.33) \qquad\qquad (f, g) = \iint\limits_{G} \rho f g \, dv.$$

These functions are assumed to be real valued and need not satisfy the boundary conditions (4.14). Since $(f, g) = (g, f)$, the inner product is symmetric. (If the functions are complex valued, a different inner product, introduced in the next section, is defined.) If $(f, g) = 0$, the functions f and g are said to be *orthogonal*. In terms of the inner product (4.33) we define a *norm* of the function $f(x)$ as

$$(4.34) \qquad \|f\| = \sqrt{(f, f)} = \sqrt{\iint_G \rho f^2 \, dv} \, .$$

The norm $\|f\|$ which is nonnegative for real valued f is a measure of the magnitude of f. If f is continuous in G, then $\|f\| = 0$ if and only if $f = 0$. Most often one defines the inner product (4.33) with $\rho(x) = 1$, but for our discussion we require the more general definition given above.

In terms of the inner product (4.33), the equation (4.28) can be expressed in the form $(w, (1/\rho)Lu) = ((1/\rho)Lw, u)$. An operator $\hat{L}$ with the property that $(w, \hat{L}u) = (\hat{L}w, u)$ is said to be *self-adjoint*. If $\rho(x) \neq 1$, the operator $(1/\rho)L$ satisfies the self-adjointness condition in terms of the inner product (4.33), whereas if $\rho(x) = 1$, the operator L is self-adjoint. We remark that the self-adjointness of $(1/\rho)L$ or L is determined not only by the definition of L but also by the boundary conditions (4.14) that u and w must satisfy. (A general discussion of adjoint differential operators is given in Section 8.3.)

Next, we consider a function u that satisfies the boundary condition (4.14). We have

$$(4.35) \qquad \iint_G [uLu] \, dv = \iint_G u[-\nabla \cdot (p\nabla u) + qu] \, dv$$

$$= -\iint_G [\nabla \cdot (pu\nabla u) - p(\nabla u)^2] \, dv + \iint_G qu^2 \, dv$$

$$= \iint_G [p(\nabla u)^2 + qu^2] \, dv - \int_{\partial G} pu \frac{\partial u}{\partial n} \, ds,$$

where (4.29) and the divergence theorem have been used. If $\beta(x) > 0$ at a point on ∂G, we have from (4.14)

$$(4.36) \qquad \frac{\partial u}{\partial n} = -\left(\frac{\alpha}{\beta}\right)u,$$

whereas if $\beta = 0$ at a point on ∂G, we have $u = 0$ there. Consequently, (4.35) can be expressed as

$$(4.37) \qquad \iint_G [uLu] \, dv = \iint_G [p(\nabla u)^2 + qu^2] \, dv + \int_{\partial G} p\left(\frac{\alpha}{\beta}\right)u^2 \, ds,$$

where we have assumed that $\beta(x) > 0$ on ∂G. If $\beta(x) = 0$, the boundary integral vanishes. The assumptions on p, q, α, and β given previously imply that the right side of (4.37) is nonnegative if u is real valued.

In terms of the inner product (4.33), (4.37) implies that if u is real valued,

$$(4.38) \qquad \iint\limits_{G} [uLu]\, dv = \left(\frac{1}{\rho}Lu, u\right) \geq 0.$$

A self-adjoint operator $\hat{L}$ with the property that $(\hat{L}u, u) \geq 0$ is said to be a *positive operator*. Thus if $\rho \neq 1$, we see that $(1/\rho)L$ is a positive operator, whereas if $\rho = 1$, the operator L is itself positive. Again the positivity of $(1/\rho)L$ or L is based not only on its definition but also on the boundary conditions (4.14) that u must satisfy.

We now apply the results of the foregoing example to the eigenvalue problem (4.26). Let $M_k(x)$ and $M_j(x)$ be eigenfunctions corresponding to the distinct eigenvalues λ_k and λ_j, respectively. Since $M_k(x)$ and $M_j(x)$ both satisfy the boundary condition (4.14), the result (4.28) implies

$$(4.39) \qquad \iint\limits_{G} \left(M_k LM_j - M_j LM_k\right) dv = 0.$$

But $LM_j = \lambda_j \rho M_j$ and $LM_k = \lambda_k \rho M_k$, so that (4.39) yields

$$(4.40) \qquad \iint\limits_{G} \left(\lambda_j \rho M_k M_j - \lambda_k \rho M_j M_k\right) dv = (\lambda_j - \lambda_k)(M_k, M_j) = 0.$$

By assumption, $\lambda_j \neq \lambda_k$, so that (4.40) implies

$$(4.41) \qquad (M_k, M_j) = 0.$$

Consequently, eigenfunctions corresponding to different eigenvalues must be orthogonal in terms of the inner product (4.33).

Using (4.41) and the fact that the coefficients in the differential operator L are assumed to be real valued, it is easy to show that the eigenvalues for our problem must be real and that the corresponding eigenfunctions may be chosen to be real valued. (This is proven for the one-dimensional eigenvalue problem in Section 4.3 and the method of proof carries over immediately to the higher dimensional problem.)

Further, let $M_k(x)$ be a real valued eigenfunction that corresponds to the real eigenvalue λ_k. Since $LM_k = \lambda_k \rho M_k$, we obtain from the positivity property (4.38)

$$(4.42) \qquad \iint\limits_{G} [M_k LM_k]\, dv = \lambda_k \iint\limits_{G} \rho M_k^2\, dv = \lambda_k \|M_k\|^2 \geq 0.$$

Because $M_k(x)$ is real valued and nonzero by assumption, we conclude that the eigenvalue $\lambda_k \geqslant 0$.

Thus the self-adjointness property of the operator $(1/\rho)L$ implies that the eigenvalues are real and that eigenfunctions corresponding to different eigenvalues are orthogonal. The positivity property implies that the eigenvalues are nonnegative. It can be shown that there are a countably infinite number of eigenvalues λ_k ($k = 1, 2, \ldots$) whose only limit point is at infinity (i.e., $\lambda_k \to \infty$ as $k \to \infty$). We denote the corresponding set of (real valued) eigenfunctions by $M_k(x)$. Further properties of eigenvalues and eigenfunctions are presented later, and the eigenvalue problem is reexamined from a variational point of view in Section 8.1.

For each eigenvalue λ_k we obtain an equation for N_k in (4.27). Assuming for simplicity that $\lambda_k > 0$ for all k, we have for N_k

$$(4.43) \quad \begin{cases} N_k(t) = a_k\cos\left(\sqrt{\lambda_k}\,t\right) + b_k\sin\left(\sqrt{\lambda_k}\,t\right), & \text{hyperbolic case} \\ N_k(t) = a_k\exp(-\lambda_k t), & \text{parabolic case} \\ N_k(y) = a_k\exp\left(\sqrt{\lambda_k}\,y\right) + b_k\exp\left(-\sqrt{\lambda_k}\,y\right), & \text{elliptic case} \end{cases}$$

The a_k and b_k are arbitrary constants that must be determined.

Forming the product $M_k N_k$, we obtain

$$(4.44) \qquad\qquad u_k = M_k N_k$$

which for each value of k satisfies the appropriate equation (4.13), (4.17), or (4.19) together with the boundary condition (4.14) or (4.15).

To satisfy the additional conditions on $u(x, t)$ or $u(x, y)$ given above, we consider the formal *superposition* of the solutions u_k and construct the series

$$(4.45) \qquad\qquad u = \sum_{k=1}^{\infty} u_k = \sum_{k=1}^{\infty} M_k N_k$$

The constants a_k and b_k in the terms N_k are chosen to satisfy the additional conditions placed on u that have not as yet been accounted for by our choice of the $M_k(x)$.

In the hyperbolic case the initial conditions (4.16) must be satisfied and we obtain formally

$$(4.46) \quad \begin{cases} u(x,0) = \sum_{k=1}^{\infty} M_k(x)N_k(0) = \sum_{k=1}^{\infty} a_k M_k(x) = f(x) \\ u_t(x,0) = \sum_{k=1}^{\infty} M_k(x)N_k'(0) = \sum_{k=1}^{\infty} \sqrt{\lambda_k}\,b_k M_k(x) = g(x) \end{cases}$$

The expressions (4.46) represent what are referred to as *eigenfunction expansions* of $f(x)$ and $g(x)$. The validity of this representation is characterized by the concept of completeness of the set of eigenfunctions $M_k(x)$. If the set is *complete*, under certain conditions on the functions $f(x)$ and $g(x)$ the expansions (4.46) will converge to these functions. The more stringent the conditions, the better the convergence. Since the $M_k(x)$ are eigenfunctions for a self-adjoint operator as we have shown, even under very mild conditions on $f(x)$ and $g(x)$ the series of eigenfunctions will converge in some sense to these functions. This question is discussed more fully for the one-dimensional case in Section 4.3 and for the higher dimensional problem in Chapter 8.

For the hyperbolic case considered above and the further discussion in this section, the $M_k(x)$ are assumed to be an orthogonal set, that is, $(M_k, M_j) = 0$ for $k \neq j$. As shown above, this assumption is valid if each $M_k(x)$ corresponds to a different eigenvalue. If the eigenvalue is multiple and a finite number of linearly independent eigenfunctions correspond to a single eigenvalue, the *Gram–Schmidt orthogonalization process*, which will be discussed later in the text (see Exercise 8.2.1), can be used to orthogonalize the finite set of eigenfunctions. (For the eigenvalue problem under consideration, there cannot exist more than a finite number of linearly independent eigenfunctions for each eigenvalue.)

To determine the a_k in (4.46) we multiply the series by $\rho(x)M_j(x)$ and integrate over G. This gives

(4.47)

$$\left(f(x), M_j(x)\right) = \left(\sum_{k=1}^{\infty} a_k M_k(x), M_j(x)\right) = \sum_{k=1}^{\infty} a_k\left(M_k, M_j\right) = a_j\left(M_j, M_j\right)$$

on using (4.41) and assuming summation and integration can be interchanged in the series. Thus

(4.48)
$$a_k = \frac{\left(f(x), M_k(x)\right)}{\left(M_k(x), M_k(x)\right)},$$

and, by the same argument,

(4.49)
$$b_k = \frac{\left(g(x), M_k(x)\right)}{\sqrt{\lambda_k}\left(M_k(x), M_k(x)\right)}.$$

The formal solution of the initial and boundary value problem for the hyperbolic equation (4.13) is then given as

(4.50) $$u(x, t) = \sum_{k=1}^{\infty} \left[a_k \cos\left(\sqrt{\lambda_k}\, t\right) + b_k \sin\left(\sqrt{\lambda_k}\, t\right)\right] M_k(x).$$

For the parabolic case, the initial condition (4.18) must be satisfied. This gives

$$(4.51) \qquad u(x,0) = \sum_{k=1}^{\infty} M_k(x) N_k(0) = \sum_{k=1}^{\infty} a_k M_k(x) = f(x).$$

Thus a_k is specified as in (4.48) and the formal solution of the initial and boundary value problem for the parabolic equation (4.17) is

$$(4.52) \qquad u(x,t) = \sum_{k=1}^{\infty} a_k \exp[-\lambda_k t] M_k(x).$$

In the elliptic case, the boundary conditions (4.20) must be satisfied. This yields the eigenfunction expansions

$$(4.53) \quad \begin{cases} u(x,0) = \sum_{k=1}^{\infty} [a_k + b_k] M_k(x) = f(x) \\ u(x,\hat{l}) = \sum_{k=1}^{\infty} \left[a_k \exp\left(\sqrt{\lambda_k}\,\hat{l}\right) + b_k \exp\left(-\sqrt{\lambda_k}\,\hat{l}\right) \right] M_k(x) = g(x) \end{cases}$$

Applying the technique used in the hyperbolic case yields

$$(4.54) \quad \begin{cases} a_k + b_k = \dfrac{(f(x), M_k(x))}{(M_k(x), M_k(x))} \\ a_k \exp\left(\sqrt{\lambda_k}\,l\right) + b_k \exp\left(-\sqrt{\lambda_k}\,\hat{l}\right) = \dfrac{(g(x), M_k(x))}{(M_k(x), M_k(x))} \end{cases}$$

Unique solutions for the a_k and b_k can be determined from the foregoing system. Then the formal solution of the boundary value problem for the elliptic equation (4.19) is

$$(4.55) \qquad u(x,y) = \sum_{k=1}^{\infty} \left[a_k \exp\left(\sqrt{\lambda_k}\, y\right) + b_k \exp\left(-\sqrt{\lambda_k}\, y\right) \right] M_k(x).$$

The above solutions are often termed *classical* solutions of the given equations if the formal operations carried out are valid and the series expansions of u can be differentiated term by term as often as required by the equations and the initial and boundary data. However, even if term by term differentiability is not valid, the sum u may still be characterized as a *generalized* solution if certain conditions specified later are met.

We have shown that the separation of variables method reduces each of the initial and boundary value problems given in this section to the study of an eigenvalue problem. Unless the eigenvalues λ_k and the eigenfunctions $M_k(x)$ can be determined, the formal series solutions obtained remain unspecified. For many of the problems considered in this text the eigenvalues and eigenfunctions can be determined exactly, as we demonstrate in this and later chapters. When exact results are not available, approximation methods such as the Rayleigh–Ritz method discussed in Section 8.2 yield effective approximations to the leading and most significant eigenvalues and eigenfunctions.

4.3. THE STURM–LIOUVILLE PROBLEM AND FOURIER SERIES

The one-dimensional version of the eigenvalue problem introduced in the preceding section

$$(4.56) \quad L[v(x)] = -\frac{d}{dx}\left[p(x)\frac{dv}{dx}\right] + q(x)v(x) = \lambda\rho(x)v(x),$$

where $0 < x < l$ and $v(x)$ satisfies the boundary conditions (4.15), that is,

$$(4.57) \qquad \begin{cases} \alpha_1 v(0) - \beta_1 v'(0) = 0 \\ \alpha_2 v(l) + \beta_2 v'(l) = 0 \end{cases},$$

is known as the *Sturm–Liouville problem*. We require that $p(x) > 0$, $\rho(x) > 0$, and $q(x) \geqslant 0$, and that $p(x)$, $\rho(x)$, $q(x)$, and $p'(x)$ be continuous in the closed interval $0 \leqslant x \leqslant l$. Also, we must have $\alpha_i \geqslant 0$, $\beta_i \geqslant 0$, and $\alpha_i + \beta_i > 0$ for $i = 1, 2$. Then we have a *regular* Sturm–Liouville problem. If one or more of the conditions on the coefficients in (4.56) are relaxed, say, if $p(x)$ or $\rho(x)$ or both are permitted to vanish at either or both of the endpoints $x = 0$ and $x = l$, we have *singular* Sturm–Liouville problem. Both cases are of interest in applications and examples of each type are considered.

Before discussing and deriving some of the important properties of the eigenvalues and eigenfunctions of the Sturm–Liouville problem, we introduce several relevant definitions and concepts. Some of these have already been given in the preceding section but are repeated here in their one-dimensional form. Unless otherwise specified, the functions considered in the following discussion will be real valued.

The *inner product* of two functions $\varphi(x)$ and $\psi(x)$ (bounded and integrable over the interval $0 \leqslant x \leqslant l$) is defined as

$$(4.58) \qquad (\varphi, \psi) = \int_0^l \rho(x)\varphi(x)\psi(x)\,dx,$$

with the weight function $\rho(x) > 0$ in $0 < x < l$. [Most often one uses a unit weight function $\rho(x) = 1$; however, for our purposes the above definition is more appropriate.] The inner product is clearly symmetric; that is, $(\varphi, \psi) =$

(ψ, φ). The *norm* of a function $\varphi(x)$ defined in terms of the inner product (4.58) (or induced by it) is

$$(4.59) \qquad \|\varphi\| = \sqrt{(\varphi, \varphi)} = \sqrt{\int_0^l \rho \varphi^2 \, dx}$$

The norm, which is clearly nonnegative, is a measure of the magnitude of the function $\varphi(x)$ over the interval $0 \leqslant x \leqslant l$. If $\varphi(x)$ is continuous in that interval, then $\|\varphi\| = 0$ if and only if $\varphi(x) = 0$. Any function $\varphi(x)$ with a finite norm (i.e., $\|\varphi\| < \infty$) is said to be *square integrable* over the interval $0 < x < l$. If $\varphi(x)$ is such that $\|\varphi\| = 1$, then $\varphi(x)$ is said to be *normalized* to unity. Any square integrable function can be normalized by defining a new function $\hat{\varphi}(x) = \varphi(x)/\|\varphi\|$ for which it is immediately seen that $\|\hat{\varphi}\| = 1$. [All square integrable functions will be required to be integrable as well (see Exercise 4.3.1)].

Two functions φ and ψ for which

$$(4.60) \qquad (\varphi, \psi) = 0,$$

are said to be *orthogonal* over the interval $0 < x < l$. A set of functions $\{\varphi_k(k)\}$, $k = 1, 2, \ldots$, for which

$$(4.61) \qquad (\varphi_k, \varphi_j) = 0, \qquad k \neq j$$

is said to be an *orthogonal set*. If, in addition, $\|\varphi_k\| = 1$ for all φ_k, the set is said to be *orthonormal*. Any orthogonal set can be orthonormalized by normalizing each of the φ_k in the manner shown above.

If the functions φ, ψ or φ_k can assume complex values, we replace the inner product (4.58) by the (weighted) *Hermitian inner product*

$$(4.62) \qquad (\varphi, \psi) = \int_0^l \rho(x) \varphi(x) \overline{\psi(x)} \, dx,$$

where the bar denotes complex conjugation. Since we now have $(\varphi, \psi) = \overline{(\psi, \varphi)}$, the inner product has Hermitian symmetry. The induced norm is

$$(4.63) \qquad \|\varphi\| = \sqrt{(\varphi, \varphi)} = \sqrt{\int_0^l \rho(x) |\varphi(x)|^2 \, dx} \quad,$$

so that $\|\varphi\| \geqslant 0$.

We now return to the consideration of real valued functions. Given the orthonormal set of square integrable functions $\{\varphi_k(x)\}$, $k = 1, 2, \ldots$, and the square integrable function $\varphi(x)$ over the interval $0 < x < l$, the set of numbers (φ, φ_k) are called the *Fourier coefficients* of $\varphi(x)$. The formal series

$$(4.64) \qquad \varphi(x) = \sum_{k=1}^{\infty} (\varphi, \varphi_k) \varphi_k(x),$$

is called the *Fourier series* of $\varphi(x)$. Even though we have equated $\varphi(x)$ to its Fourier series in (4.64), we have yet to specify in what sense the Fourier series converges to the function $\varphi(x)$, if it converges at all. Further, although we use the term "Fourier series" to denote any expansion of a function in terms of a set of orthonormal (or more generally orthogonal) functions, in our later discussions, as well as in the literature, this term is often used to denote an expansion in a series of trigonometric functions.

To discuss the convergence properties of the series (4.64) we note that for the finite sum $\sum_{k=1}^{N}(\varphi, \varphi_k)\varphi_k(x)$, we have

(4.65)

$$\left\| \varphi(x) - \sum_{k=1}^{N}(\varphi, \varphi_k)\varphi_k(x) \right\|^2 = \left(\varphi - \sum_{k=1}^{N}(\varphi, \varphi_k)\varphi_k, \varphi - \sum_{k=1}^{N}(\varphi, \varphi_k)\varphi_k \right)$$

$$= (\varphi, \varphi) - 2\sum_{k=1}^{N}(\varphi, \varphi_k)^2 + \sum_{k=1}^{N}(\varphi, \varphi_k)^2 = \|\varphi\|^2 - \sum_{k=1}^{N}(\varphi, \varphi_k)^2 \geq 0,$$

where we have used elementary properties of the inner product, the orthonormality of the set $\{\varphi_k(x)\}$, and the nonnegativity of the norm. Since $\|\varphi\| < \infty$ and (4.65) is valid for all N, we obtain in the limit as $N \to \infty$

(4.66)
$$\sum_{k=1}^{\infty}(\varphi, \varphi_k)^2 \leq \|\varphi\|^2,$$

which is known as *Bessel's inequality*. This shows that the sum of squares of the Fourier coefficients of any square integrable function $\varphi(x)$ converges.

A sequence of (square integrable) functions $\{\psi_N(x)\}$, $N = 1, 2, \ldots$, is said to converge to a function $\varphi(x)$ in the mean if

(4.67)
$$\lim_{N \to \infty} \|\varphi(x) - \psi_N(x)\| = 0.$$

This type of convergence, which differs from and does not imply pointwise convergence, is called *mean square convergence*. If the partial sums of the series (4.64) are denoted by

(4.68)
$$\psi_N(x) = \sum_{k=1}^{N}(\varphi, \varphi_k)\varphi_k(x),$$

we see that if *Parseval's equality*

(4.69)
$$\sum_{k=1}^{\infty}(\varphi, \varphi_k)^2 = \|\varphi\|^2$$

is satisfied, (4.65) implies that

$$(4.70) \quad \lim_{N \to \infty} \|\varphi(x) - \psi_N(x)\|^2 = \lim_{N \to \infty} \left\{ \|\varphi\|^2 - \sum_{k=1}^{N} (\varphi, \varphi_k)^2 \right\} = 0.$$

Thus the Fourier series (4.64) converges in the mean square sense to the function $\varphi(x)$ in the interval $0 < x < l$, if Parseval's equality is satisfied.

A set of square integrable functions $\{\varphi_k(x)\}$ is said to be *complete* (with respect to mean square convergence) if, for any square integrable function $\varphi(x)$, its Fourier series (4.64) converges to it in the mean. It follows from (4.70) that if Parseval's equation (4.69) is satisfied for all square integrable functions $\varphi(x)$, the set of functions $\{\varphi_k(x)\}$, $k = 1, 2, \ldots$, is complete. Although few restrictions need to be placed on a function $\varphi(x)$ in order to achieve mean square convergence for its Fourier series, in applications of Fourier series to the solution of boundary value problems, stronger forms of convergence such as uniform convergence are generally required.

Returning to our consideration of the Sturm–Liouville problem, we now list the basic properties of the eigenvalues and eigenfunctions and derive the simplest of these properties.

1. Eigenfunctions corresponding to different eigenvalues are orthogonal. Let λ_i and λ_j be two distinct eigenvalues and $v_i(x)$ and $v_j(x)$ two corresponding eigenfunctions for the problem (4.56)–(4.57). Then

$$(4.71) \quad \int_0^l \left[v_i L v_j - v_j L v_i \right] dx = \int_0^l \left[v_j (p v_i')' - v_i (p v_j')' \right] dx$$

$$= \int_0^l \frac{d}{dx} \left[p v_j v_i' - p v_i v_j' \right] dx = 0,$$

as is easily verified on using the boundary conditions (4.57) for v_i and v_j. But we also have

$$(4.72) \quad \int_0^l \left[v_i L v_j - v_j L v_i \right] dx = (\lambda_i - \lambda_j) \int_0^l \rho v_i v_j \, dx = (\lambda_i - \lambda_j)(v_i, v_j).$$

Since $\lambda_i \neq \lambda_j$, by assumption, we conclude that

$$(4.73) \quad (v_i(x), v_j(x)) = 0,$$

which implies orthogonality. [This result should be compared with (4.39)–(4.41) and follows from the self-adjointness of the operator L.]

2. The eigenvalues are real and nonnegative and the eigenfunctions may be chosen to be real valued. Suppose an eigenvalue λ_i is complex valued. Then $\bar{\lambda}_i = \lambda_j$ represents a second distinct eigenvalue. Because the coefficients in the operator L are real valued, eigenfunctions corresponding to λ_i and $\lambda_j = \bar{\lambda}_i$ are

$v_i(x)$ and $v_j(x) = \bar{v}_i(x)$ where $v_j(x)$ is obtained by complex conjugation in the equation for $v_i(x)$. Then since $\lambda_i \neq \lambda_j$, (4.73) shows that

$$(4.74) \quad \big(v_i(x), v_j(x)\big) = \big(v_i(x), \bar{v}_i(x)\big) = \int_0^l \rho(x)|v_i(x)|^2 \, dx = 0.$$

But $v_i(x) \neq 0$ so that the integral in (4.76) must be nonzero, and we have a contradiction. Consequently, we must have $\lambda_j = \bar{\lambda}_i$ and λ_i is real valued. Since the coefficients and the eigenvalues in the equation (4.56) are now real valued, the real and imaginary parts of any solution (i.e., eigenfunction) must also satisfy (4.56). Therefore, the eigenfunctions can be chosen to be real valued and we assume them to be such.

The nonnegativity of the eigenvalues can be shown as follows. Let $v(x)$ be an eigenfunction. Then

$$(4.75) \qquad \left(v, \frac{1}{\rho} Lv\right) = -\int_0^l v \frac{d}{dx}(pv') \, dx + \int_0^l q v^2 \, dx$$

$$= -pvv'\big|_0^l + \int_0^l pv'^2 \, dx + \int_0^l qv^2 \, dx \geq 0,$$

since, say, at $x = l$ we have

$$(4.76) \qquad -p(l)v(l)v'(l) = \begin{cases} \left(\dfrac{\alpha_2}{\beta_2}\right)p(l)v^2(l) \geq 0; & \beta_2 > 0 \\[2mm] 0; & \beta_2 = 0 \end{cases}$$

with a similar result valid at $x = 0$ (the last two integrals in (4.75) are clearly nonnegative). But we also have

$$(4.77) \qquad \left(v, \frac{1}{\rho} Lv\right) = (v, \lambda v) = \lambda(v, v) = \lambda \|v\|^2,$$

with $\|v\| > 0$, by assumption. Combining (4.75) and (4.77) gives

$$(4.78) \qquad \lambda = \frac{\left(v, \dfrac{1}{\rho} Lv\right)}{\|v\|^2} \geq 0,$$

which proves that the eigenvalue λ is nonnegative. (Note that the positivity of the operator $(1/\rho)L$ has been used to prove this result. This may be compared with the discussion in the preceding section.)

The above discussion may be used to show that if $q(x) > 0$ in the interval $0 \leq x \leq l$, then $\lambda = 0$ cannot be an eigenvalue of the Sturm–Liouville problem. If $q(x) = 0$, then $\lambda = 0$ is an eigenvalue if and only if $\alpha_1 = \alpha_2 = 0$. This is shown in the exercises.

3. Each eigenvalue is simple. Since (4.56) is a second order equation, there can be at most two linearly independent eigenfunctions for each eigenvalue λ. However, for the Sturm–Liouville problem, as shown in the exercises, there is only one linearly independent eigenfunction for each eigenvalue. That is, each eigenvalue is simple.

4. There is a countable infinity of eigenvalues having a limit point at infinity. The set of eigenvalues can be arranged as follows: $0 \leqslant \lambda_1 < \lambda_2 < \lambda_3 < \ldots$ with $\lambda_k \to \infty$ as $k \to \infty$. The set of eigenvalues is called the *spectrum* of the operator L, and we see that the spectrum is discrete, nonnegative, and has a limit point at infinity.

5. The set of eigenfunctions $\{v_k(x)\}$, $k = 1, 2, \ldots$ forms a complete orthonormal set of square integrable functions on the interval $0 < x < l.$ As a result, if the function $v(x)$ is square integrable over the interval $0 < x < l$, the Fourier series or, equivalently, the eigenfunction expansion

$$(4.79) \qquad v(x) = \sum_{k=1}^{\infty} (v, v_k) v_k(x)$$

converges to $v(x)$ in the mean. It can be shown, however, that if $v(x)$ is continuous and, in addition, has a piecewise continuous first derivative in $0 \leqslant x \leqslant l$, and $v(x)$ satisfies the boundary conditions (4.57), the Fourier series (4.79) converges absolutely and uniformly to $v(x)$ in the given interval. If $v(x)$ has a jump discontinuity at some interior point x_0 in the interval, the Fourier series converges to the value $(1/2)v(x_0 -) + (1/2)v(x_0 +)$ at that point, where $v(x_0 -)$ and $v(x_0 +)$ represent one-sided limits of $v(x)$ as x approaches x_0 from the left and the right, respectively.

Proofs of properties 4 and 5 are not given here. However, the discussion in Section 8.1 which deals with higher dimensional eigenvalue problems can be adapted to yield a proof of these properties for the Sturm–Liouville problem.

The eigenvalues and eigenfunctions of the Sturm–Liouville problem can be formally obtained as follows. Let $V(x; \lambda)$ and $W(x; \lambda)$ be solutions of the initial value problems

$$(4.80) \qquad V(0; \lambda) = 1; \qquad V'(0; \lambda) = 0,$$

$$(4.81) \qquad W(0; \lambda) = 0; \qquad W'(0; \lambda) = 1,$$

respectively for the equation (4.56). Then, the function

$$(4.82) \qquad v(x; \lambda) = \beta_1 V(x; \lambda) + \alpha_1 W(x; \lambda)$$

is a solution of (4.56) that satisfies the boundary condition (4.57) at $x = 0$. To satisfy the condition at $x = l$ we must have

$$(4.83) \quad \alpha_2\beta_1 V(l; \lambda) + \alpha_2\alpha_1 W(l; \lambda) + \beta_2\beta_1 V'(l; \lambda) + \beta_2\alpha_1 W'(l; \lambda) = 0,$$

which results from the substitution of (4.82) into (4.57). The eigenvalues of the Sturm–Liouville problem $\lambda = \lambda_k$ ($k = 1, 2, \dots$) are determined as the roots of the equation (4.83). The corresponding eigenfunctions are given as

$$(4.84) \quad v_k(x) = v(x; \lambda_k) = \beta_1 V(x; \lambda_k) + \alpha_1 W(x; \lambda_k), \qquad k = 1, 2, \dots.$$

In the following examples we consider a number of regular and singular Sturm–Liouville problems that lead to trigonometric, Bessel, and Legendre eigenfunctions.

EXAMPLE 4.3.   TRIGONOMETRIC EIGENFUNCTIONS

We set $p(x) = \rho(x) = 1$ and $q(x) = 0$ in (4.56) and retain the general boundary conditions (4.57). The resulting equation for the eigenfunction $v(x)$ is

$$(4.85) \qquad\qquad -v''(x) = \lambda v(x).$$

The solutions of the initial value problems (4.80)–(4.81) for (4.85) are found to be

$$(4.86) \qquad V(x; \lambda) = \cos(\sqrt{\lambda}\, x); \qquad W(x; \lambda) = \frac{\sin(\sqrt{\lambda}\, x)}{\sqrt{\lambda}},$$

where we have assumed $\lambda > 0$. Then

$$(4.87) \qquad\qquad v(x; \lambda) = \beta_1 \cos(\sqrt{\lambda}\, x) + \alpha_1 \frac{\sin(\sqrt{\lambda}\, x)}{\sqrt{\lambda}},$$

and the eigenvalues are determined from the transcendental equation

$$(4.88) \quad \sqrt{\lambda}\,(\alpha_1\beta_2 + \beta_1\alpha_2)\cos(\sqrt{\lambda}\, l) + (\alpha_2\alpha_1 - \lambda\beta_2\beta_1)\sin(\sqrt{\lambda}\, l) = 0.$$

We do not discuss the solutions of this equation but consider two special cases. Once the eigenvalues λ_k ($k = 1, 2, \dots$) are determined from (4.88), we obtain the eigenfunctions

$$(4.89) \quad v_k(x) = \beta_1 \cos\left(\sqrt{\lambda_k}\, x\right) + \alpha_1 \frac{\sin\left(\sqrt{\lambda_k}\, x\right)}{\sqrt{\lambda_k}}; \qquad k = 1, 2, \dots.$$

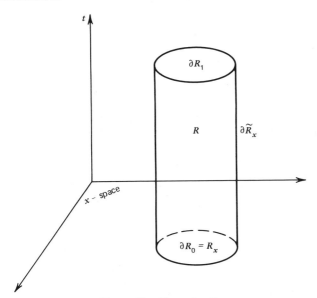

Figure 6.1. The region R.

(6.6) can, therefore, be written as

$$(6.7) \qquad \int_{R_x} [\rho(x) u_t(x, t_1) - \rho(x) u_t(x, t_0)] \, dx$$

$$= \int_{t_0}^{t_1} \int_{\partial R_x} p \frac{\partial u}{\partial n_x} \, ds_x \, dt - \int_{t_0}^{t_1} \int_{R_x} (qu - \rho F) \, dx \, dt$$

where $\partial u / \partial n_x$ is the exterior normal derivative of u on the surface ∂R_x. In one space dimension with R_x equal to the interval (x_0, x_1), (6.7) has the form

$$(6.8) \qquad \int_{x_0}^{x_1} [\rho(x) u_t(x, t_1) - \rho(x) u_t(x, t_0)] \, dx$$

$$= \int_{t_0}^{t_1} [p(x_1) u_x(x_1, t) - p(x_0) u_x(x_0, t)] \, dt$$

$$- \int_{t_0}^{t_1} \int_{x_0}^{x_1} [q(x) u(x, t) - \rho(x) F(x, t)] \, dx \, dt.$$

To consider a concrete example, we assume that (6.1) describes the longitudinal vibration of a rod and that the problem is one-dimensional. Then the integral relation (6.8) characterizes the change in momentum of a segment of a rod (x_0, x_1) in the time interval (t_0, t_1). The displacement of the rod at the time t is given by $u(x, t)$, the density is $\rho(x)$, and the tension is $T(x, t) = p(x) u_x(x, t)$—as required by Hooke's law—where $p(x)$ equals Young's modulus at the point x. The momentum equals $\rho(x)u_t(x, t)$ and we put $q(x) = 0$ in (6.8). Also $\rho(x)F(x, t)$ represents the external force density. Thus (6.8) equates the change in momentum to the forces acting on the segment (x_0, x_1) of the rod. In the limit as $x_0 \to x_1$ and $t_0 \to t_1$, we obtain the partial differential equation (6.1) with $q = 0$ and ∇ replaced by $\partial/\partial x$. For the limiting procedure to be valid, $u(x, t)$ must have continuous second derivatives and ρ, p', and F must be nonsingular. This enables the mean value theorems for derivatives and integrals to be applied to (6.8). At points where the limit process is not valid, it does yield matching conditions.

For completeness we present the appropriate integral relations for the elliptic and parabolic equations of Section 4.1. Their derivations are similar to those given in the hyperbolic case and are not presented. For the parabolic equation (4.4) we have

$$(6.9) \qquad \int_{\partial R} [p\nabla u, -\rho u] \cdot \mathbf{n} \, ds = \iint_R (qu - \rho F) \, dv$$

and, in a form equivalent to that in (6.7),

$$(6.10) \quad \int_{R_x} [\rho(x)u(x, t_1) - \rho(x)u(x, t_0)] \, dx$$

$$= \int_{t_0}^{t_1} \int_{\partial R_x} p \frac{\partial u}{\partial n_x} \, ds_x \, dt - \int_{t_0}^{t_1} \int_{R_x} (qu - \rho F) \, dx \, dt.$$

For the elliptic case, given the equation

$$(6.11) \qquad -\nabla \cdot (p\nabla u) + qu = \rho F,$$

we have the equivalent integral relation

$$(6.12) \qquad \int_{\partial R} p(\nabla u \cdot \mathbf{n}) \, ds = \int_{\partial R} p \frac{\partial u}{\partial n} \, ds = \iint_R (qu - \rho F) \, dv.$$

The regions R and R_x and their boundaries are defined as in the foregoing hyperbolic problem for the parabolic case, whereas $R = R_x$ in the elliptic case. We do not write down the one-dimensional forms of (6.9)–(6.12).

In each of the above integral relations in two or three-dimensional x-space, if we choose R_x to be a rectangular region (i.e., $x_0 < x < x_1$, $y_0 < y < y_1$, $z_0 < z < z_1$ in three dimensions), it is an easy matter to retrieve the partial differential equations from the integral relations in the limit as R_x shrinks to a point and $t_0 \to t_1$. However, the function u and the coefficients, as well as F, must be sufficiently smooth for this limiting process to be valid. This implies the equivalence of the differential and integral representations of the given equations in regions where everything is smooth. In the following sections we show how the integral relations are to be used in cases where the foregoing limit process breaks down in some region, so that the given differential equation is not valid there.

6.2. COMPOSITE MEDIA: DISCONTINUOUS COEFFICIENTS

We consider any one of the three integral relations (6.7), (6.10), or (6.12) and assume that a Cauchy, an initial and boundary value or a strict boundary value problem, whichever is appropriate for the equation, is given over a region G in x-space. (The region G may be bounded or unbounded.) Let G be divided into two subregions G_1 and G_2 with S_0 as the boundary region separating G_1 and G_2. We assume that the properties of the given medium vary discontinuously across S_0 (for all time t if the problem has a time dependence), so that one or more of the functions $\rho(x)$, $p(x)$, or $q(x)$ is discontinuous across S_0 (they are permitted to have, at most, jump discontinuities). For example, if two strings of different densities are attached at a point x_0 and we consider the equation for the vibration of the composite string, the density $\rho(x)$ will be discontinuous at the point x_0.

At points x or (x, t) that do not lie in the region of discontinuity and where the coefficients and the solution are assumed to be smooth functions, we may go to the limit in the integral relations and derive the appropriate partial differential equation. Thus at points x or (x, t) such that x is interior G_1 or G_2, the partial differential equation is valid. Let us denote the solution in G_1 by u_1 and the solution in G_2 by u_2. It is meaningful to require that the solution of the problem be continuous across S_0. Thus if u is a temperature distribution or represents the displacement of a string, we expect it to be continuous in the interior of the region G. Therefore, the first *matching condition* is

$$(6.13) \qquad u_1\big|_{S_0} = u_2\big|_{S_0}.$$

To obtain a second matching condition, we apply the appropriate integral relation over a region R_x that contains a portion of S_0 in its interior, as pictured in Figure 6.2. The region R_x is extended parallel to itself from t_0 to t_1 in the hyperbolic and parabolic cases (as was done above). We now consider the integral relations (6.7), (6.10), and (6.12) [with $R = R_x$ in (6.12)]. Let the region R_x collapse onto S_0. Since u (as well as u_t) is continuous across S_0 and p, ρ, q, and, possibly, F have, at most, jump discontinuities across S_0, the only contribution that results in the limit comes from the normal derivative terms that may have different values on both sides of S_0. Because this result is valid over any portion of S_0, we conclude on the basis of the duBois–Reymond lemma (see Exercise 8.1.9) that the second *matching condition* is

$$(6.14) \qquad p_1 \frac{\partial u_1}{\partial n}\bigg|_{S_0} = p_2 \frac{\partial u_2}{\partial n}\bigg|_{S_0}$$

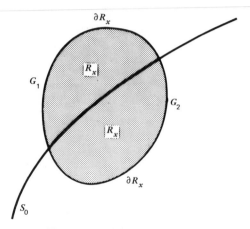

Figure 6.2. The composite region.

where $p_1(x)$ and $p_2(x)$ represent the limiting values of $p(x)$ as x approaches a point in S_0 from the subregions G_1 and G_2, respectively. Both normal derivatives (in S_0) in (6.14) are taken in the same direction. We remark that (6.13) and (6.14) state that both u and $p\,\partial u/\partial n$ are continuous across S_0, and these conditions are valid for all time if u is time dependent.

In the one-dimensional case, if S_0 corresponds to the point x_0, and G_1 and G_2 correspond to values of x less than and greater than x_0, respectively, (6.13)–(6.14) are replaced by

$$(6.15) \qquad u_1\big|_{x_0} = u_2\big|_{x_0}$$

$$(6.16) \qquad p_1(x)\frac{\partial u_1}{\partial x}\bigg|_{x_0} = p_2(x)\frac{\partial u_2}{\partial x}\bigg|_{x_0}.$$

For the sake of concreteness we formulate an initial and boundary value problem for a hyperbolic equation in a composite medium with S_0 as the region of discontinuity. We have, with G equal to the union of G_1, S_0, and G_2

$$(6.17) \quad \begin{cases} \rho_1(x)\dfrac{\partial^2 u_1}{\partial t^2} - \nabla \cdot (p_1 \nabla u_1) + q_1 u_1 = \rho_1 F_1, & x \in G_1, t > 0 \\[2mm] \rho_2(x)\dfrac{\partial^2 u_2}{\partial t^2} - \nabla \cdot (p_2 \nabla u_2) + q_2 u_2 = \rho_2 F_2, & x \in G_2, t > 0 \end{cases}$$

The initial conditions are

$$(6.18) \quad u(x,0) = f(x); \qquad u_t(x,0) = g(x); \qquad u = \begin{cases} u_1, & x \in G_1 \\ u_2, & x \in G_2 \end{cases}.$$

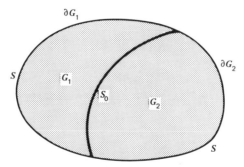

Figure 6.3. The region G.

The boundary conditions given on the boundary S of the region G are

$$(6.19) \qquad \alpha u + \beta \frac{\partial u}{\partial n}\bigg|_S = B(x, t); \qquad u = \begin{cases} u_1, & x \in \partial G_1 \\ u_2, & x \in \partial G_2 \end{cases}.$$

and the matching conditions are

$$(6.20) \qquad u_1\big|_{S_0} = u_2\big|_{S_0}$$

$$(6.21) \qquad p_1 \frac{\partial u_1}{\partial n}\bigg|_{S_0} = p_2 \frac{\partial u_2}{\partial n}\bigg|_{S_0}.$$

The subscripts 1 and 2 indicate functions evaluated in the regions G_1 and G_2, respectively.

In case we consider the Cauchy problem for the preceding hyperbolic equation, the region G is of infinite extent so we drop the boundary conditions (6.19) and retain all other equations. In either case, we can solve the problem by considering (6.17) in the regions G_1 and G_2 separately. We apply the appropriate initial and boundary conditions for each subregion with the additional conditions $u_1 = \tilde{u}_1$ and $u_2 = \tilde{u}_2$ on S_0 where $\tilde{u}_1$ and $\tilde{u}_2$ are as yet unknown. Once the solutions u_1 and u_2 are determined, the functions $\tilde{u}_1$ and $\tilde{u}_2$ are then specified by applying the matching conditions (6.20)–(6.21).

For the parabolic case we replace $\partial^2 u_i / \partial t^2$, $i = 1, 2$ by $\partial u_i / \partial t$, $i = 1, 2$ in (6.17) and drop the second initial condition $u_t(x, 0) = g(x)$ in (6.18). In the elliptic case, the time derivatives and the initial conditions in (6.17) and (6.18) are dropped. However, the other conditions remain for both cases.

EXAMPLE 6.1. VIBRATION OF AN INFINITE COMPOSITE ROD

We consider the longitudinal vibration of a composite rod of infinite extent. The rod is made up of two homogeneous rods joined at $x = 0$, each of which has a constant density ρ and a constant Young's modulus p (see the foregoing discussion). The region G is the infinite interval $-\infty < x < \infty$ with G_1 equal to $-\infty < x < 0$ and G_2 equal to $0 < x < \infty$ and the discontinuity region S_0 is the point $x = 0$. We set q and F equal to zero in (6.17) and replace ∇ by $\partial/\partial x$. Also, ρ_1, ρ_2 and p_1, p_2 are constants that are given for the problem.

The appropriate wave equations in the regions G_1 and G_2 are

$$(6.22) \qquad \frac{\partial^2 u_i}{\partial t^2} - c_i^2 \frac{\partial^2 u_i}{\partial x^2} = 0, \qquad i = 1, 2$$

where $c_i^2 = p_i/\rho_i$, $i = 1, 2$. Thus there are different speeds of wave propagation c_1 and c_2 in the two regions G_1 and G_2, respectively.

We assume that a wave $f(t - x/c_1)$ is approaching the junction point $x = 0$ as t tends to zero (from negative values) and we want to determine the resulting reflected and transmitted waves. To do so, we formulate this as a Cauchy problem with the initial data

$$(6.23) \quad u_1(x,0) = f\left(\frac{-x}{c_1}\right); \qquad \frac{\partial u_1}{\partial t}(x,0) = f'\left(\frac{-x}{c_1}\right); \qquad x < 0, (x \in G_1);$$

$$(6.24) \quad u_2(x,0) = 0; \qquad \frac{\partial u_2}{\partial t}(x,0) = 0; \qquad x > 0, (x \in G_2).$$

Note that a solution of (6.22) with $i = 1$ and the initial data (6.23) is just $u_1(x, t) = f(t - x/c_1)$—that is, a wave traveling to the right with speed c_1. For small t, this is assumed to be the only disturbance in the region G_1. As t increases, the wave reaches the junction point $x = 0$ and gives rise to a reflected and transmitted wave. This is the case if we assume that $f(-x) \equiv 0$ for $x > x_1$ where $x_1 < 0$. In addition to the conditions (6.23)–(6.24), the solutions u_1 and u_2 must satisfy the matching conditions (6.15)–(6.16) where $x_0 = 0$.

The general solutions of the wave equations (6.22) are

$$(6.25) \qquad u_i(x, t) = f_i\left(t - \frac{x}{c_i}\right) + g_i\left(t + \frac{x}{c_i}\right), \qquad i = 1, 2.$$

From (6.23)–(6.24), we have

$$(6.26) \qquad f_1\left(\frac{-x}{c_1}\right) + g_1\left(\frac{x}{c_1}\right) = f\left(\frac{-x}{c_1}\right), \qquad x < 0,$$

$$(6.27) \qquad f_1'\left(\frac{-x}{c_1}\right) + g_1'\left(\frac{x}{c_1}\right) = f'\left(\frac{-x}{c_1}\right), \qquad x < 0,$$

$$(6.28) \qquad f_2\left(\frac{-x}{c_2}\right) + g_2\left(\frac{x}{c_2}\right) = 0, \qquad x > 0.$$

$$(6.29) \qquad f_2'\left(\frac{-x}{c_2}\right) + g_2'\left(\frac{x}{c_2}\right) = 0; \qquad x > 0.$$

On differentiating (6.26) and (6.28) we easily conclude that $g_1(z) = 0$ for $z < 0$, $g_2(z) = 0$ for $z > 0$, $f_1(z) = f(z)$ for $z > 0$, and $f_2(z) = 0$ for $z < 0$. The matching conditions

$$(6.30) \qquad f_1(t) + g_1(t) = f_2(t) + g_2(t), \qquad t > 0;$$

$$(6.31) \quad p_1\left[\frac{-1}{c_1}f_1'(t) + \frac{1}{c_1}g_1'(t)\right] = p_2\left[\frac{-1}{c_2}f_2'(t) + \frac{1}{c_2}g_2'(t)\right], \qquad t > 0$$

yield a simultaneous system of equations for $f_2(t)$ and $g_1(t)$ with $t > 0$. Solving these equations and introducing the results into (6.25) yields the solution as

$$(6.32) \qquad u_1(x, t) = f\left(t - \frac{x}{c_1}\right) + Rf\left(t + \frac{x}{c_1}\right), \qquad x < 0, t > 0$$

$$(6.33) \qquad u_2(x, t) = Tf\left(t - \frac{x}{c_2}\right), \qquad x > 0, t > 0$$

where the reflection coefficient R and the transmission coefficient T are given as

$$(6.34) \qquad R = \frac{\sqrt{\rho_1 p_1} - \sqrt{\rho_2 p_2}}{\sqrt{\rho_1 p_1} + \sqrt{\rho_2 p_2}}; \qquad T = \frac{2\sqrt{\rho_1 p_1}}{\sqrt{\rho_1 p_1} + \sqrt{\rho_2 p_2}}.$$

Near $t = 0$ we have $u_1 = f(t - x/c_1)$ and $u_2 = 0$, with $f(t - x/c_1)$ equal to a wave traveling to the right (i.e., towards the juncture point $x = 0$). This wave is called the incident wave. When the incident wave reaches the interface $x = 0$, two additional waves arise: the reflected wave $Rf(t + x/c_1)$ which travels to the left and the transmitted wave $Tf(t - x/c_2)$ which travels to the right. Since apart from a constant factor each wave has the same waveform $f(x)$, we call R and T the reflection and transmission coefficients. These coefficients are a measure of the amplitudes of the waveforms. We observe that $1 + R = T$ so that there is an equal distribution of amplitudes (in a sense) in both the regions $x < 0$ and $x > 0$. If $p_1\rho_1 = p_2\rho_2$ we have $R = 0$ and $T = 1$, so that the incident wave is transmitted without undergoing reflection. The solution is described in Figure 6.4.

___

We have indicated that to solve initial and boundary value problems in composite media, the problems should first be solved in the two subregions G_1 and G_2 with (as yet) unspecified boundary data on the discontinuity region S_0. The solutions are then fully determined by applying the matching conditions. This procedure was used in the foregoing example. There is an alternative approach that we now discuss for problems in bounded spatial regions. In the absence of a discontinuity region, such problems were solved in Chapter 4 with

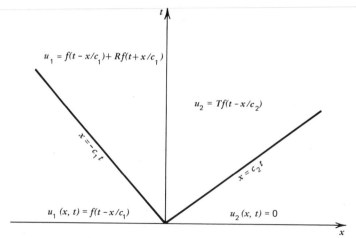

Figure 6.4.   Vibration of a composite rod.

the use of eigenfunction expansions. The eigenfunctions were determined from
the eigenvalue problem associated with the given (homogeneous) equation and
boundary conditions. We now demonstrate that it is also possible to construct
eigenvalue problems and solve the problems considered for a composite
medium by means of eigenfunction expansions.

Again considering the same region G with S_0 as a discontinuity region for
the coefficients, we introduce the following *eigenvalue problem* for the function
$v(x)$:

$$(6.35) \quad L[v] \equiv -\nabla \cdot (p\nabla v) + qv = \lambda\rho v; \qquad x \in G, x \notin S_0,$$

$$(6.36) \qquad\qquad \alpha v + \beta\frac{\partial v}{\partial n}\bigg|_S = 0,$$

$$(6.37) \qquad\qquad v \text{ and } p\frac{\partial v}{\partial n} \text{ are continuous across } S_0.$$

The coefficients p, q, and ρ may have jump discontinuities across S_0. All the
significant properties of eigenvalues and eigenfunctions given in Chapter 4
remain valid for this eigenvalue problem.

We demonstrate the orthogonality of eigenfunctions corresponding to different eigenvalues in the one-dimensional case. With the operator L defined as
in (4.56) and v_i and v_j equal to eigenfunctions corresponding to different
eigenvalues λ_i and λ_j ($\lambda_i \neq \lambda_j$), we have from (4.71):

$$(6.38) \quad \int_0^l (v_i L v_j - v_j L v_i)\, dx = \int_0^{x_0} \frac{d}{dx}\left(pv_j v_i' - pv_i v_j'\right) dx$$

$$+ \int_{x_0}^l \frac{d}{dx}\left(pv_j v_i' - pv_i v_j'\right) dx = 0,$$

where x_0 is the point of discontinuity and $0 < x_0 < l$. The contribution to each integral vanishes at the boundary points $x = 0$ and $x = l$ as before. At the point $x = x_0$, the functions $pv_i v_j'$ and $pv_j v_i'$ both occur with opposite signs and their contribution vanishes since v_i, v_j, pv_i', and pv_j' are assumed to be continuous at $x = x_0$. The inner product for this eigenvalue problem is defined as

$$(6.39) \qquad (v_i, v_j) = \int_0^l \rho v_i v_j \, dx = \int_0^{x_0} \rho v_i v_j \, dx + \int_{x_0}^l \rho v_i v_j \, dx,$$

and the norm is

$$(6.40) \qquad \|v_i\| = \sqrt{(v_i, v_i)} \, .$$

Then using (4.72) and (6.38) we conclude that $v_i(x)$ and $v_j(x)$ are orthogonal. The other properties listed and proven in Section 4.3 for the Sturm–Liouville problem can be proven for the present problem as well.

In the following example we solve the problem heat conduction in an insulated composite, finite rod with the ends kept at zero temperature in terms of the eigenvalue problem that results on solving this problem by separation of variables.

EXAMPLE 6.2.   HEAT CONDUCTION IN A FINITE COMPOSITE ROD

We consider heat conduction in an insulated piecewise homogeneous rod of length l, with the ends kept at zero temperature for all time. With the rod situated in the interval $0 \leqslant x \leqslant l$ and the juncture of the two homogeneous portions occurring at the point $x = x_0$, we have the following initial and boundary value problem for the temperature $u(x, t)$ with $u = u_1(x, t)$ for $x < x_0$ and $u = u_2(x, t)$ for $x > x_0$:

$$(6.41) \qquad \begin{cases} \rho_1 \dfrac{\partial u_1}{\partial t} - p_1 \dfrac{\partial^2 u_1}{\partial x^2} = 0, & 0 < x < x_0, t > 0 \\[2mm] \rho_2 \dfrac{\partial u_2}{\partial t} - p_2 \dfrac{\partial^2 u_2}{\partial x^2} = 0, & x_0 < x < l, t > 0 \end{cases}.$$

The coefficients ρ_1, ρ_2, p_1, and p_2 are all constant. The boundary conditions are

$$(6.42) \qquad u_1(0, t) = 0, \, u_2(l, t) = 0, \qquad t > 0.$$

The initial condition is

$$(6.43) \qquad u(x, 0) = f(x), \qquad 0 < x < l.$$

The matching conditions at $x = x_0$ are

(6.44)
$$\begin{cases} u_1(x_0, t) = u_2(x_0, t), & t > 0 \\ p_1 \dfrac{\partial u_1(x_0, t)}{\partial x} = p_2 \dfrac{\partial u_2(x_0, t)}{\partial x}, & t > 0 \end{cases}.$$

Using separation of variables we are led to the following eigenvalue problem for the function $v(x)$ with $v = v_1(x)$ for $x < x_0$ and $v = v_2(x)$ for $x > x_0$:

(6.45)
$$\begin{cases} p_1 v_1''(x) + \lambda \rho_1 v_1(x) = 0, & 0 < x < x_0 \\ p_2 v_2''(x) + \lambda \rho_2 v_2(x) = 0, & x_0 < x < l \end{cases}.$$

The homogeneous boundary conditions are

(6.46)
$$v_1(0) = 0, \qquad v_2(l) = 0$$

and the matching conditions at $x = x_0$ are

(6.47)
$$\begin{cases} v_1(x_0) = v_2(x_0) \\ p_1 \dfrac{\partial v_1(x_0)}{\partial x} = p_2 \dfrac{\partial v_2(x_0)}{\partial x} \end{cases}.$$

To determine the eigenfunctions we solve (6.45) for $v_1(x)$ and $v_2(x)$ applying the boundary conditions (6.46) to obtain with arbitrary constants A and B,

(6.48) $v_1(x) = A \sin\left(\sqrt{\dfrac{\lambda \rho_1}{p_1}}\, x \right); \qquad v_2(x) = B \sin\left[\sqrt{\dfrac{\lambda \rho_2}{p_2}}\, (x - l) \right].$

Applying the matching conditions gives

(6.49)
$$\begin{cases} A \sin\left(\sqrt{\dfrac{\lambda \rho_1}{p_1}}\, x_0 \right) - B \sin\left[\sqrt{\dfrac{\lambda \rho_2}{p_2}}\, (x_0 - l) \right] = 0 \\ A \sqrt{\lambda \rho_1 p_1}\, \cos\left(\sqrt{\dfrac{\lambda \rho_1}{p_1}}\, x_0 \right) - B \sqrt{\lambda \rho_2 p_2}\, \cos\left[\sqrt{\dfrac{\lambda \rho_2}{p_2}}\, (x_0 - l) \right] = 0 \end{cases}.$$

We require a nonzero solution of this homogeneous system for A and B so that the determinant of the coefficients must vanish. This can be written as

(6.50) $c_1 \rho_1 \cot\left(\dfrac{\sqrt{\lambda}}{c_1} x_0 \right) + c_2 \rho_2 \cot\left[\dfrac{\sqrt{\lambda}}{c_2} (l - x_0) \right] = 0,$

where $c_1 = \sqrt{p_1/\rho_1}$ and $c_2 = \sqrt{p_2/\rho_2}$. This equation determines the eigenvalues λ_k ($k = 1, 2, \ldots$). We do not solve this equation (which can be done graphically and numerically) but note that the λ_k are real and countably infinite and positive according to the general theory. The eigenfunctions $v^{(k)}(x)$ corresponding to the λ_k are

(6.51)

$$v^{(k)}(x) = \begin{cases} \dfrac{\sin\left(\sqrt{\lambda_k}\, x/c_1\right)}{\sin\left(\sqrt{\lambda_k}\, x_0/c_1\right)}; & 0 < x < x_0 \\[2em] \dfrac{\sin\left[\sqrt{\lambda_k}\,(l - x)/c_2\right]}{\sin\left[\sqrt{\lambda_k}\,(l - x_0)/c_2\right]}; & x_0 < x < l \end{cases} \quad ; \quad k = 1, 2, 3, \ldots .$$

For the $v^{(k)}(x)$ we have

$$(6.52) \quad \|v^{(k)}(x)\|^2 = \left(v^{(k)}(x), v^{(k)}(x)\right)$$

$$= \int_0^{x_0} \rho_1 v_1^{(k)}(x)\, dx + \int_{x_0}^l \rho_2 v_2^{(k)}(x)\, dx$$

$$= \frac{\rho_1 x_0}{2 \sin^2\left(\sqrt{\lambda_k}\, x_0/c_1\right)} + \frac{\rho_2(l - x_0)}{2 \sin^2\left[\sqrt{\lambda_k}\,(l - x_0)/c_2\right]}.$$

The orthogonality of the set of eigenfunctions $v^k(x)$ may be verified directly, and using (6.52), the set may be orthonormalized. The completeness property of the eigenfunctions asserts that a function $\varphi(x)$ under suitable conditions can be expanded in a series of eigenfunctions.

Applying the foregoing results to the initial boundary value problem for the heat equation (6.41)–(6.44), we obtain on separating variables or using finite Fourier transforms, the series representation for the solution $u(x, t)$ in the form

$$(6.53) \qquad u(x, t) = \sum_{k=1}^{\infty} \alpha_k e^{-\lambda_k t} v^{(k)}(x),$$

where the λ_k and $v^{(k)}(x)$ are the eigenvalues and eigenfunctions of the problem (6.45)–(6.47). The α_k are determined as the Fourier coefficients of the initial temperature distribution $f(x)$; that is,

$$(6.54) \qquad u(x, 0) = \sum_{k=1}^{\infty} \alpha_k v^{(k)}(x) = f(x)$$

and they are given as

$$(6.55) \quad \alpha_k = \frac{1}{\|v^{(k)}(x)\|^2}\left[\rho_1 \int_0^{x_0} f(x) v_1^{(k)}(x)\, dx + \rho_2 \int_{x_0}^l f(x) v_2^{(k)}(x)\, dx\right]$$

for $k = 1, 2 \ldots$ and $\|v^{(k)}(x)\|^2$ given in (6.52).

The series (6.53) represents the formal solution of the problem. The nonhomogeneous version of the foregoing heat conduction problem can be solved using the preceding eigenfunctions and the finite Fourier transform. Similarly, related problems for hyperbolic and elliptic equations also can be solved in terms of the preceding eigenfunctions.

6.3. SOLUTIONS WITH DISCONTINUOUS FIRST DERIVATIVES

It has been shown in Section 3.2 that discontinuities in second derivatives for second order partial differential equations must occur across characteristic curves or surfaces. A large class of second order partial differential equations were shown in Section 6.1 to have equivalent integral relations in the case where the coefficients and the unknown functions were smooth. Since the integral relations contain, at most, first derivatives of the unknown functions, they can be used to attach a meaning to solutions of differential equations that are continuous and have only piecewise continuous first derivatives.

We consider the integral form (6.5) of the hyperbolic equation (6.1) in three dimensions. Assume that a solution $u(x, t)$ of (6.5) is continuous across a surface $\varphi(x, t) = 0$ but has jump discontinuities in its first derivatives across that surface. Let the surface be denoted by S_0 and assume that it divides the given region R into subregions R_1 and R_2. The solution $u(x, t)$ is assumed to be smooth in R_1 and R_2 and the functions $\rho, p, q,$ and F in (6.1) and (6.5) are assumed to be smooth throughout R. A unit normal vector $\mathbf{n}$ to S_0 can be given in terms of the gradient vector $\tilde{\nabla}\varphi$ as

$$(6.56) \qquad \mathbf{n} = \frac{\tilde{\nabla}\varphi}{|\tilde{\nabla}\varphi|} = \frac{[\nabla\varphi, \varphi_t]}{\sqrt{(\nabla\varphi)^2 + \varphi_t^2}}.$$

We apply the integral relation (6.5) over R_1 and R_2 and allow these regions to collapse onto S_0. With the exception of u_t and ∇u, all functions in the integral relation (6.5) including $\mathbf{n}(x, t)$ are continuous across S_0. Therefore, we easily obtain

$$(6.57) \qquad \int_{S_0} \frac{p[\nabla u] \cdot \nabla\varphi - \rho[u_t]\varphi_t}{|\tilde{\nabla}\varphi|} \, ds = 0$$

where $[V]$ denotes the jump in V across S_0. Since the same expression can be obtained over any arbitrary subregion of S_0, we conclude on the basis of the duBois–Reymond lemma (see Exercise 8.1.9) that the integrand itself must vanish and obtain

$$(6.58) \qquad p[\nabla u] \cdot \nabla\varphi - \rho[u_t]\varphi_t = 0.$$

Now $u(x, t)$ is continuous across S_0 and is continuously differentiable in the interior of S_0—that is, directional derivatives of $u(x, t)$ in tangential directions in S_0 are continuous. Thus we can obtain additional equations for the jumps in u_t and ∇u across S_0 (note that there are four derivatives to be determined in the three-dimensional case so that four equations are needed). We observe that the vectors $\mathbf{n}_1 = [\varphi_t, 0, 0, -\varphi_x]$, $\mathbf{n}_2 = [0, \varphi_t, 0, -\varphi_y]$, and $\mathbf{n}_3 = [0, 0, \varphi_t, -\varphi_z]$ are linearly independent and are all orthogonal to the normal vector $\mathbf{n}$. Thus the scalar product of the (space-time) gradient vector $\tilde{\nabla} u$ into any of the vectors, $\mathbf{n}_1$, $\mathbf{n}_2$, and $\mathbf{n}_3$ yields an interior derivative in S_0. Denoting $u(x, t)$ by $u_1(x, t)$ in R_1 and by $u_2(x, t)$ in R_2, we differentiate u_1 and u_2 in each of the directions $\mathbf{n}_1$, $\mathbf{n}_2$, and $\mathbf{n}_3$. The resulting interior derivatives of u are continuous across S_0 so that the difference of the results for u_1 and u_2 must vanish. This yields the following three additional equations:

$$(6.59) \qquad \left[\tilde{\nabla} u \cdot \mathbf{n}_1\right] = [u_x]\varphi_t - [u_t]\varphi_x = 0,$$

$$(6.60) \qquad \left[\tilde{\nabla} u \cdot \mathbf{n}_2\right] = [u_y]\varphi_t - [u_t]\varphi_y = 0,$$

$$(6.61) \qquad \left[\tilde{\nabla} u \cdot \mathbf{n}_3\right] = [u_z]\varphi_t - [u_t]\varphi_z = 0,$$

for the jumps in ∇u and u_t across S_0.

The four equations (6.58)–(6.61) are a homogeneous linear system for the jumps $[u_x]$, $[u_y]$, $[u_z]$, and $[u_t]$ across S_0. Since we assume that one or more of these jumps is nonzero, the determinant of the coefficients of this system must vanish to guarantee a nonzero solution. Evaluating the determinant gives the equation

$$(6.62) \qquad \varphi_t^2\left[\rho\varphi_t^2 - p(\nabla\varphi)^2\right] = 0.$$

Equating the bracketed term in (6.62) to zero yields the result

$$(6.63) \qquad \rho\varphi_t^2 - p(\nabla\varphi)^2 = 0.$$

On comparing with (3.76) we note that (6.63) is the characteristic equation for the hyperbolic equation (6.1). Thus $\varphi(x, t) = 0$ must be a characteristic surface and discontinuities in first derivatives can occur only across characteristic surfaces. It is shown in the exercises that $\varphi_t = 0$, which is also a consequence of (6.62), does not lead to a surface of discontinuity. We conclude that a continuous, piecewise continuously differentiable solution of (6.1) or, more precisely, of (6.5), satisfies (6.1) in regions where it is smooth and the jumps in the first derivatives must satisfy (6.58) on all characteristic discontinuity surfaces.

Without exhibiting the calculations, we note that for the parabolic case (6.9)—where u_t does not appear in the integral relation—we obtain instead of (6.62) the equation

$$(6.64) \qquad p(\nabla\varphi)^2 = 0,$$

which characterizes the surface S_0 across which u_t and ∇u can have jump discontinuities. This is again the characteristic equation for the parabolic equation (4.4), and we conclude that $\varphi = \varphi(t)$, so that the surfaces $t = $ constant are the characteristics. As a result, ∇u is continuous across the characteristics, since it represents interior differentiation on the surfaces $t = $ constant. Only u_t can have a jump across the characteristics, but because this is the highest order t-derivative in (4.4), the integral relation (6.9) does not extend the class of solutions in the parabolic case.

For the elliptic equation (6.12) we also obtain the result (6.64). Since the surface S_0 is given as $\varphi(x, y, z) = 0$ in this case (i.e., there is no t-dependence), we again conclude that there are no real characteristics or discontinuity surfaces for the elliptic equation (6.11).

EXAMPLE 6.3.   THE CAUCHY PROBLEM FOR THE WAVE EQUATION

We consider a Cauchy problem for the one-dimensional wave equation with discontinuous initial data. Let $u(x, t)$ satisfy

$$(6.65) \qquad u_{tt} - c^2 u_{xx} = 0, \qquad -\infty < x < \infty, t > 0$$

with the initial conditions

$$(6.66) \qquad u(x, 0) = 0, \qquad -\infty < x < \infty,$$

$$(6.67) \qquad u_t(x, 0) = \begin{cases} 0, & x < 0 \\ 1, & x \geqslant 0 \end{cases}.$$

Since $u(x, t)$ is initially continuous and $u_t(x, 0)$ has a jump discontinuity, we expect the solution to have discontinuous first derivatives on the characteristics that issue from the point $(x, t) = (0, 0)$.

A formal application of d'Alembert's solution (2.29) yields

$$(6.68) \qquad u(x, t) = \frac{1}{2c} \int_{x-ct}^{x+ct} u_t(s, 0) \, ds,$$

with $u_t(x, 0)$ given as in (6.67). We divide the half-plane $t > 0$ into three sectors as shown in Figure 6.5. The sectors I, II, and III are separated by the characteristic curves $x = \pm ct$ that issue from the initial discontinuity point $(0, 0)$ of the velocity $u_t(x, t)$. The integration in (6.6.8) is easily carried out and yields

$$(6.69) \quad u(x, t) = \begin{cases} t; & x - ct > 0: & \text{I} \\ \dfrac{1}{2}\left(t + \dfrac{x}{c}\right); & x - ct \leqslant 0 \leqslant x + ct: & \text{II} \\ 0; & x + ct < 0: & \text{III} \end{cases}.$$

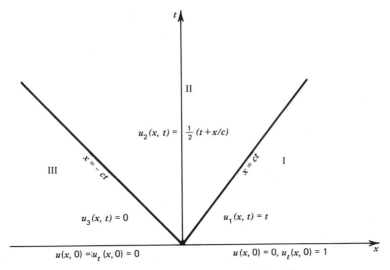

Figure 6.5. The solution of the Cauchy problem.

Clearly, $u(x, t)$ satisfies the wave equation (6.65) in the interior of each sector I, II, and III. Also, the solution $u(x, t)$ is continuous across the characteristics $x = \pm ct$. In sector I we have $\partial u/\partial x = 0$ and $\partial u/\partial t = 1$. In sector II, $\partial u/\partial x = 1/2c$ and $\partial u/\partial t = 1/2$, whereas in sector III, $\partial u/\partial x = \partial u/\partial t = 0$. Thus the derivatives of $u(x, t)$ have jump discontinuities across the characteristics $x = \pm ct$. Therefore, $u(x, t)$ is not a strict solution of (6.65) since it lacks the required number of derivatives.

To show that (6.69) satisfies the integral form (6.5) of the wave equation, we show that u_t and u_x satisfy the one-dimensional form of the jump condition (6.58) across the characteristics $x = \pm ct$. We have already seen that $u(x, t)$ is a solution of the wave equation (6.65) away from these characteristics. For the characteristic $x = ct$, we have $\varphi(x, t) = x - ct = 0$ so that $\varphi_x = 1$ and $\varphi_t = -c$. Also, $\rho = 1$ and $p = c^2$ so that (6.58) becomes

$$(6.70) \qquad c^2[u_x] + c[u_t] = c^2\left\{\frac{\partial u_2}{\partial x} - \frac{\partial u_1}{\partial x}\right\} + c\left\{\frac{\partial u_2}{\partial t} - \frac{\partial u_1}{\partial t}\right\}$$

$$= \frac{1}{2}c - \frac{1}{2}c = 0.$$

On the characteristic $x = -ct$, we have $\varphi = x + ct = 0$ with $\varphi_x = 1$ and $\varphi_t = c$. Thus (6.58) becomes

$$(6.71) \qquad c^2[u_x] - c[u_t] = c^2\left\{\frac{\partial u_3}{\partial x} - \frac{\partial u_2}{\partial x}\right\} - c\left\{\frac{\partial u_3}{\partial t} - \frac{\partial u_2}{\partial t}\right\}$$

$$= -\frac{c}{2} + \frac{c}{2} = 0.$$

We have demonstrated that $u(x, t)$ is a twice differentiable solution of the wave equation (6.65) away from the characteristics and that it satisfies the jump condition (6.58) across the characteristics $x = \pm ct$. Thus (6.69) satisfies the integral form (6.5) of the wave equation.

6.4. WEAK SOLUTIONS

The results of the preceding section have extended the concept of the solution of second order partial differential equations to the case of discontinuous first derivatives. Had we introduced initial displacements $u(x, 0)$ with jump discontinuities in Example 6.3, a formal application of d'Alembert's solution would have shown that the solution has discontinuities across characteristics for the wave equation. Such "solutions" cannot be discussed on the basis of the foregoing methods where the continuity of solutions was assumed. To deal with such problems, we weaken the concept of solution even further and obtain a new integral expression for each of the hyperbolic, parabolic, and elliptic equations considered. Again, this expression is equivalent to the given differential equation when the solutions are smooth. However, it remains valid even if the solution has jump discontinuities.

We discuss the hyperbolic equation (6.1) in detail and then state the results for the parabolic and elliptic equations of Section 4.1. Given the region R, we consider the smooth function $v(x, t)$ that is assumed to vanish identically near the boundary ∂R of R if R is of finite extent. If R is of infinite extent, we assume $v(x, t)$ vanishes outside some bounded region as well as near any finite boundary of R. We define the operator $\tilde{L}$ as

$$(6.72) \qquad \tilde{L}[u] = \rho u_{tt} - \nabla \cdot (p\nabla u) + qu,$$

where we are again considering two or three-dimensional spatial regions. Then we readily obtain

$$(6.73) \qquad v\tilde{L}[u] - u\tilde{L}[v]$$

$$= \rho[vu_{tt} - uv_{tt}] - [v\nabla \cdot (p\nabla u) - u\nabla \cdot (p\nabla v)]$$

$$= \frac{\partial}{\partial t}[\rho v u_t - \rho u v_t] - \nabla \cdot [pv\nabla u - pu\nabla v],$$

since $\rho = \rho(x)$. Using the space-time gradient vector $\tilde{\nabla} = [\nabla, \partial/\partial t]$ we have

$$(6.74) \qquad v\tilde{L}[u] - u\tilde{L}[v] = -\tilde{\nabla} \cdot [pv\nabla u - pu\nabla v, -\rho v u_t + \rho u v_t].$$

Integrating over the region R we obtain

$$(6.75) \qquad \iint\limits_{R} \left\{ v\left(\tilde{L}[u] - \rho F \right) - \left(u\tilde{L}[v] - v\rho F \right) \right\} dV$$

$$= -\iint\limits_{R} \tilde{\nabla} \cdot \left[pv\nabla u - pu\nabla v, -\rho vu_t + \rho uv_t \right] dV$$

$$= -\int_{\partial R} \left[pv\nabla u - pu\nabla v, -\rho vu_t + \rho uv_t \right] \cdot \mathbf{n} \, ds = 0$$

where $\mathbf{n}$ is the exterior unit normal to ∂R. The last integral in (6.75) results on using the divergence theorem, and it equals zero since v vanishes identically near ∂R. If R is unbounded, the result is still zero since $v(x, t)$ vanishes for sufficiently large x and t. The divergence theorem is then applied to a region bounded by ∂R and by portions of circles or spheres whose radius is subsequently allowed to tend to infinity.

We conclude from (6.75) that

$$(6.76) \qquad \iint\limits_{R} v\left(\tilde{L}[u] - \rho F \right) dV = \iint\limits_{R} \left(u\tilde{L}[v] - v\rho F \right) dV.$$

If $u(x, t)$ is a smooth solution of (6.1), the integral on the left in (6.76) vanishes and we conclude that for any $v(x, t)$ satisfying these conditions we have

$$(6.77) \qquad \iint\limits_{R} \left(u\tilde{L}[v] - v\rho F \right) dV = 0.$$

Conversely, if $u(x, t)$ satisfies the integral expression (6.77) for all admissible $v(x, t)$ and $u(x, t)$ is twice continuously differentiable, we obtain from the equation (6.76)

$$(6.78) \qquad \iint\limits_{R} v\left(\tilde{L}[u] - \rho F \right) dV = 0.$$

Using the *fundamental lemma of the calculus of variations* (see Exercise 8.1.8), we conclude from the arbitrariness of $v(x, t)$ that $\tilde{L}[u] - \rho F = 0$ so that $u(x, t)$ satisfies (6.1).

If $u(x, t)$ is a twice continuously differentiable solution of the hyperbolic equation (6.1), we shall say that $u(x, t)$ is a *classical solution* of the differential equation. If $u(x, t)$ is a solution of (6.77) for all admissible $v(x, t)$ but does not have the required number of derivatives to be a classical solution of the differential equation (6.1), we say that $u(x, t)$ is a *weak* or *generalized solution* of the differential equation. Note that the solutions of the integral relation (6.5) may also be termed weak solutions if they are not twice differentiable. We do not concern ourselves here with the question of the assumption of initial and boundary conditions by the weak solution as the additional discussion required would take us too far afield.

It can again be shown that if $u(x, t)$ is a classical solution of (6.1) in two subregions R_1 and R_2 of R and has a jump discontinuity across the region S_0 separating R_1 and R_2, then S_0 must be a characteristic. This is done by applying (6.76) to the two regions R_1 and R_2 and considering the nonzero contributions-the integrals over the region S_0. We do not carry out the discussion which is similar to that given in Section 6.4.

We now define weak or generalized solutions for the parabolic and elliptic equations of Section 4.1. We say that $u(x, t)$ is a classical solution of the parabolic equation

$$(6.79) \qquad\qquad \rho u_t - \nabla \cdot (p\nabla u) + qu = \rho F$$

if $u(x, t)$ has a continuous time derivative and two continuous x-derivatives. With the region R and the smooth function $v(x, t)$ as previously defined, we easily conclude by adapting the foregoing procedure for classical solutions $u(x, t)$, that

$$(6.80) \qquad \iint\limits_{R} \left[u\{-\rho v_t - \nabla \cdot (p\nabla v) + qv\} - v\rho F \right] dV = 0.$$

Note that ρu_t is replaced by $-\rho v_t$ in (6.80). If $u(x, t)$ satisfies (6.80) for all admissible $v(x, t)$ and is not a classical solution of (6.79), we say that it is a weak or generalized solution of the differential equation (6.79).

Similarly, if $u(x)$ is a twice continuously differentiable solution of the elliptic equation

$$(6.81) \qquad\qquad - \nabla \cdot (p\nabla u) + qu = \rho F,$$

u is said to be a classical solution of (6.81) and it is easily shown to satisfy the integral relation

$$(6.82) \qquad \iint\limits_{R} \left[u\{\nabla \cdot (p\nabla v) - qv\} + v\rho F \right] dV = 0,$$

for all admissible $v(x, t)$. If $u(x)$ satisfies (6.82) but is not a classical solution, then it is said to be a weak or generalized solution of (6.81). In view of the smoothness properties of solutions of elliptic equations, there is generally no distinction between weak and classical solutions in the elliptic case.

We now consider some examples. In the first example we consider a weak solution of the wave equation (6.65). In the second example we examine the Fourier series solutions obtained in Chapter 4 and show how they may be interpreted as weak solutions.

EXAMPLE 6.4. WEAK SOLUTIONS OF THE WAVE EQUATION

The Cauchy problem for the wave equation

(6.83) $$u_{tt} - c^2 u_{xx} = 0, \qquad -\infty < x < \infty, t > 0$$

with the initial data

(6.84) $$u(x,0) = \begin{cases} 0, & x < 0 \\ 1, & x \geqslant 0 \end{cases}$$

and

(6.85) $$u_t(x,0) = 0,$$

has the formal d'Alembert's solution

(6.86) $$u(x,t) = \begin{cases} 1, & x - ct \geqslant 0 \\ \frac{1}{2}, & x - ct < 0 < x + ct. \\ 0, & x + ct \leqslant 0 \end{cases}$$

Since $u(x,t)$ is discontinuous across the characteristics $x = \pm ct$, we must interpret the solution in a generalized sense.

We consider the rectangular region R displayed in Figure 6.6 and assume that the smooth function $v(x,t)$ vanishes near the boundary ∂R of R. The region R is divided into sectors I, II, and III, and we now show that the solution (6.86) satisfies (6.77) for all admissible $v(x,t)$ so that it is a weak

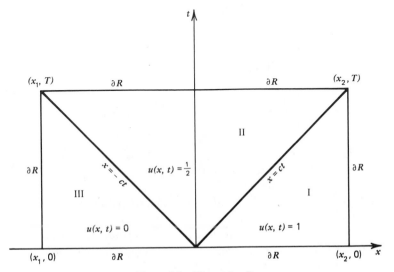

Figure 6.6. The region R.

solution of (6.83). We have

(6.87) $\qquad \iint\limits_{R} u\left[v_{tt} - c^2 v_{xx}\right] dV = \iint\limits_{I} 1\left[v_{tt} - c^2 v_{xx}\right] dx\, dt$

$$+ \iint\limits_{II} \frac{1}{2}\left[v_{tt} - c^2 v_{xx}\right] dx\, dt$$

$$+ \iint\limits_{III} 0\left[v_{tt} - c^2 v_{xx}\right] dx\, dt.$$

For the integral over the region II we have

(6.88) $\quad \dfrac{1}{2} \iint\limits_{II} \left[v_{tt} - c^2 v_{xx}\right] dx\, dt$

$$= -\frac{1}{2} \iint\limits_{II} \left[\frac{\partial}{\partial x}, \frac{\partial}{\partial t}\right] \cdot \left[c^2 v_x, -v_t\right] dx\, dt$$

$$= -\frac{1}{2} \int_{x=-ct} \left[c^2 v_x, -v_t\right] \cdot \mathbf{n}\, ds - \frac{1}{2} \int_{x=ct} \left[c^2 v_x, -v_t\right] \cdot \mathbf{n}\, ds,$$

on using the divergence theorem and the fact that $v = 0$ on ∂R. On $x = -ct$, the exterior unit normal vector is $\mathbf{n} = [-1, -c]/\sqrt{1 + c^2}$ and on $x = ct$, $\mathbf{n} = [1, -c]/\sqrt{1 + c^2}$. Further, on $x = -ct$, $v(x, t) = v(-ct, t)$ so that $dv/dt = -cv_x + v_t$. Similarly, on $x = ct$ we have $v(x, t) = v(ct, t)$ so that $dv/dt = cv_x + v_t$. Also, on each line $x = \pm ct$ we have $ds = \sqrt{dx^2 + dt^2} = \sqrt{1 + c^2}\, |dt|$.

Combining these results and noting that the direction of integration in the application of the divergence theorem (in the two-dimensional form) is counterclockwise, we have

(6.89) $\quad \dfrac{1}{2} \iint\limits_{II} \left[v_{tt} - c^2 v_{xx}\right] dx\, dt = \dfrac{c}{2} \int_T^0 (-cv_x + v_t)\, dt$

$$- \frac{c}{2} \int_0^T (cv_x + v_t)\, dt = \frac{c}{2} \int_T^0 \frac{dv}{dt}\, dt - \frac{c}{2} \int_0^T \frac{dv}{dt}\, dt = 0,$$

since $v(x, t)$ vanishes at $t = 0$ and $t = T$. A similar calculation shows that the integral in (6.87) over the sector I also vanishes.

We have shown that the integral in (6.87) over the region R vanishes for all admissible $v(x, t)$. Since T is arbitrary, the foregoing result is valid for $-\infty < x < \infty$ and $t > 0$. Thus $u(x, t)$ is a weak solution of the wave equation (6.83). We have not considered the initial conditions for the problem, but have merely shown that $u(x, t)$ as defined by (6.86) is a weak solution of the differential equation (6.83).

EXAMPLE 6.5. FOURIER SERIES AND WEAK SOLUTIONS

The series solutions obtained in Chapter 4 by the method of separation of variables or by finite Fourier transforms were shown to require substantial conditions on the data for the given problems to render them classical solutions. This was especially the case for the hyperbolic equations that were considered. It was indicated in Chapter 4 how the series may be understood as generalized solutions in case they are not classical solutions, and we now use the preceding results to show in a more precise manner how the series solutions are to be interpreted as weak or generalized solutions.

Our discussion is restricted to the hyperbolic equation (4.10) with the homogeneous boundary conditions (4.14) and the initial conditions (4.16). Proceeding as in Section 4.6 we expand $u(x, t)$ in a series of eigenfunctions.

$$(6.90) \qquad u(x, t) = \sum_{k=1}^{\infty} N_k(t) M_k(x),$$

with $N_k(t)$ defined as in (4.207) except that $B_k = 0$ for all k since the boundary conditions are homogeneous. We assume that each of the eigenfunctions $M_k(x)$ is twice continuously differentiable and that so are the $N_k(t)$. The partial sum

$$(6.91) \qquad u_n(x, t) = \sum_{k=1}^{n} N_k(t) M_k(x)$$

is easily seen to be a classical solution of the equation

$$(6.92) \qquad \rho \frac{\partial^2 u_n}{\partial t^2} - \nabla \cdot (p \nabla u_n) + q u_n = \rho F_n$$

where $F_n(x, t)$ is the nth partial sum of the Fourier series of $F(x, t)$. The initial conditions for the $u_n(x, t)$ are

$$(6.93) \qquad u_n(x, 0) = \sum_{k=1}^{n} N_k(0) M_k(x) = \sum_{k=1}^{n} (f, M_k) M_k(x),$$

$$(6.94) \qquad \frac{\partial u_n(x, 0)}{\partial t} = \sum_{k=1}^{n} N_k'(0) M_k(x) = \sum_{k=1}^{n} (g, M_k) M_k(x),$$

and the boundary conditions are

$$(6.95) \qquad \alpha u_n + \beta \frac{\partial u_n}{\partial n} \bigg|_{\partial G} = 0.$$

We consider the time interval $0 \leqslant t \leqslant T$ and the region G on which the initial data are assigned together with its boundary ∂G. Let region G and its boundary ∂G be extended parallel to themselves from $t = 0$ to $t = T$ and denote the resulting region by R. Let $v(x, t)$ be a smooth function that vanishes near ∂R. Then since each u_n is a classical solution of (6.92), we have from (6.77)

$$(6.96) \qquad \iint_R u_n \{ \rho v_{tt} - \nabla \cdot (p \nabla v) + q v \} \, dV = \iint_R \rho F_n v \, dV.$$

We now assume that as $n \to \infty$ we have $u_n(x, 0) \to f(x)$, $\partial u_n(x, 0)/\partial t \to g(x)$, $\nabla u_n(x, 0) \to \nabla f$ and $F_n \to F$ where f and g are the initial data and F is the inhomogeneous term of the hyperbolic equation. The convergence of $u_n(x, 0)$ to $f(x)$ is assumed to be uniform, whereas all other cases represent mean square convergence. It can then be shown that the partial sums $u_n(x, t)$ converge to a function $u(x, t)$ in the region R in the mean square sense.

If we proceed to the limit as $n \to \infty$ in (6.96), since $u_n \to u$ and $F_n \to F$, we conclude that

$$(6.97) \qquad \iint_R u \{ \rho v_{tt} - \nabla \cdot (p \nabla v) + q v \} \, dV = \iint_R \rho F v \, dV,$$

(in the sense of weak convergence) so that $u(x, t)$ is a weak solution (see Section 7.2).

6.5. THE INTEGRAL WAVE EQUATION

The integral relation (6.5) that corresponds to the one-dimensional wave equation (6.83) leads to a number of interesting and useful expressions that can be applied to obtain the solution of various initial and boundary value problems for the wave equation as we now demonstrate.

If $u(x, t)$ is a solution of the one-dimensional wave equation (6.83), then (6.5) takes the form of a line integral

$$(6.98) \qquad \int_{\partial R} [c^2 u_x, -u_t] \cdot \mathbf{n} \, ds = \int_{\partial R} u_t \, dx + c^2 u_x \, dt = 0,$$

since $\mathbf{n} \, ds = [dt, -dx]$. When the wave equation (6.83) has a nonhomogeneous term ρF, this term must be integrated over the region R. We shall call (6.98) the integral wave equation, and we now choose various regions R and integrate (6.98) over their boundaries.

First we let R be a region in (x, t)-space bounded on four sides by characteristic lines $x \pm ct = $ constant as pictured in Figure 6.7. Thus ∂R is a

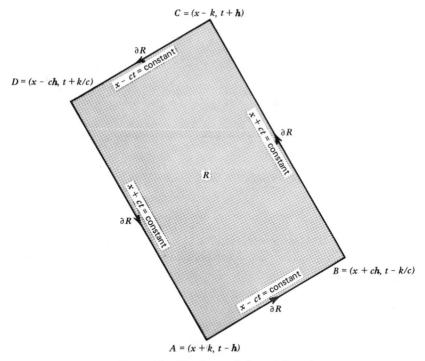

Figure 6.7. The characteristic quadrilateral.

characteristic quadrilateral. Now on the lines $x \pm ct = $ constant we have $dx = \mp c \, dt$ and $u_t \, dx + c^2 u_x \, dt = \mp c \, du$. The integral (6.98) takes the form

$$(6.99) \quad \int_{\partial R} u_t \, dx + c^2 u_x \, dt = \int_A^B c \, du - \int_B^C c \, du + \int_C^D c \, du - \int_D^A c \, du$$

$$= 2c[u(B) + u(D) - u(A) - u(C)] = 0$$

where $u(C)$, for example, is given as $u(C) = u(x - k, t + h)$, where k and h are given positive constants. We, therefore, obtain the difference equation

$$(6.100) \quad u(x - k, t + h) + u(x + k, t - h)$$

$$= u\left(x + ch, t - \frac{k}{c}\right) + u\left(x - ch, t + \frac{k}{c}\right).$$

Using Taylor's expansion it can be shown that any twice continuously differentiable solution of the difference equation must satisfy the wave equation in view of the arbitrariness of h and k (see Exercises 6.5.1 and 6.5.2).

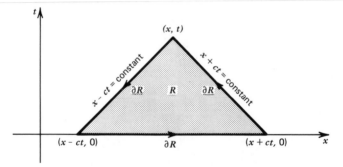

Figure 6.8.   The characteristic triangle.

As our next choice for the region R, we choose a characteristic triangle with base on the x-axis and the two other sides equal to characteristic line segments as shown in Figure 6.8. Then ∂R is the boundary of the characteristic triangle, and using these results we obtain

$$(6.101) \quad \int_{\partial R} u_t\, dx + c^2 u_x\, dt$$

$$= -\int_{(x+ct,0)}^{(x,t)} c\, du + \int_{(x,t)}^{(x-ct,0)} c\, du + \int_{(x-ct,0)}^{(x+ct,0)} u_t(x,0)\, dx$$

$$= -2cu(x,t) + cu(x+ct,0)$$

$$+ cu(x-ct,0) + \int_{x-ct}^{x+ct} u_t(x,0)\, dx = 0.$$

Given the Cauchy problem for the wave equation (6.83) with initial data

$$(6.102) \qquad u(x,0) = f(x), \qquad u_t(x,0) = g(x)$$

we obtain from (6.101),

$$(6.103) \quad u(x,t) = \frac{1}{2}[f(x-ct) + f(x+ct)] + \frac{1}{2c}\int_{x-ct}^{x+ct} g(x)\, dx,$$

which is just d'Alembert's solution. For the inhomogeneous wave equation with a term ρF on the right of (6.83), we must add the integral of ρF over the characteristic triangle R to the solution (6.103).

A third choice for the region of integration R in (6.98) is the region pictured in Figure 6.9 with the boundary ∂R, three of whose sides are characteristics and whose fourth side is an interval on the x-axis. We apply (6.98) over the

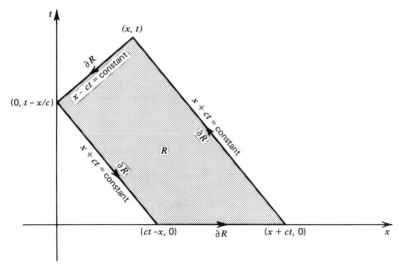

Figure 6.9.   The quadrilateral.

boundary of the region R to obtain

(6.104)

$$\int_{\partial R} u_t \, dx + c^2 u_x \, dt = -\int_{(x+ct,0)}^{(x,t)} c \, du + \int_{(x,t)}^{(0, t-x/c)} c \, du$$

$$-\int_{(0, t-x/c)}^{(ct-x,0)} c \, du + \int_{(ct-x,0)}^{(x+ct,0)} u_t(x,0) \, dx$$

$$= -2cu(x,t) + cu(x+ct,0) + 2cu\left(0, t - \frac{x}{c}\right) - cu(ct-x,0)$$

$$+ \int_{(ct-x,0)}^{(x+ct,0)} u_t(x,0) \, dx = 0.$$

With the initial conditions $u(x,0) = f(x)$, $u_t(x,0) = g(x)$ and the boundary condition $u(0, t) = h(t)$, we obtain

(6.105) $$u(x,t) = \frac{1}{2}\left[f(x+ct) - f(ct-x)\right]$$

$$+ \frac{1}{2c} \int_{ct-x}^{ct+x} g(x) \, dx + h\left(t - \frac{x}{c}\right).$$

Differentiating (6.105) with respect to x and evaluating the result at $x = 0$

gives

$$(6.106) \qquad \frac{\partial u(0, t)}{\partial x} = f'(ct) + \frac{1}{c}g(ct) - \frac{1}{c}h'(t)$$

$$= u_x(ct, 0) + \frac{1}{c}u_t(ct, 0) - \frac{1}{c}u_t(0, t).$$

Using these results, any initial and boundary value problem for the wave equation over a finite, semi-infinite, or infinite interval can be solved and domains of dependence for the solutions can be established as we now show.

EXAMPLE 6.6.   THE WAVE EQUATION IN A SEMI-INFINITE INTERVAL

We consider the initial and boundary value problem for the inhomogeneous wave equation in a semi-infinite interval; that is,

$$(6.107) \qquad u_{tt} - c^2 u_{xx} = F(x, t), \qquad 0 < x < \infty, t > 0,$$

with initial conditions

$$(6.108) \qquad u(x, 0) = f(x); \qquad u_t(x, 0) = g(x); \qquad 0 < x < \infty,$$

and the boundary condition

$$(6.109) \qquad \alpha u(0, t) - \beta u_x(0, t) = H(t), \qquad t > 0$$

where $\alpha \geqslant 0$, $\beta \geqslant 0$, and $\alpha + \beta > 0$ (with α and β both equal to constants) and F, f, g, and H are given functions.

To solve this problem we break up the first quadrant in the (x, t)-plane into two regions. In region I, $x > ct$ and in region II, $x < ct$. These regions are separated by the characteristic curve $x = ct$ which issues from the origin as shown in Figure 6.10. For a point (x, t) in region I, we integrate over a characteristic triangle and obtain d'Alembert's solution (6.103) plus an integral of $F(x, t)$; that is,

$$(6.110) \quad u(x, t) = \frac{1}{2}[f(x - ct) + f(x + ct)]$$

$$+ \frac{1}{2c} \int_{x-ct}^{x+ct} g(s)\, ds + \frac{1}{2c} \int_0^t \int_{x-c(t-\tau)}^{x+c(t-\tau)} F(\sigma, \tau)\, d\sigma\, d\tau,$$

with the last integral taken over the characteristic triangle. If the point (x, t) is in region II, we integrate over the quadrilateral pictured in Figure 6.10, three of whose sides are characteristics, and obtain from (6.105)

$$(6.111) \quad u(x, t) = \frac{1}{2}[f(x + ct) - f(ct - x)] + \frac{1}{2c} \int_{ct-x}^{ct+x} g(s)\, ds$$

$$+ h\left(t - \frac{x}{c}\right) + \frac{1}{2c} \iint F(\sigma, \tau)\, d\sigma\, d\tau,$$

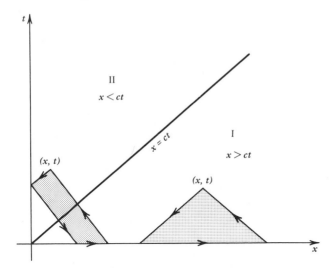

Figure 6.10.   The first quadrant.

where $h(t) = u(0, t)$ which is as yet unspecified if $\beta \neq 0$ in (6.109) and $F(x, t)$ is integrated over the interior of the quadrilateral. From (6.106) we obtain

$$(6.112) \qquad u_x(0, t) = f'(ct) + \frac{1}{c}g(ct) - \frac{1}{c}h'(t).$$

Substituting (6.112) into the boundary condition (6.109) gives

$$(6.113) \quad \alpha u(0, t) - \beta u_x(0, t)$$

$$= \alpha h(t) - \beta\left[f'(ct) + \frac{1}{c}g(ct) - \frac{1}{c}h'(t)\right] = H(t),$$

which is an ordinary differential equation for $u(0, t) = h(t)$. The initial condition is $h(0) = \lim_{t \to 0} u(0, t) = u(0,0)$, assuming $u(x, t)$ is continuous at $(0,0)$. In that case $u(0,0) = \lim_{x \to 0} u(x, 0) = \lim_{x \to 0} f(x) = f(0)$.

Now if $\beta = 0$ and $\alpha = 1$ in (6.109), then $u(0, t) = h(t) = H(t)$ so that (6.111) is already completely specified. However, if $H(0) \neq f(0)$, the solution (6.110)–(6.111) is discontinuous across the characteristic $x = ct$. As the characteristic is approached from points in region I, the limit of $u(x, t)$ is $f(0) + constant$ and the limit of $u(x, t)$ as $x = ct$ is approached from region II is $H(0) + constant$, where the constant is identical for both regions. We remark that even if $H(0) = f(0)$, the derivatives u_x and u_t may be discontinuous across the characteristic $x = ct$, say, if $H'(0) = \lim_{t \to 0} u_t(0, t) \neq g(0) = \lim_{x \to 0} u_t(x, 0)$. In these cases, the solution of the initial and boundary value problem must be interpreted in the generalized sense. If $\beta \neq 0$ in (6.109), $u(0, t)$ is unspecified and we may put $u(0,0) = f(0)$ to make the solution continuous across the characteristic line $x = ct$. The solution of (6.113) with $h(0) = f(0)$ is

$$h(t) = f(ct) + \int_0^t \exp\left[\frac{c\alpha}{\beta}(\tau - t)\right]\left\{\frac{c}{\beta}H(\tau) - \frac{c\alpha}{\beta}f(c\tau) + g(c\tau)\right\} d\tau,$$

when $\beta \neq 0$.

In the special case where $f = g = F = 0$ we obtain the following solution of the initial and boundary value problem (6.107)–(6.109):

$$(6.114) \quad u(x, t) = \begin{cases} 0; & x > ct \\ \dfrac{c}{\beta}\displaystyle\int_0^{t - \frac{x}{c}} \exp\left[\dfrac{c\alpha}{\beta}\left(\tau - t + \dfrac{x}{c}\right)\right] H(\tau)\, d\tau; & 0 < x < ct \end{cases}$$

if $\beta \neq 0$. If $\alpha = 0$ and $\beta = 1$, we have

$$(6.115) \qquad u(x, t) = \begin{cases} 0; & x > ct \\ c\displaystyle\int_0^{t - \frac{x}{c}} H(\tau)\, d\tau; & 0 < x < ct \end{cases}$$

whereas if $\beta = 0$ and $\alpha = 1$, we obtain

$$(6.116) \qquad u(x, t) = \begin{cases} 0; & x > ct \\ H\left(t - \dfrac{x}{c}\right); & 0 < x < ct \end{cases}.$$

In each case, the boundary condition (6.109) gives rise to a wave of the form $h(t - x/c)$ that travels to the right with speed c. The foregoing problem is often referred to as a *signaling problem*.

The domain of dependence for a point (x, t) in the first quadrant is given by the characteristic triangle in region I and by the quadrilateral in region II. This follows from the solutions (6.110)–(6.111).

EXAMPLE 6.7. THE WAVE EQUATION IN A FINITE INTERVAL

The initial and boundary value problem for the wave equation in a finite interval can be solved by separation of variables and by the use of finite transform methods in terms of standing waves as was shown in Chapter 4. We now use the foregoing results to construct a solution that yields a useful small time description in terms of propagating waves. The two forms of the solution may be compared to the results obtained for the heat or diffusion equation for which two different representations of the solution useful for small and large times were given (see Section 5.6).

Given the interval $0 < x < l$ we consider the problem

$$(6.117) \qquad u_{tt} - c^2 u_{xx} = F(x, t), \qquad 0 < x < l, t > 0$$

with the initial data

$$(6.118) \qquad \begin{cases} u(x, 0) = f(x) \\ u_t(x, 0) = g(x) \end{cases}, \qquad 0 < x < l$$

and the boundary data

$$(6.119) \qquad \begin{cases} \alpha u(0, t) - \beta u_x(0, t) = H_1(t) \\ \alpha u(l, t) + \beta u_x(l, t) = H_2(t) \end{cases},$$

where $\alpha \geqslant 0$, $\beta \geqslant 0$, and $\alpha + \beta > 0$.

To solve the problem we break up the strip $0 \leqslant x \leqslant l$, $t \geqslant 0$ into a collection of regions $R_1, R_2, R_3, \ldots,$ as shown in Figure 6.11. The regions R_k ($k = 1, 2, \ldots$) are bounded by portions of the initial line $t = 0$, the boundary lines $x = 0, l$, and portions of characteristic lines $x \pm ct = $ constant. The problem can be solved successively by starting with the solutions in $R_1, R_2,$ and R_3. In R_1, $u(x, t)$ is given by d'Alembert's solution (6.110). In R_2 and R_3 the solution is obtained as in Example 6.6 from (6.111) or a slight modification thereof. In R_4 we find $u(x, t)$ by using a characteristic quadrilateral and (6.100) which gives $u(x, t)$ in terms of its (known) values in $R_1, R_2,$ and R_3. Similarly, it is possible to construct quadrilaterals and to use (6.100) or (6.111) to determine the solution in all the regions R_k with $k > 4$, on proceeding step by step from R_5 to R_6 to R_7 and so on. The values of $u(x, t)$ on the characteristic lines that separate the regions R_k depend on the compatibility of the data at $(0, 0)$ and $(l, 0)$ which determines whether the solution is continuous.

The hatched strips in Figure 6.11 indicate how the solution $u(x, t)$ propagates, if the initial or boundary data are concentrated in a small x or a small t interval. Two cases are depicted. In one case we assume that $u(x, 0) = f(x)$ vanishes outside the interval (x_0, x_1) and that the other data $g(x)$, $F(x, t)$, $H_1(t)$, and $H_2(t)$ all vanish. This results in two waves with sharply defined wave fronts if $f(x)$ is discontinuous at x_0 and x_1, traveling to the right and to the left with speed c. As they hit the boundaries $x = 0$ and $x = l$, they are reflected and move in the opposite direction, as shown, again with sharp wave fronts. This process continues indefinitely.

In the second case, we assume $H_1(t) = 0$ everywhere outside the interval (t_0, t_1) and all the other data vanish. This yields a wave traveling to the right with speed c until it reaches $x = l$. It is then reflected and reverses direction until it is again reflected from the boundary $x = 0$ and so on. In both cases if $\beta \neq 0$ in (6.119) the waves do not have sharp trailing edges.

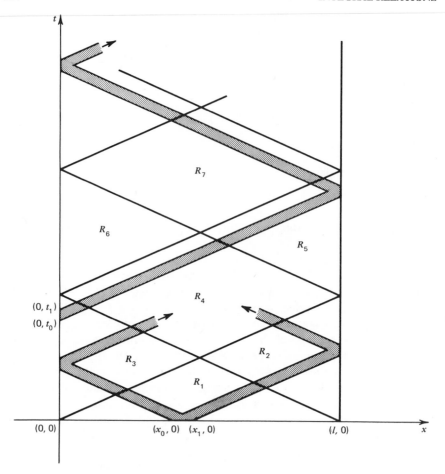

Figure 6.11. The regions R_k.

For the general case, it may be possible to detect at the initial stages of the wave motion (if the data are concentrated in small intervals or regions), propagating waves resulting from the effect of the data. However, as time increases, the interference from all waves makes it impossible to distinguish individual waves and the representation of the solution in terms of standing waves given in Chapter 4 becomes more useful.

EXAMPLE 6.8. MOVING BOUNDARIES

We consider the problem of a piston initially located at $x = 0$ moving into a gas located in the region $x > 0$. The equation of motion of the piston is $x = h(t)$ with $h'(t) > 0$ and $h(0) = 0$. Assuming we are dealing with small disturbances, we find that $u(x, t)$, which may represent the density of the gas, satisfies the wave equation (see Example 8.10)

$$(6.120) \qquad u_{tt} - c^2 u_{xx} = 0, \qquad h(t) < x < \infty, t > 0;$$

with the initial conditions

$$(6.121) \qquad u(x,0) = f(x); \qquad u_t(x,0) = g(x); \qquad 0 < x < \infty.$$

On the surface of the piston we have

$$(6.122) \qquad u(x,t)|_{x=h(t)} = u(h(t),t) = H(t), \qquad t > 0.$$

If the speed of the moving piston $dx/dt = h'(t)$ exceeds the sound speed c of the gas, the problem (6.120)–(6.122) is not well posed. This is seen by examining Figure 6.12. The solution $u(x, t)$ at a point $(h(t), t)$ on the path of the piston is uniquely determined (by means of d'Alembert's solution) by the initial data $f(x)$ and $g(x)$. Thus unless the data $u(h(t), t) = H(t)$ are such that the values of u equal those resulting from the initial data, a solution does not exist. Consequently, the problem is not well posed in this case. This difficulty arises because when the piston speed exceeds the (local) sound speed, the linearized version of the gas dynamics equations is not valid and the nonlinear equations presented in Example 8.10 must be used.

When the piston speed $dx/dt = h'(t) < c$, the sound speed, the problem (6.120)–(6.122) is well posed and the solution can be obtained in the manner indicated in Figure 6.13. We consider the regions R_1 and R_2 separated by the

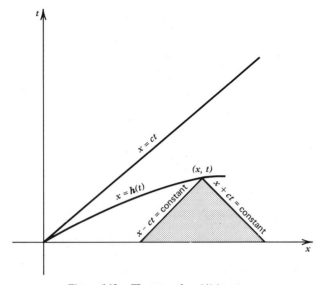

Figure 6.12. The case when $h'(t) > c$.

characteristic line $x = ct$. At a point (x, t) in R_1, d'Alembert's solution is valid and it expresses $u(x, t)$ in terms of the initial data. At a point (x, t) in R_2, we use the quadrilateral indicated in Figure 6.13 to obtain the result,

(6.123)
$$u(x, t) = u(x_1, t_1) + \frac{1}{2}[f(x + ct) - f(x_1 + ct_1)] + \frac{1}{2c}\int_{x_1 + ct_1}^{x + ct} g(s)\, ds.$$

The point (x_1, t_1) is where the characteristic $x - ct = $ constant through (x, t) intersects the curve $x = h(t)$. It is determined from the equations

(6.124)
$$\begin{cases} x_1 - ct_1 = x - ct \\ h(t_1) - x_1 = 0 \end{cases}.$$

Once (x_1, t_1) is determined, we may express $u(x, t)$ in R_2 as

(6.125)
$$u(x, t) = H(t_1) + \frac{1}{2}[f(x + ct) - f(x_1 + ct_1)] + \frac{1}{2c}\int_{x_1 + ct_1}^{x + ct} g(s)\, ds.$$

As $(x, t) \to (0, 0)$, we also have $(x_1, t_1) \to (0, 0)$ so that with

(6.126)
$$\lim_{t \to 0} u(h(t), t) = \lim_{t \to 0} H(t) = H(0),$$

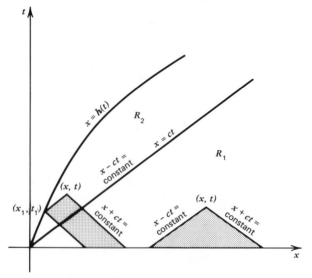

Figure 6.13.   The case when $h'(t) < c$.

we find that unless $\lim_{x \to 0} u(x, 0) = \lim_{x \to 0} f(x) = f(0) = H(0)$, the solution is discontinuous across the characteristic $x = ct$ and must be interpreted in a generalized sense.

As a concrete example we consider the piston curve

$$(6.127) \qquad x = h(t) = c_0 t, \qquad 0 < c_0 < c$$

and the boundary condition

$$(6.128) \qquad u(h(t), t) = u(c_0 t, t) = A \cos \omega t$$

where A and ω are constants. We assume that $f(x) = g(x) = 0$. It is then easily seen that

(6.129)

$$u(x, t) = \begin{cases} H\left[\dfrac{ct - x}{c - c_0}\right] = A \cos\left[\left(\dfrac{c\omega}{c - c_0}\right)\left(t - \dfrac{x}{c}\right)\right]; & c_0 t < x < ct \\ 0; & ct < x \end{cases}.$$

This result has the following interesting interpretation. The wave $u(x, t) = A \cos[(c\omega/(c - c_0))(t - x/c)]$, which results from the effect of the piston motion—which may be thought of as a moving energy source—has a frequency of oscillation $\omega_1 = [c/(c - c_0)]\omega$ which exceeds the frequency of oscillation ω of the source term (i.e., $A \cos \omega t$). This phenomenon is known as the *Doppler effect*.

Generally speaking, a curve $x = h(t)$ in the (x, t)-plane for which $|h'(t)| < c$ where c is the characteristic speed of the wave equation $u_{tt} = c^2 u_{xx}$ is called a *time-like curve*. If $|h'(t)| > c$ it is called a *space-like curve*. Any space-like curve can be used as an initial curve for the wave equation on which u and $\partial u / \partial n$, the normal derivative on the curve, must be specified. Any time-like curve can play the role of a boundary curve on which one condition on $u(x, t)$ can be assigned, as is the case for the time-like line $x = 0$. We now demonstrate how the foregoing results can be used to solve an initial and boundary value problem with initial data given on a space-like curve and boundary data prescribed on a time-like curve.

Let $u(x, t)$ be a solution of the wave equation (6.120) and suppose that $x = h_1(t)$ is a space-like curve and $x = h_2(t)$ is a time-like curve that bound the region R depicted in Figure 6.14. Let $u(x, t)$ and $\partial u(x, t) / \partial n$, the normal derivative, be specified on $x = h_1(t)$ as

$$(6.130) \qquad u(h_1(t), t) = f(t); \qquad \frac{\partial u(h_1(t), t)}{\partial n} = g(t);$$

and let $u(x, t)$ be prescribed on $x = h_2(t)$ as

$$(6.131) \qquad\qquad u(h_2(t), t) = H(t).$$

It follows from (6.130) that

$$(6.132) \quad \begin{cases} \dfrac{du(h_1(t), t)}{dt} = h_1'(t)u_x + u_t = f'(t) \\[2mm] \dfrac{\partial u(h_1(t), t)}{\partial n} = \dfrac{[-1, h_1'(t)]}{\sqrt{1 + h_1'(t)^2}} \cdot \nabla u = \dfrac{-u_x + h_1'(t)u_t}{\sqrt{1 + h_1'(t)^2}} = g(t). \end{cases}$$

It is possible to determine u_x and u_t on $x = h_1(t)$ in terms of $f(t)$ and $g(t)$ from (6.132). In the region R_1 of Figure 6.14 the solution at an arbitrary point (x, t) can be specified in terms of $f(t)$ and $g(t)$ by applying the formula (6.98) to the characteristic triangle shown in that figure. In the region R_2 of Figure 6.14 the solution at a point (x, t) can be specified in terms of $f(t)$, $g(t)$, and $H(t)$ by using (6.98) and integrating over the quadrilateral indicated in that figure.

Another problem that is of great interest for the wave equation is the *characteristic initial value problem*. In this case $u(x, t)$ is specified on the two characteristics issuing from the point $(x_0, 0)$ and we look for a solution of the

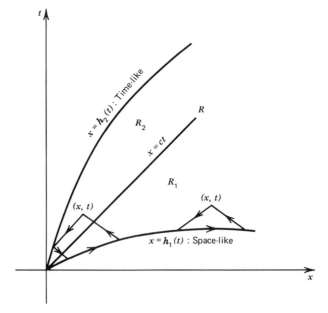

Figure 6.14. Time-like and space-like curves.

wave equation in the interior of the sector bounded by the characteristics. We assume that

$$(6.133) \quad u_{tt} - c^2 u_{xx} = F(x, t), \qquad x_0 - ct < x < x_0 + ct, t > 0$$

with the characteristic data

$$(6.134) \qquad \begin{cases} u(x, t)|_{x - ct = x_0} = H_1(t) \\ u(x, t)|_{x + ct = x_0} = H_2(t) \end{cases} ; \qquad t > 0$$

and the compatibility condition $H_1(0) = H_2(0)$. Using the difference equation (6.100) we obtain the solution $u(x, t)$ as

$$(6.135) \quad u(x, t) = H_1\left[\frac{1}{2}\left(t + \frac{x - x_0}{c}\right)\right] + H_2\left[\frac{1}{2}\left(t + \frac{x_0 - x}{c}\right)\right] - H_1(0)$$

$$+ \frac{1}{2c} \iint_{\diamondsuit} F(\sigma, \tau) \, d\sigma \, d\tau$$

where the integration is carried out over the characteristic quadrilateral depicted in Figure 6.15.

It may be noted that in contrast to the results for first order equations, where characteristic initial data resulted in nonunique solutions, the solution (6.135) is unique. This can be attributed to the fact that the data are given on two intersecting characteristics and are compatible at the point of intersection.

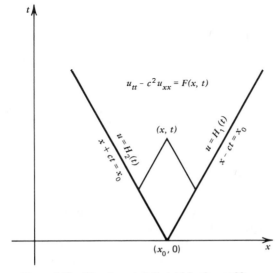

Figure 6.15. The characteristic initial value problem.

The characteristic initial value problem plays a role in Riemann's method of solution of the Cauchy problem for the second order hyperbolic equation in two variables discussed in Section 8.3. The equation is assumed to be in the canonical form for hyperbolic equations (apart from the fact that $c \neq 1$),

$$(6.136) \qquad u_{tt} - c^2 u_{xx} = au_x + bu_t + du + F(x, t),$$

where a, b, and d may be functions of x and t. We assume that the characteristic initial data for $u(x, t)$ are given as in (6.134) and are compatible. By treating the right side of (6.136) as an inhomogeneous term, we obtain from (6.135) the solution in the form of an integral equation

$$(6.137) \quad u(x, t) = H_1\left[\frac{1}{2}\left(t + \frac{x - x_0}{c}\right)\right] + H_2\left[\frac{1}{2}\left(t + \frac{x_0 - x}{c}\right)\right] - H_1(0)$$

$$+ \frac{1}{2c} \iint_\diamond F(\sigma, \tau)\, d\sigma\, d\tau + \frac{1}{2c} \iint_\diamond (au_\sigma + bu_\tau + du)\, d\sigma\, d\tau$$

$$\equiv R(x, t) + T[u, u_x, u_t],$$

where R represents the terms in H_1, H_2, and F, and T represents the integral terms in u_x, u_t, and u.

The integral equation (6.137) can be solved by iteration. We select an arbitrary initial approximation $u^{(0)}(x, t)$ [we can set $u^{(0)} = 0$ in this case] and insert $u^{(0)}$ into the right side of (6.137). In the next approximation we have

$$(6.138) \qquad u^{(1)}(x, t) = R(x, t) + T\left[u^{(0)}, u_x^{(0)}, u_t^{(0)}\right].$$

Iterating this process gives

$$(6.139) \qquad u^{(n)}(x, t) = R(x, t) + T\left[u^{(n-1)}, u_x^{(n-1)}, u_t^{(n-1)}\right].$$

It can be shown if the data and the coefficients in (6.136) are smooth, that the sequence of functions $u^{(n)}(x, t)$ converges to a unique solution of the characteristic initial value problem; that is, $u^{(n)}(x, t) \to u(x, t)$ as $n \to \infty$. We also have $\partial u^{(n)}/\partial x \to \partial u/\partial x$ and $\partial u^{(n)}/\partial t \to \partial u/\partial t$, which implies that $u(x, t)$ satisfies (6.137). The details of the proof are not presented here.

6.6. CONCENTRATED SOURCE OR FORCE TERMS

It has been generally assumed in the preceding sections and chapters that the inhomogeneous term ρF which occurs in the equations of Section 4.1, and which represents a source of force term, is fairly smoothly distributed throughout the region where the problem is to be considered. Often it happens, however, that the source or force term is effectively concentrated near some lower dimensional region, such as a curve for a problem in two-dimensional space-time. In such cases it is convenient to idealize the inhomogeneous term as having infinite density in the region of concentration and vanishing outside that region. Thus ρF must be singular in the region where it is concentrated, and the given differential equation must be interpreted in terms of the integral relations given in Section 6.1. We assume that although ρF is singular, the integral of ρF over the region of its concentration is finite, and we now show how to characterize the effect of the inhomogeneous term on the solution.

We consider the hyperbolic equation (6.1) and its integral relation (6.5). If the equation is defined over $(n + 1)$-dimensional space-time, the inhomogeneous term ρF is assumed to be concentrated over an n-dimensional region S_0 and to have the property that

$$(6.140) \qquad \iint_R \rho F \, dv = \int_{S_0} \rho F_0 \, ds$$

where ds is a surface or line element. The result (6.140) is a consequence of the fact that $F(x, t) = 0$ for (x, t) not on S_0, but $F(x, t)$ is concentrated on S_0 such that the integral over R collapses to an integral over S_0. The function ρF_0 represents the surface distribution of the force or source term over S_0. The solution $u(x, t)$ of (6.5) is assumed to be continuous across S_0.

To determine the variation of u across S_0, we let the region R collapse onto S_0 so that in the limit ∂R coincides with S_0 and, in fact, covers it twice as shown in Figure 6.16. The limiting normal vectors $\mathbf{n}$ of ∂R [in the integral relation (6.5)] have opposite direction at common points on S_0 so that they differ only in sign. This implies that

$$(6.141) \qquad \int_{S_0} [[p\nabla u, -\rho u_t]]_{S_0} \cdot \mathbf{n} \, ds = - \int_{S_0} \rho F_0 \, ds,$$

with the term involving qu vanishing in the limit since ρ, p, and q are assumed

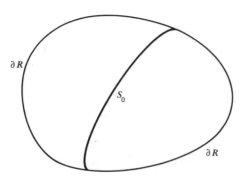

Figure 6.16. The region R.

to be smooth functions. The bracket $[\ldots]_{S_0}$ in the integral represents the jump in the quantity across S_0. Since (6.141) is also valid if the integrals are taken over an arbitrary portion of S_0, we conclude from the duBois–Reymond lemma (see Exercise 8.1.9) that the jump condition

$$(6.142) \qquad [[\,p\nabla u, -\rho u_t]]_{S_0} \cdot \mathbf{n} = -\rho F_0$$

is valid at any point (x, t) on S_0. The continuity of $u(x, t)$ across S_0 yields a second jump condition

$$(6.143) \qquad [u]_{S_0} = 0.$$

Applying interior differentiation to (6.143) in S_0 yields an expression for $[u_t]_{S_0}$ in terms of the components of $[\nabla u]_{S_0}$ and the condition (6.142) can be simplified somewhat. In the special case that S_0 is a cylindrical surface with generators parallel to the t-axis, we conclude that $[u_t]_{S_0} = 0$ since $[u]_{S_0} = 0$.

We only consider problems for (6.1) in which (each) S_0 divides the region R over which the problem is defined into two disjoint regions. Away from S_0, $u(x, t)$ satisfies a homogeneous version of (6.1) since $F(x, t)$ vanishes there. To solve the given problem, we can solve for u in each of the two regions, using the given data for the problem and matching these solutions across S_0 with the aid of (6.141)–(6.143). This approach is used in Example 6.10. A more direct approach to the solution of such problems is presented in Example 6.9.

In the case of one space-dimension if ρF is concentrated at the point $x = x_0$ for all time in (6.1), we have instead of (6.142)–(6.143),

$$(6.144) \qquad [u]_{x_0} = 0,$$

$$(6.145) \qquad \left[\frac{\partial u}{\partial x}\right]_{x_0} = -\frac{\rho(x_0)F_0(t)}{p(x_0)}.$$

With appropriate modifications the foregoing matching conditions are valid for the elliptic and parabolic equations of Section 4.2.

EXAMPLE 6.9.   VIBRATION OF A LOADED STRING

We consider the vibration of a finite string with fixed endpoints and with concentrated masses m_i $(i = 1, 2, \ldots n)$ placed at points x_i $(i = 1, \ldots, n)$ on the string, with $0 < x_i < l$. As the string vibrates, the masses m_i exert concentrated forces $F_i(t)$ at the points x_i, and these are determined from Newton's second law of motion to be

$$(6.146) \qquad F_i(t) = -m_i u_{tt}(x_i, t), \qquad i = 1, 2, \ldots, n, \qquad t > 0$$

where $u(x, t)$ is the displacement of the string. Since the acceleration of the mass point m_i is just the acceleration of the point x_i on the string and this is given by $u_{tt}(x_i, t)$, we obtain the formulas in (6.146).

The problem to be solved for an arbitrary, not necessarily homogeneous, string requires that $u(x, t)$ satisfy

$$(6.147) \quad \rho(x)u_{tt} = \frac{\partial}{\partial x}\left(p(x)\frac{\partial u}{\partial x}\right), \qquad 0 < x < l, t > 0, x \neq x_i,$$

with the homogeneous boundary conditions

$$(6.148) \qquad u(0, t) = 0; \qquad u(l, t) = 0; \qquad t > 0,$$

and the initial conditions

$$(6.149) \qquad u(x, 0) = f(x); \qquad u_t(x, 0) = g(x); \qquad 0 < x < l.$$

The jump conditions at the points $x = x_i$ are

$$(6.150) \quad \begin{cases} [u]_{x_i} = 0 \\ p(x_i)\left[\dfrac{\partial u}{\partial x}\right]_{x_i} = -F_i(t) = m_i u_{tt}(x_i, t); \qquad i = 1, 2, \ldots, n \end{cases}$$

where (6.146) was used and ρF is replaced by F in this case.

The initial and boundary value problem (6.147)–(6.150) can be solved as follows. Let

$$(6.151) \qquad\qquad u(x, t) = M(x)N(t),$$

and we obtain, on separating variables,

$$(6.152) \qquad\qquad N''(t) + \lambda N(t) = 0, \qquad t > 0,$$
$$(6.153) \quad (p(x)M'(x))' + \lambda \rho(x)M(x) = 0, \qquad 0 < x < l, \quad x \neq x_i$$

with the boundary conditions

$$(6.154) \qquad\qquad M(0) = 0, \qquad M(l) = 0$$

and the jump conditions

$$(6.155) \quad \begin{cases} [M(x)]_{x_i} = 0 \\ m_i M(x_i)N''(t) = p(x_i)[M'(x)]_{x_i}N(t) \end{cases}; \qquad i = 1, 2, \ldots, n.$$

Using (6.152) we can replace the second matching condition in (6.155) by

$$(6.156) \quad p(x_i)[M'(x)]_{x_i} + \lambda m_i M(x_i) = 0, \qquad i = 1, 2, \ldots, n.$$

The equations (6.153)–(6.156) constitute an *eigenvalue problem* for $M(x)$, with the interesting aspect that λ, the eigenvalue parameter, enters not only in the differential equation (6.153) but also in the jump conditions. It can be shown that the eigenvalues and eigenfunctions for this problem have essentially the same properties as those occurring in the Sturm–Liouville problem. In the exercises a problem of this type with a load (or mass) placed at one end of the string is considered.

Once the eigenfunctions are determined, the solution of the initial and boundary value problem is found in the usual manner by using eigenfunction expansions.

EXAMPLE 6.10.   MOVING CONCENTRATED FORCES OR SOURCES

Let us consider the one-dimensional versions of the hyperbolic or parabolic equations of Section 4.1 and assume the force or source term moves and is concentrated along the curve $x = h(t)$ which plays the role of S_0 in the foregoing discussion. In the hyperbolic case we use the one-dimensional form of the matching or jump conditions (6.142) and (6.143) on $x = h(t)$. Since $u(x, t)$ is assumed to be continuous across $x = h(t)$, a relationship between the jumps $[u_x]$ and $[u_t]$ across the curve can be established by differentiating along $x = h(t)$. Also, an expression relating the jumps $[u_x]$ and $[u_t]$ is determined from (6.142). In the parabolic case it is easily seen that the jump condition corresponding to (6.142) does not involve a time derivative of u. The matching conditions in both the hyperbolic and parabolic cases can be given in terms of a known function $f_0(t)$ as

$$(6.157) \qquad \begin{cases} [u]_{x=h(t)} = 0 \\ \left[\dfrac{\partial u}{\partial x} \right]_{x=h(t)} = -f_0(t) \end{cases}.$$

An interesting problem of this type is the Cauchy problem for the wave equation with a concentrated source term moving along the line $x = c_0 t$ with $|c_0| < c$, where c is the speed of wave propagation. We assume that the source term $f_0(t)$ is oscillatory and obtain a result similar to that for the piston problem considered in Example 6.8. The problem is formulated as follows. The function $u(x, t)$ satisfies the wave equation

$$(6.158) \quad u_{tt} - c^2 u_{xx} = 0, \qquad -\infty < x < c_0 t, \quad c_0 t < x < \infty, \quad c_0 \geqslant 0$$

where

$$(6.159) \qquad u = \begin{cases} u_1(x, t), & -\infty < x < c_0 t \\ u_2(x, t), & c_0 t < x < \infty \end{cases}; \qquad t > 0.$$

The matching conditions are

$$(6.160) \qquad u_2(c_0 t, t) - u_1(c_0 t, t) = 0, \qquad t > 0,$$

$$(6.161) \qquad \frac{\partial u_2}{\partial x}(c_0 t, t) - \frac{\partial u_1}{\partial x}(c_0 t, t) = -f_0(t) = -A \cos \omega t,$$

where A and ω are constants. The initial conditions are

(6.162)
$$\begin{cases} u_1(x,0) = \dfrac{\partial u_1(x,0)}{\partial t} = 0, & -\infty < x < 0 \\[2mm] u_2(x,0) = \dfrac{\partial u_2(x,0)}{\partial t} = 0, & 0 < x < \infty \end{cases}.$$

In the regions $\tilde{R}_1$ and $\tilde{R}_2$ of Figure 6.17 we have $u_1 = u_2 = 0$ since the domains of dependence of points in these regions contain only initial data that are all zero. In R_1 we look for a solution in the form $u_1(x, t) = g(x + ct)$ and in R_2 $u(x, t)$ has in the form $u_2(x, t) = f(x - ct)$ since the moving source generates waves moving to the right and to the left. (A more systematic approach based on the integral wave equation of Section 6.5 can also be used.) Applying the matching conditions (6.160)–(6.161) we readily obtain

(6.163)
$$\begin{cases} u_1(x, t) = \dfrac{c^2 - c_0^2}{2\omega c} A \sin\left[\dfrac{\omega}{c + c_0}(x + ct)\right], & -ct < x < c_0 t \\[3mm] u_2(x, t) = \dfrac{c_0^2 - c^2}{2\omega c} A \sin\left[\dfrac{\omega}{c - c_0}(x - ct)\right], & c_0 t < x < ct \end{cases}.$$

We again observe the Doppler effect in (6.163). The wave $u_1(x, t)$ that moves in a direction opposite to that of the source if $c_0 > 0$ has the frequency

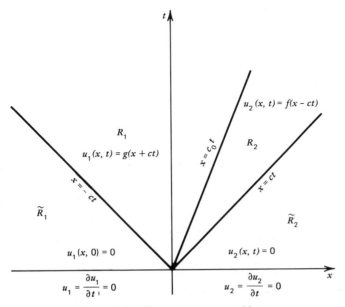

Figure 6.17. The moving source problem.

$\omega_1 = [c/(c + c_0)]\omega$ which is smaller than the frequency ω of the source term. The wave $u_2(x, t)$ that travels in the same direction as the source has the frequency $\omega_2 = [c/(c - c_0)]\omega$ which exceeds the frequency ω of the source term.

6.7. POINT SOURCES AND FUNDAMENTAL SOLUTIONS

In this section we assume that the inhomogeneous term ρF in the differential equations of Section 4.1 that represents a source or force term is effectively concentrated at a point P_0. We consider two cases. To begin, we deal with higher dimensional hyperbolic and parabolic equations and assume that ρF is concentrated at the point $P_0 = x_0$ for all time $t \geqslant 0$. The one-dimensional version of this problem was considered in Section 6.6. Next we assume that the point of concentration in the hyperbolic or parabolic case is $P_0 = (x_0, t_0)$ and in the elliptic case, $P_0 = x_0$. For the time dependent problems this constitutes an instantaneous point source.

In both cases ρF can be expressed in terms of the Dirac δ-function briefly discussed in Chapter 1. Since the theory of the δ-function and other generalized functions is not presented until Chapter 7, we discuss the foregoing problems, as far as possible, on the basis of the integral relations of Section 6.1. However, the theory of Green's functions developed in Chapter 7 is closely related to the point source problem.

We start with the integral relations (6.7) and (6.10) for the hyperbolic and parabolic cases, respectively. We assume that ρF is concentrated at the point $P_0 = x_0$ in the region R_x for all $t > 0$ and that it has the property

$$(6.164) \qquad \iint\limits_{R_x} \rho F \, dx = f(t)$$

where $f(t)$ is a (smooth) function that characterizes the source strength for $t > 0$ and vanishes for $t < 0$. In the limit as R_x shrinks down to the point P_0, we obtain from (6.7) and (6.10)

$$(6.165) \qquad \lim_{\partial R_x \to P_0} \int_{\partial R_x} p \frac{\partial u}{\partial n_x} \, ds_x = -f(t).$$

Although (6.165) is, in fact, integrated over the interval $t_0 < t < t_1$ in (6.7) and (6.10), our result is valid since the interval of integration is arbitrary. The integrals in (6.7) and (6.10) that involve u or u_t vanish in the limit as $R_x \to P_0$ since these terms are not more singular than x-derivatives in u and the relevant integrals are higher dimensional than the integral in (6.165).

The result (6.165) is valid for an arbitrary region R_x and its boundary ∂R_x. It takes on a simpler form if R_x is chosen to be a specific region. In the two-dimensional case let ∂R_x be the circle of radius r with center at $P_0 =$

(x_0, y_0) and introduce polar coordinates with the pole at P_0. Then (6.165) becomes

(6.166)

$$\lim_{\partial R_x \to (x_0, y_0)} \int_{\partial R_x} p \frac{\partial u}{\partial n_x} \, ds_x = \lim_{r \to 0} \int_0^{2\pi} p \frac{\partial u}{\partial r} r \, d\theta = \lim_{r \to 0} \left(2\pi pr \frac{\partial u}{\partial r} \right) = -f(t)$$

where the integral is taken over the circle of radius r. In the three-dimensional case let ∂R_x be a sphere of radius r with center at $P_0 = (x_0, y_0, z_0)$ and introduce spherical coordinates with the pole at P_0. Then (6.165) easily yields

$$(6.167) \qquad \lim_{\partial R_x \to (x_0, y_0, z_0)} \int_{\partial R_x} p \frac{\partial u}{\partial n_x} \, ds_x = \lim_{r \to 0} \left(4\pi pr^2 \frac{\partial u}{\partial r} \right) = -f(t)$$

on integrating over the sphere of radius r. In (6.166) and (6.167) we have used the fact that $\partial u / \partial n_x = \partial u / \partial r$ on ∂R_x and assumed that the mean value theorem for integrals can be applied in both cases.

The point source problem can now be formulated as follows. The function $u(x, t)$ satisfies the homogeneous form of the relevant hyperbolic or parabolic equation for $x \ne x_0$ and $t > 0$, since pF vanishes away from the source point P_0. At P_0, $u(x, t)$ must satisfy the condition (6.166) or (6.167). Initially, at $t = 0$ it is required that $u(x, t)$, as well as $u_t(x, t)$ in the hyperbolic case, vanish. If the problem is given over a bounded spatial region, $u(x, t)$ must satisfy prescribed boundary conditions.

The foregoing problem requires a different approach from that used in solving the concentrated source problem of the preceding section or, for that matter, all the problems of the previous sections. In the earlier problems it was possible to divide the region R over which the problems were formulated into two (or more) subregions separated by the curves or surfaces on which the sources or singularities were concentrated. Consequently, these curves or surfaces could be treated as additional boundaries for the problems, on which boundary conditions (i.e., matching or jump conditions) could be assigned. For the preceding point source problem and the one treated in the following, such a subdivision is not a priori possible because the source is concentrated either on a line or at a point and, as such, cannot represent a boundary for the region, given the space-time or space dimension of the problems we are considering. (However, if the problem has certain symmetries, such as axial symmetry, it may be possible to reduce the number of independent variables and treat the source points as boundary points.)

The nature of the singularity of the solution $u(x, t)$ of this problem at the source point may be determined by treating (6.166) and (6.167) as approximate equations near $x = x_0$. Integrating (6.166) yields

$$(6.168) \quad u(x, y, t) \approx \frac{-f(t)}{2\pi p(x_0, y_0)} \log \left[(x - x_0)^2 + (y - y_0)^2 \right]^{1/2},$$

$$(x, y) \approx (x_0, y_0)$$

in the two-dimensional case. In three dimensions (6.167) gives

(6.169)

$$u(x, y, z, t) \approx \frac{f(t)}{4\pi p(x_0, y_0, z_0)} \left[(x - x_0)^2 + (y - y_0)^2 + (z - z_0)^2 \right]^{-1/2},$$

$$(x, y, z) \approx (x_0, y_0, z_0).$$

In the following example we solve the point source problem for the wave equation and show that the behavior of the solution at the source point agrees with the preceding results. The solution of the point source problem for the heat equation is considered in the exercises.

EXAMPLE 6.11. THE POINT SOURCE PROBLEM FOR THE WAVE EQUATION

Let u satisfy the two or three-dimensional wave equation

(6.170) $$u_{tt} - c^2 \nabla^2 u = 0, \qquad P \neq P_0, t > 0$$

everywhere except at the source point P_0. At $t = 0$ we assume that $u = u_t = 0$ and at P_0, the condition (6.166) or (6.167) is satisfied.

To solve this problem we let r be the (spatial) distance from the source point P_0 and look for a solution of (6.170) in the form $u = u(r, t)$. The approximate results (6.168) and (6.169) suggest that the solution depends only on these two variables locally, even in the case of variable coefficients. Then (6.170) becomes

(6.171) $$u_{tt} = c^2 \left[u_{rr} + \frac{n-1}{r} u_r \right], \qquad r > 0, t > 0, n = 2, 3$$

in the case of two or three (space) dimensions.

We begin with the three-dimensional case since it is easier to handle. Let $n = 3$ in (6.171) and put $v = ru$. Then we easily verify that v satisfies the one-dimensional wave equation $v_{tt} - c^2 v_{rr} = 0$ which has the general solution $v(r, t) = F[t - (r/c)] + G[t + (r/c)]$. Thus $u(r, t) = (1/r)F[t - (r/c)] + (1/r)G[t + (r/c)]$. Since u must vanish at $t = 0$ and u satisfies the source point condition (6.167) where $f(t) = 0$ for $t < 0$, we readily conclude that $G[t + (r/c)] = 0$ and the solution is

(6.172) $$u(r, t) = \frac{1}{4\pi c^2 r} f\left(t - \frac{r}{c} \right).$$

Because $f(t)$ vanishes for negative t, $u(r, t)$ vanishes at $t = 0$ when $r > 0$. Near $r = 0$, (6.172) behaves like (6.169) where we must set $p = c^2$.

The solution (6.172) represents a *propagating spherical wave* that spreads out from the source point P_0 at the time $t = 0$ with speed c. The spherical wave fronts are given as $r = ct$ and $u = 0$ when $r > ct$ since $f(t)$ vanishes when its argument is negative.

In the two-dimensional case (6.171) cannot be simplified as nicely as was done for $n = 3$. However, the point source problem for $n = 2$ can be solved in terms of Hankel transforms in which case the condition (6.166) at the source can effectively be treated as a boundary condition at $r = 0$ (see Examples 5.10 and 5.9).

A simpler method for solving this problem uses the three-dimensional result (6.172) and proceeds as follows. We think of the point source at $P_0 = (x_0, y_0)$ as representing a line source for the three-dimensional problem located on the line $x = x_0$, $y = y_0$, $z = z$ in space. By summing (i.e., integrating) over all source points (x_0, y_0, z_0) along this line, we obtain the solution to the two-dimensional problem. Thus we integrate (6.172) with respect to z_0 from $-\infty$ to $+\infty$ and obtain

$$(6.173) \quad u(x, y, t) = \frac{1}{4\pi c^2} \int_{-\infty}^{\infty} \frac{1}{r} f\left(t - \frac{r}{c}\right) dz_0 = \frac{1}{2\pi c^2} \int_0^{\infty} \frac{1}{r} f\left(t - \frac{r}{c}\right) d\hat{z}$$

$$= \frac{1}{2\pi c} \int_0^{t - (\hat{r}/c)} \left[c^2(t - \tau)^2 - \hat{r}^2\right]^{-1/2} f(\tau) \, d\tau,$$

where $\hat{z} = z - z_0$, $\hat{r}^2 = (x - x_0)^2 + (y - y_0)^2$, $\tau = t - (r/c)$ and we have used the fact that $f(t) = 0$ for $t < 0$. It can be shown that (6.173) is a solution of the two-dimensional wave equation and satisfies the condition (6.166) but we do not verify this. Also, (6.173) vanishes at $t = 0$ since $f(t)$ vanishes for negative values of its argument. In fact, (6.173) vanishes when $\hat{r} > ct$ so that the circles $\hat{r} = ct$ are circular (or cylindrical) wave fronts for the two-dimensional point source problem.

Next we consider *point source problems* for elliptic equations and *instantaneous point source problems* for parabolic and hyperbolic equations. In the elliptic case the source point is $P_0 = x_0$, whereas in the hyperbolic and parabolic cases it is $P_0 = (x_0, t_0)$. The source is assumed to have unit strength, and the concentration of ρF at the point P_0 implies that

$$(6.174) \qquad \iint_R \rho F \, dv = 1$$

for any region R in space or space-time that contains P_0, whereas ρF vanishes when $P \neq P_0$. Thus ρF is identical with the Dirac δ-function with singular point at P_0.

Since $\rho F = 0$ for $P \neq P_0$, the solution $u(x)$ or $u(x, t)$ of the point source problem satisfies the homogeneous form of the relevant differential equation or the equivalent integral relation away from P_0. The behavior of u at P_0 must be determined from the integral relation since u is not expected to be smooth at P_0.

Proceeding formally, we consider the integral relation (6.5) for the hyperbolic case and assume the region R contains the source point P_0. Then as R shrinks down to P_0 we obtain the formal limit

$$(6.175) \qquad \lim_{\partial R \to P_0} \int_{\partial R} [\, p\nabla u, \, -\rho u_t \,] \cdot \mathbf{n} \, ds = -1$$

on using (6.174) and assuming the other terms in the integral relation vanish in the limit. Similarly, the integral relation (6.9) for the parabolic case yields, as $R \to P_0$,

$$(6.176) \qquad \lim_{\partial R \to P_0} \int_{\partial R} [\, p\nabla u, \, -\rho u \,] \cdot \mathbf{n} \, ds = -1.$$

In the elliptic case the integral relation (6.12) gives

$$(6.177) \qquad \lim_{\partial R \to P_0} \int_{\partial R} p \frac{\partial u}{\partial n} \, ds = -1.$$

In obtaining (6.176) and (6.177) we have used (6.174) and assumed that all other terms in the integral relations (6.9) and (6.12) vanish in the limit as $R \to P_0$.

Now for the hyperbolic and parabolic problems the singularity of $u(x, t)$ does not remain isolated at the source point P_0 but is transmitted along the characteristics that contain that point. Thus $u(x, t)$ may not possess the derivatives near P_0 required to make the integral relations (6.5) and (6.9) and their limits (6.175) and (6.176) valid. In that case we must replace the foregoing integral relations by those obtained in Section 6.4 for which the solutions $u(x, t)$ were not required to be differentiable. In the hyperbolic case, the relevant integral relation is (6.77) and in terms of that expression it is easy to see that (6.175) should be replaced by

$$(6.178) \qquad \lim_{\partial R \to P_0} \iint_R u [\, \rho v_{tt} - \nabla \cdot (p\nabla v) + qv \,] \, dV = v(P_0).$$

This condition must be satisfied for every smooth function $v(x, t)$ that vanishes near the boundary of R. A similar condition based on (6.80) can be obtained for the parabolic case.

In the hyperbolic problem, the family of characteristic curves or surfaces containing a given point $P_0 = (x_0, t_0)$ envelops a (singular) characteristic known as the *characteristic conoid* which has a vertex at P_0. This conoid reduces to the two characteristic curves that pass through the point P_0 in the case of one space dimension. For the wave equation the conoid coincides with the characteristic cone (3.80). The solution of the point source problem is thus expected to be singular along the entire characteristic conoid and it appears that (6.178) rather than (6.175) is the relevant condition at P_0 for the hyperbolic case.

In fact, if we formally set $f(t) = \delta(t - t_0)$ [where $\delta(t)$ is the Dirac δ-function and $t_0 > 0$] in the point source problem for the wave equation considered in Example 6.11, the problem reduces to that for an instantaneous point source at (x_0, t_0). It has the formal solution $u = [1/(4\pi c^2 r)]\delta[t - t_0 - (r/c)]$ as follows from (6.172). This shows the solution to be singular not just at P_0 but along the entire forward characteristic cone $r = c(t - t_0)$. We remark that in the continuous source problem treated in the beginning of this section, we assumed that $f(t)$ is smooth for $t > 0$ so that once the wavefront $r = ct$ passes, the solution is smooth for $r > 0$ and the integral relation can be used for $t > 0$ without encountering the foregoing difficulty.

In the parabolic problem, the characteristic curves or surfaces that contain the point $P_0 = (x_0, t_0)$ are the lines or planes $t = t_0$, as follows from the results of Chapter 3. Thus we expect $u(x, t)$ to be singular along $t = t_0$ and not just at P_0. Since ∇u represents interior differentiation on $t = t_0$, the singularity of $u(x, t)$ is not enhanced by differentiation across the characteristic in the surface integral in (6.176), as was the case for the hyperbolic problem considered earlier. Consequently, it may not be necessary to replace the condition (6.176) by a (weaker) condition based on (6.80) as was required in the hyperbolic case.

The source point condition (6.177) for the elliptic case poses the fewest problems. Since elliptic equations have no real characteristics, the singularity of $u(x)$ must be confined to the source point $P_0 = x_0$, and u and its derivatives are expected to be smooth on ∂R in (6.177). Effectively, the elliptic point source problem can be treated in the same way as the continuous point source problem for hyperbolic and parabolic equations discussed at the beginning of this section, as shown in the following.

As a result of the foregoing observations, we discuss the elliptic point source problem on the basis of the condition (6.177), although the point source problems for the hyperbolic and parabolic cases are dealt with in a different way. We remark that it is generally easier to deal with the latter problems by expressing ρF as a Dirac δ-function and using the theory of generalized functions to solve the problem. This approach is indicated in Chapter 7. Nevertheless, a substantial body of results for the point source and related problems was achieved before the theory of generalized functions was developed.

Solutions of the foregoing point source problems are called *fundamental solutions* of the given differential equation. As such they are merely required to satisfy the homogeneous form of the equation away from the source point P_0 and the appropriate condition (6.175), (6.176), (6.177), or (6.178) at P_0. Fundamental solutions are, therefore, not uniquely determined since any smooth solution of the homogeneous differential equation can be added to them. Here we are mostly concerned with fundamental solutions that are required to satisfy additional conditions that serve to specify them uniquely.

For the hyperbolic and parabolic problems we require that the fundamental solution $u(x, t)$ satisfy a *causality condition*. This states that the solution $u(x, t)$ vanishes identically for $t < t_0$ where $P_0 = (x_0, t_0)$ is the source point. Thus we are considering the effect of the source point P_0 on the solution $u(x, t)$ in the region $t > t_0$. This is referred to as the *causal fundamental solution*. The causality assumption translates into the condition that u and u_t vanish at $t = t_0$ for all $P \neq P_0$. That is, we effectively have an initial value problem for the given equations with zero data away from P_0 and the data at P_0 determined from the source point conditions. Since $t = t_0$ is a characteristic for the parabolic problem, we can arbitrarily set $u(x, t) = 0$ for $t < t_0$ since singularities in the solution can occur only on characteristics. For the hyperbolic problem (as we have already observed in a number of special cases), the domain of influence of a point $P_0 = (x_0, t_0)$ is the forward characteristic conoid—that is, that part of it for which $t > t_0$—with vertex at P_0. Then with $u(x, t) = 0$ for $t < t_0$, the causal fundamental solution vanishes identically in the exterior of the forward characteristic conoid. It is shown in Chapter 7 how the causal fundamental solutions for hyperbolic and parabolic problems can be determined as solutions of homogeneous equations for $t > t_0$ with appropriate (singular) initial data at $t = t_0$.

In the elliptic case we require that the fundamental solution have a specified behavior at infinity. Then the fundamental solution is known as the *free space Green's function*. In this section we only consider fundamental solutions in unbounded regions. If the point source problem is given for a bounded region, the fundamental solution must satisfy the conditions prescribed on the boundary. Solutions of such problems are generally referred to as *Green's functions* or *influence functions*. Such problems are discussed in Chapter 7.

Since elliptic problems are easiest to deal with, we begin our discussion of fundamental solutions with the elliptic case. In the one-dimensional case, the source point condition (6.177) is easily seen to reduce to a jump condition for u_x at the point x_0. In two and three dimensions the condition (6.177) is essentially of the form (6.165) if we set $f(t) = 1$. Consequently, the simplified forms of (6.165) given in (6.166) and (6.167) are valid for the present case as well if we set $f(t) = 1$ in these equations. With n equal to the number of dimensions, we conclude that the source point condition (6.177) can be expressed as

$$(6.179) \qquad \begin{cases} \left[\dfrac{\partial u}{\partial x}\right]_{x=x_0} = -\dfrac{1}{p(x_0)}, & n = 1 \\[2ex] \lim_{r \to 0}\left(2\pi p r \dfrac{\partial u}{\partial r}\right) = -1, & n = 2 \\[2ex] \lim_{r \to 0}\left(4\pi p r^2 \dfrac{\partial u}{\partial r}\right) = -1, & n = 3 \end{cases}$$

The bracket in the case where $n = 1$ represents the jump in u_x at $x = x_0$. For $n = 2$ we have $r^2 = (x - x_0)^2 + (y - y_0)^2$ and for $n = 3$, $r^2 = (x - x_0)^2 + (y - y_0)^2 + (z - z_0)^2$.

Away from the source point P_0, the fundamental solution u satisfies the homogeneous equation

$$(6.180) \qquad\qquad -\nabla \cdot (p\nabla u) + qu = 0, \qquad P \neq P_0$$

where ∇ is replaced by d/dx in the one-dimensional case. We note that for the one-dimensional problem, the source condition (6.179) can be treated as a matching condition for two solutions of (6.180) in the intervals $x < x_0$ and $x > x_0$. This is not the case, however, for the higher dimensional problems. It is also seen that the conditions (6.179) are insufficient to determine solutions of (6.180) uniquely.

Fundamental solutions can be found by looking for singular solutions of (6.180) that have the appropriate behavior (6.179) at the source point. This approach is used when we discuss fundamental solutions of equations with constant coefficients in Example 6.12. Even if these solutions do not satisfy the auxiliary conditions placed on the problem, they are important because we can add smooth solutions of (6.180) to the problem to account for the additional conditions. (This technique will be used in our discussion of Green's functions in Chapter 7.)

To examine the form of the singularity at the source point P_0, we may consider (6.179) to represent approximate equations near P_0, just as was done in obtaining (6.168) and (6.169). On solving these (approximate) equations we obtain

$$(6.181) \qquad u \approx \begin{cases} -\dfrac{1}{2p(x_0)}|x - x_0|, & n = 1 \\[2ex] -\dfrac{1}{2\pi p(x_0, y_0)}\log r, & n = 2 \\[2ex] \dfrac{1}{4\pi p(x_0, y_0, z_0)r}, & n = 3 \end{cases}$$

which is valid for points P near P_0 and where r is defined as in (6.179) for $n = 2$ and $n = 3$. The one-dimensional form of $u(x)$ in (6.181) was chosen to make the solution symmetric with respect to x_0. In all cases only the most singular terms of the solution were retained. We note that the singularity of u is localized at the source point P_0 in all cases, and that the strength of the singularity increases as the dimension n increases.

In the following example we construct fundamental solutions of the elliptic equation (6.180) in the case of constant coefficients.

EXAMPLE 6.12.   ELLIPTIC EQUATIONS WITH CONSTANT COEFFICIENTS

We consider the elliptic equation

$$(6.182) \qquad\qquad p\nabla^2 u - qu = 0$$

where $p > 0$ and q are constants. Guided by (6.181) we look for a fundamental solution of (6.182) that depends only on the distance r from the source point P_0. In one dimension (6.182) is an ordinary differential equation and $r = |x - x_0|$. In two and three dimensions r is the radial variable in polar and spherical coordinates, respectively, with the pole at P_0.

Let $u = \tilde{u}(r)$ in (6.182) and we obtain

$$(6.183) \qquad p\left[\tilde{u}_{rr} + \frac{n-1}{r}\tilde{u}_r\right] - q\tilde{u} = 0, \qquad r > 0,\, n = 1, 2, 3$$

with n equal to the number of dimensions in the problem. Note that for $n = 2$ and $n = 3$ we are assuming the fundamental solution is independent of the angular variables.

To begin, we set $q = 0$ in (6.182) which reduces it to Laplace's equation when $n = 2$ and $n = 3$. For each n we require a singular solution of (6.183), and a set of such solutions is easily found to be given by

(6.184)

$$u = \tilde{u}(r) = \begin{cases} cr = c|x - x_0|, & n = 1 \\ c\log r = c\log\sqrt{(x - x_0)^2 + (y - y_0)^2}, & n = 2 \\ \dfrac{c}{r} = c\left[(x - x_0)^2 + (y - y_0)^2 + (z - z_0)^2\right]^{-1/2}, & n = 3 \end{cases}$$

with the constant c to be determined in each case. On using the conditions (6.179) at P_0 or, more simply, by comparing (6.184) with (6.181) we find that

$$(6.185) \qquad\qquad c = \begin{cases} -\dfrac{1}{2p}, & n = 1 \\[2mm] -\dfrac{1}{2\pi p}, & n = 2 \\[2mm] \dfrac{1}{4\pi p}, & n = 3 \end{cases}.$$

Other solutions of (6.183) exist that also satisfy the conditions (6.179) at P_0. However, the fundamental solutions (6.184)–(6.185) are in the standard form associated with (6.182) when $q = 0$, particularly when $n = 2$ or $n = 3$, in which case it reduces to Laplace's equation.

If $q \neq 0$ in (6.182)–(6.183) and $n = 1$, we obtain

$$(6.186) \quad u = \tilde{u}(r) = \begin{cases} c_1\exp\left[\sqrt{\dfrac{q}{p}}\,r\right] + c_2\exp\left[-\sqrt{\dfrac{q}{p}}\,r\right]; & q > 0 \\[2em] c_1\exp\left[i\sqrt{-\dfrac{q}{p}}\,r\right] + c_2\exp\left[-i\sqrt{-\dfrac{q}{p}}\,r\right]; & q < 0 \end{cases}$$

with the constants c_1 and c_2 to be determined. This is done so that we obtain the free space Green's functions for these equations by specifying the behavior of the fundamental solutions at infinity as well as at the source point.

For $q > 0$ we require that (6.186) vanish at infinity—that is, as $|x - x_0| \to \infty$—and that it satisfy (6.179) at x_0. Since $r = |x - x_0|$ in (6.186) we conclude that

$$(6.187) \quad u(x) = \frac{1}{2}(pq)^{-1/2}\exp\left[-\sqrt{\frac{q}{p}}\,|x - x_0|\right]; \qquad q > 0$$

is the (uniquely determined) free space Green's function for this case.

With $q < 0$ and $n = 1$, (6.182) has the form of the Helmholtz or reduced wave equation in one dimension. Given the wave equation $v_{tt} = pv_{xx}$, if we set $v(x, t) = u(x)\exp[-i\sqrt{-q}\,t]$ with $q < 0$, we find that $u(x)$ satisfies the one-dimensional form of (6.182). The free space Green's function for the Helmholtz equation requires that the fundamental solution represent a wave traveling away from the source point. (This requirement is related to the *Sommerfeld radiation condition* discussed in Chapters 7 and 9.) In combination with the source point condition (6.179) it implies that (6.186) has the form

$$(6.188) \quad u(x) = \frac{i}{2}(-pq)^{-1/2}\exp\left[i\sqrt{-\frac{q}{p}}\,|x - x_0|\right]; \qquad q < 0.$$

When multiplied by $\exp[-i\sqrt{-q}\,t]$, (6.188) is a plane wave traveling away from the source point $x = x_0$ with speed $\sqrt{p}$.

In both of the foregoing problems, the presence of exponentially growing solutions at infinity when $q > 0$ and the existence of waves traveling from infinity to the source point when $q < 0$ is physically unreasonable. A physical interpretation of the case $q > 0$ is given in the following.

With $n = 2$ and $q < 0$ (6.183) has the form of Bessel's equation of zero order. We require a singular solution of this equation and this is given by

$$(6.189) \quad u = \tilde{u}(r) = c_0 Y_0\left[\sqrt{-\frac{q}{p}}\,\sqrt{(x - x_0)^2 + (y - y_0)^2}\right]; \qquad q < 0$$

where c_0 is a constant and $Y_0(z)$ is the *Neumann function* of zero order. With $n = 2$ and $q > 0$ in (6.183) we have the modified Bessel equation of zero order and a singular solution is

$$(6.190) \quad u = \tilde{u}(r) = c_1 K_0 \left[\sqrt{\frac{q}{p}} \sqrt{(x - x_0)^2 + (y - y_0)^2} \right]; \quad q > 0$$

where c_1 is a constant and $K_0(z)$ is the *modified zero order Bessel function* of the second kind. For $z \approx 0$ these functions have the following behavior:

$$(6.191) \quad \begin{cases} Y_0(z) \approx \dfrac{2}{\pi} \log z, & z \approx 0 \\ K_0(z) \approx -\log z, & z \approx 0 \end{cases}.$$

Comparing (6.189)–(6.190) with (6.181) and using (6.191), we conclude that

$$(6.192) \quad \begin{cases} c_0 = -\dfrac{1}{4p} \\ c_1 = \dfrac{1}{2\pi p} \end{cases}.$$

With $n = 3$ and $q \neq 0$ in (6.182) we easily obtain

$$(6.193) \quad u(x, y, z) = \begin{cases} \dfrac{1}{4\pi pr} \exp\left[-\sqrt{\dfrac{q}{p}}\, r \right], & q > 0 \\ \dfrac{1}{4\pi pr} \exp\left[i\sqrt{-\dfrac{q}{p}}\, r \right], & q < 0 \end{cases}$$

where $r^2 = (x - x_0)^2 + (y - y_0)^2 + (z - z_0)^2$. As $r \to 0$ these fundamental solutions have the required behavior (6.181).

The foregoing two and three-dimensional fundamental solutions were specified somewhat arbitrarily without regard to their behavior at infinity. To obtain the free space Green's functions for these problems we require that the fundamental solutions have a prescribed behavior at infinity. If $q > 0$, (6.182) represents a stationary form of the diffusion equation $v_t + qv = p\nabla^2 v$, which is obtained on setting $v(x, t) \equiv u(x)$. We require that $v(x, t)$ and, consequently, $u(x)$ tend to zero as the distance r from the source point P_0 tends to infinity. If $q < 0$, (6.182) is again the reduced wave equation or Helmholtz's equation. If $v(x, t)$ is a solution of the wave equation $v_{tt} = p\nabla^2 v$, we require that $v(x, t) = u(x)\exp[-i\sqrt{-q}\, t]$ represent a diverging wave traveling away from the source point P_0 as $r \to \infty$.

To apply the conditions at infinity we must know the behavior of the Bessel functions Y_0 and K_0 for large values of the argument. We have

(6.194)
$$\begin{cases} Y_0(z) \approx \left[\dfrac{2}{\pi z} \right]^{1/2} \sin\left(z - \dfrac{\pi}{4} \right), & z \to \infty \\[2ex] K_0(z) \approx \left[\dfrac{\pi}{2z} \right]^{1/2} \exp(-z), & z \to \infty \end{cases}.$$

We conclude that for $q > 0$ the fundamental solutions (6.190) and (6.193) both decay exponentially as $r \to \infty$ and are, therefore, the free space Green's functions in two and three dimensions for this case.

In the three-dimensional case with $q < 0$, the fundamental solution (6.193) is easily seen to represent a diverging spherical wave traveling away from the source point P_0 with speed $\sqrt{p}$. Thus it is the free space Green's function for this problem. However, since $\sin z = (1/2i)[e^{iz} - e^{-iz}]$, we find that the fundamental solution (6.189) contains both diverging and converging cylindrical waves for large r in view of (6.194). Thus (6.189) is not the free space Green's function for this problem.

The correct form for the free space Green's function in two dimensions is given by

(6.195) $\quad u(x, y) = \dfrac{i}{4p} H_0^{(1)} \left[\sqrt{-\dfrac{q}{p}} \sqrt{(x - x_0)^2 + (y - y_0)^2} \right], \qquad q < 0$

where $H_0^{(1)}(z)$ is the zero order *Hankel function* of the first kind. $H_0^{(1)}(z)$ is a singular solution of the Bessel equation of zero order with the following behavior:

(6.196) $\qquad H_0^{(1)}(z) \approx \dfrac{2i}{\pi} \log z, \qquad z \approx 0$

(6.197) $\qquad H_0^{(1)}(z) \approx \left(\dfrac{2}{\pi z} \right)^{1/2} \exp\left[i\left(z - \dfrac{\pi}{4} \right) \right], \qquad z \to \infty.$

Consequently, (6.195) exhibits the correct behavior (6.181) at the source point and represents a diverging cylindrical wave as the distance r from the source point tends to infinity.

We now turn to the consideration of fundamental solutions for hyperbolic and parabolic equations. Because of the aforementioned complications with the interpretations of these solutions and the determination of their behavior at the source point, we restrict our discussion to one-dimensional problems for equations with constant coefficients. Higher dimensional problems are considered in Chapter 7.

The solution of the inhomogeneous wave equation

$$(6.198) \qquad u_{tt} - c^2 u_{xx} = F(x, t), \qquad -\infty < x < \infty, t > 0$$

with homogeneous initial data

$$(6.199) \qquad u(x, 0) = u_t(x, 0) = 0, \qquad -\infty < x < \infty$$

was found to be (see Section 6.5)

$$(6.200) \qquad u(x, t) = \frac{1}{2c} \int_0^t \int_{x-c(t-\tau)}^{x+c(t-\tau)} F(\sigma, \tau) \, d\sigma \, d\tau.$$

As has been noted, the foregoing Cauchy problem reduces to the instantaneous point source problem for the wave equation with source point at $P_0 = (x_0, t_0)$ if we set

$$(6.201) \qquad F(x, t) = \delta(x - x_0)\delta(t - t_0)$$

where the two-dimensional Dirac delta function $\delta(x - x_0)\delta(t - t_0)$ is defined as in (1.106)–(1.107). The integral in (6.200) is taken over a characteristic triangle with vertices at (x, t) and at the points $(x - ct, 0)$ and $(x + ct, 0)$ on the x-axis. From (6.201) and the property of the delta function we find that the integral has the value unity if (x_0, t_0) lies within the characteristic triangle, whereas it vanishes if (x_0, t_0) lies outside the characteristic triangle. The causality condition for the problem requires that the fundamental solution vanish for $t < t_0$, so that the conditions (6.199) are appropriate since $t_0 > 0$ by assumption.

It is apparent from Figure 6.18 that any point (x, t) in the forward characteristic sector (or the domain of influence) of the source point P_0 contains P_0 in its characteristic triangle (or domain of dependence). Consequently, the causal fundamental solution for the wave equation (i.e., the solution of the instantaneous point source problem) is given as

$$(6.202) \qquad u(x, t) = \begin{cases} \dfrac{1}{2c}, & |x - x_0| < c(t - t_0) \\ 0, & |x - x_0| > c(t - t_0) \end{cases}.$$

Since $u(x, t)$ is discontinuous across the characteristic lines $x - x_0 = \pm c(t - t_0)$, $u(x, t)$ must be interpreted as a weak solution in the sense of Section 6.4. By proceeding as in Example 6.4 it may be shown that (6.202) is a weak solution of the wave equation that satisfies the condition (6.178) at the source point P_0. This shows why it is not possible to apply the source point condition (6.175) to determine the behavior of the fundamental solution at P_0 unless the theory of generalized functions is used to obtain derivatives of $u(x, t)$ across the characteristics.

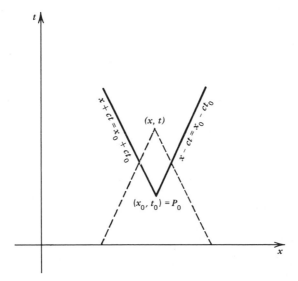

Figure 6.18. The forward characteristic sector.

The solution of the inhomogeneous heat equation

(6.203) $u_t - c^2 u_{xx} = F(x, t), \qquad -\infty < x < \infty, t > 0$

with homogeneous initial data

(6.204) $u(x, 0) = 0, \qquad -\infty < x < \infty,$

was given in (5.46) as

(6.205) $u(x, t) = \int_0^t \int_{-\infty}^{\infty} G(x - s, t - \tau) F(s, \tau) \, ds \, d\tau$

where

(6.206) $G(x, t) = \dfrac{1}{\sqrt{4\pi c^2 t}} \exp\left[-\dfrac{x^2}{4c^2 t} \right].$

Putting $F(x, t)$ equal to the two-dimensional Dirac delta function as in (6.201) we find the causal fundamental solution for the heat equation:

(6.207) $u(x, t) = \begin{cases} G(x - x_0, t - t_0), & t > t_0 \\ 0, & t < t_0 \end{cases}.$

We have set $u(x, t) = 0$ for $t < t_0$ because of the causality condition and we observe that $u(x, t)$ is singular on the characteristic line $t = t_0$. The function $G(x - x_0, t - t_0)$ is itself a fundamental solution of the heat equation and has already been referred to as such in Chapter 5. We do not discuss how (6.207) is to be interpreted as a weak or generalized solution of the heat equation and how the condition (6.176) at the source point is satisfied. The higher dimensional fundamental solutions for the heat equation are similar in form to $G(x, t)$ and are considered in the exercises and in Chapter 7.

The construction of the foregoing fundamental solutions for the wave and heat equations by specializing the results obtained for the general inhomogeneous equation to the case of a point source must be considered somewhat unsatisfactory; first, because such an approach was not necessary for the elliptic point source problem whose solution was obtained in a more direct manner; second, because fundamental solutions and the related Green's functions are expected to play a role in the solution of more general problems as shown in Chapter 7, so that we would like to determine the fundamental solutions independently of having to solve more general problems. Although such direct determinations of fundamental solutions are given in Chapter 7 by using the theory of generalized functions and transform methods, we now present two alternative methods for obtaining fundamental solutions for hyperbolic and parabolic equations with constant coefficients.

We begin with the hyperbolic equation

$$(6.208) \qquad\qquad \rho u_{tt} - p u_{xx} + q u = 0$$

with constant coefficients and present a method for determining fundamental solutions of (6.208) similar to that given for elliptic equations. Let $P_0 = (x_0, t_0)$ be the source or singular point for the fundamental solution. Then the *characteristic cone* for (6.208) with vertex at P_0 is given as

$$(6.209) \qquad\qquad c^2 |t - t_0|^2 - |x - x_0|^2 = 0$$

where $c^2 = p/\rho$. In fact, (6.209) is just the two characteristic lines $x - x_0 = \pm c(t - t_0)$ that pass through the point P_0. As we have seen for the wave equation, the causal fundamental solution is singular not only at the source point (x_0, t_0) but also along the forward characteristic cone (6.209).

Therefore, we look for a singular solution of (6.208) that depends only on the variable r defined as

$$(6.210) \qquad\qquad r = \sqrt{c^2 (t - t_0)^2 - (x - x_0)^2} \, .$$

In a hyperbolic (nonEuclidean) geometry appropriate for the hyperbolic equation (6.208), r represents distance from the point (x_0, t_0). In this sense our approach to the hyperbolic problem is similar to that for the elliptic problem. In the interior of the cone (6.209)—that is, when $(x - x_0)^2 < c^2 (t - t_0)^2$—$r$ is real, whereas r becomes imaginary in the cone's exterior. Since the domains of dependence and influence of the point P_0 are determined by the interior of the cone, the points (x, t) where r is real are of greatest interest to us.

Let $u = \tilde{u}(r)$ in (6.208) with r given as in (6.210) and we easily obtain

$$(6.211) \qquad p\left[\tilde{u}_{rr} + \frac{1}{r}\tilde{u}_r\right] + qu = 0.$$

Apart from the sign in front of q, this equation is identical with (6.183) obtained for the elliptic equation (6.182) in the case of two dimensions. Proceeding as in the elliptic problem, we look for a singular solution of (6.211) that satisfies the source point condition (6.178) at $r = 0$.

First we set $q = 0$ in (6.211) and find, as in the elliptic case, the singular solution

$$(6.212) \qquad u(x, t) = \tilde{u}(r) = \alpha \log\sqrt{c^2(t - t_0)^2 - (x - x_0)^2},$$

with the constant α to be determined. We note that in contrast to the fundamental solution for the two-dimensional Laplace's equation, (6.212) is singular not only at the source point (x_0, t_0) but along the entire characteristic cone $(x - x_0)^2 = c^2(t - t_0)^2$ through that point. It can be shown, but we do not demonstrate this, that the source point condition requires that we set

$$(6.213) \qquad \alpha = \frac{1}{2c\rho}.$$

The fundamental solution (6.212)–(6.213) is quite different from the causal fundamental solution (6.202) and does not reduce to that form even if we set it equal to zero outside the forward characteristic cone or sector. Yet Hadamard used (6.212)–(6.213) and more general fundamental solutions for equations with variable coefficients to solve initial value problems for these equations. To overcome difficulties with the singularities of the fundamental solutions when integrations are to be performed, Hadamard invented the *method of finite parts* for integrals. A more modern approach to fundamental solutions and their use in solving more general problems for hyperbolic equations is based on the theory of generalized functions. In terms of the present problem this means that we would work with a causal fundamental solution in the form (6.202) rather than one constructed on the basis of (6.212)–(6.213). The problems resulting from differentiating and integrating such fundamental solutions are dealt with in the context of the theory of generalized functions.

We now demonstrate that a fundamental solution of the form (6.202) can also be obtained as a solution of (6.211). Although this requires that we use some properties of generalized functions that are not discussed until Chapter 7, our derivation is sufficiently elementary that it can be given here.

Recalling from Example 1.2 the (substitution) property of the Dirac δ-function that gives $f(x)\delta(x - \xi) = f(\xi)\delta(x - \xi)$ if $f(x)$ is continuous, we have

$x\delta(x) = 0 \cdot \delta(x) = 0$. Assuming a meaning can be assigned to $[d/dx]\delta(x) = \delta'(x)$, we obtain on differentiating $x\delta(x) = 0$,

(6.214) $x\delta'(x) + \delta(x) = 0.$

Now with $q = 0$, (6.211) takes the form

(6.215) $r\tilde{u}_{rr} + \tilde{u}_r = 0$

so that a (generalized) first integral of (6.215) is

(6.216) $\tilde{u}_r = \alpha\delta(x)$

in view of (6.214) where α is a constant. Formally integrating (6.216) gives

(6.217) $\tilde{u}(r) = \alpha\displaystyle\int_{-\infty}^{r} \delta(r)\, dr = \begin{cases} 0, & r < 0 \\ \alpha, & r > 0 \end{cases}.$

This follows since $\delta(r) = 0$ for $r \neq 0$ so that the integral vanishes for $r < 0$ and the integral equals unity for $r > 0$ because of the property of the delta function.

Since $r > 0$ represents the interior of the characteristic cone (6.209) and $r = 0$ is the cone itself, we conclude that the function $\tilde{u}(r)$ in (6.217) is constant within and may be equated to zero outside the characteristic cone. To obtain the causal fundamental solution we apply the condition (6.178) from which we determine that

(6.218) $\alpha = \dfrac{1}{2c\rho}.$

Therefore, the causal fundamental solution of (6.208) with $q = 0$ is

(6.219) $u(x, t) = \begin{cases} \dfrac{1}{2c\rho}, & |x - x_0| < c(t - t_0) \\ 0, & |x - x_0| > c(t - t_0) \end{cases}$

where the causality condition has been used so that $u(x, t) = 0$ for $t < t_0$.

A more satisfying analysis of the foregoing problem can be given if we replace the variable r defined in (6.210) by the squared variable $\sigma = r^2$. Then $\sigma > 0$ is the interior of the characteristic cone (6.209) and $\sigma < 0$ is its exterior. Next, with $u(x, t) = \hat{u}(\sigma)$ we find that $\hat{u}$ satisfies (6.215) with r replaced by σ. Continuing as before, we conclude that $\hat{u}(\sigma)$ is given as in (6.217)—with r replaced by σ. Now, however, we determine that $u(x, t) = \hat{u} = \alpha$ when $\sigma > 0$ (i.e., in the interior of the characteristic cone) and $u(x, t) = \hat{u} = 0$ when $\sigma < 0$ (i.e., in the cone's exterior). Consequently, (6.219) follows from the causality condition.

When $q < 0$ in (6.211) we have a modified Bessel equation of zero order which has the singular solution

$$(6.220) \qquad \tilde{u}(r) = \alpha K_0 \left[\sqrt{-\frac{q}{p}} \, r \right]$$

where $K_0(z)$ is the zero order *modified Bessel function of the second kind* and the constant α is to be determined. Now $K_0(z)$ has the form

$$(6.221) \qquad K_0(z) = -I_0(z)\log z + M_0(z)$$

where $I_0(z)$ is the zero order *modified Bessel function of the first kind* and $M_0(z)$ is given as a power series that converges for all z. The function $I_0(z)$ is a regular solution of the Bessel equation with $I_0(0) = 1$ and $I_0'(0) = 0$.

The behavior of $K_0(z)$ near $z = 0$ is dominated by the $\log z$ term in (6.221). Consequently, (6.220) corresponds to the fundamental solution (6.212) and if we set $\alpha = -1/2c\rho$ so that

$$(6.222) \quad u(x, t) = \tilde{u}(r) = -\frac{1}{2c\rho} K_0 \left[\sqrt{-\frac{q}{p}} \sqrt{c^2(t - t_0)^2 - (x - x_0)^2} \, \right],$$

we find that (6.222) and (6.212)–(6.213) have the same behavior near $r = 0$, as they indeed must.

The causal fundamental solution of (6.208) with $q < 0$ and source point at $P_0 = (x_0, t_0)$ vanishes outside the forward characteristic sector $|x - x_0| < c(t - t_0)$ and may be obtained by replacing the singular solution K_0 of (6.211) by the regular solution I_0. It has the form

$$(6.223) \quad u(x, t) = \begin{cases} \dfrac{1}{2c\rho} I_0 \left[\sqrt{-\frac{q}{p}} \sqrt{c^2(t - t_0)^2 - (x - x_0)^2} \, \right]; & \\ & |x - x_0| < c(t - t_0) \cdot \\ 0; & \\ & |x - x_0| > c(t - t_0) \end{cases}$$

This corresponds to the fundamental solution (6.219) and reduces to it as $q \to 0$ since $I_0(0) = 1$. In fact, as $r \to 0$, (6.223) behaves like (6.219) since $I_0(0) = 1$ so that the condition (6.178) at the source point P_0 is satisfied by (6.223). It may be verified that (6.223) is a weak solution of (6.208) but we do not show this. We remark that $(1/2c\rho)I_0$ is the coefficient of the log term in the fundamental solution (6.222) just as $1/2c\rho$ in (6.219) is the coefficient of the log term in (6.212). We defer a discussion of the fundamental solution of (6.208) with $q > 0$ to Chapter 7.

The causal fundamental solution for the one-dimensional heat or diffusion equation was given in (6.207) where it was determined as a special case of a more general result. In the exercises we show how fundamental solutions for the heat equation can be constructed by looking for solutions that depend on the space and time variables in a special way. We now show how the causal fundamental solution (6.207) for the heat equation can be obtained as a limit of the causal fundamental solution for the telegrapher's equation. This result is of some interest in connection with the Brownian motion problems discussed in Sections 1.1 and 1.2.

The (hyperbolic) telegrapher's equation

$$(6.224) \qquad \epsilon u_{tt} - c^2 u_{xx} + u_t = 0$$

where $\epsilon > 0$ formally reduces to the heat equation if we set $\epsilon = 0$. We might expect that as $\epsilon \to 0$, the causal fundamental solution of (6.224) corresponding to a source point at $P_0 = (x_0, t_0)$ reduces to the causal fundamental solution of the heat equation. Indeed, the causality condition for both problems requires that $u(x, t) = 0$ for $t < t_0$. Furthermore, the causal fundamental solution of (6.224) vanishes outside the forward characteristic sector $|x - x_0| < \hat{c}(t - t_0)$ where $\hat{c} = c/\sqrt{\epsilon}$. As $\epsilon \to 0$, the boundary $\sqrt{\epsilon}|x - x_0| = c(t - t_0)$ of this sector tends to the line $t = t_0$, which is a characteristic line for the heat equation. In the limit, therefore, the fundamental solution of (6.224) is expected to be nonzero in the region $t > t_0$, a result appropriate for the heat equation.

Let

$$(6.225) \qquad u(x, t) = \exp\left[-\frac{1}{2\epsilon}(t - t_0)\right] v(x, t)$$

in (6.224) and we obtain

$$(6.226) \qquad \epsilon v_{tt} - c^2 v_{xx} - \frac{1}{4\epsilon} v = 0.$$

At the source point $P_0 = (x_0, t_0)$, (6.225) shows that $u(x_0, t_0) = v(x_0, t_0)$ so that the fundamental solution of (6.226) should yield the fundamental solution of (6.224) by means of (6.225). With $\rho = \epsilon$, $p = c^2$, and $q = -1/4\epsilon$, we obtain the causal fundamental solution of (6.224) from that of (6.208) in the form

$$(6.227)$$

$$u(x, t) = \begin{cases} \dfrac{1}{\sqrt{4c^2\epsilon}} \exp\left[-\dfrac{(t - t_0)}{2\epsilon}\right] I_0\left[\dfrac{1}{\sqrt{4c^2\epsilon}} \sqrt{\dfrac{c^2}{\epsilon}(t - t_0)^2 - (x - x_0)^2}\right]; \\ \qquad\qquad\qquad\qquad\qquad\qquad\qquad |x - x_0| < \dfrac{c}{\sqrt{\epsilon}}(t - t_0) \\ \qquad\qquad\qquad 0; \\ \qquad\qquad\qquad\qquad\qquad\qquad\qquad |x - x_0| > \dfrac{c}{\sqrt{\epsilon}}(t - t_0) \end{cases}$$

where (6.225) was used in connection with the result (6.223). As $z \to \infty$, the modified Bessel function $I_0(z)$ has the asymptotic behavior

$$(6.228) \qquad I_0(z) \approx \frac{1}{\sqrt{2\pi z}} e^z; \qquad z \to \infty.$$

Also, for $t > t_0$ and $0 < \epsilon \ll 1$ we have

(6.229)

$$\sqrt{\frac{c^2}{\epsilon}(t - t_0)^2 - (x - x_0)^2} = \frac{c}{\sqrt{\epsilon}}(t - t_0)\left[1 - \frac{\epsilon}{2c^2}\frac{(x - x_0)^2}{(t - t_0)^2} + O(\epsilon^2)\right],$$

which follows from the binomial expansion. Thus

(6.230)

$$(4c^2\epsilon)^{-1/2}\sqrt{\frac{c^2}{\epsilon}(t - t_0)^2 - (x - x_0)^2} = \frac{t - t_0}{2\epsilon} - \frac{(x - x_0)^2}{4c^2(t - t_0)} + O(\epsilon).$$

As $\epsilon \to 0$ in (6.227), the argument of I_0 tends to infinity. Using (6.228) and (6.230), we easily obtain in the limit as $\epsilon \to 0$,

$$(6.231) \quad u(x, t) = \begin{cases} \dfrac{1}{\sqrt{4\pi c^2(t - t_0)}} \exp\left[-\dfrac{(x - x_0)^2}{4c^2(t - t_0)}\right]; & t > t_0 \\[2mm] 0; & t < t_0, \end{cases}$$

which is identical to the causal fundamental solution of the heat equation given in (6.207).

We conclude our discussion of fundamental solutions and point source problems by noting that Duhamel's principle, introduced in Section 4.5, implies that causal fundamental solutions for parabolic and hyperbolic equations can be obtained as solutions of (equivalent) initial value problems. We have further occasion to discuss and use fundamental solutions later in the text.

6.8. ENERGY INTEGRALS

Until now we have been concerned with the construction of solutions of partial differential equations but not with the question of uniqueness and continuous dependence on the data. As was stated in Section 3.4, existence, uniqueness, and continuous dependence on data are required if an initial and/or boundary value problem is to be well posed. Using the concept of energy integrals, we now briefly discuss the uniqueness and continuous dependence on the data of the solutions of the second order differential equations we have considered in the last three chapters.

We begin with the hyperbolic equation

$$(6.232) \qquad \rho u_{tt} - \nabla \cdot (p\nabla u) + qu = \rho F; \qquad x \in G, \qquad t > 0$$

where G is a region in x-space with the boundary ∂G, $p(x)$ and $\rho(x)$ are positive, and $q(x)$ is nonnegative in G. Initial and boundary data are given as in (4.16) and (4.11). The coefficients and the data for (6.232) are assumed to be such that classical solutions can be constructed. We show that these solutions are unique and depend continuously on the data.

To construct an *energy integral* for (6.232) we multiply across by u_t and integrate over G. Since

$$(6.233) \qquad u_t \nabla \cdot (p\nabla u) = \nabla \cdot (pu_t \nabla u) - p\nabla u \cdot \nabla u_t,$$

we obtain

(6.234)

$$\iint_G u_t [\rho u_{tt} - \nabla \cdot (p\nabla u) + qu] \, dv = \frac{\partial}{\partial t} \iint_G \left[\frac{1}{2}\rho u_t^2 + \frac{1}{2}p(\nabla u)^2 + \frac{1}{2}qu^2 \right] dv$$

$$- \int_{\partial G} pu_t \frac{\partial u}{\partial n} \, ds = \iint_G \rho F u_t \, dv,$$

where we have used the divergence theorem and the fact that the region G and the coefficients in (6.232) are time-independent.

If $\beta = 0$ in the boundary condition $\alpha u + \beta \partial u/\partial n|_{\partial G} = B(x, t)$, the integral

$$(6.235) \qquad E(t) = \frac{1}{2} \iint_G \left[\rho u_t^2 + p(\nabla u)^2 + qu^2 \right] dv$$

represents the sum of the kinetic and potential energies in the (vibrating) system represented by (6.232) at the time t. If $\beta \neq 0$ in the boundary condition $\alpha u + \beta \partial u/\partial n|_{\partial G} = B(x, t)$, the energy of the system is given as

$$(6.236) \quad E(t) = \frac{1}{2} \iint_G \left[\rho u_t^2 + p(\nabla u)^2 + qu^2 \right] dv + \frac{1}{2} \int_{\partial G} p\frac{\alpha}{\beta} u^2 \, ds,$$

where the surface integral term comes from substituting for $\partial u/\partial n$ in (6.234) on using the boundary condition. Note that based on the assumptions on ρ, p, and q, as well as α and β, all of which are nonnegative, each of the integrals (6.235) and (6.236) is nonnegative.

Thus if $\beta = 0$, on using (6.235) in (6.234) we have

$$(6.237) \qquad \frac{dE(t)}{dt} = \iint_G \rho F u_t \, dv + \int_{\partial G} \frac{p}{\alpha} B_t \frac{\partial u}{\partial n} \, ds,$$

whereas if $\beta \neq 0$, we obtain

(6.238) $$\frac{dE(t)}{dt} = \iint\limits_{G} \rho F u_t \, dv + \int\limits_{\partial G} \frac{p}{\beta} B u_t \, ds.$$

The terms on the right in (6.237)–(6.232) represent changes in the energy of the system due to internal and boundary sources.

To prove that the initial and boundary value problem for (6.232) has a unique solution, we show that on assuming that two solutions exist, we conclude that they must be identical. Let $u_1(x, t)$ and $u_2(x, t)$ be two solutions of the initial and boundary value problem for (6.232). Since the problem is linear, the difference $u(x, t) = u_1 - u_2$ must be a solution of the homogeneous form of (6.232), that is, with $F = 0$ and with zero initial and boundary data. Since $F = B = 0$, we find from (6.237)–(6.238) that $E'(t) = 0$. Thus $E(t) = $ constant and energy is conserved. In particular, if $\beta = 0$, we have

(6.239)
$$E(t) = E(0) = \frac{1}{2} \iint\limits_{G} \left[\rho u_t^2(x, 0) + p(\nabla u(x, 0))^2 + q u^2(x, 0) \right] dv = 0,$$

since the initial data $u(x, 0)$ and $u_t(x, 0)$ vanish. Note that $E(t) = $ constant implies that $E(t) = E(0)$. Then for any $t > 0$ we have

(6.240) $$E(t) = \frac{1}{2} \iint\limits_{G} \left[\rho u_t^2 + p(\nabla u)^2 + q u^2 \right] dv = 0.$$

Since ρ, p are positive and q is nonnegative in G, the integrand in (6.240) is nonnegative and we conclude that

(6.241) $$u_t = 0 \quad \text{and} \quad \nabla u = \mathbf{0}.$$

The vanishing of all the first derivatives of $u(x, t)$ implies that $u(x, t) = $ constant. But $u(x, 0) = 0$ so that $u(x, t) = 0$ and we have $u_1(x, t) = u_2(x, t)$. Thus we have shown that the (classical) solutions of the initial and boundary value problem are unique if $\beta = 0$ in the boundary condition. A similar conclusion follows if $\beta > 0$ since the additional surface integral that results in that case is nonnegative. In obtaining (6.241) we have admitted the possibility that $q = 0$. If $q > 0$ in G, we conclude directly from (6.240) that $u(x, t) = 0$.

If the coefficient q in (6.232) is permitted to be negative, this result fails to prove uniqueness since we no longer conclude from the vanishing $E(t)$ in (6.240) that the integrand and, consequently, u_t and ∇u vanish. To obtain uniqueness for this case we employ the following device. Let

(6.242) $$u(x, t) = e^{rt} v(x, t),$$

where r is a positive constant to be specified. Inserting (6.242) into (6.232) yields

(6.243) $\rho v_{tt} + 2r\rho v_t - \nabla \cdot (p\nabla v) + (r^2\rho + q)v = e^{-rt}\rho F.$

We assume that ρ and q are uniformly bounded in the region G with $|q| < M$ where M is some constant. Then we choose r such that $r^2\rho > M$ (recall that $\rho > 0$), and this implies that $r^2\rho + q$, the coefficient of v in (6.243), is positive.

We multiply across by v_t in (6.243) and integrate over G. Proceeding as before, we easily obtain

(6.244) $\dfrac{\partial}{\partial t}\left\{ \dfrac{1}{2} \iint_G \left[\rho v_t^2 + p(\nabla v)^2 + (r^2\rho + q)v^2 \right] dV \right\}$

$$+ 2r \iint_G \rho v_t^2\, dV = \int_{\partial G} \frac{p}{\alpha}(e^{-rt}B)_t \frac{\partial v}{\partial n}\, ds + \iint_G e^{-rt}\rho F v_t\, dV,$$

in the case $\alpha u|_{\partial G} = B$. Let $\tilde{E}(t)$ represent the term in curly brackets in (6.244). Assuming there are two solutions u_1 and u_2 of the initial and boundary value problem for (6.232), we consider the difference $u = u_1 - u_2$. This leads to a homogeneous initial and boundary value problem for the solution of the homogeneous form of (6.243). Then since $F = B = 0$ in (6.244) we have

(6.245) $\tilde{E}'(t) = -2r \iint_G \rho v_t^2\, dV \leqslant 0$

inasmuch as ρ and r are positive. From (6.245) we determine that $\tilde{E}(t)$ is nonincreasing in time. But $\tilde{E}(0) = 0$ since $v = v_t = 0$ at the time $t = 0$, and $\tilde{E}(t) \geqslant 0$ for all t since ρ, p, and $r^2\rho + q$ are positive in G. Since $\tilde{E}(t)$ cannot increase from its value $\tilde{E}(0) = 0$ as t increases, we conclude that $\tilde{E}(t) = 0$. Again the nonnegativity of the integral implies that $v = 0$, so that $v_1 = v_2$ and, consequently, $u_1 = u_2$ and uniqueness is proven.

The energy integral method can also be used to prove continuous dependence of the solution of the initial and boundary value problem for (6.232) on the data. For simplicity we assume that $F = B = 0$; that is, the equation and the boundary data are homogeneous (to consider the case with nonzero F and B we would need additional estimates of the given integrals in F and B). Let $u(x,0) = f(x)$ and $u_t(x,0) = g(x)$. We assume there are two sets of initial data $f_1(x), f_2(x)$ and $g_1(x), g_2(x)$ with corresponding solutions $u_1(x, t)$ and $u_2(x, t)$. The data are assumed to be close to each other in the sense that

(6.246) $\|f_1 - f_2\| < \epsilon_1; \qquad \|\nabla f_1 - \nabla f_2\| < \epsilon_2; \qquad \|g_1 - g_2\| < \epsilon_3$

where the norm $\| \cdots \|$ represents the absolute value of the maximum difference between the indicated functions over the region G and the constants ϵ_1,

ϵ_2, and ϵ_3 are small. We consider the case where $q > 0$ and $\beta = 0$ in the boundary condition. Thus the energy $E(t)$ [i.e., (6.235)] is conserved.

Let $u(x, t) = u_1(x, t) - u_2(x, t)$. The energy integral for u yields

(6.247)
$$E(t) = E(0) = \frac{1}{2} \iint_G \left\{ \rho(g_1 - g_2)^2 + p(\nabla f_1 - \nabla f_2)^2 + q(f_1 - f_2)^2 \right\} dv$$

since, for example, $u_t(x, 0) = g_1(x) - g_2(x)$. Using (6.246) we can bound $E(t)$ in terms of ϵ_1, ϵ_2, and ϵ_3. We write this as

(6.248) $E(t) \leqslant M(\epsilon_1, \epsilon_2, \epsilon_3),$

where $M(\epsilon_1, \epsilon_2, \epsilon_3) \to 0$ as ϵ_1, ϵ_2, and $\epsilon_3 \to 0$. Thus we can make $E(t)$ arbitrarily small, say, less than ϵ, if the ϵ_i $(i = 1, 2, 3)$ are small. This yields

(6.249)
$$E(t) = \frac{1}{2} \iint_G \left[\rho \left(\frac{\partial u_1}{\partial t} - \frac{\partial u_2}{\partial t} \right)^2 + p(\nabla u_1 - \nabla u_2)^2 + q(u_1 - u_2)^2 \right] dv < \epsilon.$$

The positivity of ρ, p, and q implies that each term in the integral is small, so that u_1 and its first derivatives are close to u_2 and its first derivatives. Thus if the data f, ∇f, and g undergo small perturbations, the solutions are only perturbed slightly. This signifies continuous dependence on the data. We remark that if F and B are not zero, slight variations in their values lead to variations of the corresponding solutions that increase linearly in time. However, in a fixed time interval $0 \leqslant t \leqslant T$ the difference between the original and perturbed solutions can be kept uniformly small.

If the domain G is of infinite extent and we are considering the Cauchy problem for (6.232) with initial data in G, we must assume that the data and the solution $u(x, t)$ vanish for sufficiently large $|x|$. Then uniqueness can be proven by the foregoing energy integral approach. The surface integrals are absent and we can conclude that zero data implies that $u = 0$. In the same fashion, continuous dependence on the data can be shown. In addition, if G is a semi-infinite region, say, a half-space, the preceding arguments also carry through.

We have seen that in the absence of interior or boundary source terms F and B, the energy $E(t)$—that is, (6.235) or (6.236)—is conserved if $q \geqslant 0$. However, for (6.243), which is essentially a dissipative hyperbolic equation with $2r\rho v_t$ representing the dissipative effect, the energy $\tilde{E}(t)$ decreases in time as seen from (6.245). That is, the kinetic and potential energy of the system is dissipated as t increases and is converted into another energy form such as heat energy. This is consistent with our results in Chapter 1 relating the telegrapher's equation to the diffusion or heat equation.

We conclude our discussion of the hyperbolic case by noting that more precise results could have been obtained for (6.232) that would have given domains of dependence for both the Cauchy and the initial and boundary value problems for (6.232). For the sake of simplicity we did not present the more complicated analysis required to obtain these sharper results.

Turning now to the parabolic equation

$$(6.250) \qquad \rho u_t - \nabla \cdot (p\nabla u) + qu = \rho F; \qquad x \in G, t > 0$$

with the initial condition $u(x,0) = f(x)$ and the boundary data (4.11), we again assume that ρ, p are positive, q is nonnegative in G, and $u(x,t)$ is a classical solution of the initial and boundary value problem.

To obtain an energy integral for (6.250) we multiply by u and integrate over G. Since

$$(6.251) \qquad u\nabla \cdot (p\nabla u) = \nabla \cdot (pu\nabla u) - p(\nabla u)^2,$$

we obtain on using the divergence theorem

$$(6.252)$$

$$\frac{\partial}{\partial t}\left[\frac{1}{2}\iint_G \rho u^2 \, dv\right] + \iint_G \left[p(\nabla u)^2 + qu^2\right] dv = \int_{\partial G} pu\frac{\partial u}{\partial n}\,ds + \iint_G \rho \, Fu \, dv.$$

If $\beta \neq 0$ in the boundary condition (4.11), we have

$$(6.253) \qquad \int_{\partial G} pu\frac{\partial u}{\partial n}\,ds = -\int_{\partial G}\frac{\alpha}{\beta}pu^2\,ds + \int_{\partial G}\frac{p}{\beta}uB\,ds,$$

and if $\beta = 0$,

$$(6.254) \qquad \int_{\partial G} pu\frac{\partial u}{\partial n}\,ds = \int_{\partial G}\frac{p}{\alpha}B\frac{\partial u}{\partial n}\,ds.$$

We define the energy term for (6.250) as

$$(6.255) \qquad E(t) = \tfrac{1}{2}\iint_G \rho u^2\,dv.$$

Assuming there are two solutions u_1 and u_2 for the initial and boundary value problem for (6.250), the difference $u = u_1 - u_2$ satisfies the homogeneous form of (6.250) with homogeneous initial and boundary data. Thus we obtain from (6.252) in the case $\beta \neq 0$,

$$(6.256) \qquad E'(t) = -\iint_G \left[p(\nabla u)^2 + qu^2\right] dv - \int_{\partial G}\frac{\alpha}{\beta}pu^2\,ds \leqslant 0,$$

because p, q, ρ, α, and β are nonnegative. Since $E(t) \geqslant 0$, $E(0) = \tfrac{1}{2}\int\int_G \rho u(x,0)^2\,dv = 0$, and $E(t)$ is nonincreasing in view of (6.256), we conclude that $E(t) = 0$ for all t. Thus $u(x, t) = 0$ and $u_1 = u_2$ so that uniqueness is established. Continuous dependence on data follows by an argument similar to that in the hyperbolic case. If the region G is of infinite extent and $u(x, t)$ vanishes as $|x| \to \infty$, we can also prove uniqueness and continuous dependence on the data.

Again if $q < 0$, this method fails but if q is uniformly bounded in G, the transformation (6.242) reduces (6.250) to a form for which this approach proves uniqueness.

The equation (6.256) shows that the energy $E(t)$ is not conserved and this is the case even if $u = 0$ on the boundary and $q = 0$. There is, however, another conserved quantity if $q = 0$ and $\partial u/\partial n$ vanishes on the boundary (i.e., there is no flux of u through the boundary). With $F = q = 0$ in (6.250) we obtain on integrating the equation over G,

(6.257)

$$\iint_G [\rho u_t - \nabla \cdot (p\nabla u)] \, dv = \frac{\partial}{\partial t} \iint_G \rho u \, dv - \int_{\partial G} p \frac{\partial u}{\partial n} \, ds = \frac{\partial}{\partial t} \iint_G \rho u \, dv = 0.$$

Thus the quantity $\iint_G \rho u \, dv$ is conserved and if, initially, $u(x, 0) = f(x)$, we have

(6.258)
$$\iint_G \rho u(x, t) \, dv = \iint_G \rho f(x) \, dv.$$

The integrals in (6.258) represent the total amount of heat, say, in the region G. Even if G is of infinite extent and $u(x, t)$ vanishes as $|x| \to \infty$, we still obtain (6.258).

For the elliptic equation

(6.259)
$$-\nabla \cdot (p\nabla u) + qu = \rho F; \qquad x \in G$$

with the boundary condition (4.11), we multiply across by $u(x)$ and integrate over G. This gives the energy integral, on using the divergence theorem and (6.251),

(6.260)
$$\iint_G [p(\nabla u)^2 + qu^2] \, dv = \int_{\partial G} pu \frac{\partial u}{\partial n} \, ds + \iint_G \rho F u \, dv.$$

If $\beta \neq 0$ in the boundary condition, the energy term is

(6.261)
$$E = \iint_G [p(\nabla u)^2 + qu^2] \, dv + \int_{\partial G} \frac{\alpha}{\beta} pu^2 \, ds,$$

with the same expression for the energy, with the surface integral absent, if $\beta = 0$. Proceeding as before, we easily prove uniqueness of classical solutions if $\alpha \neq 0$ in the boundary condition $\alpha u + \beta \partial u/\partial n|_{\partial G} = B$. However, if α and q

are zero, the solution is unique only up to an additive constant since $u =$ constant is a solution of $\nabla \cdot (p\nabla u) = \rho F$ with the boundary condition $\beta \partial u/\partial n|_{\partial G} = B$. Furthermore, uniqueness can only be demonstrated if $q > 0$ in (6.259). If $q < 0$, the energy integral method cannot be used to prove uniqueness. In fact, as shown in Chapter 8, the solution need not be unique in that case.

The uniqueness theorem can also be extended to exterior boundary value problems for (6.259) where G is a semi-infinite region or a region of infinite extent outside a closed and bounded region with the conditions (4.11) assigned on the finite boundaries. The solution is required to satisfy appropriate conditions at infinity. We do not discuss this problem. Also, the continuous dependence on the data for solutions of the boundary value problem for (6.259) can be proven by the energy integral method.

The energy integral method can be applied to systems of equations. It can also be used to prove existence of solutions, not only uniqueness and continuous dependence on data, but we do not consider this here.

To conclude our discussion, we note that for parabolic and elliptic equations there exist *maximum principles* whereby uniqueness and continuous dependence on data can be proven. The appropriate maximum principles are considered in Chapter 8.

EXERCISES FOR CHAPTER 6

Section 6.1

6.1.1. Derive the one-dimensional form of (6.1) from the integral relation (6.8) by using mean value theorems for derivatives and integrals, assuming the functions are sufficiently smooth.

6.1.2. Obtain the integral relation (6.9) from the parabolic equation (4.4).

6.1.3. Obtain the integral relation (6.10) from the parabolic equation (4.4).

6.1.4. Specialize (6.10) to the case of one space dimension and derive the (one-dimensional) parabolic equation (4.5) from it using appropriate mean value theorems.

6.1.5. Derive the integral relation (6.12) from the elliptic equation (6.11).

6.1.6. Consider a two-dimensional form of (6.12) and let R be the rectangle $x_0 < x < x_1$ and $y_0 < y < y_1$. Assuming the functions are smooth, derive (6.11) from (6.12) by using appropriate mean value theorems.

Section 6.2

6.2.1. Formulate, in the manner of (6.17)–(6.21), an initial and boundary value problem for the parabolic equations (4.4) and (6.9) if S_0 is a discontinuity region.

6.2.2. Formulate, in the manner of (6.17)–(6.21), a boundary value problem for the elliptic equations (6.11)–(6.12) if S_0 is a discontinuity region.

6.2.3. (a) Show that if $p_2 \approx 0$ in Example 6.1, then $R \approx 1$ and $T \approx 2$ in (6.34). Demonstrate that the (approximate) solution to this problem (i.e., with $p_2 \approx 0$) can be obtained by solving for $u_1(x, t)$ in $x < 0$ with the boundary condition $\partial u_1(0, t)/\partial x = 0$, and then obtaining $u_2(x, t)$ in $x > 0$, by using the boundary condition $u_2(0, t) = u_1(0, t)$.
(b) Show that for $p_2 \to \infty$ we have $R \to -1$ and $T \to 0$ so that there is no transmitted wave. Introduce the assumption that $p_2 \gg p_1$ into the matching condition for Example 6.1, set up (approximate) boundary values for the two regions $x < 0$ and $x > 0$, and determine that their solutions are consistent with the results obtained from (6.32)–(6.33) for large p_2.

6.2.4. Consider the problem of heat conduction in an infinite composite rod composed of two homogeneous rods connected at $x = 0$. Put $u = u_1(x, t)$ for $x < 0$ and $u = u_2(x, t)$ for $x > 0$ and assume they satisfy the equations:

$$\rho_1 \frac{\partial u_1}{\partial t} - p_1 \frac{\partial^2 u_1}{\partial x^2} = 0, \qquad -\infty < x < 0, t > 0$$

$$\rho_2 \frac{\partial u_2}{\partial t} - p_2 \frac{\partial^2 u_2}{\partial x^2} = 0, \qquad 0 < x < \infty, t > 0.$$

Let $u_1(x, 0) = A$ and $u_2(x, 0) = B$ where A and B are constants and apply the matching conditions (6.44) at $x_0 = 0$. Determine the temperature $u(x, t)$ for $t > 0$. [Hint: Let

$$u_1(x, t) = \alpha_1 + \beta_1 \Phi\left[-\frac{x}{2c_1\sqrt{t}}\right]; \qquad u_2(x, t) = \alpha_2 + \beta_2 \Phi\left[\frac{x}{2c_2\sqrt{t}}\right]$$

where α_i and β_i are constants and $c_i^2 = p_i/\rho_i$, for $i = 1, 2$. The function $\Phi(z)$ is the error integral defined in (5.105)].

6.2.5. Reconsider the problem of Exercise 6.2.4 and replace the constant initial data by the arbitrary initial conditions $u_1(x, 0) = f_1(x)$ and $u_2(x, 0) = f_2(x)$. Solve the respective problems for $u_1(x, t)$ and $u_2(x, t)$ by assuming that

$$\rho_1 \frac{\partial u_1(0, t)}{\partial x} = \rho_2 \frac{\partial u_2(0, t)}{\partial x} = g(t)$$

is a known function. The solution of the initial and boundary value problems for u_1 and u_2 may then be obtained from the result (5.115). Use the matching condition $u_1(0, t) = u_2(0, t)$ to show that $g(t)$ is the solution of the *Abel integral equation*

$$G(t) = \int_0^t \frac{g(\tau)}{\sqrt{t - \tau}} d\tau$$

where $G(t)$ is a known function. The solution of this equation is

$$g(t) = \frac{1}{\pi} \frac{d}{dt} \int_0^t \frac{G(\tau)}{\sqrt{t - \tau}} d\tau.$$

6.2.6. Show that the solution of the problem in Exercise 6.2.4 may be obtained by applying the Laplace transform in the time variable.

6.2.7. Adapt the discussion of Example 4.2 to show that the operator L associated with the eigenvalue problem (6.35)–(6.37) is a positive operator. Conclude that the eigenvalues for the problem are nonnegative.

6.2.8. Discuss the eigenvalue problem (6.45)–(6.47) if the boundary conditions (6.46) are replaced by $v_1'(0) = v_2'(l) = 0$. Obtain an equation for the determination of the eigenvalues and obtain an orthonormal set of eigenfunctions. Show that $\lambda = 0$ is an eigenvalue and $v(x) = 1$ is an eigenfunction for this problem.

6.2.9. Expand the function $f(x) = 1$ in a series of the eigenfunctions $v^{(k)}(x)$ given in (6.51) and use this result to obtain a solution of the problem (6.41)–(6.44) if $u(x, 0) = 1$.

6.2.10. Consider the initial and boundary value problem for the hyperbolic equations

$$\rho_1 \frac{\partial^2 u_1}{\partial t^2} - p_1 \frac{\partial^2 u_1}{\partial x^2} = 0, \qquad 0 < x < x_0, t > 0$$

$$\rho_2 \frac{\partial^2 u_2}{\partial t^2} - p_2 \frac{\partial^2 u_2}{\partial x^2} = 0, \qquad x_0 < x < l, t > 0$$

in a composite medium. Let

$$u(x, 0) = f(x); \qquad u_t(x, 0) = g(x), \qquad 0 < x < l$$

where $u = u_1$ for $x < x_0$ and $u = u_2$ for $x > x_0$. Assuming u_1 and u_2 satisfy the boundary conditions (6.42) and the matching conditions (6.44), apply separation of variables and show how the solution can be expressed in terms of the eigenfunctions obtained in Example 6.2.

6.2.11. Develop a finite Fourier transform approach for the eigenfunctions obtained in Example 6.2 for the purpose of solving the problem (6.41)–(6.44) if the equation and the boundary conditions are inhomogeneous.

6.2.12. Show how Laplace transforms can be used to solve the problem (6.41)–(6.44).

6.2.13. Let $u_1(x, y)$ and $u_2(x, y)$ satisfy the Helmholtz equations

$$\nabla^2 u_1 + k_1^2 u_1 = 0, \qquad -\infty < y < \infty, \, x < 0,$$
$$\nabla^2 u_2 + k_2^2 u_2 = 0, \qquad -\infty < y < \infty, \, x > 0$$

with k_1 and k_2 as constants and let u_1 and u_1 be related across $x = 0$ by the matching conditions

$$u_1(0, y) = u_2(0, y); \qquad \frac{\partial u_1(0, y)}{\partial x} = \frac{\partial u_2(0, y)}{\partial x}.$$

Represent u_1 and u_2 as

$$u_1(x, y) = \exp[ik_1(x \cos \theta + y \sin \theta)]$$
$$+ R \exp[ik_1(x \cos \hat{\theta} + y \sin \hat{\theta})]$$
$$u_2(x, y) = T \exp[ik_2(x \cos \phi + y \sin \phi)]$$

where $-\pi < \theta$, $\phi < \pi$, and $\pi < \hat{\theta} < 2\pi$. The angle θ is assumed to be specified. Determine the constants $\hat{\theta}$, ϕ, R, and T such that the Helmholtz equations and the matching conditions are satisfied. This problem characterizes the scattering of a plane wave at an interface. A general discussion of scattering problems is given in Chapter 9. The restrictions on the angles $\hat{\theta}$ and ϕ are required to guarantee a unique solution to this problem and to correspond to physically motivated *radiation conditions*. The relationships between the angle of incidence θ and the angles of reflection and refraction $\hat{\theta}$ and ϕ are known as *Snell's laws*.

Section 6.3

6.3.1. Show that for the one-dimensional wave equation (6.65) the jump conditions (6.58)–(6.59) reduce to the single condition

$$c[u_x] \pm [u_t] = 0$$

across the characteristics $\phi = x \mp ct = $ constant.

6.3.2. Determine that the function

$$u(x, t) = \begin{cases} f(x - ct), & x - ct < 0 \\ 0, & x - ct > 0 \end{cases}$$

with $f(0) = 0$ and $f'(0) \neq 0$ satisfies the jump conditions across the characteristic $\phi = x - ct = 0$ for the wave equation given in Exercise 6.3.1.

6.3.3. Show that

$$u(x, t) = \begin{cases} e^{-t}f(x - t), & x - t < 0 \\ 0, & x - t > 0 \end{cases}$$

where f is a smooth function with $f(0) = 0$ and $f'(0) \neq 0$ is a solution of the hyperbolic equation

$$u_{tt} - u_{xx} + 2u_t + u = 0$$

with discontinuous first derivatives across $x - t = 0$. Verify that the jump conditions across the characteristic $x - t = 0$ are satisfied.

6.3.4. Determine that the function

$$u(x, y, t) = \begin{cases} \sin\phi, & \phi < 0 \\ 0, & \phi > 0 \end{cases}$$

where $\phi = x\cos\theta + y\sin\theta - ct$ with $\theta = $ constant is a solution of the two-dimensional wave equation

$$u_{tt} - c^2[u_{xx} + u_{yy}] = 0,$$

whose derivatives are discontinuous across the characteristic $\phi = 0$. Show that the jump conditions at $\phi = 0$ are satisfied.

6.3.5. Show that

$$u(x, y, z, t) = \begin{cases} \dfrac{1}{r}\sinh(r - ct), & r - ct < 0 \\ 0, & r - ct > 0 \end{cases}$$

where $r^2 = x^2 + y^2 + z^2$ satisfies the three-dimensional wave equation

$$u_{tt} - c^2\nabla^2 u = 0$$

and that its first derivatives satisfy the jump conditions across the characteristic $r - ct = 0$.

6.3.6. Show that if we have $\phi_t = 0$ in (6.62), we conclude from (6.59)–(6.61) that $[u_t] = 0$ across the surface $\phi(x, y, z) = $ constant, and find from (6.58) that the normal derivative of u must be continuous across $\phi = $ constant. Therefore, both u and its first derivatives must be continuous across $\phi = $ constant.

6.3.7. Obtain (6.64) from the integral relation (6.9) in the parabolic case.

6.3.8. Show that the solution of the initial and boundary value problem for the heat equation,

$$u_t - c^2 u_{xx} = 0, \qquad 0 < x < \infty, t > 0$$
$$u(x, 0) = 0, \qquad x > 0$$
$$u(0, t) = \begin{cases} 0, & 0 < t < t_0 \\ 1, & t_0 < t \end{cases}$$

has the form

$$u(x, t) = \begin{cases} 0, & 0 < x < \infty, 0 < t < t_0 \\ 1 - \Phi\left[\dfrac{x}{2c\sqrt{t - t_0}}\right], & 0 < x < \infty, t_0 < t \end{cases}$$

where $\Phi(z)$ is the error integral defined in (5.105). Verify that even though $u(0, t)$ has a discontinuity at $t = t_0$, $u(x, t)$ and $u_x(x, t)$ have a zero jump across $t = t_0$ for $x > 0$. Note that $t = t_0$ is a characteristic for the heat equation.

Section 6.4

6.4.1. Show that

$$u(x, y, t) = \begin{cases} 0, & x - ct < 0 \\ 1, & x - ct > 0 \end{cases}$$

is a weak solution of the two-dimensional wave equation.

$$u_{tt} - c^2(u_{xx} + u_{yy}) = 0$$

[Hint: Use (6.75) and apply it to regions on both sides of the characteristic surface $\phi = x - ct = 0$. Show that the boundary integral over the surface $\phi = 0$, which results from the integration over the region where $\phi > 0$, vanishes by showing that the terms in v can be expressed as a directional derivative.]

6.4.2. Verify that

$$u(x, t) = \begin{cases} \cos(x + ct), & x + ct < 0 \\ \sin(x + ct), & x + ct < 0 \end{cases}$$

is a weak solution of the one-dimensional wave equation.

6.4.3. Obtain the integral (6.80) that characterizes weak solutions for the parabolic equation (6.79).

6.4.4. Using the convergence properties of Fourier sine series discussed in Chapter 4, set up an initial and boundary value problem for the one-dimensional homogeneous wave equation in the finite interval $0 < x < l$ with $u(0, t) = u(l, t) = 0$ such that its separation of variables result is a weak solution and not a classical solution. That is, choose the initial data $f(x)$ and $g(x)$ such that the conditions given in Example 6.5 are met, but the series solution is not twice differentiable term by term.

6.4.5. Show, by using separation of variables, that the solution of the steady state problem for the heat equation

$$u_t - c^2 u_{xx} = 0, \qquad -\infty < t < \infty, x > 0$$

with the boundary condition

$$u(0, t) = \begin{cases} \sin \omega t, & t > 0 \\ 0, & t < 0 \end{cases}$$

where ω = constant, is given as

$$u(x, t) = \begin{cases} \exp\left[-\sqrt{\dfrac{\omega}{2c^2}}\, x\right] \sin\left[-\sqrt{\dfrac{\omega}{2c^2}}\, x + \omega t\right]; & t > 0 \\ 0; & t < 0 \end{cases}$$

if we require that $u(x, t) \to 0$ as $x \to \infty$. Verify that $u(x, t)$ is a weak solution of the heat equation.

Section 6.5

6.5.1. Use Taylor's series to show that a solution of the difference equation (6.100) must be a solution of the wave equation (6.83) in view of the arbitrariness of h and k.

6.5.2. Show that the general solution of the wave equation (6.83), that is,

$$u(x, t) = F(x - ct) + G(x + ct),$$

is also a solution of the difference equation (6.100).

6.5.3. Obtain the solution $u(x, t)$ for each of the formulas (6.114)–(6.116) if $H(t) = \sin \omega t$. Verify that in each case the solution obtained satisfies the boundary condition.

6.5.4. Let $f(x) = 1$ and $g = \beta = H_1 = H_2 = F = 0$ in the problem (6.117)–(6.119). Use the method of Example 6.7 to obtain the solution $u(x, t)$ of this problem in the regions $R_1, R_2, \ldots, R_7$ indicated in Figure 6.11.

6.5.5. Show that the problem

$$\begin{aligned} u_{tt} - u_{xx} &= 0, & 2t < x < \infty, t > 0 \\ u(x, 0) &= x, & u_t(x, 0) = 0; & 0 < x < \infty \\ u(2t, t) &= 1, & t > 0 \end{aligned}$$

has no solution.

6.5.6. Solve the initial and (moving) boundary value problem (6.120)–(6.122) if $f(x) = g(x) = 0$, $h(t) = -c_0 t$, $0 < c_0 < c$, and $H(t) = A \cos \omega t$. Obtain the Doppler effect and show that the solution oscillates at a frequency below the input frequency ω.

6.5.7. Solve the characteristic initial value problem (6.133)–(6.134) if $F(x, t) = \cos \omega t$, $H_1(t) = e^{-t}$, and $H_2(t) = 1$.

6.5.8. Let $x = h_1(t) = 2t$ be a space-like curve on which the data (6.130) where $f(t) = \sin t$ and $g(t) = 1$ are given. Solve the Cauchy problem for the wave equation $u_{tt} = u_{xx}$ in the region $2t - x > 0$ with the Cauchy data on $x = 2t$ given as in the foregoing.

6.5.9. Solve the *Goursat problem* for the wave equation $u_{tt} = c^2 u_{xx}$ in the region $0 < x < ct$, $t > 0$ with the data

$$u(0, t) = g(t); \qquad u(ct, t) = f(t); \qquad t > 0$$

assuming that $f(0) = g(0)$. (A problem in which the data are given on a characteristic curve and on an intersecting time-like curve is called a Goursat problem.)

6.5.10. Solve the characteristic initial value problem for the hyperbolic equation

$$u_{tt} - u_{xx} + 2u_t + u = 0, \qquad -t < x < t, t > 0$$

with the data

$$u(t, t) = e^{-t}, \qquad u(-t, t) = e^t, \qquad t > 0,$$

by using the iteration procedure given in (6.137)–(6.139). [Hint: The solution is $u(x, t) = e^{-x}$.]

6.5.11. Show that the initial and (moving) boundary value problem for the heat equation

$$u_t - c^2 u_{xx} = 0, \qquad c_0 t < x < \infty, t > 0$$
$$u(x, 0) = f(x), \qquad 0 < x < \infty$$
$$u(c_0 t, t) = g(t), \qquad t > 0$$

can be solved by introducing the moving coordinate system

$$\sigma = x - c_0 t, \qquad \tau = t.$$

In the new variables σ and τ we obtain a problem for $\hat{u}(\sigma, \tau)$ in the region $\sigma > 0$, $\tau > 0$ with data given at $\sigma = 0$ and $\tau = 0$. Introduce a change of the dependent variable that yields an initial and boundary value problem for a heat equation with data given at $\sigma = 0$ and $\tau = 0$.

6.5.12. Solve the problem in Exercise 6.5.11 if $c_0 = 1$, $f(x) = 1$, and $g(t) = e^{-t}$.

Section 6.6

6.6.1. Adapt the method presented in (4.71)–(4.72) to determine that the orthogonality condition for the eigenvalue problem (6.153)–(6.156) is given as

$$\int_0^l M_j(x) M_k(x) \rho(x) \, dx + \sum_{i=1}^n m_i M_j(x_i) M_k(x_i) = 0$$

where M_j and M_k correspond to the eigenvalues λ_j and λ_k, with $\lambda_j \neq \lambda_k$. Conclude that the appropriate inner product associated with this eigenvalue problem is

$$(\phi, \psi) = \int_0^l \phi(x)\psi(x)\rho(x)\,dx + \sum_{i=1}^n m_i \phi(x_i)\psi(x_i).$$

The (induced) norm is given as

$$\|\phi\|^2 = (\phi, \phi) = \int_0^l \phi^2 \rho\,dx + \sum_{i=1}^n m_i [\phi(x_i)]^2$$

Determine a formula for the Fourier coefficients c_k in an expansion of a function $f(x)$ in a series of the eigenfunctions $M_k(x)$.

6.6.2. Let $\rho(x) = p(x) = 1$, $l = \pi$, $n = 1$ and $x_1 = 1$ in the problem (6.153)–(6.156). Determine an equation for the eigenvalues and obtain the corresponding eigenfunctions.

6.6.3. Expand $f(x) = 1$ in a series of eigenfunctions determined in Exercise 6.6.2.

6.6.4. Assume that the load in the vibrating string problem of Example 6.9 is placed at the end of the string at $x = l$. The boundary condition $u(l, t) = 0$ is then replaced by

$$m_1 u_{tt}(l, t) = -p(l)u_x(l, t).$$

Show why this follows from (6.150). Use separation of variables to obtain the eigenvalue problem for this case that replaces the one given in Example 6.9.

6.6.5. Let $p = \rho = 1$ in the problem of Exercise 6.6.4. Show that the eigenvalues for the resulting eigenvalue problem are determined from the equation

$$\cot[l\sqrt{\lambda}] = m_1\sqrt{\lambda}.$$

Demonstrate graphically that there are infinitely many eigenvalues $\lambda = \lambda_k$. Show that the eigenfunctions $M_k(x)$ are given as

$$M_k(x) = \frac{\sin\left(\sqrt{\lambda_k}\,x\right)}{\sin\left(\sqrt{\lambda_k}\,l\right)}, \qquad k = 1, 2, \ldots$$

and that they are orthogonal with respect to the inner product of Exercise 6.6.1 where $n = 1$ and $x_1 = l$. Calculate the norm of these eigenfunctions.

6.6.6. Obtain the limit of the solutions (6.163) as $c_0 \to 0$ and discuss the resulting problem and its solution in that case.

6.6.7. Use the matching conditions (6.157) to solve the (heat) equation

$$\rho u_t - p u_{xx} = 0, \qquad -\infty < x < \infty, t > 0, x \neq 0$$

where ρ and p are constants, if $u(x, 0) = 0$ and $h(t) = 0$ in (6.157). Assume that $f_0(t) = e^{-t}$ and apply the Laplace transform.

Section 6.7

6.7.1. Use (6.179) to determine that the constants c in (6.184) are given as in (6.185).

6.7.2. Show that (6.202) is a weak solution of the wave equation.

6.7.3. Show that if $u(x, t)$ is a solution of the heat equation $u_t = c^2 u_{xx}$, so is the function $u(ax, a^2 t)$ where a is a constant. Put $a = 1/\sqrt{t}$ and show that the heat equation has a *similarity solution* of the form $u = F(x/\sqrt{t})$ where $F(z)$ satisfies the equation

$$F''(z) + \frac{z}{2c^2} F'(z) = 0$$

and that $F'(z) = \tilde{a} \exp[-z^2/4c^2]$. Noting that u_x is a solution of the heat equation if $u(x, t)$ is a solution, conclude that

$$u_x(x, t) = \frac{\partial}{\partial x}\left(F\left(\frac{x}{\sqrt{t}}\right)\right) = \frac{1}{\sqrt{t}} F'\left(\frac{x}{\sqrt{t}}\right) = \frac{\tilde{a}}{\sqrt{t}} \exp\left[-\frac{x^2}{4c^2 t}\right]$$

(with $\tilde{a} = 1/\sqrt{4\pi c^2}$ this is the fundamental solution of the heat equation).

6.7.4. Let $r^2 = x^2 + y^2$ and $u(r, t)$ be a solution of the heat equation in two dimensions. Show that $u(ar, a^2 t)$ is also a solution and construct a *similarity solution* of the form $u = F(r/\sqrt{t})$. Noting that $u_t(r, t)$ is also a solution of the heat equation if $u(r, t)$ is a solution, conclude that

$$u_t = \frac{\partial}{\partial t}\left(F\left(\frac{r}{\sqrt{t}}\right)\right) = \frac{\tilde{a}}{t} \exp\left[-\frac{r^2}{4c^2 t}\right]$$

satisfies the heat equation $u_t = c^2 \nabla^2 u$ (with $\tilde{a} = 1/4\pi c^2$ this is the fundamental solution of the heat equation).

6.7.5. Let $r^2 = x^2 + y^2 + z^2$ and show that the heat equation $u_t = c^2 \nabla^2 u$ in three dimensions has a *similarity solution* of the form $u = F(r/\sqrt{t})$. Show that

$$u_{tt} = \frac{\partial^2}{\partial t^2}\left(F\left(\frac{r}{\sqrt{t}}\right)\right) = \frac{\tilde{a}}{t^{3/2}} \exp\left[-\frac{r^2}{4c^2 t}\right]$$

satisfies the heat equation. With $\tilde{a} = (1/\sqrt{4\pi c^2})^3$ this is the fundamental solution of the heat equation in three dimensions.

6.7.6. Verify that (6.208) reduces to (6.211) if $u = \tilde{u}(r)$.

6.7.7. Obtain the solution of the (continuous) point source problem for the two-dimensional heat equation discussed in the text, by constructing a superposition of appropriately modified fundamental solutions as given in Exercise 6.7.4.

6.7.8. Proceeding as in Exercise 6.7.7. and using Exercise 6.7.5, solve the (continuous) point source problem for the three-dimensional heat equation.

Section 6.8

6.8.1. Use the energy integral method to prove uniqueness for the solution of the Cauchy problem for the hyperbolic equation (6.232).

6.8.2. Construct the appropriate energy integral for the one-dimensional form

$$\rho u_{tt} - (pu_x)_x + qu = \rho F$$

of the hyperbolic equation (6.232) and prove uniqueness and continuous dependence on the data for the initial and boundary value problem for this equation.

6.8.3. Prove that the solution of the initial and boundary value problem for the parabolic equation (6.250) is continuously dependent on the data.

6.8.4. Determine conditions at infinity that would guarantee uniqueness for the exterior Dirichlet problem for Laplace's and Poisson's equations on the basis of the energy integral method. (Hint: Apply the energy integral method to the region bounded by the curve or surface where the data are assigned and by a circle or sphere that completely contains the boundary on which that data are given. Allow the radius of the circle or sphere to tend to infinity and obtain conditions on the solution that make the integral over the circle or sphere tend to zero as the radius tends to infinity.)

6.8.5. Show that $u(x, y) = \log[\sqrt{x^2 + y^2}/a]$ and $u = 0$ are both solutions of $u_{xx} + u_{yy} = 0$ for $x^2 + y^2 > a$ with $u = 0$ on $x^2 + y^2 = a^2$. Use the conditions at infinity determined in Exercise 6.8.4 to eliminate one of these solutions and to obtain uniqueness for the foregoing exterior Dirichlet problem.

6.8.6. Show that $u(x, y, z) = a/\sqrt{x^2 + y^2 + z^2}$ and $u = 1$ are both solutions of the exterior Dirichlet problem for $u_{xx} + u_{yy} + u_{zz} = 0$ in the region $x^2 + y^2 + z^2 > a^2$ with the boundary condition $u = 1$ on $x^2 + y^2 + z^2 = a^2$. Use the conditions at infinity obtained in Exercise 6.8.4 to eliminate one of the above solutions so that this problem has a unique solution.

6.8.7. Develop an energy integral approach for the first order linear equation

$$u_t + au_x + bu = 0, \qquad -\infty < x < \infty, t > 0$$

by multiplying across by u, using the identity $auu_x = \left(\frac{1}{2}au^2\right)_x - \frac{1}{2}a_x u^2$ and integrating with respect to x from $-\infty$ to $+\infty$. Assuming all coefficients have the necessary behavior at $\pm\infty$ and using the change of variable (6.242), show how the energy integral can be used to prove uniqueness for the Cauchy problem for the given equation.

6.8.8. Consider the first order hyperbolic system with constant coefficients

$$\mathbf{u}_t + A\mathbf{u}_x + B\mathbf{u} = 0$$

where A is a real symmetric matrix. Multiply across by the vector $\mathbf{u}^T$ and integrate from $-\infty$ to $+\infty$ with respect to x. Assuming that $\mathbf{u}(x, t)$ vanishes as $|x| \to \infty$, show that the energy integral $E(t) = \int_{-\infty}^{\infty} |\mathbf{u}|^2 \, dx \leqslant 0$ if B is a nonnegative matrix with the property that $\mathbf{u}^T B\mathbf{u} \geqslant 0$ for all vectors $\mathbf{u}$. If B is not nonnegative, the transformation $\mathbf{u} = e^{rt}\mathbf{v}$ with $r > 0$ yields a system for $\mathbf{v}$ in which the coefficient $\hat{B}$ of $\mathbf{v}$ can be made nonnegative for an appropriate choice of r. Use the energy integral $E(t)$ to prove uniqueness for the Cauchy problem for the given system.

6.8.9. Use the method of Exercise 6.8.8 to show that the Cauchy problem for the system (2.2), which is equivalent to the wave equation, has a unique solution.

6.8.10. Apply the method of Exercise 6.8.8 to prove uniqueness for the solution of the Cauchy problem for the system (1.73)–(1.74), which is equivalent to the telegrapher's equation.

6.8.11. Use an energy integral to show that the initial and boundary value problem for the equation of a vibrating rod (see Section 8.5)

$$u_{tt} + u_{xxxx} = 0, \qquad 0 < x < l, t > 0$$

with $u(x, 0)$ and $u_t(x, 0)$ specified and $u(x, t)$ and $u_x(x, t)$ given at $x = 0$ and $x = l$, has a unique solution. [Hint: Use the following identity:

$$u_t u_{xxxx} = \left(u_t u_{xxx}\right)_x - \left(u_{tx} u_{xx}\right)_x + \frac{1}{2}\left(u_{xx}^2\right)_t.]$$

6.8.12. Show that the boundary value problem for the biharmonic equation (see Section 8.5)

$$\nabla^2 \nabla^2 u = 0; \qquad u = u(x, y)$$

in a bounded region R, with u and $\partial u / \partial n$ specified on ∂R has a unique solution by the use of an appropriate energy integral. [Hint:

$$u\nabla^2\nabla^2 u = u\nabla \cdot \left(\nabla\nabla^2 u\right) = \nabla \cdot \left(u\nabla\nabla^2 u\right)$$
$$- \nabla \cdot \left(\nabla u\nabla^2 u\right) + \left(\nabla^2 u\right)^2.$$

Deduce that if $u = \partial u / \partial n = 0$ on ∂R, we must have $\nabla^2 u = 0$, and conclude from the uniqueness theorem for Laplace's equation that $u = 0$.]

6.8.13. Combine the methods of Exercises 6.8.11 and 6.8.12 to develop an energy integral that yields uniqueness for the initial and boundary value problem for the equation of a vibrating plate (see Section 8.5)

$$u_{tt} + c^2 \nabla^2 \nabla^2 u = 0; \qquad u = u(x, y, t)$$

in a bounded region R where $u(x, y, 0)$ and $u_t(x, y, 0)$ are prescribed and u together with $\partial u / \partial n$ is specified on ∂R.

7

GREEN'S FUNCTIONS

The method of *Green's functions* is an important technique for solving boundary value, initial and boundary value, and Cauchy problems for partial differential equations. It is most commonly identified with the solution of boundary value problems for Laplace's equation and a Green's function has already been introduced in that context in Chapter 1. In Example 1.6 it was shown by way of an application of Green's second theorem that the Dirichlet or first boundary value problem for Laplace's equation can be solved if the Green's function for that problem can be determined. It was also seen in Section 1.3 and in our study of point source problems in Section 6.7, that the Green's function is often worthwhile determining in its own right rather than as a tool to be used only for solving another problem.

In this chapter we begin by constructing generalizations of Green's second theorem that are appropriate for the second order differential equations introduced in Chapter 4. These integral theorems are then used to show how boundary value, initial and boundary value, and Cauchy problems can be solved in terms of appropriately defined Green's functions for each of these problems. Even though the construction of Green's functions requires that a problem similar to the original (given) problem must be solved, it is often easier to solve the Green's function problem in a number of important cases as we shall see. In this regard the fundamental solutions, of which Green's functions are a special case, considered in Section 6.7 play an important role. Since the determination and use of Green's functions require the use of generalized functions such as the Dirac delta function, a brief discussion of the theory of generalized functions will be given in this chapter. Most of the chapter, however, is devoted to the construction and use of Green's functions for problems involving equations of elliptic, hyperbolic, and parabolic types.

7.1. INTEGRAL THEOREMS AND GREEN'S FUNCTIONS

In this section we construct integral theorems appropriate for the elliptic, hyperbolic, and parabolic equations introduced in Section 4.1. Each of these theorems follows from an application of the divergence theorem and represents a generalization of Green's second theorem. These theorems form the basis for the construction of the Green's functions we consider in this chapter. Technically, the theorems are valid only if the functions occurring in the integrals are sufficiently smooth and, as we have seen in Section 6.7, this is generally not the case for Green's functions. Nevertheless, we shall assume these theorems are formally valid in all cases and rely on the theory of generalized functions presented in Section 7.2 to form a basis for their validity, even though this is not demonstrated. We begin our discussion with problems in two or three space dimensions and present the one-dimensional results at the end of this section.

We start with the elliptic equation

$$(7.1) \qquad Lu = -\nabla \cdot (p\nabla u) + qu = \rho F$$

in two or three dimensions given over a bounded region G with the boundary conditions

$$(7.2) \qquad \alpha u + \beta \left. \frac{\partial u}{\partial n} \right|_{\partial G} = B.$$

The conditions on the coefficients in (7.1) and (7.2) given in Section 4.1 are assumed to remain in effect. Introducing a function $w(x)$ whose properties are to be specified and proceeding as in Example 4.2, we obtain

$$(7.3) \qquad \iint_G (wLu - uLw)\, dv = -\iint_G \nabla \cdot (pw\nabla u - pu\nabla w)\, dv$$

$$= -\int_{\partial G} p(w\nabla u - u\nabla w) \cdot \mathbf{n}\, ds$$

$$= \int_{\partial G} p\left(u\frac{\partial w}{\partial n} - w\frac{\partial u}{\partial n} \right) ds,$$

on applying the divergence theorem with $\mathbf{n}$ equal to the exterior unit normal on ∂G. Equation (7.3) is the basic integral theorem from which the Green's function method proceeds in the elliptic case.

The function $w(x)$ is now determined such that (7.3) expresses u at an arbitrary point ξ in the region G in terms of w and known functions in (7.1) and (7.2). Let $w(x)$ be a solution of

$$(7.4) \qquad Lw = \delta(x - \xi)$$

where $\delta(x - \xi)$ is a two or three-dimensional Dirac delta function. The substitution property of the delta function then yields

$$(7.5) \qquad \iint_G uLw\, dv = \iint_G u\delta(x - \xi)\, dv = u(\xi).$$

In view of (7.1) we also have

$$(7.6) \qquad \iint_G wLu \, dv = \iint_G \rho w F \, dv.$$

It now remains to choose boundary conditions for $w(x)$ on ∂G so that the boundary integral in (7.3) involves only $w(x)$ and known functions. This can be accomplished by requiring $w(x)$ to satisfy the homogeneous version of the boundary condition (7.2); that is,

$$(7.7) \qquad \alpha w + \beta \left. \frac{\partial w}{\partial n} \right|_{\partial G} = 0.$$

If $\alpha \neq 0$ in (7.7), we have

$$(7.8) \qquad u\frac{\partial w}{\partial n} - w\frac{\partial u}{\partial n} = \frac{1}{\alpha}\left(\alpha u \frac{\partial w}{\partial n} - \alpha w \frac{\partial u}{\partial n}\right)$$

$$= \frac{1}{\alpha}\frac{\partial w}{\partial n}\left(\alpha u + \beta \frac{\partial u}{\partial n}\right) = \frac{1}{\alpha}B\frac{\partial w}{\partial n},$$

in view of (7.2). If $\alpha = 0$ in (7.7), we have

$$(7.9) \qquad u\frac{\partial w}{\partial n} - w\frac{\partial u}{\partial n} = -\frac{1}{\beta}Bw.$$

The function $w(x)$ is called the Green's function for the boundary value problem (7.1)–(7.2). To indicate its dependence on the point ξ, we denote the *Green's function* by

$$(7.10) \qquad w = K(x; \xi)$$

as in Section 1.3. In terms of the Green's function $K(x; \xi)$ the foregoing results imply that (7.3) takes the form

$$(7.11) \qquad u(\xi) = \iint_G \rho KF \, dv - \int_{\partial G} \frac{p}{\alpha}B\frac{\partial K}{\partial n} \, ds$$

in the case where $\alpha \neq 0$ and

$$(7.12) \qquad u(\xi) = \iint_G \rho KF \, dv + \int_{\partial G} \frac{p}{\beta}BK \, ds$$

when $\alpha = 0$.

The Green's function $K(x; \xi)$ thus satisfies the equation

$$(7.13) \qquad -\nabla \cdot (p\nabla K) + qK = \delta(x - \xi), \qquad x, \xi \in G$$

and the boundary condition

$$(7.14) \qquad\qquad \alpha K + \beta \left.\frac{\partial K}{\partial n}\right|_{\partial G} = 0$$

with the derivatives taken in the x-variable. It follows from (7.13) and our discussion in Section 6.7 that the Green's function is a fundamental solution of (7.1). This fact will be exploited in the construction of certain Green's functions. It must be noted that not all problems (7.13)–(7.14) have solutions. In certain cases that are considered later, a generalized or modified Green's function must be constructed that satisfies an equation that differs from (7.13) or boundary conditions that differ from (7.14). However, once the Green's function has been determined, the formulas (7.10) and (7.11) or slightly modified ones in the generalized case yield the solution $u(x)$ of the boundary value problem (7.1)–(7.2) at any point in G. By introducing appropriate assumptions on the behavior of the solutions at infinity, the Green's function technique can also be applied to problems over unbounded regions. We construct Green's functions for specific elliptic equations of the form (7.1) over bounded and unbounded regions in this chapter.

Turning now to a consideration of hyperbolic problems, we examine the initial and boundary value problem for the hyperbolic equation

$$(7.15) \qquad\qquad \rho u_{tt} + Lu = \rho F, \qquad x \in G, \qquad t > 0,$$

where the operator L is defined as in (7.1) and G is a bounded region in two or three-dimensional space. The initial conditions for $u(x, t)$ are

$$(7.16) \qquad u(x,0) = f(x); \qquad u_t(x,0) = g(x); \qquad x \in G.$$

The boundary conditions for $u(x, t)$ are given on ∂G for all $t > 0$ in the form

$$(7.17) \qquad\qquad \alpha u + \beta \left.\frac{\partial u}{\partial n}\right|_{\partial G} = B(x, t); \qquad t > 0.$$

The integral theorem appropriate for the problem is given over the bounded (cylindrical) region R in (x, t)-space obtained by extending the region G parallel to itself from $t = 0$ to $t = T$, as shown in Figure 7.1 ($T > 0$ is an arbitrary number). The lateral boundary of R is denoted by ∂R_x and the two caps of the cylinder, which are portions of the planes $t = 0$ and $t = T$, are denoted by ∂R_0 and ∂R_T, respectively. We remark that ∂R_0 is identical to the region G and the initial conditions for $u(x, t)$ are assigned on it. The boundary conditions for $u(x, t)$ are assigned on ∂R_x. The exterior unit normal $\mathbf{n}$ to ∂R has the form $\mathbf{n} = [\mathbf{n}_x, 0]$ on ∂R_x, where $\mathbf{n}_x$ is the exterior unit normal to ∂G. On ∂R_0, $\mathbf{n}$ has the form $\mathbf{n} = [\mathbf{0}, -1]$, and on ∂R_T, it has the form $\mathbf{n} = [\mathbf{0}, 1]$.

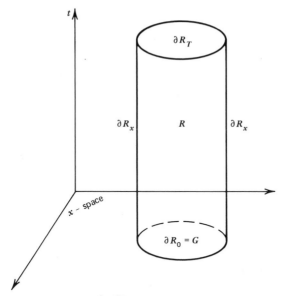

Figure 7.1. The region R.

It follows from the results at the beginning of Section 6.4 that

$$(7.18) \quad \iint_R \left[w(\rho u_{tt} + Lu) - u(\rho w_{tt} + Lw) \right] dv$$

$$= \iint_R \tilde{\nabla} \cdot \left[-pw\nabla u + pu\nabla w, \rho wu_t - \rho uw_t \right] dv$$

$$= \int_{\partial R} \left[-pw\nabla u + pu\nabla w, \rho wu_t - \rho uw_t \right] \cdot \mathbf{n} \, ds$$

$$= \int_{\partial R_x} \left(-pw\nabla u + pu\nabla w \right) \cdot \mathbf{n}_x \, ds + \int_{\partial R_T} \left(\rho wu_t - \rho uw_t \right) dx$$

$$- \int_{\partial R_0} \left(\rho wu_t - \rho uw_t \right) dx$$

where $\tilde{\nabla} = [\nabla, \partial/\partial t]$ is the gradient operator in space-time, and we have used the divergence theorem as well as the foregoing results concerning the exterior unit normal to the boundary ∂R. The integral relation (7.18) forms the basis for the Green's function method for solving the initial and boundary value problem (7.15)–(7.17).

We now show how $w(x, t)$ is determined so that the solution $u(x, t)$ of (7.15)–(7.17) can be specified at an arbitrary point (ξ, τ) in the region R from (7.18). First we require that $w(x, t)$ be a solution of

$$(7.19) \quad \rho w_{tt} + Lw = \delta(x - \xi)\delta(t - \tau), \qquad \xi \in G, 0 < \tau < T.$$

The product of the Dirac delta functions in (7.19) has the effect

$$(7.20) \quad \iint_R u(\rho w_{tt} + Lw)\, dv = \iint_R u\delta(x - \xi)\delta(t - \tau)\, dv = u(\xi, \tau).$$

In addition, we obtain from (7.15)

$$(7.21) \qquad\qquad \iint_R w(\rho u_{tt} + Lu)\, dv = \iint_R \rho wF\, dv$$

so that this term is known once $w(x, t)$ is specified.

Since

$$(7.22) \quad \int_{\partial R_x} p(-w\nabla u + u\nabla w) \cdot \mathbf{n}_x\, ds = \int_{\partial R_x} p\left(-w\frac{\partial u}{\partial n} + u\frac{\partial w}{\partial n}\right)\, ds,$$

we see that if we require, as in the elliptic case, that

$$(7.23) \qquad\qquad \alpha w + \beta \left.\frac{\partial w}{\partial n}\right|_{\partial R_x} = 0,$$

we obtain

$$(7.24) \quad \int_{\partial R_x} p\left(-w\frac{\partial u}{\partial n} + u\frac{\partial w}{\partial n}\right)\, ds = \begin{cases} \int_{\partial R_x} (p/\alpha)B\dfrac{\partial w}{\partial n}\, ds, & \alpha \neq 0 \\[4mm] -\int_{\partial R_x} (p/\beta)Bw\, ds, & \alpha = 0 \end{cases}.$$

To complete the determination of $w(x, t)$ we expect that initial conditions must be assigned to it for some value of t. If w and w_t are specified at $t = 0$, the integral over ∂R_0 in (7.18) is completely determined since u and u_t are given at $t = 0$. However, u and u_t at $t = T$ (i.e., on ∂R_T) are not known. If we specify w and w_T at $t = T$, it must be done in such a way that the unknown values of u and u_t play no role in the integral over ∂R_T. The only possible choice is to set

$$(7.25) \qquad\qquad w(x, T) = 0; \qquad w_t(x, T) = 0$$

so that the entire integral over ∂R_T vanishes.

The equation (7.19) together with the boundary condition (7.23) and the conditions (7.25) at $t = T$ constitutes a *backward initial and boundary value problem* for the function $w(x, t)$. It differs from the types of problems considered previously for hyperbolic equations where initial conditions were assigned at $t = 0$ and the problem was solved for $t > 0$. Here we assign end conditions at $t = T$ and solve the problem for $t < T$. The problem for $w(x, t)$

is well posed because if t is replaced by $-t$ in the hyperbolic equation in (7.19) (i.e., reversing the direction of time), the equation is not altered. We refer to problems with initial conditions or end conditions as initial value problems. The function $w(x, t)$ determined from (7.19), (7.23), and (7.25) is called the *Green's function* for the initial and boundary value problem (7.15)–(7.17) for $u(x, t)$. It is denoted as $w(x, t) = K(x, t; \xi, \tau)$. Once the initial and boundary value problem for K is solved, the values of K and K_t at $t = 0$ are known. Then the foregoing results yield the solution u at an (arbitrary) point (ξ, τ) as

$$(7.26) \qquad u(\xi, \tau) = \iint_R \rho K F \, dv + \int_{\partial R_0} (\rho K g - \rho K_t f) \, dx$$

$$- \int_{\partial R_x} \frac{p}{\alpha} B \frac{\partial K}{\partial n} \, ds$$

for the case when $\alpha \neq 0$, with a modified result obtained from (7.24) valid in the case where $\alpha = 0$.

The parabolic equation

$$(7.27) \qquad \rho u_t + L u = \rho F, \qquad x \in G, t > 0$$

with the initial condition

$$(7.28) \qquad u(x, 0) = f(x), \qquad x \in G$$

and the boundary condition

$$(7.29) \qquad \alpha u + \beta \left. \frac{\partial u}{\partial n} \right|_{\partial G} = B(x, t), \qquad t > 0,$$

can be treated in the same way as the hyperbolic problem (7.15)–(7.17). The operator L, the regions R and G, and their boundaries are defined as in the foregoing hyperbolic problem.

We introduce the function $w(x, t)$ and consider the integral relation

$$(7.30) \quad \iint_R \left[w(\rho u_t + L u) - u(-\rho w_t + L w) \right] dv$$

$$= \iint_R \tilde{\nabla} \cdot \left[-pw\nabla u + pu\nabla w, \rho w u \right] dv$$

$$= \int_{\partial R_x} \left(-pw \frac{\partial u}{\partial n} + pu \frac{\partial w}{\partial n} \right) ds + \int_{\partial R_T} \rho u w \, dx - \int_{\partial R_0} \rho u w \, dx.$$

Again, $\tilde{\nabla} = [\nabla, \partial/\partial t]$ is the gradient operator in space-time and the region R and its boundaries are as shown in Figure 7.1. The result (7.30) is a consequence of the divergence theorem, but it differs from the preceding integral theorems for the elliptic and hyperbolic problems in the following respect. The operator $\rho(\partial/\partial t) + L$ that occurs in the parabolic equation (7.27) is not self-adjoint as was the case in the preceding problems. Its adjoint operator is given as $-\rho(\partial/\partial t) + L$. With this choice for the adjoint operator we find that $w(\rho u_t + Lu) - u(-\rho w_t + Lw)$ is a divergence expression as is shown in (7.30). A discussion of adjoint differential operators is given in Section 8.3.

If we require $w(x, t)$ to be a solution of

$$(7.31) \qquad -\rho w_t + Lw = \delta(x - \xi)\delta(t - \tau), \qquad \xi \in G, 0 < \tau < T,$$

with the end condition

$$(7.32) \qquad\qquad\qquad w(x, T) = 0$$

and the boundary condition

$$(7.33) \qquad\qquad\qquad \alpha w + \beta \left.\frac{\partial w}{\partial n}\right|_{\partial R_x} = 0,$$

we immediately obtain from (7.30)

$$(7.34) \qquad u(\xi, \tau) = \iint_R \rho K F \, dv + \int_{\partial R_0} \rho f K \, dx - \int_{\partial R_x} \frac{p}{\alpha} B \frac{\partial K}{\partial n} \, ds,$$

where we have set $w(x, t) = K(x, t; \xi, \tau)$ and considered the case where $\alpha \neq 0$ in (7.33). When $\alpha = 0$, the last integral in (7.34) is replaced by the one given in (7.24). $K(x, t; \xi, \tau)$ is the Green's function for the initial and boundary value problem (7.27)–(7.29).

The equation (7.31) satisfied by the *Green's function K* is a *backward parabolic equation* that results on reversing the direction of time in the (forward) parabolic equation (7.27). However, since the problem for the Green's function is to be solved backwards in time, the initial and boundary value problem (7.31)–(7.33) for K is well posed. (That is, we must determine K for $t < T$ with an end condition given at $t = T$.) Once K has been determined, all the terms on the right side of (7.34) are known and the solution $u(x, t)$ of the initial and boundary value problem (7.27)–(7.29) is completely specified.

It follows from our discussion in Section 6.7 that the Green's function constructed for the hyperbolic equation (7.15) is a fundamental solution of that equation. However, for the parabolic problem the Green's function is not a fundamental solution of the parabolic equation (7.27) but rather of its adjoint equation (7.31). Indeed, the Green's function is always given as a fundamental solution of the adjoint of the given equation, as we see from our construction of Green's functions for the special equations considered. It is only because the operators in the elliptic equation (7.1) and the hyperbolic equation (7.15) are self-adjoint that the Green's functions for these equations are also fundamental solutions of the same equations.

The Green's function method can also be used to solve Cauchy problems for hyperbolic and parabolic equations. In the hyperbolic case we assume that $u(x, t)$ is a solution of (7.15) with initial data (7.16) and in the parabolic case $u(x, t)$ is a solution of (7.27) that satisfies the initial condition (7.28). Both problems are given over the entire two or three-dimensional space. The integral theorems (7.18) and (7.30) can be used for these problems if we assume that the (spatial) boundary ∂R_x tends to infinity and the solution u and the Green's function K are such that the contributions from these integrals vanish in the limit.

The Green's function K for the hyperbolic case is taken to be the solution of the backward Cauchy problem (7.19) and (7.25) and the solution of the Cauchy problem (7.15)–(7.16) is given as

$$(7.35) \qquad u(\xi, \tau) = \int_0^T \int_{-\infty}^{\infty} \rho KF \, dx \, dt + \int_{-\infty}^{\infty} \left[\rho Kg - \rho K_t f \right]_{t=0} dx,$$

as is easily seen from the (modified) integral relation (7.18). In the parabolic case the Green's function K is chosen to satisfy the backward Cauchy problem (7.31)–(7.32) and it then follows from the (modified) integral theorem (7.30) that the solution of the Cauchy problem (7.27)–(7.28) takes the form

$$(7.36) \qquad u(\xi, \tau) = \int_0^T \int_{-\infty}^{\infty} \rho KF \, dx \, dt + \int_{-\infty}^{\infty} \left[\rho Kf \right]_{t=0} dx.$$

The foregoing results are easily modified to yield Green's functions and solution formulas for initial and boundary value problems for hyperbolic and parabolic equations given over semi-infinite spatial regions.

There is an alternative approach to the construction of Green's functions that applies in the hyperbolic and parabolic cases. Instead of having $w(x, t)$ satisfy the inhomogeneous equations (7.19) and (7.31), we require that they be solutions of the homogeneous equations

$$(7.37) \qquad\qquad \rho w_{tt} + Lw = 0, \qquad x \in G, t < T$$

and

$$(7.38) \qquad\qquad - \rho w_t + Lw = 0, \qquad x \in G, t < T,$$

in the hyperbolic and parabolic cases, respectively. The homogeneous initial conditions (7.25) and (7.32) are replaced by

$$(7.39) \qquad w(x, T) = 0; \qquad w_t(x, T) = - \frac{\delta(x - \xi)}{\rho}; \qquad \xi \in G$$

and

$$(7.40) \qquad\qquad w(x, T) = \frac{\delta(x - \xi)}{\rho}; \qquad \xi \in G,$$

respectively. The boundary conditions (7.23) and (7.33) for $w(x, t)$ are retained.

In this formulation we obtain

$$(7.41) \qquad \int_{\partial R_T} [\rho w u_t - \rho u w_t] \, dx = \int_{\partial R_T} u \delta(x - \xi) \, dx = u(\xi, T)$$

and

$$(7.42) \qquad \int_{\partial R_T} \rho u w \, dx = \int_{\partial R_T} u \delta(x - \xi) \, dx = u(\xi, T),$$

for the hyperbolic and parabolic cases, respectively, when (7.39) and (7.40) are used. The solutions $u(\xi, T)$ then have the form (7.26) and (7.34) in the hyperbolic and parabolic cases as is easily seen. The only difference is that τ is replaced by T. Since both τ and T are chosen arbitrarily to be greater than zero, this difference is of no significance. In fact, if we set $T = \tau$ in (7.37)–(7.40), the Green's functions determined from each of the foregoing methods in the hyperbolic and parabolic cases are identical in the region $t < T = \tau$, as demonstrated later. The relation between these two approaches is connected with Duhamel's principle (see Section 4.5) which relates inhomogeneous equations with homogeneous initial conditions to homogeneous equations with inhomogeneous initial conditions.

The preceding results are valid in two or three space dimensions. The one-dimensional case for the elliptic equation (7.1) leads to the Green's function for a boundary value problem for an ordinary differential equation. The Green's function $K(x; \xi)$ satisfies the equation

$$(7.43) \qquad LK = -\frac{\partial}{\partial x}\left(p \frac{\partial K}{\partial x} \right) + qK = \delta(x - \xi)$$

in the interval $0 < x < l$ with the boundary conditions

$$(7.44) \qquad \begin{cases} \alpha_1 K(0; \xi) - \beta_1 \dfrac{\partial K(0; \xi)}{\partial x} = 0 \\[2mm] \alpha_2 K(l; \xi) + \beta_2 \dfrac{\partial K(l; \xi)}{\partial x} = 0 \end{cases}$$

where $\alpha_1, \alpha_2, \beta_1, \beta_2$ satisfy the conditions given in Chapter 4.

The one-dimensional versions of the hyperbolic and parabolic equations (7.15) and (7.27) lead to the consideration of the region R given as $0 < x < l$ and $0 < t < T$. The boundary ∂R is made up of the portion ∂R_x that comprises the lines $x = 0$ and $x = l$ with $0 \leqslant t \leqslant T$ and ∂R_0 and ∂R_T which represent the lines $t = 0$ and $t = T$, respectively, with $0 < x < l$. The region R is depicted in Figure 7.2.

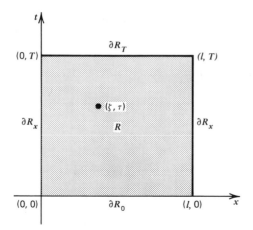

Figure 7.2. The region R.

For the hyperbolic case we have

$$(7.45) \qquad \rho u_{tt} - \frac{\partial}{\partial x}\left(p\frac{\partial u}{\partial x}\right) + qu = \rho F, \qquad 0 < x < l, t > 0$$

with the initial conditions

$$(7.46) \qquad u(x,0) = f(x); \qquad u_t(x,0) = g(x); \qquad 0 < x < l$$

and the boundary conditions

$$(7.47) \qquad \begin{cases} \alpha_1 u(0,t) - \beta_1 u_x(0,t) = g_1(t) \\ \alpha_2 u(l,t) + \beta_2 u_x(l,t) = g_2(t) \end{cases}.$$

With the operator L defined as in (4.7) we have:

$$(7.48) \qquad \iint_R \left[w(\rho u_{tt} + Lu) - u(\rho w_{tt} + Lw)\right]\,dx\,dt$$

$$= \iint_R \tilde{\nabla} \cdot \left[-pwu_x + puw_x, \rho wu_t - \rho uw_t\right]\,dx\,dt$$

$$= -\int_0^l \left[\rho wu_t - \rho uw_t\right]_{t=0}\,dx + \int_0^T \left[-pwu_x + puw_x\right]_{x=l}\,dt$$

$$+ \int_0^l \left[\rho wu_t - \rho uw_t\right]_{t=T}\,dx - \int_0^T \left[-pwu_x + puw_x\right]_{x=0}\,dt,$$

where $\tilde{\nabla} = [\partial/\partial x, \partial/\partial t]$ is the space-time gradient operator and we have used Green's theorem in the plane, which is just the planar version of the divergence theorem.

In a similar fashion for the one-dimensional parabolic case we consider the equation

$$(7.49) \qquad \rho u_t - \frac{\partial}{\partial x}\left(p\frac{\partial u}{\partial x}\right) + qu = \rho F, \qquad 0 < x < l, t > 0$$

with the initial condition

$$(7.50) \qquad\qquad\qquad u(x,0) = f(x)$$

and the boundary conditions (7.47). Proceeding as in the higher dimensional case we obtain

$$(7.51) \qquad \iint_R \left[w(\rho u_t + Lu) - u(-\rho w_t + Lw)\right] dx\, dt$$

$$= \iint_R \tilde{\nabla} \cdot \left[-pwu_x + puw_x, \rho wu\right] dx\, dt$$

$$= -\int_0^l [\rho uw]_{t=0}\, dx + \int_0^T [-pwu_x + puw_x]_{x=l}\, dt$$

$$+ \int_0^l [\rho uw]_{t=T}\, dx - \int_0^T [-pwu_x + puw_x]_{x=0}\, dt.$$

With (ξ, τ) equal to an interior point in the region R, we choose $w(x, t)$ to be a solution of

$$(7.52) \qquad\qquad \rho w_{tt} + Lw = \delta(x - \xi)\delta(t - \tau), \qquad t < T$$

and

$$(7.53) \qquad\qquad -\rho w_t + Lw = \delta(x - \xi)\delta(t - \tau), \qquad t < T$$

with $0 < x, \xi < l$, and $0 < t, \tau < T$ in the hyperbolic and parabolic problems, respectively. In addition, we have the end conditions

$$(7.54) \qquad\qquad w(x, T) = 0; \qquad w_t(x, T) = 0$$

in the former case and

$$(7.55) \qquad\qquad\qquad w(x, T) = 0$$

in the latter case. In both cases $w(x, t)$ is required to satisfy the homogeneous version of (7.47). If we have $\beta_1 = \beta_2 = 0$ in the boundary conditions, we obtain

$$(7.56) \qquad \int_0^T [-pwu_x + puw_x]_{x=l}\, dt = \int_0^T \left(\frac{p(l)}{\alpha_2}\right) g_2(t) \frac{\partial w(l, t)}{\partial x}\, dt$$

and

$$(7.57) \quad \int_0^T [-pwu_x + puw_x]_{x=0} \, dt = \int_0^T \left(\frac{p(0)}{\alpha_1} \right) g_1(t) \frac{\partial w(0, t)}{\partial x} \, dt.$$

Using these results, we easily obtain the following expression for the solution u at the arbitrary point (ξ, τ) in the hyperbolic case:

(7.58)

$$u(\xi, \tau) = \iint_R \rho F w \, dx \, dt + \int_0^l [\rho(x)g(x)w(x,0) - \rho(x)f(x)w_t(x,0)] \, dx$$
$$+ \int_0^T \left(\frac{p(0)}{\alpha_1} \right) g_1(t) \frac{\partial w(0, t)}{\partial x} \, dt - \int_0^T \left(\frac{p(l)}{\alpha_2} \right) g_2(t) \frac{\partial w(l, t)}{\partial x} \, dt,$$

when $\beta_1 = \beta_2 = 0$ in the boundary conditions. For the parabolic case we have

(7.59)

$$u(\xi, \tau) = \iint_R \rho w F \, dx \, dt + \int_0^l \rho(x)f(x)w(x,0) \, dx$$
$$+ \int_0^T \left(\frac{p(0)}{\alpha_1} \right) g_1(t) \frac{\partial w(0, t)}{\partial x} \, dt - \int_0^T \left(\frac{p(l)}{\alpha_2} \right) g_2(t) \frac{\partial w(l, t)}{\partial x} \, dt,$$

with $\beta_1 = \beta_2 = 0$ in the boundary conditions. If β_1 and β_2 are not zero, a somewhat different expression for the solution is readily obtained. Also, if the problem is given over a semi-infinite interval or we are dealing with the Cauchy problem over the infinite interval, appropriate expressions for the solution are easily found in a manner similar to that previously used in the higher dimensional problems. Further, $w(x, t)$ can be characterized in an alternative manner as was done in (7.37)–(7.40) for the higher dimensional case. These Green's functions are all denoted as $K(x, t; \xi, \tau)$.

Before proceeding to obtain Green's functions for various problems, we present the theory of generalized functions in the following section since these functions play an important role in Green's function theory.

7.2. GENERALIZED FUNCTIONS

Throughout the text and especially in connection with Green's functions, we have used the Dirac delta function in various calculations. Though the delta function is neither integrable nor differentiable in the conventional sense, we have integrated and differentiated this function. In this section, the Dirac delta and other functions are characterized as *generalized functions* and it is shown how the formal operations carried out on these functions are to be interpreted. Only the elementary and basic ideas of the theory of generalized functions are presented. Our discussion is restricted mostly to the one-dimensional case, as we shall see that the results for higher dimensional problems can often be characterized in terms of the one-dimensional case. The approach we use was developed by Laurent Schwartz and defines generalized functions in terms of what are known as linear functionals.

We begin by considering the collection of *test functions* $\varphi(x)$ that are assumed to be infinitely differentiable (i.e., they have derivatives of all orders) and vanish identically outside a bounded region. (Such functions are said to have compact support and in the exercises it is shown that such functions exist.)

The generalized function $f(x)$ is defined with respect to the set of test functions $\varphi(x)$ as the *linear functional* (f, φ). For each $\varphi(x)$ the functional has a uniquely defined numerical value. Since $f(x)$ is, therefore, technically a function of a function, assuming a (numerical) value for each $\varphi(x)$, we call (f, φ) a *functional*. Strictly speaking, the generalized function $f(x)$ is not a function of x (i.e., it does not assume values for each x) but rather of each test function $\varphi(x)$. The $\varphi(x)$ are called test functions since f is determined by specifying its effect on the test function $\varphi(x)$. However, for the cases we consider, generalized functions can be characterized as ordinary functions of x for almost all values of x.

The functional (f, φ) is assumed to be linear; that is, if $\varphi(x)$ and $\psi(x)$ are two test functions and a and b are constants,

$$(7.60) \qquad\qquad (f, a\varphi + b\psi) = a(f, \varphi) + b(f, \psi).$$

In addition, (f, φ) is required to be continuous in the following sense. If $\langle \varphi_k(x) \rangle$ is a sequence of test functions, all of which vanish outside a common region and $\langle \varphi_k \rangle$ converges (say, uniformly) to a function $\varphi(x)$ as $k \to \infty$, then we have

$$(7.61) \qquad\qquad \lim_{k \to \infty} (f, \varphi_k) = (f, \varphi).$$

The collection of generalized functions $f(x)$ is a linear (vector) space; that is, they can be added together and multiplied by constants. Given a sequence of generalized functions $\langle f_k(x) \rangle$ for which

$$(7.62) \qquad\qquad \lim_{k \to \infty} (f_k, \varphi) = (f, \varphi),$$

we say that $f_k(x)$ converges to the (generalized) function $f(x)$ as $k \to \infty$. This type of convergence is called *weak convergence* and has been considered earlier in the text in a different context.

A concrete general representation of the linear functional (f, φ) is given in the case where $f(x)$ is an (ordinary) integrable function over any finite region.

We then associate the functional (f, φ) with an integral given as

$$(7.63) \qquad (f, \varphi) = \int f(x)\varphi(x)\, dx,$$

where the integration is carried out over the region where $\varphi(x)$ is nonzero. The linearity of the functional follows from the linearity of the integral and the continuity of the functional is not hard to show. In view of the representation (7.63) the reason for the use of the inner product notation (f, φ) for the functional becomes clear.

If $f(x)$ and $g(x)$ are two integrable functions such that

$$(7.64) \qquad (f, \varphi) = (g, \varphi)$$

for all $\varphi(x)$, we must have $f(x) = g(x)$ for almost all x. This follows since

$$(7.65) \qquad (f, \varphi) - (g, \varphi) = \int (f - g)\varphi \, dx = 0.$$

The vanishing of the integral for all test functions $\varphi(x)$ implies (say, by using the fundamental lemma of the calculus of variations) that $f = g$ for almost all x. This shows that the functional representation (7.63) of $f(x)$ essentially specifies $f(x)$ uniquely if $f(x)$ is integrable. Thus with each integrable function a unique generalized function (f, φ) can be identified. Such generalized functions are often called *regular generalized functions*.

Generalized functions that are not ordinary integrable functions are often called *singular generalized functions*. The basic example is given by the *Dirac delta function* $\delta(x)$ which is defined as

$$(7.66) \qquad (\delta, \varphi) = \int \delta(x)\varphi(x)\, dx = \varphi(0).$$

Although we have formally carried over the integral representation of (f, φ) given in (7.63) to this case, the delta function is not an integrable function. We recall that $\delta(x)$ was defined to be zero for all $x \neq 0$. Thus the improper integral in (7.66)—since $\delta(x)$ is assumed to be singular at zero—must vanish for all x. Consequently, since $\varphi(0)$ need not vanish for all $\varphi(x)$, we find that $\delta(x)$ is a singular generalized function. It may be noted that the integral representation (7.63) for singular generalized functions is not strictly valid but it is generally used.

The generalized function $f(x)$ can be identified with (the values of) an ordinary function $g(x)$ in a region G if we have

$$(7.67) \qquad (f, \varphi) = (g, \varphi)$$

for all test functions $\varphi(x)$ that vanish outside G. We then say that $f(x) = g(x)$

in the region G. In particular, if $g(x) = 0$ in G, we say that $f(x)$ vanishes in the region G. As an example, we now show that $\delta(x)$ vanishes for all $x \neq 0$. We consider all test functions $\varphi(x)$ that vanish in a neighborhood of $x = 0$ so that, in particular, $\varphi(0) = 0$. Then using the definition (7.66) of $\delta(x)$ we find that $(\delta, \varphi) = \varphi(0) = 0 = (0, \varphi)$. Thus $\delta(x) = 0$ for all $x \neq 0$.

In Example 1.1 we characterized the delta function as a limit of the sequence of integrable functions $f_k(x)$,

$$f_k(x) = \begin{cases} \dfrac{k}{2}, & -\dfrac{1}{k} < x < \dfrac{1}{k} \\ 0, & |x| > \dfrac{1}{k} \end{cases}.$$

We have

$$(7.68) \qquad (f_k, \varphi) = \int_{-\infty}^{\infty} f_k(x)\varphi(x)\, dx$$

$$= \frac{k}{2} \int_{-1/k}^{1/k} \varphi(x)\, dx = \frac{k\varphi(\hat{x})}{2} \int_{-1/k}^{1/k} dx = \varphi(\hat{x}),$$

on using the mean value theorem for integrals with $-1/k \leqslant \hat{x} \leqslant 1/k$. Since $\varphi(x)$ is continuous at $x = 0$ we obtain

$$(7.69) \qquad \lim_{k \to \infty} (f_k, \varphi) = \lim_{k \to \infty} \varphi(\hat{x}) = \varphi(0) = (\delta, \varphi).$$

Recalling our definition of the convergence of sequences of generalized functions, we conclude that the functions $f_k(x)$ converge to the delta function $\delta(x)$ as $k \to \infty$.

The properties of generalized functions are usually derived from results valid for the functional (f, φ) if $f(x)$ is a regular generalized function. These results are then defined as properties of generalized functions. Thus in the one-dimensional case we have

(7.70)

$$(f(cx), \varphi(x)) = \int_{-\infty}^{\infty} f(cx)\varphi(x)\, dx = \begin{cases} \dfrac{1}{c} \displaystyle\int_{-\infty}^{\infty} f(x)\varphi\left(\dfrac{x}{c}\right) dx, & c > 0 \\[2mm] -\dfrac{1}{c} \displaystyle\int_{-\infty}^{\infty} f(x)\varphi\left(\dfrac{x}{c}\right) dx, & c < 0 \end{cases}$$

and

(7.71)

$$(f(x + \alpha), \varphi(x)) = \int_{-\infty}^{\infty} f(x + \alpha)\varphi(x)\, dx = \int_{-\infty}^{\infty} f(x)\varphi(x - \alpha)\, dx,$$

where c and α are constants. Therefore, we define for the generalized function $f(x)$

(7.72)
$$(f(cx), \varphi(x)) = \frac{1}{|c|}\left(f(x), \varphi\left(\frac{x}{c}\right)\right)$$

and

(7.73)
$$(f(x + \alpha), \varphi(x)) = (f(x), \varphi(x - \alpha)).$$

As a consequence of these definitions we have for the delta function $\delta(x)$,

(7.74) $\quad (\delta(cx), \varphi(x)) = \frac{1}{|c|}\left(\delta(x), \varphi\left(\frac{x}{c}\right)\right) = \frac{1}{|c|}\varphi(0) = \left(\frac{1}{|c|}\delta(x), \varphi(x)\right)$

so that $\delta(cx) = (1/|c|)\delta(x)$ formally. In particular, if $c = -1$, we obtain $\delta(-x) = \delta(x)$ so that $\delta(x)$ may be said to be an *even function*. Similarly, we have

(7.75)
$$(\delta(x - y), \varphi(x)) = (\delta(x), \varphi(x + y)) = \varphi(y),$$

which is known as the *substitution property* of the delta function and is generally written as

(7.76)
$$\int \delta(x - y)\varphi(x)\,dx = \varphi(y).$$

If $f(x)$ and $g(x)$ are regular functions, we have

(7.77)
$$(g(x)f(x), \varphi(x)) = \int (g(x)f(x))\varphi(x)\,dx = \int f(x)(g(x)\varphi(x))\,dx$$
$$= (f(x), g(x)\varphi(x)).$$

Thus if $g(x)$ is infinitely differentiable, $g(x)\varphi(x)$ is also a test function and we therefore define the product of an infinitely differentiable function $g(x)$ and the generalized function $f(x)$ as

(7.78)
$$(gf, \varphi) = (f, g\varphi).$$

An example of this result is given by

(7.79)
$$g(x)\delta(x - y) = g(y)\delta(x - y),$$

which follows from

(7.80) $\quad (g(x)\delta(x - y), \phi(x)) = (\delta(x - y), g(x)\varphi(x)) = g(y)\varphi(y)$
$$= (g(y)\delta(x - y), \varphi(x)),$$

in view of (7.75). A special case of the above is

$$(7.81) \qquad\qquad x\delta(x) = 0 \cdot \delta(x) = 0.$$

If $f(x)$ and $f'(x)$ are regular generalized functions, we obtain the result

$$(7.82) \qquad (f'(x), \varphi(x)) = \int_{-\infty}^{\infty} f'(x)\varphi(x)\, dx = \int_{-\infty}^{\infty} \varphi\, df$$

$$= \varphi f|_{-\infty}^{+\infty} - \int_{-\infty}^{\infty} f\varphi'\, dx = -(f, \varphi'),$$

since $\varphi(x)$ vanishes as $|x| \to \infty$. If $f(x)$ has n derivatives, repeated integration by parts yields

$$(7.83) \qquad \left(\frac{d^n f(x)}{dx^n}, \varphi(x) \right) = (-1)^n \left(f(x), \frac{d^n \varphi(x)}{dx^n} \right); \qquad n \geqslant 1.$$

We take (7.83) to be the definition of the nth derivative of the generalized function $f(x)$. Since the test functions $\varphi(x)$ are assumed to be infinitely differentiable, we conclude that any generalized function $f(x)$ has derivatives of all orders. In higher dimensions a similar result is valid for partial derivatives of $f(x)$.

As an example of the preceding we consider the function

$$(7.84) \qquad\qquad f(x) = \begin{cases} 0, & x < 0 \\ x, & x \geqslant 0. \end{cases}$$

This function is continuous at $x = 0$ but is not differentiable there. To obtain the *generalized derivative* we have

(7.85)

$$(f'(x), \varphi(x)) = -(f(x), \varphi'(x)) = -\int_{-\infty}^{\infty} f(x)\varphi'(x)\, dx$$

$$= -\int_{0}^{\infty} x\varphi'(x)\, dx$$

$$= -\int_{0}^{\infty} x\, d\varphi = -x\varphi|_{0}^{\infty} + \int_{0}^{\infty} \varphi(x)\, dx = \int_{0}^{\infty} \varphi(x)\, dx.$$

Now the functional representing the *Heaviside function* $H(x)$ defined as

$$(7.86) \qquad\qquad H(x) = \begin{cases} 0, & x < 0 \\ 1, & x \geqslant 0 \end{cases}$$

is given as

$$(7.87) \qquad (H(x), \varphi(x)) = \int_{-\infty}^{\infty} H(x)\varphi(x)\, dx = \int_{0}^{\infty} \varphi(x)\, dx.$$

Thus (7.85) shows that

(7.88) $$f'(x) = H(x),$$

with $f(x)$ defined as in (7.84).

For the derivative of the Heaviside function we have

(7.89)
$$(H'(x), \varphi(x)) = - (H(x), \varphi'(x)) = - \int_0^\infty \varphi'(x)\,dx = \varphi(0) = (\delta, \varphi)$$

so that

(7.90) $$H'(x) = \delta(x).$$

Further, we have

(7.91) $$(\delta'(x), \varphi(x)) = - (\delta(x), \varphi'(x)) = -\varphi'(0)$$

and

(7.92) $$(\delta''(x), \varphi(x)) = (\delta(x), \varphi''(x)) = \varphi''(0).$$

An additional result that has already been used in Chapter 6 is the following:

(7.93)
$$(x\delta'(x), \varphi(x)) = (\delta', x\varphi) = - (\delta, (x\varphi)') = - (\delta, x\varphi' + \varphi)$$
$$= - (x\varphi'(x) + \varphi(x))|_{x=0} = -\varphi(0) = - (\delta(x), \varphi(x)).$$

This may be formally expressed as the equation

(7.94) $$x\delta'(x) + \delta(x) = 0.$$

EXAMPLE 7.1.   THE DERIVATIVE OF A DISCONTINUOUS FUNCTION

Let the function $f(x)$ be continuously differentiable everywhere except at $x = \alpha$ where it has a jump discontinuity. Consequently, $f(x)$ does not have an ordinary derivative at $x = \alpha$, but the generalized function associated with $f(x)$ is differentiable everywhere and we now obtain its first derivative.

Using (7.82) we have

(7.95)
$$(f'(x), \varphi(x)) = - (f, \varphi') = - \int_{-\infty}^\alpha f(x)\varphi'(x)\,dx - \int_\alpha^\infty f(x)\varphi'(x)\,dx$$
$$= -f(x)\varphi(x)|_{-\infty}^\alpha + \int_{-\infty}^\alpha f'(x)\varphi(x)\,dx - f(x)\varphi(x)|_\alpha^\infty$$
$$+ \int_\alpha^\infty f'(x)\,\varphi(x)\,dx$$
$$= [f(x)]_{x=\alpha}\varphi(\alpha) + \int_{-\infty}^\infty f'(x)\varphi(x)\,dx,$$

where $[f(x)]_{x=\alpha}$ is the jump in $f(x)$ at $x = \alpha$ and the derivative in the last integral is defined at all x except $x = \alpha$. Since $f'(x)$ is assumed to have a finite limit as x approaches α from the left and from the right, the value of $f'(x)$ at $x = \alpha$ plays no role in the integral. We can express $[f(x)]_{x=\alpha}\varphi(\alpha)$ in the form $([f(x)]_{x=\alpha}\delta(x - \alpha), \varphi(x))$ and the last integral in (7.95) can be written as $(f'(x), \varphi(x))$. Thus the *generalized derivative* of $f(x)$ can be expressed as

$$(7.96) \qquad f' = [f(x)]_{x=\alpha}\delta(x - \alpha) + f'(x)|_{x \neq \alpha}.$$

It may be noted that (7.90) is a special case of this result.

An application of (7.96) may be made to the Green's function $K(x; \xi)$ determined from the ordinary differential equation (7.43). We have

$$(7.97) \qquad \frac{\partial}{\partial x}\left[p(x)\frac{\partial K(x; \xi)}{\partial x}\right] = -\delta(x - \xi) + q(x)K(x; \xi).$$

Using (7.96) we conclude that qK is the ordinary derivative of pK_x for $x \neq \xi$, and -1 is the jump of pK_x at $x = \xi$. Since $p(x)$ is continuous at $x = \xi$, the jump in $\partial K(x; \xi)/\partial x$ at $x = \xi$ is

$$(7.98) \qquad \left[\frac{\partial K(x; \xi)}{\partial x}\right]_{x=\xi} = -\frac{1}{p(\xi)},$$

a result that agrees with (6.179).

In determining Green's functions for differential equations it is often necessary to take transforms of the delta function. Therefore, we now discuss how Fourier transforms of generalized functions are to be defined. Since we are mostly concerned with transforms of the delta function and possibly its derivatives, we begin by considering generalized functions that vanish outside a bounded region.

For the one-dimensional problem, let the generalized function $f(x)$ vanish outside the interval $[-R, R]$ (i.e., it has compact support). It is possible to construct test functions $\hat{\varphi}(x)$ that equal unity in $[-R, R]$ and vanish outside an interval containing $[-R, R]$. Assuming for the moment that $f(x)$ is a regular function, the Fourier transform $F(\lambda)$ of $f(x)$ is given as

$$(7.99) \quad F(\lambda) = \frac{1}{\sqrt{2\pi}}\int_{-\infty}^{\infty} e^{i\lambda x}f(x)\,dx = \frac{1}{\sqrt{2\pi}}\int_{-R}^{R} e^{i\lambda x}f(x)\,dx$$

$$= \frac{1}{\sqrt{2\pi}}\int_{-R}^{R} e^{i\lambda x}f(x)\hat{\varphi}(x)\,dx$$

$$= \frac{1}{\sqrt{2\pi}}\int_{-\infty}^{\infty} e^{i\lambda x}f(x)\hat{\varphi}(x)\,dx = \left(f, \frac{1}{\sqrt{2\pi}}e^{i\lambda x}\hat{\varphi}(x)\right).$$

Since $(1/\sqrt{2\pi})e^{i\lambda x}\hat{\varphi}(x)$ is again a test function, we define the *Fourier transform* of the generalized function $f(x)$ to be

(7.100) $$F(\lambda) = \left(f, \frac{1}{\sqrt{2\pi}} e^{i\lambda x}\hat{\varphi}(x) \right).$$

For example, the Fourier transform of the delta function $\delta(x)$ is given as

(7.101) $$F(\lambda) = \left(\delta, \frac{1}{\sqrt{2\pi}} e^{i\lambda x}\hat{\varphi}(x) \right) = \frac{1}{\sqrt{2\pi}},$$

since $\hat{\varphi}(0) = 1$ by assumption. This result is formally equivalent to the following

(7.102) $$F(\lambda) = \frac{1}{\sqrt{2\pi}} \int_{-\infty}^{\infty} e^{i\lambda x}\delta(x)\, dx = \frac{1}{\sqrt{2\pi}}.$$

All the properties of Fourier transforms given in Section 5.2 are valid for this case. Higher dimensional Fourier transforms are defined in a similar way.

To define Fourier transforms for generalized functions without compact support we proceed as follows. If $f(x)$ and $g(x)$ are regular functions whose Fourier transforms are $F(\lambda)$ and $G(\lambda)$, respectively, we obtain from (5.12) on setting $x = 0$, the relation

(7.103) $$\int_{-\infty}^{\infty} F(\lambda)\overline{G}(\lambda)\, d\lambda = \int_{-\infty}^{\infty} f(x)\overline{g}(x)\, dx.$$

Noting the formulas (5.10)–(5.11) relating the transform $F(\lambda)$ and its inverse transform $f(x)$, we see that $\overline{g}(x)$ may be considered to be the transform of $\overline{G}(\lambda)$. Then if $\overline{G}(\lambda)$ is assumed to be a test function $\varphi(\lambda)$ and we denote its transform by $\Phi(x)$, (7.103) may be written as

(7.104) $$\int_{-\infty}^{\infty} F(\lambda)\varphi(\lambda)\, d\lambda = \int_{-\infty}^{\infty} f(x)\Phi(x)\, dx.$$

Clearly, both sides of the equation are linear functionals of the form (7.63) and we can write (7.104) as

(7.105a) $$(F, \varphi) = (f, \Phi).$$

This is taken as the definition of the Fourier transform $F(\lambda)$ of the generalized function $f(x)$.

Unfortunately, even though every test function φ has a Fourier transform, the transform is not itself a test function unless φ vanishes identically. Thus the right side of (7.105a) does not define a generalized function in general since Φ

is not necessarily a test function, so that (7.105a) does not result in a meaningful definition of the Fourier transform for all generalized functions. To obtain a useful definition of the Fourier transform on the basis of (7.105a) we must introduce a new class of test functions whose properties are preserved under Fourier transformation and, correspondingly, a new class of generalized functions.

Accordingly, we define the class of test functions of *rapid decay* that are required to be infinitely differentiable and to vanish, together with all their derivatives, more rapidly than any negative power of $|x|$ as $|x|$ tends to infinity. The function $\exp(-x^2)$ belongs to this class of test functions but not to the previously defined collection of test functions since it does not have compact support. However, every test function with compact support belongs to this class of test functions. It is not hard to show that the Fourier transform of the present class of test functions is again a test function of the same class.

If $\varphi(x)$ is a test function of rapid decay, the linear functional (f, φ) determines a generalized function of *slow growth*. The basic definitions and properties given for the previously defined generalized functions carry over to this class of generalized functions. Thus they are infinitely differentiable. (We do not discuss convergence of test functions and weak convergence of generalized functions for this case.) Any regular function $f(x)$ that is integrable over any finite interval and does not grow more rapidly than any power of $|x|$ as $|x|$ tends to infinity, determines a (regular) generalized function of slow growth by means of the formula (7.63). Since our main interest lies in the definition of the Fourier transform for the class of generalized functions of slow growth, we do not discuss all of their properties. We conclude with the observation that if $f(x)$ is a generalized function of slow growth and $\varphi(x)$ is a test function of rapid decay, then the linear functional on the right side of (7.105a) determines a generalized function of slow growth that we define to be the Fourier transform $F(\lambda)$ of $f(x)$.

As an example, we consider the Fourier transform of the (generalized) function of slow growth $f(x) = 1$. The conventional Fourier transform is not defined for this function and it does not have compact support so that the definition (7.100) cannot be used. Using (7.105a) with $F(\lambda)$ equal to the Fourier transform of $f(x) = 1$, we have

$$(7.105b) \quad (F, \varphi) = (1, \Phi) = \int_{-\infty}^{\infty} \Phi(x) \, dx$$

$$= \frac{1}{\sqrt{2\pi}} \int_{-\infty}^{\infty} \int_{-\infty}^{\infty} e^{i\lambda x} \varphi(\lambda) \, d\lambda \, dx$$

$$= \sqrt{2\pi} \left[\frac{1}{2\pi} \int_{-\infty}^{\infty} \int_{-\infty}^{\infty} e^{-i\lambda(t-x)} \varphi(\lambda) \, dx \, d\lambda \right]_{t=0} = \sqrt{2\pi} \, \varphi(0)$$

where we have used the Fourier integral formula (5.9) after interchanging the order of integration. Thus

$$(7.106) \qquad (F, \varphi) = \sqrt{2\pi}\, \varphi(0) = \sqrt{2\pi}\, (\delta, \varphi)$$

so that the Fourier transform of $f(x) = 1$ equals $\sqrt{2\pi}\, \delta(x)$. This result can be obtained formally from the inversion formula for the Fourier transform (5.11), where the use of (7.101) gives

$$(7.107) \qquad \delta(x) = \frac{1}{\sqrt{2\pi}} \int_{-\infty}^{\infty} e^{-i\lambda x} \frac{1}{\sqrt{2\pi}}\, d\lambda.$$

Multiplying across by $\sqrt{2\pi}$ and applying complex conjugation on both sides yields the required relationship.

In connection with the application of Fourier series methods for the solution of partial differential equations, there is a further property of generalized functions that assigns a significance to such series even if they do not converge everywhere. Let $\sum_{n=1}^{\infty} g_n(x)$ be an infinite series of functions uniformly convergent in any bounded region and let the sequence of partial sums of the series be defined as

$$(7.108) \qquad f_k(x) = \sum_{n=1}^{k} g_n(x), \qquad k = 1, 2 \ldots .$$

Then if $f(x)$ is the limit of the sequence $\{f_k(x)\}$, we have

(7.109)

$$\lim_{k \to \infty} (f_k(x), \varphi(x)) = \lim_{k \to \infty} \int f_k(x)\varphi(x)\, dx$$

$$= \int \lim_{k \to \infty} f_k(x)\varphi(x)\, dx = \int f(x)\varphi(x)\, dx = (f, \varphi),$$

since the uniform convergence of the sequence $f_k(x)$ permits the interchange of the limit process and integration. Note that the vanishing of the test function $\varphi(x)$ outside a bounded region implies that the integration is carried out only over that region so that the convergence is uniform.

The result (7.109) shows that the sequence $\{f_k(x)\}$ and, consequently, the given series converges to the function $f(x)$ in the sense of weak convergence. Thus $f(x)$ and the $f_k(x)$ can be interpreted as generalized functions. If each term $g_n(x)$ in the series has m derivatives, so does each term in the sequence $f_k(x)$. Using (7.83) we have

$$(7.110) \qquad \left(\frac{d^m f_k(x)}{dx^m}, \varphi(x) \right) = (-1)^m \left(f_k(x), \frac{d^m \varphi(x)}{dx^m} \right).$$

Since the right side of (7.110) converges as k tends to infinity (for the term

$d^m\varphi/dx^m$ vanishes outside a bounded interval), we conclude that the sequence of derivatives $\{d^m f_k/dx^m\}$ ($k = 1, 2, \dots$) converges (weakly) as $k \to \infty$, to the mth derivative of $f(x)$. In fact, we have

$$(7.111) \quad \lim_{k \to \infty} \left(\frac{d^m f_k(x)}{dx^m}, \varphi(x) \right) = (-1)^m \lim_{k \to \infty} \left(f_k, \frac{d^m \varphi}{dx^m} \right)$$

$$= (-1)^m \left(f(x), \frac{d^m \varphi}{dx^m} \right) = \left(\frac{d^m f}{dx^m}, \varphi \right).$$

As an example of the use of these results we consider the Fourier sine series

$$(7.112) \qquad\qquad F(x) = \sum_{n=1}^{\infty} a_n \sin nx,$$

where we assume that $|a_n| < M < \infty$ for all n. By formally differentiating the series

$$(7.113) \qquad\qquad G(x) = - \sum_{n=1}^{\infty} \frac{a_n}{n^2} \sin nx$$

twice term by term, we obtain the series (7.112). Since the terms in the series $G(x)$ are majorized by M/n^2, the series is uniformly convergent. Thus even though the series for $F(x)$ may not converge pointwise everywhere, we conclude on the basis of the foregoing results that it converges in the generalized sense to the generalized function $F(x)$ which equals the second (generalized) derivative of $G(x)$.

EXAMPLE 7.2. GENERALIZED SOLUTIONS OF THE WAVE EQUATION

It was shown in Example 6.6 that the solution of the wave equation

$$(7.114) \qquad\qquad u_{tt} - c^2 u_{xx} = 0, \qquad x > 0, t > 0$$

in the semi-infinite interval $x > 0$, with homogeneous initial data

$$(7.115) \qquad u(x,0) = 0; \qquad u_t(x,0) = 0; \qquad x > 0$$

and the boundary value

$$(7.116) \qquad\qquad u(0, t) = 1, \qquad t > 0,$$

is given as [see (6.116)]

$$(7.117) \qquad\qquad u(x, t) = \begin{cases} 0, & x > ct \\ 1, & 0 < x < ct \end{cases}.$$

In terms of the Heaviside function $H(x)$ defined in (7.86), the solution (7.117) may be written in the form of a generalized function

(7.118)
$$u(x, t) = H\left(t - \frac{x}{c}\right).$$

Since $H(x)$ is infinitely differentiable in the generalized sense, we have

(7.119)
$$u_{tt} = H''\left(t - \frac{x}{c}\right) = \delta'\left(t - \frac{x}{c}\right)$$

and

(7.120)
$$u_{xx} = \frac{1}{c^2}H''\left(t - \frac{x}{c}\right) = \frac{1}{c^2}\delta'\left(t - \frac{x}{c}\right).$$

Thus (7.118) is formally a solution of the wave equation if we admit generalized functions as solutions. This interpretation of a generalized solution of the wave equation is closely related to the concept of weak solutions introduced in Section 6.4.

We next consider the wave equation (7.114) in the finite interval $0 < x < l$ with the boundary conditions

(7.121)
$$u(0, t) = u(l, t) = 0, \qquad t > 0$$

and the initial conditions

(7.122)
$$u(x,0) = 1; \qquad u_t(x,0) = 0; \qquad 0 < x < l.$$

Using the results of Example 4.9 we obtain the (formal) Fourier sine series solution.

(7.123)
$$u(x, t) = \sqrt{\frac{2}{l}} \sum_{k=1}^{\infty} a_k \cos\left(\frac{\pi kc}{l}t\right)\sin\left(\frac{\pi k}{l}x\right).$$

The Fourier coefficients a_k are given as

(7.124)

$$a_k = \sqrt{\frac{2}{l}} \int_0^l \sin\left(\frac{\pi k}{l}x\right) dx = \frac{\sqrt{2l}}{\pi k}\left(1 - (-1)^k\right) = \begin{cases} \dfrac{2\sqrt{2l}}{\pi k}; & k = 1,3,\ldots \\ 0; & k = 2,4,\ldots \end{cases}.$$

Thus (7.123) can be written as

(7.125)
$$u(x, t) = \frac{4}{\pi} \sum_{k=1}^{\infty}{}' \frac{1}{k} \cos\left(\frac{\pi kc}{l}t\right)\sin\left(\frac{\pi k}{l}x\right),$$

with the summation taken only over the odd numbers $k = 1, 3, 5 \ldots$, as indicated by the prime.

In view of (4.148) we can write (7.125) as

(7.126)

$$u(x, t) = \frac{2}{\pi} \sum_{k=1}^{\infty} {}' \frac{1}{k} \sin\left[\frac{\pi k}{l}(x + ct)\right] + \frac{2}{\pi} \sum_{k=1}^{\infty} {}' \frac{1}{k} \sin\left[\frac{\pi k}{l}(x - ct)\right],$$

where each term in the series represents a propagating or progressive wave. As was noted in Section 4.3, the Fourier sine series (7.123) evaluated at $t = 0$ represents the odd periodic extension of the function $f(x) = 1$ defined in the given interval $0 < x < l$. The graph of the extended function is shown in Figure 7.3.

Let $F(x)$ denote the extended function of $f(x)$. It is seen that $F(x)$ has jumps of magnitude 2 at the points $x = nl$, $n = 0, \pm 1, \pm 2, \ldots$. Then (7.126) shows that $u(x, t)$ has the form

(7.127) $$u(x, t) = \tfrac{1}{2}F(x + ct) + \tfrac{1}{2}F(x - ct).$$

The Fourier series does not converge   pointwise to $F(x)$ at $x \pm ct = nl$, $n = 0, \pm 1, \pm 2, \ldots$, and $F(x)$ is certainly not differentiable there. Consequently, (7.127) cannot be interpreted as a classical solution of the initial and boundary value problem for the wave equation.

However, each of the Fourier sine series in (7.126) converges in the generalized sense to the functions $\tfrac{1}{2}F(x \pm ct)$ in view of our discussion. In addition, they can be differentiated twice to show that (7.127) is a generalized

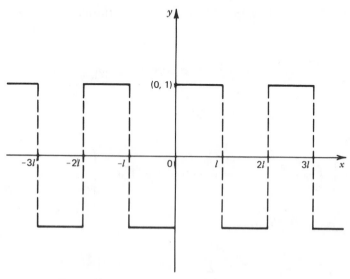

Figure 7.3.   The odd periodic extension of $f(x) = 1$.

solution of the wave equation. For example, we have for $F(x + ct)$,

$$(7.128) \qquad F(x + ct) = \frac{4}{\pi} \sum_{k=1}^{\infty} {}' \frac{1}{k} \sin\left[\frac{\pi k}{l}(x + ct)\right],$$

and on using Example 7.1 we obtain,

$$(7.129)$$

$$\frac{\partial}{\partial x} F(x + ct) = \sum_{n=-\infty}^{\infty} 2\delta(x + ct - nl) = \frac{4}{l} \sum_{k=1}^{\infty} {}' \cos\left[\frac{\pi k}{l}(x + ct)\right],$$

since $F(x + ct) = 1$ at $x + ct \neq nl$ and has jumps of magnitude 2 at $x + ct = nl$. A second derivative yields

$$(7.130)$$

$$\frac{\partial^2 F(x + ct)}{\partial x^2} = \sum_{n=-\infty}^{\infty} 2\delta'(x + ct - nl) = -\frac{4\pi}{l^2} \sum_{k=1}^{\infty} {}' k \sin\left[\frac{\pi k}{l}(x + ct)\right].$$

Again, this interpretation of the solution is closely related to that given in Example 6.5 where such Fourier series were characterized as weak solutions.

We have defined the product of an ordinary function and a generalized function in (7.78). However, there does not seem to exist a useful definition of the ordinary product of two generalized functions. Thus we cannot define the product $\delta(x)\delta(x)$ or $H(x)H(x)$ in a consistent manner (see Exercise 7.2.11). Yet $\delta(x)\delta(x - a)$, with $a \neq 0$ is well defined since the singular points of both delta functions do not coincide. Thus at $x = 0$, $\delta(x - a)|_{x=0} = \delta(-a) = 0$ and at $x = a$, $\delta(x)|_{x=a} = \delta(a) = 0$. Consequently, we are effectively considering the product of an ordinary function and the delta function and we have $\delta(x)\delta(x - a) = 0$.

In a similar fashion it is possible to consider the product of two generalized functions if they involve different variables. For example, the two-dimensional delta function $\delta(x, y)$ can be written as

$$(7.131) \qquad \delta(x, y) = \delta(x)\delta(y).$$

Using the definition of the delta function (7.66) which is valid in any number of dimensions, we have

$$(7.132)$$

$$\iint_{-\infty}^{\infty} \delta(x, y)\varphi(x, y)\, dx\, dy = \int_{-\infty}^{\infty} \int_{-\infty}^{\infty} \delta(x)\delta(y)\varphi(x, y)\, dx\, dy$$

$$= \int_{-\infty}^{\infty} \delta(y)\varphi(0, y)\, dy = \varphi(0, 0) = (\delta, \varphi),$$

so that the two expressions in (7.131) are seen to be equivalent. Similarly, we obtain in three dimensions

$$(7.133) \qquad \delta(x, y, z) = \delta(x)\delta(y)\delta(z).$$

By changing variables in the integral in (7.132) it is possible to obtain delta function representations in polar or other coordinate systems. This can also be done in the three-dimensional case.

In addition, it is possible to define generalized composite functions such as $\delta[g(x)]$ where $g(x)$ is assumed to be an ordinary function. The interpretation of these functions is based on the integral representation (7.63) and a change of variables, in the manner used in (7.70)–(7.71).

For example, if $g(x)$ is a smooth monotonic function that vanishes at $x = x_0$ and is such that $g'(x) > 0$ for all x, we have formally

$$(7.134) \quad (\delta[g(x)], \varphi(x)) = \int_{-\infty}^{\infty} \delta[g(x)]\varphi(x)\, dx$$

$$= \int_{-\infty}^{\infty} \delta(\sigma)\varphi[x(\sigma)] \frac{d\sigma}{g'[x(\sigma)]} = \frac{\varphi(x_0)}{g'(x_0)}$$

$$= \left(\frac{1}{g'(x_0)}\delta(x - x_0), \varphi(x) \right)$$

where we have used the transformation

$$(7.135) \qquad \sigma = g(x)$$

with $\sigma = 0$ corresponding to $x = x_0$. Thus

$$(7.136) \qquad \delta[g(x)] = \frac{1}{g'(x_0)}\delta(x - x_0).$$

If $g'(x) < 0$ for all x, we find that the change of variables (7.135) reverses the order of integration in (7.134). Then for both $g'(x) > 0$ or $g'(x) < 0$ for all x with $g(x_0) = 0$ (and $x = x_0$ as the only zero) we obtain

$$(7.137) \qquad \delta[g(x)] = \frac{1}{|g'(x_0)|}\delta(x - x_0).$$

If $g(x)$ has more than one zero and $g'(x)$ does not vanish at each of the zeros, we have, on applying the preceding argument in the neighborhood of each zero (which we assume to be isolated),

$$(7.138) \qquad \delta[g(x)] = \sum_{n} \frac{1}{|g'(x_n)|}\delta(x - x_n)$$

where we sum over all zeros.

EXAMPLE 7.3.   SOME PROPERTIES OF THE DELTA FUNCTION

We show that

$$(7.139) \qquad \delta(ax + by)\delta(cx + dy) = \frac{1}{|ad - bc|}\delta(x)\delta(y).$$

Using a formal approach, we would consider the two-dimensional integral representation of the functional $(\delta(ax + by)\delta(cx + dy), \varphi(x, y))$ and change the variables to obtain (7.139). However, we use a more straightforward approach that treats the problem as one involving the product of two delta functions with different arguments.

We first assume that $a = 0$ but b and c are not zero. Since $\delta(ax + by)|_{a=0} = \delta(by) = 0$ for $y \neq 0$, on using (7.79) we find that

$$(7.140) \qquad \delta(by)\delta(cx + dy) = \delta(by)\delta(cx).$$

Then (7.74) implies that

$$(7.141) \qquad \delta(by)\delta(cx) = \frac{1}{|bc|}\delta(x)\delta(y)$$

and this yields (7.139) in the case $a = 0$. If $a \neq 0$, we have

(7.142)

$$\delta(ax + by)\delta(cx + dy) = \delta(ax + by)\delta\left[\left(-\frac{bc}{a} + d\right)y\right]$$

$$= \delta(ax + by)\frac{|a|}{|ad - bc|}\delta(y) = \frac{|a|}{|ad - bc|}\delta(ax)\delta(y)$$

$$= \frac{1}{|ad - bc|}\delta(x)\delta(y)$$

where we have used (7.79) and (7.74) several times.

In a similar fashion we may consider the generalized function $\delta[g(x)]$ in the case where $g(x) = 0$ only at $x = x_0$ and $g'(x) \neq 0$ for all x. We have [since $g'(x_0) \neq 0$], on using Taylor's theorem with remainder,

$$(7.143) \quad \delta[g(x)] = \delta\left[g(x_0) + g'(x_0)(x - x_0) + \frac{1}{2}g''(\xi)(x - x_0)^2\right]$$

$$= \delta\left\{g'(x_0)(x - x_0)\left[1 + \frac{1}{2}\frac{g''(\xi)}{g'(x_0)}(x - x_0)\right]\right\}$$

Since $\delta[g(x)]$ is singular only at $x = x_0$ we may write it as

$$(7.144) \qquad \delta[g(x)] = \delta[g'(x_0)(x - x_0)] = \frac{1}{|g'(x_0)|}\delta(x - x_0),$$

which is in agreement with (7.137).

Both results obtained show that taking account of the localized (point) singularity of the delta function can lead quickly to simplified expressions in problems that involve the Dirac delta function. The basic property we have used is that the delta function vanishes everywhere except where its argument equals zero.

We conclude our discussion of generalized functions by noting that generalized functions are often referred to as *distributions*. This is because generalized functions may be thought of as densities of masses, charges, or other physical quantities. For example, the delta function $\delta(x)$ has infinite density at $x = 0$ but

$$(7.145) \qquad (\delta(x), 1) = \int_{-\infty}^{\infty} \delta(x) \, dx = 1$$

so that the total amount of mass, say, is unity. Similarly, the derivative of the delta function $\delta'(x)$ may be thought of as the density of a dipole located at $x = 0$. The total charge is given as

$$(7.146) \qquad (\delta'(x), 1) = - \left(\delta(x), \frac{d(1)}{dx} \right) = - (\delta(x), 0) = 0;$$

that is, the total charge vanishes as expected. However, the moment of the dipole around $x = 0$ is given as

(7.147)

$$\int_{-\infty}^{\infty} x\delta'(x) \, dx = (\delta'(x), x) = - \left(\delta(x), \frac{dx}{dx} \right) = - (\delta(x), 1) = -1.$$

Thus $-\delta'(x)$ is a dipole distribution located at $x = 0$ with a unit moment.

Further properties of one-dimensional or higher dimensional generalized functions are considered as needed in our discussion.

7.3. GREEN'S FUNCTIONS FOR BOUNDED REGIONS: EIGENFUNCTION EXPANSIONS

A general procedure for determining Green's functions for problems given over bounded (spatial) regions is the method of finite Fourier transforms presented in Section 4.6. For each of the problems considered in Section 7.1, the Green's function is expanded in a series of eigenfunctions of the elliptic operator L defined in (7.1). The coefficients in the series of eigenfunctions are specified in the manner shown in Section 4.6. Although this procedure is identical to that given in Section 4.6, the solutions of the given problems expressed in terms of the Green's functions have a somewhat different form than that given earlier.

However, the uniqueness theorems guarantee that there can be only one solution for each of these problems. We do not demonstrate in the general case the equivalence of the various solution forms.

We begin by constructing the Green's function $K(x; \xi)$ for the elliptic problem. As shown in Section 7.1, the function $K(x; \xi)$ satisfies the equation.

$$(7.148) \qquad LK \equiv -\nabla \cdot (p\nabla K) + qK = \delta(x - \xi), \qquad x \in G$$

and the boundary condition

$$(7.149) \qquad \alpha K + \beta \left. \frac{\partial K}{\partial n} \right|_{\partial G} = 0$$

where derivatives are taken with respect to the variable x. Let $M_k(x)$ be the orthonormalized set of eigenfunctions of the operator L; that is,

$$(7.150) \qquad LM_k = \lambda_k \rho M_k, \qquad k = 1, 2, \dots .$$

where the λ_k are the eigenvalues of L and $\rho(x)$ is a given weight function. The boundary condition for the M_k is (7.149) with K replaced by M_k.

We express $K(x; \xi)$ as a series of eigenfunctions

$$(7.151) \qquad K(x; \xi) = \sum_{k=1}^{\infty} N_k(\xi) M_k(x),$$

as in (4.195) with the (Fourier) coefficients $N_k(\xi)$ to be determined. Proceeding as in Section 4.6, we multiply (7.148) by $M_k(x)$, ($k = 1, 2, \dots$) and integrate over the region G. (Note that the operator K introduced in Section 4.6 is unrelated to the Green's function K of the present chapter.) Using the results (4.197)–(4.200) and noting that both $K(x; \xi)$ and $M_k(x)$ satisfy homogeneous boundary conditions of the form (7.149), we obtain

$$(7.152) \quad \lambda_k N_k(\xi) = \iint_G \delta(x - \xi) M_k(x) \, dx = M_k(\xi), \qquad k = 1, 2 \dots$$

since $\xi \in G$. Then if all the $\lambda_k > 0$, we have

$$(7.153) \qquad N_k(\xi) = \frac{1}{\lambda_k} M_k(\xi),$$

and this yields the *bilinear expansion* of the Green's function $K(x; \xi)$,

$$(7.154) \qquad K(x; \xi) = \sum_{k=1}^{\infty} \frac{M_k(\xi) M_k(x)}{\lambda_k}.$$

We notice that $K(x; \xi) = K(\xi; x)$, so that the Green's function for the elliptic problem is *symmetric*. This result can be proven directly for the Green's function without the use of the bilinear expansion (see Exercise 7.1.1). The symmetry is a consequence of the fact that the operator L taken together with the boundary conditions is self-adjoint. It implies that the interchange of the *source point* ξ and the *observation point* x does not alter the solution.

It has already been shown in Section 4.2 that the eigenvalues λ_k for the eigenvalue problem (7.150) with the boundary condition (7.149) [with K replaced by $M_k(x)$] are nonnegative. It can be shown that $\lambda_0 = 0$ is an eigenvalue for this problem if and only if $q = 0$ in the operator L and $\alpha = 0$ in the boundary condition. The corresponding eigenfunction $M_0(x)$ is clearly a constant in that case. Consequently, unless zero is an eigenvalue, the bilinear expansion (7.154) yields a formal solution of the boundary value problem (7.148)–(7.149) for the Green's function $K(x; \xi)$.

To verify that (7.154) is a (generalized) solution of (7.148)–(7.149), we apply the operator L to the series (7.154) term by term. We have

$$(7.155) \qquad LK = \sum_{k=1}^{\infty} \frac{M_k(\xi) L M_k(x)}{\lambda_k} = \sum_{k=1}^{\infty} \rho(x) M_k(\xi) M_k(x),$$

in view of (7.150). The eigenfunction expansion of $\delta(x - \xi)$ is given as

$$(7.156)$$

$$\delta(x - \xi) = \sum_{k=1}^{\infty} \big(\delta(x - \xi), M_k(x)\big) M_k(x) = \sum_{k=1}^{\infty} \rho(\xi) M_k(\xi) M_k(x),$$

since the Fourier coefficients are

$$(7.157) \quad \big(\delta(x - \xi), M_k(x)\big) = \iint_G \rho(x) \delta(x - \xi) M_k(x) \, dx = \rho(\xi) M_k(\xi).$$

From (7.156) we obtain, on using (7.79),

$$(7.158) \quad \frac{\delta(x - \xi)}{\rho(\xi)} = \frac{\delta(x - \xi)}{\rho(x)} = \sum_{k=1}^{\infty} M_k(\xi) M_k(x) = \frac{1}{\rho(x)} LK,$$

so that $LK = \delta(x - \xi)$. Since each eigenfunction $M_k(x)$ satisfies the boundary condition (7.149), the bilinear series in the generalized sense also satisfies the boundary condition (7.149). We have, therefore, shown that if $K(x; \xi)$ is interpreted as a generalized function so that term by term differentiation in the above series is possible, its representation as a bilinear series satisfies the problem (7.148)–(7.149).

For certain elliptic problems it is possible to construct the Green's function in terms of eigenfunctions for lower dimensional problems. Proceeding as in Section 4.2 we consider the function $u = u(x, y)$ where x is a point in the (bounded) region G and y is a scalar variable defined over the interval $0 < y < \hat{l}$. In place of (7.1) we now consider the equation

$$(7.159) \quad \rho(x)u_{yy} - Lu = -\rho(x)F(x, y), \qquad x \in G, 0 < y < \hat{l}$$

and the boundary conditions (7.2) and (4.20) if G is a two-dimensional region. If G is one-dimensional, $u(x, y)$ satisfies the boundary conditions (7.47) where t is replaced by y and l by $\hat{l}$.

The Green's function for the problem satisfies the equation

$$(7.160) \quad \rho \frac{\partial^2 K}{\partial y^2} - LK = -\delta(x - \xi)\delta(y - \eta), \qquad x, \xi \in G, \quad 0 < y, \eta < \hat{l}$$

and the boundary condition (7.149)—if G is two-dimensional—as well as a homogeneous form of the boundary condition (4.20). The Green's function is expressed in the form $K = K(x, y; \xi, \eta)$ and the derivatives in (7.160) are taken with respect to the variables x and y. If the region G is one-dimensional, K satisfies (7.47) with u replaced by K and $g_1 = g_2 = 0$.

To determine $K(x, y; \xi, \eta)$ we consider the eigenfunctions of the operator L in (7.160)—that is, the set $M_k(x)$ ($k = 1, 2 \ldots$)—and construct the eigenfunction expansion

$$(7.161) \quad K(x, y; \xi, \eta) = \sum_{k=1}^{\infty} N_k(y)M_k(x).$$

The $N_k(y)$ are the Fourier coefficients of K given as

$$(7.162) \quad N_k(y) = (K, M_k) = \iint_G \rho K(x, y; \xi, \eta)M_k(x) \, dx, \qquad k \geqslant 1,$$

assuming G is two-dimensional. We use the finite Fourier transform procedure of Section 4.6 and multiply (7.160) by $M_k(x)$ and integrate over the region G. Again using the results (4.197)–(4.200), we obtain

$$(7.163) \quad N_k''(y) - \lambda_k N_k(y) = -M_k(\xi)\delta(y - \eta), \qquad k = 1, 2, \ldots.$$

Since $K(x, 0; \xi, \eta) = K(x, \hat{l}; \xi, \eta) = 0$, we must have

$$(7.164) \qquad\qquad N_k(0) = 0; \qquad N_k(\hat{l}) = 0$$

as the boundary conditions for $N_k(y)$. To determine the $N_k(y)$ we must essentially obtain a Green's function for an ordinary differential equation in a finite interval. Although this problem may also be solved using eigenfunction expansions, we solve it instead in a more concrete manner in the following example.

EXAMPLE 7.4.   GREEN'S FUNCTION FOR AN ORDINARY DIFFERENTIAL
EQUATION

Using the notation given in Example 7.1, we consider the Green's function
$K(x; \xi)$ defined over the interval $0 < x < l$ and satisfying the equation

(7.165) $\qquad \dfrac{\partial^2 K(x; \xi)}{\partial x^2} - c^2 K(x; \xi) = -\delta(x - \xi), \qquad 0 < x, \xi < l$

with the homogeneous boundary conditions

(7.166) $\qquad\qquad\qquad K(0; \xi) = 0; \qquad K(l; \xi) = 0.$

From our discussion in Example 7.1 we find that $K(x; \xi)$ satisfies the
homogeneous equation

(7.167) $\qquad\qquad \dfrac{\partial^2 K(x; \xi)}{\partial x^2} - c^2 K(x; \xi) = 0, \qquad x \neq \xi$

and the matching conditions at $x = \xi$,

(7.168) $\qquad\qquad\qquad K(x; \xi) \text{ continuous at } x = \xi,$

(7.169) $\qquad\qquad\qquad \left[\dfrac{\partial K(x; \xi)}{\partial x} \right]_{x=\xi} = -1,$

where the square bracket represents the jump in the first derivative of $K(x; \xi)$
across $x = \xi$.

To solve this problem we denote K by K_1 for $x < \xi$ and K_2 for $x > \xi$. Both
K_1 and K_2 satisfy the homogeneous equation (7.167) and K_1 vanishes at $x = 0$,
whereas K_2 vanishes at $x = l$. We easily conclude that

(7.170) $\qquad\qquad \begin{cases} K_1(x; \xi) = a_1 \sinh(cx) \\ K_2(x; \xi) = a_2 \sinh[c(l - x)], \end{cases}$

where a_1 and a_2 are as yet unspecified constants (note that c is assumed to be a
constant). The continuity condition (7.168) implies that

(7.171) $\qquad\qquad a_1 \sinh(c\xi) = a_2 \sinh[c(l - \xi)],$

and the jump condition (7.166) requires that

(7.172) $\qquad\qquad -a_2 c \cosh[c(l - \xi)] - a_1 c \cosh(c\xi) = -1$

The solution of the system (7.171)–(7.172) for a_1 and a_2 is readily found to be

(7.173) $\qquad\qquad\qquad a_1 = \dfrac{\sinh[c(l - \xi)]}{c \sinh[cl]}$

and

$$(7.174) \qquad a_2 = \frac{\sinh(c\xi)}{c\sinh(cl)}.$$

Inserting these expressions into (7.170) we obtain the Green's function $K(x; \xi)$ as

$$(7.175) \quad K(x; \xi) = \frac{1}{c\sinh[cl]} \begin{cases} \sinh[c(l-\xi)]\sinh[cx]; & 0 < x < \xi \\ \sinh[c\xi]\sinh[c(l-x)]; & \xi < x < l \end{cases}.$$

Notice that this Green's function is symmetric since $K(x; \xi) = K(\xi; x)$.

In the limit as $c \to 0$ we obtain the Green's function $K(x; \xi)$ satisfying the equation

$$(7.176) \qquad \frac{\partial^2 K(x; \xi)}{\partial x^2} = -\delta(x - \xi); \qquad 0 < x, \xi < l$$

and the boundary conditions (7.166). It is given as

$$(7.177) \qquad K(x; \xi) = \frac{1}{l} \begin{cases} x(l-\xi); & 0 < x < \xi \\ \xi(l-x); & \xi < x < l \end{cases},$$

since $\sinh x \approx x$ as $x \to 0$.

We continue our discussion of (7.160) and assume the eigenvalues λ_k of (7.150) are all positive. Then on using the results of the preceding example, we find that the solution of (7.163)–(7.164) is given as

$$(7.178)$$

$$N_k(y) = \frac{M_k(\xi)}{\sqrt{\lambda_k}\sinh\left[\sqrt{\lambda_k}\,\hat{l}\right]} \begin{cases} \sinh\left[\sqrt{\lambda_k}(\hat{l}-\eta)\right]\sinh\left[\sqrt{\lambda_k}\,y\right]; & 0 < y < \eta \\ \sinh\left[\sqrt{\lambda_k}\,\eta\right]\sinh\left[\sqrt{\lambda_k}(\hat{l}-y)\right]; & \eta < y < \hat{l} \end{cases}.$$

Inserting (7.178) into (7.161) completes the formal solution of the problem (7.159) for the Green's function $K(x, y; \xi, \eta)$ for the given special region.

To see that the series (7.161) represents a generalized solution of (7.160), we find as in (7.155) that

$$(7.179) \qquad LK = \rho(x) \sum_{k=1}^{\infty} \lambda_k N_k(y) M_k(x).$$

Also, on using (7.163) we have

$$(7.180) \quad \rho(x)\frac{\partial^2 K}{\partial y^2} = \rho(x) \sum_{k=1}^{\infty} N_k''(y)M_k(x) = \rho(x) \sum_{k=1}^{\infty} \lambda_k N_k(y)M_k(x)$$

$$-\rho(x)\delta(y-\eta) \sum_{k=1}^{\infty} M_k(\xi)M_k(x).$$

Then

(7.181)

$$\rho\frac{\partial^2 K}{\partial y^2} - LK = - \left\{ \rho(x) \sum_{k=1}^{\infty} M_k(\xi)M_k(x) \right\} \delta(y-\eta) = -\delta(x-\xi)\delta(y-\eta)$$

in view of (7.158).

The preceding technique can be generalized to deal with elliptic equations of the form

$$(7.182) \qquad\qquad \rho(x)\hat{L}u - Lu = -\rho(x)F(x, y)$$

where $u = u(x, y)$ and

$$(7.183) \qquad\qquad \hat{L}u = a(y)\frac{\partial^2 u}{\partial y^2} + b(y)\frac{\partial u}{\partial y} + c(y)u,$$

with a, b, and c specified. The (x, y)-region may be defined as in the foregoing, and the boundary conditions in the y-variable may be of the more general form (7.47) where t is replaced by x, x is replaced by y, and l by $\hat{l}$.

The Green's function $K(x, y; \xi, \eta)$ satisfies the equation

$$(7.184) \qquad\qquad \rho\hat{L}K - LK = -\delta(x-\xi)\delta(y-\eta),$$

and K satisfies a homogeneous version of the boundary conditions for $u(x, y)$. Expanding K as in (7.161), we conclude that $N_k(y)$ satisfies

$$(7.185) \quad \hat{L}N_k(y) - \lambda_k N_k(y) = -M_k(\xi)\delta(y-\eta), \qquad k = 1, 2, \ldots,$$

and $N_k(y)$ satisfies appropriate boundary conditions at $y = 0$ and $y = \hat{l}$. Thus we again obtain a Green's function problem for an ordinary differential equation. A simple case of the Green's function problem was considered in Example 7.4, and further cases are studied in the exercises. We have further occasion to consider the technique of reducing the Green's function problem to a lower dimensional one later in this chapter.

In the following example we construct the Green's function appropriate for the first boundary value problem for Laplace's equation in a rectangle, using each of the given methods.

EXAMPLE 7.5. LAPLACE'S EQUATION: GREEN'S FUNCTION FOR A RECTANGLE

We consider the rectangle G, defined as $0 < x < l$ and $0 < y < \hat{l}$, and construct the Greens function $K(x, y; \xi, \eta)$ that satisfies the equation

$$(7.186) \quad \nabla^2 K = \frac{\partial^2 K}{\partial x^2} + \frac{\partial^2 K}{\partial y^2} = -\delta(x - \xi)\delta(y - \eta), \quad (x, y) \in G$$

with $0 < \xi < l$ and $0 < \eta < \hat{l}$ and the boundary condition

$$(7.187) \quad K(x, y; \xi, \eta) = 0, \quad (x, y) \in \partial G.$$

Applying the first method presented, we must solve the following eigenvalue problem in the rectangle G:

$$(7.188) \quad -\nabla^2 M(x, y) = \lambda M(x, y), \quad (x, y) \in G,$$

with the boundary condition

$$(7.189) \quad M(x, y) = 0, \quad (x, y) \in \partial G.$$

The eigenvalues and eigenfunctions are determined by using separation of variables.

Let $M(x, y) = F(x)G(y)$ and insert this expression into (7.188). We have

$$(7.190)$$
$$M_{xx} + M_{yy} + \lambda M = F''(x)G(y) + F(x)G''(y) + \lambda F(x)G(y) = 0.$$

Dividing by $F(x)G(y)$ and separating variables gives

$$(7.191) \quad \frac{F''(x)}{F(x)} + \lambda = -\frac{G''(y)}{G(y)} = k^2,$$

where k^2 is the separation constant. The equations for F and G are

$$(7.192) \quad F''(x) + (\lambda - k^2)F(x) = 0$$
$$(7.193) \quad G''(y) + k^2 G(y) = 0.$$

The boundary condition (7.189) implies that

$$(7.194) \quad F(0) = F(l) = 0$$

and

$$(7.195) \quad G(0) = G(\hat{l}) = 0.$$

Consequently, we are led to consider one-dimensional eigenvalue problems for $F(x)$ and $G(y)$ of a type studied in Section 4.3.

For the eigenvalue problem (7.193) and (7.195) we obtain as the eigenvalues

$$(7.196) \qquad k^2 = \left(\frac{\pi m}{\hat{l}}\right)^2, \qquad m = 1, 2, \ldots$$

and the eigenfunctions

$$(7.197) \qquad G_m(y) = \sin\left[\frac{\pi m}{\hat{l}}\right], \qquad m = 1, 2, \ldots .$$

For each of these eigenvalues, we have an eigenvalue problem (7.192) and (7.194) for $F(x)$. As in Section 4.3, we find the eigenvalues

$$(7.198) \qquad \lambda - k^2 = \left(\frac{\pi n}{l}\right)^2, \qquad n = 1, 2, 3, \ldots$$

and the eigenfunctions

$$(7.199) \qquad F_n(x) = \sin\left[\frac{\pi n}{l}x\right], \qquad n = 1, 2, 3, \ldots .$$

Combining the results obtained, we conclude that the eigenvalues λ for the problem (7.188)–(7.189) are given as

$$(7.200) \qquad \lambda_{nm} = \left(\frac{\pi n}{l}\right)^2 + \left(\frac{\pi m}{\hat{l}}\right)^2, \qquad n, m = 1, 2, 3, \ldots,$$

and the corresponding eigenfunctions are

$$(7.201) \qquad \hat{M}_{nm} = \sin\left(\frac{\pi n}{l}x\right)\sin\left(\frac{\pi m}{\hat{l}}y\right), \qquad n, m = 1, 2, 3, \ldots .$$

We denote the eigenfunctions as $\hat{M}_{nm} = F_n G_m$ since they are not as yet normalized. It may be noted that the eigenvalues λ_{nm} are positive, infinite in number, and tend to ∞ as n and m tend to infinity.

The appropriate inner product for the rectangular region G is given as

$$(7.202) \qquad (f, g) = \int_0^{\hat{l}}\int_0^l fg \, dx \, dy.$$

If we consider the two pairs (n, m) and (j, k) with $(n, m) \neq (j, k)$, we have

(7.203)

$$\left(\hat{M}_{nm}, \hat{M}_{jk}\right) = \int_0^{\hat{l}}\int_0^l \sin\left(\frac{\pi n}{l}x\right)\sin\left(\frac{\pi m}{\hat{l}}y\right)\sin\left(\frac{\pi j}{l}x\right)\sin\left(\frac{\pi k}{\hat{l}}y\right) dx \, dy$$

$$= \left[\int_0^{\hat{l}}\sin\left(\frac{\pi m}{\hat{l}}y\right)\sin\left(\frac{\pi k}{\hat{l}}y\right) dy\right]\left[\int_0^l \sin\left(\frac{\pi n}{l}x\right)\sin\left(\frac{\pi j}{l}x\right) dx\right] = 0$$

since $\{\sin[(\pi n/l)x]\}$ and $\{\sin[(\pi m/l)y]\}$ are orthogonal sets and $m \neq k$ and/or $n \neq j$. This shows that the set of eigenfunctions $\{\hat{M}_{nm}(x, y)\}$ is an orthogonal set. To orthonormalize the set we proceed as follows. The square of the norm of $\hat{M}_{nm}$ is given as

$$(7.204) \quad \|\hat{M}_{nm}\|^2 = (\hat{M}_{nm}, \hat{M}_{nm}) = \int_0^{\hat{l}}\int_0^l \sin^2\left(\frac{\pi n}{l}x\right)\sin^2\left(\frac{\pi m}{\hat{l}}y\right) dx\, dy$$

$$= \left[\int_0^{\hat{l}}\sin^2\left(\frac{\pi m}{\hat{l}}y\right) dy\right]\left[\int_0^l \sin^2\left(\frac{\pi n}{l}x\right) dx\right] = \left(\frac{\hat{l}}{2}\right)\left(\frac{l}{2}\right),$$

on using the results of Section 4.3. Thus an orthonormal set of eigenfunctions is given to be

$$(7.205) \quad M_{nm}(x, y) = \frac{2}{\sqrt{l\hat{l}}}\sin\left(\frac{\pi n}{l}x\right)\sin\left(\frac{\pi m}{\hat{l}}y\right); \qquad n, m = 1, 2, \ldots.$$

It should be noted that different sets of values of (n, m) do not necessarily yield distinct eigenvalues λ_{nm}. For example, if $l = \hat{l}$, we see that $\lambda_{12} = (\pi/l)^2 + (2\pi/l)^2 = \lambda_{21}$. Nevertheless, the set of eigenvalues λ_{nm} can be arranged in a sequence that corresponds to the positive integers $k = 1, 2, 3, \ldots$ with an equivalent arrangement for the eigenfunctions M_{nm}. It is then possible to speak of the set of eigenvalues λ_k and the eigenfunctions $M_k(x, y)$, with $k = 1, 2, 3, \ldots$. This notation was used in our earlier discussions of multidimensional eigenvalue problems. Here we retain the double subscript notation for the eigenvalues and eigenfunctions.

Given a function $f(x, y)$ defined in the rectangle $0 < x < l$ and $0 < y < \hat{l}$ and satisfying certain smoothness conditions we have the expansion

$$(7.206) \qquad f(x, y) = \frac{2}{\sqrt{l\hat{l}}} \sum_{n=1}^{\infty} \sum_{m=1}^{\infty} c_{nm}\sin\left(\frac{\pi n}{l}x\right)\sin\left(\frac{\pi m}{\hat{l}}y\right)$$

with the Fourier coefficients c_{nm} given as

$$(7.207) \quad c_{nm} = (f, M_{nm}) = \frac{2}{\sqrt{l\hat{l}}} \int_0^{\hat{l}}\int_0^l f(x, y)\sin\left(\frac{\pi n}{l}x\right)\sin\left(\frac{\pi m}{\hat{l}}y\right) dx\, dy$$

The expansion (7.206) is known as a double Fourier sine series.

Having determined the eigenfunctions and eigenvalues for the problem (7.188)–(7.189) we are now in a position to construct the Green's function $K(x, y; \xi, \eta)$. In view of (7.154), we obtain

$$(7.208) \quad K(x, y; \xi, \eta)$$
$$= \frac{4}{l\hat{l}} \sum_{n=1}^{\infty} \sum_{m=1}^{\infty} \frac{\sin[(\pi n/l)x]\sin[(\pi n/l)\xi]\sin[(\pi m/\hat{l})y]\sin[(\pi m/\hat{l})\eta]}{(\pi n/l)^2 + (\pi m/\hat{l})^2}$$

as the eigenfunction expansion of the Green's function for the rectangle G.

The alternative approach for the construction of the preceding Green's function has better convergence properties than those of the series (7.208), as we now demonstrate. The equation (7.186) for K is written as

$$(7.209) \quad \nabla^2 K = \frac{\partial^2 K}{\partial y^2} - \left(-\frac{\partial^2 K}{\partial x^2} \right) = \frac{\partial^2 K}{\partial y^2} - LK = -\delta(x - \xi)\delta(y - \eta),$$

with the operator L given as $L = -(\partial^2/\partial x^2)$. Thus the appropriate eigenvalue problem is

$$(7.210) \qquad LM(x) = -M''(x) = \lambda M(x), \qquad 0 < x < l$$

with the boundary conditions

$$(7.211) \qquad\qquad\qquad M(0) = M(l) = 0.$$

This represents a Sturm–Liouville problem which was solved in Example 4.4. The eigenvalues are

$$(7.212) \qquad\qquad\qquad \lambda_k = \left(\frac{\pi k}{l} \right)^2, \qquad k = 1, 2, \ldots,$$

and the orthonormal set of eigenfunctions is

$$(7.213) \qquad\qquad M_k(x) = \sqrt{\frac{2}{l}} \sin\left(\frac{\pi k}{l} x \right), \qquad k = 1, 2, \ldots.$$

With $N_k(y)$ defined as in (7.178) the Green's function is given by the eigenfunction expansion

$$(7.214) \qquad\qquad K(x, y; \xi, \eta) = \sum_{k=1}^{\infty} N_k(y) M_k(x).$$

For y different from η, the hyperbolic functions that occur in $N_k(y)$ can be approximated by exponentials as was done in Example 4.11, and the series can be shown to converge fairly rapidly if $|y - \eta|$ is not small.

Both expressions (7.208) and (7.214) can be used in solving the boundary value problem of the first kind for Laplace's equation in a rectangle. We have already seen in Example 4.11 how a direct separation of variables approach can be used to solve that problem with quite satisfactory results.

The eigenfunction expansion method for constructing Green's functions for the foregoing elliptic equations fails if $\lambda_0 = 0$ is an eigenvalue of the associated eigenvalue problem. As indicated, this occurs for the Green's function problem

(7.148)–(7.149) if and only if $q(x) = 0$ in (7.148) and $\alpha = 0$ in the boundary condition (7.149).

Let $\lambda_0 = 0$ be an eigenvalue of the operator L [see (7.150)] and $M_0(x)$ be the eigenfunction corresponding to λ_0. Then (7.152) implies that $\lambda_0 N_0 = 0 = M_0(\xi)$, and this is not possible since M_0 cannot vanish identically. Clearly then, the Green's function cannot be constructed in the given manner if $\lambda_0 = 0$ is an eigenvalue. There are two methods whereby a *modified Green's function* can be constructed in the special case that $\lambda_0 = 0$ is an eigenvalue. For each of the two modified Green's functions it is possible to solve the corresponding boundary value problem for $u(x)$ in a manner similar to that presented for the (ordinary) Green's function $K(x; \xi)$ in Section 7.1.

Now if $\lambda_0 = 0$ is an eigenvalue of the operator L (so that $q = 0$ in L and $\alpha = 0$ in the boundary condition), we find from (4.200) that

$$(7.215) \qquad \lambda_0 N_0 = 0 = \iint_G \rho F M_0 \, dx + \int_{\partial G} \frac{p}{\beta} M_0 B \, ds.$$

Since the eigenfunction $M_0(x)$ must be a constant (see Exercise 4.2.5), we conclude that

$$(7.216) \qquad \iint_G \rho F \, dx + \int_{\partial G} \frac{p}{\beta} B \, ds = 0.$$

Unless F and B are specified such that (7.216) is satisfied, the boundary value problem (7.1)–(7.2) for $u(x)$—with $\alpha = q = 0$—has no solution. We assume that (7.216) is satisfied and proceed with the construction of the modified Green's function.

In the first method the modified Green's function, which we denote by $\hat{K}(x; \xi)$, is expanded in a series of eigenfunctions of the operator L:

$$(7.217) \qquad \hat{K}(x; \xi) = \sum_{k=1}^{\infty} N_k(\xi) M_k(x)$$

where the $M_k(x)$, $k = 1, 2, \ldots$ correspond to the positive eigenvalues of L. This expansion differs from that given in (7.151) since the eigenfunction $M_0(x)$ is absent from the series, even though $\lambda_0 = 0$ is an eigenvalue for this problem. We have removed the term $N_0(\xi) M_0(x)$ from the expansion so as to avoid obtaining the contradictory result $\lambda_0 N_0 = 0 = M_0(\xi)$ that was derived in the foregoing.

Since the complete set of eigenfunctions is $\{M_k(x)\}$, $k = 0, 1, 2, \ldots$, the eigenfunction expansion of $\delta(x - \xi)$ is

(7.218)

$$\delta(x - \xi) = \sum_{k=0}^{\infty} \left(\delta(x - \xi), M_k(x) \right) M_k(x) = \rho(\xi) \sum_{k=0}^{\infty} M_k(\xi) M_k(x).$$

As shown, this series can be expressed as

$$(7.219) \qquad \delta(x - \xi) = \rho(x) \sum_{k=0}^{\infty} M_k(\xi) M_k(x).$$

Further, we have

$$(7.220) \qquad L\hat{K} = \sum_{k=1}^{\infty} N_k(\xi) L M_k(x) = \rho(x) \sum_{k=1}^{\infty} \lambda_k N_k(\xi) M_k(x).$$

Comparing (7.218) with (7.220) shows that if we set

$$(7.221) \qquad N_k(\xi) = \frac{M_k(\xi)}{\lambda_k}, \qquad k = 1, 2, \ldots,$$

the modified Green's function satisfies the equation

$$(7.222) \quad L\hat{K} = \sum_{k=1}^{\infty} \rho(x) M_k(\xi) M_k(x) = \delta(x - \xi) - \rho(x) M_0(\xi) M_0(x).$$

Noting the foregoing discussion, we define the modified Green's function $\hat{K}(x; \xi)$ to be a solution of the equation

$$(7.223)$$
$$L\hat{K} = -\nabla \cdot (p\nabla \hat{K}) = \delta(x - \xi) - \rho(x) M_0(\xi) M_0(x), \qquad x, \xi \in G$$

with the boundary condition

$$(7.224) \qquad \left. \frac{\partial \hat{K}}{\partial n} \right|_{\partial G} = 0.$$

The bilinear series of eigenfunctions for $\hat{K}(x; \xi)$ has the form

$$(7.225) \qquad \hat{K}(x; \xi) = \sum_{k=1}^{\infty} \frac{M_k(\xi) M_k(x)}{\lambda_k}.$$

Given the boundary value problem for $u(x)$,

$$(7.226) \qquad -\nabla \cdot (p\nabla u) = \rho F, \qquad x \in G$$

with the boundary condition

$$(7.227) \qquad \left. \beta \frac{\partial u}{\partial n} \right|_{\partial G} = B,$$

we now construct a solution formula for $u(x)$ in terms of the modified Green's function $\hat{K}(x; \xi)$. Following the procedure given in Section 7.1, we set $w = \hat{K}$ and note that (7.5) gives

$$(7.228) \qquad \iint_G uL\hat{K}\, dv = \iint_G u[\delta(x - \xi) - \rho(x)M_0(\xi)M_0(x)]\, dv$$

$$= u(\xi) - M_0(\xi) \iint_G \rho(x)u(x)M_0(x)\, dv.$$

The last integral in (7.228) is the Fourier coefficient of $u(x)$ with respect to the eigenfunction $M_0(x)$, and it can be expressed as (u, M_0). Then we obtain from (7.12) the solution formula

$$(7.229) \qquad u(\xi) = \iint_G \rho\hat{K}F\, dv + \int_{\partial G} \frac{p}{\beta}\hat{K}B\, ds + (u, M_0)M_0(\xi).$$

The solution exists only if the compatibility condition (7.216) is satisfied, and even then it is determined only up to an arbitrary constant multiple of the constant eigenfunction $M_0(x)$ as seen from the last term in (7.229). Consequently, as has already been noted in Section 6.8, the solution to the boundary value problem (7.226)–(7.227) is not unique. Any constant can be added to it.

The modified Green's function $\hat{K}(x; \xi)$ differs from the ordinary Green's function $K(x; \xi)$ in that $L\hat{K}$ does not equal $\delta(x - \xi)$, but is given as in (7.223). However, $\hat{K}(x; \xi)$ does satisfy the homogeneous boundary condition (7.224). There is an alternative construction of a modified Green's function, which we denote by $\tilde{K}(x; \xi)$, associated with the boundary value problem (7.226)–(7.227), for which we have

$$(7.230) \qquad L\tilde{K} = -\nabla \cdot (p\nabla\tilde{K}) = \delta(x - \xi), \qquad x, \xi \in G.$$

An application of the divergence theorem shows that

(7.231)

$$\iint_G L\tilde{K}\, dv = -\iint_G \nabla \cdot (p\nabla\tilde{K})\, dv = -\int_{\partial G} p\frac{\partial\tilde{K}}{\partial n}\, ds = \iint_G \delta(x - \xi)\, dv = 1.$$

Consequently, if we set $L\tilde{K} = \delta(x - \xi)$ we cannot have $\partial\tilde{K}/\partial n = 0$ on ∂G, as was the case for $\hat{K}(x; \xi)$. Instead, we must set

$$(7.232) \qquad\qquad\qquad \left.\frac{\partial\tilde{K}}{\partial n}\right|_{\partial G} = b$$

where $b(x)$ is any function for which

$$(7.233) \qquad\qquad\qquad \int_{\partial G} pb\, ds = -1,$$

in view of (7.231). (Note that b may be chosen to equal a constant.) The solution formula for $u(x)$ that satisfies (7.226)–(7.227) is then obtained from the results given at the beginning of Section 7.1. With $w = \tilde{K}$ we obtain in place of (7.9):

$$(7.234) \qquad u\frac{\partial \tilde{K}}{\partial n} - \tilde{K}\frac{\partial u}{\partial n} = ub - \frac{1}{\beta}\tilde{K}B,$$

and the solution is obtained in the form

$$(7.235) \qquad u(\xi) = \iint_G \rho\tilde{K}F\,dv + \int_{\partial G}\frac{p}{\beta}B\tilde{K}\,ds - \int_{\partial G}pub\,ds,$$

on using (7.12). The last integral in (7.235) is an arbitrary constant.

The expansion of $\tilde{K}(x;\xi)$ in a series of eigenfunctions is not as straightforward as that for $\hat{K}(x;\xi)$ since $\tilde{K}$ does not satisfy a homogeneous boundary condition. Nevertheless, it can be carried out using the finite Fourier transform techniques of Section 4.6. We do not present this expansion here. It may be noted that $\hat{K}$ and $\tilde{K}$ may be related to each other by way of the procedure developed in Chapter 4, whereby inhomogeneous boundary conditions can be transformed to homogeneous conditions. This is considered in the exercises.

EXAMPLE 7.6.   LAPLACE'S EQUATION: THE MODIFIED GREEN'S FUNCTION IN A RECTANGLE

Given the rectangle G defined as $0 < x < l$ and $0 < y < \hat{l}$, we construct the modified Green's function $\hat{K}$ that satisfies the equation [(see (7.223)]:

(7.236)

$$\nabla^2\hat{K}(x,y;\xi,\eta) = \frac{\partial^2\hat{K}}{\partial x^2} + \frac{\partial^2\hat{K}}{\partial y^2} = -\delta(x-\xi)\,\delta(y-\eta) + M_0(\xi)M_0(x)$$

with $(\xi,\eta) \in G$ and the Neumann condition

$$(7.237) \qquad \frac{\partial\hat{K}}{\partial n}(x,y;\xi,\eta) = 0, \qquad (x,y) \in \partial G,$$

where $\partial\hat{K}/\partial n$ is an exterior normal derivative.

To solve for $\hat{K}$, we must determine the eigenvalues and eigenfunctions of the problem

$$(7.238) \qquad -\nabla^2 M(x,y) = \lambda M(x,y), \qquad (x,y) \in G$$

where $M(x,y)$ satisfies the boundary condition

$$(7.239) \qquad \left.\frac{\partial M(x,y)}{\partial n}\right|_{\partial G} = 0.$$

Using separation of variables, we set $M(x, y) = F(x)G(y)$ in (7.238) and obtain

(7.240)
$$M_{xx} + M_{yy} + \lambda M = F''(x)G(y) + F(x)G''(y) + \lambda F(x)G(y) = 0.$$

Dividing by $F(x)G(y)$ and separating variables gives

(7.241)
$$\frac{F''(x)}{F(x)} + \lambda = -\frac{G''(y)}{G(y)} = k^2,$$

with k^2 as the separation constant. Then $F(x)$ satisfies the boundary value problem

(7.242)
$$F''(x) + (\lambda - k^2)F(x) = 0, \qquad 0 < x < l,$$

with

(7.243)
$$F'(0) = 0, \qquad F'(l) = 0.$$

Also, $G(y)$ satisfies

(7.244)
$$G''(y) + k^2 G(y) = 0, \qquad 0 < y < \hat{l}$$

with

(7.245)
$$G'(0) = 0; \qquad G'(\hat{l}) = 0.$$

Each of these eigenvalue problems for F and G was studied in Section 4.3.
The eigenvalues for (7.244)–(7.245) are

(7.246)
$$k^2 = \left(\frac{\pi m}{\hat{l}}\right)^2, \qquad m = 0, 1, 2, \ldots,$$

and the eigenfunctions are

(7.247)
$$G_m(y) = \cos\left[\frac{\pi m}{\hat{l}} y\right], \qquad m = 0, 1, 2, \ldots.$$

For $F(x)$ we obtain the eigenvalues

(7.248)
$$\lambda - k^2 = \left(\frac{\pi n}{l}\right)^2, \qquad n = 0, 1, 2, \ldots$$

and the eigenfunctions

(7.249)
$$F_n(x) = \cos\left[\frac{\pi n}{l} x\right], \qquad n = 0, 1, 2, \ldots.$$

Then the eigenvalues for (7.238)–(7.239) are

$$(7.250) \qquad \lambda_{nm} = \left(\frac{\pi n}{l} \right)^2 + \left(\frac{\pi m}{\hat{l}} \right)^2, \qquad n, m = 0, 1, 2, \ldots,$$

and the eigenfunctions are

$$(7.251) \quad \hat{M}_{nm}(x, y) = \cos\left(\frac{\pi n}{l} x \right) \cos\left(\frac{\pi m}{\hat{l}} y \right), \qquad n, m = 0, 1, 2, \ldots.$$

The inner product for the region G is given as in (7.202), and it is easily shown that $\{\hat{M}_{nm}\}$ is an orthogonal set using the results of Section 4.3. Also, the square of the norm of $\hat{M}_{nm}$ is

$$(7.252) \qquad \|\hat{M}_{nm}\|^2 = \left(\hat{M}_{nm}, \hat{M}_{nm} \right) = \frac{l\hat{l}}{4}, \qquad n, m = 1, 2, 3 \ldots.$$

For $\hat{M}_{00}$ we have

$$(7.253) \qquad \qquad \|\hat{M}_{00}\|^2 = \int_0^{\hat{l}} \int_0^l dx \, dy = l\hat{l},$$

and for $\hat{M}_{n0}$ and $\hat{M}_{0m}$ we obtain

$$(7.254) \qquad \qquad \|\hat{M}_{n0}\|^2 = \frac{l\hat{l}}{2} = \|M_{0m}\|^2.$$

Consequently, the orthonormal set of eigenfunctions is given by

$$(7.255)$$

$$M_{nm}(x, y) = \begin{cases} \dfrac{1}{\sqrt{l\hat{l}}}; & m = n = 0 \\[3mm] \sqrt{\dfrac{2}{l\hat{l}}} \cos\left(\dfrac{\pi n}{l} x \right); & m = 0, n = 1, 2, \ldots \\[3mm] \sqrt{\dfrac{2}{l\hat{l}}} \cos\left(\dfrac{\pi m}{\hat{l}} y \right); & n = 0, m = 1, 2, \ldots \\[3mm] \sqrt{\dfrac{4}{l\hat{l}}} \cos\left(\dfrac{\pi n}{l} x \right) \cos\left(\dfrac{\pi m}{\hat{l}} y \right); & n, m = 1, 2, \ldots \end{cases}$$

We note that $\lambda_{00} = 0$ is an eigenvalue for which the eigenfunction M_{00} is a constant.

Having determined the eigenfunctions $M_{nm}(x, y)$, we expand the modified Green's function in a series of these functions as in (7.225). The eigenfunction

$M_{00}(x, y)$ must be excluded from the series. We have

(7.256)

$$\hat{K}(x, y; \xi, \eta) = \frac{2}{\hat{l}l} \sum_{m=1}^{\infty} \frac{\cos[(\pi m/\hat{l})\eta]\cos[(\pi m/\hat{l})y]}{(\pi m/\hat{l})^2}$$

$$+ \frac{2}{\hat{l}l} \sum_{n=1}^{\infty} \frac{\cos[(\pi n/l)\xi]\cos[(\pi n/l)x]}{(\pi n/l)^2}$$

$$+ \frac{4}{\hat{l}l} \sum_{n=1}^{\infty} \sum_{m=1}^{\infty} \frac{\cos[(\pi n/l)\xi]\cos[(\pi n/l)x]\cos[(\pi m/\hat{l})\eta]\cos[(\pi m/\hat{l})y]}{(\pi n/l)^2 + (\pi m/\hat{l})^2}$$

The Green's function for the hyperbolic problem considered in Section 7.1 is expressed as $K(x, t; \xi, \tau)$ and satisfies the equation

(7.257)
$$\rho \frac{\partial^2 K}{\partial t^2} + LK = \delta(x - \xi)\delta(t - \tau),$$

where x and ξ are points in the bounded region G and t and τ are less than T. In addition, $K(x, t; \xi, \tau)$ satisfies the initial conditions or, equivalently, the end conditions

(7.258)
$$K(x, T; \xi, \tau) = \frac{\partial K(x, T; \xi, \tau)}{\partial t} = 0$$

and the boundary condition

(7.259)
$$\alpha K + \beta \left. \frac{\partial K}{\partial n} \right|_{x \in \partial G} = 0; \qquad t < T.$$

If the region G is one-dimensional (i.e., $0 < x < l$), the boundary condition (7.259) is replaced by a one-dimensional form as in (7.44).

Proceeding as in Section 4.6, we expand $K(x, t; \xi, \tau)$ in a series of eigenfunctions,

(7.260)
$$K(x, t; \xi, \tau) = \sum_{k=1}^{\infty} N_k(t) M_k(x)$$

where $M_k(x)$ are the eigenfunctions of the operator L, satisfying a boundary condition of the form (7.259). To determine the $N_k(t)$, (7.257) is multiplied by $M_k(x)$ and integrated over G. This yields the equations

(7.261) $\quad N_k''(t) + \lambda_k N_k(t) = M_k(\xi)\delta(t - \tau), \qquad t, \tau < T, k = 1, 2, \ldots.$

The initial conditions for $N_k(t)$ are

(7.262)
$$N_k(T) = N_k'(T) = 0.$$

On the basis of Example 7.1 we conclude that $N_k(t)$ is continuous at $t = \tau$ and $N_k'(t)$ has a jump discontinuity

(7.263)
$$\left[N_k'(t)\right]_{t=\tau} = M_k(\xi).$$

The conditions at $t = T$ imply that

(7.264)
$$N_k(t) = 0, \qquad \tau < t \leqslant T.$$

The continuity of $N_k(t)$ and the jump condition on $N_k'(t)$ at $t = \tau$ imply that

(7.265)
$$N_k(\tau) = 0; \qquad N_k'(\tau) = -M_k(\xi).$$

These serve as initial conditions for $N_k(t)$ in the interval $t < \tau$. Since $\delta(t - \tau) = 0$ for $t < \tau$, $N_k(t)$ satisfies the equation

(7.266)
$$N_k''(t) + \lambda_k N_k(t) = 0, \qquad t < \tau.$$

The solution of (7.265)–(7.266) is easily found to be

(7.267)
$$N_k(t) = \frac{1}{\sqrt{\lambda_k}} \sin\left[\sqrt{\lambda_k}\,(\tau - t)\right] M_k(\xi), \qquad t < \tau.$$

Combining (7.264) and (7.267) we can write $N_k(t)$ in the form

(7.268)
$$N_k(t) = \frac{1}{\sqrt{\lambda_k}} \sin\left[\sqrt{\lambda_k}\,(\tau - t)\right] M_k(\xi) H(\tau - t)$$

where $H(x)$ is the Heaviside function (7.86). The Green's function $K(x, t; \xi, \tau)$ thus has the form

(7.269)
$$K(x, t; \xi, \tau) = \left[\sum_{k=1}^{\infty} \frac{1}{\sqrt{\lambda_k}} \sin\left[\sqrt{\lambda_k}\,(\tau - t)\right] M_k(\xi) M_k(x)\right] H(\tau - t).$$

Notice that K is symmetric in x and ξ but not in t and τ.

It may be verified directly that (7.269) is a generalized solution of (7.257). Let $G(x, t)$ represent the bracketed term in (7.269) and write K as

(7.270)
$$K(x, t; \xi, \tau) = G(x, t) H(\tau - t).$$

Then

(7.271) $\qquad \dfrac{\partial K}{\partial t} = \dfrac{\partial G}{\partial t} H(\tau - t) - G\delta(\tau - t) = \dfrac{\partial G}{\partial t} H(\tau - t),$

since $G(x, t)\delta(\tau - t) = G(x, \tau)\delta(\tau - t)$ and $G(x, \tau) = 0$. Further,

(7.272) $\qquad \dfrac{\partial^2 K}{\partial t^2} = \dfrac{\partial^2 G}{\partial t^2} H(\tau - t) - \dfrac{\partial G}{\partial t}\bigg|_{t=\tau} \delta(\tau - t).$

Now

(7.273)

$$\dfrac{\partial G}{\partial t}\bigg|_{t=\tau} = \left[-\sum_{k=1}^{\infty} \cos\left[\sqrt{\lambda_k}\,(\tau - t)\right] M_k(\xi) M_k(x) \right]_{t=\tau} = -\sum_{k=1}^{\infty} M_k(\xi) M_k(x)$$

$$= -\dfrac{\delta(x - \xi)}{\rho(\xi)} = -\dfrac{\delta(x - \xi)}{\rho(x)},$$

in view of (7.158). Also,

(7.274)

$$LK = LGH(\tau - t) = \left[\sum_{k=1}^{\infty} \dfrac{1}{\sqrt{\lambda_k}} \sin\left[\sqrt{\lambda_k}\,(\tau - t)\right] M_k(\xi) LM_k(x) \right] H(\tau - t)$$

$$= \left[\sum_{k=1}^{\infty} \rho\sqrt{\lambda_k} \sin\left[\sqrt{\lambda_k}\,(\tau - t)\right] M_k(\xi) M_k(x) \right] H(\tau - t)$$

$$= -\rho \dfrac{\partial^2 G}{\partial t^2} H(\tau - t).$$

Combining these results we have

(7.275) $\quad \rho\dfrac{\partial^2 K}{\partial t^2} + LK = -\rho\dfrac{\partial G}{\partial t}\bigg|_{t=\tau} \delta(\tau - t) + \left[\rho\dfrac{\partial^2 G}{\partial t^2} - \rho\dfrac{\partial^2 G}{\partial t^2} \right] H(\tau - t)$

$$= \delta(x - \xi)\delta(t - \tau).$$

Recall that $\delta(\tau - t) = \delta(t - \tau)$. It has been assumed that $\lambda_0 = 0$ is not an eigenvalue. If $\lambda_0 = 0$ is an eigenvalue, an additional term $(\tau - t)M_0(\xi)M_0(x)$ must be added to the series in (7.269), as shown in the exercises.

The Green's function $K(x, t; \xi, \tau)$ for the parabolic problem of Section 7.1 satisfies the equation

(7.276) $\quad -\rho\dfrac{\partial K}{\partial t} + LK = \delta(x - \xi)\delta(t - \tau), \qquad x \in G, \quad t, \tau < T,$

the initial condition

(7.277) $$K(x, T; \xi, \tau) = 0,$$

and the boundary condition

(7.278) $$\alpha K + \beta \left. \frac{\partial K}{\partial n} \right|_{\partial G} = 0; \qquad t < T,$$

if G is a two or three-dimensional region.

As was done for the hyperbolic problem, we expand $K(x, t; \xi, \tau)$ in a series of eigenfunctions.

(7.279) $$K(x, t; \xi, \tau) = \sum_{k=1}^{\infty} N_k(t) M_k(x).$$

Multiplying (7.276) by $M_k(x)$ and integrating over the region G we obtain

(7.280) $$-N_k'(t) + \lambda_k N_k(t) = M_k(\xi)\delta(t - \tau), \qquad k = 1, 2, \ldots .$$

Since $\delta(t - \tau) = \delta(\tau - t)$ we can write (7.280) as

(7.281) $$\frac{d}{dt}\left[e^{-\lambda_k t} N_k(t) \right] = -M_k(\xi) e^{-\lambda_k t} \delta(\tau - t)$$
$$= -M_k(\xi) e^{-\lambda_k \tau} \delta(\tau - t),$$

where (7.79) was used. Integrating (7.281) and using the initial condition $N_k(T) = 0$, we obtain

(7.282) $$N_k(t) = e^{\lambda_k(t - \tau)} M_k(\xi) H(\tau - t).$$

[We recall that $dH(x)/dx = \delta(x)$.] Thus

(7.283) $$K(x, t; \xi, \tau) = \left[\sum_{k=1}^{\infty} e^{\lambda_k(t - \tau)} M_k(\xi) M_k(x) \right] H(\tau - t).$$

It is not difficult to show directly (see the exercises) that the series (7.283) is a generalized solution of (7.276). If $\lambda_0 = 0$ is an eigenvalue, the series (7.283) must be modified as shown in the exercises. Some specific examples of eigenfunction expansions of Green's functions for hyperbolic and parabolic problems are also considered in the exercises.

We conclude this section with the observation that the Green's functions (7.269) and (7.283) are solutions of the homogeneous versions of (7.257) and (7.276), respectively, for $t < \tau$. At $t = \tau$ the Green's function (7.269) vanishes, and its time derivative equals $-\delta(x - \xi)/\rho$. The Green's function (7.283) equals $\delta(x - \xi)/\rho$ at $t = \tau$. Also, the boundary condition (7.259) or (7.278) is satisfied by both Green's functions for $t < \tau$. These results follow easily from our discussion. Consequently, we find that for $t < \tau$ the Green's functions determined by each of the methods given in Section 7.1 for the hyperbolic and parabolic cases are identical.

7.4. GREEN'S FUNCTIONS FOR UNBOUNDED REGIONS

The transforms introduced in Chapters 4 and 5 (i.e., finite and infinite transforms) can be used to determine Green's functions for problems given over unbounded spatial regions. Rather than give a general discussion for equations of different types, we consider several examples in which specific Green's functions are obtained.

EXAMPLE 7.7.   THE HEAT EQUATION IN AN UNBOUNDED REGION

We begin by considering the heat equation in one dimension over the infinite interval $-\infty < x < \infty$. The medium is assumed to be homogeneous. In view of our discussion in Section 7.1, the Green's function $K(x, t; \xi, \tau)$ is a solution of the equation

$$(7.284) \qquad \frac{\partial K}{\partial t} + c^2 \frac{\partial^2 K}{\partial x^2} = -\delta(x - \xi)\delta(t - \tau); \qquad t, \tau < T,$$

$$-\infty < x < \infty$$

and satisfies the initial condition at $t = T$

$$(7.285) \qquad\qquad K(x, T; \xi, \tau) = 0.$$

The initial value problem for K will be solved by using the one-dimensional *Fourier transform* in x. Let the Fourier transform of K be denoted by $k(\lambda, t; \xi, \tau)$. Then

$$(7.286) \qquad k(\lambda, t; \xi, \tau) = \frac{1}{\sqrt{2\pi}} \int_{-\infty}^{\infty} e^{i\lambda x} K(x, t; \xi, \tau)\, dx.$$

To obtain an equation for k, we multiply (7.284) by $(1/\sqrt{2\pi})e^{i\lambda x}$ and integrate from $-\infty$ to $+\infty$. Using (5.15) we obtain the equation

$$(7.287) \qquad \frac{\partial k}{\partial t} - (c\lambda)^2 k = \frac{-1}{\sqrt{2\pi}} e^{i\lambda\xi}\delta(t - \tau), \qquad t < T.$$

Also, k satisfies the condition

$$(7.288) \qquad\qquad k(\lambda, T; \xi, \tau) = 0.$$

The equation (7.287) has the form of equation (7.280) with N_k replaced by k, λ_k equal to $(c\lambda)^2$, and M_k given as $(1/\sqrt{2\pi})e^{i\lambda\xi}$. The initial condition at $t = T$ is identical for both functions N_k and k. Therefore, the solution of (7.287)–(7.288) is obtained from (7.282) as

$$(7.289) \quad k(\lambda, t; \xi, \tau) = \frac{1}{\sqrt{2\pi}} \exp\left[(c\lambda)^2(t - \tau) + i\lambda\xi\right] H(\tau - t).$$

The inversion formula for the Fourier transform then gives

(7.290)

$$K(x, t; \xi, \tau) = \frac{1}{\sqrt{2\pi}} \int_{-\infty}^{\infty} e^{-i\lambda x} k(\lambda, t; \xi, \tau)\, d\lambda$$

$$= \frac{1}{2\pi} H(\tau - t) \int_{-\infty}^{\infty} \exp\left[-i\lambda(x - \xi) - c^2\lambda^2(\tau - t)\right] d\lambda.$$

This integral was evaluated in Example 5.2—see (5.35)–(5.39). Using the results of that example gives the Green's function K as

$$(7.291) \qquad K(x, t; \xi, \tau) = \frac{H(\tau - t)}{\sqrt{4\pi c^2(\tau - t)}} \exp\left[-\frac{(x - \xi)^2}{4c^2(\tau - t)}\right].$$

We note that K is symmetric in x and ξ but not in t and τ. In fact, there is a reversal in time in relation to the fundamental solution $G(x - \xi, t - \tau)$ defined in (5.41). This results because K satisfies the *backward* heat equation, whereas G satisfies the *forward* heat equation.

By using the two-dimensional Fourier transform it is easy to show that the two-dimensional Green's function is

$$(7.292) \quad K(x, y, t; \xi, \eta, \tau) = \frac{H(\tau - t)}{4\pi c^2(\tau - t)} \exp\left[-\frac{(x - \xi)^2 + (y - \eta)^2}{4c^2(\tau - t)}\right].$$

Similarly, the three-dimensional Fourier transform yields

(7.293)

$$K(x, y, z, t; \xi, \eta, \zeta, \tau)$$

$$= \frac{H(\tau - t)}{\left[\sqrt{4\pi c^2(\tau - t)}\right]^3} \exp\left[-\frac{(x - \xi)^2 + (y - \eta)^2 + (z - \zeta)^2}{4c^2(\tau - t)}\right]$$

as the three-dimensional Green's function. We will refer to (7.291)–(7.293) as the *free space Green's functions* for the heat equation.

Given the initial value problem for $u(x, t)$,

(7.294) $\qquad u_t - c^2 u_{xx} = F(x, t), \qquad -\infty < x < \infty, t > 0$

with the initial condition $u(x, 0) = f(x)$, the Green's function K may be used to obtain the solution u. With $-\infty < \xi < \infty$ and $\tau > 0$, the solution formula is

(7.295) $\qquad u(\xi, \tau) = \int_0^T \int_{-\infty}^{\infty} KF \, dx \, dt + \int_{-\infty}^{\infty} K(x, 0; \xi, \tau) f(x) \, dx,$

as follows from the expression (7.36) in Section 7.1, with K given as in (7.291). Now

(7.296)

$$\int_0^T \int_{-\infty}^{\infty} KF \, dx \, dt = \int_0^T \int_{-\infty}^{\infty} \frac{H(\tau - t)}{\sqrt{4\pi c^2 (\tau - t)}} \exp\left[-\frac{(x - \xi)^2}{4c^2 (\tau - t)} \right] F(x, t) \, dx \, dt$$

$$= \int_0^\tau \int_{-\infty}^{\infty} \frac{F(x, t)}{\sqrt{4\pi c^2 (\tau - t)}} \exp\left[-\frac{(x - \xi)^2}{4c^2 (\tau - t)} \right] dx \, dt$$

since $H(\tau - t) = 0$ for $t > \tau$. Also,

(7.297)

$$\int_{-\infty}^{\infty} K(x, 0; \xi, \tau) f(x) \, dx = \frac{H(\tau)}{\sqrt{4\pi c^2 \tau}} \int_{-\infty}^{\infty} f(x) \exp\left[-\frac{(x - \xi)^2}{4c^2 \tau} \right] dx$$

$$= \frac{1}{\sqrt{4\pi c^2 \tau}} \int_{-\infty}^{\infty} f(x) \exp\left[-\frac{(x - \xi)^2}{4c^2 \tau} \right] dx,$$

since $H(\tau) = 1$ for $\tau > 0$. It is immediately seen that the solution (7.295) agrees with that given in Example 5.2.

EXAMPLE 7.8. GREEN'S FUNCTION FOR THE WAVE EQUATION

In this example we construct the Green's function $K(x, y, z, t; \xi, \eta, \zeta, \tau)$ appropriate for the solution of the Cauchy problem for the wave equation in three dimensions. Using this function, the two-dimensional Green's function is obtained. In addition, the Green's function for the one-dimensional problem is discussed.

In three (space) dimensions, the Green's function K satisfies the equation

(7.298)

$$\frac{\partial^2 K}{\partial t^2} - c^2 \left[\frac{\partial^2 K}{\partial x^2} + \frac{\partial^2 K}{\partial y^2} + \frac{\partial^2 K}{\partial z^2} \right] = \delta(x - \xi)\delta(y - \eta)\delta(z - \zeta)\delta(t - \tau)$$

with $-\infty < x, y, z, \xi, \eta, \zeta < \infty$, and $t, \tau < T$. The conditions at $t = T$ are

(7.299)
$$K\Big|_{t=T} = \frac{\partial K}{\partial t}\Big|_{t=T} = 0.$$

To solve, we use the three-dimensional Fourier transform (see Section 5.4) and define

(7.300)

$$k(\lambda_1, \lambda_2, \lambda_3, t; \xi, \eta, \zeta, \tau) = \frac{1}{(\sqrt{2\pi})^3} \int\int\int_{-\infty}^{\infty} e^{i(\lambda_1 x + \lambda_2 y + \lambda_3 z)} K \, dx \, dy \, dz.$$

To obtain an equation for k, multiply $1/(\sqrt{2\pi})^3 \exp[i(\lambda_1 x + \lambda_2 y + \lambda_3 z)]$ into (7.298) and integrate with respect to x, y, and z from $-\infty$ to $+\infty$. Using the properties of the Fourier transform (see Example 5.7) gives

(7.301)

$$\frac{\partial^2 k}{\partial t^2} + c^2 (\lambda_1^2 + \lambda_2^2 + \lambda_3^2) k = \frac{1}{(\sqrt{2\pi})^3} \exp[i(\lambda_1 \xi + \lambda_2 \eta + \lambda_3 \zeta)]\delta(t - \tau)$$

and the initial conditions

(7.302)
$$k\Big|_{t=T} = \frac{\partial k}{\partial t}\Big|_{t=T} = 0.$$

The problem (7.301)–(7.302) is equivalent to that for $N_k(t)$ given in (7.261)–(7.262). Using the result (7.267) gives the solution

(7.303) $k = \dfrac{1}{(\sqrt{2\pi})^3 c|\lambda|} \sin[c|\lambda|(\tau - t)]\exp[i(\lambda_1 \xi + \lambda_2 \eta + \lambda_3 \zeta]H(\tau - t)$

where $|\lambda| = \sqrt{\lambda_1^2 + \lambda_2^2 + \lambda_3^2}$. Inverting the Fourier transform yields

(7.304)

$$K = \frac{H(\tau - t)}{(2\pi)^3} \int\int\int_{-\infty}^{\infty} \frac{\sin[c|\lambda|(\tau - t)]}{c|\lambda|}$$

$$\times \exp\{-i[\lambda_1(x - \xi) + \lambda_2(y - \eta) + \lambda_3(z - \zeta)]\} \, d\lambda_1 \, d\lambda_2 \, d\lambda_3$$

This integral, which is to be interpreted in a generalized sense, can be evaluated by transforming to spherical coordinates. Let $|\lambda|$, θ, and φ be spherical coordinates with $|\lambda| \geqslant 0$, $0 \leqslant \theta \leqslant 2\pi$, and $0 \leqslant \varphi \leqslant \pi$. The polar axis (i.e., $\varphi = 0$) is chosen to coincide with the (half) line connecting the origin (i.e., $\lambda_1 = \lambda_2 = \lambda_3 = 0$) to the fixed point $(x - \xi, y - \eta, z - \zeta)$. Since $(x - \xi, y - \eta, z - \zeta)$ is on the polar axis, we have

$$(7.305) \qquad \lambda_1(x - \xi) + \lambda_2(y - \eta) + \lambda_3(z - \zeta) = |\lambda| r \cos \varphi$$

where $r = \sqrt{(x - \xi)^2 + (y - \eta)^2 + (z - \zeta)^2}$. The volume element is given as

$$(7.306) \qquad d\lambda_1 \, d\lambda_2 \, d\lambda_3 = |\lambda|^2 \sin \varphi \, d|\lambda| \, d\theta \, d\varphi,$$

and (7.304) takes the form

(7.307)

$$K = \frac{H(\tau - t)}{c(2\pi)^3} \int_0^\infty \int_0^\pi \int_0^{2\pi} \{\sin[c|\lambda|(\tau - t)] e^{-i|\lambda| r \cos \varphi} |\lambda| \sin \varphi\} \, d\theta \, d\varphi \, d|\lambda|$$

The two inner integrals are easily evaluated, and their contribution is found to be $(4\pi/|\lambda| r) \sin[|\lambda| r]$. Then (7.307) is evaluated as follows:

(7.308)

$$\begin{aligned} K &= \frac{H(\tau - t)}{2\pi^2 cr} \int_0^\infty \sin[c|\lambda|(\tau - t)] \sin[|\lambda| r] \, d|\lambda| \\ &= \frac{H(\tau - t)}{4\pi^2 cr} \int_0^\infty (\cos\{|\lambda|[c(\tau - t) - r]\} - \cos\{|\lambda|[c(\tau - t) + r]\}) \, d|\lambda| \\ &= \frac{H(\tau - t)}{8\pi^2 cr} \int_{-\infty}^\infty (\exp\{i|\lambda|[c(\tau - t) - r]\} - \exp\{i|\lambda|[c(\tau - t) + r]\}) \, d|\lambda| \\ &= \frac{H(\tau - t)}{4\pi cr} \{\delta[c(\tau - t) - r] - \delta[c(\tau - t) + r]\}, \end{aligned}$$

where (7.107) has been used. Now $\delta[c(\tau - t) + r] = 0$ since for $\tau - t > 0$ and $r > 0$, $c(\tau - t) + r \neq 0$. For the same reason, $\delta[c(\tau - t) - r]$ vanishes for $(\tau - t) < 0$ so that there is no need to include the Heaviside function in (7.308). Thus K is given as

$$(7.309) \qquad K(x, y, z, t; \xi, \eta, \zeta, \tau) = \frac{\delta[c(\tau - t) - r]}{4\pi cr}$$

with $r = \sqrt{(x - \xi)^2 + (y - \eta)^2 + (z - \zeta)^2}$.

To obtain the Green's function in two (space) dimensions, we integrate the Green's function (7.309) with respect to ζ from $-\infty$ to $+\infty$. The resulting Green's function $K = K(x, y, t; \xi, \eta, \tau)$, is independent of z and ζ and satisfies the equation

$$(7.310) \quad \frac{\partial^2 K}{\partial t^2} - c^2 \left[\frac{\partial^2 K}{\partial x^2} + \frac{\partial^2 K}{\partial y^2} \right] = \delta(x - \xi)\delta(y - \eta)\delta(t - \tau)$$

for $-\infty < x, \xi, y, \eta < \infty$, and $t, \tau < T$. In effect, we are considering a line source in three-dimensional space and because of the homogeneity of the medium (there are constant coefficients in the equation) this corresponds to a point source in two dimensions. We remark that if we formally integrate (7.298) with respect to ζ and assume K is independent of z, (7.310) results.

On integrating over ζ, the two-dimensional Green's function becomes

$$(7.311) \qquad K(x, y, t; \xi, \eta, \tau) = \frac{1}{4\pi c} \int_{-\infty}^{\infty} \frac{\delta[c(\tau - t) - r]}{r} \, d\zeta.$$

Introducing the change of variables $s = z - \zeta$ gives

$$(7.312) \quad K = \frac{1}{4\pi c} \int_{-\infty}^{\infty} \frac{\delta[c(\tau - t) - r]}{r} \, ds = \frac{1}{2\pi c} \int_{0}^{\infty} \frac{\delta[c(\tau - t) - r]}{r} \, ds$$

where $r = \sqrt{(x - \xi)^2 + (y - \eta)^2 + s^2}$ and the second integral from 0 to ∞ results since the integrand is an even function of s. A further change of variables

$$(7.313) \qquad\qquad\qquad r^2 = \rho^2 + s^2,$$

with $\rho^2 = (x - \xi)^2 + (y - \eta)^2$ yields

$$(7.314) \qquad\qquad\qquad \frac{ds}{r} = \frac{dr}{s} = \frac{dr}{\sqrt{r^2 - \rho^2}}$$

and (7.312) becomes

$$(7.315) \qquad\qquad K = \frac{1}{2\pi c} \int_{\rho}^{\infty} \frac{\delta[c(\tau - t) - r]}{\sqrt{r^2 - \rho^2}} \, dr.$$

If $c(\tau - t) < \rho$, the integral vanishes, since the argument of the delta function is negative in that case. If $c(\tau - t) > \rho$, the substitution property of the delta function yields $K = (1/2\pi c)(1/\sqrt{c^2(\tau - t)^2 - \rho^2})$. With the use of the Heaviside function both results can be combined into a single expression

$$(7.316) \qquad K(x, y, t; \xi, \eta, \tau) = \frac{1}{2\pi c} \frac{H[c(\tau - t) - \rho]}{\sqrt{c^2(\tau - t)^2 - \rho^2}},$$

where $\rho^2 = (x - \xi)^2 + (y - \eta)^2$.

In the one-dimensional problem the Green's function $K(x, t; \xi, \tau)$ satisfies the equation

(7.317) $$\frac{\partial^2 K}{\partial t^2} - c^2 \frac{\partial^2 K}{\partial x^2} = \delta(x - \xi)\delta(t - \tau)$$

with $-\infty < x, \xi < +\infty$, and $\tau < T$. Again the conditions at $t = T$ are

(7.318) $$K\Big|_{t=T} = \frac{\partial K}{\partial t}\Big|_{t=T} = 0.$$

It is a simple matter (see the exercises) to adapt the general solution of the Cauchy problem to show that K is given as

(7.319) $$K(x, t; \xi, \tau) = \begin{cases} \dfrac{1}{2c}; & |x - \xi| < c(\tau - t) \\ 0; & |x - \xi| > c(\tau - t) \end{cases}.$$

In terms of Heaviside functions K can be expressed as

(7.320) $$K(x, t; \xi, \tau) = \frac{1}{2c} H[x - ct - (\xi - c\tau)] H[\xi + c\tau - (x + ct)]$$

and we now show that K is a solution of (7.317)–(7.318).

Putting $t = T$ in (7.320) gives

(7.321)
$$K(x, T; \xi, \tau) = \frac{1}{2c} H[x - cT - (\xi - c\tau)] H[\xi + c\tau - (x + cT)] = 0$$

because if the argument of the first Heaviside function—that is, $x - \xi - c(T - \tau)$—is positive, we have $x - \xi > c(T - \tau) > 0$ since T is greater than τ. Consequently, $(\xi - x) - c(T - \tau) = -(x - \xi) - c(T - \tau) < 0$, and the second Heaviside function vanishes. A similar result holds if the argument of the second Heaviside function is positive.

Further, $\partial K / \partial t$ is given as

(7.322) $$\frac{\partial K}{\partial t} = -\frac{1}{2}\delta[x - ct - (\xi - c\tau)] H[\xi + c\tau - (x + ct)]$$

$$-\frac{1}{2} H[x - ct - (\xi - c\tau)]\delta[\xi + c\tau - (x + ct)]$$

and K_t vanishes at $t = T$ for the same reason that K vanishes there.

Finally, we easily obtain

$$(7.323) \quad K_{tt} - c^2 K_{xx} = 2c\delta[x - ct - (\xi - c\tau)]\delta[\xi + c\tau - (x + ct)]$$
$$= 2c\delta[x - ct - (\xi - c\tau)]\delta[2c(\tau - t)]$$
$$= \delta[x - ct - (\xi - c\tau)]\delta[(\tau - t)]$$
$$= \delta(x - \xi)\delta(\tau - t) = \delta(x - \xi)\delta(t - \tau)$$

where the properties of the delta function given in the preceding section have been used.

A common feature of the Green's functions (7.309), (7.316), and (7.320) obtained is that they are symmetric in x and ξ where x is the observation point and ξ is the source point in one, two, or three dimensions. We shall refer to them as the *free space Green's functions* for the wave equation. If t and τ are interchanged, we obtain the *causal fundamental solution* for each of the problems. Let the fundamental solutions be denoted by S. In the three-dimensional case we have

$$(7.324) \quad S = \frac{\delta[c(t - \tau) - r]}{4\pi cr}; \qquad r = \sqrt{(x - \xi)^2 + (y - \eta)^2 + (z - \zeta)^2}.$$

In two dimensions

$$(7.325) \quad S = \frac{1}{2\pi c} \frac{H[c(t - \tau) - \rho]}{\sqrt{c^2(t - \tau)^2 - \rho^2}}; \qquad \rho = \sqrt{(x - \xi)^2 + (y - \eta)^2},$$

whereas in one dimension

$$(7.326) \quad S = \frac{1}{2c} H[c(t - \tau) + (x - \xi)] H[c(t - \tau) - (x - \xi)].$$

Each of these functions satisfies the same equation as the Green's function K, except that for each of them we have

$$(7.327) \qquad\qquad S\Big|_{t=0} = \frac{\partial S}{\partial t}\Big|_{t=0} = 0.$$

The fundamental solutions yield a vivid distinction between the nature of wave propagation in two and three dimensions as characterized by solutions of the wave equation. (These differences have already been discussed in Example 5.7.) Given the source point (ξ, η, ζ) and the time τ, the forward characteristic cone for the wave equation is

$$(7.328) \quad c(t - \tau) = \sqrt{(x - \xi)^2 + (y - \eta)^2 + (z - \zeta)^2}; \qquad t \geq \tau.$$

For the two-dimensional problem with the source point (ξ, η) and the time τ

we have

$$(7.329) \qquad c(t - \tau) = \sqrt{(x - \xi)^2 + (y - \eta)^2} \; ; \qquad t \geq \tau.$$

Noting the behavior of the delta function, (7.324) indicates that the disturbance due to a point source at (ξ, η, ζ) acting at the time τ, is concentrated on a sphere of radius $c(t - \tau)$ with center at (ξ, η, ζ) at the later time t and vanishes elsewhere in the three-dimensional case. For the two-dimensional problem, the disturbance resulting from a point source at (ξ, η) acting at the time τ, is distributed throughout a circle of radius $c(t - \tau)$ with center at (ξ, η) at the later time t. Consequently, both disturbances travel with speed c in all directions. However, in the three-dimensional case there is a sharp wave front that leaves no wake and the disturbance is only felt instantaneously at any point. In the two-dimensional problem the solution does not return to zero as soon as the wave front passes but there is a wake that decays like $1/c(t - \tau)$ as t increases. The sharp signals that occur in three-dimensional wave propagation and are lacking in the two-dimensional problem characterize what is known as *Huygens' principle*.

EXAMPLE 7.9. GREEN'S FUNCTIONS FOR THE KLEIN–GORDON AND THE MODIFIED TELEGRAPHER'S EQUATIONS

In connection with the Cauchy problem for the one-dimensional *Klein–Gordon equation* (5.239), we are led to consider the Green's function $K(x, t; \xi, \tau)$ that satisfies the equation

$$(7.330) \qquad \frac{\partial^2 K}{\partial t^2} - \gamma^2 \frac{\partial^2 K}{\partial x^2} + c^2 K = \delta(x - \xi)\delta(t - \tau),$$

with $-\infty < x, \xi < \infty$ and $t, \tau < T$. The conditions at $t = T$ are

$$(7.331) \qquad K\bigg|_{t=T} = \frac{\partial K}{\partial t}\bigg|_{t=T} = 0.$$

Adapting the methods developed in Section 6.7 for the construction of fundamental solutions (see the exercises), we conclude that the Green's function K should be given as

(7.332)

$$K(x, t; \xi, \tau) = \begin{cases} \dfrac{1}{2\gamma} J_0\left[\dfrac{c}{\gamma}\sqrt{\gamma^2(t - \tau)^2 - (x - \xi)^2}\right]; & |x - \xi| < \gamma(\tau - t) \\[2mm] 0; & |x - \xi| > \gamma(\tau - t) \end{cases}$$

where J_0 is the Bessel function of order zero. Noting (7.320) we can express

K as

$$(7.333) \quad K(x, t; \xi, \tau) = \frac{1}{2\gamma} J_0 \left[\frac{c}{\gamma} \sqrt{\gamma^2(t - \tau)^2 - (x - \xi)^2} \right]$$

$$\times H[x - \gamma t - (\xi - \gamma \tau)] H[\xi + \gamma \tau - (x + \gamma t)].$$

Since $J_0(0) = 1$ we see that (7.333) reduces to the Green's function for the one-dimensional wave equation if we set $c = 0$.

Inasmuch as we have not derived the Green's function K directly by Fourier transforms or some other means, we now show that it satisfies (7.330)–(7.331). We set

$$(7.334) \qquad\qquad\qquad K = J_0 \hat{K}$$

where $\hat{K}$ is the Green's function for the wave equation, and satisfies (7.317) with c replaced by γ. Then

$$(7.335)$$

$$\frac{\partial^2 K}{\partial t^2} - \gamma^2 \frac{\partial^2 K}{\partial x^2} + c^2 K = \left\{ \frac{\partial^2 J_0}{\partial t^2} - \gamma^2 \frac{\partial^2 J_0}{\partial x^2} + c^2 J_0 \right\} \hat{K}$$

$$+ 2 \left\{ \frac{\partial J_0}{\partial t} \frac{\partial \hat{K}}{\partial t} - \gamma^2 \frac{\partial J_0}{\partial x} \frac{\partial \hat{K}}{\partial x} \right\} + \left\{ \frac{\partial^2 \hat{K}}{\partial t^2} - \gamma^2 \frac{\partial^2 \hat{K}}{\partial x^2} \right\} J_0.$$

In view of the results of Section 6.7 [see (6.210)–(6.211)] we conclude that the first bracketed term in (7.335) vanishes. Also, we have

$$(7.336)$$

$$2 \left\{ \frac{\partial J_0}{\partial t} \frac{\partial \hat{K}}{\partial t} - \gamma^2 \frac{\partial J_0}{\partial x} \frac{\partial \hat{K}}{\partial x} \right\} = \frac{c J_0' \left[(c/\gamma) \sqrt{\gamma^2(t - \tau)^2 - (x - \xi)^2} \right]}{\sqrt{\gamma^2(t - \tau)^2 - (x - \xi)^2}}$$

$$\times \{ [-\gamma(t - \tau) + (x - \xi)] \delta[x - \gamma t - (\xi - \gamma \tau)] H[\xi + \gamma \tau - (x + \gamma t)]$$

$$+ [-\gamma(t - \tau) - (x - \xi)] H[x - \gamma t - (\xi - \gamma \tau)] \delta[\xi + \gamma \tau - (x + \gamma t)] \}.$$

The expression (7.336) vanishes, since

$$(7.337) \quad [-\gamma(t - \tau) + (x - \xi)] \delta[x - \gamma t - (\xi - \gamma \tau)]$$

$$= [-\gamma(t - \tau) + (x - \xi)] \delta[-\gamma(t - \tau) + (x - \xi)] = 0,$$

$$(7.338) \quad [-\gamma(t - \tau) - (x - \xi)] \delta[\xi + \gamma \tau - (x + \gamma t)]$$

$$= [-\gamma(t - \tau) - (x - \xi)] \delta[-\gamma(t - \tau) - (x - \xi)] = 0,$$

on using $x\delta(x) = 0$. Finally, we have

(7.339)

$$\left[\frac{\partial^2 \hat{K}}{\partial t^2} - \gamma^2 \frac{\partial^2 \hat{K}}{\partial x^2}\right] J_0 = \delta(x - \xi)\delta(t - \tau) J_0\left[\frac{c}{\gamma}\sqrt{\gamma^2(t - \tau)^2 - (x - \xi)^2}\right]$$

$$= \delta(x - \xi)\delta(t - \tau) J_0(0) = \delta(x - \xi)\delta(t - \tau)$$

since $J_0(0) = 1$ and $\hat{K}$ satisfies (7.317). Thus K is a solution of the equation (7.330). The conditions (7.331) are also satisfied because $\hat{K}$ and its time derivative vanish at $t = T$.

It may be noted that with $c = i\hat{c}$ (where $i = \sqrt{-1}$), in the Klein–Gordon equation (7.330) we obtain the *modified telegrapher's equation*

(7.340)
$$\frac{\partial^2 K}{\partial t^2} - \gamma^2 \frac{\partial^2 K}{\partial x^2} - \hat{c}^2 K = \delta(x - \xi)\delta(t - \tau).$$

This equation results when the first time derivative term is eliminated from the *telegrapher's equation*. Since $J_0(ix) = I_0(x)$, the modified Bessel function of zero order, we obtain for K in place of (7.333),

(7.341) $\quad K(x, t; \xi, \tau) = \frac{1}{2\gamma} I_0\left[\frac{\hat{c}}{\gamma}\sqrt{\gamma^2(t - \tau)^2 - (x - \xi)^2}\right]$

$$\times H[x - \gamma t - (\xi - \gamma\tau)] H[\xi + \gamma\tau - (x + \gamma t)].$$

Again, (7.333) and (7.341) are referred to as *free space Green's functions*.

Given the initial value problem for the *Klein–Gordon equation*,

(7.342) $\quad u_{tt} - \gamma^2 u_{xx} + c^2 u = F(x, t), \qquad -\infty < x < \infty, t > 0$

with the initial conditions

(7.343) $\quad u(x, 0) = f(x); \qquad u_t(x, 0) = g(x); \qquad -\infty < x < \infty,$

the solution at an arbitrary point (ξ, τ) is given in terms of (7.333) as

(7.344)

$$u(\xi, \tau) = \int_0^T \int_{-\infty}^{\infty} FK \, dx \, dt + \int_{-\infty}^{\infty} \left[gK(x, 0; \xi, \tau) - f\frac{\partial K}{\partial t}(x, 0; \xi, \tau)\right] dx,$$

where $\tau < T$ and (7.35) has been appropriately modified.

In the double integral in (7.344) we have $K = 0$ for $t > \tau$, so that the limit in the t integral extends only up to τ. Also, from (7.332) we conclude that K vanishes unless $|x - \xi| < \gamma(\tau - t)$ and this is equivalent to

(7.345) $\qquad \xi - \gamma(\tau - t) < x < \xi + \gamma(\tau - t).$

Therefore, we obtain

(7.346)

$$\int_0^T \int_{-\infty}^\infty FK \, dx \, dt = \frac{1}{2\gamma} \int_0^T \int_{\xi-\gamma(\tau-t)}^{\xi+\gamma(\tau-t)} F(x,t) J_0\left[\frac{c}{\gamma}\sqrt{\gamma^2(t-\tau)^2 - (x-\xi)^2}\right] dx \, dt.$$

Further, we have

(7.347)

$$K(x,0;\xi,\tau) = \frac{1}{2\gamma} J_0\left[\frac{c}{\gamma}\sqrt{\gamma^2\tau^2 - (x-\xi)^2}\right] H[x - (\xi - \gamma\tau)] H[\xi + \gamma\tau - x]$$

so that

(7.348)

$$\int_{-\infty}^\infty g(x) K(x,0;\xi,\tau) \, dx = \frac{1}{2\gamma} \int_{\xi-\gamma\tau}^{\xi+\gamma\tau} g(x) J_0\left[\frac{c}{\gamma}\sqrt{\gamma^2\tau^2 - (x-\xi)^2}\right] dx,$$

since the product of the Heaviside functions vanishes outside the interval $(\xi - \gamma\tau, \xi + \gamma\tau)$. Finally,

(7.349)

$$\frac{\partial K(x,0;\xi,\tau)}{\partial t}$$

$$= -\frac{c\tau}{2} \frac{J_0'\left[\frac{c}{\gamma}\sqrt{\gamma^2\tau^2 - (x-\xi)^2}\right]}{\sqrt{\gamma^2\tau^2 - (x-\xi)^2}} H[x - (\xi - \gamma\tau)] H[\xi + \gamma\tau - x]$$

$$- \frac{1}{2} J_0\left[\frac{c}{\gamma}\sqrt{\gamma^2\tau^2 - (x-\xi)^2}\right] \delta[x - (\xi - \gamma\tau)] H[\xi + \gamma\tau - x]$$

$$- \frac{1}{2} J_0\left[\frac{c}{\gamma}\sqrt{\gamma^2\tau^2 - (x-\xi)^2}\right] H[x - (\xi - \gamma\tau)] \delta[\xi + \gamma\tau - x].$$

In view of the substitution property of the delta function, the last two terms in (7.349) reduce to $-\frac{1}{2}\delta[x - (\xi - \gamma\tau)] - \frac{1}{2}\delta[x - (\xi + \gamma\tau)]$ since $J_0(0) = 1$ and $H(2\gamma\tau) = 1$. Therefore, we obtain

(7.350) $\displaystyle \int_{-\infty}^\infty f(x) \frac{\partial K}{\partial t}(x,0;\xi,\tau) \, dx$

$$= -\frac{c\tau}{2} \int_{\xi-\gamma\tau}^{\xi+\gamma\tau} f(x) \frac{J_0'\left[(c/\gamma)\sqrt{\gamma^2\tau^2 - (x-\xi)^2}\right]}{\sqrt{\gamma^2\tau^2 - (x-\xi)^2}} dx$$

$$- \frac{1}{2}f(\xi - \gamma\tau) - \frac{1}{2}f(\xi + \gamma\tau).$$

Combining these results and noting that $-J_0'(x) = J_1(x)$, the Bessel function of order one, gives the solution $u(x, t)$ of the initial value problem (7.342)–(7.343) as

(7.351)

$$
u(x, t) = \frac{f(x - \gamma t) + f(x + \gamma t)}{2}
$$

$$
+ \frac{1}{2\gamma} \int_{x-\gamma t}^{x+\gamma t} g(\xi) J_0 \left[\frac{c}{\gamma} \sqrt{\gamma^2 t^2 - (x - \xi)^2} \right] d\xi
$$

$$
- \frac{ct}{2} \int_{x-\gamma t}^{x+\gamma t} f(\xi) \frac{J_1 \left[\frac{c}{\gamma} \sqrt{\gamma^2 t^2 - (x - \xi)^2} \right]}{\sqrt{\gamma^2 t^2 - (x - \xi)^2}} d\xi
$$

$$
+ \frac{1}{2\gamma} \int_0^t \int_{x-\gamma(t-\tau)}^{x+\gamma(t-\tau)} F(\xi, \tau) J_0 \left[\frac{c}{\gamma} \sqrt{\gamma^2 (\tau - t)^2 - (x - \xi)^2} \right] d\xi \, d\tau.
$$

This solution formula reduces to that for the Cauchy problem for the inhomogeneous wave equation if we set $c = 0$. Also, if we set $c = i\hat{c}$ in (7.351) and note that $J_0(iz) = I_0(z)$ and $J_1(iz) = i I_1(z)$, we obtain as the solution of the *modified telegrapher's equation*

$$
(7.352) \quad u_{tt} - \gamma^2 u_{xx} - \hat{c}^2 u = F(x, t), \qquad -\infty < x < \infty, t > 0
$$

with the initial conditions (7.343),

(7.353)

$$
u(x, t) = \frac{f(x - \gamma t) + f(x + \gamma t)}{2}
$$

$$
+ \frac{1}{2\gamma} \int_{x-\gamma t}^{x+\gamma t} g(\xi) I_0 \left[\frac{\hat{c}}{\gamma} \sqrt{\gamma^2 t^2 - (x - \xi)^2} \right] d\xi
$$

$$
+ \frac{\hat{c}t}{2} \int_{x-\gamma t}^{x+\gamma t} f(\xi) \frac{I_1 \left[\frac{\hat{c}}{\gamma} \sqrt{\gamma^2 t^2 - (x - \xi)^2} \right]}{\sqrt{\gamma^2 t^2 - (x - \xi)^2}} d\xi
$$

$$
+ \frac{1}{2\gamma} \int_0^t \int_{x-\gamma(t-\tau)}^{x+\gamma(t-\tau)} F(\xi, \tau) I_0 \left[\frac{\hat{c}}{\gamma} \sqrt{\gamma^2 (t - \tau)^2 - (x - \xi)^2} \right] d\xi \, d\tau.
$$

Both solutions (7.351) and (7.353) exhibit the *domains of dependence* and *influence* for the *Klein–Gordon* and *modified telegrapher's equation*. It has already been indicated that these domains, which characterize the maximum speed at which disturbances or signals travel, are determined by the principal parts of the given equations (i.e., the second derivative terms) and do not depend on the lower order terms. Our results show that these equations and the wave equation have identical domains of dependence and influence.

Before considering an example dealing with the Green's function for an elliptic equation, we examine the relationship between the Green's functions for hyperbolic and parabolic equations defined in terms of the inhomogeneous equations (7.19) and (7.31), and those defined in terms of the homogeneous equations (7.37) and (7.38). The relationship between these two determinations of Green's functions for the case of bounded regions was discussed at the end of Section 7.3.

Let the Green's functions w be defined in terms of the (backwards) initial value problems (7.37)–(7.40) for the hyperbolic and parabolic cases with the initial data given at $t = \tau$ (instead of at $t = T$) and the problem defined over the entire space. Then if we set $K = wH(\tau - t)$ with $\tau < T$, it is easy to verify from the properties of w and by direct differentiation that K is indeed the solution of the Green's function problem given in terms of the inhomogeneous equations (7.19) and (7.31) for the hyperbolic and parabolic cases, respectively. Thus for $t < \tau$ both of these formulations lead to the same Green's functions.

In a similar fashion, it can be shown that the *causal fundamental solutions* for hyperbolic and parabolic equations can be characterized in terms of initial value problems. If the instantaneous source point is at $(x, t) = (\xi, \tau)$, the initial conditions for the homogeneous form of the hyperbolic equation (7.15) are $u = 0$ and $u_t = \delta(x - \xi)/\rho$ at $t = \tau$. For the homogeneous form of the parabolic equation (7.27) with the same source point, the initial condition is $u = \delta(x - \xi)/\rho$ at $t = \tau$. Then the causal fundamental solutions for these problems, which we denote by S, can be expressed in terms of solutions u of the aforementioned initial value problems in the form $S = uH(t - \tau)$. The verification of this result is almost identical to that for the Green's function problems discussed in the preceding paragraph.

Consequently, instantaneous point source problems and the related causal fundamental solutions need not necessarily be characterized strictly in terms of inhomogeneous differential equations as has been done in Section 6.7 and Example 7.8, for instance. We can determine the solutions of these problems in terms of homogeneous equations with appropriate Dirac delta function initial conditions. This approach is used in Section 9.5 when we discuss methods for analyzing the propagation of singularities for hyperbolic equations.

EXAMPLE 7.10.   GREEN'S FUNCTION FOR THE REDUCED WAVE EQUATION

In this problem we construct a Green's function for a boundary value problem for the reduced wave equation which is of interest in the theory of ocean acoustics.

The ocean which we assume to be homogeneous, is taken to be of infinite extent in the x and y directions and is assumed to have a constant depth $h > 0$.

Thus $-\infty < x, y < \infty$, and $-h < z < 0$. A point source is located at $x = y = 0$ and $z = \zeta$ with $-h < \zeta < 0$. Then the Green's function $K = K(x, y, z; \zeta)$ satisfies the equation

(7.354)

$$\nabla^2 K + k^2 K = \frac{\partial^2 K}{\partial x^2} + \frac{\partial^2 K}{\partial y^2} + \frac{\partial^2 K}{\partial z^2} + k^2 K = -\delta(x)\delta(y)\delta(z - \zeta),$$

and the boundary conditions

(7.355) $$K(x, y, 0; \zeta) = 0,$$

(7.356) $$\frac{\partial K}{\partial z}(x, y, -h; \zeta) = 0.$$

The real part of $Ke^{-i\omega t}$, where ω is the frequency and $k = \omega/c$, represents the time harmonic acoustic pressure and satisfies a wave equation with constant wave speed c in the absence of any source terms. Thus the equation for K is known as the *reduced wave equation*. The plane $z = 0$ represents the surface of the ocean, and the plane $z = -h$ is the (rigid) bottom. The conditions (7.355) and (7.356) are appropriate for the sound pressure at the surface and at the bottom, respectively. The homogeneity of the ocean implies that k in (7.354) is a constant. This problem for K does not have a unique solution unless a *radiation condition* is imposed. This is done in our discussion.

To solve for the Green's function K we use the method of eigenfunction expansions. We consider an eigenvalue problem for the function $M(z)$ over the interval $-h < z < 0$. The equation for $M(z)$ is

(7.357) $$-M''(z) - k^2 M(z) = \lambda M(z), \qquad -h < z < 0$$

where λ is the eigenvalue parameter. The boundary conditions are

(7.358) $$M(0) = 0; \qquad M'(-h) = 0.$$

On comparing (7.357)–(7.358) with the Sturm–Liouville problem of Section 4.3 we find that, apart from the fact that the coefficient $-k^2$ of $M(z)$ is negative, all conditions are the same. All properties of the eigenvalues and eigenfunctions given in that section are valid for (7.357)–(7.358) except that a finite number of the eigenvalues are negative as shown.

The general solution of (7.357) is

(7.359) $$M(z) = c_1 \sin\left[z\sqrt{k^2 + \lambda}\right] + c_2 \cos\left[z\sqrt{k^2 + \lambda}\right].$$

The condition $M(0) = 0$ implies that $c_2 = 0$. The condition $M'(-h) = 0$ implies that

(7.360) $$\cos\left[h\sqrt{k^2 + \lambda}\right] = 0,$$

so that λ is specified in terms of the zeros of the cosine function. The eigenvalues λ_n are given as

$$(7.361) \qquad \lambda_n = \left(n + \frac{1}{2}\right)^2 \left(\frac{\pi}{h}\right)^2 - k^2, \qquad n = 0, 1, 2, \ldots,$$

and the eigenfunctions are

$$(7.362) \qquad \hat{M}_n(z) = \sin\left[\left(n + \frac{1}{2}\right)\left(\frac{\pi}{h}\right)z\right], \qquad n = 0, 1, 2, \ldots.$$

Note that some of the λ_n may be negative.

The inner product appropriate for this eigenvalue problem is

$$(7.363) \qquad (\varphi, \psi) = \int_{-h}^{0} \varphi(z)\psi(z)\, dz$$

and $(\hat{M}_n, \hat{M}_m) = 0$ for $m \neq n$ as can easily be shown. In addition,

$$(7.364) \qquad (\hat{M}_n, \hat{M}_n) = \int_{-h}^{0} \sin^2\left[\left(n + \frac{1}{2}\right)\left(\frac{\pi}{h}\right)z\right] dz$$

$$= \int_{-h}^{0} \left\{\frac{1}{2} - \frac{1}{2}\cos\left[(2n + 1)\left(\frac{\pi}{h}\right)z\right]\right\} dz = \frac{h}{2}.$$

Thus the normalized set of eigenfunctions are

$$(7.365) \qquad M_n(z) = \sqrt{\frac{2}{h}}\, \sin\left[\left(n + \frac{1}{2}\right)\left(\frac{\pi}{h}\right)z\right], \qquad n = 0, 1, 2, \ldots.$$

In terms of the $M_n(z)$ we construct the eigenfunction expansion of the Green's function K:

$$(7.366) \qquad K(x, y, z; \zeta) = \sum_{n=0}^{\infty} N_n(x, y) M_n(z).$$

The coefficients N_n are determined by multiplying (7.354) by $M_n(z)$ and integrating from $-h$ to 0. Using the procedures of Section 4.6 we obtain

$$(7.367) \qquad \frac{\partial^2 N_n}{\partial x^2} + \frac{\partial^2 N_n}{\partial y^2} + \left[k^2 - \left(n + \frac{1}{2}\right)^2 \left(\frac{\pi}{h}\right)^2\right] N_n = -M_n(\zeta)\delta(x)\delta(y)$$

for $n = 0, 1, 2, \ldots$.

To obtain a unique solution of (7.367) we must specify the behavior of N_n as x and y tend to infinity. Apart from the factor $M_n(\zeta)$, we are essentially interested in constructing a free space Green's function for (7.367). If $k^2 >$

$[n + (1/2)]^2(\pi/h)^2$, (7.367) is an inhomogeneous Helmholtz or reduced wave equation, whereas if $k^2 < [n + (1/2)]^2(\pi/h)^2$, we have a modified Helmholtz equation.

With $r = \sqrt{x^2 + y^2}$, a function $\varphi(x, y)$ is said to satisfy the *Sommerfeld radiation condition* at infinity if

$$(7.368) \qquad \lim_{r \to \infty} \sqrt{r}\left[\frac{\partial \varphi}{\partial r} - i\omega\varphi\right] = 0$$

and φ satisfies the equation

$$(7.369) \qquad \varphi_{xx} + \varphi_{yy} + \omega^2\varphi = 0$$

for large values of r. With $\omega_n^2 = -\lambda_n$ and λ_n defined in (7.361), we write (7.367) as

$$(7.370) \qquad \frac{\partial^2 N_n}{\partial x^2} + \frac{\partial^2 N_n}{\partial y^2} + \omega_n^2 N_n = -M_n(\zeta)\delta(x)\delta(y); \qquad n = 0, 1, 2, \ldots.$$

With $\omega_n^2 > 0$ we use the results of Example 6.12 to conclude, in view of (6.195), that

$$(7.371) \qquad N_n(x, y) = \left(\frac{i}{4}\right)M_n(\zeta)H_0^{(1)}(\omega_n r), \qquad n = 0, 1, \ldots$$

where $H_0^{(1)}(z)$ is the (zero-order) Hankel function of the first kind. With the use of (6.197) it is easy to see that (7.371) satisfies the radiation condition (7.366) if $\omega_n^2 > 0$. If $\omega_n^2 < 0$ (i.e., ω_n is complex valued), the appropriate solution of (7.370) must be given in terms of the modified Bessel function $K_0(z)$ as in (6.190). However, since $H_0^{(1)}(iz) = (2/\pi i)K_0(z)$ for $z > 0$, we conclude that (7.371) is a suitable solution for all values of n. Since $N_n(x, y)$ decays exponentially at infinity, in view of (6.194), when $\omega_n^2 < 0$, the radiation condition is certainly satisfied.

Collecting our results, we obtain the Green's function K in the form

$$(7.372) \quad K(x, y, z; \zeta) = \left(\frac{i}{2h}\right)\sum_{n=0}^{\infty} \sin\left[\left(n + \frac{1}{2}\right)\left(\frac{\pi}{h}\right)\zeta\right]\sin\left[\left(n + \frac{1}{2}\right)\left(\frac{\pi}{h}\right)z\right]$$

$$\times H_0^{(1)}\left[\omega_n\sqrt{x^2 + y^2}\right]$$

where $\omega_n^2 = k^2 - [n + (1/2)]^2(\pi/h)^2$. As n increases, the argument of the Hankel function eventually becomes imaginary and the terms in the series decay exponentially for large n.

7.5. THE METHOD OF IMAGES

Given any of the partial differential equations of Section 7.1, the Green's function K can be expressed in the form

$$(7.373) \qquad\qquad K = K_F + K_G,$$

where K_F is the free space Green's function. We recall that the free space Green's function satisfies the same differential equation as the Green's function K. In addition, K_F satisfies (backward) causality conditions in the hyperbolic and parabolic cases, and appropriate conditions at infinity in the elliptic case. Consequently, K_G satisfies an homogeneous differential equation with homogeneous end conditions at $t = T$ if these are relevant. The boundary conditions for K_G are no longer homogeneous, however. For example, if K is required to vanish on ∂G, then $K_G = -K_F$ on ∂G. Although it may be possible to use eigenfunction expansions or transform methods to determine K_G, we do not use these approaches since K itself can just as easily be determined in the same fashion. Instead we shall use the *method of images* to specify K_G in case the equations of Section 7.1 have constant coefficients and the boundary ∂G has a special form as described in the following.

We assume that the boundaries for the given problem are made up of (portions of) lines or planes, or (portions of) circles or spheres. Given the singular point of the delta function in the equation for K, we consider all possible image points obtained by reflection through lines and planes and inversion through circles and spheres. (The inversion process is defined later.) If none of the resulting image points lies in the interior of the region in which the problem is specified and certain additional conditions are met, the Green's function K can be specified in a simple manner. We do not describe the most general regions and equations for which the method of images works. Instead we consider a number of examples and exercises that exhibit the basic features of the method.

To apply the method of images, it is necessary to know the free space Green's functions for the given equations and most of the relevant ones have already been determined.

EXAMPLE 7.11. LAPLACE'S EQUATION IN A HALF - SPACE

We consider Laplace's equation in the half-space $z > 0$. The Green's function $K = K(x, y, z; \xi, \eta, \zeta)$ satisfies the equation

$$(7.374) \quad \nabla^2 K = \frac{\partial^2 K}{\partial x^2} + \frac{\partial^2 K}{\partial y^2} + \frac{\partial^2 K}{\partial z^2} = -\delta(x - \xi)\delta(y - \eta)\delta(z - \zeta)$$

in the region G defined as the half-space $z > 0$. On the boundary ∂G—that is, the (x, y)-plane $z = 0$—we have

(7.375) $$\alpha K + \beta \frac{\partial K}{\partial n}\Big|_{\partial G} = \alpha K - \beta \frac{\partial K}{\partial z}\Big|_{z=0} = 0.$$

We assume that α and β are constants and consider three cases. For the Dirichlet problem (i.e., the first boundary value problem) we set $\alpha = 1$ and $\beta = 0$; for the Neumann problem (i.e., the second boundary value problem) we put $\alpha = 0$ and $\beta = -1$; for the third boundary value problem we set $\alpha = h$ and $\beta = 1$ with $h > 0$.

To begin, we recall that the *free space Green's function* for Laplace's equation is

(7.376) $$K_F = \frac{1}{4\pi}\left[(x - \xi)^2 + (y - \eta)^2 + (z - \zeta)^2\right]^{-1/2}$$

as was shown in Example 6.12. It follows from our discussion in that example that $\nabla^2 K_F = -\delta(x - \xi)\delta(y - \eta)\delta(z - \zeta)$. The Green's function K is then expressed as

(7.377) $$K = K_F + K_G.$$

Since we must have

(7.378) $$\nabla^2 K = \nabla^2 K_F + \nabla^2 K_G = -\delta(x - \xi)\delta(y - \eta)\delta(z - \zeta),$$

we conclude that

(7.379) $$\nabla^2 K_G = 0,$$

so that K_G is an *harmonic function* (i.e., a solution of the homogeneous Laplace's equation). In view of (7.375), the boundary condition for K_G is

(7.380) $$\alpha K_G - \beta \frac{\partial K_G}{\partial z}\Big|_{z=0} = -\alpha K_F + \beta \frac{\partial K_F}{\partial z}\Big|_{z=0}.$$

The point (ξ, η, ζ) is the source (or singular) point for the Green's function K and K_F. Now K_F is given in terms of the distance from the observation point (x, y, z) to the source point (ξ, η, ζ). As shown in Figure 7.4, if we introduce the image (source) point $(\xi, \eta, -\zeta)$—that is, the reflection of (ξ, η, ζ) in the plane $z = 0$—then as the observation point (x, y, z) tends to a boundary point $(x, y, 0)$, its distance from (ξ, η, ζ) equals its distance from $(\xi, \eta, -\zeta)$.

Consequently, if we introduce the function

(7.381) $$\hat{K}_G = \frac{1}{4\pi}\left[(x - \xi)^2 + (y - \eta)^2 + (z + \zeta)^2\right]^{-1/2}$$

we have in effect a free space Green's function corresponding to the image

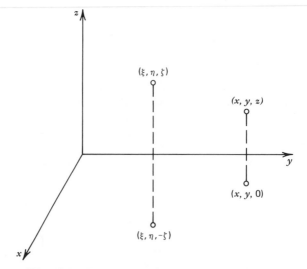

Figure 7.4.   The source point (ξ, η, ζ) and its image point.

source point $(\xi, \eta, -\zeta)$. At the boundary $z = 0$, $\hat{K}_G = K_F$ and

$$(7.382) \quad \left.\frac{\partial \hat{K}_G}{\partial z}\right|_{z=0} = \left.\frac{-(z+\zeta)}{4\pi}\left[(x-\xi)^2 + (y-\eta)^2 + (z+\zeta)^2\right]^{-3/2}\right|_{z=0}$$

$$= -\left.\frac{\partial K_F}{\partial z}\right|_{z=0}.$$

Furthermore,

$$(7.383) \qquad\qquad \nabla^2 \hat{K}_G = -\delta(x-\xi)\delta(y-\eta)\delta(z+\zeta),$$

and the right hand side vanishes in the half-space $z > 0$ since $\delta(z+\zeta) = 0$ there. Therefore, if we set $K_G = -\hat{K}_G$ for the Dirichlet problem and $K_G = \hat{K}_G$ for the Neumann problem, the Green's function $K = K_F + K_G$ satisfies (7.374) in the half-space $z > 0$ and the boundary conditions $K|_{z=0} = 0$ and $\partial K/\partial z|_{z=0} = 0$ which are appropriate for the Dirichlet and Neumann problems, respectively.

The Green's function for the third boundary value problem cannot be obtained solely in terms of an image source point at $(\xi, \eta, -\zeta)$. Instead we must introduce an entire line of image sources on the line $x = \xi$ and $y = \eta$ with z extending from $z = -\zeta$ to $z = -\infty$ and a source density function to be determined. Let

$$(7.384) \quad K_G = \frac{1}{4\pi}\left[(x-\xi)^2 + (y-\eta)^2 + (z+\zeta)^2\right]^{-1/2}$$

$$+ \frac{1}{4\pi}\int_{-\infty}^{-\zeta}\frac{\rho(s)\,ds}{\left[(x-\xi)^2 + (y-\eta)^2 + (z-s)^2\right]^{1/2}}$$

where $\rho(s)$ is the source density. When $h = 0$ in the boundary condition

$\partial K/\partial z - hK|_{z=0} = 0$, we expect (7.384) to reduce to the form appropriate for the Neumann problem so that $\rho = 0$ when $h = 0$. For this reason we have added the free space Green's function corresponding to an image source at $(\xi, \eta, -\zeta)$ to the integral term in (7.384).

It is assumed that $\rho(s)$ decays sufficiently rapidly at infinity that the integral in (7.384) converges and that differentiation under the integral sign is permitted. We then find that $\nabla^2 K_G = 0$ for $z > 0$, since all the singular points in (7.384) occur in the lower half-space $z < 0$. Applying the boundary condition at $z = 0$ gives in view of (7.380)

(7.385)

$$\frac{\partial K_G}{\partial z} - hK_G\bigg|_{z=0} = -\frac{\zeta}{4\pi}\left[(x - \xi)^2 + (y - \eta)^2 + \zeta^2\right]^{-3/2}$$

$$- \frac{1}{4\pi}\int_{-\infty}^{-\zeta}\rho(s)\frac{\partial}{\partial s}\left[(x - \xi)^2 + (y - \eta)^2 + s^2\right]^{-1/2} ds$$

$$- \frac{h}{4\pi}\left[(x - \xi)^2 + (y - \eta^2) + \zeta^2\right]^{-1/2}$$

$$- \frac{h}{4\pi}\int_{-\infty}^{-\zeta}\frac{\rho(s)\,ds}{\left[(x - \xi)^2 + (y - \eta)^2 + s^2\right]^{1/2}}$$

$$= -\frac{\partial K_F}{\partial z} + hK_F\bigg|_{z=0} = -\frac{\zeta}{4\pi}\left[(x - \xi)^2 + (y - \eta)^2 + \zeta^2\right]^{-3/2}$$

$$+ \frac{h}{4\pi}\left[(x - \xi)^2 + (y - \eta)^2 + \zeta^2\right]^{-1/2}.$$

The operator $\partial/\partial z$ equals $-\partial/\partial s$ at $z = 0$, and the use of $\partial/\partial s$ in the integral term enables us to integrate by parts. We have

(7.386) $\displaystyle\int_{-\infty}^{-\zeta}\rho(s)\frac{\partial}{\partial s}\left[(x - \xi)^2 + (y - \eta)^2 + s^2\right]^{-1/2} ds$

$$= \rho(-\zeta)\left[(x - \xi)^2 + (y - \eta)^2 + \zeta^2\right]^{-1/2}$$

$$- \int_{-\infty}^{-\zeta}\rho'(s)\left[(x - \xi)^2 + (y - \eta)^2 + s^2\right]^{-1/2} ds.$$

Combining results gives

(7.387) $\displaystyle\left[-\frac{h}{2\pi} - \frac{\rho(-\zeta)}{4\pi}\right]\left[(x - \xi)^2 + (y - \eta)^2 + \zeta^2\right]^{-1/2}$

$$+ \frac{1}{4\pi}\int_{-\infty}^{-\zeta}\frac{[\rho'(s) - h\rho(s)]\,ds}{\left[(x - \xi)^2 + (y - \eta)^2 + s^2\right]^{1/2}} = 0.$$

Therefore, the boundary condition is satisfied if we set

(7.388) $\qquad\qquad\qquad \rho'(s) - h\rho(s) = 0, \qquad s < -\zeta,$

(7.389) $\qquad\qquad\qquad\qquad \rho(-\zeta) = -2h.$

The solution of the initial value problem (7.388)–(7.389) is

$$(7.390) \qquad\qquad \rho(s) = -2he^{h(s+\zeta)}.$$

We note that $\rho(s)$ vanishes for $h = 0$ and that it decays exponentially as $s \to -\infty$.

The Green's function for the third boundary value problem thus has the form

$$(7.391)$$

$$K(x, y, z; \xi, \eta, \zeta) = \frac{1}{4\pi}\left[(x - \xi)^2 + (y - \eta)^2 + (z - \zeta)^2\right]^{-1/2}$$

$$+ \frac{1}{4\pi}\left[(x - \xi)^2 + (y - \eta)^2 + (z + \zeta)^2\right]^{-1/2}$$

$$- \frac{h}{2\pi}\int_{-\infty}^{-\zeta} \frac{e^{h(s+\zeta)}\, ds}{\left[(x - \xi)^2 + (y - \eta)^2 + (z - s)^2\right]^{1/2}}.$$

With $h = 0$ this reduces to Green's function for the Neumann problem. The Green's function for the Dirichlet problem is

$$(7.392) \quad K(x, y, z; \xi, \eta, \zeta) = \frac{1}{4\pi}\left[(x - \xi)^2 + (y - \eta)^2 + (z - \zeta)^2\right]^{-1/2}$$

$$- \frac{1}{4\pi}\left[(x - \xi)^2 + (y - \eta)^2 + (z + \zeta)^2\right]^{-1/2}.$$

On using the free space Green's function for Laplace's equation in two dimensions, it is easy to obtain the Green's function for the half-plane problem in two dimensions. Furthermore, on using the formulas given in Section 7.1, one can readily obtain the solutions of boundary value problems in the half-space or the half-plane for Laplace's equation, as seen in the exercises.

EXAMPLE 7.12.   HYPERBOLIC EQUATIONS IN A SEMI - INFINITE INTERVAL

The free space Green's functions corresponding to the one-dimensional hyperbolic equation (7.45) with constant coefficients were obtained in our discussion in Examples 7.8 and 7.9.

For the *Klein–Gordon equation*, the free space Green's function K_F satisfies (7.330) and is given as

$$(7.393) \quad K_F(x, t; \xi, \tau) = \frac{1}{2\gamma} J_0\left[\frac{c}{\gamma}\sqrt{\gamma^2(t - \tau)^2 - (x - \xi)^2}\right]$$

$$\times H[x - \gamma t - (\xi - \gamma\tau)] H[\xi + \gamma\tau - (x + \gamma t)]$$

in view of (7.333). For the *modified telegrapher's equation* (7.340) we have

$$(7.394) \quad K_F(x, t; \xi, \tau) = \frac{1}{2\gamma} I_0 \left[\frac{\hat{c}}{\gamma} \sqrt{\gamma^2(t - \tau)^2 - (x - \xi)^2} \right]$$

$$\times H[x - \gamma t - (\xi - \gamma \tau)] H[\xi + \gamma \tau - (x + \gamma t)].$$

The free space Green's function for the one-dimensional *wave equation* (7.317) is

$$(7.395) \quad K_F(x, t; \xi, \tau) = \frac{1}{2c} H[x - ct - (\xi - ct)] H[\xi + c\tau - (x + ct)].$$

It should be noted that for the wave equation K_F can also be expressed as

$$(7.396) \quad K_F(x, t; \xi, \tau) = \frac{1}{2c} H(\tau - t) H\left[c^2(t - \tau)^2 - (x - \xi)^2 \right].$$

It is readily seen that (7.396) is consistent with (7.319) and that the product of the Heaviside functions given in (7.393)–(7.394), can also be expressed in the form (7.396) with appropriately modified constants. In connection with the application of the method of images, it is apparent from the form of K_F in (7.396) that an image source can be introduced at $x = -\xi$ if the Green's function in the interval $x > 0$ is to be obtained.

We now show by direct substitution that (7.396) is a solution of (7.317). We have

$$(7.397) \quad \frac{\partial K_F}{\partial t} = -\frac{1}{2c} \delta(\tau - t) H\left[c^2(t - \tau)^2 - (x - \xi)^2 \right]$$

$$+ c(t - \tau) H(\tau - t) \delta\left[c^2(t - \tau)^2 - (x - \xi)^2 \right].$$

The first term on the right side of (7.397) vanishes since $\delta(\tau - t) H[-(x - \xi)^2] = \delta(\tau - t) H[c^2(t - \tau)^2 - (x - \xi)^2]$ and $H(x)$ vanishes for negative values of the argument. Then

(7.398)

$$\frac{\partial^2 K_F}{\partial t^2} - c^2 \frac{\partial^2 K_F}{\partial x^2}$$

$$= -c(t - \tau) \delta(\tau - t) \delta\left[c^2(t - \tau)^2 - (x - \xi)^2 \right]$$

$$+ 2cH(\tau - t) \left\{ \left[c^2(t - \tau)^2 - (x - \xi)^2 \right] \delta'\left[c^2(t - \tau)^2 - (x - \xi)^2 \right] \right.$$

$$\left. + \delta\left[c^2(t - \tau)^2 - (x - \xi)^2 \right] \right\}.$$

The second term on the right of (7.398) vanishes in view of (7.94). Using

(7.138) gives

$$(7.399) \quad \delta\left[c^2(t-\tau)^2 - (x-\xi)^2\right]$$

$$= \frac{1}{2c|t-\tau|}\{\delta[c(t-\tau)-(x-\xi)] + \delta[c(t-\tau)+(x-\xi)]\}.$$

In the region $t < \tau$, which is of greatest interest, we have $|t-\tau| = -(t-\tau)$. Thus

$$(7.400) \quad -c(t-\tau)\delta(\tau-t)\delta\left[c^2(t-\tau)^2 - (x-\xi)^2\right]$$

$$= -\frac{c(t-\tau)}{2c|t-\tau|}\delta(\tau-t)\{\delta[c(t-\tau)-(x-\xi)]$$

$$+ \delta[c(t-\tau)+(x-\xi)]\}$$

$$= \frac{1}{2}\delta(\tau-t)\{\delta[c(t-\tau)-(x-\xi)]$$

$$+ \delta[c(t-\tau)+(x-\xi)]\}$$

$$= \frac{1}{2}\delta(\tau-t)\{\delta[-(x-\xi)] + \delta(x-\xi)\}$$

$$= \delta(t-\tau)\delta(x-\xi)$$

since $\delta(-z) = \delta(z)$ and $\delta(\tau-t)\delta[c(t-\tau)\pm(x-\xi)] = \delta(\tau-t)\delta[\pm(x-\xi)]$.

For each of the hyperbolic equations considered we now obtain Green's functions for the semi-infinite interval $x > 0$. In the case of Dirichlet boundary conditions at $x = 0$ (i.e., $K = 0$ at $x = 0$), the Green's function is given as

$$(7.401) \quad K(x,t;\xi,\tau) = K_F(x,t;\xi,\tau) - K_F(x,t;-\xi,\tau),$$

with K_F equal to the appropriate free space Green's function (7.393), (7.394), or (7.395). If a Neumann boundary condition is given at $x = 0$ (i.e., $\partial K/\partial x|_{x=0} = 0$), the Green's function is

$$(7.402) \quad K(x,t;\xi,\tau) = K_F(x,t;\xi,\tau) + K_F(x,t;-\xi,\tau).$$

Finally, if a boundary condition of the third kind is given (i.e., $\partial K/\partial x - hK|_{x=0} = 0$ with $h > 0$), the Green's function is

$$(7.403) \quad K(x,t;\xi,\tau) = K_F(x,t;\xi,\tau) + K_F(x,t;-\xi,\tau)$$

$$- 2h\int_{-\infty}^{-\xi} e^{h(s+\xi)}K_F(x,t;s,\tau)\,ds.$$

By comparison with the methods used in the preceding example or by direct verification it can be determined that (7.401)–(7.403) give the required Green's functions for each of the boundary value problems considered.

EXAMPLE 7.13. THE HEAT EQUATION IN A FINITE INTERVAL

In this example we construct the Green's function $K(x, t; \xi, \tau)$ for the equation of heat conduction in a finite interval $0 < x < l$. Thus K satisfies the equation

$$(7.404) \qquad \frac{\partial K}{\partial t} + c^2 \frac{\partial^2 K}{\partial x^2} = -\delta(x - \xi)\delta(t - \tau); \qquad 0 < x, \xi < l, t, \tau < T.$$

It is assumed that K vanishes at the endpoints so that

$$(7.405) \qquad K(0, t; \xi, \tau) = K(l, t; \xi, \tau) = 0.$$

In addition, we have the end condition

$$(7.406) \qquad K(x, T; \xi, \tau) = 0.$$

The free space Green's function K_F for the one-dimensional *heat equation* was found in Example 7.7 to be [see (7.291)],

$$(7.407) \qquad K_F(x, t; \xi, \tau) = \frac{H(\tau - t)}{\sqrt{4\pi c^2(\tau - t)}} \exp\left[-\frac{(x - \xi)^2}{4c^2(\tau - t)}\right].$$

We express K in the form (7.373), (i.e., $K = K_F + K_G$) and use the method of images to specify K_G. The source point $x = \xi$ must have an image with respect to $x = 0$ and $x = l$, and each of the image sources, in turn, must also have images with respect to $x = 0$ and $x = l$. Consequently, we are led to consider an infinite sequence of source points at the points $\xi_n = \pm\xi \pm 2nl$, with $n = 0, 1, 2, 3, \ldots$. Some of these points are shown in Figure 7.5.

The Green's function K can then be written as an infinite series

(7.408)

$$K(x, t; \xi, \tau) = \frac{H(\tau - t)}{\sqrt{4\pi c^2(\tau - t)}}$$

$$\times \sum_{n=-\infty}^{\infty} \left\{ \exp\left[-\frac{(x - \xi - 2nl)^2}{4c^2(\tau - t)}\right] - \exp\left[-\frac{(x + \xi - 2nl)^2}{4c^2(\tau - t)}\right] \right\}.$$

The term with $n = 0$ and a positive coefficient corresponds to K_F. Clearly, $K = 0$ at $t = T$ since $\tau < T$. It can be shown that the series can be differenti-

Figure 7.5. The source point and the image sources.

ated term by term. Inasmuch as each of the functions in the series except that term corresponding to K_F has it source point ξ_n outside the interval $0 < x < l$, we see that (7.408) satisfies (7.404). Also, it is not difficult to see that at $x = 0$ and $x = l$ there corresponds to each term in the series with a positive coefficient an identical term with a negative coefficient. For example, Figure 7.5 shows that $x = \xi$ and $x = -\xi$ are images with respect to $x = 0$, and $x = \xi$ and $x = 2l - \xi$ are images with respect to $x = l$. The terms in the series (7.408) that correspond to $x = \xi$ and $x = -\xi$ are $\exp[-(x - \xi)^2/4c^2(\tau - t)] - \exp[-(x + \xi)^2/4c^2(\tau - t)]$, and this difference vanishes at $x = 0$. Similarly, for $x = \xi$ and $x = 2l - \xi$ we have $\exp[-(x - \xi)^2/4c^2(\tau - t)] - \exp[-(x + \xi - 2l)^2/4c^2(\tau - t)]$, and this difference vanishes when $x = l$.

The solution of the initial and boundary value problem for the heat equation in a finite interval when the formula (7.36) of Section 7.1 is used has a form similar to that given in Example 5.12. As was shown there, this result is expected to be useful for small values of t, whereas that given by the finite Fourier transform method is more useful for t large.

EXAMPLE 7.14.   GREEN'S FUNCTION FOR LAPLACE'S EQUATION IN A SPHERE

As shown in the preceding examples, the method of images can be applied to equations with constant coefficients of all three types if there are linear or planar boundaries. For Laplace's equation it is possible to extend the image method to problems that involve circular or spherical boundaries as demonstrated in this example.

We now construct the Green's function for the Dirichlet problem for *Laplace's equation* in the interior of a sphere using inversion with respect to the sphere. The Green's function $K = K(x, y, z; \xi, \eta, \zeta)$ satisfies the equation

$$(7.409) \quad \nabla^2 K = \frac{\partial^2 K}{\partial x^2} + \frac{\partial^2 K}{\partial y^2} + \frac{\partial^2 K}{\partial z^2} = -\delta(x - \xi)\delta(y - \eta)\delta(z - \zeta)$$

and the boundary condition $K = 0$ on the sphere of radius a with center at the origin. Let the observation point P be denoted by $P = (x, y, z)$. The source point is $P_0 = (\xi, \eta, \zeta)$ and the origin of coordinates (i.e., the center of the sphere) is $O = (0, 0, 0)$.

As shown in Figure 7.6, we introduce the image source point $P_0' = (\xi', \eta', \zeta')$. The point P_0' lies on the radial line extending from the origin O through the source point P_0. Its distance from the origin equals ρ_1, and the distance of P_0

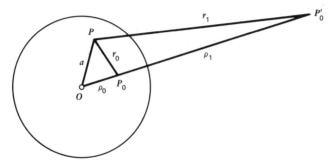

Figure 7.6. Inversion with respect to the sphere.

from the origin is ρ_0. These distances are related by the equation

$$(7.410) \qquad \rho_0 \rho_1 = a^2$$

where a is the radius of the sphere. The foregoing process of obtaining P_0' from P_0 is called *inversion with respect to the sphere*. Note that if the radius $a = 1$, we have $\rho_1 = 1/\rho_0$, so that the distances ρ_0 and ρ_1 are inverse to one another. The observation point P is assumed to lie on the sphere in Figure 7.6.

The triangles $\Delta OP P_0$ and $\Delta OP P_0'$ in the figure are similar since they have a common angle $\sphericalangle POP_0$ and proportional sides $\overline{OP_0}/\overline{OP} = \overline{OP}/\overline{OP_0'}$. This follows since $\overline{OP} = a$ (the radius of the sphere), $\overline{OP_0} = \rho_0$, $\overline{OP_0'} = \rho_1$, and $(\overline{OP_0})(\overline{OP_0'}) = \rho_0 \rho_1 = (\overline{OP})^2 = a^2$ in view of (7.410). The similarity of the triangles implies that all three sides are proportional and we have

$$(7.411) \qquad \frac{\rho_0}{a} = \frac{a}{\rho_1} = \frac{r_0}{r_1}.$$

To complete the solution of the problem we set $K = K_F + K_G$ where K_F is the free space Green's function (7.376) and K_G is a constant multiple of the free space Green's function with source point at $P_0' = (\xi', \eta', \zeta')$. That is,

$$(7.412) \qquad K = \frac{1}{4\pi}\left\{ \left[(x - \xi)^2 + (y - \eta)^2 + (z - \zeta)^2 \right]^{-1/2} \right.$$
$$\left. + c\left[(x - \xi')^2 + (y - \eta')^2 + (z - \zeta')^2 \right]^{-1/2} \right\}$$

with the constant c to be specified. Since P_0' lies outside the sphere, the second term in (7.412) is a solution of (the homogeneous) Laplace's equation within the sphere. Thus (7.412) is a solution of (7.409). On the sphere—that is, when $P = (x, y, z)$ lies on the sphere—we have, in the notation of Figure 7.6,

$$(7.413) \qquad K = \frac{1}{4\pi}\left[\frac{1}{r_0} + \frac{c}{r_1} \right] = \frac{1}{4\pi r_0}\left[1 + \frac{\rho_0 c}{a} \right],$$

in view of (7.411). Thus if $c = -a/\rho_0$, the Green's function K vanishes on the sphere. Therefore, we obtain the Green's function K as

$$(7.414) \qquad K = \frac{1}{4\pi}\left[\frac{1}{r_0} - \frac{a}{\rho_0 r_1}\right].$$

Let G represent the interior of the sphere $x^2 + y^2 + z^2 = a^2$ with ∂G equal to the sphere itself. The Dirichlet boundary value problem for *Poisson's equation*

$$(7.415) \quad \nabla^2 u = u_{xx} + u_{yy} + u_{zz} = -F(x, y, z), \qquad (x, y, z) \in G$$

with the boundary condition

$$(7.416) \qquad u|_{\partial G} = B(x, y, z),$$

has the solution

$$(7.417) \qquad u(\xi, \eta, \zeta) = \int\int_G\int KF\, dv - \int\int_{\partial G} \frac{\partial K}{\partial n} B\, ds,$$

with (ξ, η, ζ) equal to an arbitrary point within the sphere as follows from (7.11).

To determine (7.417) more explicitly it is necessary to evaluate the exterior normal derivative $\partial K/\partial n$ on the surface of the sphere. Now

$$(7.418) \qquad \frac{\partial K}{\partial n} = \frac{1}{4\pi}\left[\frac{\partial}{\partial n}\left(\frac{1}{r_0}\right) - \frac{a}{\rho_0}\frac{\partial}{\partial n}\left(\frac{1}{r_1}\right)\right]$$

$$= \frac{1}{4\pi}\left[\frac{\partial}{\partial r_0}\left(\frac{1}{r_0}\right)\frac{\partial r_0}{\partial n} - \frac{a}{\rho_0}\frac{\partial}{\partial r_1}\left(\frac{1}{r_1}\right)\frac{\partial r_1}{\partial n}\right]$$

$$= \frac{1}{4\pi}\left[-\frac{1}{r_0^2}\frac{\partial r_0}{\partial n} + \frac{a}{\rho_0}\frac{1}{r_1^2}\frac{\partial r_1}{\partial n}\right].$$

We recall that $r_0^2 = (x - \xi)^2 + (y - \eta)^2 + (z - \xi)^2$ and that $r_1^2 = (x - \xi')^2 + (y - \eta')^2 + (z - \zeta')^2$, with (x, y, z) given as a point on the sphere in the present discussion. On introducing coordinate systems with the origins at (ξ, η, ζ) and (ξ', η', ζ'), we immediately conclude that

$$(7.419) \qquad\qquad\qquad \frac{\partial r_0}{\partial n} = \cos\theta_0,$$

$$(7.420) \qquad\qquad\qquad \frac{\partial r_1}{\partial n} = \cos\theta_1,$$

respectively. The angle θ_0 is the angle between the exterior unit normal vector to the sphere at the point $P = (x, y, z)$ and the vector extending from the source point $P_0 = (\xi, \eta, \zeta)$ to P. Similarly, θ_1 is the angle between the normal vector at P and the vector from $P_0' = (\xi', \eta', \zeta')$ to P. Referring to Figure 7.6 and using the law of cosines we obtain

$$(7.421) \qquad \cos \theta_0 = \frac{a^2 + r_0^2 - \rho_0^2}{2ar_0},$$

$$(7.422) \qquad \cos \theta_1 = \frac{a^2 + r_1^2 - \rho_1^2}{2ar_1}.$$

Using (7.411) to replace ρ_1 and r_1 by ρ_0 and r_0 in (7.472) gives

$$(7.423) \qquad \cos \theta_1 = \frac{a^2 + \left(a^2/\rho_0^2\right)r_0^2 - \left(a^4/\rho_0^2\right)}{2a(a/\rho_0)r_0} = \frac{\rho_0^2 + r_0^2 - a^2}{2\rho_0 r_0}.$$

Combining these results we conclude that

$$(7.424)$$

$$\left. \frac{\partial K}{\partial n} \right|_{\partial G} = \frac{1}{4\pi} \left[-\frac{1}{r_0^2} \left(\frac{a^2 + r_0^2 - \rho_0^2}{2ar_0} \right) + \frac{\rho_0^2}{a^2 r_0^2} \left(\frac{a}{\rho_0} \right) \left(\frac{\rho_0^2 + r_0^2 - a^2}{2\rho_0 r_0} \right) \right]$$

$$= -\frac{1}{4\pi a} \frac{a^2 - \rho_0^2}{r_0^3}.$$

Thus the solution formula (7.417) reduces to

$$(7.425)$$

$$u(\xi, \eta, \zeta) = \frac{1}{4\pi} \int \int \int_G \left[\frac{1}{r_0} - \frac{a}{\rho_0 r_1} \right] F \, dv + \frac{1}{4\pi a} \int \int_{\partial G} \left[\frac{a^2 - \rho_0^2}{r_0^3} \right] B \, ds.$$

By transforming to spherical coordinates with center at the origin, the second integral in (7.425) can be expressed as, with $\gamma = \sphericalangle POP_0$,

$$(7.426) \qquad \frac{1}{4\pi a} \int \int_{\partial G} \left[\frac{a^2 - \rho_0^2}{r_0^3} \right] B \, ds = \frac{a}{4\pi} \int_0^{2\pi} \int_0^{\pi} \frac{\left(a^2 - \rho_0^2\right) B \sin \varphi \, d\varphi \, d\theta}{\left[a^2 - 2a\rho_0 \cos \gamma + \rho_0^2\right]^{3/2}}.$$

This expression is known as *Poisson's integral* for the sphere and can be obtained by using separation of variables for Laplace's equation in the sphere. It represents the solution of the Dirichlet problem for Laplace's equation, that is, (7.415)–(7.416) with $F = 0$.

If we put $F = 0$ in (7.425) and evaluate u at the origin, we obtain

$$(7.427) \qquad u(0, 0, 0) = \frac{1}{4\pi a^2} \int \int_{\partial G} B \, ds,$$

since $\rho_0 = 0$ and $r_0 = a$ in that case. Now $4\pi a^2$ equals the area of the sphere and B is the value of u on the surface of the sphere. Thus u at the center of the sphere equals the average of its values on the surface of the sphere. This *mean value property* is valid for harmonic functions (i.e., solutions of Laplace's equation) in both two and three dimensions. It also arises in discrete formulations of Laplace's equation as was seen in Section 1.3. This property can be derived in a more direct manner using Green's theorem as shown in the exercises.

The preceding method of inversion used to obtain Green's function for the Dirichlet problem does not work for the case of Neumann's problem or the third boundary value problem for the sphere. Similarly, it cannot be applied to other elliptic equations or, for that matter, to hyperbolic or parabolic equations. For example, we consider the *reduced wave equation* in three dimensions

$$(7.428) \qquad \nabla^2 u + k^2 u = u_{xx} + u_{yy} + u_{zz} + k^2 u = 0$$

in a sphere and try to use inversion to find the Green's function for the first boundary value problem. As shown in Example 6.12, the free space Green's function for this problem is

$$(7.429) \qquad K_F = \frac{1}{4\pi r} e^{ikr}.$$

with $r^2 = (x - \xi)^2 + (y - \eta)^2 + (z - \xi)^2$. This function satisfies the (three-dimensional) *radiation condition of Sommerfeld*

$$(7.430) \qquad \lim_{r \to \infty} r\left[\frac{\partial K_F}{\partial r} - ikK_F\right] = 0.$$

The Green's function K is sought in the form

$$(7.431) \qquad K = \frac{1}{4\pi}\left[\frac{1}{r_0} e^{ikr_0} + \frac{c}{r_1} e^{ikr_1}\right]$$

with r_0 and r_1 defined as in the preceding example and the constant c to be determined. Clearly, c must have the same value given in the foregoing, that is, $c = -a/\rho_0$. However, on the sphere $r_1 = (a/\rho_0)r_0$ so that the exponentials have different arguments and no cancellation occurs. It is not possible to replace e^{ikr_1} by $e^{ik(\rho_0/a)r_1}$ in which case both exponentials would be equal on the sphere, since $(1/r_1)\exp[ik(\rho_0/a)r_1]$ is not a solution of the reduced wave equation.

There are other techniques for obtaining Green's functions that we have not discussed. For example, conformal mapping methods are extremely useful in obtaining Green's functions for Laplace's equation in two dimensions. However, they require a knowledge of the theory of complex variables for their application. In addition, it is often possible to use known Green's functions to convert a differential equation to an integral equation and thereby construct other Green's functions. Again, it is rarely a simple matter to solve the integral equations that result, and most often this must be done approximately. Finally, there exist perturbation and asymptotic methods for finding certain Green's functions. An indication of how to use such methods is given in Chapter 9, which is devoted to approximation methods.

EXERCISES FOR CHAPTER 7

Section 7.1

7.1.1. Show that for the elliptic case the Green's function $K(x; \xi)$ determined from (7.13)–(7.14) is symmetric—that is, $K(x; \xi) = K(\xi; x)$. [Hint: Let $u = K(x; \hat{\xi})$ and $w = K(x; \xi)$ in (7.3).]

7.1.2. Show that the Green's function for the hyperbolic problem (7.15)–(7.17) satisfies the equation $K(x, t; \xi, \tau) = K(\xi, -\tau; x, -t)$. [Hint: Let $u = K(x, -t; \hat{\xi}, -\hat{\tau})$ and $w = K(x, t; \xi, \tau)$ in (7.18).]

7.1.3. Show that the Green's function for the parabolic problem (7.27)–(7.29) satisfies the equation $K(x, t; \xi, \tau) = K(\xi, -\tau; x, -t)$. [Hint: Let $u = K(x, -t; \hat{\xi}, -\hat{\tau})$ and $w = K(x, t; \xi, \tau)$ in (7.30).]

7.1.4. Let $F = 0$ in (7.11)–(7.12) and let B have delta function behavior with the singular point at $x = \hat{x}$ on the boundary so that (7.11) reduces to

$$u(\xi) = -\frac{p}{\alpha} \frac{\partial K(\hat{x}; \xi)}{\partial n}$$

and (7.12) becomes

$$u(\xi) = \frac{p}{\beta} K(\hat{x}; \xi).$$

Use these results to show that not only can the solution of (7.1)–(7.2) with $B = 0$ and $F \neq 0$ be expressed as the superposition of the solutions of point source or singularity problems, but the same can be done for (7.1)–(7.2) if $F = 0$ and $B \neq 0$. In the latter case the point sources lie on the boundary.

7.1.5. Use the expression (7.26) to characterize the solution of the initial and boundary value problem for the hyperbolic equation (7.15) as a superposition of solutions of point source problems, in the manner of Exercise 7.1.4.

7.1.6. Use the expression (7.34) to characterize the solution of the initial and boundary value problem for the parabolic equation (7.27) as a superposition of solutions of point source problems.

7.1.7. Obtain an expression for the solution $u(\xi, T)$ of the hyperbolic equation (7.15) based on the Green's function obtained from (7.37) and (7.39).

7.1.8. Using the Green's function determined from (7.38) and (7.40) obtain an expression for the solution $u(\xi, T)$ of the parabolic equation (7.27).

7.1.9. Verify that $u(\xi, \tau)$ is given as in (7.58) if $\beta_1 = \beta_2 = 0$ and find the appropriate form for $u(\xi, \tau)$ if $\alpha_1 = \alpha_2 = 0$.

7.1.10. Verify the expression for $u(\xi, \tau)$ in (7.59) if $\beta_1 = \beta_2 = 0$ and obtain the correct form for $u(\xi, \tau)$ if $\alpha_1 = \alpha_2 = 0$.

7.1.11. Let

$$Mu = u_{tt} + c^2 u_{xxxx}.$$

Express $wMu - uMw$ in divergence form and use this result in the manner done in the text for the hyperbolic equation (7.15) to determine how a Green's function $K(x, t; \xi, \tau)$ should be constructed for the initial and boundary value problem for $Mu = F$. Here u and u_t are specified at $t = 0$ and, say, u and u_x are given on the boundary of the interval $0 < x < l$.

7.1.12. Generalize the result of Exercise 7.1.11 to the equation

$$u_{tt} + c^2 \nabla^2 \nabla^2 u = F.$$

[Hint: Use (8.322).]

7.1.13. Consider the hyperbolic system

$$\mathbf{u}_t + A\mathbf{u}_x + B\mathbf{u} = \mathbf{c}; \qquad A, B \text{ constant}$$

where A is a (real) symmetric matrix and B and $\mathbf{c}$ are real valued. With B^T equal to the transpose of B show that

$$\mathbf{w}^T[\mathbf{u}_t + A\mathbf{u}_x + B\mathbf{u}] - \mathbf{u}^T[-\mathbf{w}_t - A\mathbf{w}_x + B^T\mathbf{w}]$$

has the form of a divergence expression. Use this fact to construct a Green's function $\mathbf{w}$ suitable for solving the Cauchy problem for the given equation for $\mathbf{u}(x, t)$ with $\mathbf{u}(x, 0) = \mathbf{f}(x)$.

Section 7.2

7.2.1. Let the sequence of functions $f_k(x, y, z)$ be defined as

$$f_k(x, y, z) = \begin{cases} \dfrac{3k^3}{4\pi}; & r < \dfrac{1}{k} \\[2mm] 0; & r > \dfrac{1}{k} \end{cases}$$

where $r^2 = x^2 + y^2 + z^2$. Let $\phi(x, y, z)$ be a test function. Show that

$$\lim_{k \to \infty} (f_k, \phi) = \lim_{k \to \infty} \int\!\!\!\int\!\!\!\int_{-\infty}^{\infty} f_k \phi \, dx \, dy \, dz = \phi(0,0,0),$$

so that the sequence f_k converges to the delta function $\delta(x, y, z)$ in three dimensions.

7.2.2. Show that the functions

$$\phi_a(x) = \begin{cases} \exp\left[-\dfrac{a^2}{a^2 - |x|^2}\right]; & |x| < a \\ 0; & |x| \geq a \end{cases}$$

where x is a one-, two- or three-dimensional variable, are test functions. That is, they are infinitely differentiable and vanish outside the region $|x| \leq a$.

7.2.3. Use (7.82) to show that generalized function

$$f(x; a) = \begin{cases} x^a; & x \geq 0 \\ 0; & x < 0 \end{cases}$$

where $a \geq 0$ has the generalized derivative

$$\frac{d}{dx}[f(x; a)] = \begin{cases} ax^{a-1}; & x > 0 \\ 0; & x < 0 \end{cases}.$$

7.2.4. Determine the generalized derivative of the following functions:
   (a) $f(x) = H(x - 1)\sin x$.
   (b) $f(x) = x^2 \delta(x)$.
   (c) $f(x) = e^x \delta'(x + 3) - H(x)\cos^2 x$.
   (d) $f(x) = H(x - 1)\log x$.
   (e) $f(x) = |x|$.

7.2.5. Obtain the formal Fourier sine and cosine series of the delta function $\delta(x - x_0)$,

$$\delta(x - x_0) = \frac{2}{l} \sum_{k=1}^{\infty} \sin\left(\frac{\pi k x_0}{l}\right) \sin\left(\frac{\pi k x}{l}\right)$$

and

$$\delta(x - x_0) = \frac{1}{l} + \frac{2}{l} \sum_{k=1}^{\infty} \cos\left(\frac{\pi k x_0}{l}\right) \cos\left(\frac{\pi k x}{l}\right)$$

where $0 < x_0 < l$. Show that these series are convergent in the generalized sense.

7.2.6. Express $\delta(x^2 - a^2)$ as a sum of delta functions in the manner of (7.138).

7.2.7. Express $\delta[\sin x]$ as a series of delta functions by using (7.138).

7.2.8. Show that the sequence of functions

$$f_\epsilon(x) = \frac{1}{\sqrt{4\pi\epsilon}} \exp\left[-\frac{x^2}{4\epsilon}\right]$$

tends to $\delta(x)$ as $\epsilon \to 0$.

7.2.9. Let $f(x)$ be defined as $\delta(x)$ in the interval $-\pi < x < \pi$ and extend it as a periodic function of period 2π. Show that the extended function $\hat{f}(x)$ can be expressed as

$$\hat{f}(x) = \sum_{k=-\infty}^{\infty} \delta[x - 2k\pi]$$

and that it has the generalized (complex) Fourier series

$$\hat{f}(x) = \frac{1}{2\pi} \sum_{k=-\infty}^{\infty} \exp[ikx].$$

7.2.10. Show that the function $f(x) = 1/x$ can be defined as a generalized function in terms of (7.63) as

$$\left(\frac{1}{x}, \phi\right) = \text{P.V.} \int_{-\infty}^{\infty} \frac{1}{x} \phi(x)\, dx,$$

where P.V. refers to the principal value of the integral. Express the principal value integral as

$$\left(\frac{1}{x}, \phi\right) = \int_0^{\infty} \frac{\phi(x) - \phi(-x)}{x}\, dx$$

and show that the generalized derivative of $g(x) = \log|x|$ is given by the generalized function $f(x) = 1/x$, as defined above.

7.2.11. Noting that $x\delta(x) = 0$, show that

$$0 = (0, \phi) = \left(x\delta(x)\frac{1}{x}, \phi\right) = (\delta(x), \phi) = \phi(0)$$

where $f(x) = 1/x$ is the generalized function defined in Exercise 7.2.10. Conclude from this that if the product of the generalized functions $x\delta(x)$ and $1/x$ were defined, we would have the result $0 = \delta(x)$.

7.2.12. Verify the following results.

(a) $\dfrac{1}{2\pi}\displaystyle\int_{-\infty}^{\infty} \exp[i\lambda x]\exp[-i\lambda y]\,d\lambda = \delta(x-y).$

(b) $\dfrac{2}{\pi}\displaystyle\int_{0}^{\infty} \cos\lambda x \cos\lambda y\,d\lambda = \delta(x-y), \qquad x, y \geqslant 0.$

(c) $\dfrac{2}{\pi}\displaystyle\int_{0}^{\infty} \sin\lambda x \sin\lambda y\,d\lambda = \delta(x-y), \qquad x, y \geqslant 0.$

Show that these results can be interpreted as orthogonality conditions for the functions arising in the Fourier transform, the Fourier cosine transform, and the Fourier sine transform. In part (a) the Hermitian inner product must be used.

Section 7.3

7.3.1. Obtain the Green's function (7.177) for the problem (7.176) and (7.166) directly—that is, not as a limit of the function (7.175). Use this Green's function to solve the problem

$$u''(x) = -f(x), \qquad 0 < x < l$$
$$u(0) = u(l) = 0.$$

7.3.2. Solve the boundary value problem for the ordinary differential equation

$$u''(x) - c^2 u(x) = -f(x), \qquad 0 < x < l$$
$$u(0) = u(l) = 0$$

using the Green's function (7.175).

7.3.3. Determine the Green's function $K(x; \xi)$ that satisfies the equation

$$\frac{\partial^2 K(x; \xi)}{\xi x^2} - c^2 K(x; \xi) = -\delta(x - \xi), \qquad 0 < x, \xi < l$$

and the boundary conditions

$$K(0; \xi) = 0; \qquad \frac{\partial K(l; \xi)}{\partial x} = 0.$$

7.3.4. Obtain the Green's function $K(x; \xi)$ that satisfies the following problem.

$$\frac{\partial^2 K(x; \xi)}{\partial x^2} + c^2 K(x; \xi) = -\delta(x - \xi), \qquad 0 < x, \xi < l,$$

$$K(0; \xi) = K(l; \xi) = 0.$$

7.3.5. Find the Green's function $K(x; \xi)$ determined from the following problem.

$$x\frac{\partial^2 K}{\partial x^2} + \frac{\partial K}{\partial x} = -\delta(x - \xi), \qquad 0 < x, \xi < 1,$$

$$K(1; \xi) = 0; \qquad K(x; \xi) \text{ bounded at } x = 0.$$

7.3.6. Solve for the Green's function $K(x; \xi)$:

$$\frac{\partial}{\partial x}\left[(1 + x)^2 \frac{\partial K}{\partial x}\right] - K = -\delta(x - \xi), \qquad 0 < x, \xi < 1,$$

$$K(0; \xi) = K(1; \xi) = 0.$$

7.3.7. Obtain a one-dimensional version of the formulas (7.11)–(7.12) for the solution of the boundary value problem for $u(x)$,

$$-(pu')' + qu = \rho F, \qquad 0 < x < l,$$
$$\alpha_1 u(0) - \beta_1 u'(0) = a_1,$$
$$\alpha_2 u(l) + \beta_2 u'(l) = a_2$$

where $\alpha_1, \alpha_2, \beta_1, \beta_2$ are nonnegative, $\alpha_1 + \beta_1 > 0$, $\alpha_2 + \beta_2 > 0$, $p > 0$, $q \geqslant 0$, $\rho > 0$, and F is a given function. [Hint: The solution formulas are given in terms of the Green's function $K(x; \xi)$ determined from (7.43)–(7.44).]

7.3.8. Use the results of Exercises 7.3.3 and 7.3.7 to solve the following problem.

$$u''(x) - c^2 u(x) = e^x, \qquad 0 < x < l$$
$$u(0) = 3; \, u'(l) = 10.$$

7.3.9. Obtain an eigenfunction expansion for the Green's function $K(x; \xi)$ determined from (7.176) and (7.166) and show that the result obtained is the Fourier sine series representation of the function $K(x; \xi)$ given in (7.177).

7.3.10. Show that the following Green's function problem has no solution.

$$\frac{\partial}{\partial x}\left[p(x)\frac{\partial K(x; \xi)}{\partial x}\right] = -\delta(x - \xi), \qquad 0 < x, \xi < l$$

$$\frac{\partial K(0; \xi)}{\partial x} = \frac{\partial K(l; \xi)}{\partial x} = 0.$$

(Hint: Integrate from $x = 0$ to $x = l$.)

7.3.11. Obtain the modified Green's function $\hat{K}(x; \xi)$ that satisfies the problem

$$\frac{\partial}{\partial x}\left[p(x)\frac{\partial \hat{K}}{\partial x}\right] = -\delta(x - \xi) + \frac{1}{l}; \qquad 0 < x, \xi < l,$$

$$\frac{\partial \hat{K}(0; \xi)}{\partial x} = \frac{\partial \hat{K}(l; \xi)}{\partial x} = 0,$$

in the form

$$\hat{K}(x; \xi) = \begin{cases} \dfrac{1}{l}\displaystyle\int_\xi^x \frac{s}{p(s)}\,ds; & 0 < x < \xi \\[2ex] \dfrac{1}{l}\displaystyle\int_\xi^x \frac{s-l}{p(s)}\,ds; & \xi < x < l. \end{cases}$$

(Note that $\hat{K}$ is determined up to an arbitrary constant.) Show how $\hat{K}(x; \xi)$ may be used to solve the boundary value problem for $u(x)$:

$$-(pu')' = f(x), \qquad 0 < x < l$$
$$u'(0) = a, \qquad u'(l) = b,$$

assuming this problem has a solution. Integrate the equation for $u(x)$ from 0 to l to determine a condition for the solution to exist.

7.3.12. Construct a modified Green's function $\tilde{K}(x; \xi)$ that satisfies the problem

$$\frac{\partial}{\partial x}\left[p(x)\frac{\partial \tilde{K}}{\partial x}\right] = -\delta(x - \xi); \qquad 0 < x, \xi < l,$$

$$\frac{\partial \tilde{K}(0; \xi)}{\partial x} = A; \qquad \frac{\partial \tilde{K}(l, \xi)}{\partial x} = B.$$

Choose the constants A and B such that the problem for $\tilde{K}$ has a solution. Then show how $\tilde{K}$ can be used to solve the boundary value problem for $u(x)$ given in Exercise 7.3.11.

7.3.13. Verify that (7.268) is the solution of (7.261)–(7.262).

7.3.14. Verify that (7.282) is the solution of (7.280) with $N_k(T) = 0$.

7.3.15. Show that if the Green's function $K(x; \xi)$ determined from (7.148)–(7.149) has the eigenfunction expansion (7.154), then the Green's function $K(x; \xi)$ determined from the problem

$$LK - \hat{\lambda}K = \delta(x - \xi)$$

with the boundary condition (7.149) has the expansion

$$K(x; \xi) = \sum_{k=1}^{\infty} \frac{M_k(\xi) M_k(x)}{\lambda_k - \hat{\lambda}}.$$

(Note that this Green's function is not defined in the preceding form if $\hat{\lambda} = \lambda_{\hat{\imath}}$.)

7.3.16. Use the result of Exercise 7.3.15 to obtain the Green's function $K(x, y; \xi, \eta)$ for the elliptic equation

$$\nabla^2 K + \hat{\lambda} K = -\delta(x - \xi)$$

in the rectangle $0 < x < l$, $0 < y < \hat{l}$ with the Dirichlet boundary condition (7.187) from the result (7.208).

7.3.17. Construct the Green's function $K(x, y; \xi, \eta)$ for Laplace's equation in the rectangle $0 < x < l$, $0 < y < \hat{l}$, where K satisfies (7.186) and the boundary conditions

$$K(x, 0; \xi, \eta) = K(x, \hat{l}; \xi, \eta) = 0, \qquad 0 < x < l,$$

$$K(0, y; \xi, \eta) = \frac{\partial K(l, y; \xi, \eta)}{\partial x} = 0; \qquad 0 < y < \hat{l}.$$

Use the method presented in the discussion following (7.159).

7.3.18. Obtain the Green's function K for Laplace's equation in a disk $0 \leqslant r \leqslant R$ with the Dirichlet boundary condition $K = 0$ on $r = R$. Express the Laplacian in polar coordinates and obtain the equation for $K(r, \theta; \rho, \phi)$,

$$\frac{\partial^2 K}{\partial r^2} + \frac{1}{r} \frac{\partial K}{\partial r} + \frac{1}{r^2} \frac{\partial^2 K}{\partial \theta^2} = -\frac{\delta(r - \rho)\delta(\theta - \phi)}{r}$$

where $0 < r, \rho < R$ and $0 \leqslant \theta, \phi \leqslant 2\pi$, and the right side of the equation is the two-dimensional delta function in polar coordinates. Construct an eigenfunction expansion of K based on the eigenvalue problem for the operator $-\partial^2/\partial\theta^2$ with periodic boundary conditions—that is, $-M''(\theta) = \lambda M(\theta)$ $(-\pi < \theta < \pi)$ with $M(-\pi) = M(\pi)$ and $M'(-\pi) = M'(\pi)$. Obtain an equation of the general form (7.185) in the radial variable and solve it subject to the conditions that the solution is bounded at $r = 0$ and vanishes at $r = R$.

7.3.19. Obtain the Green's function in the disk $0 \leqslant r \leqslant R$ for the reduced wave equation with Dirichlet boundary condition. That is, the Green's function K satisfies the equation

$$\nabla^2 K + k^2 K = -\delta, \qquad r < R$$

and the boundary condition $K = 0$ on $r = R$ (here δ is the two-dimensional delta function). Proceed as in Exercise 7.3.18, the only difference being that the equation in the radial variable now has the form of Bessel's equation. Two independent solutions of Bessel's equation of order n are $J_n(z)$ and $Y_n(z)$, the Bessel and Neumann functions, respectively.

7.3.20. Consider the modified Green's functions $\hat{K}$ and $\tilde{K}$ for the Neumann problem for Laplace's equation in the disk $0 \leqslant r \leqslant R$. Construct a function $f(r)$ that is smooth in the disk and satisfies the condition $f'(R) = -1/2\pi R$. Use $f(r)$ to transform the problem for $\tilde{K}$ to a new problem with an homogeneous boundary condition. Discuss the relationship between this new problem and that for the function $\hat{K}$.

7.3.21. Obtain the form of the eigenfunction expansion (7.269) for $K(x, t; \xi, \tau)$ if $\lambda_0 = 0$ is an eigenvalue.

7.3.22. Obtain the form of the expansion (7.283) if $\lambda_0 = 0$ is an eigenvalue.

7.3.23. Determine the Green's function for the wave equation $u_{tt} - u_{xx} - u_{yy} = 0$ in the rectangle $0 < x < l$, $0 < y < \hat{l}$ with the boundary condition $u = 0$ on the rectangle. That is, find the solution of (7.257)–(7.259) with $\rho = p = 1$ and $q = 0$ in (7.257) and $\alpha = 1$, $\beta = 0$ in the boundary condition (7.259).

7.3.24. Construct the Green's function for the one-dimensional wave equation $u_{tt} - u_{xx} = 0$ in the interval $0 < x < l$ with the following boundary conditions:

    (a) $u(0, t) = u(l, t) = 0$, $t > 0$.
    (b) $u(0, t) = u_x(l, t) = 0$, $t > 0$.
    (c) $u_x(0, t) = u_x(l, t) = 0$, $t > 0$.

7.3.25. Use the Green's functions determined in Exercise 7.3.24 to write down the solution of the wave equation $u_{tt} = u_{xx}$ in the interval $0 < x < l$, with the initial data $u(x, 0) = \sin x$, $u_t(x, 0) = 1$, and the boundary conditions:

    (a) $u(0, t) = e^{-t}$; $u(l, t) = 0$; $t > 0$.
    (b) $u(0, t) = 0$; $u_x(l, t) = 1$; $t > 0$.
    (c) $u_x(0, t) = \sin t$; $u_x(l, t) = 0$; $t > 0$.

7.3.26. Verify that (7.283) is a generalized solution of (7.276).

7.3.27. Construct the Green's function for the one-dimensional heat equation $u_t = u_{xx}$ in the interval $0 < x < l$ with the following boundary conditions:

    (a) $u(0, t) = u(l, t) = 0$, $t > 0$.
    (b) $u(0, t) = u_x(l, t) = 0$, $t > 0$.
    (c) $u_x(0, t) = u_x(l, t) = 0$, $t > 0$.

7.3.28. Use the Green's functions obtained in Exercise 7.3.27 to solve the initial and boundary value problems for the heat equation $u_t = u_{xx}$ in the interval $0 < x < l$ with $u(x, 0) = 1$, and the boundary data:

    (a) $u(0, t) = e^{-t}$; $u(l, t) = 0$; $t > 0$.
    (b) $u(0, t) = \sin t$; $u_x(l, t) = 0$; $t > 0$.
    (c) $u_x(0, t) = 1$; $u_x(l, t) = 0$; $t > 0$.

7.3.29. Determine the Green's function for Laplace's equation in the cube $0 < x < l$, $0 < y < l$, $0 < z < l$ with Dirichlet conditions on the boundary. Use the eigenfunctions for the rectangle obtained in Example 7.5.

Section 7.4

7.4.1. Use the two-dimensional Fourier transform to obtain the Green's function (7.292).

7.4.2. Use the three-dimensional Fourier transform to obtain the Green's function (7.293).

7.4.3. Apply the Fourier sine transform to obtain the Green's function for the one-dimensional heat equation $u_t = c^2 u_{xx}$ in the semi-infinite interval $0 < x < \infty$ with $u(x, t)$ specified at $x = 0$ and $t = 0$.

7.4.4. Apply the Fourier cosine transform to obtain the Green's function for $u_t = c^2 u_{xx}$ in $0 < x < \infty$, $t > 0$ with $u(x, 0)$ and $u_x(0, t)$ specified.

7.4.5. Use the general solution of the Cauchy problem for the wave equation in one dimension to obtain the Green's function K as given in (7.319).

7.4.6. Apply the formula (7.35) to obtain the solution of the Cauchy problem for the wave equation $u_{tt} = c^2 \nabla^2 u$ in three dimensions if the initial data are $u(x, y, z, 0) = 0$ and $u_t(x, y, z, 0) = f$. The solution has the form given in (5.151).

7.4.7. Obtain the result (5.159) for the Cauchy problem for the two-dimensional wave equation $u_{tt} = c^2 \nabla^2 u$ if $u(x, y, 0) = 0$ and $u_t(x, y, 0) = f$, using the Green's function (7.316) and the formula (7.35).

7.4.8. Use the Green's functions for the wave equation in one, two, and three dimensions obtained in Example 7.8 to determine domains of dependence for the solutions of the Cauchy problem for the inhomogeneous wave equation in each of the three cases.

7.4.9. Show that the Klein–Gordon equation $u_{tt} - \gamma^2 u_{xx} + c^2 u = 0$ has the solution $u = J_0[(c/\gamma)\sqrt{\gamma^2 t^2 - x^2}]$. Noting that the Green's function for the Klein–Gordon equation should reduce to the Green's function (7.319) for the wave equation as $c \to 0$, argue that the Green's function should have the form (7.332).

7.4.10. The Green's function for the Klein–Gordon equation $u_{tt} - \gamma^2 \nabla^2 u + c^2 u = 0$ in three dimensions is given as

$$K(x, y, z, t; \xi, \eta, \zeta, \tau) = \frac{\delta[\gamma(\tau - t) - r]}{4\pi\gamma r}$$

$$- \left(\frac{c}{\gamma^2}\right) \frac{J_1\left[(c/\gamma)\sqrt{\gamma^2(\tau - t)^2 - r^2}\right]}{\sqrt{\gamma^2(\tau - t)^2 - r^2}} H[\gamma(\tau - t) - r]$$

where $r^2 = (x - \xi)^2 + (y - \eta)^2 + (z - \zeta)^2$, as can be shown by using the Fourier transform. Let $\hat{r}^2 = \gamma^2(\tau - t)^2 - r^2$ in the Klein–Gordon equation and obtain the equation

$$\hat{u}'' + \frac{3}{\hat{r}}\hat{u}' + \left(\frac{c}{\gamma}\right)^2 \hat{u} = 0$$

where $\hat{u} = u(\hat{r})$. Show that this equation has the solutions

$$\hat{u}_1 = \frac{1}{\hat{r}}J_1\left[\left(\frac{c}{\gamma}\right)\hat{r}\right]$$

and

$$\hat{u}_2 = \frac{1}{\hat{r}}Y_0'\left[\left(\frac{c}{\gamma}\right)\hat{r}\right]$$

where J_1 and Y_0 are the Bessel and Neumann functions, respectively and that the coefficient of the log term in the solution $\hat{u}_2$ is a constant times the solution $\hat{u}_1$. Noting the discussion following (6.223), conclude that the preceding expression for the Green's function K is reasonable in view of the fact that it reduces to the Green's function for the wave equation when $c = 0$.

7.4.11. Apply the discussion given in Exercise 7.4.10 to obtain the Green's function K for the hyperbolic equation $u_{tt} - \gamma^2\nabla^2 u - c^2 u = 0$ in three dimensions. (Hint: It has the same form as the Green's function for Exercise 7.4.10 except that the Bessel function J_1 is replaced by I_1, the modified Bessel function.)

7.4.12. Obtain the Green's function $K(x, y, t; \xi, \eta, \tau)$ for the two-dimensional Klein–Gordon equation $u_{tt} - \gamma^2\nabla^2 u + c^2 u = 0$ in the form

$$K = \frac{1}{2\pi\gamma}\frac{\cos\left[(c/\gamma)\sqrt{\gamma^2(\tau - t)^2 - \rho^2}\right]}{\sqrt{\gamma^2(\tau - t)^2 - \rho^2}}H[\gamma(\tau - t) - \rho]$$

where $\rho^2 = (x - \xi)^2 + (y - \eta)^2$, by following the procedure given in Exercise 7.4.10.

7.4.13. Obtain the Green's function $K(x, y, t; \xi, \eta, \tau)$ for the two-dimensional hyperbolic equation $u_{tt} - \gamma^2\nabla^2 u - c^2 u = 0$ in the form

$$K = \frac{1}{2\pi\gamma}\frac{\cosh\left[(c/\gamma)\sqrt{\gamma^2(\tau - t)^2 - \rho^2}\right]}{\sqrt{\gamma^2(\tau - t)^2 - \rho^2}}H[\gamma(\tau - t) - \rho]$$

where $\rho^2 = (x - \xi)^2 + (y - \eta)^2$, by following the method presented in Exercise 7.4.10.

7.4.14. Obtain the Green's function appropriate for the Cauchy problem for Schrödinger's equation

$$i\hbar \frac{\partial u}{\partial t} + \frac{\hbar^2}{2m} \frac{\partial^2 u}{\partial x^2} = 0,$$

where $\hbar$ is Planck's constant, m is the mass, and $i = \sqrt{-1}$. (Hint: Use the Fourier transform.)

7.4.15. Show that the solution $K(x; \xi)$ of the problem

$$\frac{\partial^2 K}{\partial x^2} - k^2 K = -\delta(x - \xi), \qquad -\infty < x, \xi < \infty,$$

$$K(x; \xi) \to 0 \text{ as } |x| \to \infty$$

is given as

$$K(x; \xi) = \frac{1}{2k} \exp[-k|x - \xi|]$$

(see Examples 5.1 and 6.12).

7.4.16. Obtain the solution $K(x; \xi)$ of the problem

$$\frac{\partial^2 K}{\partial x^2} + k^2 K = -\delta(x - \xi), \qquad -\infty < x, \xi < \infty,$$

$$\lim_{|x| \to \infty} \left[\frac{\partial K}{\partial |x|} - ikK \right] = 0,$$

—that is, $K(x; \xi)$ satisfies the radiation condition at infinity—in the form

$$K(x; \xi) = \frac{i}{2k} \exp[ik|x - \xi|]$$

(see Example 6.12).

7.4.17. Solve the following problems for $K(x, \xi)$:

$$\frac{\partial^2 K}{\partial x^2} - k^2 K = -\delta(x - \xi), \qquad 0 < x, \xi < \infty.$$

(a) $K(0; \xi) = 0$; $K(x; \xi) \to 0$ as $x \to \infty$.

(b) $\dfrac{\partial K(0; \xi)}{\partial x} = 0$; $K(x; \xi) \to 0$ as $x \to \infty$.

7.4.18. Obtain the Green's function for the following problem.

$$\frac{\partial^2 K}{\partial x^2} + k^2 K = -\delta(x - \xi), \qquad 0 < x, \xi < \infty;$$

$$K(0; \xi) = 0; \qquad \lim_{x \to \infty}\left[\frac{\partial K}{\partial x} - ikK\right] = 0.$$

7.4.19. Determine the Green's function $K(x, y; \xi, \eta)$ that satisfies the following problem.

$$\frac{\partial^2 K}{\partial x^2} + \frac{\partial^2 K}{\partial y^2} = -\delta(x - \xi)\delta(y - \eta); \qquad \begin{cases} 0 < x, \xi < l \\ -\infty < y, \eta < \infty, \end{cases}$$

$$K = 0 \text{ at } x = 0 \text{ and } x = l,$$
$$K \text{ is bounded as } |y| \to \infty.$$

(Hint: Use the eigenfunctions for the interval $0 < x < l$ with zero boundary conditions.)

7.4.20. Obtain the solution of the following problem.

$$\frac{\partial^2 K}{\partial x^2} + \frac{\partial^2 K}{\partial y^2} = -\delta(x - \xi)\delta(y - \eta); \qquad \begin{cases} 0 < x, \xi < l \\ -\infty < y, \eta < \infty, \end{cases}$$

$$\frac{\partial K}{\partial x} = 0 \text{ at } x = 0 \text{ and } x = l,$$

$$K \to 0 \text{ as } |y| \to \infty.$$

[Hint: Use the cosine eigenfunctions for $0 < x < l$.]

7.4.21. Obtain the Green's function $K(x, y; \xi, \eta)$.

$$\nabla^2 K - c^2 K = -\delta(x - \xi)\delta(y - \eta); \qquad 0 < x, \xi < l, -\infty < y, \eta < \infty,$$
$$K = 0 \text{ on } x = 0, x = l,$$
$$K \to 0 \text{ as } |y| \to \infty.$$

(Hint: Use the eigenfunctions for the interval $0 < x < l$)

7.4.22. Set up a two-dimensional version of the problem in Example 7.10—that is, drop the y dependence in the problem—and solve for the Green's function $K(x, z; \zeta)$.

7.4.23. Solve for the Green's function $K(x, y, z; \xi, \eta, \zeta)$:

$$\nabla^2 K - c^2 K = -\delta(x - \xi)\delta(y - \eta)\delta(z - \zeta); \qquad \begin{cases} 0 < x, \xi < l \\ -\infty < y, \eta, z, \zeta < \infty, \end{cases}$$

$$K = 0 \text{ on } x = 0, x = l$$
$$K \to 0 \text{ as } |y| \to \infty, |z| \to \infty.$$

(Hint: Use the free space Green's function found in Example 6.12.)

Section 7.5

7.5.1. Use the Green's functions obtained in Example 7.11 and appropriate forms of the solution formulas obtained in Section 7.1 to construct the solution of Laplace's equation $\nabla^2 u = 0$ in the half-space $z > 0$ if the following boundary conditions are given:

(a) $u(x, y, 0) = f(x, y)$, $\qquad -\infty < x, y < \infty$.

(b) $\dfrac{\partial u(x, y, 0)}{\partial z} = f(x, y)$, $\qquad -\infty < x, y < \infty$.

(c) $\dfrac{\partial u(x, y, 0)}{\partial z} - hu(x, y, 0) = f(x, y)$, $\qquad -\infty < x, y < \infty$.

7.5.2. Use the method of images to obtain the Green's function for $\nabla^2 u = 0$ in the half-plane $y > 0$ for the Dirichlet problem in the form,

$$K(x, y; \xi, \eta) = -\frac{1}{2\pi} \log \frac{\rho}{\hat\rho}; \qquad 0 < y, \eta < \infty, -\infty < x, \xi < \infty$$

where $\rho^2 = (x - \xi)^2 + (y - \eta)^2$ and $\hat\rho^2 = (x - \xi)^2 + (y + \eta)^2$. Use this Green's function to obtain the solution (5.67) of the boundary value problem (5.58)–(5.60).

7.5.3. Apply the method of images to obtain the Green's function for $\nabla^2 u = 0$ in the half-plane $y > 0$ for the Neumann problem in the form

$$K(x, y; \xi, \eta) = -\frac{1}{2\pi} \log \rho\hat\rho; \qquad 0 < y, \eta < \infty, -\infty < x, \xi < \infty$$

where ρ and $\hat\rho$ are defined as in Exercise 7.5.2. Use this Green's function to obtain the solution (5.75) of the problem (5.58), (5.68).

7.5.4. Obtain the Green's function that corresponds to (7.391) in the two-dimensional case.

7.5.5. Use the method of images to obtain the Green's function for Laplace's equation in a composite medium in three dimensions. That is, determine the function $K(x, y, z; \xi, \eta, \zeta)$ that satisfies the equation

$$p\nabla^2 K = -\delta(x - \xi)\delta(y - \eta)\delta(z - \zeta), \qquad z \neq 0, \qquad \zeta > 0$$

with $-\infty < x, \xi, y, \eta, z < \infty$ ($z \neq 0$), the condition

$$K \to 0 \text{ as } |z| \to \infty,$$

as well as the jump conditions at $z = 0$,

$$K \text{ continuous at } z = 0,$$
$$p\frac{\partial K}{\partial z} \text{ continuous at } z = 0$$

where $p = p_1$ for $z < 0$, $p = p_2$ for $z > 0$, and p_1 and p_2 are constants. [Hint:

Let $K = K_1$ for $z < 0$ and $K = K_2$ for $z > 0$. Since the source point lies in $z > 0$ use the free space Green's function and its image for $z > 0$ and the free space Green's function with a source at (ξ, η, ζ) in $z < 0$. Then apply the matching conditions.]

7.5.6. Use the procedure of Example 7.13 to construct a Green's function for Laplace's equation $\nabla^2 u = 0$ in the region $-\infty < x < \infty$, $-\infty < y < \infty$, $0 < z < l$ for the case of Dirichlet boundary conditions.

7.5.7. Apply the method of images to construct the Green's function for the Helmholtz equation $\nabla^2 u + k^2 u = 0$ in the half-space $z > 0$, in the case where $K(x, y, z; \xi, \eta, \zeta)$ satisfies one of the following boundary conditions:

(a) $K = 0$ at $z = 0$,

(b) $\dfrac{\partial K}{\partial z} = 0$ at $z = 0$,

(c) $\dfrac{\partial K}{\partial z} - hK = 0$ at $z = 0$,

with $-\infty < x, y, \xi, \eta < \infty$ and where K satisfies the radiation condition at infinity in all cases. (Hint: Use Example 6.12.)

7.5.8. Construct the two-dimensional forms of the Green's functions obtained in Exercise 7.5.7 for the Helmholtz equation $\nabla^2 u + k^2 u = 0$ in the half-plane $y > 0$.

7.5.9. Construct the Green's function for $\nabla^2 u - c^2 u = 0$ in the half-space $z > 0$ if the Green's function $K(x, y, z; \xi, \eta, \zeta)$ satisfies the following boundary conditions;

(a) $K = 0$ at $z = 0$,

(b) $\dfrac{\partial K}{\partial z} = 0$ at $z = 0$,

for all $-\infty < x, y < \infty$ and $K \rightarrow 0$ as $z \rightarrow \infty$. (Hint: Use Example 6.12 and the method of images.)

7.5.10. Carry out the solution of the two-dimensional version of Exercise 7.5.9 in the half-plane $y > 0$.

7.5.11. Use the Green's function (7.401) for the wave equation in the semi-infinite interval $0 < x < \infty$, to solve the following initial and boundary value problem.

$$u_{tt} - c^2 u_{xx} = 0, \qquad 0 < x < \infty, t > 0$$
$$u(x, 0) = u_t(x, 0) = 0, \qquad 0 < x < \infty$$
$$u(0, t) = g(t), \qquad t > 0.$$

7.5.12. Solve the following problem by use of the Green's function (7.402):

$$u_{tt} - \gamma^2 u_{xx} + c^2 u = 0, \qquad x > 0, t > 0$$
$$u(x, 0) = u_t(x, 0) = 0, \qquad x > 0$$
$$\dfrac{\partial u(0, t)}{\partial x} = h(t), \qquad t > 0.$$

7.5.13. Verify that (7.402) is indeed the appropriate Green's function for the modified telegrapher's equation (7.394) in the case of Neumann boundary conditions.

7.5.14. Noting the result of Exercise 7.2.9,

$$2\pi \sum_{k=-\infty}^{\infty} \delta[x - 2k\pi] = \sum_{k=-\infty}^{\infty} \exp[ikx],$$

multiply both sides by a Fourier transformable function $\phi(x)$ and conclude upon integrating that

$$\sqrt{2\pi} \sum_{k=-\infty}^{\infty} \phi[2k\pi] = \sum_{k=-\infty}^{\infty} F[k]$$

where $F[k]$ is the Fourier transform—see (5.10)—of $\phi(x)$. This result is known as *Poisson's summation formula*.

7.5.15. Use the Poisson summation formula (Exercise 7.5.14) to show the equivalence of the Green's function expression (7.408) for the heat equation and that obtained by the method of eigenfunction expansions.

7.5.16. Apply the method of images to construct a Green's function for the heat equation in the semi-infinite interval $0 < x < \infty$ with Dirichlet boundary conditions and use it to solve the following problem.

$$u_t - c^2 u_{xx} = 0, \qquad 0 < x < \infty, t > 0$$
$$u(x,0) = f(x), \qquad 0 < x < \infty$$
$$u(0,t) = g(t), \qquad 0 < t < \infty.$$

7.5.17. Use the inversion process of Example 7.14 to obtain the Green's function for the Dirichlet problem for Laplace's equation in a disk $x^2 + y^2 < a^2$.

7.5.18. Verify the result (7.425).

7.5.19. Obtain Poisson's integral for the circle—see (7.426)—by using the Green's function obtained in Exercise 7.5.17.

7.5.20. Use the method of images to determine the appropriate Green's function for Laplace's equation $\nabla^2 u = 0$ in the quadrant $0 < x < \infty$, $0 < y < \infty$ with the following boundary conditions for $K(x, y; \xi, \eta)$:

(a) $K = 0$ on $x = 0$, $y > 0$ and $y = 0$, $x > 0$.
(b) $K = 0$ on $x = 0$, $y > 0$; $\partial K/\partial y = 0$ on $y = 0$, $x > 0$.
(c) $\partial K/\partial x = 0$ on $x = 0$, $y > 0$, $\partial K/\partial y = 0$ on $y = 0$, $x > 0$.

7.5.21. Apply the method of images to construct the Green's function for Laplace's equation within the hemisphere $x^2 + y^2 + z^2 < a^2$, $z \geq 0$, with

Dirichlet boundary conditions. (Hint: Use inversion in the sphere and reflection in the plane $z = 0$.)

7.5.22. Obtain the Green's function for the Dirichlet problem for Laplace's equation within the quarter-circle $x^2 + y^2 < a^2$, $x > 0$, $y > 0$. (Hint: Reflect in the x and y-axis as in the quarter-plane problem and then invert each source and image source in the circle.)

7.5.23. Let u be a solution of Laplace's equation $\nabla^2 u = 0$ in the region G and let R be the interior of a circle or a sphere centered at P_0 and completely contained within G.

(a) Use Green's theorem or the divergence theorem to show that

$$\int_{\partial R} \frac{\partial u}{\partial n} \, ds = 0.$$

(b) Let K be the free space Green's function for Laplace's equation with singular point at P_0. Apply Green's theorem in the region R to the harmonic function u and the Green's function K. Deduce the mean value theorems

$$u(P_0) = \frac{1}{4\pi a^2} \int_{\partial R} u \, ds$$

and

$$u(P_0) = \frac{1}{2\pi a} \int_{\partial R} u \, ds,$$

in three and two dimensions, respectively, where a is the radius of the sphere and the circle with center at P_0.

8

VARIATIONAL AND OTHER METHODS

In this chapter we present a number of methods and results that either apply to problems studied previously or to a new class of problems. We begin with a presentation of a variational characterization of the eigenvalue problems introduced in Chapter 4, and show how the variational approach enables us to prove some of the properties of eigenvalues and eigenfunctions given in that chapter. An important consequence of this approach is the Rayleigh–Ritz method that yields approximate determinations of eigenvalues and eigenfunctions and this method is considered next.

We continue with a discussion of adjoint differential operators and this leads to a discussion of the classical method of Riemann for integrating linear hyperbolic equations of second order.

This is followed by a presentation of maximum and minimum principles for the diffusion and Laplace's equations, and their consequences for uniqueness and continuous dependence on data for problems for these equations. In addition, we consider a positivity property for the telegrapher's equation.

We conclude with a discussion of some basic equations of mathematical physics. These are the equations of fourth order governing the vibration of rods and plates and the equations of fluid dynamics, Maxwell's equations of electromagnetic theory, and the equations of elasticity theory, which are systems of equations. Techniques for simplifying and solving these equations will be considered.

8.1. VARIATIONAL PROPERTIES OF EIGENVALUES AND EIGENFUNCTIONS

It has been demonstrated in the preceding chapters that eigenvalue problems play an important role in various methods for solving problems for partial differential equations. Up to now we have determined the eigenvalues and eigenfunctions by using separation of variables for partial differential equations and general solutions of ordinary differential equations for the Sturm–Liouville problem. In this section we show how the eigenvalues and eigenfunctions can be specified by means of a *variational principle*. Thereby, it is possible to prove some of the properties of the eigenvalues and eigenfunctions given at the beginning of Chapter 4. Furthermore, the variational approach leads to a useful approximate method for determining the first few eigenvalues and eigenfunctions. This method, known as the *Rayleigh–Ritz method*, is discussed in Section 8.2.

To motivate the use of the variational principle, we begin by considering the eigenvalue problem

$$(8.1) \qquad LM = - \nabla \cdot (p\nabla M) + qM = \lambda \rho M$$

for the function $M(x)$ defined in the bounded region G. As in Sections 4.1 and 4.2, we assume that $p(x) > 0$ and $\rho(x) > 0$ in G, whereas $q(x) \geqslant 0$ in G. The eigenfunction $M(x)$ is assumed to satisfy the boundary condition on ∂G,

$$(8.2) \qquad \alpha M + \beta \frac{\partial M}{\partial n} \bigg|_{\partial G} = 0$$

where $\alpha(x) \geqslant 0$ and $\beta(x) \geqslant 0$, whereas $\alpha + \beta > 0$ on ∂G. (We use notation appropriate for the two or three-dimensional eigenvalue problem and occasionally indicate the appropriate forms for the one-dimensional Sturm–Liouville problem.)

As was discussed in Section 4.2, we assume that there are a countable infinity of eigenvalues λ_k ($k = 1, 2, \ldots$) that are real valued and nonnegative. They are numbered according to increasing values as

$$(8.3) \qquad 0 \leqslant \lambda_1 \leqslant \lambda_2 \leqslant \lambda_3 \leqslant \lambda_4 \leqslant \cdots \leqslant \lambda_k \leqslant \cdots.$$

The associated eigenfunctions $M_k(x)$ are assumed to form a complete orthonormal set. That is,

$$(8.4) \qquad (M_k, M_j) = \iint_G \rho M_k M_j \, dv = \delta_{kj},$$

where $\delta_{kj} = 1$ for $k = j$ and $\delta_{kj} = 0$ for $k \neq j$, with $k, j = 1, 2, 3, \ldots$. Furthermore, any function $u(x)$ that is sufficiently smooth and satisfies the boundary condition (8.2)—where M is replaced by u—can be expanded in a series of eigenfunctions

$$(8.5) \qquad u(x) = \sum_{k=1}^{\infty} N_k M_k(x).$$

The Fourier coefficients N_k are given as

$$(8.6) \qquad N_k = (u(x), M_k(x)) = \iint_G \rho u M_k \, dv.$$

We assume that $u(x)$ is normalized to unity so that

$$(8.7) \qquad \|u\|^2 = (u, u) = \iint_G \rho u^2 \, dx = 1.$$

To proceed, we consider the *energy integral*

$$(8.8) \qquad E(u) = \iint_G [u L u] \, dv = \iint_G \left[p(\nabla u)^2 + q u^2 \right] dv$$

$$- \int_{\partial G} p u \frac{\partial u}{\partial n} \, ds$$

where (4.35) has been used. Now $u(x)$ satisfies the boundary condition

$$(8.9) \qquad \alpha u + \beta \left. \frac{\partial u}{\partial n} \right|_{\partial G} = 0.$$

Thus if $\beta = 0$ on ∂G and $u = 0$ is the boundary condition, we have the energy integral

$$(8.10) \qquad E(u) = \iint_G \left[p(\nabla u)^2 + q u^2 \right] dv.$$

If $\alpha = 0$ on ∂G so that $\partial u / \partial n = 0$ is the boundary condition, we again have the energy integral (8.10). If both α and β are nonzero on ∂G, the energy integral becomes

$$(8.11) \qquad E(u) = \iint_G \left[p(\nabla u)^2 + q u^2 \right] dv + \int_{\partial G} \left(\frac{\alpha}{\beta} \right) p u^2 \, ds.$$

We refer to $E(u)$ as an energy integral on the basis of our discussion in Section 6.8.

On inserting (8.5) into (8.8) we have

$$(8.12) \quad E(u) = \iint\limits_G \left[\sum_{k=1}^{\infty} N_k M_k Lu \right] dv = \sum_{k=1}^{\infty} N_k \iint\limits_G [M_k Lu] \, dv$$

$$= \sum_{k=1}^{\infty} N_k \iint\limits_G \left[M_k(x) L \left\{ \sum_{j=1}^{\infty} N_j M_j(x) \right\} \right] dv$$

$$= \sum_{k=1}^{\infty} N_k \iint\limits_G \left[M_k(x) \sum_{j=1}^{\infty} N_j L M_j(x) \right] dv$$

$$= \sum_{k=1}^{\infty} N_k \iint\limits_G \left[\sum_{j=1}^{\infty} \lambda_j N_j \{ \rho(x) M_k(x) M_j(x) \} \right] dv$$

$$= \sum_{k=1}^{\infty} N_k \left\{ \sum_{j=1}^{\infty} \lambda_j N_j (M_k, M_j) \right\} = \sum_{k=1}^{\infty} \lambda_k N_k^2,$$

since $LM_j = \lambda_j \rho M_j$ and $(M_k, M_j) = \delta_{kj}$. Also,

$$(8.13) \quad 1 = (u, u) = \left(\sum_{k=1}^{\infty} N_k M_k(x), \sum_{j=1}^{\infty} N_j M_j(x) \right)$$

$$= \sum_{k=1}^{\infty} N_k \iint\limits_G \left[\rho(x) M_k(x) \sum_{j=1}^{\infty} N_j M_k(x) \right] dv$$

$$= \sum_{k=1}^{\infty} N_k \left\{ \sum_{j=1}^{\infty} N_j (M_k, M_j) \right\} = \sum_{k=1}^{\infty} N_k^2,$$

since $(M_k, M_j) = \delta_{kj}$. We have assumed that the interchanges of summation, differentiation, and integration used in (8.12)–(8.13) are valid.

Noting (8.3) and (8.13), we obtain from (8.12)

$$(8.14) \quad E(u) = \sum_{k=1}^{\infty} \lambda_k N_k^2 \geqslant \sum_{k=1}^{\infty} \lambda_1 N_k^2 = \lambda_1 \sum_{k=1}^{\infty} N_k^2 = \lambda_1.$$

Thus for all *admissible* functions $u(x)$ that satisfy (8.9) and (8.7) we have $E(u) \geqslant \lambda_1$. If $u(x)$ satisfies the additional *constraints*

$$(8.15) \quad (u, M_k) = \iint\limits_G \rho u M_k \, dx = 0, \quad k = 1, 2, \ldots, n-1,$$

the first $n - 1$ coefficients in the series (8.5)—that is, $N_1, N_2, \ldots, N_{n-1}$—all vanish. Then (8.12) yields

$$(8.16) \qquad E(u) = \sum_{k=n}^{\infty} \lambda_k N_k^2 \geqslant \lambda_n \sum_{k=n}^{\infty} N_k^2 = \lambda_n$$

on using (8.3) and (8.13). In addition,

$$(8.17) \qquad E[M_n] = \iint_{\tilde{G}} M_n L M_n \, dv = \lambda_n \iint_{\tilde{G}} \rho M_n^2 \, dv = \lambda_n,$$

since the $M_n(x)$ are normalized by assumption.

In view of these results, the energy integral $E(u)$ is greater than or equal to the smallest eigenvalue λ_1 associated with the eigenvalue problem (8.1)–(8.2). The minimum value λ_1 is assumed if $u = M_1(x)$, the first eigenfunction. If $u(x)$ satisfies the constraints (8.15), we have $E(u) \geqslant \lambda_n$ and the minimum is assumed if $u = M_n(x)$. Consequently, we may characterize the eigenvalues and eigenfunctions λ_n and $M_n(x)$ in terms of the following minimum problem. *The nth eigenvalue of the problem* (8.1)–(8.2) *is the minimum value of the energy integral $E(u)$, where the admissible functions $u(x)$ satisfy the conditions* (8.7), (8.9), *and* (8.15). *If $n = 1$, the conditions* (8.15) *are absent. Further, among all admissible functions $u(x)$, the minimum value of $E(u)$ is assumed if u is one of the eigenfunctions of* (8.1)–(8.2) *and the minimum value is equal to the corresponding eigenvalue.*

In a slight variation of this procedure, we drop the restriction that the admissible functions $u(x)$ are normalized. If $w(x)$ is a function that satisfies the boundary condition (8.9) and the constraints (8.15) (where u is replaced by w), we can set

$$(8.18) \qquad u(x) = \frac{w(x)}{\|w(x)\|}.$$

Then $u(x)$ satisfies all the conditions stated in the minimum problem presented. We have

$$(8.19) \qquad E(u) = E\left(\frac{w}{\|w\|}\right) = \frac{1}{\|w\|^2} E(w) = \frac{E(w)}{(w, w)}.$$

With $E(w)$ defined as in (8.10) or (8.11), the ratio $E(w)/(w, w)$ is known as the *Rayleigh quotient*. The minimum values of the Rayleigh quotient still determine the eigenvalues, and these minima are assumed by the eigenfunctions. However, the resulting eigenfunctions are no longer normalized but are determined up to an arbitrary multiplicative constant.

For the purpose of comparing the properties of eigenvalues arising in different eigenvalue problems, our formulation of the variational problem is unsatisfactory because it characterizes the nth eigenvalue for a given problem in terms of the first $n - 1$ eigenfunctions of that problem in view of (8.15). That is, we must first determine the eigenfunctions $M_1(x)..., M_{n-1}(x)$ in order to specify the eigenvalue λ_n. It is of interest to obtain a variational formulation that permits the determination of any eigenvalue directly and independently of first having to solve any other problem. Such an approach was developed by Courant and is known as the *maximum–minimum principle*.

Given the eigenvalue problem (8.1)–(8.2) we consider a collection of (sufficiently smooth) functions $\{m_k(x)\}$, $k = 1, 2, 3, \ldots$, defined in the region G. These functions need not satisfy any specific conditions in G or on the boundary ∂G. To determine the nth eigenvalue λ_n, we introduce the function $w(x)$ that satisfies the boundary condition (8.9) and the $n - 1$ constraints

$$(8.20) \qquad (w, m_k) = \iint_G \rho w m_k \, dv = 0, \qquad k = 1, 2, \ldots, n - 1.$$

By varying over all admissible functions $w(x)$, we minimize the Rayleigh quotient $E(w)/(w, w)$. For each set of functions $m_1(x), \ldots, m_{n-1}(x)$ we obtain a minimum for the Rayleigh quotient. We then vary these minima over all possible sets of functions $m_1(x), \ldots, m_{n-1}(x)$. The maximum of these minima is equal to the nth eigenvalue λ_n. This maximum–minimum principle can be stated as

$$(8.21) \qquad \max_{m_1, \ldots, m_{n-1}} \left[\min_w \frac{E(w)}{(w, w)} \right] = \lambda_n.$$

To prove this result, we note that on selecting $w(x)$ as

$$(8.22) \qquad w(x) = \sum_{j=1}^{n} c_j M_j(x)$$

where the $M_k(x)$ are the eigenfunctions for (8.1)–(8.2), we may specify the constants c_k so that the $n - 1$ constraints (8.20) are satisfied. In fact, we have

$$(8.23) \qquad (w, m_k) = \sum_{j=1}^{n} c_j (M_j, m_k) = 0, \qquad k = 1, \ldots, n - 1.$$

This is an underdetermined system of homogeneous linear equations for the n constants c_j, and it has a solution with at least one of the c_j remaining arbitrary. Therefore, we can satisfy the additional condition.

$$(8.24) \qquad (w, w) = \sum_{j=1}^{n} c_j^2 = 1$$

by choosing the undetermined constant(s) appropriately. Note that $w(x)$

satisfies the boundary condition (8.9) since each of the $M_j(x)$ does so. Therefore, $w(x)$ is an admissible function for the variational problem.

We have, for the $w(x)$ given above,

$$(8.25) \qquad \frac{E(w)}{(w, w)} = \sum_{j=1}^{n} \lambda_j c_j^2 \leqslant \lambda_n \sum_{j=1}^{n} c_j^2 = \lambda_n,$$

in view of (8.12) and (8.24), since

$$(8.26) \qquad 0 \leqslant \lambda_j \leqslant \lambda_n; \qquad j = 1, 2, \ldots, n - 1.$$

Therefore, for each set of functions $m_1(x), \ldots, m_{n-1}(x)$ we have found an admissible $w(x)$ for which the Rayleigh quotient is less than or equal to λ_n. Consequently, the maxima of all the minima of the Rayleigh quotient cannot exceed λ_n. However, if $w(x) = M_n(x)$, the nth eigenfunction for the problem (8.1)–(8.2), we have

$$(8.27) \qquad \frac{E(M_n)}{(M_n, M_n)} = E(M_n) = \lambda_n$$

on using (8.17), so that the maximum–minimum principle (8.21) has been verified.

EXAMPLE 8.1.   THE EIGENVALUE PROBLEM FOR A SQUARE

In this example we reconsider the eigenvalue problem for the *Laplacian operator*

$$(8.28) \qquad \nabla^2 M + \lambda M = M_{xx} + M_{yy} + \lambda M = 0$$

in the square $0 < x < \pi$ and $0 < y < \pi$, with the *Dirichlet boundary condition*

$$(8.29) \qquad M(x, y)|_{\partial G} = 0.$$

The region G is the aforementioned square, and ∂G is given by the sides of the square. The eigenvalues and eigenfunctions for the problem (8.28)–(8.29) were determined in Example 7.5. The first eigenvalue λ_{11} was found to be

$$(8.30) \qquad \lambda_{11} = 2,$$

and the associated normalized eigenfunction $M_{11}(x, y)$ is

$$(8.31) \qquad M_{11}(x, y) = \frac{2}{\pi} \sin x \sin y.$$

The appropriate energy integral for this problem is

$$(8.32) \qquad E(u) = \int_0^\pi \int_0^\pi \left[u_x^2 + u_y^2 \right] dx\, dy,$$

and an admissible function $u(x, y)$ must vanish on the boundary of the square. Also, $u(x, y)$ must be normalized so that

$$(8.33) \qquad (u, u) = \int_0^\pi \int_0^\pi u^2\, dx\, dy = 1.$$

Now if $u(x, y) = M_{11}(x, y)$, we already know that

$$(8.34) \qquad (M_{11}, M_{11}) = \frac{4}{\pi^2} \int_0^\pi \int_0^\pi \sin^2 x \sin^2 y\, dx\, dy = 1,$$

as can be verified directly. Also,

$$(8.35) \qquad E(M_{11}) = \frac{4}{\pi^2} \int_0^\pi \int_0^\pi \left[\cos^2 x \sin^2 y + \sin^2 x \cos^2 y \right] dx\, dy = 2,$$

so that $E(M_{11}) = \lambda_{11}$, as was to be expected in view of (8.17).

As was shown, if $u(x, y)$ is any admissible function other than $M_{11}(x, y)$, we must have $E(u) > \lambda_{11} = 2$. To see this in a particular case we consider the function

$$(8.36) \qquad u(x, y) = \frac{30}{\pi^5} xy(\pi - x)(\pi - y).$$

The coefficient $30/\pi^5$ has been selected so that (8.33) is satisfied, and $u(x, y)$ clearly vanishes on the boundary of the square. We readily find that

$$(8.37)$$
$$E(u) = \frac{900}{\pi^{10}} \int_0^\pi \int_0^\pi \left[x^2(\pi - x)^2(\pi - 2y)^2 + y^2(\pi - y)^2(\pi - 2x)^2 \right] dx\, dy$$
$$= 2.03.$$

Thus $E(u) = 2.03 > E(M_{11}) = 2$ as follows from (8.16). Furthermore, the value 2.03 is a remarkably good approximation to the exact eigenvalue $\lambda_{11} = 2$ given that $u(x, y)$ was chosen quite arbitrarily.

Although the given function $u(x, y)$ yields an excellent approximation to the eigenvalue λ_{11}, it does not appear to be a good approximation to the eigenfunction $M_{11}(x, y)$. In fact, if we measure the difference between $u(x, y)$ and $M_{11}(x, y)$ in the mean square norm, we obtain

$$\|M_{11} - u\|^2 = \int_0^\pi \int_0^\pi [M_{11} - u]^2\, dx\, dy = 0.73,$$

so that $\|M_{11} - u\| = 0.85$. The norm is not very small. This indicates that the variational approach may be expected to yield a much better approximation to the eigenvalues than to the eigenfunctions. This is seen in the discussion of the Rayleigh–Ritz method in Section 8.2.

We have shown that subject to appropriate constraints the minima of the Rayleigh quotient (8.19) are the eigenvalues of (8.1)–(8.2) and the minima are assumed when the minimizing functions are eigenfunctions. Reversing this process, we now formulate a variational problem for the Rayleigh quotient and show directly that the eigenvalues and eigenfunctions for (8.1)–(8.2) result from the minimization problem. The type of variational problem considered depends on the boundary conditions for the eigenvalue problem (8.1)–(8.2). It has already been seen in (8.10)–(8.11) that the form of energy integral $E(u)$ varies with the boundary conditions (8.11).

In the case of *Dirichlet boundary conditions* (i.e., $M = 0$ on ∂G) the appropriate *Rayleigh quotient* is

$$(8.38) \qquad \frac{E(w)}{\|w\|^2} = \frac{1}{\|w\|^2} \iint_G \left[p(\nabla w)^2 + qw^2 \right] dv.$$

In addition, in the case of *Neumann boundary conditions* (i.e., $\partial M/\partial n = 0$ on ∂G) the Rayleigh quotient again has the form (8.38). However, for *boundary conditions of the third kind* (i.e., $\partial M/\partial n + hM = 0$ on ∂G) we obtain, in view of (8.11),

$$(8.39) \qquad \frac{E(w)}{\|w\|^2} = \frac{1}{\|w\|^2} \left\{ \iint_G \left[p(\nabla w)^2 + qw^2 \right] dv + \int_{\partial G} hpw^2 \, ds \right\}.$$

It is assumed that $h > 0$.

The variational problem is given as follows. *The function w that yields a minimum for the variational problem*

$$(8.40) \qquad \frac{E(w)}{\|w\|^2} = \text{minimum},$$

over all functions w that vanish on ∂G and satisfy the constraints

$$(8.41) \qquad (w, M_1) = (w, M_2) = \cdots = (w, M_{n-1}) = 0,$$

is—upon normalization—the nth eigenfunction M_n of the problem (8.1)–(8.2) with $\alpha = 1$ and $\beta = 0$ in (8.2). The nth successive minimum value of the Rayleigh quotient (8.40) is the eigenvalue λ_n that corresponds to M_n. The eigenvalues λ_j are ordered as $0 \leqslant \lambda_1 \leqslant \lambda_2 \leqslant \cdots \leqslant \lambda_n$. For the eigenvalue problem (8.1)–(8.2) with boundary conditions of the second and third kinds, there are no restrictions placed on the boundary values of admissible functions $w(x)$.

To demonstrate this result we assume that a sufficiently smooth minimizing function for the variational problem exists. (This is difficult to prove, in general.) Let the minimizing function be denoted by $w_n(x)$, and the admissible functions $w(x)$ be represented as

(8.42)
$$w = w_n + \epsilon W$$

where ϵ is a constant and W satisfies the same admissibility conditions as $w(x)$.
   In terms of (8.42) the variational problem becomes

(8.43)
$$\frac{E(w)}{\|w\|^2} = \frac{E(w_n + \epsilon W)}{\|w_n + \epsilon W\|^2} = \text{minimum.}$$

As a function of the parameter ϵ, the Rayleigh quotient in (8.43) assumes a minimum value when $\epsilon = 0$, that is, when $w = w_n$. Consequently, $\epsilon = 0$ must be a critical or stationary point of the function $E(w)/\|w\|^2$, so that its derivative with respect to ϵ must vanish at $\epsilon = 0$. We have

(8.44)
$$\frac{\partial}{\partial \epsilon} \left[\frac{E(w_n + \epsilon W)}{\|w_n + \epsilon W\|^2} \right]_{\epsilon=0}$$

$$= \frac{\partial}{\partial \epsilon} \left\{ \frac{\iint\limits_G \left\{ p(\nabla[w_n + \epsilon W])^2 + q[w_n + \epsilon W]^2 \right\} dv}{\iint\limits_G \rho(w_n + \epsilon W)^2 dv} \right\}_{\epsilon=0}$$

$$+ \frac{\partial}{\partial \epsilon} \left\{ \frac{\int_{\partial G} hp[w_n + \epsilon W]^2 ds}{\iint\limits_G \rho(w_n + \epsilon W)^2 dv} \right\}_{\epsilon=0}$$

$$= \frac{2}{\|w_n\|^2} \iint\limits_G \left\{ p\nabla w_n \cdot \nabla W + q w_n W - \left[\frac{E(w_n)}{\|w_n\|^2} \right] \rho w_n W \right\} dv$$

$$+ \frac{2}{\|w_n\|^2} \int_{\partial G} hp w_n W \, ds = 0.$$

If we are dealing with the Dirichlet or Neumann problem, the surface integral is absent in (8.44). Using (4.29) and the divergence theorem gives

(8.45)
$$\iint\limits_G p\nabla w_n \cdot \nabla W \, dv = -\iint\limits_G W \nabla \cdot (p\nabla w_n) \, dv + \iint\limits_G \nabla \cdot (pW \nabla w_n) \, dv$$

$$= -\iint\limits_G W \nabla \cdot (p\nabla w_n) \, dv + \int_{\partial G} pW \frac{\partial w_n}{\partial n} \, ds.$$

Introducing (8.45) into (8.44) yields

$$(8.46) \qquad \iint\limits_G W \left\{ - \nabla \cdot (p \nabla w_n) + q w_n - \left[\frac{E(w_n)}{\|w_n\|^2} \right] \rho w_n \right\} dv$$

$$+ \int_{\partial G} W p \left\{ \frac{\partial w_n}{\partial n} + h w_n \right\} ds = 0.$$

For the eigenvalue problem with *Dirichlet boundary conditions*, the surface integral in (8.46) is zero since $W = 0$ on the boundary. From the arbitrariness of W in the region G (see Exercises 8.1.7 and 8.1.8 for some additional details regarding this point) we conclude that the bracketed term in the volume integral must vanish. Therefore, w_n satisfies the equation

$$(8.47) \qquad - \nabla \cdot (p \nabla w_n) + q w_n = \left[\frac{E[w_n]}{\|w_n\|^2} \right] \rho w_n,$$

and $w_n = 0$ on ∂G. This implies that

$$(8.48) \qquad \frac{E[w_n]}{\|w_n\|^2} = \lambda_n,$$

and $w_n(x)/\|w_n(x)\| = M_n(x)$, in view of (8.1)–(8.2).

In the case of *Neumann conditions* for the eigenvalue problem, we must set $h = 0$ in the surface integral in (8.44). The arbitrariness of W again implies that the integrals over G and ∂G in (8.46) must vanish separately. Thus $w_n(x)$ again is a solution of (8.47) but the vanishing of the surface integral leads to the *natural boundary condition*

$$(8.49) \qquad \frac{\partial w_n}{\partial n} \bigg|_{\partial G} = 0.$$

We conclude that $w_n / \|w_n(x)\| = M_n$, the nth eigenfunction for the problem (8.1) with $\alpha = 0$ and $\beta = 1$ in (8.2), and that (8.48) is again valid.

For the *third boundary value problem* (i.e., with $h \neq 0$), the arbitrariness of W implies that both integrals in (8.46) vanish separately. The *natural boundary condition* for w_n becomes

$$(8.50) \qquad \frac{\partial w_n}{\partial n} + h w_n \bigg|_{\partial G} = 0$$

and w_n satisfies (8.47). Consequently, $w_n / \|w_n(x)\| = M_n$, the nth eigenfunction for (8.1), with $\alpha = h$ and $\beta = 1$ in (8.2) and (8.48) yields the eigenvalue.

This concludes our direct demonstration that the eigenvalues and eigenfunctions for (8.1)–(8.2) can be determined from a variational problem. It has already been shown in Example 8.1 how this variational principle may be used to approximate the first eigenvalue and eigenfunction for a specific problem.

The preceding variational problem requires that the first $n - 1$ eigenfunctions must be known if we are to determine the nth eigenvalue and eigenfunction. However, *Courant's maximum–minimum principle* permits the determination of each eigenvalue independently of the other (lesser) eigenvalues. Restricting ourselves to the variational eigenvalue problem with Dirichlet boundary conditions, we now show that the eigenvalues $\lambda_n \to \infty$ as $n \to \infty$. Although the variational problem shows (see Exercise 8.1.6) that the eigenvalues can be ordered as in (8.3), it may happen that the eigenvalues have a finite limit point and that infinitely many (independent) eigenfunctions can be associated with a single eigenvalue. However, $\lambda_n \to \infty$ implies that each eigenvalue has finite multiplicity; that is, there are only a finite number of linearly independent eigenfunctions associated with each distinct eigenvalue.

The use of the maximum–minimum principle permits the comparison of variational problems with different sets of coefficients in their Rayleigh quotients and problems defined over different regions G. We assume the coefficients p, ρ, and q that occur in the Rayleigh quotient have maximum and minimum values in the region G together with its boundary ∂G and still require that $p > 0$, $\rho > 0$, and $q \geqslant 0$ in G. Let the constants p_M, ρ_M, and q_M represent the maximum values and p_m, ρ_m, and q_m represent the minimum values of the function p, ρ, and q, respectively. (We assume that p_m and ρ_m are not zero.) Further, the admissible functions $w(x)$ will be required to vanish on ∂G and to satisfy the constraints (8.20).

On examining the Rayleigh quotient appropriate for the Dirichlet problem

$$(8.51) \qquad \frac{E(w)}{\|w\|^2} = \frac{\iint\limits_G \left[p(\nabla w)^2 + qw^2 \right] dv}{\iint\limits_G \rho w^2 \, dv},$$

it is clear that if we replace the functions p, ρ, and q by the constants p_M, ρ_m and q_M, we obtain a new Rayleigh quotient, denoted by $E_M(w)/\|w\|_m^2$ for which

$$(8.52) \qquad \frac{E_M(w)}{\|w\|_m^2} \geqslant \frac{E(w)}{\|w\|^2},$$

for each admissible function w. Now the inner product that occurs in the constraints (8.20) for the new problem with constant coefficients can be expressed as

$$(8.53) \qquad (w, m_k) = \iint\limits_{G} \rho_m w m_k \, dv = \iint\limits_{G} \rho w \left(m_k \frac{\rho_m}{\rho} \right) dv.$$

Thus the constraints for the new problem must be given in terms of the set of functions $\hat{m}_k = (\rho_m/\rho) m_k$ if we want the new problem to have the same set of constraints. But in the maximum–minimum principle the set of functions $m_k(x)$ ranges over all sufficiently smooth functions and, therefore, so does the set $\hat{m}_k(x)$. Consequently, we may conclude that max–min of $E_M(w)/\|w\|_m^2$ is greater than or equal to the max–min of $E(w)/\|w\|^2$.

In a similar fashion, if we replace the functions p, ρ, and q in (8.51) by the constants p_m, ρ_M, and q_m, we obtain a new Rayleigh quotient $E_m(w)/\|w\|_M^2$ for which

$$(8.54) \qquad \frac{E_m(w)}{\|w\|_M^2} \leqslant \frac{E(w)}{\|w\|^2}.$$

The foregoing procedure may be applied to the constraints (8.20), and we conclude that the max–min of $E_m(w)/\|w\|_M^2$ does not exceed the max–min of $E(w)/\|w\|^2$. Since for the set of admissible functions w, vanishing on ∂G and satisfying the constraints (8.20), each of the aforementioned max–min's yields the nth eigenvalue for the given variational problem, we conclude that

$$(8.55) \qquad \lambda_n^{(m)} \leqslant \lambda_n \leqslant \lambda_n^{(M)},$$

where $\lambda_n^{(m)}$, λ_n, and $\lambda_n^{(M)}$ are the nth eigenvalues associated with the variational problem for the Rayleigh quotients $E_m(w)/\|w\|_M^2$, $E(w)/\|w\|^2$, and $E_M(w)/\|w\|_m^2$, respectively.

Next we consider changes in the region G. Suppose we consider a subregion $\hat{G}$ contained in G (a portion of the boundary $\partial \hat{G}$ may coincide with ∂G). Let $E(w)/\|w\|^2$ be the Rayleigh quotient for the variational problem in G, and suppose we require that all admissible functions $w(x)$ in addition to vanishing on ∂G and satisfying the constraints (8.20), must vanish on $\partial \hat{G}$ and in that part of G exterior to $\hat{G}$. Since the additional requirement on the functions $w(x)$ effectively reduces the number of admissibile functions for the variational problem, the maximum–minimum for the new problem cannot be smaller than the max–min λ_n attained when the added restriction is absent. However, the new set of admissible functions is precisely that which is appropriate for the variational problem in the subregion $\hat{G}$. The max–min for the variational problem in $\hat{G}$ is $\hat{\lambda}_n$ the nth eigenvalue for that problem. In view of the preceding observation, we obtain

$$(8.56) \qquad \lambda_n \leqslant \hat{\lambda}_n.$$

To complete our discussion we consider two rectangular regions R_M and R_m. (In three dimensions these are rectangular boxes.) The region R_M contains the region G within it, whereas the region R_m is completely contained within G. First we consider the max–min variational problem for the Rayleigh quotient $E_M(w)/\|w\|_m^2$, with the region G replaced by R_m but the problem otherwise unchanged. In view of (8.56) we conclude that

$$(8.57) \qquad \lambda_n^{(M)} \leqslant \hat{\lambda}_n^{(M)},$$

where $\lambda_n^{(M)}$ and $\hat{\lambda}_n^{(M)}$ are the nth eigenvalues for the problem associated with $E_M(w)/\|w\|_m^2$ over the regions G and R_m, respectively. Next, we consider the max–min variational problem for $E_m(w)/\|w\|_M^2$ with the region G replaced by R_M, but the problem otherwise unchanged. Using (8.56) gives

$$(8.58) \qquad \hat{\lambda}_n^{(m)} \leqslant \lambda_n^{(m)},$$

where $\hat{\lambda}_n^{(m)}$ and $\lambda_n^{(m)}$ are the nth eigenvalues for the problem associated with $E_m(w)/\|w\|_M^2$ over the regions R_M and G, respectively.

Combining (8.55), (8.57), and (8.58) yields the set of inequalities

$$(8.59) \qquad \hat{\lambda}_n^{(m)} \leqslant \lambda_n^{(m)} \leqslant \lambda_n \leqslant \lambda_n^{(M)} \leqslant \hat{\lambda}_n^{(M)}.$$

The eigenvalues $\hat{\lambda}_n^{(m)}$ and $\hat{\lambda}_n^{(M)}$ can be determined exactly be separation of variables since the equations have constant coefficients and the regions are rectangular. This has already been carried out for a special two-dimensional problem in Example 7.5, and the general case is considered in Example 8.2 for two and three dimensions. Since the eigenvalues $\hat{\lambda}_n^{(m)}$ and $\hat{\lambda}_n^{(M)}$ tend to infinity as $n \to \infty$ (as shown in Example 8.2) so do the eigenvalues λ_n.

In the following example we consider some specific problems that illustrate the results obtained in our discussion. There are many further properties of eigenvalues that may be obtained by arguments similar to those given. For example, it can be shown that the eigenvalues λ_n for the Neumann and third boundary value problem also tend to infinity as $n \to \infty$ but we do not prove this here. We use the property that $\lambda_n \to \infty$ as $n \to \infty$ to prove completeness in the mean square sense of the eigenfunctions for the Dirichlet problem in our discussion following Example 8.2.

EXAMPLE 8.2. DIRICHLET EIGENVALUE PROBLEMS FOR EQUATIONS WITH CONSTANT COEFFICIENTS

The eigenvalue problem

$$(8.60) \qquad -p\nabla^2 M + qM = \lambda\rho M$$

in the region G with the *Dirichlet boundary condition*

$$(8.61) \qquad M(x)|_{\partial G} = 0$$

where p, q, and ρ are assumed to be constants, is associated with the *variational problem*

$$(8.62) \qquad \frac{E(w)}{\|w\|^2} = \frac{\iint\limits_G \left[p(\nabla w)^2 + qw^2 \right] dv}{\iint\limits_G \rho w^2 \, dv} = \text{minimum},$$

where all admissible functions $w(x)$ must vanish on ∂G and satisfy appropriate additional constraints.

Now (8.60) can be written as

$$(8.63) \qquad \nabla^2 M + \left[\frac{\lambda \rho - q}{p} \right] M \equiv \nabla^2 M + \tilde{\lambda} M = 0,$$

that is, $\tilde{\lambda} = (\lambda \rho - q)/p$. Similarly, the variational problem (8.62) can be expressed as

$$(8.64) \qquad \frac{E(w)}{\|w\|^2} = \left(\frac{p}{\rho} \right) \frac{\iint\limits_G (\nabla w)^2 \, dv}{\iint\limits_G w^2 \, dv} + \frac{q}{\rho} = \text{minimum}.$$

Both (8.63) and (8.64) imply that if we can solve a simplified version of (8.60)–(8.61) or (8.62) with $p = \rho = 1$ and $q = 0$ and we denote the eigenvalues of that simplified problem by $\tilde{\lambda}_n$, then the eigenvalues λ_n of (8.60)–(8.61) or (8.62) are given as

$$(8.65) \qquad \lambda_n = \frac{p\tilde{\lambda}_n + q}{\rho}, \qquad n = 1, 2, 3, \ldots .$$

The relationship (8.65) between λ_n and $\tilde{\lambda}_n$ demonstrates the property discussed in the general case of variable coefficients. If the coefficients p and q are increased, the eigenvalues increase, whereas the reverse is true if p and q are decreased. Also if ρ is decreased, the eigenvalues are increased, whereas if ρ is increased, the eigenvalues decrease.

In the two-dimensional case, if we specialize G to be a *rectangle* as given in Example 7.5 with $0 < x < l$ and $0 < y < \hat{l}$, we obtain the eigenvalues for (8.63) and (8.61) as

$$(8.66a) \qquad \tilde{\lambda}_{nm} = \left(\frac{\pi n}{l} \right)^2 + \left(\frac{\pi m}{\hat{l}} \right)^2, \qquad n, m = 1, 2, 3, \ldots,$$

as shown in that example. Similarly, on considering the *rectangular box*

$0 < x < l$, $0 < y < \hat{l}$, and $0 < z < \tilde{l}$ in three dimensions and separating variables, we readily obtain the eigenvalues for (8.63) and (8.61) as

$$(8.66b) \quad \tilde{\lambda}_{nmk} = \left(\frac{\pi n}{l}\right)^2 + \left(\frac{\pi m}{\hat{l}}\right)^2 + \left(\frac{\pi k}{\tilde{l}}\right)^2, \qquad n, m, k = 1, 2, 3, \ldots.$$

In both cases the eigenvalues can be expressed as a sequence $\{\tilde{\lambda}_n\}$ and we have $\tilde{\lambda}_n \to \infty$ as $n \to \infty$, since $\tilde{\lambda}_{nm}$ and $\tilde{\lambda}_{nmk}$ both tend to infinity as their subscripts tend to infinity. Consequently, in view of (8.65), $\lambda_n \to \infty$ as $n \to \infty$ since p and ρ are positive. Furthermore, the rectangular regions R_M and R_m introduced in our earlier discussion may be assumed to have sides parallel to the coordinate lines or planes in the two or three-dimensional cases, respectively. Since the eigenvalue equation in the case of constant coefficients is invariant under the translation of axes, the eigenvalues are given as in (8.66a) or (8.66b). Therefore, we conclude that the eigenvalues $\hat{\lambda}_n^{(M)}$ and $\hat{\lambda}_n^{(m)}$ tend to infinity as $n \to \infty$ and (8.59) implies that $\lambda_n \to \infty$ as $n \to \infty$.

It is seen from (8.66a) and (8.66b) that as the dimensions of the rectangular region G are changed, the eigenvalues change in a specific manner. As the lengths of the sides are decreased, the eigenvalues $\tilde{\lambda}_{nm}$ and $\tilde{\lambda}_{nmk}$ increase, whereas the reverse is true if the lengths of the sides are increased. This result is consistent with our conclusion relating the eigenvalues of region G to those of a subregion $\hat{G}$.

To conclude this example we discuss the *Dirichlet eigenvalue problem* for the *unit circle*. We first obtain the eigenvalues and eigenfunctions by means of separation of variables and then use the inequalities (8.59) to estimate the first eigenvalue for the circle in terms of the eigenvalues for inscribed and circumscribed squares.

Expressing the Laplacian operator in polar coordinates, we consider the eigenvalue problem

$$(8.67) \qquad \nabla^2 M + \lambda M = M_{rr} + \frac{1}{r} M_r + \frac{1}{r^2} M_{\theta\theta} + \lambda M = 0$$

in the unit disk $0 \leqslant r < 1$, with the Dirichlet boundary condition for $M(r, \theta)$,

$$(8.68) \qquad\qquad\qquad M(1, \theta) = 0.$$

The solutions M are required to be single valued in the disk so that we have

$$(8.69) \qquad\qquad M(r, \theta + 2\pi) = M(r, \theta).$$

To solve for $M(r, \theta)$ we use separation of variables and set

$$(8.70) \qquad\qquad M(r, \theta) = F(r)G(\theta).$$

Inserting (8.70) in (8.67) gives

$$(8.71) \qquad r^2 \left[\frac{F''(r) + (1/r)F'(r) + \lambda F(r)}{F(r)} \right] = -\frac{G''(\theta)}{G(\theta)} = m^2,$$

where m^2 is the separation constant. The equation for $G(\theta)$ is

$$(8.72) \qquad\qquad\qquad G''(\theta) + m^2 G(\theta) = 0.$$

The equation (8.69) requires $G(\theta)$ to be periodic of period 2π, and this implies that $m = n$, an integer. The equation for $F_n(r)$ (i.e., for each n) is

$$(8.73) \qquad\qquad F_n''(r) + \frac{1}{r} F_n'(r) + \left(\lambda - \frac{n^2}{r^2} \right) F_n(r) = 0,$$

which is Bessel's equation of order n. This equation was discussed in Example 4.7. The solutions F_n that are finite at $r = 0$ are the Bessel functions and are given as

$$(8.74) \qquad\qquad\qquad F_n(r) = J_n(\sqrt{\lambda}\, r).$$

The boundary condition (8.68) requires that

$$(8.75) \qquad\qquad\qquad J_n(\sqrt{\lambda}\,) = 0$$

so that the eigenvalues λ are the squares of the zeros of the Bessel functions. These were denoted by α_{kn} ($k = 1, 2, \dots$) in Example 4.7 so that we have

$$(8.76) \qquad\qquad \lambda_{kn} = (\alpha_{kn})^2, \qquad k = 1, 2, \dots, \qquad n = 0, 1, 2, \dots.$$

Since with $m = n$ (8.72) yields the trigonometric functions $\cos n\theta$ and $\sin n\theta$, the eigenfunctions M_{kn} are obtained from the functions

$$(8.77) \qquad J_0(\alpha_{k0} r); \qquad J_n(\alpha_{kn} r)\cos n\theta; \qquad J_n(\alpha_{kn} r)\sin n\theta.$$

Using (4.116) it is not difficult to normalize these eigenfunctions. At present we are mainly concerned with the eigenvalues λ_{kn}, and the first eigenvalue is given in terms of the first zero of the Bessel function $J_0(x)$, that is,

$$(8.78) \qquad\qquad\qquad \lambda_{10} = (\alpha_{10})^2 \approx (2.40)^2 \approx 5.76.$$

To obtain an upper and lower bound for the first eigenvalue λ_{10} for the unit circle, we consider inscribed and circumscribed squares of sides $\sqrt{2}$ and 2, respectively. They are pictured in Figure 8.1. The leading eigenvalue $\tilde{\lambda}_{11}$ for the

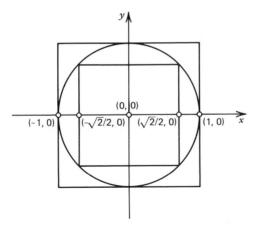

Figure 8.1.   Eigenvalues for a circle.

circumscribed square is—on using (8.66a)—

$$(8.79) \qquad \tilde{\lambda}_{11} = 2\left(\frac{\pi}{2}\right)^2 = \frac{\pi^2}{2} \approx 4.93.$$

The first eigenvalue for the inscribed square is

$$(8.80) \qquad \hat{\lambda}_{11} = 2\left(\frac{\pi}{\sqrt{2}}\right)^2 = \pi^2 \approx 9.87.$$

Thus we have

$$(8.81) \qquad \tilde{\lambda}_{11} < \lambda_{10} < \hat{\lambda}_{11},$$

a result consistent with (8.59). Clearly, (8.81) yields a very poor method for approximating the eigenvalue λ_{10}, since the squares considered do not approximate the circular region very well. However, this method can be used to provide bounds for all the zeros of the Bessel functions. In Section 8.2 we use the Rayleigh–Ritz method to approximate the first eigenvalue λ_{10} for the circle.

We are now prepared to prove the *completeness* of the orthonormalized set of eigenfunctions $\{M_k(x)\}$ that satisfy (8.1) in G and the Dirichlet conditions $M_k(x) = 0$ on ∂G. Let $u(x)$ be a smooth function defined in G that vanishes on ∂G; that is, $u(x)$ satisfies the admissibility conditions for the variational problem. We expand $u(x)$ in a series of eigenfunctions as in (8.5) with the Fourier coefficients given in (8.6). It is not required that $u(x)$ be normalized as in (8.7). Let the remainder R_n be defined as

$$(8.82) \qquad R_n(x) = u(x) - \sum_{k=1}^{n} N_k M_k(x).$$

Then we show that

$$(8.83) \qquad \lim_{n \to \infty} \| R_n(x) \| = \lim_{n \to \infty} \| u(x) - \sum_{k=1}^{n} N_k M_k(x) \| = 0$$

so that the series (8.5) converges to $u(x)$ in the *mean square sense*. Thereby, the completeness of the eigenfunctions $\{ M_k(x) \}$ is demonstrated. The norm $\| R_n \|$ is defined as

$$(8.84) \qquad \| R_n(x) \|^2 = \iint_G \rho R_n^2 \, dv.$$

Since $u(x)$ and the eigenfunctions $\{ M_k(x) \}$ are admissible for the variational problem, so is the remainder term $R_n(x)$. We have, in addition,

$$(8.85) \quad (R_n, M_j) = \left(u - \sum_{k=1}^{n} N_k M_k, M_j \right) = (u, M_j) - \sum_{k=1}^{n} N_k (M_k, M_j)$$

$$= N_j - N_j = 0, \qquad j = 1, \ldots, n,$$

in view of (8.4) and (8.6). Consequently, $R_n(x)$ satisfies the constraints for the minimum problem that determines the $(n + 1)$st eigenvalue λ_{n+1}, and this yields for the Rayleigh quotient (8.51)

$$(8.86) \qquad \frac{E(R_n)}{\| R_n \|^2} \geq \lambda_{n+1},$$

since the minimum is assumed for the eigenfunction $M_{n+1}(x)$. We rewrite (8.86) as

$$(8.87) \qquad \| R_n \|^2 \leq \frac{E(R_n)}{\lambda_{n+1}}.$$

We already know that $\lambda_{n+1} \to \infty$ as $n \to \infty$ so that we need only show that $E(R_n)$ is bounded as $n \to \infty$ to conclude that $\| R_n \| \to 0$ as $n \to \infty$.

Now, as is easily shown,

$$(8.88) \qquad E(u) = E\left[R_n + \sum_{k=1}^{n} N_k M_k \right] = E(R_n) + E\left[\sum_{k=1}^{n} N_k M_k \right]$$

$$+ 2 \sum_{k=1}^{n} N_k \iint_G [p \nabla R_n \cdot \nabla M_k + q R_n M_k] \, dv.$$

Using (8.45), the integral term in (8.88) can be expressed as

(8.89)

$$\iint_G [p\nabla R_n \cdot \nabla M_k + qR_n M_k]\, dv = \iint_G R_n[-\nabla \cdot (p\nabla M_k) + qM_k]\, dv$$

$$= \lambda_k \iint_G \rho R_n M_k\, dv = \lambda_k(R_n, M_k) = 0, \qquad k = 1, \ldots, n$$

where we have used the fact that $R_n = 0$ on ∂G, the orthogonality properties (8.85), as well as the fact that $M_k(x)$ satisfies (8.1) with $\lambda = \lambda_k$. Furthermore, we obtain from (8.12),

$$(8.90) \quad E\left[\sum_{k=1}^n N_k M_k(x)\right] = \iint_G \sum_{j=1}^n N_j M_j(x) L\left[\sum_{k=1}^n N_k M_k(x)\right]\, dv$$

$$= \sum_{j=1}^n \sum_{k=1}^n \lambda_k N_k N_j(M_k, M_j) = \sum_{k=1}^n \lambda_k N_k^2.$$

Combining these results gives

$$(8.91) \qquad E(R_n) = E(u) - \sum_{k=1}^n \lambda_k N_k^2 \leqslant E(u),$$

since $E(u)$ is finite by assumption and the eigenvalues λ_k are nonnegative. We conclude from (8.91) that $E(R_n)$ is bounded for all n. Therefore, (8.87) shows that $\|R_n\| \to 0$ as $n \to \infty$, and the *mean square convergence* of the series (8.5) and the *completeness* of the $M_k(x)$ is demonstrated. As was indicated earlier, if we can show that the eigenvalues λ_n for the second and third boundary value problems tend to infinity, we can use the given argument to show that the corresponding set of eigenfunctions is complete.

For the one-dimensional *Sturm–Liouville eigenvalue problem* discussed in Section 4.3, the appropriate Rayleigh quotient is

$$(8.92) \qquad \frac{E(w)}{\|w\|^2} = \frac{\int_0^l [p(w')^2 + qw^2]\, dx}{\int_0^l \rho w^2\, dx},$$

when the admissible functions $w(x)$ satisfy the Dirichlet conditions

$$(8.93) \qquad w(0) = w(l) = 0.$$

The appropriate Rayleigh quotients for the second and third boundary value problems are presented in the exercises. All the basic results given for the higher dimensional eigenvalue problem carry over to the Sturm–Liouville problem. Thus the completeness of the eigenfunctions (i.e., the sine functions) of Example 4.4, and of the convergence of the Fourier sine series in the mean square sense can be demonstrated.

To conclude this section we note that we can drop our assumption that the function $u(x)$ expanded in a series of eigenfunctions must vanish on the boundary ∂G. This follows since any admissible function $u(x)$ that does not vanish on ∂G can be approximated arbitrarily closely by a function that does vanish on ∂G in the mean square norm. In this norm the values of functions at specific points or lower dimensional regions do not play a significant role, in general.

8.2. THE RAYLEIGH–RITZ METHOD

The formulation of the eigenvalue problem in variational form has led to some interesting and useful conclusions concerning the general properties of eigenvalues and eigenfunctions as we have seen in the preceding section. For the most part, however, we have had to rely on separation of variables or exact solutions in the one-dimensional case to determine the eigenvalues and eigenfunctions for a given problem. The *Rayleigh–Ritz method* presented in this section is an elegant approximation method for the determination of the first few eigenvalues and eigenfunctions. It is, in fact, most effective in approximating the lowest eigenvalue and this is often the most important one in applications. As we shall see, the method yields an algebraic problem for the determination of the eigenvalues and eigenfunctions.

To begin, we must select n functions $\varphi_1(x), \ldots, \varphi_n(x)$ that are admissible for the variational problem under consideration. That is, they must be sufficiently smooth and vanish on ∂G in the case of Dirichlet boundary conditions. For the case of the Neumann or the third boundary value problem, they need not satisfy any conditions on ∂G. These functions are chosen to be linearly independent and are selected, if possible, to be good approximations to the eigenfunctions for the problem.

We form the linear combination

$$(8.94) \qquad w(x) = \sum_{k=1}^{n} c_k \varphi_k(x)$$

with as yet undetermined coefficients c_k and insert this sum into the Rayleigh quotient $E(w)/\|w\|^2$. This gives for the Rayleigh quotient (8.38),

$$(8.95) \qquad \frac{E(w)}{\|w\|^2} = \frac{\displaystyle\sum_{j,k=1}^{n} c_j c_k \iint_G \left[p\nabla\varphi_j \cdot \nabla\varphi_k + q\varphi_j\varphi_k \right] dv}{\displaystyle\sum_{j,k=1}^{n} c_j c_k \iint_G \rho\varphi_j\varphi_k \, dv}.$$

For the Rayleigh quotient (8.39) appropriate for the third boundary value problem, there is an additional term $\int_{\partial G} hp\varphi_j\varphi_k\,ds$ in the numerator in (8.95) that multiplies c_jc_k. We use the form (8.95) in our discussion.

Let the n by n matrices A and B be defined by their respective elements

$$(8.96) \qquad a_{jk} = \iint\limits_{G} \left[p\nabla\varphi_j \cdot \nabla\varphi_k + q\varphi_j\varphi_k \right] dv,$$

$$(8.97) \qquad b_{jk} = \iint\limits_{G} \rho\varphi_j\varphi_k\,dv.$$

We define the n-component column vector $\mathbf{c}$ to have the components $c_1,\ldots, c_n$ and $\mathbf{c}^T$ to be its transpose—that is, $[c_1, c_2,\ldots, c_n]$. Then the equation (8.95) can be expressed as

$$(8.98) \qquad \frac{E(w)}{\|w\|^2} = \frac{\mathbf{c}^T A \mathbf{c}}{\mathbf{c}^T B \mathbf{c}},$$

as is easily verified. We remark that both $\mathbf{c}^T A \mathbf{c}$ and $\mathbf{c}^T B \mathbf{c}$ are *quadratic forms*. Further, the (real valued) matrices A and B are *symmetric* as follows from (8.96)–(8.97). Both matrices are also *positive definite* if $q(x) > 0$, whereas B is positive definite in all cases, as shown in the exercises.

The Rayleigh quotient $E(w)/\|w\|^2$ is now a function of the vector $\mathbf{c}$ and we wish to minimize this expression. If the functions $\varphi_1,\ldots, \varphi_n$ are part of a complete set of functions, it is possible to represent any admissible function $w(x)$ in an infinite series of those functions, and the minimum problem can technically be replaced by a problem where the coefficients c_k in the expansion are chosen to achieve a minimum. We replace the infinite sum by a finite sum and expect this to yield approximate eigenvalues and eigenfunctions. We do not prove the validity of the method.

To proceed with the method we assume that $\mathbf{c} = \boldsymbol{\sigma}$ is the vector that minimizes the expression (8.98) and represent the arbitrary vector $\mathbf{c}$ in the form

$$(8.99) \qquad \mathbf{c} = \boldsymbol{\sigma} + \sum_{k=1}^{n} \epsilon_k \mathbf{e}_k,$$

where the vectors $\mathbf{e}_1,\ldots, \mathbf{e}_n$ are the standard basis vectors for n-component vectors with $\mathbf{e}_k$ having unity as its kth component and zero as its remaining components. The constants ϵ_k characterize the variation around the minimizing vector $\boldsymbol{\sigma}$. Inserting (8.99) into (8.98) gives

$$(8.100) \qquad \frac{E(w)}{\|w\|^2} = \frac{\left[\boldsymbol{\sigma} + \sum_{k=1}^{n} \epsilon_k \mathbf{e}_k \right]^T A \left[\boldsymbol{\sigma} + \sum_{k=1}^{n} \epsilon_k \mathbf{e}_k \right]}{\left[\boldsymbol{\sigma} + \sum_{k=1}^{n} \epsilon_k \mathbf{e}_k \right]^T B \left[\boldsymbol{\sigma} + \sum_{k=1}^{n} \epsilon_k \mathbf{e}_k \right]}.$$

To determine the equation satisfied by σ we differentiate (8.100) with respect to ϵ_k ($k = 1, \ldots, n$) and then set ϵ_k and the derivatives equal to zero since the minimum occurs when $\epsilon_k = 0$. This gives

$$(8.101) \qquad \frac{\partial}{\partial \epsilon_k}\left[\frac{E(w)}{\|w\|^2}\right]_{\epsilon_k = 0} = \frac{e_k^T A\sigma + \sigma^T A e_k}{\sigma^T B\sigma}$$

$$-\frac{\left[e_k^T B\sigma + \sigma^T B e_k\right]\sigma^T A\sigma}{\left(\sigma^T B\sigma\right)^2} = 0.$$

Since A and B are symmetric matrices and $\sigma^T A e_k$ and $\sigma^T B e_k$ are scalars we have

$$(8.102) \qquad \left(\sigma^T A e_k\right)^T = e_k^T A^T \sigma = e_k^T A\sigma = \sigma^T A e_k,$$

and a similar result with A replaced by B. Thus (8.101) reduces to

$$(8.103) \qquad e_k^T\left[A\sigma - \left(\frac{\sigma^T A\sigma}{\sigma^T B\sigma}\right)B\sigma\right] = 0; \qquad k = 1, \ldots, n.$$

Since the bracketed vector in (8.103) is orthogonal to the n basis vectors $e_1, \ldots, e_n$, we conclude that it must be the zero vector. As a result we obtain

$$(8.104) \qquad A\sigma = \hat{\lambda}B\sigma$$

where we have set

$$(8.105) \qquad \hat{\lambda} = \frac{\sigma^T A\sigma}{\sigma^T B\sigma}.$$

Now if the functions $\varphi_1, \ldots, \varphi_n$ are an orthonormal set, we find from (8.97) that $b_{jk} = \delta_{jk}$, the Kronecker delta (i.e., $\delta_{jk} = 0$ for $j \neq k$ and $\delta_{jk} = 1$ for $j = k$). Then B is a unit matrix, and (8.104) reduces to the standard eigenvalue problem for the matrix A with σ as the eigenvector and $\hat{\lambda}$ as the eigenvalue. If B is not a unit matrix, we also refer to σ as an eigenvector of A with respect to B and to $\hat{\lambda}$ as the corresponding eigenvalue.

The characteristic equation for the determination of the eigenvalues $\hat{\lambda}$ is

$$(8.106) \qquad \det[A - \hat{\lambda}B] = 0.$$

This is an nth degree algebraic equation for $\hat{\lambda}$, and the symmetry of A and B implies that all the eigenvalues are real. Since the matrices A and B are positive definite if $q(x) > 0$, we also find that the eigenvalues must be positive. (If

$q = 0$, they are nonnegative.) Therefore, they can be ordered as

$$(8.107) \qquad 0 \leqslant \hat{\lambda}_1 \leqslant \hat{\lambda}_2 \leqslant \cdots \leqslant \hat{\lambda}_n.$$

For each eigenvalue $\hat{\lambda}_k$ there is at least one independent eigenvector $\boldsymbol{\sigma}^{(k)}$. All together there are n linearly independent *eigenvectors* $\{\boldsymbol{\sigma}^{(k)}\}$, $k = 1, \ldots, n$. It is easy to show that eigenvectors corresponding to different eigenvalues are orthogonal with respect to the matrix B. In fact, we have from (8.104),

$$
\begin{aligned}
(8.108) \qquad 0 &= \boldsymbol{\sigma}^{(k)^T}\big[A - \hat{\lambda}_j B\big]\boldsymbol{\sigma}^{(j)} - \boldsymbol{\sigma}^{(j)^T}\big[A - \hat{\lambda}_k B\big]\boldsymbol{\sigma}^{(k)} \\
&= \boldsymbol{\sigma}^{(k)^T}A\boldsymbol{\sigma}^{(j)} - \boldsymbol{\sigma}^{(j)^T}A\boldsymbol{\sigma}^{(k)} - \hat{\lambda}_j\boldsymbol{\sigma}^{(k)^T}B\boldsymbol{\sigma}^{(j)} + \hat{\lambda}_k\boldsymbol{\sigma}^{(j)^T}B\boldsymbol{\sigma}^{(k)} \\
&= \big(\hat{\lambda}_k - \hat{\lambda}_j\big)\big[\boldsymbol{\sigma}^{(k)^T}B\boldsymbol{\sigma}^{(j)}\big]
\end{aligned}
$$

where the symmetry of A and B was used. Thus if $\hat{\lambda}_k \neq \hat{\lambda}_j$, we have

$$(8.109) \qquad \boldsymbol{\sigma}^{(k)^T}B\boldsymbol{\sigma}^{(j)} = 0.$$

If B is the identity matrix, (8.109) yields the scalar or dot product of the two vectors $\boldsymbol{\sigma}^{(k)}$ and $\boldsymbol{\sigma}^{(j)}$ and shows that they are orthogonal. If B is not a unit matrix, we say that the vectors are orthogonal with respect to B. Using a Gram–Schmidt process (see Exercise 8.2.1) it is possible to orthonormalize the set of eigenvectors with respect to B and obtain

$$(8.110) \qquad \boldsymbol{\sigma}^{(k)^T}B\boldsymbol{\sigma}^{(j)} = \delta_{kj}; \qquad k, j = 1, \ldots, n$$

where δ_{kj} is the Kronecker delta.

Each *eigenvector* $\boldsymbol{\sigma}^{(j)}$ determines an *approximate eigenfunction* $\hat{M}_j(x)$,

$$(8.111) \qquad \hat{M}_j(x) = \sum_{k=1}^{n} c_k^{(j)}\varphi_k(x); \qquad j = 1, \ldots, n$$

where the $c_k^{(j)}$ are the components of the vector $\boldsymbol{\sigma}^{(j)}$. If is of interest to observe that the inner product $(\hat{M}_j, \hat{M}_k)$ of the set of approximate eigenfunctions $\{\hat{M}_j(x)\}$ is

$$
\begin{aligned}
(8.112) \qquad \big(\hat{M}_j, \hat{M}_k\big) &= \iint\limits_{G} \rho\hat{M}_j\hat{M}_k \, dv \\
&= \sum_{l, m=1}^{n} c_l^{(j)}c_m^{(k)}b_{lm} = \boldsymbol{\sigma}^{(j)^T}B\boldsymbol{\sigma}^{(k)}.
\end{aligned}
$$

Thus approximate eigenfunctions corresponding to different approximate eigenvalues are *orthogonal*. Furthermore, if the $\boldsymbol{\sigma}^{(j)}$ are orthonormalized with respect to B, so are the $\hat{M}_j(x)$ with respect to the above inner product and its (induced) mean square norm.

By using the *Courant maximum–minimum principle* of Section 8.1 we may compare the approximate eigenvalues $\hat{\lambda}_k$ with the exact eigenvalues λ_k. In replacing the given variational problem with the Rayleigh–Ritz formulation we note that we are placing an additional constraint on the problem by requiring all admissible functions to be linear combinations of the functions $\varphi_1(x), \ldots, \varphi_n(x)$. This means that every eigenvalue obtained by means of the Rayleigh–Ritz process cannot be smaller than the exact eigenvalue and we obtain

$$(8.113) \qquad \lambda_k \leqslant \hat{\lambda}_k; \qquad k = 1, \ldots, n.$$

In Example 8.1 we chose a single admissible function $u(x, y)$ for the purpose of approximating the lowest eigenvalue and eigenfunction for the Dirichlet eigenvalue problem in the rectangle. As the preceding discussion shows, this is equivalent to using the Rayleigh–Ritz method in terms of a single function $\varphi_1(x, y)$ that vanishes on the boundary of the given region.

EXAMPLE 8.3. THE EIGENVALUE PROBLEM FOR A SPHERE

We consider a sphere of unit radius with center at the origin and the eigenvalue problem within the sphere

$$(8.114) \qquad \nabla^2 M(x, y, z) + \lambda M(x, y, z) = 0$$

with the Dirichlet boundary conditions

$$(8.115) \qquad M(x, y, z)\big|_{\partial G} = 0$$

where ∂G is the surface of the sphere. Introducing spherical coordinates (r, θ, φ) with $0 \leqslant r < \infty$, $0 \leqslant \theta \leqslant 2\pi$, and $0 \leqslant \varphi \leqslant \pi$, we have

$$(8.116) \quad \nabla^2 M + \lambda M = \frac{1}{r^2} \frac{\partial}{\partial r}\left(r^2 \frac{\partial M}{\partial r}\right) + \frac{1}{r^2 \sin\varphi} \frac{\partial}{\partial \varphi}\left[\sin\varphi \frac{\partial M}{\partial \varphi}\right]$$

$$+ \frac{1}{\sin^2\varphi} \frac{\partial^2 M}{\partial \theta^2} + \lambda M = 0,$$

where $M = M(r, \theta, \varphi)$. The boundary condition (8.115) becomes

$$(8.117) \qquad M(1, \theta, \varphi) = 0.$$

Using separation of variables, we set

$$(8.118) \qquad M(r, \theta, \varphi) = F(r) Y(\theta, \varphi)$$

and insert (8.118) into (8.116) to obtain after some simplification

$$(8.119) \quad \frac{(r^2 F'(r))'}{F(r)} + \lambda r^2 = -\frac{1}{Y(\theta, \varphi)} \left[\frac{1}{\sin \varphi} \frac{\partial}{\partial \varphi} \left(\sin \varphi \frac{\partial Y}{\partial \varphi} \right) \right.$$

$$\left. + \frac{1}{\sin^2 \varphi} \frac{\partial^2 Y}{\partial \theta^2} \right] = k^2$$

with k^2 equal to the separation constant. The equation for $Y(\theta, \varphi)$ is

$$(8.120) \quad \frac{1}{\sin \varphi} \frac{\partial}{\partial \varphi} \left[\sin \varphi \frac{\partial Y}{\partial \varphi} \right] + \frac{1}{\sin^2 \varphi} \frac{\partial^2 Y}{\partial \theta^2} + k^2 Y = 0.$$

The conditions on $Y(\theta, \varphi)$ are that it be periodic in θ [i.e., $Y(\theta + 2\pi, \varphi) = Y(\theta, \varphi)$] and that it be bounded at the poles of the sphere (i.e, when $\varphi = 0$ and $\varphi = \pi$) as the coefficients in (8.120) are singular there. These conditions determine the values of k^2 to be

$$k^2 = n(n + 1); \qquad n = 0, 1, 2, \ldots,$$

and yield the solutions $Y_n(\theta, \varphi)$ known as the *spherical harmonics*. These are specified by means of a further separation of variables in (8.120) to be products of Legendre functions and sine and cosine functions (they are discussed in the exercises).

The equation for $F(r)$ then becomes

$$(8.121) \quad \frac{1}{r^2} \frac{d}{dr} \left(r^2 \frac{dF}{dr} \right) + \left[\lambda - \frac{n(n + 1)}{r^2} \right] F = 0$$

with the conditions

$$(8.122) \qquad\qquad\qquad F(1) = 0$$

and that $F(r)$ is bounded at the origin. The substitution

$$(8.123) \qquad\qquad\qquad F(r) = \frac{f(r)}{\sqrt{r}}$$

reduces (8.121) to the Bessel equation of order $n + \frac{1}{2}$,

$$(8.124) \quad f''(r) + \frac{1}{r} f'(r) + \left[\lambda - \frac{(n + 1/2)^2}{r^2} \right] f(r) = 0.$$

The boundedness condition of $F(r)$ at $r = 0$ implies that

$$(8.125) \qquad\qquad f(r) = J_{n+1/2}(\sqrt{\lambda}\, r)$$

and the boundary condition (8.122) requires that

$$(8.126) \qquad\qquad J_{n+1/2}(\sqrt{\lambda}) = 0.$$

If α_{mn} $(m = 1, 2, \ldots)$ are the roots of the Bessel function $J_{n+1/2}(\alpha)$, we obtain the eigenvalues

$$(8.127) \qquad \lambda_{mn} = (\alpha_{mn})^2; \qquad n = 0, 1, 2, \ldots; \qquad m = 1, 2, \ldots.$$

The roots of the Bessel functions are all real and for each n are listed in increasing order. It is well known that

$$(8.128) \qquad\qquad J_{1/2}(x) = \sqrt{\frac{2}{\pi x}} \sin x$$

and, in fact, all the functions $J_{n+1/2}(x)$ can be expressed in terms of powers of x and the sine and cosine functions.

We have shown that the eigenvalues for the sphere are given by (8.127) and the associated eigenfunctions are

$$(8.129) \qquad\qquad \hat{M}_{mn} = \frac{1}{\sqrt{r}} J_{n+1/2}(\alpha_{mn} r) Y_n(\theta, \varphi).$$

(The $\hat{M}_{mn}$ are not normalized.) The lowest eigenvalue is given by $\alpha_{10}^2 = \lambda_{10}$, and α_{10} is the smallest positive zero of the Bessel function $J_{1/2}(x)$. In view of (8.128) we have

$$(8.130) \qquad\qquad \lambda_{10} = \alpha_{10}^2 = \pi^2.$$

To approximate the eigenvalue λ_{10} using the Rayleigh–Ritz method, we introduce the function

$$(8.131) \qquad\qquad \varphi_1(r, \theta, \varphi) = 1 - r,$$

which appears to be the simplest function that vanishes at $r = 1$ and is bounded at $r = 0$. Then

$$(8.132) \qquad\qquad w = c_1 \varphi_1$$

is admissible function for the variational problem associated with (8.114)–

(8.115). We have

$$
(8.133) \qquad \frac{E(w)}{\|w\|^2} = \frac{c_1^2 \int_0^\pi \int_0^{2\pi} \int_0^1 (\nabla\varphi_1)^2 r^2 \sin\varphi \, dr \, d\theta \, d\varphi}{c_1^2 \int_0^\pi \int_0^{2\pi} \int_0^1 \varphi_1^2 r^2 \sin\varphi \, dr \, d\theta \, d\varphi}
$$

$$
= \frac{\int_0^1 r^2 \, dr}{\int_0^1 (1-r)^2 r^2 \, dr} = 10.
$$

Since the exact eigenvalue is $\pi^2 = 9.87$, the approximate eigenvalue $\hat{\lambda}_{10} = 10$ given in (8.133) yields excellent agreement. We do not discuss how well the first eigenfunction is approximated nor approximations for higher eigenvalues and eigenfunctions.

The previous determination of the lowest eigenvalue for the sphere that was carried out using separation of variables and the Rayleigh–Ritz procedure is of interest in the following problem. Let $u(x, y, z, t)$ be the *concentration* of some diffusing substance and let the sources in the substance be proportional to the concentration. According to our discussion in Section 4.1, the equation for u is given as

$$
(8.134) \qquad u_t = c^2 \nabla^2 u + \gamma u,
$$

where c^2 and γ are positive constants, both of which are characteristic of the substance. Note that $-\gamma$ which corresponds to q in (4.5) is now negative rather than nonnegative. An initial and boundary value problem of interest for (8.134) occurs if we consider the *diffusion process* within a bounded region G, with the initial condition

$$
(8.135) \qquad u(x, y, z, 0) = f(x, y, z)
$$

and the boundary condition

$$
(8.136) \qquad u(x, y, z, t)|_{\partial G} = 0.
$$

Applying the separation of variables method to the problem (8.134)–(8.136) we set

$$
(8.137) \qquad u = N(t) M(x, y, z)
$$

and insert (8.137) into (8.134). This leads to

$$(8.138) \qquad -\frac{\nabla^2 M}{M} = \frac{-N' + \gamma N}{c^2 N} = \lambda,$$

where λ is the separation constant. Consequently, we must consider the eigenvalue problem

$$(8.139) \qquad \nabla^2 M + \lambda M = 0, \qquad (x, y, z) \in G,$$

$$(8.140) \qquad M(x, y, z)|_{\partial G} = 0,$$

for the function M. The eigenvalues λ_k $(k = 1, 2, \dots)$ for this problem are positive and can be arranged as

$$(8.141) \qquad 0 < \lambda_1 \leqslant \lambda_2 \leqslant \cdots \leqslant \lambda_k \leqslant \dots,$$

as has been shown previously. The associated eigenfunctions are denoted by $M_k(x, y, z)$. For each λ_k we obtain a solution $N_k(t)$ of the separated equation for N in the form

$$(8.142) \qquad N_k(t) = a_k \exp\left[(\gamma - \lambda_k c^2) t \right]; \qquad k = 1, 2, \dots,$$

with arbitrary constants a_k. Finally, the solution of (8.134)–(8.136) is given as

$$(8.143) \qquad u(x, y, z, t) = \sum_{k=1}^{\infty} a_k M_k(x, y, z) \exp\left[(\gamma - \lambda_k c^2) t \right]$$

with the a_k given as the Fourier coefficients in the eigenfunction expansion of the initial concentration $f(x, y, z)$.

An important question regarding the solution (8.143) is whether the concentration $u(x, y, z, t)$ grows without bound as the time t increases, in which case the problem (8.134)–(8.136) is unstable. Since the concentration u at any time t is a result of reactions that occur within the substance as was indicated, we say that when the concentration u increases exponentially, a *chain reaction* is taking place because of the rapid increase in u. The solution (8.143) shows that exponential growth in t occurs if $\gamma - \lambda_k c^2 > 0$. In view of (8.141) we see that the lowest eigenvalue λ_1 is the most significant and if

$$(8.144) \qquad \frac{\gamma}{c^2} > \lambda_1,$$

a chain reaction takes place in the substance. Now the eigenvalue varies as the region R is changed, and the parameters c^2 and γ can presumably be adjusted for given substances. Therefore, if we fix two of the three constants, say, γ and c^2, we may define a *critical value* of the third constant λ_1. If $\lambda_1 > \gamma/c^2$, no chain reaction takes place, whereas if $\lambda_1 < \gamma/c^2$, a chain reaction does occur. Thus the critical value of λ_1, which we denote by $\lambda_1^{(c)}$, is given as

$$\lambda_1^{(c)} = \frac{\gamma}{c^2}.$$

For a specific problem it is important to determine the first eigenvalue (i.e., the lowest) to see if it is greater or less than the critical value. We have found that for the sphere of radius one, the lowest eigenvalue equals π^2. The Rayleigh–Ritz method yielded a slightly higher value. It is clearly important to be able to approximate the lowest eigenvalue as accurately as possible.

EXAMPLE 8.4.   A STURM–LIOUVILLE PROBLEM

The eigenvalue problem

(8.145) $M''(x) + \lambda M(x) = 0, \qquad 0 < x < \pi$

with the boundary conditions

(8.146) $M'(0) = 0, \qquad M(\pi) = 0,$

has the eigenvalues

(8.147) $\lambda_k = \left(k + \tfrac{1}{2}\right)^2, \qquad k = 0, 1, 2, \ldots$

and the (unnormalized) eigenfunctions

(8.148) $\hat{M}_k(x) = \cos\left[\left(k + \tfrac{1}{2}\right)x\right], \qquad k = 0, 1, 2, \ldots,$

as is easily shown. We shall use the Rayleigh–Ritz method to approximate the lowest two eigenvalues and the associated eigenfunctions.

The appropriate Rayleigh quotient for this problem is

(8.149) $$\frac{E(w)}{\|w\|^2} = \frac{\displaystyle\int_0^\pi (w')^2 \, dx}{\displaystyle\int_0^\pi w^2 \, dx}.$$

The admissible functions $w(x)$ are required to vanish at $x = \pi$—that is, $w(\pi) = 0$. At $x = 0$ there is a natural boundary condition so that no restrictions need be placed on $w(x)$ at that point.

To apply the Rayleigh–Ritz method we select two functions $\varphi_1(x)$ and $\varphi_2(x)$ that satisfy the admissibility conditions; that is, $\varphi_1(\pi) = \varphi_2(\pi) = 0$. As in (8.94) we set

(8.150)
$$w(x) = c_1\varphi_1(x) + c_2\varphi_2(x),$$

with φ_1 and φ_2 chosen as

(8.151)
$$\varphi_1(x) = \pi^2 - x^2; \qquad \varphi_2(x) = \pi^3 - x^3.$$

We note that $\varphi_1(\pi) = \varphi_2(\pi) = 0$ and that, in addition, $\varphi_1'(0) = \varphi_2'(0) = 0$. Although admissible functions need not have vanishing derivatives at $x = 0$, we expect to get improved results with our choices for φ_1 and φ_2 since they are expected to more closely approximate the eigenfunctions (8.148) whose derivatives do vanish at $x = 0$.

Evaluating the integrals (8.96)–(8.97), with $p = \rho = 1$ and $q = 0$ and $\nabla\varphi$ replaced by φ', we obtain the matrices

(8.152)
$$A = \begin{bmatrix} \frac{4}{3}\pi^3 & \frac{3}{2}\pi^4 \\ \frac{3}{2}\pi^4 & \frac{9}{5}\pi^5 \end{bmatrix}; \qquad B = \begin{bmatrix} \frac{8}{15}\pi^5 & \frac{7}{12}\pi^6 \\ \frac{7}{12}\pi^6 & \frac{9}{14}\pi^7 \end{bmatrix}.$$

The two roots of the characteristic equation

(8.153)
$$\det(A - \hat{\lambda}B) = 0$$

are

(8.154)
$$\hat{\lambda}_1 = 0.25; \qquad \hat{\lambda}_2 = 2.39.$$

We have retained only two decimal places, and $\hat{\lambda}_1$ is, in fact, slightly larger than 0.25. Comparing the approximate eigenvalues $\hat{\lambda}_1$ and $\hat{\lambda}_2$ with the exact eigenvalues $\lambda_1 = 0.25$ and $\lambda_2 = 2.25$—see (8.147)—we observe that there is excellent agreement between $\hat{\lambda}_1$ and λ_1 but that $\hat{\lambda}_2$ is not as close to λ_2. This seems to be the norm for the Rayleigh–Ritz procedure in that the higher eigenvalues are more poorly approximated than the lower ones. We do observe that for both eigenvalues we have $\lambda_1 < \hat{\lambda}_1$ and $\lambda_2 < \hat{\lambda}_2$ as required by (8.113).

Given the approximate eigenvalues $\hat{\lambda}_1$ and $\hat{\lambda}_2$ we can determine approximate eigenfunctions by specifying two sets of coefficients c_1 and c_2 in (8.150). We do not carry out this calculation.

EXAMPLE 8.5. A STURM–LIOUVILLE PROBLEM WITH A VARIABLE COEFFICIENT

We consider the equation

(8.155)
$$M''(x) + (\lambda - \epsilon x)M(x) = 0, \qquad 0 < x < \pi$$

with the boundary conditions

(8.156)
$$M(0) = M(\pi) = 0.$$

We assume that $0 < \epsilon \ll 1$ so that the term $q(x) = \epsilon x$ is uniformly small in the interval $0 \leqslant x \leqslant \pi$ and is nonnegative. With $\epsilon = 0$, (8.155) reduces to an exactly solvable problem. In any case, (8.155) can be solved exactly in terms of *Airy functions*, which are solutions of the equation

$$(8.157) \qquad\qquad y''(x) - xy(x) = 0$$

and are tabulated. However, we do not use the exact solutions of (8.155) but use approximate techniques to determine the lowest eigenvalue for the Sturm–Liouville problem (8.155)–(8.156).

First we apply the Rayleigh–Ritz method. In view of (8.92) the Rayleigh quotient appropriate for (8.155)–(8.156) is

$$(8.158) \qquad\qquad \frac{E(w)}{\|w\|^2} = \frac{\int_0^\pi \left[(w')^2 + \epsilon x w^2 \right] dx}{\int_0^\pi w^2 \, dx}.$$

The admissible functions $w(x)$ are required to vanish at $x = 0$ and $x = \pi$. Since ϵ is small and for $\epsilon = 0$ the lowest eigenvalue for (8.155)–(8.156) is $\lambda_1 = 1$ and the corresponding eigenfunction is $M_1(x) = \sqrt{(2/\pi)} \sin x$, we pick $\varphi_1(x)$ in the Rayleigh–Ritz approximation (8.94) to be

$$(8.159) \qquad\qquad\qquad \varphi_1(x) = \sin x.$$

Clearly, $\varphi_1(x)$ is admissible for the variational problem.

With $w = c_1 \varphi_1(x)$ we have

$$(8.160) \qquad\qquad \frac{E(w)}{\|w\|^2} = \frac{\int_0^\pi \left[\cos^2 x + \epsilon x \sin^2 x \right] dx}{\int_0^\pi \sin^2 x \, dx} = 1 + \frac{\epsilon \pi}{2} = \hat{\lambda}_1$$

Now in the interval $0 < x < \pi$, the function $q(x) = \epsilon x$ satisfies the inequalities

$$(8.161) \qquad\qquad\qquad 0 \leqslant \epsilon x \leqslant \epsilon \pi,$$

with $q_m = 0$ and $q_M = \epsilon \pi$ representing the minimum and maximum value of $q(x)$, respectively. Then (8.55) implies that

$$(8.162) \qquad\qquad\qquad \lambda_1^{(m)} \leqslant \lambda_1 \leqslant \lambda_1^{(M)},$$

where $\lambda_1^{(m)} = 1$ and $\lambda_1^{(M)} = 1 + \epsilon \pi$, as is easily seen on replacing $\lambda - \epsilon x$ by λ and by $\lambda - \epsilon \pi$ to determine $\lambda_1^{(m)}$ and $\lambda_1^{(M)}$, respectively. The Rayleigh–Ritz procedure implies that

$$(8.163) \qquad\qquad\qquad \lambda_1 \leqslant \hat{\lambda}_1 = 1 + \frac{\epsilon \pi}{2},$$

and we see that $\hat{\lambda}_1 < \lambda_1^{(M)}$. The smaller ϵ is, the better the approximation to λ_1 is expected to be.

An alternative procedure that may be applied directly to the problem (8.155)–(8.156) is a perturbation method that we now present. The solutions M and the eigenvalues λ are functions of ϵ; that is, $M = M(x; \epsilon)$ and $\lambda = \lambda(\epsilon)$. Consequently, we expand both M and λ in a power series in ϵ in the form

$$(8.164) \qquad M(x; \epsilon) = M^{(0)}(x) + \epsilon M^{(1)}(x) + \cdots$$

and

$$(8.165) \qquad \lambda(\epsilon) = \lambda^{(0)} + \epsilon \lambda^{(1)} + \cdots .$$

We insert (8.164)–(8.165) into (8.155)–(8.156) to obtain

$$(8.166) \quad M'' + (\lambda - \epsilon x) M = \left[\frac{d^2 M^{(0)}}{dx^2} + \lambda^{(0)} M^{(0)} \right] + \epsilon \left[\frac{d^2 M^{(1)}}{dx^2} \right.$$
$$\left. + \lambda^{(0)} M^{(1)} + (\lambda^{(1)} - x) M^{(0)} \right] + \cdots = 0$$

and

$$(8.167) \quad M^{(0)}(0) + \epsilon M^{(1)}(0) + \cdots = M^{(0)}(\pi) + \epsilon M^{(1)}(\pi) + \cdots = 0.$$

Equating coefficients of like powers of ϵ to zero in (8.166)–(8.167) leads to the following problems. For $M^{(0)}(x)$ we have

$$(8.168) \qquad \frac{d^2 M^{(0)}}{dx^2} + \lambda^{(0)} M^{(0)} = 0; \qquad M^{(0)}(0) = M^{(0)}(\pi) = 0.$$

The leading eigenvalue and eigenfunction for this problem are

$$(8.169) \qquad \lambda_1^{(0)} = 1; \qquad M_1^{(0)}(x) = \sqrt{\frac{2}{\pi}} \sin x.$$

The equation for $M^{(1)}(x)$ is

$$(8.170) \qquad \frac{d^2 M^{(1)}}{dx^2} + \lambda^{(0)} M^{(1)} = -(\lambda^{(1)} - x) M^{(0)}$$

with the boundary conditions

$$(8.171) \qquad M^{(1)}(0) = M^{(1)}(\pi) = 0.$$

With $\lambda^{(0)} = \lambda_1^{(0)} = 1$ the first eigenvalue for (8.168), the inhomogeneous equation (8.170) has no solution unless the inhomogeneous term is orthogonal to the eigenfunction $M_1^{(0)}(x)$. This follows on multiplying (8.170) by $M_1^{(0)}(x)$ and integrating from 0 to π. We have

(8.172)

$$\int_0^\pi M_1^{(0)}\left[\frac{d^2M^{(1)}}{dx^2} + \lambda_1^{(0)}M^{(1)}\right]dx = \int_0^\pi M^{(1)}\left[\frac{d^2M_1^{(0)}}{dx^2} + \lambda_1^{(0)}M_1^{(0)}\right]dx = 0$$

on integrating by parts and using (8.168) and (8.171). This implies that

(8.173)
$$\int_0^\pi [\lambda^{(1)} - x]\sin^2 x\, dx = 0,$$

from which we conclude that

(8.174)
$$\lambda^{(1)} = \lambda_1^{(1)} = \frac{\pi}{2}.$$

Having determined $\lambda_1^{(1)}$ we can specify $M_1^{(1)}(x)$ but this is not carried out.
Inserting (8.169) and (8.174) in (8.165) gives

(8.175)
$$\lambda_1(\epsilon) \approx 1 + \frac{\epsilon\pi}{2}.$$

The approximation (8.175) to the lowest eigenvalue is identical to that given by the Rayleigh–Ritz method. This appears to be due to our choice of $\varphi_1(x)$ in the Rayleigh–Ritz method to equal the leading eigenfunction for (8.155) with $\epsilon = 0$. However, the perturbation method readily yields approximations to the higher eigenvalues, whereas this is not so easily accomplished by way of the Rayleigh–Ritz method as we have seen (see also Section 9.2.).

8.3. ADJOINT DIFFERENTIAL OPERATORS AND RIEMANN'S METHOD

In our discussion of second order partial differential equations and their solution in this and previous chapters we have dealt for the most part with equations that contain the elliptic operator L defined as [see (4.6)]

(8.176)
$$Lu = -\nabla \cdot (p\nabla u) + qu.$$

The one-dimensional form of L is given in (4.7). The operator L has the important property

(8.177)
$$wL[u] - uL[w] = \nabla \cdot [-pw\nabla u + pu\nabla w],$$

as was shown in Example 4.2 That is, $wL[u] - uL[w]$ is a divergence expression and under suitable boundary conditions for a bounded region G, it was shown in Example 4.2 that

$$(8.178) \qquad \iint_G \{wL[u] - uL[w]\} \, dv = 0,$$

and the operator L was said to be *self-adjoint* with respect to the inner product $(f, g) = \iint_G fg \, dv$. As was stated in that example, the self-adjointness is tied to the fact that in (8.177) the operators L acting on u and w are identical and that the boundary conditions for the problem lead to the result (8.178) which can be written as $(w, Lu) = (Lw, u)$ in terms of the aforementioned inner product. The choice of a specific inner product does not play a role in our discussion in this section.

In Section 6.4 we defined the hyperbolic operator $\tilde{L}$ as

$$(8.179) \qquad \tilde{L}[u] = \rho u_{tt} + Lu,$$

with L given as in (8.176) and it was shown that [see (6.74)]

$$(8.180) \quad w\tilde{L}[u] - u\tilde{L}[w] = \tilde{\nabla} \cdot [-pw\nabla u + pu\nabla w, \rho wu_t - \rho uw_t]$$

where $\tilde{\nabla} = [\nabla, \partial/\partial t]$ is the space-time gradient operator. Since $w\tilde{L}[u] - u\tilde{L}[w]$ is a divergence expression and the operators $\tilde{L}$ acting on u and w are identical, we say that the operator $\tilde{L}$ is *formally self-adjoint*. We use the term formally self-adjoint since one often reserves the term self-adjoint for operators in the context of a particular boundary (or initial and boundary) value problem as was done for the preceding elliptic operator. In this section we call operators for which expressions of the form (8.177) or (8.180) are valid, self-adjoint operators.

For the parabolic operator $\hat{L}$ defined as

$$(8.181) \qquad \hat{L}[u] = \rho u_t + Lu,$$

with L given by (8.176), we have [see (7.30)]

$$(8.182) \qquad w\hat{L}[u] - u\hat{L}^*[w] = \tilde{\nabla} \cdot [-pw\nabla u + pu\nabla w, \rho wu].$$

The operator $\hat{L}^*$ is defined as

$$(8.183) \qquad \hat{L}^*[u] = -\rho u_t + Lu$$

and $\tilde{\nabla} = [\nabla, \partial/\partial t]$. Thus in order to obtain a divergence expression in (8.182) for the parabolic operator $\hat{L}$, it is necessary to introduce a new operator $\hat{L}^*$ that differs from $\hat{L}$. We call $\hat{L}^*$ the *adjoint operator* of $\hat{L}$, and (8.182) shows that $\hat{L}$ is not a self-adjoint differential operator.

We have dealt with self-adjoint elliptic operators of the special form L in this text since they have very nice properties, especially in connection with eigenvalue problems and eigenfunction expansions, and because they occur naturally in many physical contexts. Rather than speak in general terms about nonself-adjoint operators in this section, we consider the adjoint operator for the general second order linear partial differential operator in the following example. Then we discuss, for a special case, *Riemann's method* for solving the Cauchy problem for hyperbolic second order linear equations in two variables. Although this method does not yield closed form solutions except in a few problems, it does provide useful information about domains of dependence and influence for solutions in general. It also has some relation to the Green's function method presented in Chapter 7.

EXAMPLE 8.6.   ADJOINT OPERATORS FOR SECOND ORDER PARTIAL
DIFFERENTIAL EQUATIONS

The general *linear second order differential operator M* can be written as

$$(8.184) \qquad M[u] = \sum_{i=1}^{n} \sum_{j=1}^{n} a_{ij} \frac{\partial^2 u}{\partial x_i \, \partial x_j} + \sum_{i=1}^{n} b_i \frac{\partial u}{\partial x_i} + cu$$

where the coefficients a_{ij}, b_i, and c may be functions of $x_1, x_2, \ldots, x_n$. The *adjoint operator M** is given as

$$(8.185) \qquad M^*[w] = \sum_{i=1}^{n} \sum_{j=1}^{n} \frac{\partial^2 (a_{ij} w)}{\partial x_i \, \partial x_j} - \sum_{i=1}^{n} \frac{\partial (b_i w)}{\partial x_i} + cw.$$

It can be verified directly that

$$(8.186) \qquad wM[u] - uM^*[w] = \sum_{i=1}^{n} \frac{\partial P_i}{\partial x_i}$$

where

$$(8.187) \qquad P_i = \sum_{j=1}^{n} \left[a_{ij} w \frac{\partial u}{\partial x_j} - u \frac{\partial (a_{ij} w)}{\partial x_j} \right] + b_i u w.$$

The expression on the right of (8.186) is a *divergence expression*, and the n-dimensional form of the divergence theorem yields

$$(8.188) \qquad \iint_G \{ wM[u] - uM^*[w] \} \, dv = \int_{\partial G} \mathbf{P} \cdot \mathbf{n} \, ds,$$

where $\mathbf{P}$ is a vector with n components $P_1, \ldots, P_n$ and $\mathbf{n}$ is the exterior unit normal vector to the boundary ∂G. The possibility of using the divergence theorem as in (8.188) is the main reason for the importance of the adjoint operator M^*. This has already been demonstrated a number of times in previous discussions in the text—especially in section 7.1—and is again seen in our presentation of Riemann's method.

To see how the adjoint operator M^* is constructed we consider a typical term. We have

$$(8.189) \quad wa_{11}\frac{\partial^2 u}{\partial x_1^2} = wa_{11}\frac{\partial}{\partial x_1}\left[\frac{\partial u}{\partial x_1}\right] = \frac{\partial}{\partial x_1}\left[wa_{11}\frac{\partial u}{\partial x_1}\right] - \frac{\partial(a_{11}w)}{\partial x_1}\frac{\partial u}{\partial x_1}$$

$$= \frac{\partial}{\partial x_1}\left[wa_{11}\frac{\partial u}{\partial x_1}\right] - \frac{\partial}{\partial x_1}\left[u\frac{\partial(a_{11}w)}{\partial x_1}\right] + u\frac{\partial^2(a_{11}w)}{\partial x_1^2}.$$

This yields

$$(8.190) \quad wa_{11}\frac{\partial^2 u}{\partial x_1^2} - u\frac{\partial^2(a_{11}w)}{\partial x_1^2} = \frac{\partial}{\partial x_1}\left[a_{11}w\frac{\partial u}{\partial x_1} - u\frac{\partial(a_{11}w)}{\partial x_1}\right].$$

If $M^* = M$, the operator M is *self-adjoint*. By carrying out all the differentiations in (8.185) and comparing M^* with M, it is easy to determine conditions that the coefficients in the operator M must satisfy to render M self-adjoint. The operators L and $\tilde{L}$ defined previously are self-adjoint. If the operator M has constant coefficients, all the b_i must vanish for M^* to equal M. Thus if M contains first derivative terms, even if it has constant coefficients a_{ij} and b_i, it is not self-adjoint, as is the case for the operator $\hat{L}$.

We now present *Riemann's method* for solving the Cauchy problem for second order hyperbolic equations in two independent variables. According to the results of Section 3.1.a, such equations can be brought into one of the two canonical forms (3.19) or (3.21), and the Riemann method is generally applied to one of these forms.

Rather than present the method in its full generality, we consider the equation

$$(8.191) \quad u_{tt} - \gamma^2(x)u_{xx} + c(x)u = F(x,t), \qquad -\infty < x < \infty, t > 0,$$

with coefficients γ^2 and c that depend only on x. The initial conditions given at $t = 0$ are

$$(8.192) \qquad u(x,0) = f(x); \qquad u_t(x,0) = g(x),$$

with $-\infty < x < \infty$. Transforming (8.191) into canonical form simplifies the principal part but the initial line $t = 0$ assumes a more complicated form under the transformation if (3.19) is obtained, while it remains unchanged if (3.21) is used. In either case, the transformed equation contains first derivative terms. Therefore, we indicate how to solve the problem (8.191)–(8.192) in the given form.

With M defined as

$$(8.193) \qquad M[u] = u_{tt} - \gamma^2(x)u_{xx} + c(x)u,$$

the adjoint operator M^* is given as

$$(8.194) \qquad M^*[w] = w_{tt} - (\gamma^2 w)_{xx} + cw,$$

in view of (8.185). It is seen that $M \neq M^*$ if $\gamma \neq$ constant so that M is not self-adjoint since γ is assumed to vary with x. Introducing a bounded but as yet arbitrary region G in (x, t)-space with boundary ∂G, we obtain on using (8.186)–(8.188),

$$(8.195) \quad \iint_G \{wM[u] - uM^*[w]\} \, dx \, dt$$

$$= \int_{\partial G} \left[-\gamma^2 wu_x + u(\gamma^2 w)_x \right] dt + \left[uw_t - wu_t \right] dx,$$

with integration over ∂G carried out in the positive direction. We assume that the two-dimensional divergence theorem or Green's theorem in the plane is valid for the region G.

The characteristic curves for (8.191) are solutions of the equations [see (3.8)]

$$(8.196) \qquad \frac{dt}{dx} = \mp \frac{1}{\gamma(x)},$$

so that they are given by

$$(8.197) \qquad t \pm \int^x \frac{ds}{\gamma(s)} = \text{constant.}$$

Let (ξ, τ) be a point in the (x, t)-plane with $\tau > 0$, and select the two characteristic curves (8.197) that pass through that point. They are the curves

$$(8.198) \qquad t \pm \int_\xi^x \frac{ds}{\gamma(s)} = \tau.$$

If the curves are extended backwards in t until they intersect the x-axis, we obtain a *characteristic triangle* whose base is a segment of the x-axis and whose other sides are characteristic curves. The interior of this triangle is denoted as the region G and the triangle itself by ∂G, as shown in Figure 8.2. The

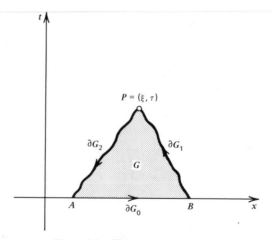

Figure 8.2. The characteristic triangle.

boundary ∂G is divided into three parts ∂G_0, ∂G_1, and ∂G_2, and we apply (8.195) to this region with the positive direction of integration displayed in the figure.

On the line segment ∂G_0 the line integral in (8.195) becomes

$$(8.199) \qquad \int_{\partial G_0} [uw_t - wu_t]\, dx = \int_A^B [f(x)w_t(x,0) - g(x)w(x,0)]\, dx,$$

since $dt = 0$, $u(x,0) = f(x)$, and $u_t(x,0) = g(x)$. On the characteristic segment ∂G_1 we have, since we assume that $\gamma(x) > 0$,

$$(8.200) \qquad dt = -\frac{1}{\gamma}\, dx$$

and for any function $v(x,t)$ defined and differentiable along the characteristic,

$$(8.201) \qquad \frac{dv}{dx} = v_x + \frac{dt}{dx} v_t = v_x - \frac{1}{\gamma} v_t.$$

Thus

$$(8.202) \qquad \int_{\partial G_1} \left[-\gamma^2 wu_x + u(\gamma^2 w)_x \right] dt + \left[uw_t - wu_t \right] dx$$

$$= \int_{\partial G_1} \left[\gamma wu_x - \frac{1}{\gamma} u(\gamma^2 w)_x + uw_t - wu_t \right] dx$$

$$= \int_{\partial G_1} \left\{ \frac{d}{dx}(\gamma uw) - \gamma u\left[2\frac{dw}{dx} + 3\frac{\gamma'(x)}{\gamma(x)} w \right] \right\} dx$$

$$= \gamma uw \Big|_B^P - \int_{\partial G_1} \gamma u\left[2\frac{dw}{dx} + \frac{3\gamma'}{\gamma} w \right] dx$$

On the characteristic segment ∂G_2 we have

(8.203)
$$dt = \frac{1}{\gamma} dx$$

and a derivative of a function $v(x, t)$ on the characteristic has the form

$$\frac{dv}{dx} = v_x + \frac{1}{\gamma} v_t.$$

Consequently,

(8.204) $\displaystyle \int_{\partial G_2} \left[-\gamma^2 wu_x + u(\gamma^2 w)_x \right] dt + \left[uw_t - wu_t \right] dx$

$$= \int_{\partial G_2} \left[-\gamma wu_x + \frac{1}{\gamma} u(\gamma^2 w)_x + uw_t - wu_t \right] dx$$

$$= \int_{\partial G_2} \left\{ -\frac{d}{dx}(\gamma uw) + \gamma u \left[2\frac{dw}{dx} + \frac{3\gamma'}{\gamma} w \right] \right\} dx$$

$$= -\gamma uw \big|_P^A + \int_{\partial G_2} \gamma u \left[2\frac{dw}{dx} + \frac{3\gamma'}{\gamma} w \right] dx.$$

Combining these results gives

(8.205)

$\gamma uw|_P = \dfrac{1}{2}\{\gamma fw|_A + \gamma fw|_B\}$

$$- \frac{1}{2} \int_A^B \left[f(x)w_t - g(x)w \right] dx + \frac{1}{2} \int_B^P \gamma u \left[2\frac{dw}{dx} + \frac{3\gamma'}{\gamma} w \right] dx$$

$$+ \frac{1}{2} \int_A^P \gamma u \left[2\frac{dw}{dx} + \frac{3\gamma'}{\gamma} w \right] dx + \frac{1}{2} \iint_G \{ wM[u] - uM^*[w] \} \, dx \, dt.$$

To obtain an expression for the solution $u(x, t)$ at the point $P = (\xi, \tau)$ we require that $w(x, t)$ satisfy the following conditions. First we set

(8.206)
$$M^*[w] = 0$$

in the characteristic triangle G. To eliminate the integrals over the characteristics ∂G_1 and ∂G_2 we set

(8.207)
$$2\frac{dw}{dx} + \frac{3\gamma'(x)}{\gamma(x)} w = 0$$

on the characteristics from B to P and from A to P. Finally, to obtain an expression for $u(x, t)$ at P we set

(8.208)
$$\gamma(x)w(x, t)|_P = \gamma(\xi)w(\xi, \tau) = 1.$$

The equations (8.207)–(8.208) represent an initial value problem for w on the characteristic curves. The solution is easily found to be

$$(8.209) \qquad\qquad w = \left[\frac{\gamma(\xi)}{\gamma^3(x)} \right]^{1/2},$$

where $w(x, t)$ is evaluated on the characteristics. The equation (8.206) for w and the conditions (8.208)–(8.209) comprise a characteristic initial value problem for $w(x, t)$. (The problem is to be solved backwards rather than forwards in time.) Such a problem is well posed and has been discussed at the end of Section 6.5 where an iteration method for solving the characteristic initial value problem for the canonical form of second order hyperbolic equations was presented. We assume a solution can be found for the present problem as well and obtain explicit solutions for two simple cases in Example 8.7.

The solution w is called the *Riemann function*. Since it depends on the point $P = (\xi, \tau)$, we express it as

$$(8.210) \qquad\qquad w(x, t) = R(x, t; \xi, \tau).$$

Then we have from (8.205), since $M[u] = F$,

$$(8.211) \qquad u(\xi, \tau) = \frac{1}{2} \{ \gamma f R|_A + \gamma f R|_B \} + \frac{1}{2} \iint_G FR \, dx \, dt$$

$$- \frac{1}{2} \int_A^B [f(x) R_t - g(x) R] \, dx.$$

The solution formula (8.211) shows that the domain of dependence of the solution $u(x, t)$ of (8.191)–(8.192) is the interior of the backward characteristic triangle with vertex at the point (x, t). Reversing the argument shows that the domain of influence of a point $(x, 0)$ is the interior of the forward characteristic sector formed by the two characteristics issuing from the point $(x, 0)$.

To establish a connection between the *Riemann function*, the *Green's function*, and the *fundamental solution* we set

$$(8.212) \qquad\qquad F(x, t) = \delta(x - \hat{\xi}) \delta(t - \hat{\tau})$$

in (8.191) and (8.211) and assume the initial data $f(x)$ and $g(x)$ vanish. Then (8.211) yields, with (ξ, τ) replaced by (x, t),

$$(8.213) \quad u(x, t) = \frac{1}{2} \iint_G R(\tilde{x}, \tilde{t}; x, t) \delta(\tilde{x} - \hat{\xi}) \delta(\tilde{t} - \hat{\tau}) \, d\tilde{x} \, d\tilde{t}$$

$$= \begin{cases} \frac{1}{2} R(\hat{\xi}, \hat{\tau}; x, t); & (\hat{\xi}, \hat{\tau}) \in G \\ 0; & (\hat{\xi}, \hat{\tau}) \notin G \end{cases}.$$

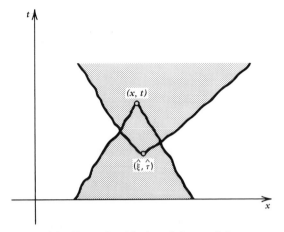

Figure 8.3. Forward and backward characteristic sectors.

This shows that $\frac{1}{2}R$ represents the effect of a point source located at the point $(\hat{\xi}, \hat{\tau})$. As shown in Figure 8.3, if the point (x, t) lies in the forward characteristic sector with vertex at $(\hat{\xi}, \hat{\tau})$, the solution is nonzero. Otherwise, $u(x, t)$ vanishes since disturbances cannot move faster than the characteristic speed. We note that the role of the variables in R as given in (8.213) has been reversed in comparison with those in (8.210). In terms of the original variables in (8.210) we see that both the *Green's function* and the *Riemann function* are effectively solutions of *backward initial value problems* resulting from point sources.

In Section 6.7 we called the solution of the *point source problem* a *causal fundamental solution*. Thus (8.213) yields the fundamental solution for (8.191). It should be compared with (6.202) and (6.223).

In the following example we obtain explicit representations for the Riemann function in the case where (8.191) has constant coefficients.

EXAMPLE 8.7. RIEMANN'S FUNCTION FOR EQUATIONS WITH CONSTANT COEFFICIENTS

With γ = constant and $c = 0$ in (8.191) we obtain the inhomogeneous wave equation

$$(8.214) \qquad u_{tt} - \gamma^2 u_{xx} = F(x, t).$$

In this case the operator M as given in (8.193) is self-adjoint. The Riemann function equals $1/\gamma$ on the characteristics and, in fact, we immediately conclude that

$$(8.215) \qquad R(x, t; \xi, \tau) = \frac{1}{\gamma}$$

in the entire characteristic triangle. The causal fundamental solutions for this problem has the form (6.202) which agrees with (8.213) and it is easily shown that (8.211) reduces to d'Alembert's solution of the initial value problem for (8.214).

If γ = constant, c = constant, and $c < 0$, we obtain for (8.191),

$$(8.216) \qquad u_{tt} - \gamma^2 u_{xx} + cu = F(x, t).$$

The operator M in (8.193) is self-adjoint since γ = constant and we have $R = 1/\gamma$ on the characteristic curves. By looking for a solution of the equation for R that depends on the variable r defined in (6.210) with c replaced by γ we find as in Section 6.7 that the Riemann function R is given as

$$(8.217) \qquad R(x, t; \xi, \tau) = \frac{1}{\gamma} I_0 \left[\sqrt{\frac{-c}{\gamma^2}} \sqrt{\gamma^2(t - \tau)^2 - (x - \xi)^2} \right],$$

since the modified Bessel function $I_0(z)$ has the property that $I_0(0) = 1$. The causal fundamental solution (8.213) agrees with (6.223) and (8.211) agrees with (7.353), as shown in the exercises.

8.4. MAXIMUM AND MINIMUM PRINCIPLES

In Chapter 1 the diffusion, telegrapher's, and Laplace's equations were derived as limiting forms of certain random walk problems. It was indicated that some of the properties of the discrete random walk models carry over to the limiting partial differential equations. For example, a mean value property for the discrete form of Laplace's equation was shown in Example 7.14 to be valid for solutions of Laplace's equation. In this section we show that the maximum and minimum properties valid for discrete forms of the diffusion and Laplace's equation carry over to their limiting forms. These properties can be used to prove uniqueness and continuous dependence on data of the solutions of these equations. In addition, we will show for the telegrapher's equation that if the initial data for the Cauchy problem are positive, then the solution must be positive. This is consistent with the interpretation of the solution in Section 1.2 as a probability density function that must be nonnegative.

We begin by discussing the *diffusion* or *heat equation*

$$(8.218) \qquad u_t - c^2 u_{xx} = 0$$

in the closed region $0 \leqslant x \leqslant l$ and $0 \leqslant t \leqslant T$. We assume that c^2 is a constant A more general equation of parabolic type with constant coefficients in the form

$$(8.219) \qquad v_t - c^2 v_{xx} = av_x + bv$$

can be brought into the form (8.218) by way of the transformation

$$v = \exp[\alpha x + \beta t]u$$

with $\alpha = -a/2c^2$ and $\beta = b - a^2/4c^2$. The *maximum and minimum principle* for (8.218) is given as follows. Let $u(x, t)$ be a solution of (8.218) in $0 < x < l$ and $0 < t \le T$ which is continuous in the closed region $0 \le x \le l$ and $0 \le t \le T$. Then the maximum and minimum values of $u(x, t)$ are assumed on the initial line $t = 0$ or at points on the boundary $x = 0$ and $x = l$.

The proof of this principle is based on showing that if a maximum or minimum occurs at an (interior) point (x_0, t_0) with $0 < x_0 < l$ and $0 < t_0 \le T$, we are led to a contradiction. Now since $u(x, t)$ is continuous in a closed and bounded region it must assume its maximum there. At the assumed maximum point (x_0, t_0) we must have

(8.220)
$$\begin{cases} \dfrac{\partial^2 u(x_0, t_0)}{\partial x^2} \le 0 \\[2mm] \dfrac{\partial u(x_0, t_0)}{\partial t} \ge 0 \end{cases}$$

as is known from calculus. If $t_0 < T$, then $u_t = 0$ at (x_0, t_0) since both u_x and u_t vanish at an interior maximum point. However, we may have $t_0 = T$ in which case $(x_0, t_0) = (x_0, T)$ is on the boundary of the closed region and we can merely assert that $u_t \ge 0$ at that point since $u(x, t)$ may be increasing at that point.

The combined inequalities (8.220) when inserted into (8.218) do not imply a contradiction since u_{xx} and u_t may vanish at (x_0, t_0). To obtain a contradiction and complete the proof we introduce the auxiliary function

(8.221)
$$w(x, t) = u(x, t) + \epsilon x^2$$

where $\epsilon > 0$ is a constant. Now, $w(x, t)$ is continuous in $0 \le x \le l$ and $0 \le t \le T$ so that it has a maximum at some point (x_1, t_1) in the region. If we assume that $0 < x_1 < l$ and $0 < t_1 \le T$, we again conclude that

(8.222)
$$w_t(x_1, t_1) \ge 0; \qquad w_{xx}(x_1, t_1) \le 0.$$

But

(8.223)
$$w_t - c^2 w_{xx} = u_t - c^2 u_{xx} - 2c^2\epsilon = -2c^2\epsilon < 0$$

since u satisfies (8.218). Inserting (8.222) into (8.223) now leads to a contradiction since the left side is nonnegative and the right side is strictly negative. Therefore, $w(x, t)$ assumes its maximum on the initial line or on the boundary.

Let M be the maximum value of $u(x, t)$ on $t = 0$, $x = 0$, and $x = l$ (i.e., on the initial and boundary lines). Then

$$(8.224) \qquad w(x, t) = u(x, t) + \epsilon x^2 \leqslant M + \epsilon l^2$$

for $0 \leqslant x \leqslant l$ and $0 \leqslant t \leqslant T$ since w has its maximum on $t = 0$, $x = 0$, or $x = l$. Consequently,

$$(8.225) \qquad u = w - \epsilon x^2 \leqslant w \leqslant M + \epsilon l^2.$$

Since ϵ is arbitrary we can let $\epsilon \to 0$ in (8.225) and conclude that

$$(8.226) \qquad u(x, t) \leqslant M$$

throughout the closed region and the proof is completed.

To obtain a *minimum principle*, we consider the function $-u$ where $u(x, t)$ is a solution of (8.218). Clearly, $-u$ is also a solution of (8.218) and the maximum values of $-u$ correspond to the minimum values of u. Since $-u$ satisfies the maximum principle, we conlude that u assumes its minimum values on the initial line or on the boundary lines. In particular, this implies that if the initial and boundary data for the problem are nonnegative, then the solution must be nonnegative. This result is consistent with the interpretation of the solution as a probability density or as a temperature.

It can also be shown for the Cauchy problem for (8.218) over the initial line $t = 0$ with $-\infty < x < \infty$ that the solution $u(x, t)$ is nonnegative if the initial values $u(x, 0) = f(x)$ are nonnegative. This follows directly from the representation of the solution given in Example 5.2. We have

$$(8.227) \qquad u(x, t) = \frac{1}{\sqrt{4\pi c^2 t}} \int_{-\infty}^{\infty} \exp\left[-\frac{(x - s)^2}{4 c^2 t} \right] f(s) \, ds.$$

Since the exponential term and the algebraic coefficient in (8.227) are positive, we conclude that $u(x, t)$ is nonnegative as long as the initial data $f(x)$ are nonnegative.

The maximum and minimum principles may be used to prove uniqueness and continuous dependence on the data for solutions of initial and boundary value problems for the diffusion equation. They can also be extended to higher space dimensions by using similar arguments and to equations with variable coefficients. These matters are considered in the exercises.

Before discussing the maximum and minimum principle for Laplace's equation, we discuss a corresponding *maximum* and *minimum principle* for *Poisson's equation*

$$(8.228) \qquad \nabla^2 u(x, y) = -F(x, y)$$

in two dimensions. We show that if we consider a bounded region G and its boundary ∂G, then the maximum values of a solution $u(x, y)$ of (8.228) are attained on ∂G if $F(x, y) < 0$ in G and the minimum values of $u(x, y)$ are assumed on ∂G if $F(x, y) > 0$ in G.

To prove this result we note that since $u(x, y)$ is continuous in a closed and bounded region by assumption, it must assume its maximum in G or in ∂G. Let us suppose the maximum is assumed at a point (x_0, y_0) in G and consider the case where $F(x, y) < 0$ in G. Then at the interior maximum point (x_0, y_0) we must have

$$(8.229) \qquad u_{xx}(x_0, y_0) \leqslant 0; \qquad u_{yy}(x_0, y_0) \leqslant 0.$$

But since $F < 0$, (8.228) yields at (x_0, y_0),

$$(8.230) \qquad\qquad u_{xx} + u_{yy} > 0,$$

which is a contradiction. Consequently, the maximum of $u(x, y)$ must occur on ∂G.

To show that the minimum of $u(x, y)$ is attained on ∂G if $F(x, y) > 0$ in G we replace u by $-u$ in the preceding argument. This is equivalent to replacing F by $-F$ in (8.228). Since $F > 0$, we obtain $-F < 0$ and conclude that $-u$ assumes its maximum on ∂G. Therefore, u assumes its minimum on ∂G and the minimum property is proven.

Now the *Green's function* $K(x, y; \xi, \eta)$ for Laplace's equation with Dirichlet boundary conditions satisfies [see (1.109)–(1.110)]

$$(8.231) \quad K_{xx} + K_{yy} = -\delta(x - \xi)\delta(y - \eta); \qquad (x, y), (\xi, \eta) \in G$$

and the boundary condition $K = 0$ on ∂G. We would like to conclude that $K \geqslant 0$ in G in view of its probabilistic interpretation in Section 1.3. Since the delta function is not an ordinary function, it is not obvious that the preceding theorem can be applied. However, we may obtain the delta function in two dimensions as a limit of sequences of ordinary nonnegative functions in the manner of Example 1.2 and Exercise 7.2.1 as is easily shown. For each of these functions we obtain a nonnegative solution of (8.231) where the delta function is replaced by a member of the sequence of functions. This follows since K and each of its approximations vanishes on ∂G and the right side of (8.231) is nonpositive. Technically, the minimum principle proved in the foregoing requires the right side of (8.228) to be strictly negative in G, and the members of the (delta) sequence are such that $-F$ in (8.228) can only be said to be nonpositive. However, the minimum principle is valid in that case also, as easily follows from the proof of that principle given for Laplace's equation. In the limit as the sequence tends to the delta function, the sequence of approximate Green's functions tends to the Green's function K and this, in turn, must also be nonnegative.

The *maximum principle* for the (homogeneous) *Laplace's equation* is proven by following the procedure used for the heat equation. We consider a solution

$u(x, y)$ of Laplace's equation

$$(8.232) \qquad\qquad u_{xx} + u_{yy} = 0$$

in the region G with boundary ∂G. Let M be the maximum value of $u(x, y)$ on ∂G and suppose that the (bounded) region G can be enclosed in a square of side $2l$ with center at the origin. Then, the function

$$(8.233) \qquad\qquad w(x, y) = u(x, y) + \epsilon x^2$$

with $\epsilon > 0$, satisfies the equation

$$(8.234) \qquad\qquad \nabla^2 w = \nabla^2[u + \epsilon x^2] = 2\epsilon > 0.$$

Based on our considerations for Poisson's equation with $F < 0$, we conclude that $w(x, y)$ attains its maximum on ∂G. Then for $(x, y) \in G$, we have

$$(8.235) \qquad\qquad w(x, y) \leqslant M + \epsilon l^2,$$

since $l^2 \geqslant x^2$ for all x on ∂G by our assumption. Finally,

$$(8.236) \qquad\qquad u = w - \epsilon x^2 \leqslant w \leqslant M + \epsilon l^2.$$

Since ϵ is arbitrary, we conclude that

$$(8.237) \qquad\qquad u(x, y) \leqslant M$$

on letting $\epsilon \to 0$, so that the maximum is assumed on the boundary. By considering $-u$ in place of u we conclude, as was done earlier, that the minimum is also assumed on the boundary.

These results can be used to prove uniqueness and continuous dependence on data for the Dirichlet problem for Laplace's and Poisson's equations. They can be extended to three dimensions and to certain problems with variable coefficients of the type considered in this text. This is done in the exercises.

The *minimum principle* for Laplace's equation again shows that nonnegative boundary data implies nonnegative solutions as required by our random walk formulation in Section 1.3. We have already derived the mean value property for solutions of Laplace's equation (see Example 7.14 for the three-dimensional case). This property can be used to prove the maximum and minimum principles for solutions of Laplace's equation. It can be used to show that whenever the maximum or minimum is attained in the interior of the region (as well as on the boundary), the solution u must be identically constant. This is discussed in the exercises.

To conclude this section, we consider the *telegrapher's equation* (1.79) of Section 1.2, that is,

$$(8.238) \qquad\qquad v_{tt} - \gamma^2 v_{xx} + 2\lambda v_t = 0$$

and express it in the form of a system

$$(8.239) \qquad \alpha_t + \gamma\alpha_x = -\lambda\alpha + \lambda\beta$$

$$(8.240) \qquad \beta_t - \gamma\beta_x = \lambda\alpha + \lambda\beta,$$

as in (1.73)–(1.74), with $v = \alpha + \beta$. Since α and β are interpreted as probability densities, we require that α and β be nonnegative. In particular, if the initial data

$$(8.241) \quad \alpha(x,0) = f(x); \qquad \beta(x,0) = g(x); \qquad -\infty < x < \infty$$

are nonnegative, we must have $\alpha(x, t) \geqslant 0$ and $\beta(x, t) \geqslant 0$ for $t > 0$. In terms of $v(x, t)$ we should have $v(x, t) \geqslant 0$ if $v(x, 0) \geqslant 0$ when solving a Cauchy problem for v. We will show that if the data $f(x)$ and $g(x)$ in (8.241) are strictly positive, the solutions $\alpha(x, t)$ and $\beta(x, t)$ must also be positive. The case of nonnegative data is considered in the exercises.

To demonstrate these results, we consider the characteristic curves for (8.238)–(8.240), which are given as

$$(8.242) \qquad \qquad x \pm \gamma t = \text{constant.}$$

On the curves $x \pm \gamma t = \text{constant}$ we have for any function $w(x, t)$:

$$(8.243) \qquad \qquad \frac{dw}{dt} = w_x \frac{dx}{dt} + w_t = \mp \gamma w_x + w_t.$$

This yields for (8.239);

$$(8.244) \qquad \alpha_t + \gamma\alpha_x = \frac{d\alpha}{dt} = -\lambda\alpha + \lambda\beta, \qquad x - \gamma t = \text{constant,}$$

and for (8.240):

$$(8.245) \qquad \beta_t - \gamma\beta_x = \frac{d\beta}{dt} = \lambda\alpha - \lambda\beta, \qquad x + \gamma t = \text{constant.}$$

These equations can be written as

$$(8.246) \qquad \qquad \frac{d}{dt}[e^{\lambda t}\alpha] = \lambda e^{\lambda t}\beta, \qquad x - \gamma t = \text{constant}$$

and

$$(8.247) \qquad \qquad \frac{d}{dt}[e^{\lambda t}\beta] = \lambda e^{\lambda t}\alpha, \qquad x + \gamma t = \text{constant.}$$

Now if $\alpha(x, 0) > 0$ and $\beta(x, 0) > 0$, (8.246) and (8.247) show that $(d/dt)[e^{\lambda t}\alpha]|_{t=0} > 0$ and $(d/dt)[e^{\lambda t}\beta]|_{t=0} > 0$ since $\lambda > 0$. Initially, therefore,

$e^{\lambda t}\alpha$ and $e^{\lambda t}\beta$ must be increasing functions of time. In view of the fact that $\alpha(x, t)$ and $\beta(x, t)$ are initially positive, it is therefore not possible that they become negative at any later time t since (8.246)–(8.247) are valid for all time. Consequently, $v(x, t) = \alpha(x, t) + \beta(x, t)$ is also positive.

8.5. SOLUTION METHODS FOR HIGHER ORDER EQUATIONS AND SYSTEMS

With the exception of Chapter 2 which deals with first order equations, we have mostly studied second order equations in this text. In this section we show by means of a number of detailed examples how to solve problems that involve higher order equations and systems. In a number of cases the techniques used previously for second order equations can be applied whereas in other cases special methods need to be introduced. We begin by considering the higher order equations that govern the vibrations of rods and plates. Then we discuss the equations of fluid dynamics, Maxwell's equations of electromagnetic theory, and the equations of elasticity theory. Both linear and nonlinear problems are considered.

EXAMPLE 8.8.   THE VIBRATION OF RODS

The *lateral vibration of a thin homogeneous rod* is governed by the fourth order equation

$$(8.248) \qquad\qquad u_{tt} + c^2 u_{xxxx} = 0$$

where $u(x, t)$ is the displacement or deflection of the rod at the time t. The constant c^2 depends on the physical properties of the rod. We do not give the derivation of (8.248) but note that in contrast to the wave equation which contains the term u_{xx} proportional to the curvature, we have the fourth derivative term u_{xxxx}. That is, because of the extreme rigidity of the rod, the internal force on a portion of the rod is proportional to a fourth rather than a second derivative of the displacement u. Invoking the results of Section 3.3, we find that (8.248) is an equation of parabolic type and the lines $t =$ constant are the characteristics.

The Cauchy problem with data $u(x, 0)$ and $u_t(x, 0)$ assigned on the x-axis is well posed. To see this we will follow the procedure of Section 3.5 and look for normal mode solutions

$$(8.249) \qquad\qquad u(x, t) = a(k) \exp[ikx + \lambda(k)t].$$

This yields

$$(8.250) \qquad\qquad \lambda^2(k) + c^2 k^4 = 0$$

and

$$(8.251) \qquad\qquad \lambda(k) = \pm ick^2$$

for all real k. The stability index $\Omega = 0$, so that the Cauchy problem is well posed. Furthermore, with

$$(8.252) \qquad\qquad \omega(k) = ck^2$$

we have

$$(8.253) \qquad\qquad \lambda(k) = \pm i\omega(k)$$

so that (8.248) is conservative as well as being of dispersive type. The dispersion relation is $\omega(k) = ck^2$ and the group velocity $d\omega(k)/dk = 2ck$ exceeds the phase velocity $\omega(k)/k = ck$ in magnitude. It is worth noting that even though (8.248) is not of hyperbolic type it is, nevertheless, an equation of *dispersive type*.

The Cauchy problem for (8.248) with initial data

$$(8.254) \quad u(x,0) = f(x); \quad u_t(x,0) = g(x); \quad -\infty < x < \infty$$

may be solved by Fourier transform methods. Let the Fourier transform of $u(x, t)$ be

$$(8.255) \qquad\qquad U(\lambda, t) = \frac{1}{\sqrt{2\pi}} \int_{-\infty}^{\infty} e^{i\lambda x} u(x, t)\, dx.$$

Proceeding as in Section 5.2, we obtain the Fourier transformed problem

$$(8.256) \qquad\qquad \frac{\partial^2 U(\lambda, t)}{\partial t^2} + c^2 \lambda^4 U(\lambda, t) = 0; \quad t > 0$$

with

$$(8.257) \qquad\qquad U(\lambda, 0) = F(\lambda),$$

$$(8.258) \qquad\qquad \frac{\partial U(\lambda, 0)}{\partial t} = G(\lambda),$$

where F and G are the Fourier transforms of f and g, respectively. It has the solution

$$(8.259) \qquad U(\lambda, t) = \left[\frac{1}{2} F(\lambda) + \frac{1}{2ic\lambda^2} G(\lambda) \right] e^{ic\lambda^2 t}$$
$$+ \left[\frac{1}{2} F(\lambda) - \frac{1}{2ic\lambda^2} G(\lambda) \right] e^{-ic\lambda^2 t}.$$

Inverting the transform yields

$$(8.260) \quad u(x, t) = \frac{1}{\sqrt{2\pi}} \int_{-\infty}^{\infty} e^{-i\lambda x} U(\lambda, t)\, d\lambda$$

$$= \frac{1}{\sqrt{2\pi}} \int_{-\infty}^{\infty} H_+(\lambda) \exp[i(\omega(\lambda)t - \lambda x)]\, d\lambda$$

$$+ \frac{1}{\sqrt{2\pi}} \int_{-\infty}^{\infty} H_-(\lambda) \exp[i(-\omega(\lambda)t - \lambda x)]\, d\lambda,$$

where $\omega(\lambda) = c\lambda^2$ and $H_\pm(\lambda)$ are the coefficients of $e^{\pm ic\lambda^2 t}$ in (8.259).

Rather than invert (8.260) for specific choices of $f(x)$ and $g(x)$ we use the *method of stationary phase* to analyze the behavior of the solution $u(x, t)$ as t gets large. This method was presented in Example 5.13. The integrals in (8.260) may be written as

$$(8.261) \qquad I_\pm(x, t) = \frac{1}{\sqrt{2\pi}} \int_{-\infty}^{\infty} H_\pm(\lambda) \exp[i\varphi_\pm(x, t; \lambda)]\, d\lambda$$

with

$$(8.262) \qquad \varphi_\pm = \pm\omega(\lambda)t - \lambda x = \pm c\lambda^2 t - \lambda x.$$

We assume that x and t are large as was done for the solution (5.247) of the Klein–Gordon equation in Example 5.13. Then the stationary points are given as

$$(8.263) \qquad \frac{d\varphi_\pm}{d\lambda} = \pm 2c\lambda t - x = 0,$$

so that

$$(8.264) \qquad \lambda = \lambda_\pm = \pm\frac{x}{2ct}.$$

Thus, say, if $x > 0$, we have $\lambda_+ = x/2ct$ and $\lambda_- = -x/2ct$ as positive and negative stationary points for the integrals I_+ and I_-, respectively. Also

$$(8.265) \qquad \varphi_\pm(x, t, \lambda_\pm) = \pm c\left(\frac{x^2}{4c^2 t^2}\right)t \mp \frac{x^2}{2ct} = \mp\frac{x^2}{4ct}$$

and $[d^2\varphi_\pm(\lambda_\pm)]/d\lambda^2 = \pm 2ct$ so that

$$(8.266) \qquad \frac{(d^2\varphi_\pm)/d\lambda^2}{|(d^2\varphi_\pm)/d\lambda^2|} = \pm 1.$$

Furthermore,

$$(8.267) \qquad\qquad \omega''(\lambda_\pm) = 2c,$$

so that on using an appropriate modification of (5.238) we have

$$(8.268) \qquad u(x, t) = I_+(x, t) + I_-(x, t) \approx \frac{H_+(\lambda_+)}{\sqrt{2ct}} \exp\left[-\frac{ix^2}{4ct} + i\frac{\pi}{4}\right]$$

$$+ \frac{H_-(\lambda_-)}{\sqrt{2ct}} \exp\left[\frac{ix^2}{4ct} - i\frac{\pi}{4}\right], \quad |x|, t \to \infty$$

If the initial data $f(x)$ and $g(x)$ are real valued, it is easy to see, on using the definition of the Fourier transform, that $H_+(\lambda)$ and $H_-(-\lambda)$ are complex conjugates. Since $\lambda_+ = -\lambda_-$, we conclude that (8.268) is real valued.

Thus near the (group) lines $x/2ct = $ constant, the solution of the initial value problem for (8.248) for large t is approximated by (8.268) and the solution decays like $1/\sqrt{t}$ as $t \to \infty$.

If the rod is of finite extent, say, of length l, we must introduce boundary conditions at $x = 0$ and $x = l$. The boundary conditions

$$(8.269) \qquad\qquad u(0, t) = u_x(0, t) = 0$$

and

$$(8.270) \qquad\qquad u(l, t) = u_x(l, t) = 0$$

imply that the rod is *clamped* at both edges so that u and the slope u_x both vanish. The initial and boundary value problem for $u(x, t)$ in the interval $0 < x < l$ with u satisfying (8.248), the initial conditions (8.254) in the finite interval, and the boundary conditions (8.269)–(8.270) can be solved by separation of variables.

We set

$$(8.271) \qquad\qquad u(x, t) = F(x)G(t)$$

in (8.248) and obtain

$$(8.272) \qquad\qquad \frac{G''(t)}{c^2 G(t)} = -\frac{F''''(x)}{F(x)} = -\lambda^2,$$

with $-\lambda^2$ as the separation constant. This gives

$$(8.273) \qquad\qquad G''(t) + (\lambda c)^2 G(t) = 0$$

and

$$(8.274) \qquad\qquad F''''(x) - \lambda^2 F(x) = 0.$$

The conditions (8.269)–(8.270) imply that

$$(8.275) \qquad\qquad F(0) = F'(0) = F(l) = F'(l) = 0,$$

so that (8.274)–(8.275) is an eigenvalue problem for $F(x)$. We do not discuss general properties of higher order eigenvalue problems but solve the preceding problem directly.

The general solution of (8.274) is

$$(8.276) \quad F(x) = a\cos(\sqrt{\lambda}\,x) + b\cosh(\sqrt{\lambda}\,x) + c\sin(\sqrt{\lambda}\,x) + d\sinh(\sqrt{\lambda}\,x).$$

The conditions $F(0) = F'(0) = 0$ imply that $b = -a$ and $d = -c$. Thus

$$(8.277) \quad F(x) = a\big[\cos(\sqrt{\lambda}\,x) - \cosh(\sqrt{\lambda}\,x)\big] + c\big[\sin(\sqrt{\lambda}\,x) - \sinh(\sqrt{\lambda}\,x)\big].$$

At $x = l$ we obtain

$$(8.278) \quad a\big[\cos(\sqrt{\lambda}\,l) - \cosh(\sqrt{\lambda}\,l)\big] + c\big[\sin(\sqrt{\lambda}\,l) - \sinh(\sqrt{\lambda}\,l)\big] = 0$$

$$(8.279) \quad a\big[-\sin(\sqrt{\lambda}\,l) - \sinh(\sqrt{\lambda}\,l)\big] + c\big[\cos(\sqrt{\lambda}\,l) - \cosh(\sqrt{\lambda}\,l)\big] = 0$$

For a and c to be nonzero, the determinant of the coefficients of the system (8.278)–(8.279) must vanish. That is,

$$
\begin{aligned}
(8.280) \quad & \begin{vmatrix} \cos(\sqrt{\lambda}\,l) - \cosh(\sqrt{\lambda}\,l) & \sin(\sqrt{\lambda}\,l) - \sinh(\sqrt{\lambda}\,l) \\ -\sin(\sqrt{\lambda}\,l) - \sinh(\sqrt{\lambda}\,l) & \cos(\sqrt{\lambda}\,l) - \cosh(\sqrt{\lambda}\,l) \end{vmatrix} \\
& = \cos^2(\sqrt{\lambda}\,l) + \sin^2(\sqrt{\lambda}\,l) + \cosh^2(\sqrt{\lambda}\,l) - \sinh^2(\sqrt{\lambda}\,l) \\
& - 2\cos(\sqrt{\lambda}\,l)\cosh(\sqrt{\lambda}\,l) = 2\big[1 - \cos(\sqrt{\lambda}\,l)\cosh(\sqrt{\lambda}\,l)\big] = 0.
\end{aligned}
$$

The eigenvalue equation (8.280) can be written as

$$(8.281) \qquad\qquad \cosh(\sqrt{\lambda}\,l) = \sec(\sqrt{\lambda}\,l).$$

As the graph in Figure 8.4 shows, the curves for $\cosh(\sqrt{\lambda}\,l)$ and $\sec(\sqrt{\lambda}\,l)$ have infinitely many intersections. The solution $\lambda = 0$ does not lead to an eigenfunction since $F(x)$ vanishes for $\lambda = 0$ as is seen from (8.277). Clearly, the eigenvalues tend to infinity. Since $\cosh(\sqrt{\lambda}\,l)$ grows exponentially with $\sqrt{\lambda}$, the large eigenvalues are given approximately by the zeros of $\cos(\sqrt{\lambda}\,l)$. Let $\lambda_k(k = 1, 2, \dots)$ denote the eigenvalues. Then a set of eigenfunctions is given

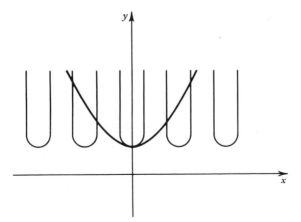

Figure 8.4. The graphs of cosh x and sec x.

by

$$(8.282) \quad F_k(x) = \left[\sin\left(\sqrt{\lambda_k}\, l\right) - \sinh\left(\sqrt{\lambda_k}\, l\right)\right]\left[\cos\left(\sqrt{\lambda_k}\, x\right) - \cosh\left(\sqrt{\lambda_k}\, x\right)\right]$$

$$- \left[\cos\left(\sqrt{\lambda_k}\, l\right) - \cosh\left(\sqrt{\lambda_k}\, l\right)\right]\left[\sin\left(\sqrt{\lambda_k}\, x\right) - \sinh\left(\sqrt{\lambda_k}\, x\right)\right].$$

The $F_k(x)$ are not normalized. By direct verification it can be shown that

$$(8.283) \qquad \int_0^l F_k(x) F_j(x)\, dx = 0$$

if $\lambda_k \neq \lambda_j$. Furthermore, it can be shown that the set $\{F_k(x)\}$ is complete and any smooth function in $0 < x < l$ can be expanded in a series of the $\{F_k(x)\}$.

On solving for $G_k(t)$ from (8.273) we have

$$(8.284) \qquad G_k(t) = \alpha_k \cos(\lambda_k ct) + \beta_k \sin(\lambda_k ct),$$

with α_k and β_k as arbitrary constants. The formal solution of (8.248) is obtained by superposition as

$$(8.285) \qquad u(x, t) = \sum_{k=1}^{\infty} F_k(x) G_k(t).$$

Formally applying the initial conditions to the series (8.285) leads to the specification of α_k and β_k in terms of the (Fourier) coefficients of the functions $f(x)$ and $g(x)$ when expanded in a series of eigenfunctions $F_k(x)$.

We consider other types of boundary conditions for (8.248) and properties of the corresponding eigenvalue problems for $F(x)$ in the exercises.

EXAMPLE 8.9.   THE VIBRATION OF PLATES

The *vibration of a thin plate* is governed by the equation

$$(8.286) \qquad u_{tt} + c^2 \nabla^2 \nabla^2 u = 0$$

where $u = u(x, y, t)$ is the transverse displacement of the point (x, y) on the plate at the time t, and ∇^2 is the two-dimensional Laplacian. The constant c^2 depends on the physical properties of the plate. The wave equation that governs the vibration of a membrane is given as $u_{tt} - c^2 \nabla^2 u = 0$ and contains only the Laplacian operator, whereas (8.286) contains the iterated Laplacian because of the greater rigidity of the plate. If we assume that the plate occupies the region G in the (x, y)-plane and is clamped at the boundary or edge ∂G, we have the boundary conditions

$$(8.287) \qquad u\Big|_{\partial G} = 0; \qquad \frac{\partial u}{\partial n}\Big|_{\partial G} = 0.$$

Also, $u(x, y, t)$ must satisfy the initial conditions

$$(8.288) \qquad u(x, y, 0) = f(x, y); \qquad u_t(x, y, 0) = g(x, y).$$

Because (8.286) is of fourth order in the spatial variables, we require two boundary conditions. However, since there are only two time derivatives in (8.286) we have two initial conditions. The problem (8.286)–(8.288) may be shown to be well posed.

We attempt to solve this initial and boundary value problem by using separation of variables. Let

$$(8.289) \qquad u(x, y, t) = F(x, y)G(t).$$

This yields on inserting (8.289) into (8.286),

$$(8.290) \qquad \frac{G''(t)}{c^2 G(t)} = -\frac{\nabla^2 \nabla^2 F(x, y)}{F(x, y)} = -\lambda,$$

where $-\lambda$ is the separation constant. The equation for $G(t)$ is

$$(8.291) \qquad G''(t) + \lambda c^2 G(t) = 0,$$

and $F(x, y)$ must satisfy

$$(8.292) \qquad \nabla^2 \nabla^2 F - \lambda F = 0$$

for $(x, y) \in G$ with the boundary conditions

$$(8.293) \qquad F(x, y)\Big|_{\partial G} = \frac{\partial F(x, y)}{\partial n}\Big|_{\partial G} = 0.$$

That is, we obtain an eigenvalue problem for $F(x, y)$.

To make any further progress with this problem we must be able to determine the eigenvalues λ and the eigenfunctions F for specific regions. It turns out, however, that (8.292) is only separable in polar coordinates so that more complicated or, possibly, approximation methods are needed if the problem does not involve a circular clamped plate. Indeed, in rectangular coordinates (8.292) takes the form

$$(8.294) \qquad F_{xxxx} + 2F_{xxyy} + F_{yyyy} - \lambda F = 0.$$

With

$$(8.295) \qquad F(x, y) = A(x)B(y)$$

we have

$$(8.296) \qquad A''''B + 2A''B'' + B''''A - \lambda AB = 0$$

and it is not possible to separate the variables x and y in (8.296).

For the case of a circular plate it is possible to obtain the eigenfunctions $F(x, y)$ by proceeding as follows. Let $\lambda = k^4$ for simplicity of notation and factor (8.292) as

$$(8.297) \qquad (\nabla^2\nabla^2 - k^4)F = (\nabla^2 + k^2)(\nabla^2 - k^2)F = 0.$$

This factorization can be carried out irrespective of the choice of coordinates, but only for the polar coordinate problem considered does it lead to a solution of the eigenvalue problem (8.292)–(8.293). We express F in polar coordinates as $F = F(r, \theta)$, and assume the region G is the interior of the circle $r < R$ with $r = R$ as the boundary ∂G. Then $F(r, \theta)$ must be bounded at $r = 0$ and must be periodic of period 2π in θ—that is, $F(r, \theta + 2\pi) = F(r, \theta)$—since F must be single valued in G. Thus if $F(r, \theta)$ satisfies either of the equations

$$(8.298) \qquad (\nabla^2 \pm k^2)F = F_{rr} + \frac{1}{r}F_r + \frac{1}{r^2}F_{\theta\theta} \pm k^2F = 0,$$

it is a solution of (8.292) since the order of the operators in (8.297) can be interchanged.

The equations (8.298) can be solved by separation of variables. Let

$$(8.299) \qquad F(r, \theta) = V(r)W(\theta),$$

and (8.298) can be expressed as

$$(8.300) \qquad V''(r) + \frac{1}{r}V'(r) \pm k^2V(r) - \frac{n^2}{r^2}V(r) = 0$$

and

(8.301) $W''(\theta) + n^2 W(\theta) = 0,$

after separating the variables with n^2 equal to the separation constant.

The periodicity of $F(r, \theta)$ implies that $W(\theta + 2\pi) = W(\theta)$ so that n must be a (positive) integer or zero (i.e., $n = 0, 1, 2, \dots$). Consequently, we have $W_n(\theta) = \cos n\theta$ and $W_n(\theta) = \sin n\theta$ as the appropriate solutions of (8.301). For each value of n, (8.300) is a Bessel equation of order n if the plus sign (i.e., $+k^2$) is chosen. When the minus sign (i.e., $-k^2$) is selected in (8.300), we have the modified Bessel equation of order n.

The product functions $F_n(r, \theta) = V_n(r)W_n(\theta)$ must be bounded at $r = 0$, so that the appropriate solution of (8.300) with the term $+k^2$ is $V_n = J_n(kr)$, the Bessel function of nth order. For the case $-k^2$ we obtain $V_n = I_n(kr)$, the modified Bessel function of nth order. Consequently, we find that for each integer n we have the solutions

(8.302) $F_n(r, \theta) = \left[\alpha_n J_n(kr) + \gamma_n I_n(kr)\right]\cos n\theta$
$$+ \left[\beta_n J_n(kr) + \delta_n I_n(kr)\right]\sin n\theta,$$

of (8.297) where α_n, β_n, γ_n, and δ_n are constants. On applying the boundary conditions (8.293) to each of the F_n, we determine the eigenvalues $\lambda = k^4$. For the circular region G, the normal derivative $\partial F/\partial n$ becomes $\partial F/\partial r$ on $r = R$. Since the boundary conditions must be valid for all θ, we see that the coefficients of $\cos n\theta$ and $\sin n\theta$ must vanish separately once the boundary conditions are applied. The coefficient of $\cos n\theta$ yields the system

(8.303) $\begin{cases} \alpha_n J_n(kR) + \gamma_n I_n(kR) = 0 \\ \alpha_n J_n'(kR) + \gamma_n I_n'(kR) = 0 \end{cases}.$

A second system with α_n and γ_n replaced by β_n and δ_n, respectively, results from the coefficient of $\sin n\theta$. Since the coefficients α_n and γ_n must not vanish, we conclude that the determinant of the coefficients in (8.303) vanishes and this yields the transcendental equation

(8.304) $J_n(kR)I_n'(kR) - I_n(kR)J_n'(kR) = 0$

for the determination of the eigenvalues. (The second system yields the same result.) The appropriate values of k must be specified approximately and we do not consider this matter here. Once the full set of eigenvalues has been found, we may solve (8.291) for the function $G(t)$. Then on summing over all the product solutions (8.289) we may specify the two sets of arbitrary constants in the solution of (8.291) so as to satisfy the initial conditions (8.288).

Since the general problem is quite complicated, we consider a simplified case where the initial data (8.288) are functions of the radial variable r only, that is,

$$(8.305) \qquad u(r,\theta,0) = f(r); \qquad u_t(r,\theta,0) = g(r).$$

Then the eigenfunctions F may be chosen to be independent of θ, and we obtain instead of (8.302):

$$(8.306) \qquad F(r) = \alpha_0 J_0(kr) + \gamma_0 I_0(kr).$$

The equations (8.303)–(8.304) remain valid with $n = 0$. Let $k_m (m = 1, 2, 3, \dots)$ represent the (real) roots of (8.304) with $n = 0$. The (unnormalized) eigenfunctions $F(r)$ then have the form

$$(8.307) \qquad F_m(r) = I_0(k_m R) J_0(k_m r) - J_0(k_m R) I_0(k_m r).$$

It is not hard to show that the $\{F_m(r)\}$ are an orthogonal set over the interval $0 \leqslant r \leqslant R$ with weight function r. With $\lambda_m = k_m^4$ the solutions of (8.291) are

$$(8.308) \qquad G_m(t) = a_m \cos(k_m^2 ct) + b_m \sin(k_m^2 ct)$$

with arbitrary constants a_m and b_m. The formal series $u = \sum_{m=1}^{\infty} F_m(r) G_m(t)$ satisfies the equation (8.286) in the circle $0 \leqslant r < R$, the boundary conditions (8.287) on $r = R$, and the initial conditions (8.305) once the constants a_m and b_m in (8.308) are specified in terms of $f(r)$ and $g(r)$. Using the orthogonality properties of the eigenfunctions $\{F_m(r)\}$ this is easy to carry out. The eigenfunctions for the eigenvalue problem (8.292)–(8.293) are known to form a complete set, so that the expansions discussed are valid if the initial data satisfy certain smoothness conditions.

It is of considerable interest to study the stationary version of the vibrating plate equation (8.286). We consider displacements u that are independent of time so that $u = u(x, y)$, and find that $u(x, y)$ satisfies the *biharmonic equation*

$$(8.309) \qquad \nabla^2 \nabla^2 u = 0.$$

Clearly, any harmonic function u (i.e., a solution of Laplace's equation $\nabla^2 u = 0$) will also satisfy (8.309). This fact can occasionally be used to advantage in solving boundary value problems for the biharmonic equation.

Using direct verification it may be shown that the function

$$(8.310) \qquad u = (r^2 - R^2)v + w$$

is a solution of the biharmonic equation if v and w are both harmonic functions, $r^2 = x^2 + y^2$, and R^2 is a constant. Given the boundary value problem for (8.309) in the circle $0 \leqslant r < R$ with the boundary conditions

$$(8.311) \qquad\qquad u\Big|_{r=R} = f, \quad \frac{\partial u}{\partial r}\Big|_{r=R} = g,$$

we may solve this problem by choosing v and w in (8.310) appropriately.
The boundary condition $u(R, \theta) = f$ implies that

$$(8.312) \qquad\qquad u(R, \theta) = w(R, \theta) = f.$$

Thus w must be a solution of Laplace's equation in the circle $0 \leqslant r < R$ which
satisfies the Dirichlet condition $w = f$ on the boundary $r = R$. This problem
has been solved previously and completely specifies w. The condition $\partial u / \partial r = g$
on $r = R$ implies

$$(8.313) \qquad\qquad \frac{\partial u}{\partial r}\Big|_{r=R} = 2Rv(R, \theta) + \frac{\partial w(R, \theta)}{\partial r} = g.$$

Since w has already been specified, $\partial w(R, \theta)/\partial r$ is a known function and
(8.313) yields $v(R, \theta) = (1/2R)g - (1/2R)\partial w(R, \theta)/\partial r$ as the Dirichlet
boundary condition for the harmonic function v. Again, v is thereby uniquely
specified and we have obtained a solution of the boundary value problem
(8.309) and (8.311) in terms of the solution of two boundary value problems
for Laplace's equation in a circle.

We complete our discussion of the vibrating plate by briefly considering the
fundamental solution and *Green's function* methods for the biharmonic equation.

We proceed as in Section 6.7 and consider the inhomogeneous biharmonic
equation

$$(8.314) \qquad\qquad \nabla^2 \nabla^2 u = -F(x, y).$$

With P_0 equal to the point (x_0, y_0), we integrate over a region R with
boundary ∂R that contains P_0 in its interior and obtain

$$(8.315) \quad \iint_R \nabla^2\nabla^2 u \, dx \, dy = \iint_R \nabla \cdot (\nabla\nabla^2 u) \, dx \, dy$$

$$= \int_{\partial R} \frac{\partial}{\partial n}(\nabla^2 u) \, ds = -\iint_R F \, dx \, dy = -1$$

where we have applied the divergence theorem and chosen $F(x, y)$ to represent
a point source or, equivalently, a Dirac delta function with singularity at P_0 so
that the integral of F equals unity. Introducing polar coordinates with the pole

located at P_0 and taking the limit in (8.315) as $\partial R \to P_0$, we obtain

$$(8.316) \quad \lim_{\partial R \to P_0} \int_{\partial R} \frac{\partial}{\partial n}(\nabla^2 u)\, ds = \lim_{r \to 0} \int_0^{2\pi} \frac{\partial}{\partial r}(\nabla^2 u)\Bigg|_{\substack{x=x_0+r\cos\theta \\ y=y_0+r\sin\theta}} r\, d\theta$$

$$= \lim_{r \to 0}\left[2\pi r \frac{\partial}{\partial r}(\nabla^2 u)\right] = -1.$$

Continuing as in Section 6.7 we assume that u is a function r only and we have

$$(8.317) \qquad\qquad \nabla^2 u(r) = \frac{1}{r}\frac{\partial}{\partial r}\left(r\frac{\partial u}{\partial r}\right).$$

As a result we obtain the approximate equation

$$(8.318) \qquad\qquad 2\pi r \frac{\partial}{\partial r}\left[\frac{1}{r}\frac{\partial}{\partial r}\left(r\frac{\partial u}{\partial r}\right)\right] \approx -1.$$

Integrating out this equation and retaining only the most singular terms yields the fundamental solution

$$(8.319) \qquad\qquad u = -\frac{1}{8\pi}r^2 \log r$$

where $r^2 = (x - x_0)^2 + (y - y_0)^2$. It can be verified that (8.319) is a solution of the biharmonic equation everywhere except at $r = 0$. In fact, apart from the constant factor, the solution (8.319) may be derived by looking for a solution of (8.309) that depends only on r.

On the basis of the identity

$$(8.320) \qquad\qquad u\nabla^2\nabla^2 v - v\nabla^2\nabla^2 u = \nabla \cdot \mathbf{p}$$

where

$$(8.321) \qquad \mathbf{p} = u\nabla\nabla^2 v - v\nabla\nabla^2 u + (\nabla^2 u)\nabla v - (\nabla^2 v)\nabla u$$

we obtain the *Green's theorem*

$$(8.322) \quad \iint_G [u\nabla^2\nabla^2 v - v\nabla^2\nabla^2 u]\, dx\, dy$$

$$= \int_{\partial G}\left[u\frac{\partial}{\partial n}(\nabla^2 v) - (\nabla^2 v)\frac{\partial u}{\partial n} + (\nabla^2 u)\frac{\partial v}{\partial n} - v\frac{\partial(\nabla^2 u)}{\partial n}\right] ds$$

on using the divergence theorem.

The Green's function K for the biharmonic equation with *clamped plate* boundary conditions is defined to be a solution of

$$(8.323) \qquad\qquad \nabla^2 \nabla^2 K = -\delta$$

where the two-dimensional Dirac delta function δ is singular at the point (ξ, η) in G and K satisfies the boundary conditions

$$(8.324) \qquad\qquad K\Big|_{\partial G} = \frac{\partial K}{\partial n}\Big|_{\partial G} = 0.$$

Now if $u(x, y)$ satisfies the inhomogeneous equation (8.314) in G and assumes given boundary conditions u and $\partial u / \partial n$ on ∂G, the Green's theorem (8.322) (where we set $v = K$) yields the result

$$(8.325) \quad u(\xi, \eta) = \iint_G KF\,dx\,dy + \int_{\partial G}\left[(\nabla^2 K)\frac{\partial u}{\partial n} - u\frac{\partial(\nabla^2 K)}{\partial n} \right] ds$$

where all the terms on the right of (8.325) are known.

Using special techniques it is possible to specify the Green's function K for the case of circular clamped plate. We do not derive or exhibit this Green's function but note that in view of (8.319) K can be expressed as

$$(8.326) \qquad\qquad K(x, y; \xi, \eta) = -\frac{1}{8\pi}r^2 \log r + V$$

where V is a regular solution of the homogeneous biharmonic equation. Applying the boundary conditions (8.324) to K yields a set of boundary conditions for V and $\partial V / \partial n$ on ∂G. Since the boundary value problem for the biharmonic equation with clamped plate boundary conditions was solved previously, it is possible to determine V and, thereby, K for the circular clamped plate.

EXAMPLE 8.10.   THE EQUATIONS OF FLUID DYNAMICS

We begin by deriving the *Euler equations* of motion for a nonviscous or inviscid fluid. Let $\mathbf{u} = \mathbf{u}(x, y, z, t)$ be the velocity vector of a fluid particle situated at the point (x, y, z) at the time t. Also let $p = p(x, y, z, t)$ and $\rho = \rho(x, y, z, t)$ represent the pressure and the density, respectively, of the fluid. We assume that no external forces are acting on the fluid.

We consider a volume of the fluid occupying a region R with boundary ∂R. Assuming there are no sources or sinks for the fluid in R, the rate of increase of the fluid in R in unit time must be balanced by the flux or inflow of fluid through the boundary ∂R. Analytically, this may be expressed as

(8.327)
$$\frac{\partial}{\partial t} \iint_R \rho \, dv = - \int_{\partial R} \rho \mathbf{u} \cdot \mathbf{n} \, ds$$

where $\mathbf{n}$ is the exterior unit normal and the minus sign occurs since we are concerned with the inflow. (We note that $\rho \, dv$ is an element of mass and that $\rho \mathbf{u} \cdot \mathbf{n} \, ds$ represents the flow of mass through the boundary element ds.) Applying the divergence theorem to the integral over ∂R and bringing the time derivative inside the integral over R in (8.327)—since R is fixed in time—we obtain

(8.328)
$$\iint_R \left[\rho_t + \nabla \cdot (\rho \mathbf{u}) \right] dv = 0.$$

Since (8.328) must be valid for an arbitrary region R, we conclude that integrand vanishes identically and obtain

(8.329)
$$\rho_t + \nabla \cdot (\rho \mathbf{u}) = 0,$$

which is known as the *equation of continuity*.

Next we consider the law of conservation of momentum for the fluid. Since we are neglecting frictional forces due to the effects of viscosity and all external forces, the only force acting on the volume of fluid in the region R is due to the pressure along the boundary ∂R. The total pressure is given as

(8.330)
$$\mathbf{P} = - \int_{\partial G} p \mathbf{n} \, ds,$$

which is effectively a resultant of all the internally directed normal pressures at the boundary. According to Newton's second law of motion, the pressure force equals the mass times the acceleration of the fluid element or particle. The acceleration of a particle is given by $d\mathbf{u}/dt$, but since the particle is moving along a path $x = x(t), y = y(t), z = z(t)$, we obtain from the chain rule

(8.331)
$$\frac{d\mathbf{u}}{dt} = \frac{\partial \mathbf{u}}{\partial t} + \frac{\partial \mathbf{u}}{\partial x}\frac{dx}{dt} + \frac{\partial \mathbf{u}}{\partial y}\frac{dy}{dt} + \frac{\partial \mathbf{u}}{\partial z}\frac{dz}{dt} = \mathbf{u}_t + (\mathbf{u} \cdot \nabla)\mathbf{u}.$$

Here we have used the fact that

(8.332)
$$\mathbf{u} = [x'(t), y'(t), z'(t)]$$

and

(8.333)
$$\mathbf{u} \cdot \nabla = x'\frac{\partial}{\partial x} + y'\frac{\partial}{\partial y} + z'\frac{\partial}{\partial z}.$$

The derivative $d\mathbf{u}/dt$ in (8.331) is often referred to as a *material derivative*. The total momentum of the fluid particles in R must equal the resultant pressure forces and we obtain

(8.334)
$$\iint_R \rho \frac{d\mathbf{u}}{dt}\, dv = -\int_{\partial R} p\mathbf{n}\, ds,$$

since $\rho\, dv$ is an element of mass. Using the gradient theorem, which is a simple consequence of the divergence theorem, the integral over ∂R can be converted into an integral over R and (8.334) becomes

(8.335)
$$\iint_R \left[\rho \frac{d\mathbf{u}}{dt} + \nabla p\right] dv = 0.$$

The arbitrariness of R implies the vanishing of the integrand and we obtain the *equation of conservation of momentum*

(8.336)
$$\mathbf{u}_t + (\mathbf{u} \cdot \nabla)\mathbf{u} + \left(\frac{1}{\rho}\right)\nabla p = 0.$$

The equations (8.329) and (8.336) are incomplete as they stand since we have only four equations and five unknowns (i.e., ρ, p, and the three components of $\mathbf{u}$). To complete the system we need an *equation of state* which we give as

(8.337)
$$p = f(\rho)$$

where $f(\rho)$ is a function that is determined from thermodynamic considerations and that differs for different fluids.

The equations (8.329), (8.336), and (8.337) constitute *Euler's equations of fluid dynamics*. It may be added that flows for which these equations are valid are known as *isentropic, inviscid fluid flows*. (Isentropic means that the entropy—which is a thermodynamic quantity—remains uniform or constant throughout the fluid motion.) The equation of state (8.337) may be used to eliminate the pressure as one of the dependent variables in the Euler system. Thus

(8.338)
$$\nabla p = f'(\rho)\nabla \rho$$

so that ∇p can be removed from (8.336) and we are left with four equations in the variables ρ and $\mathbf{u}$. This system of equations is a quasilinear first order system and it is not an easy matter to solve the system in general. Consequently, certain simplifying assumptions are introduced to deal with specific types of fluid flow.

In dealing with the flow of liquids rather than gases we assume that the density ρ is identically constant; that is,

$$(8.339) \qquad\qquad \rho = \text{constant}.$$

This means that we are dealing with an *incompressible fluid* and (8.339) replaces the equation of state (8.337). In view of (8.339), the equation of continuity (8.329) reduces to

$$(8.340) \qquad\qquad \nabla \cdot \mathbf{u} = 0.$$

We further assume that the fluid flow is *irrotational* which means that the velocity field $\mathbf{u}$ has the property that its curl vanishes; that is,

$$(8.341) \qquad\qquad \nabla \times \mathbf{u} = \mathbf{0}.$$

On the basis of Stokes' theorem from vector analysis we conclude that if (8.341) is satisfied everywhere, the integral of the velocity around any closed curve vanishes; that is,

$$(8.342) \qquad\qquad \oint \mathbf{u} \cdot d\mathbf{r} = 0.$$

The integral in (8.342) defines the *circulation* of the fluid around the closed curve and irrotationality means that no (nonzero) circulation exists. Since the Euler equations describe the fluid motion as it develops in time, we cannot necessarily assume that if the flow is irrotational at one time it may not become rotational at a later time. It can be shown, however, that the fluid remains rotation free for all time (based on Euler's equations) if it was so at a given instant of time. Rather than consider this question we examine steady fluid flow for which the pressure p and the velocity $\mathbf{u}$ are independent of time.

Now it is easy to show that if (8.342) is satisfied for all closed curves, the velocity vector $\mathbf{u}$ can be derived from a *potential function* φ and can be expressed as

$$(8.343) \qquad\qquad \mathbf{u}(x, y, z) = \nabla\varphi(x, y, z).$$

Inserting (8.343) into (8.340) gives

$$(8.344) \qquad\qquad \nabla \cdot \mathbf{u} = \nabla \cdot \nabla\varphi = \nabla^2\varphi = 0,$$

so that φ satisfies *Laplace's equation*. Once Laplace's equation for the velocity potential φ is solved, we may determine the pressure p from (8.336) where we set $\mathbf{u}_t = \mathbf{0}$. We see that for *steady, incompressible, and irrotational flow*, the study of the fluid motion is reduced to the solution of Laplace's equation, subject, of course, to appropriate boundary conditions.

A different analysis of Euler's equations that results in a study of the (linear) wave equation is a linearization procedure that leads to the *theory of acoustics*. We perturb the solutions $\mathbf{u}$, p, and ρ around a (constant) equilibrium state $\mathbf{u} = \mathbf{0}$, $p = p_0$, and $\rho = \rho_0$ [with $p_0 = f(\rho_0)$] and write

$$(8.345) \qquad \begin{cases} \mathbf{u} = \epsilon\mathbf{u}_1 + \cdots \\ p = p_0 + \epsilon p_1 + \cdots \\ \rho = \rho_0 + \epsilon\rho_1 + \cdots \end{cases}$$

where $0 < \epsilon \ll 1$ with $\epsilon = $ constant and the dots refer to terms that are of higher order in ϵ. Inserting (8.345) into Euler's equations and retaining only terms of order ϵ gives

$$(8.346) \qquad \frac{\partial\rho_1}{\partial t} + \rho_0\nabla \cdot \mathbf{u}_1 = 0,$$

$$(8.347) \qquad \frac{\partial\mathbf{u}_1}{\partial t} + \frac{1}{\rho_0}\nabla p_1 = 0,$$

$$(8.348) \qquad p_1 = f'(\rho_0)\rho_1$$

as the *linearized versions* of the Euler equations.

From (8.347) we find that

$$(8.349) \qquad \mathbf{u}_1(x, y, z, t) = \mathbf{u}_1(x, y, z, 0) - \frac{1}{\rho_0}\nabla \int_0^t p_1\, dt.$$

If, initially, $\mathbf{u}_1 = 0$ or if $\mathbf{u}_1$ can be written as

$$(8.350) \qquad \mathbf{u}_1(x, y, z, 0) = -\nabla\psi(x, y, z)$$

(i.e., it can be derived from a potential), we obtain

$$(8.351) \qquad \mathbf{u}_1(x, y, z, t) = -\nabla\left[\psi + \frac{1}{\rho_0}\int_0^t p_1 dt\right] \equiv -\nabla\varphi$$

where φ equals the bracketed term in (8.351). Inserting (8.351) into (8.347) yields

$$(8.352) \qquad \nabla\left[-\varphi_t + \frac{1}{\rho_0}p_1\right] = 0,$$

and we may choose

$$(8.353) \qquad p_1 = \rho_0\varphi_t.$$

Then (8.348) implies that

$$(8.354) \qquad \rho_1 = \left[\frac{\rho_0}{f'(\rho_0)} \right] \varphi_t.$$

Inserting (8.351) and (8.354) into (8.346) yields

$$(8.355) \qquad \varphi_{tt} - [f'(\rho_0)] \nabla^2 \varphi = 0,$$

so that φ satisfies the wave equation if $f'(\rho_0) > 0$. In fact, for a *polytropic gas* we have

$$(8.356) \qquad f(\rho) = A\rho^\gamma,$$

where A and γ are positive constants with $\gamma > 1$ and given as $\gamma = c_p/c_v$ with c_p and c_v equal to the specific heats of the gas at constant pressure and constant volume, respectively. We set

$$(8.357) \qquad c^2 = f'(\rho),$$

so that $c(\rho_0) = \sqrt{f'(\rho_0)}$ in (8.355) represents the *propagation speed* for disturbances. We note that not only φ but p_1, ρ_1, and each component of $\mathbf{u}_1$ also satisfy the *wave equation* (8.355).

For the one-dimensional case we replace $\mathbf{u}$ by u and obtain the one-dimensional wave equation for $\varphi(x, t)$ with the general solution

$$(8.358) \qquad \varphi(x, t) = F(x - c_0 t) + G(x + c_0 t),$$

with $c_0 = c(\rho_0)$. The function φ must satisfy appropriate initial and boundary conditions and then u_1, p_1, and ρ_1 can be expressed in terms of φ by means of (8.351), (8.353), and (8.354). In particular, if the solution φ results in unidirectional wave motion, that is,

$$(8.359) \qquad \varphi(x, t) = F(x - c_0 t),$$

and only right moving waves occur, we find that

$$(8.360) \qquad \begin{cases} u_1 = -F'(x - c_0 t) \\ p_1 = -c_0 \rho_0 F'(x - c_0 t) \\ \rho_1 = -\dfrac{\rho_0}{c_0} F'(x - c_0 t) \end{cases}$$

so that both p_1 and ρ_1 can be expressed in terms of u_1 in the form

$$(8.361) \qquad p_1 = c_0 \rho_0 u_1; \qquad \rho_1 = \left(\frac{\rho_0}{c_0} \right) u_1.$$

This result suggests that we look for special solutions of the full nonlinear system of Euler's equations for the one-dimensional case in the form

$$(8.362) \qquad\qquad p = p(u); \qquad \rho = \rho(u)$$

where $u(x, t)$ is the velocity. Using (8.357), the one-dimensional form of Euler's equations are

$$(8.363) \qquad\qquad \rho_t + \rho u_x + u\rho_x = 0,$$

$$(8.364) \qquad\qquad u_t + uu_x + \left(\frac{c^2(\rho)}{\rho}\right)\rho_x = 0.$$

Since p has been eliminated from (8.364) we need only use $\rho = \rho(u)$. Inserting this into (8.363) and (8.364) gives

$$(8.365) \qquad\qquad u_t + uu_x = -\left(\frac{\rho}{\rho'}\right)u_x,$$

$$(8.366) \qquad\qquad u_t + uu_x = -\left(\frac{c^2\rho'}{\rho}\right)u_x$$

where $\rho' = d\rho/du$. Equating the right sides of these equations yields

$$(8.367) \qquad\qquad (c\rho')^2 = \rho^2$$

so that

$$(8.368) \qquad\qquad \rho' = \pm\frac{\rho}{c}.$$

Inserting (8.368) in (8.365) gives, with $\hat{c}(u) = c[\rho(u)]$,

$$(8.369) \qquad\qquad u_t + [u \pm \hat{c}(u)]u_x = 0,$$

a first order *quasilinear (wave) equation* studied in Chapter 2. Once $u(x, t)$ is determined from (8.369), we may use the solution of (8.368) to specify $\rho(x, t)$.

Rather than deal with the general case, we consider a polytropic gas for which (8.356) is valid and choose the plus sign in (8.368)–(8.369). Then

$$(8.370) \qquad\qquad c^2 = \frac{d}{d\rho}(A\rho^\gamma) = \gamma A\rho^{\gamma-1}$$

and (8.368) takes the form

$$(8.371) \qquad\qquad \sqrt{\gamma A}\,\rho^{(\gamma-3)/2}\rho' = 1.$$

This ordinary differential equation is separable. Assuming that at $u = 0$ we have $\rho = \rho_0$ and $c(\rho)|_{u=0} = c(\rho_0) = c_0$, the solution in terms of $\hat{c}(u)$ is

$$(8.372) \qquad \hat{c}(u) = c_0 + \left(\frac{\gamma - 1}{2}\right)u,$$

from which ρ can be determined. Then (8.369) takes the form (with the plus sign chosen)

$$(8.373) \qquad u_t + \left[c_0 + \frac{\gamma + 1}{2}u\right]u_x = 0.$$

If the initial waveform of $u(x, t)$ at $t = 0$ is given as $u(x, 0) = h(x)$, the results of Example 2.6 imply that

$$(8.374) \qquad u = h\left[x - \left(c_0 + \frac{\gamma + 1}{2}u\right)t\right]$$

is the implicit form of the solution of (8.372). This represents a wave traveling to the right for positive u and, according to the results of Section 2.3, each point x on the waveform $u(x, t)$ travels with speed

$$(8.375) \qquad \frac{dx}{dt} = c_0 + \frac{\gamma + 1}{2}u = u + \left[c_0 + \frac{\gamma - 1}{2}u\right].$$

In fact, for the general case of (8.369) we have

$$(8.376) \qquad \frac{dx}{dt} = u \pm \hat{c}.$$

We see from (8.375) and (8.376) that $\pm \hat{c}$ represent an addition to the actual velocity u of the fluid at which disturbances propagate in the fluid. Therefore, $c = \sqrt{f'(\rho)}$ is called the *(local) speed of sound* of the fluid or gas. For the linearized equations of acoustics where a perturbation around a zero velocity was carried out, $c_0 = \sqrt{f'(\rho_0)}$ represents the (constant) speed of wave propagation or speed of sound for the gas. By choosing the minus sign in (8.368)–(8.369), we may construct solutions that represent waves traveling to the left. If we assume that the velocity u is small and approximate $\hat{c}(u)$ by $\hat{c}(u_0) = c_0$, our results reduce to those for the equations of linear acoustics.

In general, solutions of (8.369) in the form

$$(8.377) \qquad u = h[x - (u \pm \hat{c}(u)t]$$

are known as *simple waves* and represent *unidirectional wave motion*. They play a role in a number of problems of fluid dynamics. A particular case is the *piston problem*, discussed in its linearized form in Example 6.8. (We do not discuss the nonlinear version of this problem.) A difficulty encountered for quasilinear equations in Chapter 2, and which plays a role for the Euler equations as well, is that of the breakdown of solutions because they become multivalued or singular. If that happens, the equations themselves are no longer valid, as the assumptions that led to their derivation are no longer met. The assumption that the flow is isentropic cannot be maintained and an equation relating to changes in entropy must be introduced as well as a new equation of state. Then the solution can be carried beyond the breaking point by the introduction of *shock waves* as was indicated in Chapter 2.

Alternatively, we may conclude that effects such as viscosity and dependence on temperature variations that were neglected in our derivation of the fluid flow equations must be included. These effects introduce higher derivative terms into the system of fluid dynamics equations and, as we have seen on a number of occasions, the inclusion of higher derivative terms may have a smoothing effect on the solutions so that no breakdown occurs at all.

Rather than discuss the general case, we restrict ourselves to one dimension. Without presenting a derivation we write down the *Navier–Stokes equations* for *viscous fluid flow* in one dimension. We choose as the unknowns the velocity u, the density ρ, the pressure p, and the temperature T. We have

$$(8.378) \qquad \rho_t + \rho u_x + u\rho_x = 0,$$

$$(8.379) \qquad u_t + uu_x + \left(\frac{1}{\rho}\right)p_x = \frac{4}{3}\left(\frac{\mu}{\rho}\right)u_{xx},$$

$$(8.380) \qquad \rho c_v\left[T_t + uT_x\right] + pu_x = kT_{xx} + \frac{4}{3}\mu(u_x)^2,$$

and

$$(8.381) \qquad p = R\rho T.$$

In these equations μ is the (constant) coefficient of viscosity, k is the coefficient of heat conduction, c_p and c_v are the specific heats at constant pressure and volume, respectively, and $R = c_p - c_v$ is the gas constant. We have assumed there are no external forces or heat sources. (We note that in the literature the Navier–Stokes equations are often written in different forms.)

The first two equations (8.378)–(8.379) are the equations of continuity and conservation of momentum. The new term in (8.379) represents forces due to viscous effects and vanishes when $\mu = 0$. The equation (8.380) characterizes the *conservation of energy*. If we neglect conductivity and viscosity (i.e., we put $k = \mu = 0$), we obtain

$$(8.382) \qquad \rho c_v\left[T_t + uT_x\right] + pu_x = 0.$$

From (8.378) we can replace u_x by $-(1/\rho)(\rho_t + u\rho_x)$ and obtain

(8.383)
$$\rho[e_t + ue_x] - \left(\frac{p}{\rho}\right)[\rho_t + u\rho_x] = 0$$

where the *internal energy e* is given as

(8.384)
$$e = c_v T.$$

Using the material derivative gives

(8.385)
$$\frac{de}{dt} - \left(\frac{p}{\rho^2}\right)\frac{d\rho}{dt} = \frac{de}{dt} + p\frac{d}{dt}\left(\frac{1}{\rho}\right) = 0.$$

Then the thermodynamic relation

(8.386)
$$T\,ds = de + p\,d\left(\frac{1}{\rho}\right)$$

where s is the *entropy* reduces (8.385) to

(8.387)
$$T\frac{ds}{dt} = 0.$$

We see that the entropy for individual fluid particles remains constant. If at the initial time we have uniform entropy, it remains so for all time and this corresponds to the isentropic flow considered.

The final equation (8.381) is an equation of state for a *perfect gas* that takes the place of (8.337). We do not discuss the relationship between these equations, say, if $\mu = k = 0$. In any case, the introduction of viscosity and heat conduction increases the order of the system of equations from first to second order by means of the addition of derivatives in x. It is shown in the exercises that for isentropic flow the *Euler equations* are a *hyperbolic system* whose characteristic curves for a given solution are (8.376) with $\hat{c}$ replaced by $c = c(\rho)$. The full *Navier–Stokes equations* are of *parabolic type* but in view of their complexity not much appears to be gained from this information. We do not discuss the solution of these equations here but in Section 9.6 we present an approximation method that enables us to draw some conclusions about the effect of additional higher derivative terms in the Navier–Stokes equations on certain solutions of the fluid flow equations.

We conclude our discussion of fluid flow by examining the Euler equations in the case of *steady* (i.e., time independent) *two-dimensional isentropic flow*. With $u(x, y)$ and $v(x, y)$ as the velocity components, $\rho(x, y)$ as the density, and $c^2 = f'(\rho)$, we have from (8.329) and (8.336),

(8.388) $\rho u u_x + \rho v u_y + c^2 \rho_x = 0,$

(8.389) $\rho u v_x + \rho v v_y + c^2 \rho_y = 0,$

(8.390) $\rho u_x + \rho v_y + u \rho_x + v \rho_y = 0.$

The system (8.388)–(8.390) can be written in matrix form as

(8.391) $A \mathbf{w}_x + B \mathbf{w}_y = \mathbf{0}$

where the matrices A and B and the vector $\mathbf{w}$ are given as

(8.392) $A = \begin{bmatrix} \rho u & 0 & c^2 \\ 0 & \rho u & 0 \\ \rho & 0 & u \end{bmatrix}; \quad B = \begin{bmatrix} \rho v & 0 & 0 \\ 0 & \rho v & c^2 \\ 0 & \rho & v \end{bmatrix}; \quad \mathbf{w} = \begin{bmatrix} u \\ v \\ \rho \end{bmatrix}.$

This first order quasilinear system may be classified according to the method of Section 3.3 for a fixed solution $\mathbf{w}$. The characteristic curves $y = y(x)$ are determined from [see (3.97)]

(8.393) $\det(B - y'A) = \begin{bmatrix} \rho(v - y'u) & 0 & -c^2 y' \\ 0 & \rho(v - y'u) & c^2 \\ -\rho y' & \rho & (v - y'u) \end{bmatrix} = 0,$

and this gives

(8.394) $[v - uy'] \big[(v - uy')^2 - (1 + y'^2) c^2 \big] = 0.$

The first factor of (8.394) yields the characteristic curves

(8.395) $y' = \dfrac{dy}{dx} = \dfrac{v}{u},$

and these represent the *streamlines* of the flow, that is, the curves tangent to the velocity vectors whose components are u and v.

The second factor in (8.394) can be written as

(8.396) $(c^2 - u^2) y'^2 + (2uv) y' + c^2 - v^2 = 0.$

The roots y' of this quadratic equation may be characterized by means of the discriminant, which is

(8.397) $D = 4c^2 [u^2 + v^2 - c^2],$

as is easy to show. If $u^2 + v^2 > c^2$, both roots y' are real and distinct and neither is equal to the third root (8.395). Thus there are three real and distinct characteristics, and the system is (totally) hyperbolic. Since c is the sound speed and $\sqrt{u^2 + v^2}$ is the speed of the fluid, this case corresponds to *supersonic flow*. If $u^2 + v^2 < c^2$, the corresponding roots y' are complex. This case corresponds to *subsonic flow*. (According to our classification scheme, since not all the y' are complex this is not an elliptic system.) If $u^2 + v^2 = c^2$, the root is real and double and the situation is complicated in that case. The flow may be termed *sonic flow* since the fluid speed and sound speed coincide.

The distinction between subsonic and supersonic flow can be brought out in another, simpler manner by linearizing the system (8.388)–(8.390). We set

(8.398)
$$\begin{cases} u = u_0 + \epsilon u_1 + \cdots \\ v = \epsilon v_1 + \cdots \\ \rho = \rho_0 + \epsilon \rho_1 + \cdots \end{cases}$$

where u_0 and ρ_0 are constant and $0 < \epsilon \ll 1$. The dots correspond to higher order terms in ϵ. In (8.398) we are perturbing around a constant state in which there is a uniform flow with velocity $u_0 > 0$ parallel to the x-axis.

Inserting (8.398) in (8.388)–(8.390) and retaining only the terms of order ϵ yields

(8.399)
$$\rho_0 u_0 \frac{\partial u_1}{\partial x} + c_0^2 \frac{\partial \rho_1}{\partial x} = 0,$$

(8.400)
$$\rho_0 u_0 \frac{\partial v_1}{\partial x} + c_0^2 \frac{\partial \rho_1}{\partial y} = 0,$$

and

(8.401)
$$\rho_0 \frac{\partial u_1}{\partial x} + \rho_0 \frac{\partial v_1}{\partial y} + u_0 \frac{\partial \rho_1}{\partial x} = 0$$

where $c_0 = c(\rho_0)$. From (8.401) we have

(8.402)
$$\rho_0 \frac{\partial v_1}{\partial y} = -u_0 \frac{\partial \rho_1}{\partial x} - \rho_0 \frac{\partial u_1}{\partial x}$$

$$= -u_0 \frac{\partial \rho_1}{\partial x} + \frac{c_0^2}{u_0} \frac{\partial \rho_1}{\partial x};$$

in view of (8.399). Since $\partial^2 v_1 / \partial x \partial y = \partial^2 v_1 / \partial y \partial x$,

(8.403)
$$\rho_0 \frac{\partial^2 v_1}{\partial y \partial x} = -\frac{c_0^2}{u_0} \frac{\partial^2 \rho_1}{\partial y^2}$$

and

$$(8.404) \qquad \rho_0 \frac{\partial^2 v_1}{\partial x \partial y} = \left[\frac{c_0^2}{u_0} - u_0 \right] \frac{\partial^2 \rho_1}{\partial x^2},$$

which follow from (8.400) and (8.402) and imply that

$$(8.405) \qquad (1 - M^2) \frac{\partial^2 \rho_1}{\partial x^2} + \frac{\partial^2 \rho_1}{\partial y^2} = 0$$

where the *Mach number* M is defined as

$$(8.406) \qquad M = \frac{u_0}{c_0};$$

that is, it is the ratio of the underlying constant velocity to the uniform sound speed $c_0 = c(\rho_0)$.

If the Mach number exceeds unity (i.e., $M > 1$), (8.405) is clearly of *hyperbolic type*. This corresponds to *supersonic flow* since $u_0 > c_0$ in the unperturbed state. If $M < 1$, (8.405) is of *elliptic type* and we have *subsonic flow* since $u_0 < c_0$. If $M = 1$, the equation is *parabolic* and we have *sonic flow* since $u_0 = c_0$. We note that the characteristics corresponding to the streamlines for the nonlinear system [i.e., (8.395)] play no role in the linearized problem. In fact, if we insert (8.398) into (8.395), we obtain to leading order $y' = \epsilon v_1 / u_0$ and from (8.396) we have $y'^2 = M^2 - 1$. Thus with $\epsilon = 0$ the characteristic equations are $y' = 0$ and $y' = \pm \sqrt{M^2 - 1}$. The solutions of $y' = 0$ (i.e., $y = $ constant) are the streamlines of the steady uniform flow, $u = u_0$, and $v = 0$. The characteristics of (8.405) (i.e., $y \pm \sqrt{M^2 - 1} \, x = $ constant) are solutions of the equation $y' = \pm \sqrt{M^2 - 1}$.

EXAMPLE 8.11.   MAXWELL'S EQUATIONS OF ELECTROMAGNETIC THEORY

The *electric* and *magnetic field* vectors $\mathbf{E}$ and $\mathbf{H}$, respectively, which are studied in electromagnetic theory, are solutions of the fundamental set of equations known as *Maxwell's equations*. Each of these equations is based on an experimentally observed property of the electric and magnetic fields. We do not present a derivation of these equations but write out the equations in mks units and discuss some of their properties.

Maxwell's equations are given as follows:

$$(8.407) \qquad \nabla \times \mathbf{E} = -\frac{\partial \mathbf{B}}{\partial t},$$

$$(8.408) \qquad \nabla \times \mathbf{H} = \frac{\partial \mathbf{D}}{\partial t} + \mathbf{J},$$

$$(8.409) \qquad \nabla \cdot \mathbf{B} = 0,$$

and

$$(8.410) \qquad\qquad \nabla \cdot \mathbf{D} = \rho.$$

In addition there are the *constitutive relations* that express $\mathbf{B}$, $\mathbf{D}$, and $\mathbf{J}$ in terms of $\mathbf{E}$ and $\mathbf{H}$. These equations are

$$(8.411) \qquad\qquad \mathbf{D} = \epsilon \mathbf{E},$$
$$(8.412) \qquad\qquad \mathbf{B} = \mu \mathbf{H},$$

and

$$(8.413) \qquad\qquad \mathbf{J} = \sigma \mathbf{E}.$$

The vectors $\mathbf{D}$, $\mathbf{B}$, and $\mathbf{J}$ are the electric displacement, the magnetic induction, and the conduction current density, respectively. Further, ρ is the density of electric charges, ϵ is the dielectric constant, and μ is the magnetic permeability.

We now introduce some simplifying assumptions that render this system more tractable. In problems of *electrostatics* we set all time derivatives equal to zero. From (8.407) we then find that

$$(8.414) \qquad\qquad \nabla \times \mathbf{E} = \mathbf{0}$$

so that $\mathbf{E}$ is an *irrotational field*. As shown in the preceding example, this implies that $\mathbf{E}$ can be derived from a potential so that

$$(8.415) \qquad\qquad \mathbf{E} = -\nabla\varphi,$$

with φ equal to the *electrostatic potential*. Further, if ϵ is a constant in (8.411), we have from (8.410):

$$(8.416) \qquad\qquad \nabla \cdot \mathbf{E} = \frac{\rho}{\epsilon} = -\nabla \cdot \nabla\varphi,$$

which shows that

$$(8.417) \qquad\qquad \nabla^2\varphi = -\frac{\rho}{\epsilon}$$

and φ is a solution of *Poisson's equation*. If the charge density ρ vanishes, φ satisfies *Laplace's equation*. If the medium is nonconducting, so that $\sigma = 0$ and $\mathbf{J} = \mathbf{0}$, and, in addition, μ is a constant, the same result is obtained for the magnetic field $\mathbf{H}$. However, since there are no magnetic charges, the *magnetostatic potential* satisfies Laplace's equation.

If the medium is homogeneous (i.e., ϵ, μ, and σ are constant) and we also have $\rho = 0$, the following simplification of Maxwell's equations results. We apply the curl operator to (8.407) and obtain

$$\nabla \times \nabla \times \mathbf{E} = -\nabla \times \frac{\partial(\mu\mathbf{H})}{\partial t} = -\mu\frac{\partial}{\partial t}(\nabla \times \mathbf{H}).$$

Using the vector identity

$$(8.418) \qquad\qquad \nabla \times \nabla \times \mathbf{E} = \nabla(\nabla \cdot \mathbf{E}) - \nabla^2\mathbf{E}$$

and (8.408) for **H** gives

$$(8.419) \qquad \nabla(\nabla \cdot \mathbf{E}) - \nabla^2\mathbf{E} = -(\mu\epsilon)\frac{\partial^2\mathbf{E}}{\partial t^2} - (\mu\sigma)\frac{\partial\mathbf{E}}{\partial t}.$$

But $\nabla \cdot \mathbf{D} = \epsilon\nabla \cdot \mathbf{E} = 0$, in view of (8.410) so that

$$(8.420) \qquad\qquad \nabla^2\mathbf{E} - (\epsilon\mu)\frac{\partial^2\mathbf{E}}{\partial t^2} - (\sigma\mu)\frac{\partial\mathbf{E}}{\partial t} = \mathbf{0},$$

with an identical equation satisfied by **H**. Thus each Cartesian component of **E** and **H** satisfies a *telegrapher's equation* of the form

$$(8.421) \qquad\qquad u_{tt} - c^2\nabla^2 u + \lambda u_t = 0$$

where $c^2 = (1/\epsilon\mu)$ and $\lambda = \sigma/\epsilon$. If the medium is nonconducting so that $\sigma = 0$, each component satisfies the *wave equation*. In either case disturbances are propagated with the speed c. We note that if $\rho \neq 0$, we obtain an inhomogeneous form of (8.420).

It is also possible to introduce scalar and vector potential functions to treat the time dependent Maxwell's equations. We assume the medium is homogeneous as well as nonconducting so that $\sigma = 0$ and $\mathbf{J} = \mathbf{0}$.

The equation (8.409), that is,

$$(8.422) \qquad\qquad \nabla \cdot \mathbf{B} = 0,$$

implies that **B** is a *solenoidal field* and it is shown in vector analysis that there exists a *vector potential* **A** such that

$$(8.423) \qquad\qquad \mathbf{B} = \nabla \times \mathbf{A}.$$

Then (8.407) yields

$$(8.424) \qquad\qquad \nabla \times \mathbf{E} = -\frac{\partial\mathbf{B}}{\partial t} = -\nabla \times \left(\frac{\partial\mathbf{A}}{\partial t}\right),$$

and we obtain

$$(8.425) \qquad \nabla \times \left[\mathbf{E} + \frac{\partial \mathbf{A}}{\partial t} \right] = \mathbf{0},$$

so that $\mathbf{E} + \mathbf{A}_t$ is an irrotational vector. Consequently, it can be derived from a potential φ and we have

$$(8.426) \qquad \mathbf{E} + \frac{\partial \mathbf{A}}{\partial t} = -\nabla\varphi.$$

From the identity $\nabla \times \nabla\psi = \mathbf{0}$, we see that if $\mathbf{A}$ is replaced by

$$(8.427) \qquad \mathbf{A} = \mathbf{A}_0 + \nabla\psi,$$

(8.423) will still be satisfied by $\mathbf{A}_0$. Further, in terms of φ we have from (8.426)

$$(8.428) \qquad \mathbf{E} + \frac{\partial \mathbf{A}_0}{\partial t} = -\nabla\psi_t - \nabla\varphi = -\nabla(\varphi + \psi_t),$$

so that a new scalar potential φ_0 can be defined as

$$(8.429) \qquad \varphi = \varphi_0 - \psi_t.$$

To remove this ambiguity we impose the *Lorentz condition*,

$$(8.430) \qquad \nabla \cdot \mathbf{A} = -(\epsilon\mu)\frac{\partial\varphi}{\partial t}.$$

With this choice $\mathbf{A}$ and φ both satisfy wave equations as we now demonstrate. From (8.426), (8.410), and (8.430) we have

$$(8.431) \qquad \nabla \cdot \left[\mathbf{E} + \frac{\partial \mathbf{A}}{\partial t} \right] = \nabla \cdot \mathbf{E} + \frac{\partial}{\partial t}(\nabla \cdot \mathbf{A})$$
$$= \frac{\rho}{\epsilon} - (\epsilon\mu)\varphi_{tt} = -\nabla^2\varphi,$$

so that

$$(8.432) \qquad c^2\nabla^2\varphi - \varphi_{tt} = -\frac{\rho}{\epsilon^2\mu},$$

which is an inhomogeneous wave equation. To obtain an equation for $\mathbf{A}$ we use (8.408) in the form

$$(8.433) \qquad \nabla \times (\mu\mathbf{H}) = \nabla \times \mathbf{B} = (\epsilon\mu)\frac{\partial\mathbf{E}}{\partial t}$$

and insert the expression (8.423) to obtain

(8.434)
$$\nabla \times \mathbf{B} = \nabla \times \nabla \times \mathbf{A} = \nabla(\nabla \cdot \mathbf{A}) - \nabla^2\mathbf{A}$$

$$= (\epsilon\mu)\frac{\partial}{\partial t}\left[-\frac{\partial\mathbf{A}}{\partial t} - \nabla\varphi\right]$$

$$= -(\epsilon\mu)\frac{\partial^2\mathbf{A}}{\partial t^2} - (\epsilon\mu)\nabla\frac{\partial\varphi}{\partial t},$$

where (8.418) and (8.426) were used. Rearranging terms and using the Lorentz condition (8.430) yields

(8.435)
$$(\epsilon\mu)\frac{\partial^2\mathbf{A}}{\partial t^2} - \nabla^2\mathbf{A} = -\nabla\left[\nabla \cdot \mathbf{A} + (\epsilon\mu)\frac{\partial\varphi}{\partial t}\right] = \mathbf{0}$$

so that $\mathbf{A}$ satisfies the homogeneous wave equation. Once $\mathbf{A}$ and φ are specified, the electric and magnetic fields are given as

(8.436)
$$\mathbf{E} = -\frac{\partial\mathbf{A}}{\partial t} - \nabla\varphi,$$

(8.437)
$$\mathbf{H} = \frac{1}{\mu}\nabla \times \mathbf{A}.$$

We have shown the relationship between *Maxwell's equations* and *Laplace's* and *Poisson's equations* in the time independent case and the *wave* and *telegrapher's equations* in the time varying case. Many problems for Maxwell's equations can be reduced to a study of these simpler equations subject to appropriate initial and boundary conditions. As a simple example we consider the *initial value problem* for *Maxwell's equations in a vacuum*. We have $\epsilon = \mu = $ constant, $\sigma = \rho = 0$, and $\mathbf{J} = 0$ in (8.407)–(8.413). The equations can then be written as

(8.438)
$$\nabla \times \mathbf{E} = -\mu\frac{\partial\mathbf{H}}{\partial t},$$

(8.439)
$$\nabla \times \mathbf{H} = \epsilon\frac{\partial\mathbf{E}}{\partial t},$$

(8.440)
$$\nabla \cdot \mathbf{E} = \nabla \cdot \mathbf{H} = 0.$$

These equations are valid throughout space and initially we have

(8.441)
$$\mathbf{E}(x,0) = \mathbf{F}(x); \qquad \mathbf{H}(x,0) = \mathbf{G}(x).$$

The procedure that led to (8.420) can be applied to (8.438)–(8.439) to yield the equations

$$(8.442) \qquad \nabla^2 \mathbf{E} - (\epsilon\mu)\mathbf{E}_{tt} = 0,$$

$$(8.443) \qquad \nabla^2 \mathbf{H} - (\epsilon\mu)\mathbf{H}_{tt} = 0.$$

Each Cartesian component of $\mathbf{E}$ and $\mathbf{H}$ satisfies a scalar wave equation. However, we have only one initial condition for $\mathbf{E}$ and $\mathbf{H}$ in (8.441) and we must also prescribe $\mathbf{E}_t$ and $\mathbf{H}_t$ initially to obtain a unique solution. From (8.438) and (8.439) we find that

$$(8.444) \qquad \begin{cases} \dfrac{\partial \mathbf{E}(x,0)}{\partial t} = \left(\dfrac{1}{\epsilon}\right)\nabla \times \mathbf{H}(x,0) = \left(\dfrac{1}{\epsilon}\right)\nabla \times \mathbf{G} \\[2mm] \dfrac{\partial \mathbf{H}(x,0)}{\partial t} = -\left(\dfrac{1}{\mu}\right)\nabla \times \mathbf{E}(x,0) = -\left(\dfrac{1}{\mu}\right)\nabla \times \mathbf{F}. \end{cases}$$

The initial value problem for the scalar wave equation has been solved previously in the text in one, two, and three space dimensions. Thus $\mathbf{E}(x,t)$ and $\mathbf{H}(x,t)$ can be determined. To complete the verification of the solution we must show that (8.440) is satisfied for $t > 0$ provided it was true at the time $t = 0$. Thus we must assume that the initial fields $\mathbf{F}$ and $\mathbf{G}$ are such that

$$(8.445) \qquad \nabla \cdot \mathbf{F} = \nabla \cdot \mathbf{G} = 0.$$

Since $\nabla \cdot \nabla \times \mathbf{A} = 0$ for any vector $\mathbf{A}$, we have

$$(8.446) \qquad 0 = \nabla \cdot \nabla \times \mathbf{E} = -\mu\nabla \cdot \mathbf{H}_t = -\mu\frac{\partial}{\partial t}(\nabla \cdot \mathbf{H}),$$

$$(8.447) \qquad 0 = \nabla \cdot \nabla \times \mathbf{H} = \epsilon\nabla \cdot \mathbf{E}_t = \epsilon\frac{\partial}{\partial t}(\nabla \cdot \mathbf{E})$$

on using (8.438)–(8.439). Thus $\nabla \cdot \mathbf{E}$ and $\nabla \cdot \mathbf{H}$ are constant in time and since they vanish at $t = 0$, they do so for all time.

For the sake of concreteness we consider a specific problem in one dimension. Let

$$(8.448) \quad \mathbf{E}(x,0) = \mathbf{F}(x) = (\sin x)\mathbf{j}; \qquad \mathbf{H}(x,0) = \mathbf{G}(x) = (\cos x)\mathbf{k}.$$

Then (8.445) is satisfied and

$$(8.449) \qquad \nabla \times \mathbf{F} = (\cos x)\mathbf{k}; \qquad \nabla \times \mathbf{G} = (\sin x)\mathbf{j}$$

so that

$$(8.450) \qquad \mathbf{E}_t(x,0) = \frac{1}{\epsilon}(\sin x)\mathbf{j}; \qquad \mathbf{H}_t(x,0) = -\frac{1}{\mu}(\cos x)\mathbf{k}.$$

Each component of $\mathbf{E}$ and $\mathbf{H}$ satisfies the one-dimensional wave equation

$$(8.451) \qquad\qquad u_{xx} - (\epsilon\mu)u_{tt} = 0.$$

The y-component of $\mathbf{E}$ has the initial conditions

$$(8.452) \qquad u(x,0) = \sin x; \qquad u_t(x,0) = \frac{1}{\epsilon}\sin x,$$

whereas the z component of $\mathbf{H}$ has the initial conditions

$$(8.453) \qquad u(x,0) = \cos x; \qquad u_t(x,0) = -\frac{1}{\mu}\cos x.$$

All other components have zero initial conditions and, therefore, vanish for all time. Using d'Alembert's solution of the wave equation we easily obtain (with $c^2 = 1/\epsilon\mu$) the standing wave representations

$$(8.454) \qquad \mathbf{E}(x,t) = (\sin x)\left(\cos ct + \frac{1}{c\epsilon}\sin ct\right)\mathbf{j},$$

$$(8.455) \qquad \mathbf{H}(x,t) = (\cos x)\left(\cos ct - \frac{1}{c\mu}\sin ct\right)\mathbf{k}.$$

Clearly, $\mathbf{E}(x,t)$ and $\mathbf{H}(x,t)$ satisfy the initial conditions (8.448). Further,

$$(8.456) \qquad \nabla \times \mathbf{E} = (\cos x)\left(\cos ct + \frac{1}{c\epsilon}\sin ct\right)\mathbf{k}$$

and

$$(8.457) \qquad -\mu\frac{\partial \mathbf{H}}{\partial t} = (\cos x)(\mu c \sin ct + \cos ct)\mathbf{k},$$

so that (8.438) is satisfied since $\mu c = \sqrt{\mu/\epsilon} = 1/c\epsilon$. It can be easily verified that (8.439)–(8.440) are also satisfied. Finally, it may be noted that although the fields $\mathbf{E}$ and $\mathbf{H}$ propagate along the x-direction, the nonzero field components lie in the transverse y and z-directions.

EXAMPLE 8.12. THE EQUATIONS OF ELASTICITY THEORY

Let $\mathbf{u}(x, y, z, t)$ be the vector representing the displacement of a point (x, y, z) in an *elastic medium* at the time t. Assuming the medium is *isotropic*, the equation of motion satisfied by $\mathbf{u}$, which is known as *Navier's equation*, is

$$(8.458) \qquad \rho\frac{\partial^2 \mathbf{u}}{\partial t^2} = (\lambda + 2\mu)\nabla(\nabla \cdot \mathbf{u}) - \mu\nabla \times \nabla \times \mathbf{u} + \mathbf{F}.$$

Here ρ is the density of the medium, λ and μ are the *Lamé constants*, which depend on the properties of the medium, and $\mathbf{F}$ represents the external forces. Using the vector identity (8.418), we can put (8.458) into the equivalent form

$$(8.459) \qquad \rho \frac{\partial^2 \mathbf{u}}{\partial t^2} = (\lambda + \mu)\nabla(\nabla \cdot \mathbf{u}) + \mu\nabla^2 \mathbf{u} + \mathbf{F}.$$

We assume in our discussion that ρ, λ, and μ are constant.

We begin our analysis of the equations of elasticity by assuming that $\mathbf{u} = \mathbf{u}(x, t)$ so that the problem is one-dimensional. Let the components of $\mathbf{u}$ be u, v, and w (i.e., $\mathbf{u} = u\mathbf{i} + v\mathbf{j} + w\mathbf{k}$). The divergence $\nabla \cdot \mathbf{u}$ has the form

$$(8.460) \qquad \nabla \cdot \mathbf{u} = \frac{\partial u}{\partial x},$$

since $\mathbf{u}$ is independent of y and z. Thus (8.459) becomes

$$(8.461) \qquad \rho \frac{\partial^2 \mathbf{u}}{\partial t^2} = (\lambda + \mu)\frac{\partial^2 u}{\partial x^2}\mathbf{i} + \mu\frac{\partial^2 \mathbf{u}}{\partial x^2} + \mathbf{F}.$$

With $\mathbf{F} = F_1\mathbf{i} + F_2\mathbf{j} + F_3\mathbf{k}$, we can uncouple the three equations (8.461) for the Cartesian components of $\mathbf{u}$ and obtain

$$(8.462) \qquad \rho u_{tt} = (\lambda + 2\mu)u_{xx} + F_1$$

and

$$(8.463) \qquad \begin{cases} \rho v_{tt} = \mu v_{xx} + F_2 \\ \rho w_{tt} = \mu w_{xx} + F_3 \end{cases}.$$

Thus each component of $\mathbf{u}$ satisfies a wave equation in one dimension. However, the speed of wave propagation for the component u equals $[(\lambda + 2\mu)/\rho]^{1/2}$, whereas the corresponding speed for the components v and w is given as $(\mu/\rho)^{1/2}$.

These results show that in the one-dimensional case there are two modes of wave propagation that can occur for elastic waves. Since the waves associated with the component u have the property that the displacement is in the direction of wave propagation, they are called *longitudinal waves* and travel with the speed $[(\lambda + 2\mu)/\rho]^{1/2}$. The second wave motion is associated with the components v and w and thus corresponds to displacements in a plane perpendicular to the direction of wave propagation. Therefore, they are called *transverse waves* and they travel with the speed $[\mu/\rho]^{1/2}$. Since λ, μ, and ρ are all positive, we see that the longitudinal waves have greater speed than the transverse waves. (The longitudinal waves are often called *compression waves* and the transverse waves are called *shear waves*.)

It may be noted that if

(8.464) $\mathbf{u} = \mathbf{u}(x, t) = u(x, t)\mathbf{i} + [v(x, t)\mathbf{j} + w(x, t)\mathbf{k}],$

            $= u\mathbf{i} + \mathbf{V},$

we have (since u and $\mathbf{V}$ are functions of x and t only)

(8.465) $\nabla \times (u\mathbf{i}) = \mathbf{0}; \qquad \nabla \cdot \mathbf{V} = 0.$

Also, as is easily verified,

$$\nabla \cdot \mathbf{u} = \nabla \cdot (u\mathbf{i}); \qquad \nabla \times \mathbf{u} = \nabla \times \mathbf{V}.$$

This suggests that we try to decompose the displacement vector $\mathbf{u}$ in the general case into two parts, one involving $\nabla \cdot \mathbf{u}$ and the other $\nabla \times \mathbf{u}$. (It is shown in vector analysis that such a decomposition is always possible for a vector field. That is, it can be decomposed into a sum of two vectors one of which has zero divergence whereas the other has zero curl.)

To achieve this decomposition we multiply (8.458) with $\nabla \cdot$ and then with $\nabla \times$; that is, we take the divergence and curl of these equations. We have

(8.466) $\rho \dfrac{\partial^2}{\partial t^2}(\nabla \cdot \mathbf{u})$

$$= (\lambda + 2\mu)\nabla^2(\nabla \cdot \mathbf{u}) - \mu\nabla \cdot (\nabla \times \nabla \times \mathbf{u}) + \nabla \cdot \mathbf{F}$$

$$= (\lambda + 2\mu)\nabla^2(\nabla \cdot \mathbf{u}) + \nabla \cdot \mathbf{F}$$

since the divergence of a curl vanishes. Also,

(8.467) $\rho \dfrac{\partial^2}{\partial t^2}(\nabla \times \mathbf{u}) = (\lambda + 2\mu)\nabla \times [\nabla(\nabla \cdot \mathbf{u})]$

$$-\mu\nabla \times [\nabla \times \nabla \times \mathbf{u}] + \nabla \times \mathbf{F}$$

$$= \mu\nabla^2(\nabla \times \mathbf{u}) + \nabla \times \mathbf{F}$$

since $\nabla \times \nabla\varphi = \mathbf{0}$ and $\nabla \times [\nabla \times \nabla \times \mathbf{u}] = -\nabla^2(\nabla \times \mathbf{u}) + \nabla(\nabla \cdot \nabla \times \mathbf{u})$
$= -\nabla^2(\nabla \times \mathbf{u})$, as $\nabla \cdot \nabla \times \mathbf{u} = 0$.

We have shown that in the general case, the elasticity equations can be reduced to the study of two wave equations for $\nabla \cdot \mathbf{u}$ and $\nabla \times \mathbf{u}$. The term $\nabla \cdot \mathbf{u}$ characterizes the compression and expansion of the body and its waves travel with the speed $[(\lambda + 2\mu)/\rho]^{1/2}$. The term $\nabla \times \mathbf{u}$ characterizes the distortion of the body and its waves travel with the speed $[\mu/\rho]^{1/2}$. In dealing with initial value problems for elastic wave propagation with no boundaries present, this decomposition can be used to great advantage since we know how to solve the initial value problem for the wave equation. However, if there are boundaries, the boundary conditions couple the components of **u** and the problems are more complicated.

We have treated a number of examples of physical problems that give rise to differential equations that are not of the general form studied previously in the text. It has been shown in some cases that the problems can be reduced to a consideration of equations studied previously. However, we have only touched upon some of the methods used for these problems and each of the theories of fluid dynamics, electromagnetics, and elasticity has a rich body of mathematical methods associated with it.

In Chapter 9 we consider approximate methods for solving partial differential equations and these techniques are useful not only for the equations studied in the main body of this text but also for the types of problems considered in this section.

EXERCISES FOR CHAPTER 8

Section 8.1

8.1.1. Given the eigenvalue problem (8.28)–(8.29) of Example 8.1, determine all the eigenvalues and the orthonormalized eigenfunctions (see Example 7.5). Show that the eigenvalues can be ordered as in (8.3). (Note that there are multiple eigenvalues for this problem.) Also show that the energy integral (8.32) satisfies the equation (8.17) for each eigenfunction and its corresponding eigenvalue.

8.1.2. Consider the eigenvalue problem of Example 8.1 with the Dirichlet boundary condition (8.29) replaced by the Neumann condition $\partial M/\partial n|_{\partial G} = 0$ (see Example 7.6). Show that the eigenvalues can be ordered as in (8.3) and that the appropriate energy integral—that is, (8.32)—satisfies (8.17).

8.1.3. Obtain the appropriate form for the energy integral (8.11) for the Sturm–Liouville eigenvalue problem

$$- (pu')' + qu = \lambda \rho u, \qquad 0 < x < l;$$
$$u'(0) - h_1 u(0) = 0, \qquad u'(l) + h_2 u(l) = 0,$$

where h_1 and h_2 are positive constants.

8.1.4. Show that the energy integral (8.11) for the eigenvalue problem

$$u''(x) + \lambda u(x) = 0, \qquad 0 < x < l,$$
$$u(0) = 0, \qquad u'(l) + hu(l) = 0, \qquad h > 0,$$

has the form

$$E(u) = \int_0^l (u')^2 \, dx + hu^2(l).$$

Let $u_n(x) = \sin(\sqrt{\lambda_n}\, x)$ be the (unnormalized) eigenfunctions for this problem with the λ_n determined from $\sqrt{\lambda_n}\cos(\sqrt{\lambda_n}\, l) + h\sin(\sqrt{\lambda_n}\, l) = 0$. Demonstrate that

$$E(u_n) = \lambda_n(u_n, u_n).$$

[Hint: Use the identity $\cos^2(\sqrt{\lambda_n}\, x) = 1 - \sin^2(\sqrt{\lambda_n}\, x)$ and the eigenvalue equation to simplify the energy integral expression].

8.1.5. Consider the function

$$u(x, y) = cxy \sin x \sin y.$$

Determine the constant c such that the norm (8.33) of this function is unity. Observe that $u(x, y)$ vanishes on the boundary of the square $0 < x < \pi$, $0 < y < \pi$ and evaluate the energy integral (8.32) for this function. Compare your result with the eigenvalue $\lambda_{11} = 2$ for the square and evaluate the norm $\|M_{11} - u\|$ as in Example 8.1.

8.1.6. Show that the eigenvalues obtained from the variational principle (8.40) with the constraints (8.41) can indeed be ordered as $0 \leqslant \lambda_1 \leqslant \lambda_2 \leqslant \cdots \leqslant \lambda_n \leqslant \cdots$, by noting that to determine λ_n an additional constraint is added to those already given for the preceding eigenvalues so that the class of admissible functions is reduced.

8.1.7. It is required in the text that the function W defined in (8.42) satisfy the constraints (8.41). Show that if we set

$$W = \hat{W} + \sum_{j=1}^{n-1} a_j M_j,$$

where $\hat{W}$ is arbitrary (apart from having to vanish on ∂G if W vanishes there), we may choose the constants $a_j = -(\hat{W}, M_j)$ so that W satisfies the constraints (8.41). Replace W by the preceding expression in the discussion following (8.44) and show that (8.46) results with W replaced by $\hat{W}$.

8.1.8. Show that if

$$\iint_G fg \, dv = 0$$

for a continuous function f in G and arbitrary functions g, then we must have $f = 0$. This result is often called the *fundamental lemma of the calculus of variations*. (Hint: Assume $f \neq 0$ at some point in G, so that it is nonzero in the neighborhood of that point because it is continuous, and then choose g appropriately.)

8.1.9. Show that if

$$\iint_G f \, dv = 0$$

for arbitrary regions G within which f is continuous, we must have $f = 0$. (Hint: Proceed as in Exercise 8.1.8) This result is known as the *du Bois–Reymond lemma* and is closely related to the lemma of Exercise 8.1.8.

8.1.10. The eigenvalues for the problem

$$(xu')' + \frac{\lambda}{x} u = 0, \qquad 1 < x < 2,$$

$$u(1) = u(2) = 0$$

(see Exercise 4.3.2), are given as

$$\lambda_n = \left(\frac{\pi n}{\ln 2} \right)^2, \qquad n = 1, 2, \dots .$$

With $p = x$, $q = 0$, $\rho = 1/x$, determine the constants p_m, p_M, ρ_m, and ρ_M for the interval $1 < x < 2$, where the subscripts m and M correspond to the minima and maxima of these functions. Obtain the eigenvalues for the appropriate constant coefficient problems obtained from this problem and verify the result (8.55) for the eigenvalues of all three problems.

8.1.11. Obtain upper and lower bounds for the eigenvalues for the following problem.

$$-\nabla^2 M + xyM = \lambda M, \qquad 0 < x < \pi, \quad 0 < y < \pi,$$
$$M(0, y) = M(\pi, y) = M(x, 0) = M(x, \pi) = 0.$$

8.1.12. Use the method of Example 8.2 to estimate the lowest eigenvalue for the following problem.

$$\nabla^2 M + \lambda M = 0; \qquad \frac{x^2}{a^2} + \frac{y^2}{b^2} \leqslant 1; \qquad a < b,$$

$$M(x, y) = 0; \qquad \frac{x^2}{a^2} + \frac{y^2}{b^2} = 1.$$

8.1.13. Obtain an upper and lower bound for the lowest eigenvalue of the following problem in a sphere.

$$\nabla^2 M + \lambda M = 0; \qquad x^2 + y^2 + z^2 < 1,$$
$$M(x, y, z) = 0; \qquad x^2 + y^2 + z^2 = 1.$$

Use the method of Example 8.2. The exact eigenvalue is found in Example 8.3 to equal π^2.

8.1.14. Use the energy integral (8.11) to show that the leading eigenvalue λ_1 for the problem (8.1)–(8.2) increases as α/β increases. (Using the Courant max–min principle, it can be shown that all the eigenvalues λ_n have this property.)

8.1.15. Use separation of variables in polar coordinates to determine the eigenvalues and eigenfunctions for the following problem.

$$-\nabla^2 M = \lambda M; \qquad 0 \leqslant r < a; \qquad 0 < \theta < \phi \,[\phi \leqslant 2\pi],$$
$$M(r, \theta) = 0 \text{ on the boundary.}$$

(Hint: Proceed as in Example 8.2. The eigenvalues are the squares of zeros of Bessel functions.)

8.1.16. Use the results of Exercise 8.1.15 and the method of Example 8.2 to obtain an upper and lower bound for the leading eigenvalue for the following problem.

$$M_{xx} + M_{yy} + \lambda M = 0; \qquad (x, y) \in G$$
$$M(x, y) = 0 \text{ on } \partial G,$$

where the G is the interior of the triangle bounded by the lines $y = 0$, $y = x$, and $x = 1$.

8.1.17. Show that the energy integral $E(w)$—as defined in (8.38)—for functions w that equal f on G is minimized by the solution u of the Dirichlet problem

$$-\nabla \cdot (p\nabla u) + qu = 0; \qquad x \in G; \quad u = f, \quad x \in \partial G.$$

[Hint: Let $w = u + W$ where u is the (assumed) solution of the problem and W is an arbitrary function that vanishes on G. Show that $E(w) = E(u) + E(W) > E(u)$, if $W \neq 0$.] If $p = 1$ and $q = 0$, this is a classical result known as *Dirichlet's principle.*

8.1.18. Adapt the procedure given in the text to show that the solution of the problem

$$E(w) - \iint_G 2wF \, dv = \text{minimum}$$

where $E(w)$ as defined in (8.38) or (8.39) is a solution of the problem

$$-\nabla \cdot (p\nabla u) + qu = F,$$

with homogeneous boundary conditions of the first, second, or third kind. In the case of Dirichlet boundary conditions the admissible functions for the variational problem must vanish on ∂G. In the other two cases the boundary conditions are natural and depend on whether the form (8.38) or (8.39) for $E(w)$ is chosen. [Hint: Let $w = u + \epsilon W$, where u is the (assumed) solution of the problem.]

8.1.19. Show that the solution of the problem

$$E(w) - \iint_G 2wF\,dv - \int_{\partial G} 2wf\,ds = \text{minimum}$$

yields the solutions of the boundary value problems

$$-\nabla \cdot (p\nabla u) + qu = F; \qquad x \in G,$$

with the boundary conditions

$$\frac{\partial u}{\partial n} = f \quad \text{or} \quad \frac{\partial u}{\partial n} + hu = f; \qquad x \in \partial G,$$

depending on whether $E(w)$ has the form (8.38) or (8.39), respectively. The boundary conditions for the minimum problem are natural. To obtain the solution of the inhomogeneous equation with the Dirichlet boundary condition $u = f$ on ∂G, we must proceed as in Exercise 8.1.18 except that all admissible functions must equal f on ∂G. [Hint: Let $w = u + \epsilon W$ where u is the (assumed) solution of the problem.]

Section 8.2

8.2.1. The *Gram–Schmidt Process.* Let (φ, ψ) represent a (weighted) inner product for either vectors or scalars. Assume the set of functions φ_k ($k = 1, 2, \ldots$) is linearly independent. Show that if $\|\varphi\|^2 = (\varphi, \varphi)$, the set of functions $\hat{\varphi}_k$ formed as follows,

$$\hat{\varphi}_1 = \frac{\varphi_1}{\|\varphi_1\|}; \qquad \tilde{\varphi}_2 = \varphi_2 - (\varphi_2, \hat{\varphi}_1)\hat{\varphi}_1,$$

$$\hat{\varphi}_2 = \frac{\tilde{\varphi}_2}{\|\tilde{\varphi}_2\|}; \qquad \tilde{\varphi}_3 = \varphi_3 - (\varphi_3, \hat{\varphi}_1)\hat{\varphi}_1 - (\varphi_3, \hat{\varphi}_2)\hat{\varphi}_2,$$

$$\hat{\varphi}_3 = \frac{\tilde{\varphi}_3}{\|\tilde{\varphi}_3\|}; \qquad \tilde{\varphi}_4 = \varphi_4 - (\varphi_4, \hat{\varphi}_1)\hat{\varphi}_1 - (\varphi_4, \hat{\varphi}_2)\hat{\varphi}_2 - (\varphi_4, \hat{\varphi}_3)\hat{\varphi}_3,$$

and so on, is an orthonormal set.

8.2.2. Let $(\varphi, \psi) = \int_0^1 \varphi(x)\psi(x)\,dx$. Let $\varphi_1(x) = 1$, $\varphi_2(x) = x$, $\varphi_3(x) = x^2$, and $\varphi_4(x) = x^3$. Construct an orthonormal set $\hat{\varphi}_1$, $\hat{\varphi}_2$, $\hat{\varphi}_3$, and $\hat{\varphi}_4$ using the Gram–Schmidt process.

8.2.3. Given the three (independent) row vectors

$$\boldsymbol{\varphi}_1^T = [1, 2, -1]; \qquad \boldsymbol{\varphi}_2^T = [0, 1, 3]; \qquad \boldsymbol{\varphi}_3^T = [-1, 1, 0],$$

form an orthonormal set of vectors $\hat{\boldsymbol{\varphi}}_1$, $\hat{\boldsymbol{\varphi}}_2$, and $\hat{\boldsymbol{\varphi}}_3$ from the preceding set using the Gram–Schmidt method and the dot product of the vectors as an inner product.

8.2.4. Conclude from the fact that $E(w)$ with $q > 0$ and $\|w\|$ are both positive if w vanishes on ∂G and is not identically zero, that the matrices A and B defined by (8.96)–(8.97) must be positive definite in view of (8.98).

8.2.5. Separate variables in Eq. (8.120) for the spherical harmonics with $k^2 = n(n+1)$, and let $Y_n(\theta, \varphi) = H(\varphi)G(\theta)$ to obtain the equations

$$\frac{1}{\sin \varphi} \frac{d}{d\varphi} \left[\sin \varphi \frac{dH}{d\varphi} \right] + \left[n(n+1) - \frac{\mu^2}{\sin^2 \varphi} \right] H = 0,$$

$$G''(\theta) + \mu^2 G(\theta) = 0$$

where μ^2 is the separation constant. Conclude, since $G(\theta)$ is periodic of period 2π, that $\mu = m =$ integer so that the $G(\theta)$ are trigonometric. Let $t = \cos \varphi$ in the equation for $H(\varphi)$ and obtain the *associated Legendre equation*

$$\frac{d}{dt} \left[(1 - t^2) \hat{H}'(t) \right] + \left[n(n+1) - \frac{m^2}{1 - t^2} \right] \hat{H}(t) = 0$$

in the interval $-1 < t < 1$ where $\hat{H}(t) = \hat{H}(\cos \varphi) = H(\varphi)$. The solutions $H(\varphi)$ must be bounded at $\varphi = 0$ and $\varphi = \pi$. Consequently, $\hat{H}(t)$ must be bounded at $t = \pm 1$. If $m = 0$, the bounded solutions are the *Legendre polynomials* $P_n(\cos \varphi)$ discussed previously. If $m \ne 0$, bounded solutions are obtained only if $m \le n$ and are given as the *associated Legendre functions* $P_n^m(\cos \varphi)$. The functions $P_n^m(t)$ can be defined in terms of the Legendre polynomials $P_n(t)$ as

$$P_n^m(t) = (-1)^m (1 - t^2)^{m/2} \frac{d^m}{dt^m} [P_n(t)].$$

8.2.6. Differentiate the Legendre equation

$$\frac{d}{dt} \left[(1 - t^2) P_n'(t) \right] + n(n+1) P_n(t) = 0,$$

m times and show that the function $H(t) = (1 - t^2)^{m/2} (d^m/dt^m)[P_n(t)]$ is a solution of the associated Legendre equation of Exercise 8.2.5. Determine that $P_n^m(t) = 0$ if $m > n$ from its definition in terms of $P_n(t)$, and show that $P_n^m(t)$ is bounded at $t = \pm 1$.

8.2.7. Determine the coefficients c_1 and c_2 corresponding to each eigenvalue $\hat{\lambda}_1$ and $\hat{\lambda}_2$ for the problem of Example 8.4 and show that the resulting functions are orthogonal.

8.2.8. Use the Rayleigh–Ritz method to approximate the leading eigenvalue for the following problem:

$$\nabla^2 M + \lambda M = 0, \qquad x^2 + y^2 < 1;$$
$$M = 0, \qquad x^2 + y^2 = 1.$$

Use the approximating function $\varphi_1 = \cos(\pi r/2)$, where $r^2 = x^2 + y^2$. Compare the result with that given in Example 8.2.

8.2.9. Construct a function $\varphi_1(x, y)$ that vanishes on the triangle given in Exercise 8.1.16 and use the Rayleigh–Ritz method to approximate the leading eigenvalue for the problem given in that exercise.

8.2.10. Approximate the lowest eigenvalue for the problem in Example 8.4 by using only the function $\varphi_1(x)$ as defined in (8.151). Compare the approximate result with that obtained in Example 8.4.

8.2.11. Using the appropriate Rayleigh quotient, approximate the first eigenvalue for the following problem:

$$M''(x) + \lambda M(x) = 0, \qquad 0 < x < 1;$$
$$M(0) = 0, \qquad M'(1) + M(1) = 0.$$

Let $\varphi_1(x) = 3x - 2x^2$.

8.2.12. Solve the problem of Example 8.5 for the leading eigenvalue by letting $w = c_1\varphi_1(x) + c_2\varphi_2(x)$ where $\varphi_1(x)$ is given by (8.159) and $\varphi_2(x) = x\cos(\frac{1}{2}x)$.

8.2.13. Explain why if the approximation functions $\varphi_k(x)$ in (8.94) are chosen to be elements of a complete set $\{\varphi_k\}$, the more terms one takes in the series (8.94), the better the approximation to the leading eigenvalue and eigenfunction.

8.2.14. Use the Rayleigh–Ritz method to approximate the leading eigenvalue for the problem in Exercise 8.1.10. Let $\varphi_1(x) = (x - 1)(x - 2)$.

8.2.15. Show how the Rayleigh–Ritz method can be applied to the variational problems given in Exercise 8.1.18. [Hint: Express $w(x)$ as in (8.94) with w required to vanish on ∂G for the Dirichlet problem, but arbitrary on ∂G for the other boundary value problems.]

8.2.16. Use the Rayleigh–Ritz method (see Exercise 8.2.15) to approximately solve the following Dirichlet problem for Poisson's equation within the square.

$$\nabla^2 u = -\sin x, \qquad 0 < x < \pi, \;\; 0 < y < \pi;$$
$$u(x, y) = 0 \text{ on the boundary.}$$

Use a one term Rayleigh–Ritz approximation.

8.2.17. Show how the Rayleigh–Ritz method can be applied to the variational problems of Exercise 8.1.19. (Hint: See Exercise 8.2.15 and for the Dirichlet problem, let $w = \varphi_0 + \Sigma c_k \varphi_k$ where $\varphi_0 = f$ on ∂G and the φ_k vanish on ∂G.)

8.2.18. Solve the Dirichlet problem,

$$\nabla^2 M = -1, \qquad x^2 + y^2 < 1;$$

$$M = x^2, \qquad x^2 + y^2 = 1,$$

using the Rayleigh–Ritz method [see Exercise 8.2.17 and let $\varphi_0 = x^2$ and $\varphi_1 = \cos(\pi r/2)$ as in Exercise 8.2.8].

Section 8.3

8.3.1. Verify the result (8.186).

8.3.2. Determine the adjoint operators for the following operators:
   (a) $M[u] = e^x u_{xx} + x^2 u_{xy} + yu_y - 10u$.
   (b) $M[u] = a(x, y)u_x + b(x, y)u_y + c(x, y)u$.
   (c) $M[u] = u_t - c^2[u_{xx} + u_{yy}]$.

8.3.3. Obtain an adjoint operator M^* for the third order operator M given as

$$M[u] = u_t - u_x + \gamma^2 u_{xxx},$$

such that $wM[u] - uM^*[w\}$ in a divergence expression.

8.3.4. Show that the operator

$$M[u] = u_{tt} + u_{xxxx}$$

is self-adjoint and express $wM[u] - uM[w]$ in divergence form.

8.3.5. Demonstrate that

$$M[u] = u_{tt} + \nabla^2\nabla^2 u$$

is self-adjoint and express $wM[u] - uM[w]$ in divergence form.

8.3.6. Show that (8.211) reduces to d'Alembert's solution of the wave equation if R is given by (8.215) and $\gamma =$ constant.

8.3.7. Verify that (8.2.11) and (7.353) are identical if R is given by (8.217) and $c = -\hat{c}^2$.

8.3.8. Obtain the Riemann function for (8.216) with $c > 0$ (i.e., the Klein–Gordon equation). Show that the solution formula for this problem agrees with (7.351).

Section 8.4

8.4.1. Use the maximum principle to show that the solution of the initial and boundary value problem for the one-dimensional heat equation is unique. That is, show that

$$u_t - c^2 u_{xx} = F(x, t), \qquad 0 < x < l, t > 0;$$
$$u(x,0) = f(x), \qquad 0 < x < l;$$
$$u(0, t) = g_1(t); \qquad u(l, t) = g_2(t); \qquad t > 0,$$

has a unique continuous solution. (Hint: Assume there are two solutions u_1 and u_2 and consider the function $\hat{u} = u_1 - u_2$.)

8.4.2. Show that if $u_1(x, t)$ and $u_2(x, t)$ are solutions of the heat equation $u_t = c^2 u_{xx}$ whose initial and boundary data at $x = 0$, $t = 0$, and $x = l$ satisfy the inequality

$$|u_1(x, t) - u_2(x, t)| < \epsilon, \qquad x = 0, \qquad t = 0, \qquad x = l,$$

then we must have $|u_1 - u_2| < \epsilon$ for all (x, t) in $0 \leqslant x \leqslant l, t \geqslant 0$. This result proves that solutions of the first boundary value problem for the heat equation depend continuously on the data. (Hint: Consider the solutions $v_1 = -\epsilon$, $u = u_1 - u_2$ and $v_2 = \epsilon$ of the heat equation and use the maximum principle to conclude that $v_1 \leqslant u \leqslant v_2$ by examining these functions two at a time.)

8.4.3. Prove that the Cauchy problem for the heat equation

$$u_t - c^2 u_{xx} = F(x, t), \qquad -\infty < x < \infty, t > 0$$
$$u(x,0) = f(x), \qquad -\infty < x < \infty$$
$$|u(x, t)| < M, \qquad -\infty < x < \infty, t \geqslant 0$$

has a unique solution. Assume the problem has two solutions $u_1(x, t)$ and $u_2(x, t)$ and consider the difference $\hat{u} = u_1 - u_2$. Consider the interval $-l \leqslant x \leqslant l$ and the function $v(x, t) = (4M/l^2)(c^2 t + x^2/2)$ which is a solution of the (homogeneous) heat equation. Show that $|\hat{u}(x, t)| \leqslant v(x, t)$ on the boundary $x = -l$, $t = 0$, $x = l$ and conclude from the maximum principle in the bounded interval $|x| \leqslant l$ that $|\hat{u}| \leqslant v$ in $|x| \leqslant l$, and $t \geqslant 0$. By letting $l \to \infty$, conclude that $\hat{u}(x, t) = 0$.

8.4.4. Determine a maximum and minimum principle for the heat equation

$$u_t - c^2 \nabla^2 u = 0$$

in a bounded region in two or three dimensions. [Hint: Proceed as in the one-dimensional case and replace w in (8.221) by $w = u + \epsilon r^2$ where $r^2 =$

$x^2 + y^2$ and $r^2 = x^2 + y^2 + z^2$ in two and three dimensions, respectively. Note that the bounded region may be assumed to be contained in a circle or sphere of sufficiently large radius.]

8.4.5. Obtain a maximum principle for the parabolic equation

$$\rho u_t - \nabla \cdot (p\nabla u) = 0$$

where $\rho > 0$ and $p > 0$ are bounded in any bounded region. [Hint: Let $w = u + \epsilon \exp[ax^2]$ and then choose a such that $\rho w_t - \nabla \cdot (p\nabla w) < 0$. Conclude that w must have its maximum at $t = 0$ or at the boundary of the given region for $t > 0$. Proceed as in the one-dimensional case in the text.]

8.4.6. Use the maximum principle of Exercise 8.4.5 to prove that the first boundary and initial value problem for $\rho u_t - \nabla \cdot (p\nabla u) = \rho F$ has a unique solution.

8.4.7. Consider the elliptic equation

$$\nabla \cdot (p\nabla u) = -F, \qquad p > 0$$

in two or three dimensions in a bounded region G with the boundary ∂G. Show that if $F < 0$ in G, the solution u assumes its maximum on ∂G and if $F > 0$ in G, the solution u assumes its minimum on ∂G. (Hint: The first partial derivatives of u vanish at an interior maximum point.)

8.4.8. Obtain a maximum principle for the elliptic equation

$$\nabla \cdot (p\nabla u) = 0$$

with $p > 0$, in the two and three-dimensional cases. (Hint: Proceed as in Exercise 8.4.5 and use Exercise 8.4.7.)

8.4.9. Use the maximum principle to prove continuous dependence on the data for the Dirichlet problem for the elliptic equation

$$\nabla \cdot (p\nabla u) = 0, \qquad p > 0,$$

in a bounded region. (Hint: Adapt the method of Exercise 8.4.2.)

8.4.10. Consider the elliptic equation

$$-\nabla \cdot (p\nabla u) + qu = 0$$

in a bounded region G where $p > 0$ and $q > 0$. Show that $u \leq 0$ at an interior maximum point and that $u \geq 0$ at an interior minimum point. Conclude from this that the Dirichlet problem for this equation with $u = f$ on ∂G must have a unique solution.

8.4.11. Let $u(x, y, z)$ satisfy $\nabla^2 u = 0$ in G, and let P_0 and the sphere S_0 centered at P_0 with radius a lie completely within G. Then the mean value theorem for harmonic functions states (see Example 7.14)

$$u(P_0) = \frac{1}{4\pi a^2} \iint\limits_{S_0} u \, ds,$$

with the integration taken over the sphere S_0. Assuming u is continuous up to ∂G so that it has a maximum in the closed region, use the mean value theorem to show that the maximum is achieved on the boundary ∂G. [Hint: Assume that the maximum point M_0 is interior to G. Apply the mean value principle to all spheres centered at M_0 extending up to the boundary ∂G. Since $u(M_0) \geqslant u(P)$ for each P in G or ∂G by assumption, conclude that if $u(P) < u(M_0)$, we would be led to the contradiction that $u(M_0) < u(M_0)$. This argument can be extended to conclude that if u does attain its maximum in G, it must be a constant function.]

8.4.12. Show that the solution $u(x, t)$ of the Cauchy problem for the wave equation

$$u_t - c^2 u_{xx} = 0, \qquad -\infty < x < \infty, t > 0$$
$$u(x,0) = f(x); \qquad u_t(x,0) = g(x); \qquad -\infty < x < \infty$$

is nonnegative if $f(x) \geqslant 0$ and $g(x) \geqslant 0$. (Hint: Use d'Alembert's solution.)

8.4.13. Use the solution formula (7.353) for the Cauchy problem for the modified telegrapher's equation (7.352) and the properties of the modified Bessel functions $I_0(z)$ and $I_1(z)$ to show that if the data $f(x)$, $g(x)$, and $F(x, t)$ are nonnegative, then so is $u(x, t)$. Show how this relates to the result for the telegrapher's equation given in the text.

Section 8.5

8.5.1. Show that if $f(x) = \delta(x)$ (the Dirac delta function) and $g(x) = 0$ in (8.254), the asymptotic solution (8.268) of (8.248) takes the form

$$u(x, t) = \frac{1}{\sqrt{4\pi ct}} \cos\left[\frac{x^2}{4ct} - \frac{\pi}{4} \right]$$

and this is the exact solution of the initial value problem.

8.5.2. Use the result of Exercise 8.5.1 to show that the solution of (8.248) with the data $u(x,0) = f(x)$, $u_t(x,0) = 0$ is

$$u(x, t) = \frac{1}{\sqrt{4\pi ct}} \int_{-\infty}^{\infty} f(s)\cos\left[\frac{(x-s)^2}{4ct} - \frac{\pi}{4} \right] ds.$$

8.5.3. A rod is said to be simply supported at $x = 0$ if $u(0, t) = u_{xx}(0, t) = 0$. Use the Fourier sine transform to solve the following problem for the vibration of a semi-infinite rod.

$$u_{tt} + c^2 u_{xxxx} = 0; \qquad\qquad 0 < x < \infty, t > 0,$$
$$u(x,0) = f(x); \qquad\qquad u_t(x,0) = g(x); \qquad 0 < x < \infty,$$
$$u(0, t) = u_{xx}(0, t) = 0; \qquad t > 0.$$

8.5.4. Apply the Fourier cosine transform to solve the following problem.

$$u_{tt} + c^2 u_{xxxx} = 0; \qquad\qquad x > 0, t > 0,$$
$$u(x,0) = f(x); \qquad\qquad u_t(x,0) = g(x); \qquad x > 0,$$
$$u_x(0, t) = u_{xxx}(0, t) = 0; \qquad t > 0.$$

8.5.5. The vibration of a finite rod simply supported at both ends is determined by the solution of the following problem:

$$u_{tt} + c^2 u_{xxxx} = 0; \qquad\qquad\qquad 0 < x < l, t > 0,$$
$$u(x,0) = f(x); \; u_t(x,0) = g(x); \qquad\qquad 0 < x < l,$$
$$u(0, t) = u_{xx}(0, t) = u(l, t) = u_{xx}(l, t) = 0; \qquad t > 0.$$

Obtain the solution by using separation of variables.

8.5.6. A rod is free or unsupported at a point if u_{xx} and u_{xxx} vanish there. Use separation of variables to solve the following problem for the vibration of a rod clamped at one end and free at the other end.

$$u_{tt} + c^2 u_{xxxx} = 0; \qquad 0 < x < l, t > 0$$
$$u(x,0) = f(x); \qquad u_t(x,0) = g(x); \qquad 0 < x < l,$$
$$u(0, t) = u_x(0, t) = u_{xx}(l, t) = u_{xxx}(l, t) = 0; \qquad t > 0.$$

8.5.7. Consider the eigenvalue problem,

$$v''''(x) + \lambda^2 v(x) = 0, \qquad 0 < x < l$$

with the following boundary conditions:

(a) $v(0) = v'(0) = v(l) = v'(l) = 0.$
(b) $v(0) = v''(0) = v(l) = v''(l) = 0.$
(c) $v'(0) = v'''(0) = v'(l) = v'''(l) = 0.$
(d) $v(0) = v'(0) = v''(l) = v'''(l) = 0.$

Let λ_n and λ_m be distinct eigenvalues for each of these problems and $v_n(x)$ and $v_m(x)$ the corresponding eigenfunctions. Use the method of Section 4.3 to show that the eigenfunctions are orthogonal with respect to the following inner product,

$$\int_0^l v_n(x) v_m(x) \, dx = 0, \qquad n \neq m.$$

8.5.8. Construct the Green's function $K(x; \xi)$ for the following problems:

$$\frac{\partial^4 K(x; \xi)}{\partial x^4} = -\delta(x - \xi), \qquad 0 < x, \xi < l$$

where $K(x; \xi)$ satisfies the boundary conditions (a), (b), (c), or (d) given in Exercise 8.5.7. (Hint: K satisfies a homogeneous equation at $x \neq \xi$ and the jump in $\partial^3 K/\partial x^3$ at $x = \xi$ is -1.)

8.5.9. Show how the finite sine transform method can be used to solve the problem in Exercise 8.5.5 if the equation and the boundary data are inhomogeneous.

8.5.10. A simply supported plate satisfies the boundary conditions $u = 0$ and $\partial^2 u/\partial n^2 = 0$ at the edge. (Here $\partial^2 u/\partial n^2$ is the second exterior normal derivative.) Use separation of variables to solve the problem of the vibrating simply supported rectangular plate. That is, solve the following problem for $u(x, y, t)$.

$$u_{tt} + c^2 \nabla^2 \nabla^2 u = 0; \qquad 0 < x < l, 0 < y < \hat{l}, t > 0,$$
$$u(x, y, 0) = f(x, y); \qquad u_t(x, y, 0) = g(x, y),$$
$$u(0, y) = u_{xx}(0, y) = u(l, y) = u_{xx}(l, y) = 0; \qquad 0 < y < \hat{l},$$
$$u(x, 0) = u_{yy}(x, 0) = u(x, \hat{l}) = u_{yy}(x, \hat{l}) = 0; \qquad 0 < x < l.$$

{Hint: Obtain (8.296) and set $A_n = \sin[(\pi n/l)x]$ to satisfy the conditions at $x = 0$ and $x = l$. Conclude that $B_m = \sin[(\pi m/\hat{l})y]$, and obtain the solution in terms of a double Fourier sine series.}

8.5.11. Use the Fourier transform to solve the Cauchy problem for the plate equation:

$$u_{tt} + c^2 \nabla^2 \nabla^2 u = 0, \qquad -\infty < x, y < \infty, t > 0$$
$$u(x, y, 0) = f(x, y), \qquad u_t(x, y, 0) = g(x, y)$$

with appropriate conditions at infinity.

8.5.12. Use the Green's theorem (8.322) to show that eigenfunctions corresponding to different eigenvalues of the problem (8.292)–(8.293) are orthogonal.

8.5.13. Verify that (8.310) satisfies the biharmonic equation.

8.5.14. Solve the boundary value problem

$$\nabla^2\nabla^2 u = 0, \qquad x^2 + y^2 < R^2,$$

$$u = f; \qquad \frac{\partial u}{\partial r} = g; \qquad x^2 + y^2 = R^2,$$

by using a Fourier series in the angular variable θ.

8.5.15. Solve the problem of Exercise 8.5.14 using the method presented in the text if the boundary conditions are

$$f = 1; \qquad g = \sin^3\theta.$$

8.5.16. Show that (8.319) satisfies the biharmonic equation if $r \neq 0$.

8.5.17. Solve the following problem by looking for a solution of the form $u = u(r)$:

$$\nabla^2\nabla^2 u = F_0, \qquad\qquad x^2 + y^2 < R^2$$

$$u = \frac{\partial u}{\partial n} = 0, \qquad x^2 + y^2 = R^2$$

where F_0 is a constant.

8.5.18. Obtain the solution of the problem

$$\nabla^2\nabla^2 u = -\delta(x)\delta(y), \qquad x^2 + y^2 < R^2$$

$$u = \frac{\partial u}{\partial n} = 0, \qquad\qquad x^2 + y^2 = R^2$$

by looking for a solution in the form

$$u = -\frac{1}{8\pi}r^2\log r + v(r)$$

where $r^2 = x^2 + y^2$.

8.5.19. Use the theory of Section 3.3 to show that Euler's equations (8.363)–(8.364) for inviscid, isentropic flow are of hyperbolic type and obtain (8.376) as the equations for the characteristic curves.

8.5.20. Multiply (8.363) by $c(\rho)$, add (8.364) to it, and then subtract (8.364) from it to obtain

$$c\big[\rho_t + (u \pm c)\rho_x\big] \pm \rho\big[u_t + (u \pm c)u_x\big] = 0.$$

Conclude from these equations that

$$c\frac{d\rho}{dt} + \rho\frac{du}{dt} = 0 \quad \text{on} \quad \frac{dx}{dt} = u + c$$

$$c\frac{d\rho}{dt} - \rho\frac{du}{dt} = 0 \quad \text{on} \quad \frac{dx}{dt} = u - c$$

where $d/dt = \partial/\partial t + (dx/dt)\,\partial/\partial x$. Since $c = c(\rho)$, show that these equations can be written as

$$\int^{\rho}\frac{c(\rho)}{\rho}\,d\rho + u = \text{constant} \quad \text{on} \quad \frac{dx}{dt} = u + c$$

$$\int^{\rho}\frac{c(\rho)}{\rho}\,d\rho - u = \text{constant} \quad \text{on} \quad \frac{dx}{dt} = u - c.$$

They are known as the *Riemann invariants*.

8.5.21. Show that the method of the text for determining simple waves leads to the same results as obtained in Exercise 8.5.20.

8.5.22. Define the *Hertz vector* $\boldsymbol{\pi}$ in terms of the scalar and vector potentials φ and $\mathbf{A}$ for Maxwell's equations as

$$\varphi = -\nabla \cdot \boldsymbol{\pi}; \qquad \mathbf{A} = \frac{1}{c^2}\frac{\partial\boldsymbol{\pi}}{\partial t}$$

where $\epsilon\mu = 1/c^2$. Show that the Lorentz condition (8.430) is thereby satisfied. Express $\mathbf{E}$ and $\mathbf{H}$ in terms of $\boldsymbol{\pi}$, and if $\rho = \sigma = 0$ show that $\boldsymbol{\pi}$ satisfies the equation

$$\boldsymbol{\pi}_{tt} - c^2\nabla^2\boldsymbol{\pi} = \mathbf{0}.$$

8.5.23. Determine the form taken by Maxwell's equations if $\rho = \sigma = 0$, ϵ, μ are constant, and $\mathbf{E}$ and $\mathbf{H}$ are given as

$$\mathbf{E} = \mathbf{e}e^{-i\omega t}; \qquad \mathbf{H} = \mathbf{h}e^{-i\omega t},$$

where $\mathbf{e}$ and $\mathbf{h}$ depend only on the spatial variables and ω is a constant.

8.5.24. Verify that $\nabla \cdot \mathbf{F} = \nabla \cdot \mathbf{G} = 0$ for the data (8.448) and show that the solution (8.454)–(8.455) also satisfies $\nabla \cdot \mathbf{E} = \nabla \cdot \mathbf{H} = 0$.

8.5.25. Formulate Exercise 8.5.22 for the Hertz vector if the problem depends only on x and t. Assume the Hertz vector $\boldsymbol{\pi}(x, t)$ is given initially as

$$\boldsymbol{\pi}(x,0) = (\cos x)\mathbf{i} + \mathbf{j}; \qquad \boldsymbol{\pi}_t(x,0) = \mathbf{k}; \qquad -\infty < x < \infty.$$

Determine $\pi(x, t)$, $E(x, t)$, and $H(x, t)$ for $t > 0$ by solving the Cauchy problem for $\pi(x, t)$.

8.5.26. Let the Hertz vector π be given as $\pi = \alpha e^{-i\omega t}$ where α depends only on the spatial variables (see Exercise 8.5.22). Assuming the conditions given in Exercise 8.5.23 apply, show that

$$\nabla^2\alpha + k^2\alpha = 0$$

where $k = \omega/c$ and

$$e = \nabla(\nabla \cdot \alpha) + k^2\alpha,$$

$$h = -\frac{ik}{\mu}\nabla \times \alpha.$$

8.5.27. To study the possibility that electromagnetic waves can propagate in a waveguide, we consider Maxwell's equations in a cylindrical region whose walls are perfectly conducting. The interior of the guide is such that ϵ and μ are constant and $\sigma = 0$. In addition, there are no sources so that $\rho = 0$ and E and H are assumed to have harmonic time dependence as in Exercise 8.5.23. The tangential component of E vanishes on the waveguide wall. Assume the cylinder has its generators parallel to the z-axis and show that if we set the Hertz vector π equal to

$$\pi = \alpha e^{-i\omega t} = ue^{-i\omega t}k,$$

the tangential component of E vanishes if

$$u = 0 \text{ on the cylinder wall}$$

(use Exercise 8.5.26). Also, conclude that u satisfies the reduced wave equation

$$\nabla^2 u + k^2 u = 0, \qquad k = \omega/c.$$

Let $u(x, y, z) = M(x, y)v(z)$ and determine $M(x, y)$ from the eigenvalue problem

$$\nabla^2 M + \lambda M = 0, \qquad (x, y) \in G$$
$$M(x, y) = 0, \qquad (x, y) \in \partial G$$

where G is a perpendicular cross-section of the cylinder. Let λ_n and $M_n(x, y)$ be the eigenvalues and eigenfunctions for this problem with $\lambda_1 \leqslant \lambda_2 \leqslant \cdots$. Then $v_n(z)$ must satisfy

$$v_n''(z) + (k^2 - \lambda_n)v_n(z) = 0.$$

Considering only waves traveling in the positive z-direction, show that the Hertz vectors $\boldsymbol{\pi}_n(x, y, z, t)$ must take the form

$$\boldsymbol{\pi}_n = a_n M_n(x, y)\exp[i(\sigma_n z - \omega t)]\mathbf{k},$$

where $\sigma_n = \sqrt{k^2 - \lambda_n}$ and a_n is a constant. Determine that no traveling waves exist in the waveguide if $k^2 < \lambda_1$ and that only a finite number of such waves exist if $k^2 < \lambda_n$. (With $k^2 < \lambda_n$, the field $\boldsymbol{\pi}_n$ decays as $z \to \infty$.)

8.5.28. Since $k = \omega/c$, show that if $\omega^2 < \lambda_1 c^2$ for a particular waveguide where λ_1 is determined as in Exercise 8.5.27, no traveling waves can exist in the waveguide. Thus $\omega = \sqrt{\lambda_1} c$ is termed a cutoff frequency. Determine the cutoff frequencies for the waveguides with the following cross-sections:

(a) $0 \leqslant x \leqslant l, \quad 0 \leqslant y \leqslant \hat{l}$.

(b) $x^2 + y^2 \leqslant a^2$.

8.5.29. Solve the initial value problem for the one-dimensional Navier equation (8.459) with $\mathbf{F} = \mathbf{0}$ and the initial data

$$\mathbf{u}(x, 0) = (\cos x)\mathbf{i} + \mathbf{j} + (\sin x)\mathbf{k},$$
$$\mathbf{u}_t(x, 0) = x\mathbf{i} + e^{-x}\mathbf{j} + \mathbf{k}.$$

8.5.30. Show that if the equation (8.466) is solved for $\nabla \cdot \mathbf{u}$ we may obtain $\mathbf{u}$ from (8.459) by treating the term involving $\nabla \cdot \mathbf{u}$ as an additional inhomogeneous term. Indicate how we may express the solution of the initial value problem for (8.459) with $\mathbf{u}(x, y, z, 0)$ and $\mathbf{u}_t(x, y, z, 0)$ specified, using the general solution formula for the wave equation.

9

PERTURBATION AND ASYMPTOTIC METHODS

9.1. INTRODUCTION

This concluding chapter deals with a collection of methods that yield approximate solutions for a large class of initial and boundary value problems for partial differential equations. Generally speaking, these methods are used when a small parameter (or a large parameter) occurs in the given equation or data for the problem. Then the (assumed) solution is expanded in a series of powers (or inverse powers) of the parameter and this expansion is inserted into the equation and data for the problem. By equating like powers of the parameter, a collection of problems results whose solution is expected to be simpler than that of the given problem.

If the series expansion of the solution converges, or is expected to converge, the aforementioned technique is often referred to as a *perturbation method*. If the series is divergent but asymptotic, so that the first few terms yield a good approximation when extreme values of the parameter are considered, the above technique is called an *asymptotic method*. The terminology is, however, not uniform in the literature and what are termed perturbation series may, in fact, be asymptotic and vice versa.

In general, it is difficult to specify all the terms in a perturbation or asymptotic series for problems involving partial differential equations. Thus only the first few terms in the series are determined and the distinction between a convergent or asymptotic series often becomes irrelevant. Although for many problems precise results are available regarding the convergent or asymptotic nature of the result, we shall concentrate on the formal aspects of constructing the series solutions and will generally only obtain the first few terms of the series.

The appropriate expansion forms for the solution of a given problem are by no means obvious in all cases. Consequently, it is often useful to study exact solutions of particular problems and expand these solutions in a series involving the relevant small or large parameter. This not only suggests appropriate expansion forms for more general problems but may also yield useful information for specifying undetermined quantities that occur in the general case. For example, in our study of the Klein–Gordon equation in Example 5.13 as a model for dispersive wave motion, it was shown that the solution for large x and t has a specific asymptotic form. By inserting that form directly into the equation it can be shown that the results of that example can be reproduced apart from an unspecified constant. In addition, related dispersive equations can be solved by using similar expansion forms. Although no large or small parameter occurred explicitly in the Klein–Gordon equation of Example 5.13, by rescaling the x and t variables such that large values are emphasized, say, $\hat{x} = kx$, $\hat{t} = kt$ with $k \gg 1$, a large parameter k can be introduced into the equation.

Perturbation techniques can also be used to replace given equations by simpler ones whose solutions contain many of the features of the solutions of the original problem. This is especially important for nonlinear equations where perturbation methods are used to linearize the problem, as has been done for Euler's equation of fluid dynamics in Example 8.10, for instance. Even if the linearization procedure breaks down in certain regions, it may still be possible to replace the given equation or system by a simpler nonlinear equation. The equation of simple wave motion that was obtained as an approximation to Euler's hydrodynamic equations is representative of this approach. This idea will be developed further in this chapter.

In our study of hyperbolic equations we have seen that discontinuities or singularities in the solutions, interpreted in the weak sense, occur across characteristics. In fact, any rapid variation in the data for these equations must be carried along the characteristics. Near the characteristics, the solutions may be described by means of series expansions which may contain a small parameter as we will show. These results are closely related to the asymptotic expansions that will be given for the Helmholtz or reduced wave equation.

It will be seen that more than one type of expansion may be necessary to completely describe the perturbation or asymptotic solution of a given problem. For example, an expansion may break down in some region or may be insufficient to satisfy the data for the problem. These difficulties signify that the given expansion is not uniformly valid over the entire region of interest. Techniques such as the *boundary layer method* or the *method of multiple scales* need to be used to remedy this problem.

Even though asymptotic equalities, asymptotic expansions, and order of magnitude symbols have often been defined and used in the preceding chapters, we now redefine these concepts in view of their particular relevance for the material in this chapter.

The order symbol O is defined as

$$(9.1) \qquad\qquad F(x) = O[G(x)]; \qquad x \to a$$

if $|F(x)/G(x)| \to A$ as $x \to a$, where A is some constant (here a may be $\pm \infty$). If (9.1) is valid, we say that F is of order G near $x = a$.

The function $f(x)$, which depends on the parameter k, has the asymptotic (power) series

$$(9.2) \qquad\qquad f(x) \approx \sum_{n=0}^{\infty} f_n(x) k^{-n}$$

valid as $k \to \infty$, if for each N we have

$$(9.3) \qquad f(x) = \sum_{n=0}^{N-1} f_n(x) k^{-n} + O[k^{-N}], \qquad k \to \infty.$$

The result is assumed to be uniformly valid for all x in some region. (Note that with $\epsilon = 1/k$, we have an asymptotic power series in ϵ, valid as $\epsilon \to 0$ through positive values.)

The order symbol o is defined as

$$(9.4) \qquad\qquad F(x) = o[G(x)], \qquad x \to a$$

if $F(x)/G(x) \to 0$ as $x \to a$. Thus we can write (9.3) as

$$(9.5) \qquad f(x) = \sum_{n=0}^{N-1} f_n(x) k^{-n} + o[k^{-N+1}], \qquad k \to \infty.$$

We shall emphasize the use of the (large) O order symbol in our discussion. The basic feature of an asymptotic series is that the remainder is of lower order than the last term retained. The full series may converge and convergence is not necessary for the result to be useful.

Often more general forms of asymptotic expansions are required, such as series in fractional powers of k, and these are also defined as having remainders of lower order than the last term retained.

As stated previously, we do not prove that the formal series solutions we obtain are either convergent or asymptotic. Occasionally, therefore, strict equality rather than asymptotic equality signs are used even though the full series may diverge. Since only a few terms in the series are found, this distinction is not always significant. It is a general property of asymptotic series that finding additional terms need not improve the approximation since the series is generally divergent. As the parameter k is (9.2) increases, the approximation provided by the series gets better. Thus we are effectively assuming in our discussion that the parameter is very large (or very small if $\epsilon = 1/k$) since we obtain only the leading terms. However, it is often the case that the results are very good even for moderate values of the parameter. Finally, we note that all formal operations that we carry out on the asymptotic series, such as term by term differentiation and integration, are assumed to be valid.

9.2. REGULAR PERTURBATION METHODS

We consider a linear or nonlinear differential equation

$$(9.6) \qquad\qquad L(u, \epsilon) = 0$$

that depends (smoothly) on the small positive parameter ϵ and a problem for (9.6) given over a bounded or unbounded spatial region G. If (9.6) is of elliptic type, appropriate boundary conditions are assigned on ∂G or at infinity. If (9.6) is of hyperbolic or parabolic type, in addition to the boundary conditions assigned on ∂G or at infinity for all $t > 0$, initial data are given in G at the time $t = 0$. The boundary or initial data may depend on ϵ, but the boundary ∂G is for the present assumed to be specified independently of ϵ.

The *reduced* or *unperturbed problem* associated with the problem for (9.6) is obtained on formally setting $\epsilon = 0$ in (9.6) and its data. That is, we consider the equation

$$(9.7) \qquad\qquad L(v, 0) = 0$$

with the *reduced data* obtained from the data for the given problem for (9.6). If the reduced problem has a unique solution, then the given problem is called a *regular perturbation problem*. If this is not the case, we have a *singular perturbation problem*. Problems of the latter type are studied in Section 9.3.

Generally speaking, if the reduced equation is of different type or order than the given equation, we have a singular perturbation problem. It may happen, however, that the reduced problem can be solved even if the order or type of the given equation is changed.

For example, the *signaling problem* for the *hyperbolic equation*

$$(9.8) \qquad \epsilon u_{tt} - c^2 u_{xx} + u_t = 0, \qquad x > 0, -\infty < t < \infty$$

with the boundary condition

$$(9.9) \qquad\qquad u(0, t) = f(t), \qquad -\infty < t < \infty$$

reduces to the *parabolic problem*

$$(9.10) \qquad c^2 v_{xx} - v_t = 0, \qquad x > 0, -\infty < t < \infty$$

with the boundary condition

$$(9.11) \qquad\qquad v(0, t) = f(t), \qquad -\infty < t < \infty.$$

Both the given and reduced problem can be solved in this case. However, the signaling problem should be interpreted as a large time limit of an initial and boundary value problem with zero initial data and in that case the initial value problem for (9.8) has an excess of initial conditions in relation to the reduced problem (9.10).

Similarly, the *hyperbolic equation*

$$(9.12) \qquad \epsilon\left(u_{tt} - c^2 u_{xx}\right) + u_x = 0, \qquad x > 0, -\infty < t < \infty$$

with the boundary condition (9.9) reduces to

$$(9.13) \qquad\qquad v_x = 0, \qquad x > 0, -\infty < t < \infty$$

with the condition (9.11). The (unique) solution of the reduced problem for $v(x, t)$ is $v = f(t)$. Since (9.13) is a first order equation and (9.12) is of second order, the comments relating to the prescription of initial values apply to (9.12) and (9.13).

In connection with boundary conditions that depend on the parameter ϵ, it may happen that the conditions for the reduced problem render it unsolvable. For example, we consider the boundary value problem for the *biharmonic equation*

$$(9.14) \qquad\qquad \nabla^2 \nabla^2 u = 0, \qquad x \in G$$

with the boundary conditions

$$(9.15) \qquad\qquad u = f; \qquad \epsilon \frac{\partial u}{\partial n} + u = g; \qquad x \in \partial G.$$

Since (9.14) does not depend explicitly on ϵ, the reduced equation for u is also the biharmonic equation, but the reduced boundary conditions require that $u = f$ and $u = g$ on ∂G. Thus unless $f = g$, the reduced problem for u has no solution. Further, even if $f = g$, the solution of the reduced problem is not unique since a boundary condition is lost. Thus (9.14)–(9.15) is, in fact, a singular perturbation problem.

Proceeding with our discussion of the *regular perturbation method*, we expand the solution u of (9.6) in the perturbation series

$$(9.16) \qquad\qquad u = \sum_{n=0}^{\infty} u_n \epsilon^n$$

The difference between u and u_0 (i.e., $u - u_0$), is referred to as a *perturbation* on the solution u_0 of the reduced or unperturbed problem. Inserting this expansion into (9.6) gives

$$(9.17) \qquad L(u, \epsilon) = L\left(\sum_{n=0}^{\infty} u_n \epsilon^n, \epsilon \right) = 0.$$

We assume that $L(u, \epsilon)$ can be expanded in a power series in u and ϵ. As a result, (9.17) can be expressed in the form of a series

$$(9.18) \qquad L(u, \epsilon) = \sum_{n=0}^{\infty} L_n(u_n, u_{n-1}, \ldots, u_1, u_0) \epsilon^n = 0$$

where the L_n represent differential operators which may be linear or nonlinear, and which act on the functions $u_0, u_1, \ldots, u_n$. The series (9.16) is also inserted into the given initial and/or boundary conditions for the problem.

To solve the given problem by means of the perturbation method, we put the coefficients of ϵ^n in (9.18) equal to zero and obtain

$$(9.19) \qquad L_n(u_n, u_{n-1}, \ldots, u_0) = 0, \qquad n = 0, 1, \ldots .$$

Similarly, we equate coefficients of like powers of ϵ in the initial and/or boundary data. This yields the system of equations (9.19) with appropriate data that we solve recursively.

That is, we first solve the reduced equation

$$(9.20) \qquad L_0(u_0) = 0$$

with the relevant data. Once u_0 is specified, the equation for u_1,

$$(9.21) \qquad L_1(u_1, u_0) = 0$$

with its data, is solved and then the equations for $u_2, u_3, \ldots$ with their data are solved successively.

If the given equation (9.6) is linear, the equations $L_n(u_n, \ldots, u_0) = 0$ are generally nonhomogeneous versions of the equation $L_0(u_0) = 0$. However, $L_0(u_0) = 0$ may itself be a nonhomogeneous equation. Even if the given problem is nonlinear but the reduced problem is linear, all the equations (9.19) are homogeneous or nonhomogeneous equations of the same form.

For example, if we consider the equation

$$(9.22) \qquad L(u, \epsilon) = u_t + uu_x - \epsilon u = 0,$$

we find that

$$(9.23) \qquad L_0(u_0) = \frac{\partial u_0}{\partial t} + u_0 \frac{\partial u_0}{\partial x} = 0,$$

whereas

(9.24) $\qquad L_1(u_1, u_0) = \dfrac{\partial u_1}{\partial t} + u_0 \dfrac{\partial u_1}{\partial x} + u_1 \dfrac{\partial u_0}{\partial x} - u_0 = 0.$

However, the problem

(9.25) $\qquad\qquad\qquad L(u, \epsilon) = u_t + \epsilon u u_x - u_{xx} = 0$

yields

(9.26) $\qquad\qquad\qquad L_0(u_0) = \dfrac{\partial u_0}{\partial t} - \dfrac{\partial^2 u_0}{\partial x^2} = 0$

and

(9.27) $\qquad L_1(u_1, u_0) = \dfrac{\partial u_1}{\partial t} - \dfrac{\partial^2 u_1}{\partial x^2} + u_0 \dfrac{\partial u_0}{\partial x} = 0.$

For (9.22) the reduced equation is nonlinear and (9.24) for u_1 is linear, whereas for (9.25) the reduced equation as well as all the higher order equations for the u_n are linear and of the same form.

We now consider a number of examples of the perturbation method. Each of the examples emphasizes a different aspect of the method. In the first example we discuss a problem for which a closed form exact solution is easily obtained and compare the results of perturbation theory with the exact result.

EXAMPLE 9.1. HELMHOLTZ'S EQUATION WITH A SMALL PARAMETER

We consider the two-dimensional *Helmholtz equation*

(9.28) $\qquad\qquad\qquad u_{xx} + u_{yy} + \epsilon^2 u = 0$

in the unit disk $x^2 + y^2 < 1$ with the *Dirichlet boundary condition*

(9.29) $\qquad\qquad\qquad u(x, y) = 1; \qquad x^2 + y^2 = 1.$

The parameter ϵ^2 is assumed to be small, such that the solution of (9.28)–(9.29) is unique.

To solve, we introduce the perturbation series

(9.30) $\qquad\qquad\qquad u(x, y) = \sum_{n=0}^{\infty} u_n(x, y) \epsilon^{2n}$

where the expansion is in powers of ϵ^2. We insert (9.30) into (9.28) and (9.29) to obtain

$$(9.31) \qquad \nabla^2 u + \epsilon^2 u = \nabla^2 \left[\sum_{n=0}^{\infty} u_n \epsilon^{2n} \right] + \sum_{n=0}^{\infty} u_n \epsilon^{2n+2}$$

$$= \nabla^2 u_0 + \sum_{n=1}^{\infty} \left[\nabla^2 u_n + u_{n-1} \right] \epsilon^{2n} = 0$$

and

$$(9.32) \quad u(x, y) = u_0(x, y) + \sum_{n=1}^{\infty} u_n(x, y) \epsilon^{2n} = 1; \qquad x^2 + y^2 = 1.$$

We have interchanged summation and differentiation in (9.31) and have collected like powers of ϵ^2. On equating like powers of ϵ^2 to zero in (9.31) and to unity or zero in (9.32) we obtain

$$(9.33) \qquad\qquad\qquad \nabla^2 u_0 = 0,$$

$$(9.34) \qquad\qquad\qquad \nabla^2 u_n = -u_{n-1}; \qquad n \geqslant 1,$$

and the boundary conditions

$$(9.35) \quad u_0(x, y) = 1; \qquad u_n(x, y) = 0; \qquad n \geqslant 1; \qquad x^2 + y^2 = 1.$$

The equations for the u_n can be solved recursively starting with (9.33) and using the boundary conditions (9.35). The perturbation method has replaced the *Helmholtz equation* (9.28) by the system of *Laplace* and *Poisson equations* (9.33)–(9.34). Introducing polar coordinates r and θ, the Helmholtz as well as the Laplace and Poisson equations can be solved by looking for solutions independent of θ since the problem has no angular dependence.

With $u = u(r)$, (9.28) becomes

$$(9.36) \qquad\qquad\qquad u_{rr} + \frac{1}{r} u_r + \epsilon^2 u = 0$$

on expressing the Laplacian operator in polar coordinate form and dropping the θ-derivative. Now (9.36) is just Bessel's equation of zero order. The solution of (9.36) that is bounded at $r = 0$ and that satisfies the boundary condition $u(1) = 1$ is clearly

$$(9.37) \qquad\qquad\qquad u = \frac{J_0(\epsilon r)}{J_0(\epsilon)}.$$

To solve for the u_n we again assume that $u_n = u_n(r)$ and obtain for $u_0(r)$,

$$(9.38) \qquad \frac{\partial^2 u_0}{\partial r^2} + \frac{1}{r}\frac{\partial u_0}{\partial r} = 0; \qquad u_0 = 1 \qquad \text{at} \qquad r = 1.$$

The bounded solution of (9.38) is

$$(9.39) \qquad\qquad\qquad u_0 = 1.$$

The equation for $u_1(r)$ is

$$(9.40) \qquad \frac{\partial^2 u_1}{\partial r^2} + \frac{1}{r}\frac{\partial u_1}{\partial r} = -u_0 = -1; \qquad u_1 = 0 \qquad \text{at} \qquad r = 1$$

and a simple integration yields the bounded solution

$$(9.41) \qquad\qquad\qquad u_1 = \frac{1 - r^2}{4}.$$

Thus to leading orders the solution is

$$(9.42) \qquad\qquad u = 1 + \frac{\epsilon^2(1 - r^2)}{4} + O(\epsilon^4).$$

Since $J_0(z)$ has the expansion

$$(9.43) \qquad\qquad J_0(z) = 1 - \frac{z^2}{4} + O(z^4),$$

we have

$$(9.44) \qquad \frac{J_0(\epsilon r)}{J_0(\epsilon)} = \frac{1 - (\epsilon r)^2/4 + O(\epsilon^4)}{1 - \epsilon^2/4 + O(\epsilon^4)}$$

$$= 1 + \epsilon^2(1 - r^2)/4 + O(\epsilon^4).$$

The series obtained from (9.44) converges for sufficiently small ϵ, and it agrees to leading orders with the perturbation result (9.42). Note that ϵ must be smaller than the first zero of $J_0(z)$ which occurs at $z \approx 2.4$. Thus the perturbation solution yields a good approximation to the exact solution throughout the unit circle, even if we retain only the first few terms, as long as ϵ is small.

Although the perturbation method yielded an approximate solution valid over the entire unit circle in the preceding example, the perturbation result in the following example is not uniformly valid over the entire region of interest.

We consider heat conduction with a slow radiation loss and show that the perturbation solution is not valid for all time. The main reason for the distinction between the two cases is that the region of interest is bounded in the preceding example and unbounded in the following example.

EXAMPLE 9.2.   HEAT CONDUCTION WITH SLOW RADIATION

We consider the *Cauchy problem* for the *parabolic equation*

$$(9.45) \qquad u_t + \epsilon u = u_{xx}, \qquad -\infty < x < \infty, t > 0$$

with initial data

$$(9.46) \qquad u(x,0) = f(x), \qquad -\infty < x < \infty.$$

The change of variables

$$(9.47) \qquad u(x,t) = e^{-\epsilon t} v(x,t)$$

yields

$$(9.48) \qquad v_t = v_{xx}$$

and $v(x,0) = f(x)$, so that v satisfies the *heat equation*. The equation (9.45) describes heat conduction in a rod in which there is heat loss due to radiation on the surface. The radiative effect gives rise to the term ϵu in (9.45) and (9.47) shows that this yields a slow heat loss since $0 < \epsilon \ll 1$.

The solution of (9.45)–(9.46) is immediately given in terms of the solution of the Cauchy problem for the heat equation (9.48) studied previously. However, we want to apply the perturbation method to solve (9.45)–(9.46) in order to study the effect of the unboundedness of the (x, t) domain for the problem on the perturbation result.

Let

$$(9.49) \qquad u(x,t) = \sum_{n=0}^{\infty} u_n(x,t)\epsilon^n$$

and insert (9.49) into (9.45)–(9.46). On equating like powers of ϵ we have

$$(9.50) \qquad \frac{\partial u_0}{\partial t} - \frac{\partial^2 u_0}{\partial x^2} = 0,$$

$$(9.51) \qquad \frac{\partial u_n}{\partial t} - \frac{\partial^2 u_n}{\partial x^2} = -u_{n-1}, \qquad n \geqslant 1.$$

The initial conditions are

(9.52) $$u_0(x,0) = f(x),$$

(9.53) $$u_n(x,0) = 0, \qquad n \geqslant 1.$$

Putting

(9.54) $$u_0(x,t) = v(x,t)$$

where $v(x,t)$ is the solution of (9.48) with the initial condition $v(x,0) = f(x)$ [see (5.40)], we easily conclude that

(9.55) $$u_n(x,t) = \frac{(-t)^n}{n!} v(x,t), \qquad n \geqslant 0$$

on using mathematical induction. We verify that

(9.56)
$$\frac{\partial u_n}{\partial t} - \frac{\partial^2 u_n}{\partial x^2} = \frac{(-t)^n}{n!}[v_t - v_{xx}] - \frac{(-t)^{n-1}}{(n-1)!}v$$
$$= -u_{n-1}$$

since $v(x,t)$ satisfies (9.48).

The full perturbation solution is

(9.57) $$u(x,t) = \sum_{n=0}^{\infty}\left[\frac{(-t)^n}{n!}v(x,t)\right]\epsilon^n = \sum_{n=0}^{\infty}\frac{(-\epsilon t)^n}{n!}v(x,t)$$
$$= e^{-\epsilon t}v(x,t),$$

and this is identical with the exact solution (9.47). However, the purpose of the perturbation approach is to approximate the exact solution by determining and retaining only the first few terms in the perturbation series. If we retain only the first two terms in the series (9.57) and write

(9.58) $$u(x,t) = v(x,t) - \epsilon t v(x,t) + O(\epsilon^2),$$

we conclude on comparing with the exact solution that for $\epsilon t \ll 1$, (9.58) yields a good approximation.

But if $\epsilon t = O(1)$ or, equivalently, $t = O(1/\epsilon)$, we find that the first perturbation $\epsilon u_1 = -\epsilon t v(x,t)$ is of the same order of magnitude in ϵ as the leading term u_0. In fact, (9.57) shows that every term $\epsilon^n u_n$ in the series is of the same order in ϵ as the leading term u_0 if $t = O(1/\epsilon)$. Consequently, no matter how small ϵ is, there is a time t at which all terms in the perturbation series are of the same order in ϵ and cannot be neglected on the basis that they constitute small corrections for small ϵ. Thus even though the leading terms of the perturbation series yield a good approximation for $\epsilon t \ll 1$, the result is not uniformly valid for all time. Terms of the form $\epsilon t v(x, t)$ are called *secular terms* and the difficulty caused by the occurrence of such terms in a perturbation series is known as *secular behavior*.

There are several methods for remedying the difficulties caused by secular behavior in a perturbation series. One approach involves the summation of all the secular terms in the series and thereby assigning equal importance to all of them. This process is called *renormalization* and can be carried out in one form or another for a number of problems. In our example each term in the perturbation series is a secular term, so that the summation of the series yields the exact solution (9.47).

Another approach is known as the *method of multiple scales* and takes note of the fact that the perturbation solution appears to exhibit two time scales for our problem. There is a *slow* time scale ϵt and a (comparatively) *rapid* time scale t, as shown by (9.57). Therefore, we look for a solution of the form

$$(9.59) \qquad u(x, t) = \hat{u}(x, t, \tau); \qquad \tau = \epsilon t$$

that depends on x, t, and $\tau = \epsilon t$. That is, we treat the fast and slow time scales as independent variables.

Since

$$(9.60) \qquad \frac{\partial u}{\partial t} = \frac{\partial \hat{u}}{\partial t} + \frac{\partial \hat{u}}{\partial \tau} \frac{d\tau}{dt} = \frac{\partial \hat{u}}{\partial t} + \epsilon \frac{\partial \hat{u}}{\partial \tau},$$

we have, instead of (9.45),

$$(9.61) \qquad \hat{u}_t + \epsilon(\hat{u}_\tau + \hat{u}) = \hat{u}_{xx}.$$

Expanding $\hat{u}$ in a perturbation series

$$(9.62) \qquad \hat{u} = \sum_{n=0}^{\infty} \hat{u}_n \epsilon^n$$

yields the recursive system

$$(9.63) \qquad \frac{\partial \hat{u}_0}{\partial t} - \frac{\partial^2 \hat{u}_0}{\partial x^2} = 0,$$

$$(9.64) \qquad \frac{\partial \hat{u}_n}{\partial t} - \frac{\partial^2 \hat{u}_n}{\partial x^2} = -\left[\frac{\partial \hat{u}_{n-1}}{\partial \tau} + \hat{u}_{n-1}\right]; \qquad n \geqslant 1.$$

Since $\hat{u}_0 = f(x)$ at $t = 0$, we find that

$$(9.65) \qquad \hat{u}_0(x, t, \tau) = c(\tau)v(x, t)$$

where $v(x, t)$ is defined as before and $c(\tau)$ is an arbitrary function of τ that need only satisfy the condition $c(0) = 1$, since $t = 0$ implies that $\tau = \epsilon t$ vanishes. The equation for u_1 is

$$(9.66) \qquad \frac{\partial \hat{u}_1}{\partial t} - \frac{\partial^2 \hat{u}_1}{\partial x^2} = -\left[\frac{\partial \hat{u}_0}{\partial \tau} + \hat{u}_0 \right] = -\left[c'(\tau) + c(\tau) \right] v(x, t)$$

with the initial condition $\hat{u}_1 = 0$ at $t = 0$. Since the terms $c'(\tau)$ and $c(\tau)$ are constants, as far the operator on the left side of (9.66) is concerned, we obtain the solution

$$(9.67) \qquad \hat{u}_1(x, t, \tau) = -t\left[c'(\tau) + c(\tau) \right] v(x, t) + d(\tau)$$

where $d(\tau)$ vanishes at $\tau = 0$ but is otherwise arbitrary. Consequently, we still obtain a secular term that grows with t in the solution (9.67). However, we are now in a position to remove the secularity by specifying $c(\tau)$ such that

$$(9.68) \qquad\qquad c'(\tau) + c(\tau) = 0, \qquad c(0) = 1$$

which yields

$$(9.69) \qquad\qquad\qquad c(\tau) = e^{-\tau} = e^{-\epsilon t}.$$

We also set $d(\tau) = 0$, since otherwise a further secular term would arise at the next level of approximation. Consequently, $\hat{u}_1 = 0$ and, as a result, $\hat{u}_n = 0$ for all $n > 1$. The series (9.62) thus terminates with the first term and yields the exact solution (9.47).

In contrast to the preceding examples where exactly solvable problems were solved by the perturbation method, we now consider a nonlinear problem that cannot be solved explicitly. The perturbation method reduces it to a class of solvable linear problems.

EXAMPLE 9.3.   A NONLINEAR KLEIN–GORDON EQUATION

The hyperbolic equation

$$(9.70) \qquad\qquad w_{tt} - \gamma^2 w_{xx} + c^2 w - \sigma w^3 = 0$$

reduces to the *Klein–Gordon equation* when the parameter σ is equated to zero and arises in a number of physical contexts. The coefficients in (9.70) are assumed to be constants. We consider the initial value problem for (9.70) over the infinite line $-\infty < x < \infty$ with the data

$$(9.71) \qquad\qquad w(x, 0) = \epsilon \cos kx; \qquad w_t(x, 0) = 0$$

where $0 < \epsilon \ll 1$ and k is a specified constant.

Since the initial data (9.71) are uniformly small in magnitude, we look for a solution of (9.70)–(9.71) in the form

$$(9.72) \qquad\qquad w = \epsilon u$$

and apply the perturbation method to the problem for $u(x, t)$. We have

$$(9.73) \qquad\qquad u_{tt} - \gamma^2 u_{xx} + c^2 u - \epsilon^2 \sigma u^3 = 0$$

with the data

$$(9.74) \qquad\qquad u(x,0) = \cos kx; \qquad u_t(x,0) = 0.$$

Expanding $u(x, t)$ as

$$(9.75) \qquad\qquad u(x, t) = \sum_{n=0}^{\infty} u_n(x, t) \epsilon^n$$

and inserting (9.75) into (9.73)–(9.74) yields as the leading order equations

$$(9.76) \qquad L_n(u_n) = \frac{\partial^2 u_n}{\partial t^2} - \gamma^2 \frac{\partial^2 u_n}{\partial x^2} + c^2 u_n = 0; \qquad n = 0, 1,$$

$$(9.77) \qquad L_2(u_2, u_1, u_0) = \frac{\partial^2 u_2}{\partial t^2} - \gamma^2 \frac{\partial^2 u_2}{\partial x^2} + c^2 u_2 - \sigma u_0^3 = 0,$$

$$(9.78) \qquad L_3(u_3, u_2, u_1, u_0) = \frac{\partial^2 u_3}{\partial t^2} - \gamma^2 \frac{\partial^2 u_3}{\partial x^2} + c^2 u_3 - 3\sigma u_0^2 u_1 = 0.$$

The initial conditions are

$$(9.79) \qquad u_0(x,0) = \cos kx; \qquad u_n(x,0) = 0; \qquad n \geqslant 1,$$

$$(9.80) \qquad \frac{\partial u_n(x,0)}{\partial t} = 0, \qquad n \geqslant 0.$$

The solution of the initial value problem for $u_0(x, t)$ is

$$(9.81) \qquad\qquad u_0(x, t) = \cos \omega t \cos kx$$

where

$$(9.82) \qquad\qquad \omega^2 = \gamma^2 k^2 + c^2,$$

which is the *dispersion relation* for the linear Klein–Gordon equation. The equation for u_1 is uncoupled from that for u_0 and has homogeneous data. Thus

$$(9.83) \qquad u_1(x, t) = 0.$$

Since

$$(9.84) \qquad \cos^3 kx = \tfrac{3}{4}\cos kx + \tfrac{1}{4}\cos 3kx,$$

the equation for $u_2(x, t)$ may be written as

$$(9.85) \qquad \frac{\partial^2 u_2}{\partial t^2} - \gamma^2 \frac{\partial^2 u_2}{\partial x^2} + c^2 u_2 = \frac{3\sigma}{4}\cos^3\omega t \cos kx + \frac{\sigma}{4}\cos^3\omega t \cos 3kx.$$

On setting

$$(9.86) \qquad u_2(x, t) = F_1(t)\cos kx + F_2(t)\cos 3kx$$

and inserting (9.86) into (9.85), we obtain the ordinary differential equations

$$(9.87) \qquad F_1''(t) + (\gamma^2 k^2 + c^2) F_1(t) = \frac{3\sigma}{4}\cos^3\omega t,$$

$$(9.88) \qquad F_2''(t) + (9\gamma^2 k^2 + c^2) F_2(t) = \frac{\sigma}{4}\cos^3\omega t.$$

The initial conditions for $u_2(x, t)$ require that F_1, F_1', F_2, and F_2' all vanish at $t = 0$. On using the identity (9.84), the equations for F_1 and F_2 are easily solved by the method of undetermined coefficients.

Using these results, the leading terms of the perturbation series solution of (9.73)–(9.74) are found to be

$$(9.89)$$

$$u(x, t) = \cos \omega t \cos kx + \epsilon^2 \left[\frac{9\sigma}{32\omega} t \sin \omega t + \frac{3\sigma}{128\omega^2}(\cos \omega t - \cos 3\omega t) \right] \cos kx$$

$$+ \epsilon^2 \left[\frac{3\sigma}{128\gamma^2 k^2}(\cos \omega t - \cos \lambda t) \right.$$

$$\left. + \frac{\sigma}{128 c^2}(\cos \lambda t - \cos 3\omega t) \right] \cos 3kx + O(\epsilon^3)$$

where $\lambda^2 = 9\gamma^2 k^2 + c^2$. The secular term

$$(9.90) \qquad f_2(x, t) = \frac{9\epsilon^2\sigma}{32\omega} t \sin \omega t \cos kx$$

in the series shows that when $t = O(\epsilon^{-2})$, this term is of the same order of magnitude as the leading term. As a result, the perturbation series becomes invalid at those values of t. In constructing the perturbation series we are assuming that terms containing higher powers of ϵ are much smaller than those with lower powers. We have already indicated in the preceding example that the conventional perturbation expansion is generally not uniformly valid over unbounded regions.

The result (9.89) is valid and useful only for times t such that $t < O(\epsilon^{-2})$. To extend the validity of the result we note that

$$(9.91) \qquad \cos \omega t + \frac{9\epsilon^2\sigma}{32\omega} t \sin \omega t = \cos\left[\left(\omega - \frac{9\epsilon^2\sigma}{32\omega}\right)t\right] + O(\epsilon^4),$$

as is easily seen by expanding the right side of (9.91) for small ϵ. Using (9.91) to replace the first two terms on the right in (9.89) yields the improved result

$$(9.92) \qquad u(x, t) = \cos\left[\left(\omega - \frac{9\epsilon^2\sigma}{32\omega}\right)t\right]\cos kx + \epsilon^2[\cdots] + O(\epsilon^3),$$

in which the secular term is no longer present. There may be other secular terms in the series but they occur for higher order terms in ϵ and the validity of the perturbation solution is presumably extended in time when written in the form (9.92).

The method used to remove the secular term is essentially the method of renormalization discussed in Example 9.2. Secular terms in the perturbation series are removed by summing a part of that series. In Example 9.2 a full perturbation series was summed, whereas in the present example, since only the leading terms of the perturbation series were determined, the renormalization method could only be carried out in an approximate manner.

In the following example we consider a linear equation with nonconstant, but slowly varying, coefficients. As we see, the perturbation method appropriately modified yields a simple approximate solution.

EXAMPLE 9.4. THE WAVE EQUATION WITH A SLOWLY VARYING WAVE SPEED

The *hyperbolic equation*

$$(9.93) \qquad u_{tt} - c^2(\epsilon x)u_{xx} = 0$$

where $c^2(\epsilon x) > 0$ is assumed to have the expansion

$$(9.94) \qquad c^2(\epsilon x) = c_0^2 + \sum_{n=1}^{\infty} c_n[\epsilon x]^n$$

with constant c_n, reduces to the *wave equation* with wave speed c_0 if $\epsilon = 0$. Since $dc^2/dx = O(\epsilon)$ and ϵ is assumed to be small, we may approximate c^2 by c_0^2 and state that the function $c^2(\epsilon x)$ represents the slowly varying wave speed for (9.96). We note that (9.93) is not expressed in self-adjoint form and that the maximum speed of disturbances as determined by the domain of influence for concentrated initial data for (9.93) (see Section 8.3) is not given by c^2. We apply the perturbation method to an initial value problem for (9.93) to determine the effect of the slow variation of $c^2(\epsilon x)$ around c_0^2.

Let the initial conditions for (9.93) be

$$(9.95) \qquad u(x,0) = f(x); \qquad u_t(x,0) = 0; \qquad -\infty < x < \infty.$$

We expand $u(x, t)$ as

$$(9.96) \qquad u(x, t) = \sum_{n=0}^{\infty} u_n(x, t)\epsilon^n$$

and insert (9.96) into (9.93) and (9.95). Equating like powers of ϵ yields as the leading order equations

$$(9.97) \qquad \frac{\partial^2 u_0}{\partial t^2} - c_0^2 \frac{\partial^2 u_0}{\partial x^2} = 0,$$

$$(9.98) \qquad \frac{\partial^2 u_1}{\partial t^2} - c_0^2 \frac{\partial^2 u_1}{\partial x^2} = xc_1 \frac{\partial^2 u_0}{\partial x^2}$$

where (9.94) has been used. The initial conditions for u_0 and u_1 are

$$(9.99) \qquad u_0(x,0) = f(x); \qquad \frac{\partial u_0(x,0)}{\partial t} = 0;$$

$$(9.100) \qquad u_1(x,0) = \frac{\partial u_1(x,0)}{\partial t} = 0.$$

The solution of (9.97) and (9.99) is

$$(9.101) \qquad u_0(x, t) = \tfrac{1}{2}f(x - c_0 t) + \tfrac{1}{2}f(x + c_0 t).$$

Then the equation for u_1 becomes

$$(9.102) \qquad \frac{\partial^2 u_1}{\partial t^2} - c_0^2 \frac{\partial^2 u_1}{\partial x^2} = \frac{1}{2}xc_1 f''(x - c_0 t) + \frac{1}{2}xc_1 f''(x + c_0 t).$$

Using the method of undetermined coefficients, we easily solve (9.102) with the data (9.100) and obtain

$$(9.103) \qquad u_1(x, t) = \frac{c_1 t}{8c_0}[f(x - c_0 t) - f(x + c_0 t)]$$

$$- \frac{c_1 xt}{4c_0}[f'(x - c_0 t) - f'(x + c_0 t)]$$

$$+ \frac{c_1 t^2}{8}[f'(x - c_0 t) + f'(x + c_0 t)].$$

We note that when $t = O(\epsilon^{-1/2})$ or $xt = O(\epsilon^{-1})$, the term ϵu_1 of the same order in ϵ as the leading term u_0, assuming f and its derivatives are uniformly bounded. Since both the x and the t intervals relevant to this problem are unbounded, the perturbation series would appear to be invalid, in general, since values of x for which $x = O[(\epsilon t)^{-1}]$ must occur if $-\infty < x < \infty$. However, if f vanishes outside a bounded interval, so does $u_1(x, t)$ and large values of x do not play a role as long as t is small. In that case, the perturbation theory solution may be valid for a substantial length of time.

The general difficulty with the preceding perturbation series may be traced to the fact that the solutions (9.101) and (9.103) are expressed in terms of the characteristics

(9.104) $$\hat{\Phi}_{\pm}(x, t) = x \pm c_0 t = \text{constant}$$

of the reduced equation—that is, (9.93) with $\epsilon = 0$—rather than the exact characteristics of (9.93) that are given as solutions of

(9.105) $$\frac{dt}{dx} = \pm \frac{1}{c(\epsilon x)}.$$

Given a point $(x_0, 0)$ on the x-axis, the difference between the *exact* and *reduced characteristics* issuing from that point becomes appreciable for large values of x and t. Consequently, we may expect that the perturbation solution becomes invalid for x and t large.

To resolve the difficulties with the secular terms that arise in the perturbation result (9.101) and (9.103), we introduce the independent families of exact characteristic curves $\Phi(x, t) = \text{constant}$ and $\Psi(x, t) = \text{constant}$, both of which satisfy (9.105). They reduce to the curves $x - c_0 t = \text{constant}$ and $x + c_0 t = \text{constant}$, respectively, when we set $\epsilon = 0$ in (9.105). Noting the results (9.101) and (9.103) we look for a solution of (9.93) in the form

(9.106) $$u(x, t) \approx g(\epsilon x, \epsilon t) f(\Phi) + h(\epsilon x, \epsilon t) f(\Psi).$$

Inserting (9.106) into (9.93) yields

(9.107) $$u_{tt} - c^2 u_{xx} = \left[\Phi_t^2 - c^2 \Phi_x^2\right] g f''(\Phi)$$
$$+ \left[\Psi_t^2 - c^2 \Psi_x^2\right] h f''(\Psi) + \left[2\epsilon\left(\Phi_t g_\tau - c^2 \Phi_x g_\sigma\right)\right.$$
$$+ \left(\Phi_{tt} - c^2 \Phi_{xx}\right) g\right] f'(\Phi) + \left[2\epsilon\left(\Psi_t h_\tau - c^2 \Psi_x h_\sigma\right)\right.$$
$$+ \left(\Psi_{tt} - c^2 \Psi_{xx}\right) h\right] f'(\Psi) + \epsilon^2 \left[g_{\tau\tau} - c^2 g_{\sigma\sigma}\right] f(\Phi)$$
$$+ \epsilon^2 \left[h_{\tau\tau} - c^2 h_{\sigma\sigma}\right] f(\Psi) = 0$$

where $\sigma = \epsilon x$ and $\tau = \epsilon t$. Since $\Phi = \text{constant}$ and $\Psi = \text{constant}$ are characteristics, they satisfy the characteristic equations for (9.93) and we have

(9.108) $$\Phi_t^2 - c^2 \Phi_x^2 = \Psi_t^2 - c^2 \Psi_x^2 = 0.$$

Also, since Φ is a solution of

(9.109) $$\Phi_t + c\Phi_x = 0,$$

we find that

(9.110) $$\Phi_{tt} = -c\Phi_{xt} = -c\Phi_{tx} = c^2\Phi_{xx} + \epsilon cc'\Phi_x$$

with a similar result valid for Ψ. This gives

(9.111) $$2\epsilon\left[\Phi_t g_\tau - c^2\Phi_x g_\sigma\right] + \left[\Phi_{tt} - c^2\Phi_{xx}\right]g = -2\epsilon c\Phi_x\left[g_\tau + cg_\sigma - \frac{c'}{2}g\right]$$

and a similar expression for h, where c and c' are replaced by $-c$ and $-c'$.

The coefficients of $f''(\Phi)$ and $f''(\Psi)$ in (9.107) vanish in view of (9.108). The leading terms of order ϵ are obtained by equating the coefficients of $f'(\Phi)$ and $f'(\Psi)$ to zero. It follows from (9.111) and the corresponding result for h that

(9.112) $$g_\tau + cg_\sigma - \frac{c'}{2}g = 0,$$

(9.113) $$h_\tau - ch_\sigma + \frac{c'}{2}h = 0.$$

The initial conditions (9.95) imply that

(9.114) $\quad u(x,0) \approx g(\epsilon x,0)f(\Phi) + h(\epsilon x,0)f(\Psi) = f(x),$
(9.115) $\quad u_t(x,0) \approx g(\epsilon x,0)\Phi_t f'(\Phi) + h(\epsilon x,0)\Psi_t f'(\Psi) + O(\epsilon) = 0.$

We conclude that the initial values for the characteristics $\Phi(x, t)$ and $\Psi(x, t)$ are

(9.116) $$\Phi(x,0) = \Psi(x,0) = x.$$

Since $\Phi_t = -c\Phi_x$ and $\Psi_t = c\Psi_x$, we find that at $t = 0$,

(9.117) $$\Phi_t(x,0) = -\Psi_t(x,0) = -c.$$

Inserting (9.116)–(9.117) into (9.114)–(9.115) gives

(9.118) $$\begin{cases} g + h = 1 \\ -cg + ch = 0 \end{cases}$$

so that

(9.119) $$g(\epsilon x,0) = h(\epsilon x,0) = \tfrac{1}{2}.$$

Using the methods of Chapter 2, the initial value problems for Φ, Ψ, g, and h

are readily solved. Rather than present the general solution, we restrict our attention to a specific problem that is exactly solvable.

Consider (9.93) in the interval $0 < x < \infty$ with

$$(9.120) \qquad c(\epsilon x) = c_0 + \epsilon x,$$

and $u(0, t) = 0$ for $t > 0$ and the data (9.95) for $x > 0$. Comparing (9.94) with (9.120) we conclude that $c_1 = 2c_0$, $c_2 = 1$, and $c_n = 0$ for $n \geqslant 2$. The characteristics Φ and Ψ corresponding to this choice of $c(\epsilon x)$ and the data (9.116) are

$$(9.121) \qquad \Phi(x, t) = xe^{-\epsilon t} + \frac{c_0}{\epsilon}(e^{-\epsilon t} - 1),$$

$$(9.122) \qquad \Psi(x, t) = xe^{\epsilon t} + \frac{c_0}{\epsilon}(e^{\epsilon t} - 1).$$

We discuss the solution in the region where $\Phi(x, t) > 0$. The equations (9.112)–(9.113) for g and h with the data (9.119) can be solved by looking for solutions that are independent of σ since $c' = 1$ in this case. We obtain

$$(9.123) \qquad g = \frac{1}{2}\exp\left[\frac{\epsilon t}{2}\right]; \qquad h = \frac{1}{2}\exp\left[-\frac{\epsilon t}{2}\right].$$

Then the approximate solution (9.106) is

$$(9.124) \qquad u(x, t) \approx \frac{1}{2}\exp\left[\frac{\epsilon t}{2}\right]f\left[xe^{-\epsilon t} + \frac{c_0}{\epsilon}(e^{-\epsilon t} - 1)\right]$$
$$+ \frac{1}{2}\exp\left[-\frac{\epsilon t}{2}\right]f\left[xe^{\epsilon t} + \frac{c_0}{\epsilon}(e^{\epsilon t} - 1)\right].$$

On expanding (9.124) in powers of ϵ it is easily shown that the first two terms in the expansion agree with (9.101) and (9.103) in the region $\Phi > 0$. Although the secular terms found in the conventional perturbation expansion are absent in (9.124), we do observe a slow exponential growth in t, say, if $f(x)$ is uniformly bounded. Though it would appear that we have used the method of multiple scales to obtain (9.124), this problem is somewhat more complicated since we not only introduced ϵx and ϵt as variables but also the characteristics Φ and Ψ.

We have seen that the separation of variables and transform methods for the solution of initial and/or boundary value problems for partial differential equations are useful only for problems given over special regions such as squares, circles, spheres, or half-spaces. If the boundary of the given region varies only slightly from a boundary for which the foregoing methods can successfully be applied, it is possible to use perturbation theory to effect an approximate solution to the problem. We now present a brief discussion of *boundary perturbation methods* for a two-dimensional problem and consider an example.

Let the given problem be specified in a two-dimensional region $\hat{G}$ with $\partial\hat{G}$ as its boundary. We assume that $\partial\hat{G}$ can be expressed in parametric form as

$$(9.125) \qquad x = g(s) + \epsilon\hat{g}(s); \qquad y = h(s) + \epsilon\hat{h}(s)$$

where $0 < \epsilon \ll 1$, s is a parameter, and the functions g, $\hat{g}$, h, and $\hat{h}$ are given. We could also consider a more general ϵ dependence such as $x = g(x, \epsilon)$ and $y = h(s, \epsilon)$, where g and h can be expanded in powers of ϵ, but do not do so for the sake of simplicity. Assuming that the function $u(x, y)$ is specified on the boundary curve $\partial\hat{G}$, we have

$$(9.126) \qquad u(x, y)|_{\partial\hat{G}} = u(g + \epsilon\hat{g}, h + \epsilon\hat{h}) = f(s)$$

where $f(s)$ is prescribed. Further, $u(x, y)$ is assumed to satisfy a linear partial differential equation with no explicit ϵ dependence

$$(9.127) \qquad Lu = 0, \qquad (x, y) \in \hat{G}.$$

Using the perturbation method we expand $u(x, y)$ as

$$(9.128) \qquad u = \sum_{n=0}^{\infty} u_n \epsilon^n$$

and insert (9.128) into (9.126)–(9.127). Since L is a linear operator we find that

$$(9.129) \qquad Lu_n = 0, \qquad n \geqslant 0.$$

In (9.126) we first expand u in powers of ϵ as

$$(9.130) \quad u(g + \epsilon\hat{g}, h + \epsilon\hat{h}) = u(g, h)$$
$$+ \epsilon\left[\hat{g}u_x(g, h) + \hat{h}u_y(g, h)\right] + O(\epsilon^2)$$

and then obtain from (9.128)

$$(9.131) \qquad u_0(g, h) = f,$$

$$(9.132) \qquad u_1(g, h) = -\hat{g}\frac{\partial u_0(g, h)}{\partial x} - \hat{h}\frac{\partial u_0(g, h)}{\partial y}.$$

Thus the perturbation method reduces the given problem to a collection of problems given over the unperturbed region G with the boundary curve ∂G whose equations are $x = g(s)$ and $y = h(s)$. It is assumed that the boundary value problems for the functions $u_n(x, y)$ in the region G can be solved. It is a straightforward matter to extend these results to three-dimensional problems and to cases with other types of boundary conditions. We now consider an example.

EXAMPLE 9.5.   BOUNDARY PERTURBATIONS FOR LAPLACE'S EQUATION

The solution of the *Dirichlet problem* for *Laplace's equation* in a rectangle was presented in Example 4.11 where the method of separation of variables was used. We now consider a small perturbation of the rectangular region $0 < x < l$ and $0 < y < L$. Let the region $\hat{G}$ be given as the interior of the trapezoid defined as $0 < x < l$ and $\epsilon x < y < L$, so that the boundary line $y = 0$ is replaced by $y = \epsilon x$ where $0 < \epsilon \ll 1$ (it is assumed that $\epsilon l < L$).

The Dirichlet problem is given as

$$(9.133) \qquad u_{xx} + u_{yy} = 0; \qquad 0 < x < l; \qquad \epsilon x < y < L$$

with the boundary conditions

$$(9.134) \qquad\qquad u(0, y) = 0, \qquad 0 < y < L,$$

$$(9.135) \qquad\qquad u(x, L) = 0, \qquad 0 < x < l,$$

$$(9.136) \qquad\qquad u(l, y) = 0, \qquad \epsilon l < y < L,$$

and

$$(9.137) \qquad\qquad u(x, \epsilon x) = f(x); \qquad 0 < x < l$$

where $f(x)$ is a prescribed function.

To solve (9.133)–(9.137) we first expand $u(x, \epsilon x)$ as

$$(9.138) \qquad u(x, \epsilon x) = u(x, 0) + \epsilon x u_y(x, 0) + O(\epsilon^2) = f(x).$$

Then the expansion (9.128) of $u(x, y)$ yields

$$(9.139) \quad u_0(x, 0) + \epsilon \left[u_1(x, 0) + x \frac{\partial u_0(x, 0)}{\partial y} \right] + O(\epsilon^2) = f(x).$$

Each of the $u_n(x, y)$ satisfies Laplace's equation. For $u_0(x, y)$ we have the boundary conditions

$$(9.140) \qquad u_0(x, y) = 0; \qquad x = 0; \qquad y = L; \qquad x = l,$$

and

$$(9.141) \qquad\qquad u_0(x, 0) = f(x); \qquad 0 < x < l,$$

in view of (9.139), with

$$(9.142) \qquad \nabla^2 u_0 = 0; \qquad 0 < x < l, 0 < y < L.$$

The function $u_1(x, y)$ also vanishes on the three sides of the rectangle (i.e., $x = 0$, $y = L$ and $x = l$), whereas on the fourth side

$$(9.143) \qquad u_1(x, 0) = -x \frac{\partial u_0(x, 0)}{\partial y}, \qquad 0 < x < l,$$

and satisfies Laplace's equation

$$(9.144) \qquad \nabla^2 u_1 = 0, \qquad 0 < x < l, 0 < y < L.$$

Using the results of Example 4.11, we find the solution of the problem for $u_0(x, y)$ to be

$$(9.145) \qquad u_0(x, y) = \sqrt{\frac{2}{l}} \sum_{k=1}^{\infty} b_k \sinh\left[\frac{\pi k}{l}(y - L)\right] \sin\left[\frac{\pi k x}{l}\right]$$

where b_k is defined in (4.161) with $\hat{l}$ replaced by L. To determine the boundary condition (9.143) for $u_1(x, y)$ we differentiate (9.145) term by term with respect to y and evaluate it at $y = 0$ to obtain

$$(9.146) \qquad u_1(x, 0) = -\sqrt{\frac{2}{l}} \sum_{k=1}^{\infty} x b_k \left(\frac{\pi k}{l}\right) \cosh\left[\frac{\pi k L}{l}\right] \sin\left[\frac{\pi k x}{l}\right].$$

Again the results of Example 4.11 can be used to specify $u_1(x, y)$ by putting $f(x)$ equal to the right side of (9.146) and setting $g(x) = 0$ in that example.

For the sake of concreteness we set

$$(9.147) \qquad f(x) = \sin\frac{\pi x}{l}.$$

in (9.137). Then $u_0(x, y)$ has the form

$$(9.148) \qquad u_0(x, y) = \frac{\sinh[\pi(L - y)/l]}{\sinh(\pi L/l)} \sin\left[\frac{\pi x}{l}\right]$$

and the boundary condition (9.146) for $u_1(x, y)$ is

$$(9.149) \qquad u_1(x, 0) = \frac{\pi x}{l} \coth\left[\frac{\pi L}{l}\right] \sin\left[\frac{\pi x}{l}\right].$$

We note that $u_1(x, 0)$ vanishes at $x = 0$ and $x = l$, and is a smooth function in the interval $0 < x < l$. Therefore, its Fourier sine series converges uniformly in the given interval and, as a result, the solution $u_1(x, y)$ obtained by the methods of Example 4.11 is continuous and, therefore, bounded in the given region. Thus ϵu_1 represents a small perturbation around u_0 for small ϵ. Similarly, if $u_1(x, 0)$ in (9.146) is a smooth function in $0 < x < l$, we conclude that ϵu_1 is a small perturbation around u_0 since $u_1(0, 0) = u_1(l, 0) = 0$.

Eigenvalue problems can also be treated by perturbation methods. A special case has already been considered in Example 8.5. We use the notation of Chapters 4 and 8 in our discussion and consider the problem

$$(9.150) \qquad LM \equiv -\nabla \cdot (p\nabla M) + qM = (\lambda + \epsilon r)\rho M$$

in the bounded region G with the boundary condition

$$(9.151) \qquad \alpha M + \beta \left.\frac{\partial M}{\partial n}\right|_{\partial G} = 0.$$

The parameter ϵ is small (i.e., $0 < \epsilon \ll 1$) and r is a given function. It is assumed that the reduced eigenvalue problem with $\epsilon = 0$ is solvable and that the eigenvalues are simple so that there is one linearly independent eigenfunction for each eigenvalue.

To determine the eigenvalues and eigenfunctions of (9.150)–(9.151) we expand both M and λ in powers of ϵ and set

$$(9.152) \qquad M(x; \epsilon) = \sum_{n=0}^{\infty} M^{(n)}(x)\epsilon^n$$

and

$$(9.153) \qquad \lambda(\epsilon) = \sum_{n=0}^{\infty} \lambda^{(n)}\epsilon^n.$$

Inserting (9.152)–(9.153) into (9.150)–(9.151) gives

$$(9.154) \qquad LM^{(0)} = \lambda^{(0)}\rho M^{(0)},$$

$$(9.155) \qquad LM^{(1)} = \lambda^{(0)}\rho M^{(1)} + r\rho M^{(0)} + \lambda^{(1)}\rho M^{(0)},$$

and

$$(9.156) \qquad LM^{(2)} = \lambda^{(0)}\rho M^{(2)} + r\rho M^{(1)} + \lambda^{(1)}\rho M^{(1)} + \lambda^{(2)}\rho M^{(0)}$$

as the equations for the first three $M^{(n)}$. The boundary conditions for each of the $M^{(n)}$ are

(9.157)
$$\alpha M^{(n)} + \beta \left.\frac{\partial M^{(n)}}{\partial n}\right|_{\partial G} = 0, \qquad n \geqslant 0.$$

Let $\lambda_k^{(0)} = \lambda_k$ and $M_k^{(0)}(x) \equiv M_k(x)$ $(k = 1, 2, \ldots)$ represent the (simple) eigenvalues and orthonormalized eigenfunctions of the unperturbed problem (9.154) and (9.157) where $n = 0$ (these are assumed to be known). The equation (9.155) is an inhomogeneous version of (9.154), with $M^{(1)}(x)$ required to satisfy the homogeneous boundary condition (9.157). This problem has no solution unless the inhomogeneous term satisfies a compatibility condition, as we now show. We put $M^{(n)}$ and $\lambda^{(n)}$ equal to $M_k^{(n)}$ and $\lambda_k^{(n)}$ in (9.154)–(9.157). Then (9.155) takes the form

(9.158)
$$LM_k^{(1)} - \lambda_k \rho M_k^{(1)} = \left(r + \lambda_k^{(1)}\right)\rho M_k$$

since $\lambda_k^{(0)} \equiv \lambda_k$ and $M_k^{(0)} \equiv M_k$.

Applying the results of Example 4.2, we multiply (9.158) by M_j and integrate over G. Since both M_j and $M_k^{(1)}$ satisfy the boundary condition (9.157) we obtain, in view of (4.28),

(9.159)
$$\iint_G \left[M_j LM_k^{(1)} - \lambda_k \rho M_j M_k^{(1)} \right] dv$$
$$= \iint_G \left[M_k^{(1)} LM_j - \lambda_k \rho M_j M_k^{(1)} \right] dv$$
$$= \left(\lambda_j - \lambda_k\right)\left(M_k^{(1)}, M_j \right)$$
$$= \iint_G r\rho M_j M_k \, dv + \lambda_k^{(1)}\left(M_j, M_k \right)$$

where the inner product (f, g) is defined as

$$(f, g) = \iint_G \rho fg \, dv.$$

Now, if $j = k$, so that $\lambda_j = \lambda_k$, the left side of (9.159) must vanish and this implies that $\lambda_k^{(1)}$ must be specified as

(9.160)
$$\lambda_k^{(1)} = -\iint_G r\rho M_k^2 \, dv$$

since $(M_k, M_k) = 1$. For $j \neq k$, the Fourier coefficients $(M_k^{(1)}, M_j)$ of the function $M_k^{(1)}$ are given as

(9.161)
$$\left(M_k^{(1)}, M_j \right) = \frac{\left(rM_k, M_j\right)}{\lambda_j - \lambda_k}, \qquad j \neq k$$

where we have used the fact that $(M_j, M_k) = 0$ for $j \neq k$.

The eigenfunctions $M_j(x)$ are assumed to form a complete set and $M_k^{(1)}(x)$ can be expanded in the series

$$(9.162) \qquad M_k^{(1)}(x) = \sum_{j=1}^{\infty} \left(M_k^{(1)}, M_j \right) M_j(x), \qquad k = 1, 2, \ldots.$$

Each of the Fourier coefficients is specified as in (9.161) except for $\left(M_k^{(1)}, M_k \right)$ which remains undetermined. If we normalize each of the eigenfunctions $M_k(x; \epsilon)$ we have

$$(9.163) \qquad \begin{aligned} 1 &= \left(M_k(x; \epsilon), M_k(x; \epsilon) \right) \\ &= \left(M_k^{(0)}, M_k^{(0)} \right) + 2\epsilon \left(M_k^{(1)}, M_k^{(0)} \right) + O(\epsilon^2) \\ &= 1 + 2\epsilon \left(M_k^{(1)}, M_k^{(0)} \right) + O(\epsilon^2), \end{aligned}$$

since the eigenfunctions $M_k^{(0)} \equiv M_k$ are already normalized. On equating like powers of ϵ we conclude that

$$(9.164) \qquad\qquad \left(M_k^{(1)}, M_k \right) = 0.$$

Consequently, the series (9.162) for $M_k^{(1)}(x)$ is completely specified.

For each unperturbed eigenvalue λ_k and eigenfunction $M_k(x)$ we have determined the first perturbation $\epsilon\lambda_k^{(1)}$ and $\epsilon M_k^{(1)}(x)$. Higher approximations can be obtained by applying the finite Fourier transform method to the equations for $M_k^{(n)}(x)$ and proceeding as in the foregoing. A modification of this procedure, which we do not present, is needed to deal with the case of multiple eigenvalues. We note that the Sturm–Liouville problem has been shown to have simple eigenvalues and these results can easily be expressed in a one-dimensional form. In higher dimensional problems multiple eigenvalues often occur. However, the perturbation method can be applied for every simple eigenvalue in a set of eigenvalues that may include multiple eigenvalues. The following example considers such a case.

EXAMPLE 9.6. EIGENVALUE PERTURBATIONS IN A SQUARE

We consider the square of side π, with the interior region G defined as $0 < x < \pi$ and $0 < y < \pi$, and consider the *perturbed eigenvalue problem*

$$(9.165) \qquad - \left(M_{xx} + M_{yy} \right) + \epsilon xy M = \lambda M, \qquad (x, y) \in G$$

and the boundary condition

$$(9.166) \qquad\qquad M(x, y; \epsilon)\big|_{\partial G} = 0$$

where $0 < \epsilon \ll 1$. The reduced eigenvalue problem with $\epsilon = 0$ was considered in Example 7.5. There it was found that the eigenvalues λ_{nm} are given as

$$(9.167) \qquad \lambda_{nm} = n^2 + m^2, \qquad n, m = 1, 2, \ldots$$

and the normalized eigenfunctions $M_{nm}(x, y)$ are

$$(9.168) \qquad M_{nm}(x, y) = \frac{2}{\pi} \sin nx \sin my, \qquad n, m = 1, 2, \ldots.$$

We see that λ_{11} is a simple eigenvalue with only one linearly independent eigenfunction M_{11}. However, λ_{12} equals λ_{21} and to this eigenvalue there correspond two independent eigenfunctions M_{12} and M_{21}. There are infinitely many other multiple eigenvalues.

As has been indicated previously, it is often of greatest interest to determine the lowest eigenvalue for a given problem. Since the lowest *unperturbed eigenvalue* λ_{11} is *simple*, we may apply our perturbation procedure and determine the corresponding *perturbed eigenvalue* as well as its eigenfunction. The situation is complicated somewhat because of the double subscript notation for the eigenvalues and the eigenfunctions of the reduced eigenvalue problem. Therefore, we shall only obtain the first perturbation of the eigenvalue λ_{11} but not the perturbation of the eigenfunction $M_{11}(x, y)$.

In the notation of the previous discussion we have $\rho = 1$ and

$$(9.169) \qquad r(x, y) = -xy.$$

Then with

$$(9.170) \qquad \lambda_{11}(\epsilon) = \lambda_{11} + \epsilon \lambda_{11}^{(1)} + O(\epsilon^2)$$

representing the lowest (perturbed) eigenvalue, we have from (9.160),

$$(9.171) \qquad \lambda_{11}^{(1)} = \frac{4}{\pi^2} \int_0^\pi \int_0^\pi xy \sin^2 x \sin^2 y \, dx \, dy$$

$$= \left(\frac{2}{\pi} \int_0^\pi t \sin^2 t \, dt \right)^2 = 1$$

so that

$$(9.172) \qquad \lambda_{11}(\epsilon) = 2 + \epsilon + O(\epsilon^2).$$

The *nonlinear Klein–Gordon equation*

$$(9.173) \qquad w_{tt} - \gamma^2 w_{xx} + c^2 w - \sigma w^3 = 0$$

discussed in Example 9.3 was found to have an approximate solution of the form

$$(9.174) \qquad w(x, t) = \epsilon \cos kx \cos\left[\left(\omega - \frac{9\epsilon^2\sigma}{32\omega}\right)t\right] + O(\epsilon^3)$$

as follows from (9.92) and (9.72). This can be rewritten as

$$(9.175) \qquad w(x, t) = \frac{\epsilon}{2} \cos\left[kx - \left(\omega - \frac{9\epsilon^2\sigma}{32\omega}\right)t\right]$$

$$+ \frac{\epsilon}{2} \cos\left[kx + \left(\omega - \frac{9\epsilon^2\sigma}{32\omega}\right)t\right] + O(\epsilon^3).$$

(We recall that k is a constant and $\omega^2 = \gamma^2 k^2 + c^2$.) Now each of the terms in (9.175) represents a *traveling wave* of the form

$$(9.176) \qquad w(x, t) = a \cos[kx \mp \hat{\omega}(k)t]$$

where the *amplitude* term $a = \epsilon/2$ is small and the *phase* term $\theta = kx \mp \hat{\omega}(k)t$, with

$$(9.177) \quad \hat{\omega}(k) = \omega(k) - \frac{9\epsilon^2\sigma}{32\omega(k)} = \omega(k)\left[1 - \frac{9a^2\sigma}{8(\gamma^2 k^2 + c^2)}\right],$$

has the phase velocity

$$(9.178) \qquad \frac{dx}{dt} = \pm\frac{\hat{\omega}(k)}{k} = \pm\frac{\omega(k)}{k}\left[1 - \frac{9a^2\sigma}{8(\gamma^2 k^2 + c^2)}\right].$$

If the amplitude $a = \epsilon/2$ were assumed to be infinitesimal, we would neglect the term involving a^2 in (9.178) and the phase velocity would reduce to that for the linear Klein–Gordon equation as given in Example 3.7. However, for small but finite amplitude traveling waves where a^2 is not neglected, we find that the phase speed or, correspondingly, the speed of the wave (9.176) depends not only on the wave number k but also on the amplitude a.

The dependence of the speed of the wave on its amplitude has already been observed in our discussion of nonlinear unidirectional wave motion in Chapter 2. The dependence of the speed of normal mode solutions of linear hyperbolic equations on the wave number k has been characterized as *dispersive wave motion* in Section 3.5. The preceding results show that the nonlinear Klein–Gordon equation has (approximate) traveling wave solutions that exhibit the combined effects of having the *wave speed* depend on both the *wave number* and the *amplitude*.

The theory of nonlinear dispersive wave motion has recently undergone much study, especially by Whitham. One aspect that has been investigated is whether *periodic finite amplitude traveling waves* of the form (9.176) or some more general form can be constructed for these equations. Since the amplitude of these waves is assumed to be small, it can serve as a perturbation parameter. It is assumed that when a small amplitude solution is inserted in the given equation and the equation is linearized, the result is a dispersive wave equation. The effect of the nonlinear terms is to introduce higher harmonics as in (9.89) and a dependence of the dispersion relation $\hat{\omega} = \hat{\omega}(k)$ on the amplitude as seen in (9.177).

We do not attempt to characterize the general form of nonlinear dispersive wave equations for which periodic traveling wave solutions can be found. Rather, we discuss a specific equation in the following example and consider further examples in the exercises. We refer to the literature for general results. One of the earliest applications of the perturbation method for determining periodic traveling waves for nonlinear equations was given by Stokes in his study of water waves. In the following example we construct approximate traveling wave solutions of the *nonlinear Korteweg–deVries equation* which plays an important role in the theory of water waves. Even though exact periodic traveling wave solutions can be found for this equation, we do not do so but apply the perturbation method instead. The main feature in our approach is that not only the solution but the dispersion function is expanded in powers of the small parameter. The need for this is suggested by (9.177).

EXAMPLE 9.7.   PERIODIC TRAVELING WAVE SOLUTIONS OF THE
KORTEWEG–DEVRIES EQUATION

The *Korteweg–deVries equation* is given as

$$(9.179) \qquad u_t + (c + u)u_x + \beta u_{xxx} = 0$$

where c and β are given constants. In its linearized form

$$(9.180) \qquad u_t + cu_x + \beta u_{xxx} = 0$$

the *dispersion relation* is (see Section 3.5)

$$(9.181) \qquad \omega = \omega(k) = ck - \beta k^3$$

so that (9.180) is of *dispersive type*. If we drop the third derivative term in (9.179), we have

$$(9.182) \qquad u_t + (c + u)u_x = 0$$

which is a *quasilinear first order wave equation* whose wave speed depends on the amplitude.

The linearized equation (9.180) has traveling wave solutions

$$(9.183) \qquad u(x, t) = a\cos[kx - \omega t]$$

where a = constant and ω is given by (9.181). The quasilinear equation (9.182) has the (implicit) solutions

$$(9.184) \qquad u = a\cos[kx - k(c + u)t].$$

The wave speed for (9.183) is $dx/dt = \omega/k = c - \beta k^2$ and the wave speed for (9.184) is, formally, $dx/dt = c + u$. In both cases the *wave speed* is perturbed around the constant c, in the linear case by the *wave number* and in the nonlinear case by the *amplitude*. Note that if a is small, then $|u|$ is small in view of (9.184). We may also characterize $k(c + u)$ as a nonlinear frequency term and then (9.183)–(9.184) show that the frequency of the traveling wave solution for the full problem (9.179) should depend on the wave number k and on the amplitude a.

Thus we look for a perturbation solution of (9.179) in the form

$$(9.185) \qquad u(x, t) = \sum_{n=1}^{\infty} u_n(\theta)a^n$$

where a is a small positive constant and

$$(9.186) \qquad \theta(x, t) = kx - \hat{\omega}t,$$

with $\hat{\omega}$ expanded as

$$(9.187) \qquad \hat{\omega} = \sum_{n=0}^{\infty} \omega_n(k)a^n.$$

We attempt to determine the $u_n(\theta)$ to be periodic functions, so that (9.185) is a periodic traveling wave solution of (9.179). To accomplish this it is necessary to specify the terms in series expansions of the frequency $\hat{\omega}$ appropriately. We note that we have included a dependence on k and a for the frequency $\hat{\omega}$ in (9.187).

Inserting (9.185) and (9.187) into (9.179) yields the equations

$$(9.188) \qquad (\omega_0 - ck)u_1' - \beta k^3 u_1''' = 0,$$
$$(9.189) \qquad (\omega_0 - ck)u_2' - \beta k^3 u_2''' = ku_1 u_1' - \omega_1 u_1',$$

and

$$(9.190) \quad (\omega_0 - ck)u_3' - \beta k^3 u_3''' = ku_1 u_2' + ku_2 u_1' - \omega_1 u_2' - \omega_2 u_1'$$

on equating like powers of a.

As the solution of (9.188) for u_1 we take

(9.191) $u_1(\theta) = \cos\theta$

and set

(9.192) $\omega_0(k) = ck - \beta k^3.$

This yields a periodic traveling wave solution corresponding to the linearized equation (9.180). Neglecting higher powers of a in (9.185)–(9.187) yields the result (9.183). With this choice for u_1, the right side of (9.189) becomes

(9.193) $ku_1 u_1' - \omega_1 u_1' = -k\sin\theta\cos\theta + \omega_1\sin\theta$

$$= -\frac{k}{2}\sin 2\theta + \omega_1\sin\theta.$$

The term $\omega_1\sin\theta$ yields a *secular term* in the expression for u_2 proportional to $\theta\sin\theta$. Since we seek periodic solutions, we must remove this term and this is the case if we set $\omega_1 = 0$. Then a solution of (9.189) is

(9.194) $u_2(\theta) = \frac{1}{12\beta k^2}\cos 2\theta.$

Continuing with (9.190) we obtain for the right side (since $\omega_1 = 0$)

(9.195) $ku_1 u_2' + ku_2 u_1' - \omega_2 u_1' = \left[-\frac{1}{24\beta k} + \omega_2\right]\sin\theta - \frac{1}{8\beta k}\sin 3\theta.$

Again the term involving $\sin\theta$ must be removed to avoid obtaining a secular term. Thus we set

(9.196) $\omega_2 = \frac{1}{24\beta k}$

and the solution $u_3(\theta)$ is

(9.197) $u_3(\theta) = \frac{1}{192\beta^2 k^4}\cos 3\theta.$

The full solution to order a^3 is

(9.198) $u(\theta) = a\cos\theta + \frac{a^2}{12\beta k^2}\cos 2\theta + \frac{a^3}{192\beta^2 k^4}\cos 3\theta$

$$+ O(a^4)$$

with θ defined as in (9.186) and $\hat{\omega}$ given as

$$(9.199) \qquad \hat{\omega} = \hat{\omega}(k, a) = ck - \beta k^3 + \frac{a^2}{24\beta k} + O(a^3).$$

The frequency $\hat{\omega}$ and, consequently, the wave speed are seen to depend both on the wave number k and the amplitude a.

We conclude this example by noting that on inserting the expression $u = u(\theta)$—with θ given as in (9.186)—into the Korteweg–deVries equation, it is possible to find an exact periodic traveling wave solution in terms of elliptic functions. If that solution is expanded for small amplitudes, the preceding results are reproduced. We do not carry out this discussion, which is given in the literature.

9.3. SINGULAR PERTURBATIONS AND BOUNDARY LAYER THEORY

In *singular perturbation theory* we are concerned with the study of partial differential equations that contain a small parameter that multiplies one or more of the highest derivative terms in the equations. Thus when that parameter is equated to zero either the order or the type (or both the order and type) of the given equation is changed. Generally, this means that a regular perturbation series solution proves inadequate to handle the initial and/or boundary data for the given problem. It thus becomes necessary to introduce boundary or initial layers where the solution of the given problem undergoes a rapid transition from a form that satisfies all the data given for the problem to a form represented by the perturbation series. The determination of the boundary layers and the approximate forms of the given equations in those regions forms the subject of *boundary layer theory*. The procedure whereby solutions valid in the boundary layers are identified with the perturbation series solution valid in the so-called outer region, is often called the *matching process*. By combining the perturbation and boundary layer solutions, a fairly good approximate description of the solution of the given problem can often be found for problems where the exact solution is difficult or impossible to determine or where the solution is not easy to interpret or evaluate.

It should be noted that singular perturbation theory is sometimes taken to encompass any problem where regular perturbation theory is inadequate for any reason. This may not involve the presence of a small parameter multiplying the highest derivative, but may be due to the presence of secular terms that result in a nonuniformity of the solution over an infinite region or the occurrence of a small parameter in the data for the problem (examples of these types have been given in Section 9.2). Nevertheless, we shall restrict our discussion in this section to the type of problem discussed in the preceding paragraph.

As it is somewhat complicated to present a general theory that encompasses all types of singular perturbation and boundary layer problems, we begin by considering two simple examples for a first order equation. Then we shall consider second and higher order equations from a general point of view and in a number of examples.

EXAMPLE 9.8. SINGULAR PERTURBATION OF A FIRST ORDER EQUATION

The *initial value problem* for the equation

$$(9.200) \qquad \epsilon(u_t + u_x) + u = \sin t, \qquad -\infty < x < \infty, t > 0$$

with

$$(9.201)^r \qquad\qquad\qquad u(x,0) = f(x)$$

where $0 < \epsilon \ll 1$ and $f(x)$ is a prescribed smooth function, has the solution

$$(9.202) \quad u(x,t) = \frac{1}{1+\epsilon^2}(\sin t - \epsilon \cos t) + \left[f(x-t) + \frac{\epsilon}{1+\epsilon^2} \right] e^{-t/\epsilon},$$

as is easily verified. (The solution may be obtained by the method of characteristics given in Chapter 2.)

If we attempt to solve (9.200)–(9.201) by using a conventional perturbation series, we set

$$(9.203) \qquad\qquad\qquad u(x,t) = \sum_{n=0}^{\infty} u_n(x,t)\epsilon^n.$$

Inserting (9.203) into (9.200) and equating like powers of ϵ yields the system

$$(9.204) \qquad\qquad\qquad u_0 = \sin t$$

and

$$(9.205) \qquad\qquad u_n = -\left(\frac{\partial u_{n-1}}{\partial t} + \frac{\partial u_{n-1}}{\partial x} \right), \qquad n \geq 1,$$

We easily conclude that

$$(9.206) \quad u_{2n} = (-1)^n \sin t; \qquad u_{2n+1} = (-1)^{n+1} \cos t; \qquad n \geq 0.$$

Thus the terms in the perturbation series (9.203) are uniquely specified without regard to the initial value (9.201) that $u(x,t)$ must satisfy. Furthermore, the perturbation series

$$(9.207) \quad u(x, t) = \left[\sum_{n=0}^{\infty} (-1)^n \epsilon^{2n} \right] \sin t - \epsilon \left[\sum_{n=0}^{\infty} (-1)^n \epsilon^{2n} \right] \cos t$$

$$= \frac{1}{1 + \epsilon^2} \sin t - \frac{\epsilon}{1 + \epsilon^2} \cos t$$

certainly does not satisfy the initial condition (9.201). The reason for this is that the *reduced problem* for (9.200)—where ϵ is set equal to zero—is not even a differential equation and, therefore, cannot absorb arbitrary initial values. This reduced problem characterizes the form of all equations arising from the conventional perturbation approach, as shown by (9.204)–(9.205), and none of them is a differential equation.

On comparing the perturbation solution (9.207) and the exact solution (9.202) we see that difference between both solutions is a term significant only in the region $0 \leqslant t \leqslant O(\epsilon)$. That is, for $t > O(\epsilon)$, the exponential $e^{-t/\epsilon}$ is small and can be neglected. Then the solution of (9.200)–(9.201) is well approximated by the perturbation result (9.207). However, within a layer of width $O(\epsilon)$ near the x-axis, the exponential term in (9.202) is significant and cannot be neglected. As we can see, it is this term in combination with the perturbation result (9.207) that enables the solution to satisfy the initial condition.

The existence of an initial layer of width $O(\epsilon)$ where the conventional perturbation series is not valid may be inferred from the fact that for $t \approx \epsilon$ we have $\sin t \approx \sin \epsilon \approx \epsilon$ and $\cos t \approx \cos \epsilon \approx 1$. Thus the first two terms in the perturbation series behave like

$$(9.208) \qquad \sin t - \epsilon \cos t \approx \epsilon - \epsilon = 0, \qquad t \approx \epsilon,$$

which implies that the two terms are of the same order in ϵ. Therefore, the perturbation series is not well ordered in the region where $t = O(\epsilon)$ and it is not expected to be a valid representation of the solution. Of course, we know that (9.207) is not valid near $t = 0$ since it fails to satisfy the initial condition. However, the foregoing argument determines an approximate description of the size of the region where the conventional perturbation method is invalid. This type of argument is similar to that given in Section 9.2 in connection with secular terms.

The significant conclusion that we have reached is that the perturbation result need not be discarded completely because of its failure to satisfy the initial condition. It need only be replaced by a different or modified approximation in an *initial layer* of width $O(\epsilon)$ near the x-axis. To study the equation in the initial layer we introduce the *stretching transformation*

$$(9.209) \qquad\qquad t = \epsilon^r \tau$$

where the positive constant r is to be specified. The equation (9.209) indicates that we wish to study (9.200) in a region where $t = O(\epsilon^r)$. Now we already have shown that the choice $r = 1$ is appropriate for our problem, but we wish to show this directly from the equation (9.200) by using *boundary layer* arguments.

We have

$$(9.210) \qquad \epsilon^{1-r}\hat{u}_\tau + \epsilon\hat{u}_x + \hat{u} = \sin \epsilon^r\tau = \epsilon^r\tau - \frac{\epsilon^{3r}\tau^3}{6} + O(\epsilon^{5r})$$

on using (9.209) in (9.200) with $\hat{u}(x, \tau) = u(x, \epsilon^r\tau)$. Now if $r = 1$, there is a balance between the terms $\hat{u}_\tau$ and $\hat{u}$ in (9.210), and they represent the leading terms for small ϵ in that equation. We need to retain the $\hat{u}_\tau$ term so that we have a differential equation in the initial layer that can absorb an initial condition. We would also like to retain the term $\hat{u}$ that occurs in the reduced equation for (9.200). Thereby, a smooth transition from the *initial layer* to the *outer region*, where the perturbation series (9.207) is valid, is expected to result. With $r > 1$, the term $\hat{u}_\tau$ is the leading term. Although this choice of r leads to equations for which the initial condition can be accounted, the initial layer seems to be too thin since the equations do not retain any part of the reduced equation. For $r < 1$, $\hat{u}$ is the leading term and nothing has been accomplished.

Putting $r = 1$ in (9.210) we obtain

$$(9.211) \qquad \hat{u}_\tau + \hat{u} + \epsilon\hat{u}_x = \sin \epsilon\tau = \epsilon\tau - \frac{\epsilon^3\tau^3}{6} + O(\epsilon^5).$$

The initial condition at $\tau = 0$ is

$$(9.212) \qquad\qquad\qquad \hat{u}(x, 0) = f(x).$$

We solve (9.212) by the perturbation method and set

$$(9.213) \qquad\qquad\qquad \hat{u}(x, \tau) = \sum_{n=0}^{\infty} \hat{u}_n(x, \tau)\epsilon^n.$$

Inserting (9.213) into (9.212) and equating like powers of ϵ yields

$$(9.214) \qquad\qquad\qquad \frac{\partial \hat{u}_0}{\partial \tau} + \hat{u}_0 = 0,$$

$$(9.215) \qquad\qquad\qquad \frac{\partial \hat{u}_1}{\partial \tau} + \hat{u}_1 = \tau - \frac{\partial \hat{u}_0}{\partial x},$$

and

$$(9.216) \qquad\qquad\qquad \frac{\partial \hat{u}_2}{\partial \tau} + \hat{u}_2 = -\frac{\partial \hat{u}_1}{\partial x}$$

as the leading equations. The initial conditions for these ordinary differential equations are

$$(9.217) \qquad \hat{u}_0(x,0) = f(x); \qquad \hat{u}_n(x,0) = 0; \qquad n \geqslant 1.$$

It is often the case that the boundary layer equations are ordinary differential equations.

The solutions of the initial value problems for the first two boundary layer terms $\hat{u}_n(x, \tau)$ are

$$(9.218) \qquad \hat{u}_0(x, \tau) = f(x)e^{-\tau}$$

and

$$(9.219) \qquad \hat{u}_1(x, \tau) = \tau - 1 + [1 - \tau f'(x)]e^{-\tau},$$

Thus

$$(9.220) \quad \hat{u}(x, \tau) \approx f(x)e^{-\tau} + \epsilon\{\tau - 1 + [1 - \tau f'(x)]e^{-\tau}\} + O(\epsilon^2).$$

In general, we now have to apply the matching process to specify unknown quantities that occur in the outer perturbation expansion. That is, we assume that the perturbation series (9.207) and the boundary layer (perturbation) series (9.213) have a common region of validity. Then we express both series in terms of a single set of variables and identify corresponding terms. Since the (outer) perturbation series in the present problem is completely specified, we carry out the matching to show how the solution undergoes a transition from its boundary layer form to the form valid in the outer region where $t > O(\epsilon)$. We have

$$(9.221) \qquad u(x, t) \approx \sum_{n=0}^{\infty} u_n(x, \epsilon\tau)\epsilon^n \approx \sum_{n=0}^{\infty} \hat{u}_n(x, \tau)\epsilon^n$$

in a common region of validity of both series that is assumed to exist. Thus

$$(9.222) \qquad u_0 + \epsilon u_1 = \sin(\epsilon\tau) - \epsilon\cos(\epsilon\tau) \approx \epsilon\tau - \epsilon + O(\epsilon^3)$$

and

$$(9.223) \quad \hat{u}_0 + \epsilon\hat{u}_1 = f(x)e^{-\tau} + \epsilon[\tau - 1] + \epsilon[1 - \tau f'(x)]e^{-\tau} \approx \epsilon\tau - \epsilon.$$

For the purpose of matching we assume that τ is large so that the exponential $e^{-\tau}$ can be neglected, but that $\epsilon\tau$ is small so that $\sin(\epsilon\tau)$ and $\cos(\epsilon\tau)$ can be approximated by the leading terms in their power series expansions. This procedure can be formalized by assuming, for example, that $\tau = O(\epsilon^{-1/2})$ so that τ is large for small ϵ but $\epsilon\tau = O(\epsilon^{1/2})$ which is small. Then the region where the matching is carried out corresponds to $t = \epsilon\tau = O(\epsilon^{1/2})$.

On comparing (9.222) and (9.223) we see that they both agree to the order of ϵ retained. Consequently, according to our perturbation and boundary layer results we find that the solution $u(x, t)$ of (9.200)–(9.202) is given as

$$(9.224) \quad u(x, t) \approx t - \epsilon + [f(x) - tf'(x) + \epsilon]e^{-t/\epsilon}, \qquad 0 \le t \le O(\epsilon)$$

and

$$(9.225) \quad u(x, t) \approx \frac{1}{1 + \epsilon^2}\sin t - \frac{\epsilon}{1 + \epsilon^2}\cos t \approx \sin t - \epsilon\cos t + O(\epsilon^2)$$

for $t > O(\epsilon)$. In the initial layer of width $O(\epsilon)$ the solution undergoes a rapid transition from the form (9.224) which satisfies the initial condition to the form (9.225) which is the (outer) perturbation solution. The solution (9.224) in the initial layer corresponds to the expansion of the exact solution (9.202) for small t as is easily verified. The fact that f and its derivatives are given as functions of x rather than $x - t$—corresponding to the characteristics $x - t = $ constant of (9.200)—as is the case in the exact solution (9.202), is not that important. Because of the rapid exponential decay of the term that involves f, this difference in the argument of that function is not significant.

To conclude this problem we note that if the term u is replaced by $-u$ in (9.200), we may still use the regular perturbation method to construct a series solution of the form (9.203). However, the leading order boundary layer equation will now be

$$(9.226) \qquad\qquad \frac{\partial \hat{u}_0}{\partial \tau} - \hat{u}_0 = 0$$

as is easily seen on proceeding as in the foregoing. The solution $\hat{u}_0 = f(x)e^\tau$ now grows exponentially as τ increases so that we do not obtain an initial layer within which a rapid transition of the solution to the form of the outer perturbation expansion takes place. The reason for the breakdown of the perturbation method and the boundary layer theory is that the problem with the term u replaced by $-u$ is unstable and has exponentially growing solutions. This is easily shown by finding the exact solution of the initial value problem or by the use of a stability analysis.

The general first order linear equation

$$(9.227) \qquad \epsilon(au_x + bu_t) + cu = d, \qquad -\infty < x < \infty, \quad t > 0$$

where $0 < \epsilon \ll 1$ and $a, b, c,$ and d are specified functions of x and t, with the initial condition

$$(9.228) \qquad\qquad u(x, 0) = f(x), \qquad -\infty < x < \infty,$$

yields a singular perturbation problem that may require the introduction of additional *internal boundary layers* apart from the initial layer for its solution. Thus if c vanishes on a curve in the region $t > 0$ where a and b do not vanish, the conventional perturbation series may become singular there. However, the solution of the full problem (9.227)–(9.228) is not expected to be singular along that curve. Consequently, a modified result that can be obtained by boundary layer methods must be introduced. In the following example a problem of this type is considered. We remark that the general process of determining perturbation and boundary or initial layer expansions and matching these results is sometimes referred to as the *method of matched asymptotic expansions.*

EXAMPLE 9.9. AN INTERNAL BOUNDARY LAYER

The first order equation

$$(9.229) \qquad \epsilon(u_t + u_x) + (t - 1)^2 u = 1, \qquad -\infty < x < \infty, t > 0$$

with the initial condition

$$(9.230) \qquad\qquad u(x, 0) = 0, \qquad -\infty < x < \infty$$

has the exact solution

$$(9.231) \qquad u(x, t) = \frac{1}{\epsilon} \exp\left[-\frac{(t - 1)^3}{3\epsilon} \right] \int_0^t \exp\left[\frac{(s - 1)^3}{3\epsilon} \right] ds,$$

as is easily verified. Because of the x-independent initial condition, we may consider (9.229)–(9.230) to be an ordinary differential equation problem. It is, nevertheless, of interest to study the phenomena that occur for the perturbation solution of this problem, as they occur for more general problems as well.

We begin by considering the *conventional perturbation series* solution

$$(9.232) \qquad\qquad u = \sum_{n=0}^{\infty} u_n \epsilon^n$$

of (9.229)–(9.230). We easily obtain as the leading terms

$$(9.233) \qquad u \approx u_0 + \epsilon u_1 = \frac{1}{(t - 1)^2} + \frac{2\epsilon}{(t - 1)^5}.$$

Not only does (9.233) not satisfy the initial condition (9.230) but it also is singular at $t = 1$, whereas the given equation (9.229) has no singularity there.

To deal with the initial condition we introduce the *stretching transformation*

$$(9.234) \qquad\qquad\qquad t = \epsilon \tau$$

which yields for (9.229)

$$(9.235) \qquad \hat{u}_\tau + \epsilon \hat{u}_x + \hat{u} - 2\epsilon\tau\hat{u} + \epsilon^2\tau^2\hat{u} = 1$$

where $\hat{u}(x, \tau) = u(x, \epsilon\tau)$. To obtain (9.234) we can use the approach given in Example 9.8. With

$$(9.236) \qquad \hat{u} = \sum_{n=0}^{\infty} \hat{u}_n \epsilon^n,$$

we have

$$(9.237) \qquad \frac{\partial \hat{u}_0}{\partial \tau} + \hat{u}_0 = 1$$

and

$$(9.238) \qquad \frac{\partial \hat{u}_1}{\partial \tau} + \hat{u}_1 = 2\tau\hat{u}_0 - \frac{\partial \hat{u}_0}{\partial x},$$

as the leading equations. The initial data for the $\hat{u}_n$ are

$$(9.239) \qquad \hat{u}_n(x, 0) = 0; \qquad n \geqslant 0.$$

The solutions are

$$(9.240) \qquad \hat{u}_0 = 1 - e^{-\tau},$$
$$(9.241) \qquad \hat{u}_1 = 2(\tau - 1) + (2 - \tau^2)e^{-\tau},$$

so that

$$(9.242) \qquad \hat{u} \approx 1 - e^{-\tau} + \epsilon\left[2(\tau - 1) + (2 - \tau^2)e^{-\tau}\right].$$

It is readily verified that (9.233) and (9.242) match one another if (9.233) is expanded for small t and (9.242) is expanded for large τ.

At t approaches unity, the (outer) perturbation series (9.233) breaks down. To determine the values of t where (9.233) first begins to break down, we note that when

$$(9.243) \qquad t - 1 = O(\epsilon^{1/3}),$$

the perturbation series (9.233) becomes disordered in that the second term is of the same order in ϵ as the leading term. This suggests that the width of the *internal boundary layer* region near $t = 1$ is given as $|t - 1| = O(\epsilon^{1/3})$. Further, we find that in the boundary layer

(9.244) $$u \approx O\left[(t-1)^{-2}\right] = O(\epsilon^{-2/3}).$$

Therefore, to study the solution of (9.229)–(9.230) near $t = 1$ we set

(9.245) $$t - 1 = \epsilon^{1/3}\tau; \qquad u = \epsilon^{-2/3}v.$$

This gives

(9.246) $$v_\tau + \tau^2 v + \epsilon^{1/3}v_x = 1.$$

The appropriate stretching exponents for ϵ in (9.245) could have been determined by the use of an argument similar to that given in the preceding problem, whereby a balancing of significant terms in the resulting equation (9.246) is carried out. A new feature for this problem is that not only an independent, but also the dependent variable was stretched. This is usually required for nonlinear problems and often arises when dealing with inhomogeneous linear equations.

To specify the solution $v(x, \tau)$ of (9.246) we require that it match the outer solution (9.233) as we approach the region $O(\epsilon) < t < 1 - O(\epsilon^{1/3})$ where the perturbation series is valid. That is, we require that as $\tau \to -\infty$, $v(x, \tau)$ tends to the form (9.233) expressed as a function of τ.

The boundary layer expansion for $v(x, \tau)$ is given in powers of $\epsilon^{1/3}$, in view of the form of (9.246), and we have

(9.247) $$v = \sum_{n=0}^{\infty} v_n \epsilon^{n/3}.$$

The leading equation is

(9.248) $$\frac{\partial v_0}{\partial \tau} + \tau^2 v_0 = 1,$$

whose general solution can be written as

(9.249) $$v_0(x, \tau) = \exp\left[-\frac{\tau^3}{3}\right] \int_{-\infty}^{\tau} \exp\left[\frac{\sigma^3}{3}\right] d\sigma + b \exp\left[-\frac{\tau^3}{3}\right],$$

where the arbitrary constant b is as yet unspecified. To match (9.249) with (9.233) we integrate by parts to obtain

(9.250) $$\int_{-\infty}^{\tau} \exp\left[\frac{\sigma^3}{3}\right] d\sigma = \frac{1}{\tau^2}\exp\left[\frac{\tau^3}{3}\right] + 2\int_{-\infty}^{\tau} \frac{1}{\sigma^3}\exp\left[\frac{\sigma^3}{3}\right] d\sigma$$

with τ assumed to be negative. A further integration by parts yields

$$(9.251) \quad \int_{-\infty}^{\tau} \exp\left[\frac{\sigma^3}{3}\right] d\sigma = \left(\frac{1}{\tau^2} + \frac{2}{\tau^5}\right)\exp\left[\frac{\tau^3}{3}\right] + 10 \int_{-\infty}^{\tau} \frac{1}{\sigma^6}\exp\left[\frac{\sigma^3}{3}\right] d\sigma$$

Thus

$$(9.252) \quad v_0(x, \tau) = \frac{1}{\tau^2} + \frac{2}{\tau^5} + 10\exp\left[-\frac{\tau^3}{3}\right]\int_{-\infty}^{\tau} \frac{1}{\sigma^6}\exp\left[\frac{\sigma^3}{3}\right] d\sigma,$$

where we have put $b = 0$, since the solution would otherwise grow exponentially as $\tau \to -\infty$, and that is ruled out by the form of (9.233). If we now express (9.233) in terms of τ, we have

$$(9.253) \qquad\qquad\qquad u \approx \epsilon^{-2/3}\left[\frac{1}{\tau^2} + \frac{2}{\tau^5}\right].$$

Since $u = \epsilon^{-2/3}v$—in view of (9.245)—we find that the *internal boundary layer expression* (9.252) matches the *outer solution* (9.233).

To complete our discussion of the solution we need to examine what happens in the region above the internal boundary layer, that is, when $t > 1 + O(\epsilon^{1/3})$. It is expected that the outer solution (9.233) should be valid there. However, for that to be the case, the outer solution must match the internal boundary layer solution as $\tau \to +\infty$. To show this we write $v_0(x, \tau)$ as

$$(9.254) \quad v_0 = \exp\left[-\frac{\tau^3}{3}\right]\left\{\int_{-\infty}^{a} \exp\left[\frac{\sigma^3}{3}\right] d\sigma + \int_{a}^{\tau}\exp\left[\frac{\sigma^3}{3}\right] d\sigma\right\}$$

where both a and τ are positive and $a < \tau$—note that we have put $b = 0$ in (9.249). Since $0 < a < \tau$, the first integral in (9.254) contributes an exponentially small term to the overall result. Integrating by parts in the second integral as was done previously, readily yields an expression that matches the outer solution (9.233) as $\tau \to \infty$, since the contributions from the lower limit result in exponentially small terms.

We have thus shown that a satisfactory description of the solution of (9.347)–(9.348) can be obtained by combining *perturbation* and *boundary layer methods*. Although an exact solution is available for this problem, the approximate results are generally easier to evaluate.

Next we discuss the singular perturbation of a second order linear hyperbolic equation that yields an initial layer as well as secular terms.

EXAMPLE 9.10. THE SINGULAR PERTURBATION OF A HYPERBOLIC EQUATION

The *second order hyperbolic equation*

$$(9.255) \qquad \epsilon\left(u_{tt} - c^2 u_{xx}\right) + u_t + a u_x = 0$$

with constant coefficients and $0 < \epsilon \ll 1$, has the property that with $\epsilon = 0$ it reduces to an equation of first order. To examine some of the problems that can result from this fact we consider the *Cauchy problem* for (9.255) with initial data on the *x*-axis given as

$$(9.256) \qquad u(x,0) = f(x); \qquad u_t(x,0) = g(x).$$

We attempt to solve (9.255)–(9.256) by the use of a *perturbation series*

$$(9.257) \qquad u(x,t) = \sum_{n=0}^{\infty} u_n(x,t)\epsilon^n.$$

Inserting (9.257) into (9.255) and equating like powers of ϵ yields the recursive system

$$(9.258) \qquad \begin{cases} \dfrac{\partial u_0}{\partial t} + a\dfrac{\partial u_0}{\partial x} = 0 \\[2mm] \dfrac{\partial u_n}{\partial t} + a\dfrac{\partial u_n}{\partial x} = -\dfrac{\partial^2 u_{n-1}}{\partial t^2} + c^2\dfrac{\partial^2 u_{n-1}}{\partial x^2}; \qquad n \geqslant 1 \end{cases}$$

with the initial conditions

$$(9.259) \qquad \begin{cases} u_0(x,0) = f(x); & u_n(x,0) = 0; & n \geqslant 1 \\[2mm] \dfrac{\partial u_0(x,0)}{\partial t} = g(x); & \dfrac{\partial u_n(x,0)}{\partial t} = 0; & n \geqslant 1 \end{cases}.$$

Since the equations for $u_n(x,t)$ are each of first order, only one initial condition can be assigned for each u_n at $t = 0$. Therefore, the initial value problems for the u_n cannot be solved in general. The singular nature of the perturbation is thereby brought into evidence. As a result, we cannot expect the perturbation series to be valid near the *x*-axis. Nevertheless, we expect the series (9.257) to provide an approximate description of the solution $u(x,t)$ away from the *x*-axis.

Using the methods of Chapter 2, we easily obtain as the general solutions of the first few equations (9.258) for the $u_n(x,t)$,

$$(9.260) \qquad u_0(x,t) = F(x - at),$$

$$(9.261) \qquad u_1(x,t) = t(c^2 - a^2)F''(x - at) + G(x - at),$$

and

$$(9.262) \quad u_2(x, t) = \tfrac{1}{2}t^2(c^2 - a^2)F''''(x - at) + 2at(c^2 - a^2)F'''(x - at)$$
$$+ t(c^2 - a^2)G''(x - at) + H(x - at)$$

where F, G, and H are arbitrary functions.

Assuming that $F(x)$, $G(x)$, and $H(x)$ are uniformly bounded functions together with their derivatives for all x, we see from (9.260)–(9.262) that *secular terms* with coefficients ϵt and $(\epsilon t)^2$ arise in the perturbation expansion (9.257). Thus the series (9.257) is not expected to be valid for $\epsilon t = O(1)$ or when $t = O(1/\epsilon)$.

We have demonstrated that the perturbation series (9.257) is not only invalid near the initial line but that it also breaks down after a sufficiently long time. Additionally, we observe from (9.260)–(9.262) that the terms $u_n(x, t)$ in the perturbation series are waves that travel to the right or left with speed $|a|$ (depending on the sign of a). Now we have already shown that the maximum speed at which disturbances for the hyperbolic equation (9.255) can travel is the characteristic speed. This speed equals c and if $|a| > c$ it can happen that disturbances as described by our perturbation approximation can travel at a speed exceeding the characteristic speed. Since this is theoretically not possible, it would appear that we have to reject the perturbation series completely for this reason if $|a| > c$.

An application of the *stability analysis* of Section 3.5 shows that (9.255) is unstable if $|a| > c$. If we insert the normal mode solution (3.123) into (9.255) we obtain for $\lambda(k)$,

$$(9.263) \qquad\qquad \epsilon\lambda^2 + \lambda + \epsilon c^2 k + iak = 0.$$

Noting that $0 < \epsilon \ll 1$ and assuming moderate values of k, we obtain two approximate solutions for λ,

$$(9.264) \qquad \lambda_1 \approx -\frac{1}{\epsilon}; \qquad \lambda_2 \approx -iak - \epsilon(c^2 - a^2)k^2.$$

Thus if $|a| > c$, the real part of λ_2 is positive and the problem is unstable. We shall, therefore, assume that $|a| < c$ in (9.255) so that disturbances associated with the equations (9.258) travel at slower speeds than the characteristic speed c. It can be shown that the real parts of λ_1 and λ_2 are negative for all $k \neq 0$ if $|a| < c$, so that (9.255) is an equation of dissipative type.

To specify the unknown functions in the perturbation series we must find a way to relate it to the initial data for the problem. This can be accomplished by the use of *boundary layer theory*. Since the perturbation series is expected to be valid from some time $t > 0$ on, we assume there is a t-interval where the solution of (9.255)–(9.256) undergoes a rapid transition from a form that satisfies the initial data to the perturbation series form. This interval which represents a layer near the x-axis has been called an *initial layer* in the preceding problems, but is more generally referred to as a *boundary layer*.

To determine the appropriate form of (9.255) in the boundary layer we introduce the *stretching transformation*

$$(9.265) \qquad \tau = \frac{t}{\epsilon^r}$$

where the positive constant r is to be specified. Since ϵ is small, the τ variable is large even for small or moderate values of t. Thus the region near the x-axis is stretched out. Also, since $t = \epsilon^r\tau$, (9.265) indicates that we are concerned with small values of t. We insert (9.265) into (9.255) to obtain

$$(9.266) \qquad \epsilon^{1-2r}\hat{u}_{\tau\tau} - \epsilon c^2\hat{u}_{xx} + \epsilon^{-r}\hat{u}_\tau + a\hat{u}_x = 0$$

where we have set

$$(9.267) \qquad \hat{u}(x,\tau) = u(x,\epsilon^r\tau).$$

The initial data for $\hat{u}$ at $\tau = 0$ are

$$(9.268) \qquad \hat{u}(x,0) = f(x); \qquad \hat{u}_\tau(x,0) = \epsilon^r g(x).$$

To specify the constant r in (9.265) we argue that the most significant terms in (9.265) are those with the lowest power (possibly negative) in ϵ. On solving the equation (9.266) by a perturbation method as is our intention, the most significant terms in (9.266) determine the basic form of the equations (in either homogeneous or inhomogeneous form) that must be satisfied by the terms in the perturbation series. Since each term is required to satisfy two initial conditions at $\tau = 0$, we require that $u_{\tau\tau}$ be retained as a leading term. Clearly, this requires that $1 - 2r \leqslant -r$, as is seen on comparing terms in (9.266). If we choose $r > 1$, only $\hat{u}_{\tau\tau}$ occurs as the leading term. However, with $r = 1$, $\hat{u}_{\tau\tau}$ and $\hat{u}_\tau$ are both of the same order in ϵ, and this choice also yields a balance between the term $\hat{u}_{\tau\tau}$ that is significant in the boundary layer region, and the term $\hat{u}_\tau$ that occurs (as u_t) in the reduced problem in the outer region where the perturbation series (9.257) is valid. Therefore, we set $r = 1$. Multiplying through by ϵ in (9.266) yields

$$(9.269) \qquad \hat{u}_{\tau\tau} + \hat{u}_\tau + \epsilon a\hat{u}_x - \epsilon^2 c^2\hat{u}_{xx} = 0$$

with the data

$$(9.270) \qquad \hat{u}(x,0) = f(x); \qquad \hat{u}_\tau(x,0) = \epsilon g(x).$$

To solve (9.269)–(9.270), we introduce the *boundary layer expansion*

$$(9.271) \qquad \hat{u}(x,\tau) = \sum_{n=0}^{\infty} \hat{u}_n(x,\tau)\epsilon^n$$

into (9.269)–(9.270) and obtain the recursive system of equations

(9.272) $$\frac{\partial^2 \hat{u}_0}{\partial \tau^2} + \frac{\partial \hat{u}_0}{\partial \tau} = 0,$$

(9.273) $$\frac{\partial^2 \hat{u}_1}{\partial \tau^2} + \frac{\partial \hat{u}_1}{\partial \tau} = -a \frac{\partial \hat{u}_0}{\partial x},$$

(9.274) $$\frac{\partial^2 \hat{u}_n}{\partial \tau^2} + \frac{\partial \hat{u}_n}{\partial \tau} = -a \frac{\partial \hat{u}_{n-1}}{\partial x} + c^2 \frac{\partial^2 \hat{u}_{n-2}}{\partial x^2}, \qquad n \geq 2,$$

on equating like powers of ϵ. The initial data are

(9.275) $$\hat{u}_0(x,0) = f(x); \qquad \hat{u}_n(x,0) = 0; \qquad n \geq 1,$$

(9.276) $$\frac{\partial \hat{u}_0(x,0)}{\partial \tau} = 0; \qquad \frac{\partial \hat{u}_1(x,0)}{\partial \tau} = g(x); \qquad \frac{\partial \hat{u}_n(x,0)}{\partial \tau} = 0; \qquad n \geq 2.$$

The boundary layer equations are all ordinary differential equations with initial data at $\tau = 0$ and are easily solved. We find that for $n = 1$ and 2,

(9.277) $$\hat{u}_0(x, \tau) = f(x)$$

and

(9.278) $$\hat{u}_1(x, \tau) = (g(x) + af'(x))(1 - e^{-\tau}) - a\tau f'(x).$$

The significant feature in $\hat{u}_1$ is the presence of the exponential $e^{-\tau} = e^{-t/\epsilon}$ which decays rapidly as τ or t increases and does not play a role when we consider the (outer) perturbation expansion (9.257). To specify the unknown functions in (9.257) we need to match (9.271) and (9.257), each of which are different representations of the unique solution $u(x, t)$. Although (9.271) is assumed to be valid in the boundary layer region near the x-axis whose width is of order ϵ, and the perturbation series is valid in some region away from the x-axis, it is assumed that they have a common region of validity. There are systematic procedures such as the method of intermediate limits for carrying out the matching process; however, we carry out the matching in a less formal manner.

In terms of the perturbation and boundary layer expansions for $u(x, t)$ we have

(9.279) $$u(x, t) \approx \sum_{n=0}^{\infty} u_n(x, \epsilon\tau)\epsilon^n \approx \sum_{n=0}^{\infty} \hat{u}_n(x, \tau)\epsilon^n; \qquad t = \epsilon\tau.$$

We assume that $\epsilon\tau$ is sufficiently small such that the terms $u_n(x, \epsilon\tau)$ can be well approximated by the leading terms in their series expansions in $t = \epsilon\tau$.

However, τ must be large enough that the exponentials $e^{-\tau}$ in the $\hat{u}_n(x, \tau)$ terms can be neglected. If ϵ is sufficiently small, since $t = \epsilon\tau$, these requirements are not inconsistent (see Example 9.8). Using (9.260)–(9.261) gives

$$(9.280) \qquad u_0 + \epsilon u_1 = F(x) - \epsilon a\tau F'(x) + \epsilon G(x) + O(\epsilon^2),$$

and neglecting the $e^{-\tau}$ terms gives

$$(9.281) \quad \hat{u}_0 + \epsilon \hat{u}_1 = f(x) - \epsilon a\tau f'(x) + \epsilon g(x) + \epsilon a f'(x) + O(\epsilon^2).$$

Comparing like powers of ϵ in (9.279) yields

$$(9.282) \qquad F(x) = f(x); \qquad G(x) = g(x) + af'(x)$$

on using (9.280)–(9.281). Consequently, the perturbation series (9.257) has the form

$$(9.283) \quad u(x, t) = f(x - at) + \epsilon\left[t(c^2 - a^2)f''(x - at)\right.$$
$$\left. + g(x - at) + af'(x - at)\right] + O(\epsilon^2).$$

Further terms in the perturbation series can be determined by carrying this matching procedure out to higher orders.

The perturbation series (9.283) breaks down when $t = O(1/\epsilon)$. A leading order approximation to the solution for all time may be obtained as follows. On inserting (9.257) into (9.255) we obtain

$$(9.284) \qquad \frac{\partial u_0}{\partial t} + a\frac{\partial u_0}{\partial x} + \epsilon\left[\frac{\partial u_1}{\partial t} + a\frac{\partial u_1}{\partial x} + \frac{\partial^2 u_0}{\partial t^2} - c^2\frac{\partial^2 u_0}{\partial x^2}\right] = O(\epsilon^2).$$

From (9.284) we see that

$$(9.285) \qquad \frac{\partial u_0}{\partial t} = -a\frac{\partial u_0}{\partial x} + O(\epsilon)$$

so that

$$(9.286) \qquad \frac{\partial^2 u_0}{\partial t^2} = a^2\frac{\partial^2 u_0}{\partial x^2} + O(\epsilon)$$

if we assume that the solutions $u_n(x, t)$ are smooth. Inserting (9.286) into (9.284) yields, to the same level of approximation,

$$(9.287) \qquad \frac{\partial u_0}{\partial t} + a\frac{\partial u_0}{\partial x} + \epsilon\left[\frac{\partial u_1}{\partial t} + a\frac{\partial u_1}{\partial x} + (a^2 - c^2)\frac{\partial^2 u_0}{\partial x^2}\right] = O(\epsilon^2).$$

Collecting like powers of ϵ, we obtain equations for u_0 and u_1 whose solutions are (9.260) and (9.261). Clearly, the term $(a^2 - c^2)\partial^2 u_0/\partial x^2$ is the one that gives rise to the secular term in (9.261). We may avoid this secularity by regrouping the terms in (9.287) and writing the equation as

$$(9.288) \quad \left[\frac{\partial u_0}{\partial t} + a\frac{\partial u_0}{\partial x} + \epsilon(a^2 - c^2)\frac{\partial^2 u_0}{\partial x^2}\right] + \epsilon\left[\frac{\partial u_1}{\partial t} + a\frac{\partial u_1}{\partial x}\right] = O(\epsilon^2).$$

The first bracketed term on the left is equated to zero so that $u_0(x, t)$ satisfies a *diffusion equation*. The initial data for $u_0(x, t)$ may be taken to be (9.283) for $u(x, t)$ evaluated at some time $t = t_0 \leqslant O(1/\epsilon)$. It may be noted that according to results given in Sections 1.2 and 5.7 we expect the solution of (9.255) to be characterized by a diffusion equation for large t.

We continue our consideration of singular perturbation problems for second order equations by studying the *linear elliptic equation*

$$(9.289)$$
$$\epsilon\left(Au_{xx} + 2Bu_{xy} + Cu_{yy} + Du_x + Eu_y + Fu\right) + au_x + bu_y + cu = g$$

where $0 < \epsilon \ll 1$ and $u(x, y)$ is specified on the boundary of the given region. We assume that $a^2 + b^2 > 0$ so that the *reduced equation* for (9.289),

$$(9.290) \qquad\qquad au_x + bu_y + cu = g,$$

obtained on setting $\epsilon = 0$ in (9.289) is of first order.

The difficulties that can arise out of an attempt to solve (9.289) by using a regular perturbation series (9.16) result from the fact that each term in the series satisfies a first order equation of the form (9.290). Consequently, the data for the given equation (9.289) may *overdetermine* the solutions of the equations for the terms in the perturbation series. Also, *singularities* in the data for (9.289) must be carried along the characteristics of the reduced equation (9.290) and, therefore, they occur in the terms of the regular perturbation series. The elliptic equation (9.289) has no real characteristic and its solutions must, therefore, be smooth functions. Further, it may happen that *characteristic initial value problems* can occur for the terms in the perturbation series or that these terms may become singular in the interior of the given region for the boundary value problem for (9.289). These and other difficulties can be remedied by the introduction of appropriate boundary layers near portions of the boundary or in the interior of the given region.

The construction of a perturbation solution for the boundary value problem for (9.289) over an arbitrary region is not a simple matter, in general. To appreciate some of the problems that can arise in solving (9.289), we first consider a simple example and then proceed to a more general discussion.

EXAMPLE 9.11. A SINGULAR PERTURBATION PROBLEM FOR AN ELLIPTIC EQUATION

We consider the *elliptic equation*

$$(9.291) \qquad \epsilon(u_{xx} + u_{yy}) + u_x + bu_y = 0,$$

over the semi-infinite region $0 < x < \infty$ and $0 < y < L$, with $0 < \epsilon \ll 1$ and $b = $ constant. The boundary conditions are

$$(9.292) \quad u(x,0) = f(x); \qquad u(0, y) = g(y); \qquad u(x, L) = h(x).$$

We assume that f and h vanish as $x \to \infty$ and require that $u(x, y) \to 0$ as $x \to \infty$ for $0 \leqslant y \leqslant L$.

As a first step in solving (9.291)–(9.292) by the perturbation method, we expand u as

$$(9.293) \qquad u(x, y) = \sum_{n=0}^{\infty} u_n(x, y)\epsilon^n$$

and insert (9.293) into (9.291)–(9.292). Then, on equating like powers of ϵ, we obtain for u_0,

$$(9.294) \qquad \frac{\partial u_0}{\partial x} + b\frac{\partial u_0}{\partial y} = 0,$$

and for the u_n,

$$(9.295) \qquad \frac{\partial u_n}{\partial x} + b\frac{\partial u_n}{\partial y} = -\left(\frac{\partial^2 u_{n-1}}{\partial x^2} + \frac{\partial^2 u_{n-1}}{\partial y^2}\right), \qquad n \geqslant 1.$$

The boundary conditions for u_0 are identical to those for u [i.e., (9.292)], whereas $u_n(x, y)$ vanishes on all three boundary lines if $n \geqslant 1$.

The general solution of (9.294) is

$$(9.296) \qquad u_0(x, y) = F(y - bx)$$

where $F(z)$ is an arbitrary function. For $u_1(x, y)$ we obtain

$$(9.297) \qquad u_1(x, y) = F_1(y - bx) - \left(\frac{1 + b^2}{b}\right)yF''(y - bx)$$

where $F_1(z)$ is arbitrary. Alternatively, the solution can be written as

$$(9.298) \qquad u_1(x, y) = G_1(y - bx) - (1 + b^2)xF''(y - bx),$$

where $G_1(z)$ is arbitrary.

The lines $y - bx = $ constant are the *characteristic lines* for (9.294) and (9.295). The solution $u_0 = F(y - bx)$ is constant on the characteristic lines. If $b \neq 0$, the characteristics intersect two of the boundary lines within the given region for the problem. If $b = 0$, the boundaries $y = 0$ and $y = L$ coincide with characteristic lines. In either case it is not possible to satisfy the full set of boundary conditions on $u_0(x, y)$ since the boundary values f, g, and h are assigned arbitrarily. If $b > 0$, for example, we may set

$$(9.299) \qquad\qquad u_0 = f\left(x - \frac{y}{b}\right)$$

in the region $x > y/b$ and satisfy the condition at $y = 0$ or we may set

$$(9.300) \qquad\qquad u_0 = h\left(x - \frac{y - L}{b}\right)$$

and satisfy the condition at $y = L$. We cannot determine from the perturbation series (9.293) and the resulting equations for its terms, which of these boundary conditions should be assigned. In any case, since the full set of boundary conditions for u_0 and the u_n cannot be applied, we must introduce *boundary layers* to deal with the lost boundary conditions.

Since each boundary line is a possible candidate for a boundary layer, we introduce stretching transformations that emphasize neighborhoods of each of the boundary lines. If $b \neq 0$, we set

$$(9.301) \qquad\qquad y = \epsilon\eta$$

near $y = 0$ and

$$(9.302) \qquad\qquad y - L = \epsilon\eta$$

near $y = L$. With $v = v(x, \eta)$ we then have the *boundary layer equation*

$$(9.303) \qquad\qquad v_{\eta\eta} + bv_\eta + \epsilon v_x + \epsilon^2 v_{xx} = 0,$$

which replaces (9.291). With v expanded in powers of ϵ and v_0 as its leading term, we find that

$$(9.304) \qquad\qquad \frac{\partial^2 v_0}{\partial \eta^2} + b\frac{\partial v_0}{\partial \eta} = 0$$

and

$$(9.305) \qquad\qquad v_0(x, \eta) = \alpha(x) + \beta(x)e^{-b\eta}$$

where α and β are arbitrary functions. Now if $b > 0$, the exponential term in

(9.305) decreases as η increases. Therefore, the boundary layer for this case must lie near $y = 0$. If $b < 0$, the exponential decreases as η decreases and—in view of (9.302)—the boundary layer should be located near $y = L$. In either case the boundary layer width is $O(\epsilon)$.

However, if $b = 0$ and we retain the stretching transformations (9.301)–(9.302), we find that v_0 is given as

$$(9.306) \qquad v_0(x, \eta) = \alpha(x) + \beta(x)\eta,$$

so that no boundary layer effect occurs at either $y = 0$ or $y = L$, since there is no exponential decay. In this case we easily conclude that the appropriate stretchings are

$$(9.307) \qquad y = \epsilon^{1/2}\eta$$

and

$$(9.308) \qquad y - L = \epsilon^{1/2}\eta$$

and the *boundary layer equation* becomes

$$(9.309) \qquad v_{\eta\eta} + v_x + \epsilon v_{xx} = 0.$$

Then $v_0(x, \eta)$ satisfies the *parabolic equation*

$$(9.310) \qquad \frac{\partial^2 v_0}{\partial \eta^2} + \frac{\partial v_0}{\partial x} = 0.$$

The boundary layer width is $O(\epsilon^{1/2})$, and there are boundary layers near $y = 0$ and $y = L$ as shown in the following.

Near $x = 0$ we set

$$(9.311) \qquad x = \epsilon\xi$$

in (9.291) and obtain the *boundary layer equation*

$$(9.312) \qquad w_{\xi\xi} + w_\xi + \epsilon b w_y + \epsilon^2 w_{yy} = 0,$$

with $w = w(\xi, y)$. If w expanded in powers of ϵ, we easily find that the leading term w_0 is given as

$$(9.313) \qquad w_0(\xi, y) = \gamma(y) + \delta(y)e^{-\xi},$$

so that a boundary layer can be located at $x = 0$ because of the exponentially decaying term in (9.313).

For the case $b \neq 0$, we now assume that $b > 0$ and complete the perturbation solution of the problem. Since we have shown that there can be boundary layers near $x = 0$ and $y = 0$ if $b > 0$, we specify $u_0 = F(y - bx)$ to satisfy the boundary condition at $y = L$. This gives

$$(9.314) \qquad u_0(x, y) = h\left(x - \frac{y - L}{b}\right).$$

The boundary layer solution $v_0(x, \eta)$—that is, (9.305)—near $y = 0$ is required to satisfy the boundary condition

$$(9.315) \qquad v_0(x, 0) = f(x)$$

and the matching condition

$$(9.316) \qquad \lim_{\eta \to \infty} v_0(x, \eta) = \lim_{y \to 0} u_0(x, y).$$

This states that for large η the boundary layer solution $v_0(x, \eta)$ must agree with the outer solution $u_0(x, y)$ evaluated for small y. We readily find that $v_0(x, \eta)$ is given as

$$(9.317) \qquad v_0(x, \eta) = h\left(x + \frac{L}{b}\right) + \left[f(x) - h\left(x + \frac{L}{b}\right)\right]e^{-b\eta}.$$

For the boundary layer near $x = 0$ we have the boundary condition

$$(9.318) \qquad w_0(0, y) = g(y)$$

and the matching condition

$$(9.319) \qquad \lim_{\xi \to \infty} w_0(\xi, y) = \lim_{x \to 0} u_0(x, y)$$

with w_0 defined in (9.313) and u_0 given in (9.314). Solving for w_0 gives

$$(9.320) \qquad w_0(\xi, y) = h\left(\frac{L - y}{b}\right) + \left[g(y) - h\left(\frac{L - y}{b}\right)\right]e^{-\xi}.$$

By inspection of the outer and boundary layer solutions we observe that the solution of (9.291)–(9.292) in the case where $b > 0$ can be expressed as

$$(9.321) \quad u(x, y) \approx h\left(x - \frac{y - L}{b}\right) + \left[f(x) - h\left(x + \frac{L}{b}\right)\right]\exp\left[-\frac{by}{\epsilon}\right]$$
$$+ \left[g(y) - h\left(\frac{L - y}{b}\right)\right]\exp\left[-\frac{x}{\epsilon}\right].$$

The terms containing the exponentials are significant only within their respective boundary layers. Apart from exponentially small terms, $u(x, y)$ satisfies the boundary conditions at $y = 0$, $x = 0$, and $y = L$ and vanishes as $x \to \infty$ since f and h vanish there. We do not consider effects due to possible incompatibilities of the boundary values at the points $(x, y) = (0, 0)$ and $(x, y) = (0, L)$. The uniform expression (9.321) is often termed a *composite expansion* and can be constructed in a systematic fashion. We do not pursue this matter here.

If $b < 0$ in (9.291), the form of the solution is similar to that in (9.321) except that the boundary layer is shifted to $y = L$. If $b = 0$, the leading term in the (outer) perturbation expansion is

$$(9.322) \qquad u_0(x, y) = F(y).$$

Since $F(y)$ is constant on $y = 0$ and $y = L$, it cannot satisfy the boundary conditions there. We could choose $F(y) = g(y)$ and thereby satisfy the boundary condition at $x = 0$. However, $u(x, y)$ must vanish as $x \to \infty$, and the solution

$$(9.323) \qquad u_0(x, y) = g(y)$$

remains constant as x varies and does not satisfy the condition at infinity unless $g(y) = 0$. Since $g(y)$ does not vanish, in general, and is, in fact, specified arbitrarily, we must put $F(y) = 0$ and obtain as the outer solution

$$(9.324) \qquad u \approx u_0 = 0,$$

for otherwise u would not vanish as $x \to \infty$. Consequently, we must use not only the boundary layers at $y = 0$ and $y = L$ but also the boundary layer at $x = 0$.

The boundary layer at $x = 0$ is the easiest to consider. Using (9.318)–(9.319) we conclude that

$$(9.325) \qquad w_0(\xi, y) = g(y)e^{-\xi},$$

since $u_0 = 0$. Near $y = 0$ and $y = L$ we have the *parabolic boundary layer equation* for $v_0 = v_0(x, \eta)$,

$$(9.326) \qquad \frac{\partial^2 v_0}{\partial \eta^2} + \frac{\partial v_0}{\partial x} = 0,$$

as given in (9.310). In the boundary layer near $y = 0$ the boundary condition is

$$(9.327) \qquad v_0(x, 0) = f(x),$$

whereas near $y = L$ the boundary condition is

$$(9.328) \qquad\qquad v_0(x,0) = h(x).$$

As $x \to \infty$ we require that $v_0(x, \eta) \to 0$. Also, as $\eta \to \infty$ in the boundary layer near $y = 0$ and $\eta \to -\infty$ in the boundary layer near $y = L$, we must have $v_0 \to 0$, because v_0 must match the outer solution u_0 that vanishes.

With x thought of as the time variable, (9.326) is a *backward heat equation*. Thus we cannot assign values for $v_0(x, \eta)$ at $x = 0$ and solve (9.326) for $x > 0$, for such a problem would not be well posed. Instead, we must place conditions on v_0 at some $x = x_0 > 0$ and solve for $x < x_0$. The only conditions given are at infinity, as we have shown. We may think of the problem as being given with data at $x = x_0$ and then let $x_0 \to \infty$. The solution is then given for all $x < \infty$. The result is equivalent to a *steady-state problem* for the ordinary heat equation where the effects of the initial temperature distribution have died out. We can think of that problem as being given for all $t > -\infty$. The solution to the steady-state problem for the heat equation and to our problem for (9.326) is unique if we assume the solution is uniformly bounded.

By adapting the results of Example 5.5, we conclude that the solution of (9.326)–(9.327) with $v_0(\infty, \eta) = 0$ is

$$(9.329) \qquad v_0(x, \eta) = \frac{\eta}{\sqrt{4\pi}} \int_x^\infty \exp\left[-\frac{\eta^2}{4(\sigma - x)} \right] \frac{f(\sigma)}{(\sigma - x)^{3/2}} \, d\sigma.$$

For the problem (9.326) and (9.328) the solution is

$$(9.330) \qquad v_0(x, \eta) = \frac{\eta}{\sqrt{4\pi}} \int_x^\infty \exp\left[-\frac{\eta^2}{4(\sigma - x)} \right] \frac{h(\sigma)}{(\sigma - x)^{3/2}} \, d\sigma.$$

We note that as $|\eta| \to \infty$ both solutions decay exponentially, and they both vanish as $x \to \infty$.

A *composite* expression for the solution may be given as

$$(9.331)$$

$$u(x, y) \approx g(y)e^{-x/\epsilon} + \frac{y}{\sqrt{4\epsilon\pi}} \int_x^\infty \exp\left[-\frac{y^2}{4\epsilon(\sigma - x)} \right] \frac{f(\sigma)}{(\sigma - x)^{3/2}} \, d\sigma$$

$$+ \frac{y - L}{\sqrt{4\epsilon\pi}} \int_x^\infty \exp\left[-\frac{(y - L)^2}{4\epsilon(\sigma - x)} \right] \frac{h(\sigma)}{(\sigma - x)^{3/2}} \, d\sigma.$$

We again do not consider the behavior of $u(x, y)$ at $(0, 0)$ and $(0, L)$. The solution is exponentially small when $x > O(\epsilon)$, $y > O(\sqrt{\epsilon})$ and $|y - L| > O(\sqrt{\epsilon})$.

We remark that near $x = 0$ and in the case where $b \neq 0$, we were led to consider ordinary differential equations in the boundary layer regions, as had been the case in the problems considered previously. However, with $b = 0$ the boundary layer equations valid near $y = 0$ and $y = L$ turned out to be parabolic partial differential equations. This can be attributed to the fact that these boundary lines coincided with the characteristics of the reduced equation for (9.291) with $b = 0$. The occurrence of parabolic boundary layer equations when portions of the boundary coincide with the characteristics of the reduced equation is generally the case, as shown later.

To study the singular perturbation problem for the elliptic equation (9.289), it is convenient to transform the coordinates so as to simplify the form of the reduced equation

$$(9.332) \qquad au_x + bu_y + cu = g.$$

We assume that the characteristic curves for (9.332) do not intersect in the region given for (9.289). Then the characteristics and their orthogonal trajectories may be introduced as a new set of coordinates in (9.289). The transformed equation can be written as

$$(9.333) \quad \epsilon\left(Au_{xx} + 2Bu_{xy} + Cu_{yy} + Du_x + Eu_y + Fu\right) - u_x + cu = g$$

where we have retained the original notation for the coefficients and the variables for the sake of convenience. The type of an equation is invariant under nonsingular transformations so that (9.333) is again of *elliptic type*.

We now consider the *Dirichlet problem* for (9.333) in a bounded region G, and note that our discussion applies to the transformed region obtained in going from (9.289) to (9.333). To begin, we assume that G is a *convex region* as shown in Figure 9.1.

The points P and R in the diagram are points where the tangent line to the (smooth) boundary curve is horizontal. To the left of those points the curve is given as $x = h_1(y)$, and $u(h_1(y), y) = f_1(y)$ is the boundary condition. To right of P and R we have $u(h_2(y), y) = f_2(y)$. The characteristics for the reduced equation for (9.333) are the lines $y = $ constant. Thus the characteristics passing through G intersect the boundary curve twice, and they are tangent to the boundary curve at the points P and R.

The *conventional perturbation series*

$$(9.334) \qquad u = \sum_{n=0}^{\infty} u_n \epsilon^n,$$

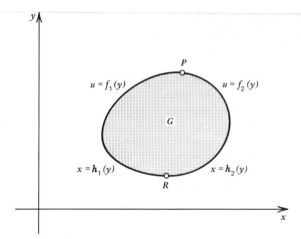

Figure 9.1. The convex region G.

when inserted into (9.418) yields for u_0,

$$(9.335) \qquad\qquad -\frac{\partial u_0}{\partial x} + cu_0 = g,$$

with similar equations satisfied by the u_n. Now u_0 must satisfy the boundary conditions

$$(9.336) \qquad\qquad \begin{cases} u_0(h_1(y), y) = f_1(y) \\ u_0(h_2(y), y) = f_2(y), \end{cases}$$

and this is not possible, in general, since u_0 is already specified uniquely either by its values on $x = h_1(y)$ or on $x = h_2(y)$. Consequently, a *boundary layer* must be introduced near $x = h_1(y)$ or $x = h_2(y)$, and we do not know in advance where to place the boundary layer.

We proceed as in Example 9.11 and *stretch* the neighborhoods of each of the boundary curves $x = h_1(y)$ and $x = h_2(y)$ to determine where the boundary layer is to be placed. The stretching transformations are given as

$$(9.337) \qquad\qquad x - h_i(y) = \epsilon\xi, \qquad i = 1, 2$$

$$(9.338) \qquad\qquad y = \eta.$$

We have anticipated that the boundary layer width is $O(\epsilon)$ as is easily shown. The *boundary layer equation* obtained from (9.333) is

$$(9.339) \qquad\qquad (A - 2h_i'B + h_i'^2C)v_{\xi\xi} - v_\xi = O(\epsilon)$$

where $v = v(\xi, \eta)$ and A, B, and C are evaluated at $\xi = 0$. Since (9.333) is elliptic, the quadratic form

$$(9.340) \qquad\qquad A - 2\lambda B + \lambda^2 C \equiv Q(\lambda)$$

is nonzero, so that the coefficient of $v_{\xi\xi}$ in (9.339) where $\lambda = h_i'$, is either strictly negative or strictly positive.

To solve the boundary layer equation, we expand v in powers of ϵ with v_0 as its leading term. Then v_0 satisfies the equation

$$(9.341) \qquad\qquad Q(h_i') \frac{\partial^2 v_0}{\partial \xi^2} - \frac{\partial v_0}{\partial \xi} = 0$$

where $h_i' = h_i'(\eta)$. The general solution of (9.341) is

$$(9.342) \qquad\qquad v_0(\xi, \eta) = \alpha(\eta) + \beta(\eta)\exp\left[\frac{\xi}{Q}\right].$$

Now ξ increases and decreases with x in view of (9.337). Thus if $Q > 0$, the boundary layer must be placed near $x = h_2(y)$, whereas if $Q < 0$, the boundary layer must be near $x = h_1(y)$. For with these choices, the exponential in (9.342) decreases as we move away from the boundary into the region G. Since $Q(\lambda)$ is either greater than zero or less than zero for all λ, we conclude from (9.340) that if A and C are positive in and on the boundary of G, then $Q > 0$, whereas if A and C are negative, then $Q < 0$. Consequently, the boundary layer is at $x = h_2(y)$ if A and C are positive, and it is at $x = h_1(y)$ if A and C are negative.

Assuming A and C are positive, we now know that a boundary layer (if it is necessary) must lie near $x = h_2(y)$. Thus the (outer) equation (9.335) must be solved subject to the boundary condition given at $x = h_1(y)$—that is, $u_0(h_1(y), y) = f_1(y)$. The solution is easily found to be

$$(9.343) \qquad u_0(x, y) = f_1(y)\exp\left[\int_{h_1(y)}^{x} c(s, y)\, ds\right]$$
$$- \int_{h_1(y)}^{x} g(\sigma, y)\exp\left[\int_{\sigma}^{x} c(s, y)\, ds\right] d\sigma.$$

For the boundary layer function $v_0(\xi, \eta)$ we require that

$$(9.344) \qquad\qquad v_0(0, \eta) = f_2(\eta)$$

since $y = \eta$, whereas the matching condition gives

$$(9.345) \qquad\qquad \lim_{\xi \to -\infty} v_0(\xi, \eta) = \lim_{x \to h_2(\eta)} u_0(x, \eta).$$

Using (9.342) yields

(9.346)

$$v_0(\xi, \eta) = u_0(h_2(\eta), \eta) + [f_2(\eta) - u_0(h_2(\eta), \eta)]\exp\left[\frac{\xi}{Q(h_2'(\eta))}\right].$$

Combining (9.343) and (9.346) we may express $u(x, y)$ in a more uniform manner as

$$(9.347) \quad u(x, y) \approx u_0(x, y) + [f_2(y) - u_0(h_2(y), y)]\exp\left[\frac{x - h_2(y)}{\epsilon Q}\right]$$

where $Q = Q(h_2'(y))$. Since $h_2'(y)$ becomes infinite at the points P and R on the boundary curve, these results are not valid near P and R. A similar result for the solution obtains if A and C are negative.

A different approach is needed if the region G has a portion of its boundary that coincides with a line $y = $ constant. A situation of this type is pictured in Figure 9.2.

The point P separates the curves $x = h_1(y)$ and $x = h_2(y)$. The transition from these curves to the line $y = y_0$ at the points R and T is assumed to be smooth. In the region $y > y_0$ within G and away from the point P, the result (9.347) again yields an approximate solution if A and C are positive. The line $y = y_0$ is a *characteristic* for the *reduced equation* (9.335) and the equation for

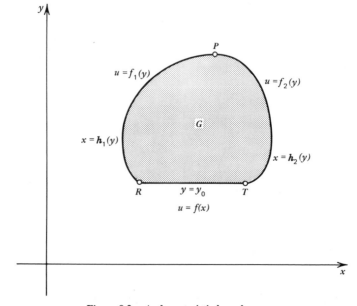

Figure 9.2. A characteristic boundary curve.

u_0 cannot be solved there subject to arbitrary data $u_0 = f(x)$. A boundary layer must be introduced near $y = y_0$.

We set

(9.348) $$y - y_0 = \epsilon^r \eta$$

to emphasize the neighborhood of $y = y_0$ and insert (9.348) into (9.333). This gives, with $w = w(x, \eta)$,

(9.349) $$\epsilon^{1-2r} C w_{\eta\eta} - w_x + cw + O(\epsilon^{1-r}) = g$$

where C, c, and g are evaluated at $y = y_0$. To achieve a balance between the reduced equation and the term $w_{\eta\eta}$, we must set $r = \frac{1}{2}$. On expanding w in powers of $\epsilon^{1/2}$ we obtain for w_0,

(9.350) $$C(x, y_0) \frac{\partial^2 w_0}{\partial \eta^2} - \frac{\partial w_0}{\partial x} + c(x, y_0) w_0 = g(x, y_0),$$

which is a *parabolic equation*. At $\eta = 0$ we have the boundary condition

(9.351) $$w_0(x, 0) = f(x),$$

whereas the matching condition requires that

(9.352) $$\lim_{\eta \to \infty} w_0(x, \eta) = \lim_{y \to y_0} u_0(x, y)$$

so that the boundary layer and the outer solution (9.343) agree. An additional condition should be placed on $w_0(x, \eta)$ regarding its behavior near $x = h_1(y)$ or $x = h_2(y)$. Thinking of x as a time-like variable in (9.350), we see that (9.350) corresponds to a forward parabolic equation. Thus it is appropriate to specify the values of $w_0(x, \eta)$ at $x = h_1(y)$ since we are considering a region where $x > h_1(y)$. However, near $x = h_1(y)$ away from the point R, the outer solution u_0 is valid, so that we may consider the (initial) condition near $x = h_1(y)$ to be equivalent to the condition (9.352). That is, $w_0(x, \eta)$ should agree with $u_0(x, y)$ away from the line $y = y_0$.

The solution of (9.350)–(9.352) is not straightforward if the coefficients C and c are not constant. We cannot use results obtained previously in the text in that case, and special techniques or possibly numerical methods must be used to solve for $w_0(x, \eta)$. If C and c are constants and $g = 0$, the solution can easily be obtained on following the procedure given in the preceding example, but we do not present it here.

We conclude our discussion of (9.333) by examining the behavior of the solution of the Dirichlet problem for (9.333) near points on the boundary where the tangent line is horizontal. As shown in Figure 9.3, we are interested

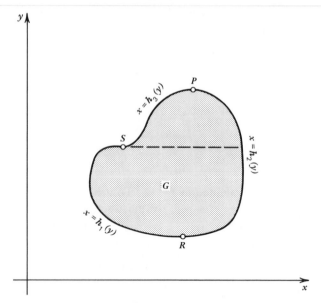

Figure 9.3. The nonconvex region G.

in points where the curve is either concave up or concave down or has a point of inflection. This contrasts with the situation considered previously where an entire portion of the boundary was horizontal.

At the point P in the figure, the curve is concave down, at R it is concave up, whereas at S the curve has a point of inflection and a zero slope.

If Dirichlet boundary conditions are given for (9.333) and A and C are both positive, we find, as was shown earlier, that a boundary layer must be introduced along $x = h_2(y)$. The outer equation (9.335) can be solved with the data given on $x = h_1(y)$ and $x = h_3(y)$. We can even solve for $u_0(x, y)$ with data given at the point of inflection. However, the derivatives of h_1, h_2, and h_3 are singular at one or more of the points P, R, and S. Thus the boundary layer solution breaks down at P and R and the second term in the ordinary perturbation series is singular at S, since it contains derivatives of h_1 or h_3. The horizontal line that extends from S in the figure indicates that the outer solution cannot be determined on that line and an internal boundary layer must be introduced along that line.

Let (x_0, y_0) represent the coordinates of the boundary point P, R, or S. We assume the boundary curve can be expressed as $y = F(x)$ near (x_0, y_0). Since the slope vanishes at (x_0, y_0) we have

$$(9.353) \qquad y - y_0 = \tfrac{1}{2}F''(x_0)(x - x_0)^2 + \cdots$$

near P or R, and

$$(9.354) \qquad y - y_0 = \tfrac{1}{6}F'''(x_0)(x - x_0)^3 + \cdots$$

near S with $y_0 = F(x_0)$. We assume that $F''(x_0)$ and $F'''(x_0)$ are not zero in these equations.

To emphasize the neighborhood of the point (x_0, y_0), we introduce a (double) *coordinate stretching*

$$(9.355) \qquad x - x_0 = \epsilon^r \xi; \qquad y - y_0 = \epsilon^s \eta$$

with r and s to be determined. Inserting (9.355) into (9.333) and with $v = v(\xi, \eta)$, we obtain

$$(9.356) \quad \epsilon^{1-2r} A v_{\xi\xi} + 2\epsilon^{1-r-s} B v_{\xi\eta} + \epsilon^{1-2s} C v_{\eta\eta} - \epsilon^{-r} v_\xi + \cdots = 0,$$

where A, B, and C are evaluated at (x_0, y_0) and only the leading terms in ϵ have been retained.

Now both r and s must be positive, otherwise the neighborhood of (x_0, y_0) is not emphasized. Also, the term v_ξ must be retained in (9.356) to maintain a balance between terms from the reduced equation and higher derivative terms. Finally, we require that r and s be such that an approximation to the boundary curve $y = F(x)$ of the form (9.353) or (9.354) be retained near (x_0, y_0). Thus if (9.355) is inserted in (9.353), we have

$$(9.357) \qquad \epsilon^s \eta = \tfrac{1}{2} \epsilon^{2r} F''(x_0) \xi^2 + \cdots,$$

and from (9.354) we obtain

$$(9.358) \qquad \epsilon^s \eta = \tfrac{1}{6} \epsilon^{3r} F'''(x_0) \xi^3 + \cdots.$$

Consequently, for points of the type P or R we must put

$$(9.359) \qquad 2r = s,$$

whereas for points of type S,

$$(9.360) \qquad 3r = s$$

in order to balance the leading terms in (9.357) and (9.358). It is then easily seen that the only way to balance v_ξ with a second derivative term in (9.356) is to set

$$(9.361) \qquad 1 - 2s = -r.$$

Thus for points of type P or R we have $r = \tfrac{1}{3}$ and $s = \tfrac{2}{3}$, whereas for points of type S, $r = \tfrac{1}{5}$ and $s = \tfrac{3}{5}$. In either case, if v is expanded in appropriate powers of ϵ, we find that the leading terms satisfy the equations

(9.362)
$$C\frac{\partial^2 v_0}{\partial \eta^2} - \frac{\partial v_0}{\partial \xi} = 0$$

and

(9.363) $$C\frac{\partial^2 v_1}{\partial \eta^2} - \frac{\partial v_1}{\partial \xi} = g - cv_0 - \xi\frac{\partial C}{\partial x}\frac{\partial^2 v_0}{\partial \eta^2} - 2B\frac{\partial^2 v_0}{\partial \xi \partial \eta}$$

where all coefficients, as well as g, are evaluated at (x_0, y_0). The last term on the right of (9.363) is absent if the point is of type S.

To leading order in ϵ, the boundary curve has the form $\eta = \frac{1}{2}F''(x_0)\xi^2$ near points of type P and R, whereas it has the form $\eta = \frac{1}{6}F'''(x_0)\xi^3$ near points of type S. To facilitate the study of the boundary layer equations (9.362)–(9.363), we introduce the new variables

(9.364) $$\alpha = \eta - \frac{1}{\kappa + 1}a_\kappa\xi^{\kappa+1}; \qquad \beta = \xi; \qquad \kappa = 1, 2$$

where $a_1 = F''(x_0)$ and $a_2 = \frac{1}{2}F'''(x_0)$. In these coordinates, $\alpha = 0$ corresponds to the boundary curve. With $w_\kappa = v_\kappa(\alpha(\xi, \eta), \beta(\xi, \eta))$ we obtain, in place of (9.362) and (9.363),

(9.365) $$C\frac{\partial^2 w_0}{\partial \alpha^2} + a_\kappa\beta^\kappa\frac{\partial w_0}{\partial \alpha} - \frac{\partial w_0}{\partial \beta} = 0, \qquad\qquad\qquad \kappa = 1, 2,$$

(9.366) $$C\frac{\partial^2 w_1}{\partial \alpha^2} + a_\kappa\beta^\kappa\frac{\partial w_1}{\partial \alpha} - \frac{\partial w_1}{\partial \beta} = g - cw_0 - \beta\frac{\partial C}{\partial x}\frac{\partial^2 w_0}{\partial \alpha^2}$$

$$+ 2Ba_\kappa\beta^\kappa\frac{\partial^2 w_0}{\partial \alpha^2} - 2B\frac{\partial^2 w_0}{\partial \alpha \partial \beta}, \qquad \kappa = 1, 2$$

where $\kappa = 1$ corresponds to a point of type P or R, $\kappa = 2$ corresponds to a point of type S, and the last two terms on the right of (9.366) are absent if $\kappa = 2$.

The solutions of (9.365)–(9.366) must be matched to the outer solution and must satisfy the boundary conditions. For definiteness we now restrict our discussion to a point of type R and assume that the coefficient C in (9.333) is positive. In that case the matching with the outer solution takes place in the region $x < x_0$, and the outer solution u_0 has the form (9.343). Also, near a point (x_0, y_0) of type R the boundary curve is given as (9.353) with $F''(x_0) > 0$.

We assume that the boundary condition near (x_0, y_0) is given as

(9.367) $$u(x, F(x)) = f(x).$$

For $x < x_0$, the boundary curve is expressed as $x = h_1(y)$, and the boundary condition is given as

(9.368) $$u(h_1(y), y) = f_1(y).$$

Consequently, $f(x)$ and $f_1(y)$ are related by

(9.369) $$f_1(y) = f(h_1(y)).$$

Since $y = F(x)$ is expressed as in (9.353) we have

(9.370)
$$x = h_1(y) = x_0 - \sqrt{\frac{2}{a_1}} \sqrt{y - y_0} + \cdots = x_0 - \epsilon^{1/3} \sqrt{\frac{2\alpha}{a_1} + \beta^2} + \cdots$$

with $a_1 = F''(x_0)$, where (9.364) with $\kappa = 1$ has been used and the minus sign follows from the fact $x - x_0 < 0$. Also,

(9.371) $$x = x_0 + (x - x_0) = x_0 + \epsilon^{1/3}\beta.$$

Expressing the outer solution $u_0(x, y)$ as given in (9.343), in terms of the boundary layer variables we obtain—to leading terms—

(9.372)
$$u_0 = f(x_0) + \epsilon^{1/3}\left[-f'(x_0)\sqrt{\frac{2\alpha}{a_1} + \beta^2} + (cf(x_0) - g)\left(\beta + \sqrt{\frac{2\alpha}{a_1} + \beta^2} \right) \right]$$

where c and g are evaluated at (x_0, y_0). We note that $\beta < 0$ in the matching region so that $\sqrt{\beta^2} = |\beta| = -\beta$. On the boundary we have

(9.373) $$u = f(x) = f(x_0) + \epsilon^{1/3}f'(x_0)\beta + O(\epsilon^{2/3}).$$

Since

(9.374) $$w = w_0 + \epsilon^{1/3}w_1 + O(\epsilon^{2/3}),$$

we find that on the boundary we must set

(9.375) $$w_0(0, \beta) = f(x_0)$$

and

$$(9.376) \qquad w_1(0, \beta) = f'(x_0)\beta.$$

We remark that $\alpha = 0$ is only an approximate leading order expression for the boundary curve $y = F(x)$. Since $w_0(\alpha, \beta)$ is shown to be identically constant, additional effects that might influence the boundary condition for w_1 do not play a role. The matching conditions require that

$$(9.377) \qquad w_0(\alpha, \beta) \to f(x_0)$$

and

$$(9.378)$$

$$w_1(\alpha, \beta) \to -f'(x_0)\sqrt{\frac{2\alpha}{a_1} + \beta^2} + (cf(x_0) - g)\left(\beta + \sqrt{\frac{2\alpha}{a_1} + \beta^2}\right),$$

as $\beta \to -\infty$ with $\alpha > 0$.

The solution of the problem for $w_0(\alpha, \beta)$ is immediately found to be

$$(9.379) \qquad w_0(\alpha, \beta) = f(x_0)$$

where, in fact, $f(x_0) = u(x_0, y_0)$. Since w_0 is a constant we may set

$$(9.380) \qquad w_1(\alpha, \beta) = (cf(x_0) - g)\beta + W(\alpha, \beta)$$

in (9.366) (with $\kappa = 1$) and conclude that $W(\alpha, \beta)$ satisfies the homogeneous equation

$$(9.381) \qquad CW_{\alpha\alpha} + a_1\beta W_\alpha - W_\beta = 0.$$

The boundary and matching condition for W must be determined from (9.376) and (9.378) with the use of (9.380). The solution of (9.381) can be expressed in terms of products of exponentials and Airy functions. The process of satisfying the boundary and matching conditions for that solution is not straightforward and is not given here.

Although we do not discuss the boundary layer equations near points of type P or S, we note that in the case of a point S there is an additional (internal) boundary layer of parabolic type. The solution of the boundary layer equation near S must be used to provide initial data for that problem.

If (9.289) is of *hyperbolic type*, perturbation and boundary layer methods may again be used in the manner given to solve initial and boundary value problems. A particular hyperbolic problem was considered in Example 9.10 where the special problem arising because of an excess of initial conditions for the reduced (perturbation) equations was examined. A general discussion of the hyperbolic problem, akin to that given for the elliptic problem, is not given here.

A common problem for both the elliptic and hyperbolic cases is where to place the boundary layer(s) if the data for the given problem result in an overdetermined problem for the reduced equation. This question can be answered, as has been done previously, by introducing boundary layer coordinates along the appropriate parts of the boundary. Those portions of the boundary along which the boundary layer equations have solutions that decay exponentially away from the boundary are the regions where boundary layers can be located. An alternative method for determining the location of the boundary layer is now presented.

To begin, (9.289) is replaced by

$$(9.382) \qquad \epsilon\left(Au_{xx} + 2Bu_{xy} + Cu_{yy}\right) + au_x + bu_y = 0$$

which may be of *elliptic* or *hyperbolic type*. The additional terms in (9.289) have been neglected for the sake of simplicity since they do not play a significant role in our discussion. Now (9.382) states that

$$(9.383) \qquad au_x + bu_y = O(\epsilon).$$

Assuming that a, b, u, and their derivatives are smooth functions, we may express u_{xx}, u_{xy}, and u_{yy} either in terms of u_{xx} and u_x or in terms of u_{yy} and u_y correct to $O(\epsilon)$. For example, if $a \neq 0$, we have

$$(9.384) \qquad u_x = -\frac{b}{a}u_y + O(\epsilon)$$

so that

$$(9.385) \qquad u_{xx} = \left(\frac{b}{a}\right)^2 u_{yy} - \left(\frac{b}{a}\right)_x u_y + \frac{b}{a}\left(\frac{b}{a}\right)_y u_y + O(\epsilon)$$

and

$$(9.386) \qquad u_{xy} = -\frac{b}{a}u_{yy} - \left(\frac{b}{a}\right)_y u_y + O(\epsilon).$$

Inserting (9.385)–(9.386) into (9.382) yields

$$(9.387) \qquad \epsilon\left[A\left(\frac{b}{a}\right)^2 - \frac{2Bb}{a} + C\right]u_{yy} + au_x + bu_y$$
$$+ \epsilon\left[\left(\frac{Ab}{a}\right)\left(\frac{b}{a}\right)_y - A\left(\frac{b}{a}\right)_x - 2B\left(\frac{b}{a}\right)_y\right]u_y = O(\epsilon^2).$$

Similarly, if $b \neq 0$, we obtain

(9.388) $\quad \epsilon \left[C \left(\dfrac{a}{b} \right)^2 - \dfrac{2Ba}{b} + A \right] u_{yy} + a u_x + b u_y$

$\quad + \epsilon \left[\dfrac{Ca}{b} \left(\dfrac{a}{b} \right)_x - C \left(\dfrac{a}{b} \right)_y - 2B \left(\dfrac{a}{b} \right)_x \right] u_x = O(\epsilon^2).$

The *parabolic equations* (9.387)–(9.388) may be solved by using a conventional perturbation series. The two leading terms in the series satisfy equations identical to those for the leading terms in the regular perturbation solution of (9.382) as is easily shown. Therefore, we expect the solutions of (9.387) or (9.388) to coincide, approximately, with the outer solution of (9.382) in regions where that solution is valid. However, as we shall see, the solutions of (9.387) and (9.388) remain valid in certain regions where the outer solution breaks down. Also, information about the location of boundary layers for (9.382) can be determined from the parabolic equations (9.387)–(9.388).

For example, if the data for the given equation (9.382) are not smooth, the discontinuities or singularities are propagated into the interior along the characteristics of the reduced equation (9.383). If the given equation is elliptic, the solution must be smooth in the interior and if it is hyperbolic, the singularities must occur across the characteristics of the given equation. These characteristics generally do not coincide with those of the reduced equation. The parabolic equations (9.387)–(9.388) generally smooth out the singularities in the data, so that its solutions are smooth across the characteristics of the reduced equation.

To determine the location of boundary layers for the elliptic or hyperbolic equation (9.382), we consider initial value problems for (9.387) and (9.388) along the boundary of the region. Those portions of the boundary for which the initial value problems for the parabolic equations (9.387) or (9.388) are not well posed, determine where the boundary layers must be located. To see how this works, we consider a simpler form of the elliptic equation (9.289) or (9.382)

(9.389) $\qquad\qquad \epsilon \left(A u_{xx} + 2 B u_{xy} + C u_{yy} \right) - u_x = 0,$

and the convex region G considered previously. Since $a = -1$ and $b = 0$ in relation to (9.382), the parabolic equation (9.387) takes the form

(9.390) $\qquad\qquad\qquad \epsilon C u_{yy} - u_x = 0$

where we have dropped the $O(\epsilon^2)$ terms. Referring to the diagram of the convex region G given in Figure 9.1, we see that if $C > 0$, the initial value problem for (9.390)—with x treated as the time variable—is well posed for data given on $x = h_1(y)$. But if $C < 0$, the data must be assigned on $x = h_2(y)$ for the initial value problem to be well posed. Thus the boundary layer must be placed along $x = h_2(y)$ if $C > 0$ and along $x = h_1(y)$ if $C < 0$. This result is consistent with that given earlier based on a direct boundary layer construction. It is not difficult to show that both the present and the earlier method for determining the location of boundary layers lead to the same result. However, this is not demonstrated here.

Given the *hyperbolic equation* with constant coefficients

$$(9.391) \qquad \epsilon\left(u_{tt} - c^2 u_{xx}\right) + u_t - au_x = 0,$$

a related *parabolic equation* is

$$(9.392) \qquad \epsilon\left(a^2 - c^2\right)u_{xx} + u_t - au_x = 0,$$

since $u_t = au_x + O(\epsilon)$. The parabolic equation (9.392) cannot absorb arbitrary initial data $u(x,0) = f(x)$ and $u_t(x,0) = g(x)$. Thus a boundary layer is generally required at $t = 0$. However, if $a^2 > c^2$, we find that even if appropriate initial values are determined for (9.392), the problem is not well posed in the region $t > 0$ since the coefficients of u_{xx} and u_t are both positive. We considered a similar problem in Example 9.10 where it was shown that if $|a| > c$, the initial value problem for (9.391) is unstable. We assume, therefore, that $|a| < c$ and note that the parabolic equation approach indicates whether the given problem is stable.

If we consider an initial and boundary value problem for (9.391) in the interval $0 < x < l$ and with $t > 0$, we find that a boundary layer is required either along $x = 0$ or $x = l$ if the conventional perturbation approach is used. Now once initial data for the parabolic equation (9.392) have been determined, it is possible to solve (9.392) with the given boundary conditions on $x = 0$ and $x = l$. That is, a boundary layer need not be introduced at all on either boundary line. Nevertheless, it is usually simpler to solve the given problem with the use of boundary layers.

Since we are concerned with the solution near the lines $x = 0$ and $x = l$, we replace (9.392) with the parabolic equation

$$(9.393) \qquad \epsilon\left(1 - \left(\frac{c}{a}\right)^2\right)u_{tt} + u_t - au_x = 0$$

where we have used $u_x = (1/a)u_t + O(\epsilon)$. If $a > 0$, the coefficients of u_{tt} and u_x have the same sign since $c > |a|$, whereas if $a < 0$, they have opposite signs. Thus if (initial) data are given at $x = 0$ and (9.393) is to be solved for $x > 0$, the problem is well posed only if $a < 0$. Consequently, a boundary layer must occur near $x = 0$ in the case $a > 0$. Similarly, if data are assigned at $x = l$ and (9.393) is solved for $x < l$, the problem is well posed only if $a > 0$, so that a boundary layer near $x = l$ occurs when $a < 0$. We note that if $a = 0$, the lines $x = 0$ and $x = l$ are characteristics of the reduced equation $u_t = 0$ and (9.392) is, in fact, the parabolic boundary layer equation. Again, these results concerning the location of the boundary layers are identical with those obtained by means of a boundary layer analysis.

In the following example we apply the *parabolic equation method* to specific equations of elliptic and hyperbolic types to demonstrate how problems can be solved by this method.

EXAMPLE 9.12.   THE PARABOLIC EQUATION METHOD

We begin by reexamining the problem of Example 9.11. That is, we consider the *elliptic equation*

$$(9.394) \qquad\qquad \epsilon(u_{xx} + u_{yy}) + u_x + bu_y = 0$$

within the region $0 < x < \infty$ and $0 < y < L$, with $b = $ constant. The boundary conditions are

$$u(x,0) = f(x); \qquad u(0, y) = g(y); \qquad u(x, L) = h(x),$$

and $u(x, y)$ is required to vanish as $x \to \infty$ in the interval $0 \leqslant y \leqslant L$.

Since $u_x = -bu_y + O(\epsilon)$ and $u_y = -(1/b)u_x + O(\epsilon)$ if $b \neq 0$, the *parabolic equations* for (9.394) are

$$(9.395) \qquad\qquad \epsilon(1 + b^2)u_{yy} + u_x + bu_y = 0$$

and

$$(9.396) \qquad\qquad \epsilon\left(1 + \frac{1}{b^2}\right)u_{xx} + u_x + bu_y = 0.$$

From (9.395) we see that a problem with initial data given at $x = 0$ and $u(x, y)$ to be determined for $x > 0$ is not well posed. Also, (9.396) shows that if $b > 0$, a problem with initial data at $y = 0$ to be solved in the region $y > 0$ is not well posed. If $b < 0$ and initial data are given at $y = L$ and (9.396) is to be solved for $y < L$, the resulting problem is not well posed. If $b = 0$, (9.396) is not valid and in this case there are parabolic boundary layers at $y = 0$ and $y = L$.

We conclude from this discussion that a boundary layer should occur at $x = 0$ in all cases. If $b > 0$, a boundary layer occurs at $y = 0$, whereas if $b < 0$, a boundary layer occurs at $y = L$. These results are consistent with those obtained in Example 9.11 by a different method.

We now suppose that $b < 0$ so that the boundary layer must be located at $y = L$. It is then possible to solve the parabolic equation (9.396) with the data given along the lines $y = 0$ and $x = 0$. Even though $x = 0$ was shown to be the site of a boundary layer, the initial and boundary value problem for (9.396) is well posed. Thus we consider the equation

$$(9.397) \qquad \epsilon\lambda^2 u_{xx} + u_x - c^2 u_y = 0$$

where we have put $\lambda^2 = 1 + (1/b)^2$ and $b = -c^2$. The initial and boundary conditions are

$$(9.398) \qquad u(x,0) = f(x); \qquad u(0, y) = g(y).$$

Though the given problem (9.291)–(9.292) was formulated in the region $0 < x < \infty$ and $0 < y < L$ so that $g(y)$ in (9.398) is unspecified for $y > L$, we may solve (9.397)–(9.398) in the first quadrant $0 < x < \infty$ and $0 < y < \infty$. The boundary function $g(y)$ may be extended arbitrarily over the interval $y > L$. Because of the causality property for (9.397), the solution $u(x, y)$ at any value of y such that $0 < y \leqslant L$ depends only on data given for smaller (nonnegative) values of y. Thus the values assigned to $g(y)$ for $y > L$ are irrelevant in the region $0 < y \leqslant L$.

To solve (9.397) we set

$$(9.399) \qquad u(x, y) = \exp\left[-\frac{x}{2\epsilon\lambda^2} - \frac{y}{4\epsilon\lambda^2 c^2}\right] v(x, y)$$

and find that $v(x, y)$ satisfies

$$(9.400) \qquad v_y = \frac{\epsilon\lambda^2}{c^2} v_{xx},$$

with the data

$$(9.401) \quad v(x,0) = f(x)\exp\left[\frac{x}{2\epsilon\lambda^2}\right]; \qquad v(0, y) = g(y)\exp\left[\frac{y}{4\epsilon\lambda^2 c^2}\right].$$

Using the results of Example 5.5 [see (5.101)], we find that $u(x, y)$ can be expressed as

(9.402)

$$u(x, y) = \frac{cx}{\sqrt{4\pi\epsilon\lambda^2}} \int_0^y \left\{ \exp\left[-\frac{(y + c^2 x - s)^2}{4\epsilon c^2\lambda^2(y - s)}\right] \frac{g(s)}{(y - s)^{3/2}} \right\} ds$$

$$+ \int_0^\infty \left[G\left(\frac{y}{c^2} + x - s, y\right) - G\left(\frac{y}{c^2} + x + s, y\right) e^{s/\epsilon\lambda^2} \right] f(s)\, ds$$

where $G(x, y)$ is given as

$$(9.403) \qquad G(x, y) = \frac{c}{\sqrt{4\pi\epsilon\lambda^2 y}} \exp\left[-\frac{c^2 x^2}{4\epsilon\lambda^2 y} \right],$$

which is the fundamental solution of (9.400).

In the first integral in (9.402) we have $y - s \geq 0$ and $x \geq 0$. Thus when $x > 0$ we see that the contribution from the entire integral is exponentially small if $\epsilon \ll 1$ since the exponential is uniformly small. This term yields the boundary layer behavior near $x = 0$. From the expression for $G(x, y)$ in (9.403) we see that as $\epsilon\lambda^2 y/c^2 \to 0$, $G(x, y) \to \delta(x)$ in view of the Dirac delta function behavior of the fundamental solution discussed previously. Thus as $\epsilon \to 0$, we find that $G[(y/c^2) + x \pm s, y] \to \delta[(y/c^2) + x \pm s]$. In the region $y > 0$, $x \geq 0$, and $s \geq 0$ the delta function $\delta[(y/c^2) + x + s]$ vanishes. Even though $\epsilon^{s/\epsilon\lambda^2}$ grows exponentially as $\epsilon \to 0$, its product with the term $G[(y/c^2) + x + s, y]$ may be shown to vanish as $\epsilon \to 0$. Thus if $\epsilon \ll 1$, we find that (9.402) reduces to

$$(9.404) \qquad u(x, y) \approx f\left(\frac{y}{c^2} + x \right), \qquad x > 0, 0 \leq y \leq L$$

to leading order in ϵ. The expression (9.404) represents the leading term in the outer solution of (9.394) that satisfies the boundary condition at $y = 0$. It agrees with the result (9.299). The term involving $g(y)$ in (9.402) is significant only in a boundary layer near $x = 0$. To satisfy the boundary condition at $y = L$, a boundary layer must be introduced as was done in Example 9.11. The solution can then be expressed in a composite form in the manner of (9.321). It may be noted that the solution form (9.402) smooths out discontinuities or singularities in the data $f(x)$ and $g(y)$.

Next we consider a *signaling problem* for the hyperbolic equation

$$(9.405) \qquad \epsilon\left(u_{tt} - c^2 u_{xx} \right) + u_t - a u_x = 0, \qquad x, t > 0$$

with the homogeneous initial data

$$(9.406) \qquad\qquad u(x, 0) = 0, \qquad x > 0$$
$$(9.407) \qquad\qquad u_t(x, 0) = 0, \qquad x > 0$$

and the boundary condition

$$(9.408) \qquad\qquad u(0, t) = g(t), \qquad t > 0.$$

It is assumed that $|a| < c$, both of which are constant, so that the problem is stable.

The relevant *parabolic approximations* for (9.405) are

(9.409) $$u_t - au_x - \epsilon(c^2 - a^2)u_{xx} = 0$$

and

(9.410) $$u_t - au_x - \epsilon\left(\left(\frac{c}{a}\right)^2 - 1\right)u_{tt} = 0.$$

Since $c^2 > a^2$, the initial and boundary value problem for (9.409) is well posed. However, in the case of (9.410) if initial data are given on $x = 0$, the problem is well posed for $a < 0$ and is not well posed for $a > 0$, as is immediately seen. This is consistent with the fact that for $a < 0$ the characteristics of the reduced equation $u_t - au_x = 0$ are the lines $x + at =$ constant. These characteristics do not simultaneously intersect the initial line $t = 0$ and the boundary line $x = 0$, so that the given initial and boundary data (9.406)–(9.408) can be assigned to the perturbation series solution of (9.405) without encountering any difficulty. However, if $a > 0$, the characteristics $x + at =$ constant intersect both the initial and boundary lines and a boundary layer at $x = 0$ is needed.

Since the initial data are homogeneous there is no need for an initial layer near $t = 0$ in this problem. We may use $u = 0$ as the initial condition when solving (9.405) either by the conventional perturbation and boundary layer approach or by the use of the parabolic equation (9.409) as we now demonstrate. We solve (9.409) which we rewrite as

(9.411) $$u_t - au_x - \epsilon\gamma^2 u_{xx} = 0, \qquad x, t > 0$$

with $\gamma^2 = c^2 - a^2$. The initial condition is

(9.412) $$u(x,0) = 0, \qquad x > 0$$

and the boundary condition

(9.413) $$u(0, t) = g(t), \qquad t > 0.$$

To solve (9.411)–(9.413) we proceed exactly as for the initial and boundary value problem for (9.397) treated earlier. Accounting for the slight change in notation we find that

(9.414) $$u(x, t) = \frac{\epsilon x}{\sqrt{4\pi\gamma^2}} \int_0^t \exp\left\{-\frac{[x + a(t - \tau)]^2}{4\epsilon\gamma^2(t - \tau)}\right\} \frac{g(\tau)}{[\epsilon(t - \tau)]^{3/2}} d\tau.$$

Now if $a > 0$, the argument of the exponential in (9.414) vanishes only when $x = 0$ in the region $x \geq 0$ and $t \geq 0$. Since $0 < \epsilon \ll 1$, we find that

$u(x, t)$ is exponentially small for $x > 0$ and $t \geq 0$, and it is only significant in a small (boundary) layer near $x = 0$. Thus the outer solution is given as

$$(9.415) \qquad\qquad u(x, t) \approx 0, \qquad x > 0, t \geq 0$$

apart from a layer of thickness $O(\epsilon)$ near $x = 0$.

If $a < 0$, the solution (9.414) is significantly different from zero for values of x and t in the first quadrant for which

$$(9.416) \qquad\qquad x = -a(t - \tau).$$

Since $t - \tau > 0$ and $t, \tau \geq 0$, (9.416) corresponds to the characteristics of the reduced equation $u_t - au_x = 0$ that occupy the sector between the line $x + at = 0$ and the t-axis. Now if $a = 0$ in (9.414), we know that as $x \to 0$ the expression (9.414) tends to $g(t)$ with the basic contribution from the integral coming from the value $\tau = t$. Similarly, we expect that as $\epsilon \to 0$ in (9.414) the main contribution to the integral comes from the value $\tau = (x/a) + t$ and that $u(x, t)$ tends to $g((x/a) + t)$. We note that (9.414) has been written in terms of $\epsilon(t - \tau)$ and ϵx, to indicate that small values of x and $t - \tau$ are equivalent to considering small values ϵ. Further, in the sector between $x + at = 0$ and the x-axis, the exponential term in (9.414) and, consequently, the integral are small. Therefore, we conclude that for small ϵ and $a < 0$ we have

$$(9.417) \qquad u(x, t) \approx \begin{cases} g\left(\dfrac{x}{a} + t\right); & x + at < 0 \\ 0; & x + at > 0 \end{cases}; \qquad x, t \geq 0.$$

This result is consistent with that obtained from the conventional perturbation method as is easily shown. However, the expression (9.414) remains valid and useful for large values of x and t, as well as in the case where the boundary function $g(t)$ is not smooth, or if $g(0) \neq 0$.

If the initial data $u(x, 0)$ and $u_t(x, 0)$ for (9.405) are not homogeneous, an initial layer of width $O(\epsilon)$ is required. The value of the solution at the edge of this initial layer may then be used as an initial condition for the parabolic equation (9.409).

This concludes our discussion of singular perturbation methods. We have treated only a few problems in some detail in order to bring out the main features and difficulties of singular perturbation methods. Some further examples and extensions are considered in the exercises. Also, singular perturbation methods play a role in some of the problems studied in later sections.

9.4. EQUATIONS WITH A LARGE PARAMETER

In the study of boundary value problems for the *reduced wave equation*

$$(9.418) \qquad \nabla^2 u + k^2 n^2 u = 0,$$

a case of greatest interest in the context of wave propagation occurs when the (constant) parameter k is large. The given function n in (9.418) is then known as the *index of refraction*. The connection between (9.418) and the wave equation is given in Exercise 2.4.4.

Assuming k is large, we apply a perturbation procedure to (9.418) and expand u in inverse powers of k as

$$(9.419) \qquad u = \sum_{j=0}^{\infty} u_j k^{-j}.$$

Inserting (9.419) into (9.418) and equating like powers of k, we immediately conclude that u_0 and all further terms in the expansion must vanish. Putting $\epsilon = 1/k$, we recognize that (9.418) is, in fact, a *singular perturbation problem* and asymptotic expansions of (9.418) must have a form different from that given in (9.419). To determine appropriate expansion forms we consider some exact solutions of (9.418) in the case where $n =$ constant.

For the two-dimensional problem, the simplest solutions of (9.418) are the *plane wave* solutions (see Example 2.10)

$$(9.420) \qquad u(x, y) = \exp[ikn(x \cos \theta + y \sin \theta)],$$

where θ and n are constants. With $r^2 = (x - \xi)^2 + (y - \eta)^2$ another (simple) solution of (9.418) in two dimensions is

$$(9.421) \qquad u(x, y) = J_0(knr),$$

where J_0 is the zero order Bessel function. For large values of k, assuming nr is not small, the asymptotic expansion of the Bessel function yields

$$(9.422)$$
$$u(x, y) = J_0(knr) \approx \left(\frac{2}{\pi knr}\right)^{1/2} \cos\left(knr - \frac{\pi}{4}\right)$$
$$= (2\pi knr)^{-1/2} \exp\left[iknr - \frac{i\pi}{4}\right] + (2\pi knr)^{-1/2} \exp\left[-iknr + \frac{i\pi}{4}\right].$$

Each of the exponential terms in (9.422) represents a *cylindrical wave* solution of (9.418).

In the three-dimensional case with $r^2 = (x - \xi)^2 + (y - \eta)^2 + (z - \zeta)^2$, a *spherical wave* solution of (9.418) is

$$(9.423) \qquad\qquad u(x, y, z) = \frac{1}{r} e^{iknr},$$

as is easily verified.

The exact and asymptotic solutions (9.420)–(9.423) of (9.418) indicate that when we expand solutions of (9.418) asymptotically for large k, they have the form of a rapidly varying exponential term multiplied by an amplitude term. On considering further terms in the asymptotic expansion of J_0 we find that the amplitude terms in (9.422) are given as series in inverse powers of k. If the exponentials in (9.422) are also expanded in powers of k, we find that the solution $u(x, y)$ contains both negative and positive powers of k in its expansion. Consequently, the series (9.419) does not lead to a useful result.

Noting these results we look for asymptotic solutions of (9.418) in the form

$$(9.424) \qquad\qquad u = v e^{ik\varphi}$$

where φ is the *phase term* and v is the *amplitude term*. Inserting (9.424) into (9.418) gives

$$(9.425) \quad \nabla^2 u + k^2 n^2 u = \left\{ k^2 \left[n^2 - (\nabla\varphi)^2 \right] v + ik \left[2\nabla\varphi \cdot \nabla v + v \nabla^2 \varphi \right] \right.$$

$$\left. + \nabla^2 v \right\} e^{ik\varphi} = 0.$$

Since k is assumed to be large, we equate the coefficient of the highest power of k in (9.425) to zero and obtain

$$(9.426) \qquad\qquad (\nabla\varphi)^2 = n^2.$$

The *eiconal equation* (9.426) was studied in Example 2.10 for the case of two dimensions. The phase term φ must be specified from (9.426) subject to appropriate conditions that result from the data given for the reduced wave equation (9.418).

Once the phase φ is specified, the amplitude v must be determined from

$$(9.427) \qquad ik \left[2\nabla\varphi \cdot \nabla v + v \nabla^2 \varphi \right] + \nabla^2 v = 0,$$

in view of (9.425)–(9.426). Dividing by ik in (9.417) and letting $k \to \infty$ reduces the order of (9.427) from second to first order. Since k is assumed to be large, we conclude that the problem of determining v from (9.427) is a singular perturbation problem. Therefore, difficulties of the type encountered in Section 9.3 are likely to occur in the present problem.

Nevertheless, we look for an asymptotic solution of (9.427) in the form

$$(9.428) \qquad\qquad v = \sum_{j=0}^{\infty} v_j (ik)^{-j}.$$

Inserting (9.428) into (9.427) and equating like powers of k yields the recursive system of equations

$$(9.429) \qquad\qquad 2\nabla\varphi \cdot \nabla v_0 + v_0 \nabla^2\varphi = 0$$

$$(9.430) \qquad\qquad 2\nabla\varphi \cdot \nabla v_j + v_j \nabla^2\varphi = -\nabla^2 v_{j-1}, \qquad j \geqslant 1.$$

These first order partial differential equations for the v_j are known as *transport equations* since they describe the variation of the amplitude terms v_j along the rays or characteristics determined from the eiconal equation, as shown later. The (initial) conditions for the v_j must be determined from the data for the reduced wave equation (9.418).

Much of the terminology associated with the large k solutions of the reduced wave equation (9.418) is drawn from the theory of *optics*. The propagation of light waves, when described in terms of (9.418), corresponds to the case where k is large. The asymptotic solution (9.424) characterizes *geometrical optics* since it is given in terms of the rays determined from the eiconal equation. Often geometrical optics is taken to mean the results determined strictly from the eiconal equation without regard to the transport equations (9.429)–(9.430). Solutions based directly on (9.418), or some approximation thereof that involves second order differential equations, yield results that characterize *wave optics*. Wave optics is required when geometrical optics fails to describe the solution of (9.418) correctly, either because the solution breaks down in the geometrical optics description or because it fails to account for certain effects due to the diffraction of light. However, a *geometrical theory of diffraction* has been developed by J. B. Keller that shows how to take into account diffraction effects within the context of a geometrical description of the solution similar to that given by geometrical optics. These asymptotic analyses of wave propagation are examined in some detail in this section.

In the following example we consider Green's functions for the two-dimensional reduced wave equation. The exact results obtained yield useful information about the asymptotic solutions of (9.418) and serve to motivate our discussion.

EXAMPLE 9.13.   GREEN'S FUNCTION FOR THE REDUCED WAVE EQUATION

The *free space Green's function* $K(x, y; \xi, \eta)$ for the reduced wave equation (9.418) in two dimensions satisfies the equation

$$(9.431) \qquad\qquad \nabla^2 K + k^2 n^2 K = -\delta(x - \xi)\delta(y - \eta),$$

and the radiation condition at infinity

$$(9.432) \qquad \lim_{r \to \infty} \sqrt{r}\left(\frac{\partial K}{\partial r} - iknK\right) = 0,$$

with $r^2 = (x - \xi)^2 + (y - \eta)^2$, and n assumed to be constant. In Example 6.12, this Green's function was found to be

$$(9.433) \qquad K = \frac{i}{4}H_0^{(1)}(knr),$$

where $H_0^{(1)}(z)$ is the zero order Hankel function of the first kind.

In view of (6.197), the leading term in the asymptotic expansion of K with $knr \gg 1$ is

$$(9.434) \qquad K(x, y; \xi, \eta) \approx \frac{i}{4}\left(\frac{2}{\pi knr}\right)^{1/2}\exp\left[iknr - \frac{i\pi}{4}\right].$$

Since with $r \neq 0$, K is a solution of the (homogeneous) reduced wave equation (9.418), we conclude that (9.434) is the leading term in an asymptotic solution of (9.418) in the form (9.424) and (9.428). To see this directly, we now construct an asymptotic solution of (9.418) that depends only on r.

Let

$$(9.435) \qquad K = e^{ik\varphi(r)}\sum_{j=0}^{\infty} v_j(r)(ik)^{-j}$$

with $r > 0$. The eiconal equation for φ takes the form

$$(9.436) \qquad (\nabla\varphi)^2 = [\varphi'(r)]^2 = n^2.$$

The solution of (9.436) for which the radiation condition (9.432) can be satisfied by the expansion (9.435) is clearly given as

$$(9.437) \qquad \varphi(r) = nr.$$

The transport equation for $v_0(r)$ is

$$(9.438) \qquad 2\nabla\varphi \cdot \nabla v_0 + v_0\nabla^2\varphi = 2\varphi'v_0' + \left(\varphi'' + \frac{1}{r}\varphi'\right)v_0$$

$$= 2nv_0' + \frac{1}{r}nv_0 = 0,$$

and its solution is

$$(9.439) \qquad v_0(r) = \frac{c_0}{\sqrt{r}}$$

where c_0 is an arbitrary constant that may depend on k. With φ and v_0 given by (9.437) and (9.439), respectively, we find that $v_0 e^{ik\varphi}$ agrees with the asymptotic form of K given in (9.434). However, the constant c_0 in (9.439) cannot be specified by the preceding (direct) asymptotic method since it depends on the behavior of the Green's function K at the source point $r = 0$. Now both the Green's function (9.433) and its asymptotic representation (9.434) are singular at $r = 0$. But the Green's function (9.433) has a logarithmic singularity at $r = 0$, whereas its asymptotic form (9.434) is algebraically singular at $r = 0$. Since the derivation of (9.434) from (9.433) required that $r > 0$, we cannot expect (9.434) nor the equivalent result obtained from the direct asymptotic method to be valid at $r = 0$.

Thus although the asymptotic expansion (9.435) generates the correct form for the Green's function K with $r > 0$, it cannot be directly related to the behavior of the Green's function at $r = 0$, so that the arbitrary constants that occur in (9.435) may be specified. An indirect approach such as the *boundary layer method* must be used to construct a modified expansion for K valid near $r = 0$. By matching this expansion with (9.435), the arbitrary constants in the latter expansion can be specified. The boundary layer method in this case yields the exact solution (9.433) as is easily shown. However, in the case of a variable index of refraction the boundary layer method yields useful results since the exact solution is not, in general, available.

The free space Green's function (9.433) is now used to construct a Green's function in the half-plane $x > 0$ with the boundary condition

$$(9.440) \qquad\qquad K(0, y; \xi, \eta) = 0.$$

The source point (ξ, η) lies in the right half-plane so that $\xi > 0$. The exact solution of this problem can be obtained by means of the *method of images* to be

$$(9.441) \qquad K(x, y; \xi, \eta) = \frac{i}{4} H_0^{(1)}(knr) - \frac{i}{4} H_0^{(1)}(kn\hat{r})$$

where $\hat{r}^2 = (x + \xi)^2 + (y - \eta)^2$. This Green's function is also required to satisfy the radiation condition at infinity.

If we expand (9.441) asymptotically for $r > 0$ in the right half-plane, we have

$$(9.442) \qquad K \approx \frac{i}{4}\left(\frac{2}{\pi knr}\right)^{1/2} \exp\left[iknr - \frac{i\pi}{4}\right]$$
$$- \frac{i}{4}\left(\frac{2}{\pi kn\hat{r}}\right)^{1/2} \exp\left[ikn\hat{r} - \frac{i\pi}{4}\right].$$

Introducing the oscillatory term $e^{-i\omega t}$ where ω is a constant, we form the function

$$(9.443) \qquad v(x, y, t) = K(x, y; \xi, \eta) e^{-i\omega t}.$$

We easily find that v satisfies the two-dimensional wave equation when $(x, y) \neq (\xi, \eta)$. We set

$$(9.444) \qquad Ke^{-i\omega t} = \frac{i}{4} H_0^{(1)}(knr) e^{-i\omega t}$$

$$- \frac{i}{4} H_0^{(1)}(kn\hat{r}) e^{-i\omega t}$$

$$\equiv u_I e^{-i\omega t} + u_S e^{-i\omega t}$$

where u_I and u_S are the *incident* and *scattered waves*, respectively.

This terminology is employed because with the introduction of the asymptotic expression (9.442) for K into (9.444), the term $v_I = u_I \exp(-i\omega t)$ represents a cylindrical wave traveling away from the source point. That part of the wave that hits the boundary $x = 0$ is moving toward the boundary. Thus it is characterized as a wave incident upon the boundary. The cylindrical wave $v_S = u_S \exp(-i\omega t)$ is a wave traveling away from the (fictitious) reflected source point and that part of the wave that intersects the boundary $x = 0$ moves away from it towards the half-plane $x > 0$. Since its existence is due to the presence of the boundary, we refer to it as a scattered wave. (Cylindrical waves are discussed in Example 2.10.)

Noting these results, we formulate the asymptotic Green's function problem as follows. We seek a function K given as

$$(9.445) \qquad K \approx u_I + u_S$$

where u_I and u_S are asymptotic solutions of the reduced wave equation (9.418). The *incident wave* u_I is completely specified—it is the free space Green's function (9.433)–(9.434)—and the scattered wave u_S is to be determined. The boundary condition (9.440) for K implies that

$$(9.446) \qquad u_S(0, y) = -u_I(0, y).$$

The scattered wave u_S is required to be an *outgoing wave* at the boundary $x = 0$. That is, $u_S e^{-i\omega t}$ must represent wave motion away from $x = 0$ into the interior region $x > 0$. The *outgoing condition* takes the place of the *radiation condition* in the asymptotic formulation of this problem.

We now demonstrate that the solution of the foregoing Green's function problem, carried out to leading terms only, agrees with (9.442). The asymptotic form of u_I is given in (9.434), and u_S is expressed as

$$(9.447) \qquad u_S \approx V e^{ik\varphi}.$$

In view of (9.446) we must have

$$(9.448) \qquad \varphi(0, y) = n\sqrt{\xi^2 + (y - \eta)^2},$$

$$(9.449) \qquad V(0, y) = -\frac{i}{4}\left(\frac{2}{\pi kn}\right)^{1/2}\left[\xi^2 + (y - \eta)^2\right]^{-1/2} e^{-i\pi/4},$$

that is, we equate both the phase and the amplitude terms in (9.446).

There are two possible solutions of the eiconal equation satisfied by the (scattered) phase term φ with the initial condition (9.448). They are

$$(9.450) \qquad \varphi_\pm(x, y) = n\sqrt{(x \pm \xi)^2 + (y - \eta)^2}.$$

We reject the minus sign in (9.450) since it does not yield an outgoing cylindrical wave for u_S. In fact, the phase will then be identical to that of the incoming wave u_I. Selecting the plus sign in (9.450) and defining $\hat{r}^2 = (x + \xi)^2 + (y - \eta)^2$, we specify the phase as $\varphi = n\hat{r}$, and readily conclude that

$$(9.451) \qquad V(x, y) = -\frac{i}{4}\left(\frac{2}{\pi kn\hat{r}}\right)^{1/2} e^{-i\pi/4}$$

on using (9.438)–(9.439) and the boundary condition (9.449). The asymptotic result (9.445) agrees with (9.442) which was obtained from the exact solution.

We now continue our discussion of the *eiconal equation* (9.426) and the *transport equations* (9.429)–(9.430). These first order partial differential equations are solved using the *method of characteristics* as developed in Chapter 2.

Our present discussion is restricted to the three-dimensional problem, but the results are presented later for the two-dimensional case as well. With $p = \varphi_x$, $q = \varphi_y$, and $r = \varphi_z$, we set

$$(9.452) \qquad F = p^2 + q^2 + r^2 - n^2,$$

so that $F = 0$ corresponds to the eiconal equation. Using the results of Exercise 2.4.7, we find that the characteristic equations for $x, y, z, \varphi, p, q,$ and r are

$$(9.453) \qquad \frac{dx}{ds} = 2p; \qquad \frac{dy}{ds} = 2q; \qquad \frac{dz}{ds} = 2r,$$

$$(9.454) \qquad \frac{d\varphi}{ds} = 2(p^2 + q^2 + r^2) = 2n^2,$$

and

$$(9.455) \qquad \frac{dp}{ds} = \frac{\partial}{\partial x}(n^2); \qquad \frac{dq}{ds} = \frac{\partial}{\partial y}(n^2); \qquad \frac{dr}{ds} = \frac{\partial}{\partial z}(n^2)$$

where s is a parameter along the characteristics and the fact that $F = 0$ along the characteristic curves was used to obtain (9.454).

The characteristic (base) curves $\mathbf{x} = \mathbf{x}(s) = [x(s), y(s), z(s)]$ are called *rays*, and the surfaces of constant phase $\varphi = $ constant are called *wave fronts*. Since the vector $[p, q, r] = \nabla\varphi$ is normal to the wave front $\varphi = $ constant, the equations (9.453) that can be written as

$$(9.456) \qquad \frac{d\mathbf{x}}{ds} = 2[p, q, r] = 2\nabla\varphi,$$

show that the rays are orthogonal to the wave fronts.

The full first order system of characteristic equations (9.453)–(9.455) must be solved simultaneously, particularly if the index of refraction n is a function of $\mathbf{x}$. However, it is possible to obtain a separate second order system of equations for the rays $\mathbf{x} = \mathbf{x}(s)$. On differentiating (9.456) with respect to s and using (9.455) we obtain

$$(9.457) \qquad \frac{d^2\mathbf{x}}{ds^2} = 2\nabla(n^2).$$

This vector equation together with

$$(9.458) \qquad \left(\frac{dx}{ds}\right)^2 + \left(\frac{dy}{ds}\right)^2 + \left(\frac{dz}{ds}\right)^2 = 4(p^2 + q^2 + r^2) = 4n^2$$

determines the rays $\mathbf{x}(s)$ and the variation of the parameter s along the rays. These equations for $\mathbf{x}(s)$ are known as the *ray equations*, and they show that the rays are specified in terms of the *index of refraction n*. In particular, if the index n is a constant, we see that $\mathbf{x}(s)$ is linear in s so that the rays are straight lines.

Now (9.458) shows that s is not generally equal to an arc length parameter on the rays. In fact, with $d\sigma^2 = dx^2 + dy^2 + dz^2$ so that σ is an arc length parameter, we have from (9.458)

$$(9.459) \qquad d\sigma = 2n\, ds.$$

In terms of σ the ray equations (9.457) become

$$(9.460) \qquad n\frac{d}{d\sigma}\left(n\frac{d\mathbf{x}}{d\sigma}\right) = \nabla\left(\frac{1}{2}n^2\right),$$

whereas (9.458) is replaced by

$$(9.461) \qquad \left(\frac{dx}{d\sigma}\right)^2 + \left(\frac{dy}{d\sigma}\right)^2 + \left(\frac{dz}{d\sigma}\right)^2 = 1,$$

as is easily shown.

Using the parameter σ in place of s and noting that $d/ds = 2nd/d\sigma$, we have for the phase φ,

$$(9.462) \qquad \frac{d\varphi}{d\sigma} = n,$$

instead of (9.454). Integrating along a ray $\mathbf{x}(\sigma)$ from σ_0 to σ we have

$$(9.463) \qquad \varphi = \varphi_0 + \int_{\sigma_0}^{\sigma} n\, d\sigma,$$

where φ and n are functions of $\mathbf{x}(\sigma)$ and φ_0 is a specified constant that may vary from ray to ray. Thus once the rays are known the phase φ can be determined.

The (standard) initial value problem for φ requires that we specify φ on a given initial surface. Let the initial surface be expressed parametrically as $\mathbf{x} = \mathbf{R}(\alpha, \beta)$ and let $\varphi = \varphi_0(\alpha, \beta)$ on that surface. To use the procedure given for determining φ we must find the rays that pass through the initial surface and then integrate along the rays as indicated in (9.463). However, since the ray equations are of second order, an initial point, as well as an initial direction, must be assigned for each ray on the initial surface $\mathbf{x} = \mathbf{R}(\alpha, \beta)$. Now (9.456) shows that the direction of a ray $\mathbf{x}$ (i.e., $d\mathbf{x}/ds$ or $d\mathbf{x}/d\sigma$) is given in terms of p, q, and r. That is, to solve for φ uniquely, we must not only specify $\mathbf{x}$ and φ initially, but also p, q, and r. This is consistent with the fact that the full set of characteristic equations (9.453)–(9.455) for $\mathbf{x}$, φ, p, q, and r is what we are effectively solving.

To determine p, q, and r initially, and thereby $d\mathbf{x}/d\sigma$, we follow the general procedure given in Section 2.4 and determine conditions equivalent to the strip conditions found in that section. From the initial condition $\varphi[\mathbf{R}(\alpha, \beta)] = \varphi_0(\alpha, \beta)$ we obtain

$$(9.464) \qquad \frac{\partial \varphi_0}{\partial \alpha} = \nabla\varphi \cdot \frac{\partial \mathbf{R}}{\partial \alpha} = \frac{1}{2} \frac{d\mathbf{x}}{ds} \cdot \frac{\partial \mathbf{R}}{\partial \alpha} = n \frac{d\mathbf{x}}{d\sigma} \cdot \frac{\partial \mathbf{R}}{\partial \alpha}$$

and, similarly,

$$(9.465) \qquad \frac{\partial \varphi_0}{\partial \beta} = \nabla\varphi \cdot \frac{\partial \mathbf{R}}{\partial \beta} = n \frac{d\mathbf{x}}{d\sigma} \cdot \frac{\partial \mathbf{R}}{\partial \beta}$$

where (9.456) and $d/ds = 2nd/d\sigma$ have been used. An additional condition, which is equivalent to setting $F = 0$ in (9.452) and which is identical with (9.461), is

$$(9.466) \qquad \frac{d\mathbf{x}}{d\sigma} \cdot \frac{d\mathbf{x}}{d\sigma} = 1.$$

This yields three equations for the three components of the vector $d\mathbf{x}/d\sigma$

evaluated on the initial surface. We assume that $\sigma = \sigma_0$ at the point where each ray intersects the surface $\mathbf{x} = \mathbf{R}(\alpha, \beta)$ so that $d\mathbf{x}/d\sigma$ evaluated at $\sigma = \sigma_0$ is determined from (9.464)–(9.466).

To simplify the discussion of the possible solutions for $d\mathbf{x}/d\sigma$ at $\sigma = \sigma_0$ we assume that α and β measure arc length in an orthogonal coordinate system on the initial surface $\mathbf{x} = \mathbf{R}(\alpha, \beta)$. Then $d\mathbf{x}/d\sigma$, $\partial \mathbf{R}/\partial \alpha$, and $\partial \mathbf{R}/\partial \beta$ are all unit vectors and, in addition, $\partial \mathbf{R}/\partial \alpha$ and $\partial \mathbf{R}/\partial \beta$ are mutually orthogonal. Let θ and ω denote the angles between the vector $d\mathbf{x}(\sigma_0)/d\sigma$ and the vectors $\partial \mathbf{R}/\partial \alpha$ and $\partial \mathbf{R}/\partial \beta$, respectively. Then (9.464)–(9.465) yield

$$(9.467) \qquad \cos \theta = \frac{1}{n} \frac{\partial \varphi_0}{\partial \alpha}; \qquad \cos \omega = \frac{1}{n} \frac{\partial \varphi_0}{\partial \beta}.$$

The preceding equations generally determine exactly two directions for each ray on the initial surface. To see this, we choose an arbitrary point P on that surface and introduce a (local) Cartesian coordinate system with its origin at that point. Let the unit vectors $\mathbf{i}$, $\mathbf{j}$, and $\mathbf{k}$ of the coordinate system correspond to the vectors $\partial \mathbf{R}/\partial \alpha$, $\partial \mathbf{R}/\partial \beta$, and a unit normal vector to the initial surface, respectively. Then our definition of θ and ω implies that

$$(9.468) \qquad \cos \theta = \frac{d\mathbf{x}}{d\sigma} \cdot \mathbf{i}; \qquad \cos \omega = \frac{d\mathbf{x}}{d\sigma} \cdot \mathbf{j}$$

at the point P on the initial surface. The angles θ and ω are uniquely specified from (9.467). If the angle γ is defined as

$$(9.469) \qquad \cos \gamma = \frac{d\mathbf{x}}{d\sigma} \cdot \mathbf{k},$$

we conclude from (9.466) that

$$(9.470) \qquad \cos^2\theta + \cos^2\omega + \cos^2\gamma = 1.$$

Now by definition we have $0 \leqslant \theta, \omega, \gamma \leqslant \pi$. If θ and ω are such that $\cos \gamma \neq 0$ in (9.470), we see that the quadratic expression (9.470) yields two possible values for $\cos \gamma$, one positive and one negative, with identical absolute values. If we choose the value of γ, say, $\gamma = \gamma_0$, for which $\cos \gamma_0 > 0$, then $0 < \gamma_0 < \pi/2$ and the vector $d\mathbf{x}(\sigma_0)/d\sigma$ points toward the same side of the initial surface as the normal vector in view of (9.469). If we set $\gamma = \gamma_1 = \pi - \gamma_0$, then $\cos \gamma_1 = -\cos \gamma_0 < 0$ and $\pi/2 < \gamma_1 < \pi$. In this case $d\mathbf{x}(\sigma_0)/d\sigma$ and the normal vector have opposite directions. The vectors $d\mathbf{x}(\sigma_0)/d\sigma$ that correspond to γ_0 and γ_1, respectively, and the normal vector to the initial surface are all coplanar at each point P.

If the initial data are such that $\cos \gamma = 0$ at a point on the initial surface, it follows from (9.469) that $d\mathbf{x}(\sigma_0)/d\sigma$ lies in the tangent plane of the surface at that point. If the rays are tangent to the initial surface at every point, there are two cases that are referred to as *characteristic initial value problems* for φ. In one case the rays are completely contained within the initial surface. Since the rays do not leave the initial surface $\mathbf{x} = \mathbf{R}(\alpha, \beta)$, the equation (9.462) that describes the variation of φ along the rays is, in fact, a condition on the variation of the initial value φ_0 along the rays on the initial surface. If the conditions on the data are met, we find that our result does not determine the phase function φ outside the initial surface.

A more interesting case occurs if the rays are merely tangent to the initial surface but do not lie within it. Then the initial surface is an envelope or a *caustic surface* of the system of rays. Again the initial value φ_0 must be such that (9.462) is satisfied when $\sigma = \sigma_0$, that is, at the point of tangency of the rays. One ray issues from each point on the initial surface and φ is determined from (9.463) along each ray. A more complicated situation that occurs often in asymptotic problems is when the rays are tangent to the initial surface only along a curve or some other subregion but are not tangent elsewhere on the surface. Although it is possible to determine the rays and the phase function φ for many of these problems, the fact that the rays intersect at the initial surface leads to difficulties with the full asymptotic solution.

In the noncharacteristic case (i.e., when $\cos \gamma \neq 0$ in the preceding) there are two possible ray directions determined at each point on the initial surface. To obtain a unique solution for the phase φ we must choose one of the two directions determined from (9.467)–(9.470). That is, we select the set of rays issuing either on one side of the initial surface or on the other side. Thereby, $\mathbf{x}(\sigma)$ and $d\mathbf{x}/d\sigma$ are uniquely specified at each point of the initial surface. Consequently, the ray equations determine $\mathbf{x}(\sigma)$ uniquely. Finally, the phase $\varphi(x, y, z)$ is found from (9.463) where integration takes place along the ray that passes through the point (x, y, z).

The *transport equations* (9.429)–(9.430) can also be solved by integrating along the rays. We have from (9.456)

$$(9.471) \qquad 2\nabla \varphi \cdot \nabla v_j = \frac{d\mathbf{x}}{ds} \cdot \nabla v_j = 2n \frac{d\mathbf{x}}{d\sigma} \cdot \nabla v_j = 2n \frac{dv_j}{d\sigma}$$

where $dv_j/d\sigma$ is a directional derivative along the ray $\mathbf{x}(\sigma)$. Thus the transport equations reduce to ordinary differential equations along the rays and are given as

$$(9.472) \qquad 2n \frac{dv_0}{d\sigma} + v_0 \nabla^2 \varphi = 0,$$

$$(9.473) \qquad 2n \frac{dv_j}{d\sigma} + v_j \nabla^2 \varphi = -\nabla^2 v_{j-1}, \qquad j \geq 1$$

where all functions are evaluated on the ray $\mathbf{x}(\sigma)$.

To solve these equations, we express $\nabla^2 \varphi$ in terms of its variation along a ray. To do so, we consider a region R bounded laterally by a *tube of rays* and capped by two segments of *wave fronts* φ = constant which we denote by S_0 and S_1 as shown in Figure 9.4. The lateral boundary is denoted by S. Let N be the exterior unit normal on the boundary of R. Applying the divergence theorem in R gives

$$(9.474) \qquad \iint\limits_R \nabla^2 \varphi \, dV = \iint\limits_R \nabla \cdot \nabla\varphi \, dv = \int_{S_0} \nabla\varphi \cdot N \, da$$

$$+ \int_S \nabla\varphi \cdot N \, da + \int_{S_1} \nabla\varphi \cdot N \, da.$$

The surface integral over S vanishes since $\nabla\varphi$ lies in the direction of the rays and N is orthogonal to that direction on S. Assuming ray parameter σ increases as we move from S_0 to S_1, we observe that $\nabla\varphi$ has the direction of N on S_1, whereas it has the direction of $-N$ on S_0, since S_0 and S_1 are portions of phase surfaces. Also, since $|\nabla\varphi| = n$, in view of (9.426), we have

$$(9.475) \qquad \iint\limits_R \nabla^2 \varphi \, dV = \int_{S_1} n \, da - \int_{S_0} n \, da.$$

To determine a simple and useful expression for the element of surface area da on a wave front, we introduce a reference wave front $\varphi = 0$ and a coordinate system (ξ, η) on that wave front. We then determine all the rays orthogonal to that wave front. Each surface φ = constant is parallel to the wave front $\varphi = 0$ and is orthogonal to the rays. With σ equal to arc length on the rays, we may introduce a (ray) coordinate system (ξ, η, σ) in space. Further, we can replace σ by the wave front coordinate φ and, in view of (9.462), we have

$$(9.476) \qquad d\varphi = n \, d\sigma$$

along the rays. In the (ξ, η, φ) coordinate system, the surfaces φ = constant are orthogonal to the rays which are given as ξ = constant and η = constant.

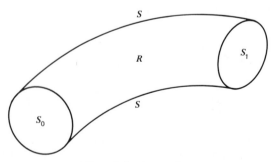

Figure 9.4.   A ray tube.

Consequently, if $\mathbf{x} = \mathbf{r}(\xi, \eta, \varphi)$ are the equations of transformation from Cartesian coordinates to (ξ, η, φ) coordinates, we find that the area element da on a wave front is given as

(9.477)
$$da = J\,d\xi\,d\eta$$

where $J = |\mathbf{r}_\xi \times \mathbf{r}_\eta|$, as is well known from vector analysis. The volume element dV in (ξ, η, φ) coordinates has the form

(9.478)
$$dV = da\,d\sigma = \frac{1}{n}J\,d\xi\,d\eta\,d\varphi,$$

since σ is arc length and the distance between two infinitesimally close wave fronts $\varphi = $ constant and $\varphi + d\varphi = $ constant is given by $d\sigma = (1/n)\,d\varphi$, in view of (9.476).

We now assume that the surface elements S_0 and S_1 considered correspond to the wave fronts $\varphi = $ constant and $\varphi + d\varphi = $ constant, respectively. Then as the region R (i.e., the ray tube) shrinks down on a particular ray, we have in (ξ, η, φ) coordinates

(9.479)
$$\iint_R \nabla^2\varphi\,dV \approx \nabla^2\varphi\frac{1}{n}J\,d\xi\,d\eta\,d\varphi$$

and

(9.480)
$$\int_{S_1} n\,da - \int_{S_0} n\,da \approx \left[nJ\big|_{\varphi+d\varphi} - nJ\big|_\varphi\right]d\xi\,d\eta,$$

so that

(9.481)
$$\nabla^2\varphi \approx \frac{n}{J}\left[\frac{nJ\big|_{\varphi+d\varphi} - nJ\big|_\varphi}{d\varphi}\right].$$

In the limit as $d\varphi \to 0$ we obtain

(9.482)
$$\nabla^2\varphi = \frac{n}{J}\frac{d}{d\varphi}(nJ) = \frac{1}{J}\frac{d}{d\sigma}(nJ)$$

on using (9.476).

We are now ready to solve the transport equations (9.472)–(9.473). Inserting (9.482) into (9.472) yields

(9.483)
$$2n\frac{dv_0}{d\sigma} + \frac{1}{J}\frac{d}{d\sigma}(nJ)v_0 = 0.$$

This can be written as

$$(9.484) \qquad \frac{2n}{\sqrt{nJ}}\frac{d\left(v_0\sqrt{nJ}\right)}{d\sigma} = 0.$$

Integrating along the ray $\mathbf{x}(\sigma)$ from σ_0 to σ with $v_0|_{\sigma_0} = V_0(\sigma_0)$, we obtain

$$(9.485) \qquad v_0 = V_0(\sigma_0)\left[\frac{n(\sigma_0)J(\sigma_0)}{n(\sigma)J(\sigma)}\right]^{1/2}$$

where the dependence of the solution on variables other than σ has been suppressed. Similarly, the transport equations (9.473) can be written as

$$(9.486) \qquad \frac{2n}{\sqrt{nJ}}\frac{d}{d\sigma}\left(v_j\sqrt{nJ}\right) = -\nabla^2 v_{j-1}, \qquad j \geq 1.$$

Again integrating along a ray with $v_j|_{\sigma_0} = V_j(\sigma_0)$ we have

$$(9.487) \qquad v_j = V_j(\sigma_0)\left[\frac{n(\sigma_0)J(\sigma_0)}{n(\sigma)J(\sigma)}\right]^{1/2}$$

$$-\frac{1}{2\sqrt{n(\sigma)J(\sigma)}}\int_{\sigma_0}^{\sigma}\sqrt{\frac{J}{n}}\,\nabla^2 v_{j-1}\,d\sigma, \qquad j \geq 1.$$

Although the derivation was carried out for the three-dimensional case, the results are valid in two dimensions as well. We need only define J appropriately. On the reference wave front $\varphi = 0$ we now introduce the single coordinate ξ, and determine all the rays orthogonal to that wave front. In the resulting (ξ, φ) coordinate system we set $\mathbf{x} = \mathbf{r}(\xi, \varphi)$ and find that the linear element ds on a wave front is given as

$$(9.488) \qquad ds = J\,d\xi$$

where $J = |\mathbf{r}_\xi|$. The area element dA then has the form

$$(9.489) \qquad dA = ds\,d\sigma = \frac{1}{n}J\,d\xi\,d\varphi,$$

since (9.462) is valid in two dimensions as well. In deriving (9.481) the integration is carried out over a plane region bounded by two wave fronts and two rays. The same expression for $\nabla^2\varphi$ as was found in three dimensions results, and the expressions for the phase function φ and the amplitude terms v_j are as given.

In both cases the function J represents an *expansion* or *contraction factor* for the rays. It is proportional to an area or linear element on the wave fronts that increases in magnitude as the rays diverge and decreases in magnitude as the rays converge. Noting the form of v_0 in (9.485) we find that, to leading order in k, the amplitude of the solution increases as the rays converge and decreases as the rays diverge. This result is consistent with a basic principle of geometrical optics according to which the field is stronger the greater the concentration of the rays.

If the rays intersect at same point or curve, the function J vanishes and the amplitude terms v_j all become infinite. The solution of the reduced wave equation of which we have found the asymptotic expansion, either remains finite in that region or has a different type of singularity, as we have seen in Example 9.13. Thus our asymptotic result is not valid in regions where the rays intersect and modified (asymptotic) results must be found in such regions.

It is of interest to observe that the dependence of v_0 on the divergence or convergence of the rays may be obtained by means of an extremely interesting alternative method. If we multiply the equation (9.429) by v_0, we have

$$(9.490) \qquad v_0\left[2\nabla\varphi \cdot \nabla v_0 + v_0\nabla^2\varphi\right] = \nabla \cdot \left(v_0^2\nabla\varphi\right) = 0.$$

Integrating this divergence expression over the ray tube R as defined and using the divergence theorem gives

$$(9.491) \qquad 0 = \iiint_R \nabla \cdot \left(v_0^2\nabla\varphi\right) dV = \int_{S_1} v_0^2 n \, da - \int_{S_0} v_0^2 n \, da,$$

as is easily seen on applying the results that lead from (9.474) to (9.475). In an approximate sense, (9.491) may be thought to represent a *conservation of energy* along the ray tubes of the solution of the reduced wave equation. If the ray tube shrinks down on a ray $x(\sigma)$, we have

$$(9.492) \qquad v_0^2 n \, da\Big|_{\sigma_1} = v_0^2 n \, da\Big|_{\sigma_0}$$

where σ_1 and σ_0 correspond to two points on the ray. In view of the relation (9.477) between da and J, we find that v_0 at some arbitrary point on a ray is given in terms of $v_0|_{\sigma_0} = V_0(\sigma_0)$ precisely as in (9.485).

In the case where the index of refraction $n = $ constant, it is possible to find some simple explicit forms for the function J and thereby for the amplitude term v_0. With $n = $ constant, the rays are straight lines and we may assume without loss of generality that $n = 1$. Then, as shown by (9.462), both σ and φ measure arc length on the rays.

If the rays are parallel lines, the area element da remains constant on all wave fronts so that $J = $ constant as well. This implies that $v_0 = $ constant on the rays. If v_0 retains the same constant value on all rays, the v_j may be chosen to vanish and the result is a *plane wave* solution of the reduced wave equation. (The wave fronts are clearly planes for this case.) If v_0 assumes different values on different rays, we obtain what is called a *general plane wave*. The wave fronts $\varphi = $ constant are again planes but the amplitude term v_0, although it is constant on each ray, may vary in magnitude from point to point on the wave fronts. A similar result applies in the two-dimensional problem.

For the case of *spherical wave fronts* we introduce spherical coordinates

$$(9.493) \qquad x = r\sin\theta\cos\omega; \qquad y = r\sin\theta\sin\omega; \qquad z = r\cos\theta,$$

where $r \geqslant 0$, $0 \leqslant \theta \leqslant \pi$, and $0 \leqslant \omega \leqslant 2\pi$. With $n = 1$, the phase term can be written as

$$(9.494) \qquad \varphi = r - r_1$$

where r_1 is a constant and the reference wave front $\varphi = 0$ is given as the sphere $r = r_1$. The ray equation (9.460) becomes

$$(9.495) \qquad \frac{d^2\mathbf{x}}{d\sigma^2} = \mathbf{0},$$

and has the solution

$$(9.496) \qquad \mathbf{x} = \mathbf{a} + \mathbf{b}\sigma$$

where $\mathbf{a}$ and $\mathbf{b}$ are constant vectors and $|\mathbf{b}| = 1$ in view of (9.461). Replacing the variable σ by r and introducing the initial data

$$(9.497) \qquad \mathbf{x}(r_1) = \mathbf{a} + \mathbf{b}r_1 = [r_1\sin\theta\cos\omega, \quad r_1\sin\theta\sin\omega, \quad r_1\cos\theta],$$

$$(9.498) \qquad \frac{d\mathbf{x}(r_1)}{dr} = \mathbf{b} = [\sin\theta\cos\omega, \quad \sin\theta\sin\omega, \quad \cos\theta],$$

we find that

$$(9.499) \qquad \mathbf{x}(r) = \mathbf{b}r = [r\sin\theta\cos\omega, \quad r\sin\theta\sin\omega, \quad r\cos\theta].$$

The data (9.497)–(9.498) imply that the rays $\mathbf{x}(r)$ pass through and are orthogonal to the sphere $r = r_1$, and (9.499) shows the rays to be radial lines issuing from the origin. The radial parameter r clearly measures arc length on the rays and the spheres $\varphi = $ constant are orthogonal to the rays.

To determine the function J for this case, we identify the coordinates ξ and η introduced previously on the wave front $\varphi = 0$ with the angular variables θ and ω. Since $n = 1$, both φ and r measure arc length on the rays. The transformation from Cartesian to ray coordinates is given as

$$(9.500) \qquad \mathbf{x} = \mathbf{r}(r, \theta, \omega) = [r\sin\theta\cos\omega, \quad r\sin\theta\sin\omega, \quad r\cos\theta]$$

as seen from (9.493). Then da is an element of area on a sphere and we have

$$(9.501) \qquad da = r^2 \sin\theta \, d\theta \, d\omega = J \, d\theta \, d\omega$$

where $J = |\mathbf{r}_\theta \times \mathbf{r}_\omega| = r^2 \sin\theta$. Thus

$$(9.502) \qquad v_0 = V_0(r_0)\left[\frac{J(r_0)}{J(r)}\right]^{1/2} = V_0(r_0)\left[\frac{r_0}{r}\right]$$

on each ray. If we assume that $V_0(r_0)$ is independent of θ and ω so that it has the same value on each ray, we easily show that the $v_j(j \geqslant 1)$ may be chosen to vanish and the solution of the reduced wave equation is the *spherical wave*

$$(9.503) \qquad u = (\text{constant})\frac{1}{r}e^{ikr}.$$

For this case, the asymptotic approach yields an exact solution of the reduced wave equation valid up to the singular point $r = 0$. (The corresponding cylindrical wave solution found in Example 9.13 is, however, not valid up to the singular point.) The point $r = 0$ represents a *focus* for the rays since they all intersect there. Even though the (asymptotic) solution may have a valid form up to $r = 0$, it does blow up there since J vanishes at that point.

More generally, the term $V_0(r_0)$ in (9.502) may vary from ray to ray; that is, it can have a nonconstant dependence on θ and ω. In that case the wave fronts are still spheres, and v_0 decays like $1/r$ along a given ray, but a more complicated expression results for the amplitude distribution of the field that involves all the v_j.

Next we consider a two-dimensional example in which the rays form a (smooth) envelope. If $\mathbf{x} = \mathbf{R}(\xi)$ is the equation of this envelope or *caustic curve* where ξ is an arc length parameter, the rays are just the tangent lines of this curve. With σ equal to arc length along the rays, an orthogonal coordinate system in ξ and σ is given by the equations of transformation

$$(9.504) \qquad \mathbf{x} = \mathbf{r}(\xi, \sigma) = \mathbf{R}(\xi) + (\sigma - \xi)\mathbf{R}'(\xi).$$

Since ξ is arc length, we have $|\mathbf{R}'(\xi)| = 1$, $\mathbf{R}' \cdot \mathbf{R}'' = 0$, and $|\mathbf{R}''| = 1/\rho$ where $\rho(\xi)$ is the radius of curvature of the caustic curve. Then

$$\mathbf{r}_\xi = (\sigma - \xi)\mathbf{R}''(\xi); \qquad \mathbf{r}_\sigma = \mathbf{R}'(\xi)$$

imply that $\mathbf{r}_\xi \cdot \mathbf{r}_\sigma = 0$, so that the (ξ, σ) coordinate system is orthogonal. The lines $\xi = \text{constant}$ are the rays, whereas the curves $\sigma = \text{constant}$ are the (orthogonal) wave fronts. This may be verified directly by noting that with

$$(9.505) \qquad h_\xi = |\mathbf{r}_\xi| = \frac{|\sigma - \xi|}{\rho}; \qquad h_\sigma = |\mathbf{r}_\sigma| = 1$$

we have

$$(9.506) \qquad (\nabla\varphi)^2 = \frac{1}{h_\xi^2}\varphi_\xi^2 + \frac{1}{h_\sigma^2}\varphi_\sigma^2 = \frac{\rho^2}{(\sigma - \xi)^2}\varphi_\xi^2 + \varphi_\sigma^2.$$

Thus $\varphi = \sigma$ is a solution of the eiconal equation $(\nabla\varphi)^2 = 1$. The function J for this system of rays is given as

$$(9.507) \qquad J = |\mathbf{r}_\xi| = \frac{|\sigma - \xi|}{\rho},$$

and the solution v_0 has the form

$$(9.508) \qquad v_0 = V_0(\sigma_0)\left[\frac{J(\sigma_0)}{J(\sigma)}\right]^{1/2} = V_0(\sigma_0)\left|\frac{\sigma_0 - \xi}{\sigma - \xi}\right|^{1/2}.$$

The amplitude term v_0 is singular when $\sigma = \xi$, and this occurs at points on the caustic $\mathbf{x} = \mathbf{R}(\xi)$ as follows from (9.504). The rays intersect at the caustic and, therefore, $J = 0$ on that curve. Although each ray intersects the caustic at $\sigma = \xi$, the set of points $\sigma \geq \xi$ and $\sigma \leq \xi$ represent two groups of rays, each of which terminates at the caustic. The asymptotic solutions given for φ and v_0, combined with the results that can be obtained for all the v_j, are valid for each group of rays away from the caustic curve. On the caustic curve the solution of the reduced wave equation of which this is the asymptotic expansion remains bounded. A *boundary layer approach* presented later yields a finite asymptotic result at the caustic.

A typical *boundary value problem* for the *reduced wave equation* (9.418), which is to be solved asymptotically for large k, is formulated in the manner indicated in Example 9.13. An *incident wave* u_I completely specified as an asymptotic series of the form (9.424) and (9.428) approaches a boundary region and is scattered by it. The resulting *scattered wave* u_S is again taken to be an asymptotic series of the type (9.424) and (9.428) and must be determined. The *total field* $u = u_I + u_S$, which is an asymptotic solution of (9.418), satisfies a homogeneous *boundary condition* of the first, second, or third kind on the boundary region. We mostly assume that $u = 0$ on the boundary. The problem is to be solved in an exterior unbounded region and we require that u_S satisfy a *radiation condition at infinity*. In our asymptotic version of the problem the radiation condition is replaced by an *outgoing condition* at the boundary. If φ is the phase term of the asymptotic expansion of u_S, then u_S is said to be outgoing at the boundary if the exterior normal derivative of φ (i.e., $\partial\varphi/\partial n = \nabla\varphi \cdot \mathbf{N}$) is positive for every exterior unit normal $\mathbf{N}$ to the boundary. The outgoing condition indicates that u_S travels away from the boundary and is, in fact, radiating towards infinity.

The specification of the incident wave u_I, the boundary condition, and the outgoing condition uniquely specify u_S and thereby the asymptotic solution u, provided the phase and the amplitude terms in the asymptotic expansion of u_S can be determined. We assume that u_I is given as

$$(9.509) \qquad u_I = e^{ik\psi} \sum_{j=0}^{\infty} w_j(ik)^{-j},$$

where the phase ψ and the amplitude terms w_j are known. The scattered wave is expanded as

$$(9.510) \qquad u_S = e^{ik\varphi} \sum_{j=0}^{\infty} v_j(ik)^{-j},$$

with φ and the v_j to be determined. If $u = u_I + u_S = 0$ on the boundary, we find that

$$(9.511) \qquad e^{ik\varphi} \sum_{j=0}^{\infty} v_j(ik)^{-j} = -e^{ik\psi} \sum_{j=0}^{\infty} w_j(ik)^{-j}$$

on the boundary. This leads to the conditions

$$(9.512) \qquad \varphi = \psi$$

and

$$(9.513) \qquad v_j = -w_j, \qquad j \geqslant 0$$

on the boundary.

As we have seen, the (initial) condition (9.512) for φ is by itself insufficient to specify the phase uniquely. In determining the rays associated with the phase φ we must also specify a direction for these rays on the (boundary) region where φ is given. We have found that there are, in general, two possible ray directions at each point where φ is specified. However, $\nabla\varphi$ has the direction of the rays as shown by (9.456). It has been demonstrated that if the rays are not tangent to the surface where φ is specified, their two possible directions lie on opposite sides of the surface. Thus the outgoing condition on the scattered wave u_S uniquely specifies a ray direction at each point as is easily seen.

Once the rays and the phase are determined, the amplitude terms v_j can be obtained by integrating along the rays as in (9.472)–(9.473) and using the data (9.513) as initial conditions.

As an example we now consider the problem of the reflection of a cylindrical wave by a parabolic cylinder. The line source of the (incident) cylindrical wave lies on the z-axis and the generators of the parabolic cylinder are parallel to the z-axis. Thus the problem may be treated as two-dimensional and we look for a solution in the form $u = u(x, y)$.

EXAMPLE 9.14.   REFLECTION OF A CYLINDRICAL WAVE BY A PARABOLA

We consider the parabola

$$x = \frac{1}{2a}y^2 - \frac{a}{2}$$

where $a > 0$ is a given constant. A *cylindrical wave* is generated by a source located at the origin, which is also the focus of the parabola. Thus the total field u satisfies the equation

$$(9.514) \qquad \nabla^2 u + k^2 u = -\delta(x)\delta(y)$$

where $u = u(x, y)$ and $\delta(x)\delta(y)$ is the two-dimensional Dirac delta function. We assume that u vanishes on the parabola; that is,

$$(9.515) \qquad u(x, y) = 0; \qquad x = \frac{1}{2a}y^2 - \frac{a}{2}$$

and u satisfies a radiation condition at infinity. An outwardly radiating wave generated by the source is

$$(9.516) \qquad u_I = \frac{i}{4}H_0^{(1)}(kr)$$

as we have shown in Example 9.13. We are concerned with the asymptotic problem for large k. Thus we replace (9.516) by its asymptotic expansion. The leading terms are given as

$$(9.517) \qquad u_I \approx \frac{i}{4}\sqrt{2/\pi kr}\left[1 + \frac{1}{8ikr}\right]\exp\left[ikr - \frac{i\pi}{4}\right],$$

as follows from the asymptotic expansion of the Hankel function for large arguments.

The wave u_I is a cylindrical wave incident upon the parabola. The asymptotic problem then looks for a solution

$$(9.518) \qquad u \approx u_I + u_S$$

where the incident wave u_I is given in (9.517) and the scattered wave u_S expanded as in (9.510) is to be determined. The boundary condition (9.515) requires that

$$(9.519) \qquad u_S = -u_I; \qquad x = \frac{1}{2a}y^2 - \frac{a}{2}$$

whereas the radiation condition implies that u_S must be an outgoing wave at the parabola.

On comparing with the notation given in our general discussion, we find that

$$(9.520) \qquad \psi = r; \qquad w_0 = \frac{c}{\sqrt{r}}; \qquad w_1 = \frac{c}{8r^{3/2}},$$

with the other terms in the expansion of u_I not having been specified in (9.517). The constant c is given as $c = (i/4)\sqrt{2/\pi k}\, e^{-i\pi/4}$. The appropriate data for the scattered or reflected wave u_S on the parabola are

$$(9.521) \qquad \varphi = r; \qquad x = \frac{1}{2a}y^2 - \frac{a}{2}$$

and φ "outgoing" on the parabola. The first two amplitude terms v_0 and v_1 satisfy

$$(9.522) \qquad v_0 = -\frac{c}{\sqrt{r}}; \qquad v_1 = -\frac{c}{8r^{3/2}}$$

on the parabola in view of (9.520).

To solve for the phase φ of the scattered wave u_S we first represent the parabola in parametric form as

$$(9.523) \qquad y = \tau; \qquad x = \frac{1}{2a}(\tau^2 - a^2); \qquad -\infty < \tau < \infty$$

(we note that τ is not an arc length parameter here). With σ as arc length along the rays we must solve the characteristic equations for $x(\sigma, \tau)$, $y(\sigma, \tau)$, $\varphi(\sigma, \tau)$, $p(\sigma, \tau)$, and $q(\sigma, \tau)$. The initial conditions are given for $\sigma = 0$. Thus

$$(9.524) \qquad y(0, \tau) = \tau; \qquad x(0, \tau) = \frac{1}{2a}(\tau^2 - a^2);$$

as follows from (9.523). That is, $\sigma = 0$ corresponds to the parabola (9.523). The condition (9.521) for φ yields

$$(9.525) \qquad \varphi(0, \tau) = \frac{1}{2a}(\tau^2 + a^2)$$

on expressing (9.521) in terms of the variable τ.

The initial conditions for p and q are determined by proceeding as in Section 2.4. The eiconal equation requires that

$$(9.526) \qquad p^2(0, \tau) + q^2(0, \tau) = 1.$$

The *strip condition* states that on the parabola,

$$(9.527) \qquad \frac{\partial \varphi}{\partial \tau} = p\frac{\partial x}{\partial \tau} + q\frac{\partial y}{\partial \tau}; \qquad \sigma = 0,$$

and this yields

(9.528)
$$\frac{\tau}{a} = p(0, \tau)\frac{\tau}{a} + q(0, \tau)$$

on using (9.524)–(9.525). On solving (9.526) and (9.528) for $p(0, \tau)$ and $q(0, \tau)$, we obtain two sets of solutions. We have

(9.529)
$$p(0, \tau) = 1; \qquad q(0, \tau) = 0;$$

and

(9.530)
$$p(0, \tau) = \frac{\tau^2 - a^2}{\tau^2 + a^2}; \qquad q(0, \tau) = \frac{2a\tau}{\tau^2 + a^2}.$$

One of these sets of initial conditions for p and q must be rejected on the basis of the outgoing condition satisfied by u_S. This states that if $\mathbf{N}$ is the unit normal pointing towards the inside of the parabola, we must have $\nabla\varphi \cdot \mathbf{N} > 0$ on the parabola. We easily find that the unit normal $\mathbf{N}$ is

(9.531)
$$\mathbf{N}(\tau) = \frac{[a, -\tau]}{\sqrt{\tau^2 + a^2}}.$$

Since $\nabla\varphi = [p, q]$ we see that on the parabola

(9.532)
$$\nabla\varphi \cdot \mathbf{N} = [p, q] \cdot \mathbf{N} = \frac{a}{\sqrt{\tau^2 + a^2}} > 0$$

if (9.529) is used, whereas

(9.533)
$$\nabla\varphi \cdot \mathbf{N} = [p, q] \cdot \mathbf{N} = \frac{-a}{\sqrt{\tau^2 + a^2}} < 0$$

if (9.530) is used. Therefore, we conclude that (9.529) are the appropriate initial conditions for this problem.

Since $n = 1$ and $d\sigma = 2\,ds$, the two-dimensional version of the characteristic equations (9.453)–(9.454) for this problem is

(9.534) $\quad \dfrac{dx}{d\sigma} = p; \qquad \dfrac{dy}{d\sigma} = q; \qquad \dfrac{d\varphi}{d\sigma} = 1; \qquad \dfrac{dp}{d\sigma} = 0; \qquad \dfrac{dq}{d\sigma} = 0.$

The last two equations in (9.536) state that p and q are constant along the characteristics. Thus they are equal to their initial values (9.529) for all σ and we have

(9.535)
$$p(\sigma, \tau) = 1; \qquad q(\sigma, \tau) = 0.$$

Inserting these results into the equations for x and y in (9.534) and using

(9.524) gives

$$(9.536) \qquad x(\sigma, \tau) = \sigma + \frac{1}{2a}(\tau^2 - a^2),$$

$$(9.537) \qquad y(\sigma, \tau) = \tau.$$

Finally we obtain for φ, in view of (9.534) and (9.525), the result

$$(9.538) \qquad \varphi(\sigma, \tau) = \sigma + \frac{1}{2a}(\tau^2 + a^2).$$

If we invert the system (9.536)–(9.537) and express σ and τ as functions of x and y, we easily obtain

$$(9.539) \qquad \sigma = x - \frac{1}{2a}(y^2 - a^2); \qquad \tau = y.$$

Inserting (9.539) into (9.538) yields

$$(9.540) \qquad \varphi(x, y) = x + a.$$

Thus the phase of the scattered wave u_S is that of a plane wave. However, since the amplitude terms of u_S are not constant, u_S is, in fact, a *general plane wave*.

To solve for the amplitude terms v_0 and v_1 we note first that $\nabla^2 \varphi = 0$ in view of (9.540). Further, we find that $r = (1/2a)(\tau^2 + a^2)$ on the parabola. Thus

$$(9.541) \qquad 2\nabla\varphi \cdot \nabla v_0 + \nabla^2\varphi v_0 = 2\frac{\partial v_0}{\partial \sigma} = 0,$$

so that

$$(9.542) \qquad v_0(\sigma, \tau) = v_0(0, \tau) = -c\sqrt{\frac{2a}{\tau^2 + a^2}}$$

on using (9.522). In terms of x and y we have

$$(9.543) \qquad v_0(x, y) = -c\sqrt{\frac{2a}{y^2 + a^2}},$$

since $\tau = y$. The equation for v_1 is

$$(9.544) \qquad 2\frac{\partial v_1}{\partial \sigma} = -\nabla^2 v_0$$

where the right side of (9.544) is a function of τ only, in view of (9.543). That is, we operate on v_0 with the Laplacian in (x, y)-coordinates and then use $y = \tau$ to convert back to (σ, τ) variables (we do not carry out this calculation). On using (9.522) we obtain

$$(9.545) \qquad v_1(\sigma, \tau) = -\frac{1}{2}(\nabla^2 v_0)\sigma - \frac{c}{8}\left(\frac{2a}{\tau^2 + a^2}\right)^{3/2}.$$

In x, y variables (9.545) has the form

$$(9.546) \qquad v_1(x, y) = V_1(y)x + V_2(y),$$

with given V_1 and V_2.

We have shown that u_S is given as

(9.547)

$$u_S(x, y) = \left\{-c\sqrt{\frac{2a}{y^2 + a^2}} + \frac{1}{ik}(V_1(y)x + V_2(y))\right\}\exp[ik(x + a)],$$

and this has the form of a *general plane wave*. We have carried out the calculation for u_S to two terms to show that the asymptotic result (9.547) is not uniformly valid. As x increases to the point where $x = O(k)$, it follows from (9.547) that the term $(1/ik)v_1$ is of the same order of magnitude as v_0. Therefore, the asymptotic expansion of u_S becomes disordered, in that our initial assumption that $(1/ik)v_1$ is of lower order in k than v_0 is not valid. The asymptotic expansion we have obtained for u_S is, consequently, useful only if $x < O(k)$.

The growth of the amplitude term v_1, as well as the growth of the further terms in the expansion of u_S, corresponds to the *secular behavior* encountered and discussed in a number of problems in the preceding sections. To analyze this question more closely we insert

$$(9.548) \qquad u_S = v e^{ik(x+a)}$$

into the reduced wave equation (9.418) with $n = 1$, and find that

$$(9.549) \qquad 2ikv_x + v_{xx} + v_{yy} = 0.$$

Inserting the expansion $v = \sum_{j=0}^{\infty}v_j(ik)^{-j}$ into (9.549) and solving for the v_j shows that $v_j = O(x^j)$ as $x \to \infty$, as we have seen for v_1.

The secularity problems encountered may be removed by using a *stretching transformation* similar to that used earlier in boundary layer theory. Since our asymptotic results must be modified only for $x \geqslant O(k)$, we set

$$(9.550) \qquad x = k\xi$$

in (9.549) to emphasize the region where x is large. This yields

$$(9.551) \qquad 2iv_\xi + v_{yy} + \frac{1}{k^2} v_{\xi\xi} = 0.$$

To leading order in k we obtain a parabolic *Schrödinger equation*

$$(9.552) \qquad 2i\hat{v}_\xi + \hat{v}_{yy} = 0.$$

We must find a solution of (9.552) that matches the asymptotic result given above for v, to leading order terms as $\xi \to 0$. Effectively, we are considering an initial value problem for (9.552) in the region $\xi > 0$ with v specified at some positive value of ξ. We remark that if the parabolic equation method of Example 9.12 is applied to (9.549), we are led to consider the same equation (9.552).

A general solution of (9.552) is

$$(9.553) \qquad \hat{v}(\xi, y) = \int_{-\infty}^{\infty} \left\{ \frac{1}{\sqrt{\xi}} \exp\left[\frac{i(y - s)^2}{2\xi} \right] \right\} f(s)\, ds.$$

The bracketed term in the integrand in (9.553) is, apart from a constant factor, a fundamental solution of (9.552). It may be verified by direct substitution into (9.552) that it satisfies the equation for $\xi > 0$. The function $f(s)$ in (9.553) is arbitrary and must be specified by matching (9.553) with the asymptotic results given previously.

To carry out the *matching procedure*, we rewrite (9.553) in terms of $x = k\xi$ and y to obtain

$$(9.554) \qquad v(x, y) = \sqrt{\frac{k}{x}} \int_{-\infty}^{\infty} \exp\left[\frac{ik(y - s)^2}{2x} \right] f(s)\, ds.$$

Now if $x \ll O(k)$, we see that the exponential in (9.554) is rapidly oscillating since k is large by assumption. We may, therefore, evaluate (9.554) by the *method of stationary phase*, as presented in Example 5.13. The phase term $\varphi(s) = (y - s)^2/2x$ in (9.554) has a stationary point where $\varphi'(s_0) = 0$, that is, at $s_0 = y$. Then the leading order result as given in (5.238) is

$$(9.555) \qquad v \approx \sqrt{2\pi}\, e^{i\pi/4} f(y).$$

This result, which is valid for small $\xi = x/k$, must be equated to the leading amplitude term v_0 given in (9.543). Thus

$$(9.556) \qquad f(y) = -\frac{c}{\sqrt{2\pi}} \sqrt{\frac{2a}{y^2 + a^2}}\, e^{-i\pi/4},$$

and $v(x, y)$—that is, (9.554)—is completely specified.

We might expect that it should be possible to generate the amplitude term v_1 from the asymptotic expansion of the integral (9.554). To see how this occurs we expand the function $f(s)$ in (9.554) around the stationary point $s_0 = y$ and obtain

$$(9.557) \quad f(s) = f(y) + (s - y)f'(y) + \tfrac{1}{2}(s - y)^2 f''(y) + \cdots .$$

Inserting (9.557) into (9.554), we find on integrating term by term that the leading term $f(y)$ yields the result (9.555). The second term $(s - y)f'(y)$ yields an exact differential and is asymptotic to zero if we neglect contributions from infinity as we do. For the third term we have, on integrating by parts,

$$(9.558) \quad \frac{1}{2}\sqrt{\frac{k}{x}}\, f''(y) \int_{-\infty}^{\infty} (y - s)^2 \exp\left[\frac{ik(y - s)^2}{2x} \right] ds$$

$$\approx -\frac{1}{2}\sqrt{\frac{k}{x}}\, f''(y)\left(\frac{x}{ik} \right) \int_{-\infty}^{\infty} \exp\left[\frac{ik(y - s)^2}{2x} \right] ds$$

$$= -\frac{1}{2}\sqrt{2\pi}\, e^{i\pi/4}\left(\frac{x}{ik} \right) f''(y) = -\frac{1}{2}\left(\frac{x}{ik} \right) \frac{\partial^2 v_0}{\partial y^2}$$

where we have neglected contributions from infinity and noted that $f(y) = (1/\sqrt{2\pi})v_0 \exp[-i\pi/4]$. On comparing (9.558) with the expression for $(1/ik)v_1$ obtained from (9.546), we find that the asymptotic expansion of the integral (9.554) agrees with the terms $v_0 + (1/ik)v_1$, provided we replace $f(s)$ in (9.554) by $f(s) + (1/ik)F(s)$ and specify $F(s)$ as to obtain full agreement between the $O(1/k)$ term obtained from the integral and the term $(1/ik)v_1$.

We have shown, formally, how to construct a solution of (9.552) that matches the first two terms v_0 and v_1 in the asymptotic expansion of the amplitude of u_S. However, the integral representation of the solution of (9.552) does not exhibit secular behavior for large x, as was the case for the (conventional) asymptotic result given previously.

The preceding examples have shown that the geometrical optics results, in which the scattered and incident fields are expanded in the form (9.424) and (9.428), are not always adequate. In regions where the rays intersect, or in the far field as in the preceding example, the geometrical optics solution is not valid and a modified expansion or result is needed. In the following example we show how to construct a valid asymptotic result near a two-dimensional caustic where the geometrical optics result fails. The boundary layer method will be used.

EXAMPLE 9.15. THE ASYMPTOTIC EXPANSION AT A CAUSTIC

We consider a two-dimensional problem for the reduced wave equation (9.418) with $n = 1$ and an asymptotic solution of the form (9.424) and (9.428). The rays for this solution are assumed to have a smooth envelope. The envelope is a *caustic curve* for this problem and the asymptotic solution breaks down on the caustic, as we have seen. We retain the notation introduced in our discussion of this problem—see (9.504)–(9.508).

The equation of the caustic is assumed to be $x = R(\xi)$ where ξ is arc length on the curve. The rays for this problem are the tangent lines to the caustic and are given as

$$(9.559) \qquad x = r(\xi, \sigma) = R(\xi) + (\sigma - \xi)R'(\xi)$$

with σ equal to arc length on the rays, as was shown following equation (9.504). The phase of the wave is given as $\varphi = \sigma$ and, in view of (9.508), the leading term in the asymptotic result is

$$(9.560) \qquad u \approx V_0(\sigma_0)\left[\frac{\xi - \sigma_0}{\xi - \sigma}\right]^{1/2} \exp[ik\sigma].$$

From (9.559) we see that when $\sigma = \xi$, the rays intersect the caustic and (9.560) shows that the asymptotic result is infinite on the caustic. We assume that σ, as well as σ_0, in (9.560) is less than ξ and that $V_0(\sigma_0)$ is a given function. Thus (9.560) represents a wave approaching the caustic $x = R(\xi)$ whose amplitude becomes infinite there.

Although the asymptotic field is singular at the caustic, the actual field is finite there. After the incoming rays pass the caustic they become outgoing rays, and the resulting outgoing field is well behaved, and we must determine its geometrical optics representation. In addition, although no rays penetrate to the concave side of the caustic in Figure 9.5 so that the geometrical optics field must vanish in that region, the actual field is, in fact, nonzero there. We wish to study the field at and near the caustic and to determine the transition undergone by the geometrical optics field on its passage through the caustic region, that is, how its amplitude and phase terms change.

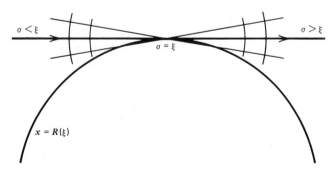

Figure 9.5. The caustic curve.

To study the field near the caustic $\mathbf{x} = \mathbf{R}(\xi)$ we express the reduced wave equation in (ξ, σ) coordinates as defined by (9.559). The coordinate system is orthogonal and we have from (9.505).

$$(9.561) \qquad\qquad h_\xi = \frac{\xi - \sigma}{\rho} ; \qquad h_\sigma = 1$$

where $\rho = \rho(\xi)$ is the radius of curvature of the caustic. We are presently only considering values of σ for which $\sigma \leqslant \xi$, since we would otherwise have a double covering of the region above the caustic curve. That is, the full tangent lines cover the region twice, so we restrict ourselves only to half-lines.

The reduced wave equation takes the form

$$(9.562) \qquad \nabla^2 u + k^2 u = \frac{1}{\xi - \sigma} \frac{\partial}{\partial \sigma} \left[(\xi - \sigma) \frac{\partial u}{\partial \sigma} \right]$$

$$+ \frac{\rho}{\xi - \sigma} \frac{\partial}{\partial \xi} \left[\frac{\rho}{\xi - \sigma} \frac{\partial u}{\partial \xi} \right] + k^2 u = 0$$

in ξ, σ coordinates. Let

$$(9.563) \qquad\qquad u = w \exp[ik\sigma]$$

and insert (9.563) into (9.562). We are retaining the phase term of the geometrical optics solution (9.560) in the expression (9.563) since it is the amplitude term of (9.560) that blows up at the caustic. We obtain for w,

$$(9.564) \qquad ik\left[2\frac{\partial w}{\partial \sigma} - \frac{1}{\xi - \sigma} w \right] + \frac{1}{\xi - \sigma} \frac{\partial}{\partial \sigma} \left[(\xi - \sigma) \frac{\partial w}{\partial \sigma} \right]$$

$$+ \frac{\rho}{\xi - \sigma} \frac{\partial}{\partial \xi} \left[\frac{\rho}{\xi - \sigma} \frac{\partial w}{\partial \xi} \right] = 0.$$

To emphasize the neighborhood of the caustic we set

$$(9.565) \qquad\qquad \xi - \sigma = k^{-r}\eta, \qquad \xi = \lambda$$

where $r > 0$ is to be determined. Since k is assumed to be large, the *stretching transformation* (9.565) implies that small values of $\xi - \sigma$ are to be considered. Now $\partial/\partial\xi = \partial/\partial\lambda + k^r\partial/\partial\eta$ and $\partial/\partial\sigma = -k^r\partial/\partial\eta$, so that (9.564) becomes

$$(9.566) \qquad -ik^{1+r}\left[2\frac{\partial w}{\partial \eta} + \frac{1}{\eta} w \right] + \frac{k^{2r}}{\eta} \frac{\partial}{\partial \eta} \left[\eta \frac{\partial w}{\partial \eta} \right]$$

$$+ \frac{k^r\rho}{\eta} \left[\frac{\partial}{\partial \lambda} + k^r \frac{\partial}{\partial \eta} \right]\left[\frac{k^r\rho}{\eta}\left(\frac{\partial w}{\partial \lambda} + k^r\frac{\partial w}{\partial \eta} \right) \right] = 0.$$

In assessing the significance of the terms in (9.566), we note that the coefficient of k^{1+r} should be retained, since in the absence of the stretching transformation it is this term that yields the geometrical optics amplitude result in (9.560). In addition, it is the absence of second derivative terms in the transport equations that invalidates the geometrical optics result at the caustic. Therefore, we must retain at least one second derivative term in (9.566). In effect, we have a *singular perturbation problem* in (9.564) and it is the neglect of the higher derivative terms that leads to difficulties in the *geometrical optics expansion*. The highest power of k that occurs in the second derivative terms in (9.566) comes from k^{4r}. Thus we choose r so that k^{1+r} and k^{4r} are equally significant. This yields

$$(9.567) \qquad\qquad 1 + r = 4r$$

so that $r = \frac{1}{3}$.

We set $r = \frac{1}{3}$ in (9.566) and expand w in a series in powers of $k^{-1/3}$. With w_0 as the leading term in the series, we find that w_0 satisfies the *boundary layer equation*

$$(9.568) \qquad -i\left[2\frac{\partial w_0}{\partial\eta} + \frac{1}{\eta}w_0\right] + \frac{\rho^2}{\eta}\frac{\partial}{\partial\eta}\left[\frac{1}{\eta}\frac{\partial w_0}{\partial\eta}\right] = 0.$$

This ordinary differential equation can be simplified by setting

$$(9.569) \qquad\qquad w_0 = \exp\left[\frac{i\eta^3}{3\rho^2}\right]W(z)$$

where

$$(9.570) \qquad\qquad z = \frac{\eta^2}{(4\rho^4)^{1/3}},$$

and $W(z)$ satisfies the equation

$$(9.571) \qquad\qquad W''(z) + zW(z) = 0.$$

With z replaced by $-z$, (9.571) becomes the *Airy equation* which has two linearly independent solutions $Ai(z)$ and $Bi(z)$, some of whose properties we now present. The Airy function $Ai(z)$ decays exponentially as $z \to +\infty$, whereas $Bi(z)$ grows exponentially as z increases. Both functions are oscillatory for $z < 0$. The asymptotic behavior of the Airy function $Ai(z)$ for complex valued z expressed as $z = |z|e^{i\theta}$, with $|z| \geqslant 0$ and $-\pi \leqslant \theta \leqslant \pi$, is given as

$$(9.572) \quad Ai(z) \approx \frac{1}{2\sqrt{\pi}}(z)^{-1/4}\exp\left[-\frac{2}{3}z^{3/2}\right], \qquad |z| \to \infty, \quad |\theta| < \pi,$$

and as $z \to -\infty$ we have

$$(9.573) \quad Ai(z) \approx \frac{1}{\sqrt{\pi}} (-z)^{-1/4} \sin\left[\frac{2}{3}(-z)^{3/2} + \frac{\pi}{4}\right], \qquad z \to -\infty.$$

From the functional relations

$$(9.574) \qquad\qquad Ai(z) = -\omega Ai(\omega z) - \omega^2 Ai(\omega^2 z)$$

and

$$(9.575) \qquad\qquad Bi(z) = i\omega Ai(\omega z) - i\omega^2 Ai(\omega^2 z)$$

where $\omega = \exp[-2\pi i/3]$, we can determine the asymptotic behavior of the solutions $Bi(z)$, $Ai(\omega z)$, and $Ai(\omega^2 z)$ of the Airy equation in terms of the results (9.572)–(9.573) for $Ai(z)$.

For the equation (9.571) for $W(z)$, a general solution is given as

$$(9.576) \qquad\qquad W(z) = c_1(\xi) Ai(-z) + c_2(\xi) Bi(-z),$$

where the functions c_1 and c_2 must be determined by matching the boundary layer result $w_0 \exp[ik\sigma]$ with the geometrical optics result (9.560). It is easily seen from (9.572) that the function $Ai(-\omega z)$ has the asymptotic behavior

$$(9.577) \quad Ai(-\omega z) = Ai[ze^{\pi i/3}] \approx \frac{1}{2\sqrt{\pi}} (z)^{-1/4} \exp\left[-i\left(\frac{2}{3}z^{3/2} + \frac{\pi}{12}\right)\right]$$

as $z \to \infty$, and (9.574)–(9.575) show that

$$(9.578) \qquad\qquad Ai(-\omega z) = \frac{1}{2i\omega}\left[Bi(-z) - iAi(-z)\right].$$

Putting $c_1 = -ic_2$ in (9.576), we express $W(z)$ as

$$(9.579) \qquad\qquad W(z) = c(\xi) Ai(-\omega z)$$

where $c(\xi)$ is to be specified. As we now demonstrate, this choice for $W(z)$ is appropriate for matching the boundary layer and geometrical optics solution.

In the (ξ, σ) variables the boundary layer solution $u_0 = w_0 \exp[ik\sigma]$ takes the form

$$(9.580) \quad u_0(\xi, \sigma) = c(\xi) Ai\left[\left(\frac{k^2 e^{\pi i}}{4\rho^4}\right)^{1/3} (\xi - \sigma)^2\right] \exp\left[ik\left(\sigma + \frac{(\xi - \sigma)^3}{3\rho^2}\right)\right].$$

To match (9.580) with the geometrical optics result (9.560) we expand the Airy function asymptotically for $\xi - \sigma > 0$ and k large. In view of (9.577) we obtain

$$(9.581) \quad u_0(\xi, \sigma) \approx c(\xi) \left[\frac{\rho^{1/3}}{\sqrt{\pi}\, k^{1/6} 2^{5/6} (\xi - \sigma)^{1/2}} \right] \exp\left[ik\sigma - \frac{i\pi}{12} \right].$$

Comparing with (9.560) shows that $c(\xi)$ must be chosen as

$$(9.582) \quad c(\xi) = k^{1/6} \sqrt{\pi}\, 2^{5/6} \rho^{-1/3} V_0(\sigma_0)(\xi - \sigma_0)^{1/2} e^{i\pi/12},$$

and thereby the boundary layer term u_0 given in (9.580) is completely specified.

This result shows that the field is not infinite at the caustic as predicted by the geometrical optics result. Instead, we see from (9.580) and (9.582) that the field has a finite value at the caustic, that is, when $\xi = \sigma$. However, the amplitude of the geometrical optics term is $O(1)$ in k away from the caustic, whereas at and near the caustic, the boundary layer term has an amplitude that is $O(k^{1/6})$. Since k is assumed to be large, we do find that there is a growth in the amplitude of the field at the caustic, but it does not become infinite there.

The preceding results are not completely satisfactory for the following reasons. First, they predict that the field below the caustic (i.e., in the region not penetrated by the rays) grows without bound as we show. The field below the caustic is expected to be weaker than the geometrical optics field. Thus it should decay rather than grow below the caustic. Second, it is not apparent from our results what role is played by the outgoing geometrical optics field that results after the incident wave passes through the caustic and how to determine that field. We now proceed to modify these results so as to remove these difficulties.

To begin, we show that the boundary layer solution (9.580) grows without bound below the caustic. To do so, an approximate expression for the distance along the *normal lines* to the caustic valid in the boundary layer region are found. The wave fronts associated with the system of rays (9.559) are the curves $\sigma = $ constant. Since the wave fronts are orthogonal to the rays and the rays are tangent to the caustic, we see that the wave fronts are orthogonal to the caustic. Now distance along the wave front $\sigma = $ constant is given as $ds = |\xi - \sigma|/\rho\, d\xi$. The boundary layer region is determined by $\xi = \sigma + O(k^{-1/3})$ in view of (9.565). On integrating the previous expression for ds on the wave front $\sigma = $ constant in the boundary layer region we obtain

$$(9.583) \quad s \approx \frac{1}{2\rho}(\xi - \sigma)^2 = k^{-2/3} \frac{\eta^2}{2\rho} = \left(\frac{\rho}{2}\right)^{1/3} k^{-2/3} z$$

where (9.565) and (9.570) are used. Within the boundary layer, the variable s yields a measure of distance along the normal line to the caustic. Positive values of s correspond to points above the caustic (i.e., in the ray region). Negative values of s correspond to points below the caustic.

Using (9.583) we express the boundary layer solution (9.580) in terms of the variables ξ and s. To leading order we have

$$(9.584) \qquad u_0 \approx c(\xi) Ai\left[e^{i\pi/3}\left(\frac{2}{\rho}\right)^{1/3} k^{2/3}s\right]\exp[ik\xi]$$

where we have set $\sigma = \xi + O(k^{-1/3})$ in the boundary layer. If (9.584) is evaluated for negative values of s in the region below the caustic, we find that $k^{2/3}s$ becomes large and negative. Then the asymptotic result (9.572) for the Airy function in (9.584) shows that $|u_0|$ grows exponentially as $k^{2/3}s \to -\infty$.

This difficulty may be resolved by considering the behavior of not only the incident field, but also of the outgoing field near the caustic. The rays for both the incident and outgoing fields are given by (9.559). On a ray of the incident field we have $\sigma < \xi$, whereas on a ray of the outgoing field we have $\sigma > \xi$. At the caustic both rays meet and $\sigma = \xi$.

The leading term of the geometrical optics form of the incident field u is given in (9.560). We denote the outgoing field by $\hat{u}$. Since the ray structure is the same for both fields, the leading term in the geometrical optics representation of $\hat{u}$ must have the form

$$(9.585) \qquad \hat{u} \approx \hat{V}_0(\xi)(\sigma - \xi)^{-1/2}\exp[ik\sigma].$$

The difference between (9.585) and (9.560) lies in the fact that $\sigma > \xi$ in (9.585), whereas $\sigma < \xi$ in (9.560). Further, the term $\hat{V}_0(\xi)$ is as yet unspecified. Both (9.585) and (9.560) are singular on the caustic (i.e., when $\sigma = \xi$).

The total field at a point (x, y) above the caustic is a sum of the incident and outgoing fields and is given as $u + \hat{u}$. We must note that even though the incident and outgoing fields are expressed in terms of the variables ξ and σ, at each point (x, y) these fields are evaluated at two different sets of values of ξ and σ. The determination of these values is indicated in Figure 9.6.

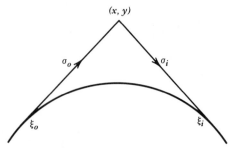

Figure 9.6.   The incoming and outgoing rays.

To determine $\hat{V}_0(\xi)$ in (9.585) and to resolve the difficulty concerning the behavior of the field below the caustic, we must consider the behavior of the total field $u + \hat{u}$ near the caustic. Now the boundary layer form of the incident field u was already given in (9.581)–(9.582). To study the boundary layer behavior of $\hat{u}$ we may carry over the preceding boundary layer results directly. The only difference is that the magnification element h_ξ has the form $h_\xi = (\sigma - \xi)/\rho$ if $\sigma > \xi$ and is not given as in (9.561).

However, (9.562) shows that the form of the reduced wave equation in the (ξ, σ) variables is unchanged if $\xi - \sigma$ is replaced by $\sigma - \xi$. Thus every result obtained up to (9.571) remains valid for the outgoing field. Therefore, if we define the leading order boundary layer term to be $\hat{w}_0$, we find that

$$(9.586) \qquad \hat{w}_0 = \exp\left[\frac{i\eta^3}{3\rho^2}\right] W(z),$$

where η, z, and W are given as in (9.565), (9.570), and (9.571), respectively.

Noting the asymptotic result (9.572), we find that

$$(9.587) \quad Ai[-\omega^2 z] = Ai[ze^{-\pi i/3}] \approx \frac{1}{2\sqrt{\pi}}(z)^{-1/4}\exp\left[i\left(\frac{2}{3}z^{3/2} + \frac{\pi}{12}\right)\right]$$

as $z \to \infty$. If we then set

$$(9.588) \qquad W(z) = \hat{c}(\xi)Ai[-\omega^2 z] = \hat{c}(\xi)Ai[ze^{-\pi i/3}]$$

in (9.586), we obtain for $\hat{u}_0 = \hat{w}_0\exp[ik\sigma]$,

$$(9.589)$$

$$\hat{u}_0(\xi, \sigma) = \hat{c}(\xi)Ai\left[\left(\frac{k^2 e^{-\pi i}}{4\rho^4}\right)^{1/3}(\xi - \sigma)^2\right]\exp\left\{ik\left[\sigma + \frac{(\xi - \sigma)^3}{3\rho^2}\right]\right\}.$$

To specify $\hat{c}(\xi)$ we expand the Airy function in (9.589) asymptotically for $\sigma - \xi > 0$ and k large. Since $\xi - \sigma$ occurs in the Airy function in squared form we must use the asymptotic form (9.587) valid for large positive z. Thereby, the matching between the boundary layer result $\hat{u}_0$ and the geometrical optics result $\hat{u}$ is carried out.

The variable z that occurs in (9.586)–(9.588) is defined in (9.570) and we have

$$(9.590) \qquad \frac{2}{3}z^{3/2} = \frac{k}{3\rho^2}|\xi - \sigma|^3 = -\frac{k}{3\rho^2}(\xi - \sigma)^3,$$

since $\xi - \sigma < 0$ so that $|\xi - \sigma| = -(\xi - \sigma)$. This result shows why it is appropriate to choose the Airy function $Ai[-\omega^2 z]$ for this case (i.e., the outgoing field), for the asymptotic form for $\hat{u}_0(\xi, \sigma)$ is then found to be

$$(9.591) \quad \hat{u}_0(\xi, \sigma) \approx \hat{c}(\xi) \left[\frac{\rho^{1/3}}{\sqrt{\pi} \, k^{1/6} 2^{5/6} (\sigma - \xi)^{1/2}} \right] \exp\left[ik\sigma + i\frac{\pi}{12} \right].$$

Comparing (9.591) with the geometrical optics result (9.585) yields

$$(9.592) \quad \hat{c}(\xi) = k^{1/6} \sqrt{\pi} \, 2^{5/6} \rho^{-1/3} \hat{V}_0(\xi) \exp\left[-\frac{i\pi}{12} \right].$$

Although (9.592) expresses $\hat{c}(\xi)$ in terms of $\hat{V}_0(\xi)$, we still do not know what form $\hat{V}_0$ has. We expect that $\hat{V}_0(\xi)$ must be expressed in terms of the (given) incident field. In order to establish this relationship we examine the total field $u_0 + \hat{u}_0$ (to leading order) in the boundary layer region. To do so it is not convenient to use the (ξ, σ) variables since they have different values for u_0 and $\hat{u}_0$ as we have indicated. Instead, we use the variables ξ and s, where s measures distance along the normal from the caustic in the boundary layer region. To the level of approximation used, it follows from (9.583) and the discussion preceding it, that the incident and outgoing fields are both evaluated at the same point (ξ, s). Thus (to leading order) the total boundary layer field $u_0 + \hat{u}_0$ is given as,

$$(9.593) \quad u_0 + \hat{u}_0 \approx k^{1/6} \sqrt{\pi} \, 2^{5/6} \rho^{-1/3} \exp[ik\xi]$$

$$\times \left\{ V_0(\sigma_0)(\xi - \sigma_0)^{1/2} e^{i\pi/12} Ai\left[e^{i\pi/3} \left(\frac{2}{\rho}\right)^{1/3} k^{2/3} s \right] \right.$$

$$\left. + \hat{V}_0(\xi) e^{-i\pi/12} Ai\left[e^{-i\pi/3} \left(\frac{2}{\rho}\right)^{1/3} k^{2/3} s \right] \right\}.$$

We require that $u_0 + \hat{u}_0$ decay in the region below that caustic, that is, when $s < 0$ in (9.593). In view of the asymptotic result (9.572), this implies that the sum of the Airy functions in the bracketed term of (9.593) add up to a multiple of $Ai[-(2/\rho)^{1/3} k^{2/3} s]$, for $Ai(-z)$ decays exponentially as $z \to -\infty$, whereas $Ai(-\omega z)$, $Ai(-\omega^2 z)$, and $Bi(-z)$ all grow exponentially as $z \to -\infty$. Using (9.574) with z replaced by $-z$ and noting that $Ai[-\omega z] = Ai[e^{i\pi/3} z]$ and $Ai[-\omega^2 z] = Ai[e^{-i\pi/3} z]$, we easily conclude that $\hat{V}_0(\xi)$ must be chosen such that

$$(9.594) \quad \hat{V}_0(\xi) e^{-i\pi/12} = \omega V_0(\sigma_0)(\xi - \sigma_0)^{1/2} e^{i\pi/12}.$$

Then (9.593) becomes

$$(9.595) \quad u_0 + \hat{u}_0 \approx k^{1/6} \sqrt{\pi} \, e^{-i\pi/4} 2^{5/6} \rho^{-1/3} V_0(\sigma_0)(\xi - \sigma_0)^{1/2}$$

$$\times Ai\left[-\left(\frac{2}{\rho}\right)^{1/3} k^{2/3} s \right] \exp[ik\xi]$$

and the field decays below the caustic when $k^{2/3} s$ gets large and negative.

Requiring that the field must decay below the caustic has enabled us to determine $\hat{V}_0(\xi)$ as given in (9.594). Inserting the result for $\hat{V}_0(\xi)$ into (9.585) yields the *outgoing geometrical optics field*

$$(9.596) \qquad \hat{u} = V_0(\sigma_0)\left[\frac{\xi - \sigma_0}{\sigma - \xi}\right]^{1/2}\exp\left[ik\sigma - \frac{i\pi}{2}\right]$$

with $\sigma > \xi$. To leading order, the incident and outgoing geometrical optics field have an amplitude term $|\sigma - \xi|^{-1/2}$ which is a consequence of the convergence and divergence of the rays as the wave approaches and leaves the caustic. However, on comparing the phase terms in (9.560) and (9.596), we find that the incident wave undergoes a *phase shift* by an amount of $\pi/2$ as it passes through the caustic and becomes the outgoing wave.

It has been noted in the preceding example that according to geometrical optics, the field in the region below the caustic must be zero since no rays penetrate into that region. However, the boundary layer result shows that there is a nonzero field below the caustic that decays exponentially. In the theory of optics when there are nonzero fields in regions where geometrical optics predicts that the field must be zero, we attribute these effects to the process of diffraction of light. In the following two examples we consider scattering problems in which diffraction effects play a significant role.

EXAMPLE 9.16.   SCATTERING BY A HALF - PLANE

We assume that the *plane wave* $u_I(x, y, z) = e^{ikz}$ is incident upon the *half-plane* $z = 0$, $y \leqslant 0$. The resulting field $u = u_I + u_S$ where u_S is the *scattered wave* satisfies the *reduced wave equation* (9.418) with a *unit index of refraction*. The boundary condition is that the *total field* $u(x, y, z)$ vanishes at $z = 0$, $y \leqslant 0$ (i.e., on the half-plane). Thus

$$(9.597) \qquad\qquad \nabla^2 u + k^2 u = 0$$

and

$$(9.598) \qquad\qquad u(x, y, 0) = 0, \qquad y \leqslant 0.$$

Following the general procedure as given, we expand u_S as

$$(9.599) \qquad\qquad u_S = e^{ik\varphi}\sum_{j=0}^{\infty} v_j(ik)^{-j}.$$

On the boundary surface $z = 0$, $y \leqslant 0$, the initial condition for $\varphi(x, y, z)$ is

$$(9.600) \qquad\qquad \varphi(x, y, 0) = 0, \qquad y \leqslant 0$$

and the initial conditions for the v_j are

(9.601) $\qquad\qquad v_0(x, y, 0) = -1, \qquad y \leqslant 0,$

(9.602) $\qquad\qquad v_j(x, y, 0) = 0, \qquad y \leqslant 0, \quad j \geqslant 1.$

These conditions follow from the fact that the incident phase term $\psi = z$ vanishes at $z = 0$, whereas the leading order amplitude term $w_0 = 1$ is the only nonzero term in the (asymptotic) expansion of the incident field u_I. We remark that this problem could be treated as being strictly two-dimensional in terms of the y, z variables, but we prefer to deal with it in three-dimensional form.

To determine the phase term $\varphi(x, y, z)$ of u_S, we require that φ satisfy the outgoing condition at the boundary $z = 0$, in addition to (9.600). Clearly, the two solutions of the eiconal equation $(\nabla \varphi)^2 = 1$ that vanish at $z = 0$ are $\varphi_+ = z$ and $\varphi_- = -z$. By requiring that u_S be an outgoing wave at the boundary, we determine the appropriate choice for φ. To apply this condition we must distinguish between the two sides of the half-plane $z = 0$, $y \leqslant 0$. Accordingly, we express the scattered wave u_S in the form $u_S = v_{\pm} e^{ik\varphi_\pm}$. On the side of the half-plane facing the half-space $z < 0$ the appropriate choice for φ is $\varphi = \varphi_- = -z$, since the wave $u_S = v_- e^{-ikz}$ travels away from the half-plane into the half-space $z < 0$. On the opposite side of the half-plane facing the region $z > 0$, we must set $\varphi = \varphi_+ = z$, since $u_S = v_+ e^{ikz}$ travels towards the half-space $z > 0$ away from the half-plane $z = 0$.

Since the scattered field in the region $z < 0$, $y < 0$ is a plane wave, we immediately conclude from the initial conditions (9.601)–(9.602) for the amplitude terms that $v_0(x, y, z) = v_0(x, y, 0) = -1$ and $v_j(x, y, z) = v_j(x, y, 0) = 0$ for all $j \geqslant 1$. Consequently, the scattered wave in this region is given as

(9.603) $\qquad\qquad u_S(x, y, z) = -e^{-ikz}, \qquad z < 0, \ y \leqslant 0.$

In the region $z > 0$, $y \leqslant 0$, u_S is also a plane wave and (9.601)–(9.602) imply that $v_0(x, y, z) = -1$ and $v_j(x, y, z) = 0$ for $j \geqslant 1$. Thus the scattered wave in this region is given as

(9.604) $\qquad\qquad u_S(x, y, z) = -e^{ikz}, \qquad z > 0, \ y \leqslant 0.$

There is no boundary surface in the half-space $y > 0$, and the scattered waves constructed previously do not penetrate into this region since the rays are orthogonal to the half-plane $z = 0$, $y \leqslant 0$. Thus the total field $u = u_I + u_S$ predicted by the *geometrical optics* approach is

(9.605) $\quad u(x, y, z) = \begin{cases} e^{ikz}; & -\infty < z < \infty; \quad y > 0 \\ e^{ikz} - e^{-ikz}; & z < 0; \qquad\qquad y < 0 \cdot \\ 0; & z > 0; \qquad\qquad y < 0 \end{cases}$

As shown in Figure 9.7, the plane $y = 0$ is a surface of discontinuity for the geometrical optics solution (9.605). Even though each of the representations of u given in (9.605) is an exact solution of the reduced wave equation (9.597) in the interior of the three indicated regions, the full expression for u is not a regular solution of (9.597). The half-plane $y = 0$, $z > 0$ is called a *shadow boundary* since it separates the (shadow) region $z > 0$, $y < 0$ where no (incident) rays penetrate, from the region $z > 0$, $y > 0$ where the incident field acts. In a similar fashion we refer to the half-plane $y = 0$, $z < 0$ as a *reflection boundary*. It separates the region where the incident and reflected field are observed from the region where there is only the incident field.

In order to obtain a *smooth transition* for the asymptotic solution of the given scattering problem across the discontinuity boundaries, we make use of the *boundary layer theory*. We consider the boundaries at $y = 0$ with $z < 0$ and $z > 0$ separately.

At the *shadow boundary* $y = 0$, $z > 0$, the field on one side is $u = 0$ whereas the field on the other side is $u = e^{ikz}$. Thus we set

$$u = we^{ikz}$$

in (9.597) and obtain

(9.606) $$2ikw_z + w_{xx} + w_{yy} + w_{zz} = 0.$$

To emphasize the neighborhood of $y = 0$, we introduce the stretching transfor-

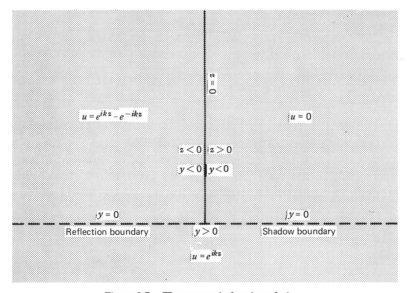

Figure 9.7. The geometrical optics solution.

mation

$$(9.607) \qquad\qquad y = k^{-r}\eta,$$

with the exponent $r > 0$ to be determined. In the x, η, z variables, (9.606) becomes

$$(9.608) \qquad\qquad 2ikw_z + k^{2r}w_{\eta\eta} + w_{xx} + w_{zz} = 0.$$

To ensure that the original leading term $2ikw_z$ remains as significant as a second derivative term in (9.608), we must set $r = \frac{1}{2}$. Then

$$(9.609) \qquad\qquad k(2iw_z + w_{\eta\eta}) + w_{xx} + w_{zz} = 0$$

and the equation for the leading order boundary layer term $\hat{w}_0$ is

$$(9.610) \qquad\qquad 2i\frac{\partial \hat{w}_0}{\partial z} + \frac{\partial^2 \hat{w}_0}{\partial \eta^2} = 0.$$

This parabolic *Schrödinger equation* has already been encountered in Example 9.14—see (9.552). We require that for large values of $\eta = k^{1/2}y$, the solution $\hat{w}_0$ of (9.610) match with the geometrical optics result (9.605).

We rewrite (9.610) in the original y and z variables and obtain

$$(9.611) \qquad\qquad 2ik\frac{\partial w_0}{\partial z} + \frac{\partial^2 w_0}{\partial y^2} = 0.$$

Noting the results (9.553)–(9.554) of Example 9.14, we consider a general solution of (9.611) in the form

$$(9.612) \qquad\qquad w_0 = \sqrt{\frac{k}{z}} \int_{-\infty}^{\infty} f(s)\exp\left[\frac{ik(y-s)^2}{2z}\right] ds$$

where $f(s)$ is to be specified by means of the matching process. Since k is large, we evaluate (9.612) asymptotically using the method of stationary phase as in Example 9.14. There is exactly one stationary point at $s = y$ and in view of (9.555) we have the asymptotic result

$$(9.613) \qquad\qquad w_0 \approx \sqrt{2\pi}\, e^{i\pi/4}f(y),$$

so that

$$(9.614) \qquad\qquad w_0 e^{ikz} \approx \sqrt{2\pi}\, f(y)\exp\left(ikz + i\frac{\pi}{4}\right).$$

The expression (9.614) must be matched with the geometrical optics field
(9.605) in the regions $z > 0$, $y > 0$ and $z > 0$, $y < 0$. Since the geometrical
optics field vanishes in the (shadow) region $z > 0$, $y < 0$ we must set $f(y) = 0$
for $y < 0$. For $z > 0$, $y > 0$, the geometrical optics field equals u_I so that
$f(y) = (1/\sqrt{2\pi})e^{-i\pi/4}$ for $y > 0$. Thus

$$(9.615) \qquad f(y) = \begin{cases} \dfrac{1}{\sqrt{2\pi}}e^{-i\pi/4}; & y > 0 \\[2mm] 0; & y < 0 \end{cases},$$

and the boundary layer result w_0 becomes

$$(9.616) \qquad w_0 = \left(\frac{k}{2\pi z}\right)^{1/2} e^{-i\pi/4} \int_0^\infty \exp\left[\frac{ik(y-s)^2}{2z}\right] ds.$$

With the change of variables

$$(9.617) \qquad \sigma = (s-y)\left(\frac{k}{2z}\right)^{1/2}$$

(9.616) takes the form

$$(9.618) \qquad w_0 = \pi^{-1/2} e^{-i\pi/4} F\left(-y\sqrt{\frac{k}{2z}}\right)$$

where the *Fresnel integral* $F(t)$ is defined as

$$(9.619) \qquad F(t) = \int_t^\infty e^{i\sigma^2} d\sigma.$$

The field near the shadow boundary $y = 0$, $z > 0$ is thus given as

$$(9.620) \qquad u \approx \pi^{-1/2} F\left(-y\sqrt{\frac{k}{2z}}\right) \exp\left[ikz - \frac{i\pi}{4}\right],$$

to leading order.

Since $F(-\infty) = \sqrt{\pi}\,e^{i\pi/4}$ and $F(+\infty) = 0$, we find that $w_0 \to 0$ as $z \to 0$
with $y < 0$, whereas $w_0 \to 1$ as $z \to 0$ with $y > 0$. Thus $w_0(z, y)$ as given in
(9.618) is, in fact, a solution of the parabolic equation (9.611) with the
discontinuous initial data $w(0, y) = 0$ for $y < 0$ and $w(0, y) = 1$ for $y > 0$.
Had we applied the parabolic equation method of Example 9.12 to this
problem, we would have been led to consider (9.611) with the aforementioned
data. Further, $F(0) = \frac{1}{2}\sqrt{\pi}\,e^{i\pi/4}$, so that the amplitude term w_0 equals $\frac{1}{2}$ on the

shadow boundary $y = 0$, $z > 0$. It equals the average of the amplitude terms of the field in the shadow and illuminated regions, respectively.

The matching procedure carried out only considered the leading term in the asymptotic expansion of the boundary layer term w_0. On evaluating the representation (9.616) of w_0 by the method of stationary phase, we find that there is only one stationary point located at $s = y$. Since the domain of integration is $0 < s < \infty$, there is no stationary point in the region $y < 0$, $z > 0$. In the region $y \geqslant 0$, $z > 0$, the stationary point coincides with the end point $s = 0$ of the integration interval if the value of y equals zero. Otherwise, the stationary point $s = y$ and the end point $s = 0$ are distinct.

In our discussion of the *stationary phase method* in Example 5.13 we assumed that the integral had infinite limits of integration. Therefore, questions about possible contributions from finite endpoints in the integration interval were avoided. In the present problem we must consider the contribution of the point $s = 0$ to the asymptotic value of the integral (9.616). It is assumed in our discussion that y is not close to zero, so that the stationary and end points of the integral (9.616) are well separated. To determine these contributions we consider a more general problem first.

We consider the integral

$$(9.621) \qquad I(k) = \int_a^b \exp[ik\varphi(t)] f(t) \, dt$$

where k is large, and $\varphi(t)$ and $f(t)$ are real valued functions. In Example 5.13 we assumed that $a = -\infty$ and $b = +\infty$. Here we assume that a or b, or possibly both, are finite. On integrating by parts in (9.621) we have

$$(9.622) \qquad I(k) = \frac{f(t)}{ik\varphi'(t)} \exp[ik\varphi(t)] \Bigg|_{t=a}^{t=b}$$
$$- \frac{1}{ik} \int_a^b \frac{d}{dt} \left[\frac{f(t)}{\varphi'(t)} \right] \exp[ik\varphi(t)] \, dt.$$

If there are no stationary points in the interval and $f(t)/\varphi'(t)$ is smooth and does not vanish at both endpoints, the main contribution to the integral $I(k)$ can be shown to come from the end points. Thus

$$(9.623) \qquad I(k) \approx \frac{f(t)}{ik\varphi'(t)} \exp[ik\varphi(t)] \Bigg|_{t=a}^{t=b},$$

and if $a = -\infty$ or $b = +\infty$, we disregard the contribution from the infinite endpoint. If there are one or more stationary points where $\varphi'(t) = 0$ in the interval of integration, we break up the interval into a collection of subintervals. Each subinterval either contains exactly one stationary point in its interior or has no stationary point but contains an endpoint of the original interval. If $\varphi''(t) \neq 0$ at each stationary point, the contribution from the corresponding subinterval is given as in (5.238). These contributions are of order $1/\sqrt{k}$. The contribution from the intervals with the endpoints of the original integral are given as in (9.623). These are seen to be of order $1/k$. Contributions from endpoints common to two subintervals clearly must vanish.

We now apply these results to the integral (9.616) for w_0. If $y < 0$, there is no stationary point in the given interval of integration. Thus the main asymptotic contribution to (9.616) comes from the endpoint $s = 0$. In view of (9.623) we have

$$(9.624) \qquad w_0 \approx \frac{1}{ik}\left(\frac{z}{y}\right)\left(\frac{k}{2\pi z}\right)^{1/2} e^{-i\pi/4} \exp\left[\frac{iky^2}{2z}\right]$$

$$= -\left(\frac{z}{y}\right)\left(\frac{1}{2\pi kz}\right)^{1/2} \exp\left[\frac{iky^2}{2z} + i\frac{\pi}{4}\right], \qquad y < 0.$$

If $y > 0$, the integral (9.616) has a stationary point contribution from the point $s = y$, as well as an endpoint contribution from $s = 0$. The stationary point result was found to be $w_0 \approx 1$, whereas the endpoint result is identical to (9.624). Thus

$$(9.625) \qquad w_0 \approx 1 - \left(\frac{z}{y}\right)\left(\frac{1}{2\pi kz}\right)^{1/2} \exp\left[\frac{iky^2}{2z} + i\frac{\pi}{4}\right], \qquad y > 0.$$

We note that both (9.624) and (9.625) are singular at $y = 0$ as well as at $z = 0$.

Since w_0 is valid only in the region near the shadow boundary $z > 0$, $y = 0$, we conclude from (9.624)–(9.625) that the total field $u(x, y, z)$ has the form

$$(9.626)$$

$$u \approx \begin{cases} -\left(\frac{z}{y}\right)\left(\frac{1}{2\pi kz}\right)^{1/2} \exp\left[ik\left(z + \frac{y^2}{2z}\right) + \frac{i\pi}{4}\right]; & z > 0, \quad y \leq 0 \\[2mm] \exp[ikz] - \left(\frac{z}{y}\right)\left(\frac{1}{2\pi kz}\right)^{1/2} \exp\left[ik\left(z + \frac{y^2}{2z}\right) + \frac{i\pi}{4}\right]; & z > 0, \quad y \geq 0 \end{cases}$$

near the edge of the shadow boundary. On comparing (9.626) with the geometrical optics result (9.605), we observe that there is no counterpart to the $O(k^{-1/2})$ term of (9.626) in (9.605). In fact, the geometrical optics result was expressed as an (asymptotic) series in integral powers of k^{-1}. Thus terms of order $k^{-1/2}$ could not arise in that series. These additional terms are due to diffraction effects. In the preceding example, the field due to diffraction below the caustic was found to decay exponentially. In this problem, the diffraction field is smaller than the geometrical optics field by a multiplicative factor of $O(k^{-1/2})$. Since it is not exponentially small, we must account for it in our discussion of the scattering problem.

An asymptotic method closely related to the geometrical optics approach and accounting for the effects of diffraction is J. B. Keller's *geometrical theory of diffraction*. In the context of the present problem this theory introduces a set of *diffracted rays* that emanate from the edge of the boundary surface (i.e., the half-plane $z = 0$, $y \leqslant 0$). The incident rays, in addition to giving rise to the reflected rays in the region $z < 0$, $y < 0$, generate a set of diffracted rays. Each incident ray that hits the edge yields a family of *edge diffracted rays* that emanate from the edge as shown in Figure 9.8. These rays correspond to those of a cylindrical wave with a focal point at the edge. If we set

$$(9.627) \qquad\qquad r = \sqrt{y^2 + z^2},$$

the resulting diffracted field u_D—which supplements the geometrical optics field—is shown in the geometrical theory of diffraction to have the form

$$(9.628) \qquad u_D(x, y, z) \approx \frac{\tilde{D}}{\sqrt{kr}} u_I(x, 0, 0)\exp[ikr].$$

Here $\tilde{D}$ is known as a *diffraction coefficient*, and $u_I(x, 0, 0)$ is the value of the incident field at the point of diffraction $(x, 0, 0)$ on the edge of the half-plane.

The diffraction coefficient $\tilde{D}$ must be determined either by solving a *canonical problem* or by using boundary layer theory near the edge. In general, a canonical problem is one in which the local geometry and other properties are the same as those for the given problem near the point of diffraction. Using either of these methods leads one to consider the problem of the scattering of a plane wave by a half-plane with the same boundary condition as for our original given problem. That is, the canonical problem is identical to the given

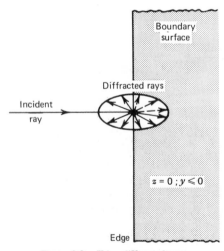

Figure 9.8.   Edge diffracted rays.

problem in the present case. (In general, of course, the canonical problem is much simpler than the given one.) The exact solution of our problem was found by Sommerfeld. On expanding his result asymptotically for large kr it is found that the diffraction coefficient $\tilde{D}$ must be chosen as

$$(9.629) \qquad \tilde{D} = -\frac{e^{i\pi/4}}{2(2\pi)^{1/2}}\left[\sec\left(\frac{1}{2}\tilde{\theta}\right) + \csc\left(\frac{1}{2}\tilde{\theta}\right)\right],$$

where $\tilde{\theta}$ is the angle that the particular diffracted ray—when projected onto the y, z-plane—makes with the negative z-axis. Also, $u_I(x, 0, 0) = 1$ for all values of x.

The terms of order $k^{-1/2}$ in (9.626) must be identified with the diffracted field u_D in the region near the shadow boundary. To effect this comparison more easily we introduce the cylindrical coordinate system

$$(9.630) \qquad x = x; \qquad y = r\cos\theta; \qquad z = r\sin\theta.$$

The angles $\tilde{\theta}$ of (9.629) and θ of (9.630) are related by

$$(9.631) \qquad \tilde{\theta} = \theta + \frac{\pi}{2}.$$

For small y with $z > 0$, we have

$$(9.632) \qquad r = \sqrt{y^2 + z^2} \approx z + \frac{y^2}{2z}.$$

Using (9.630) in (9.632) we easily conclude that we must have

$$(9.633) \qquad \sin\theta \approx 1; \qquad \cos\theta \approx 0$$

if (9.632) is valid. Also,

$$(9.634) \qquad \frac{1}{2}\left[\sec\frac{1}{2}\tilde{\theta} + \csc\frac{1}{2}\tilde{\theta}\right] = \frac{\cos(\tilde{\theta}/2) + \sin(\tilde{\theta}/2)}{2\sin(\tilde{\theta}/2)\cos(\tilde{\theta}/2)}$$

$$= \frac{\cos(\tilde{\theta}/2) + \sin(\tilde{\theta}/2)}{\sin\tilde{\theta}}$$

Noting (9.630)–(9.633) we have

$$(9.635) \qquad -\left(\frac{z}{y}\right)\left(\frac{1}{2\pi kz}\right)^{1/2}\exp\left[ik\left(z + \frac{y^2}{2z}\right) + i\frac{\pi}{4}\right]$$

$$\approx -\left(\frac{\sin\theta}{\cos\theta}\right)\left(\frac{1}{2\pi kr}\right)^{1/2}\exp\left[ikr + i\frac{\pi}{4}\right]$$

$$\approx -\left(\frac{1}{\sin\tilde{\theta}}\right)\left(\frac{1}{2\pi kr}\right)^{1/2}\exp\left[ikr + i\frac{\pi}{4}\right],$$

since $\cos\theta = \sin\tilde{\theta}$ and $\sin\theta \approx 1$. Further, (9.633) implies that $\theta \approx \pi/2$ so that $\tilde{\theta} \approx \pi$. Consequently, $\cos(\tilde{\theta}/2) \approx 0$ and $\sin(\tilde{\theta}/2) \approx 1$ and, in view of (9.634), $\tilde{D}$ has the approximate form

$$(9.636) \qquad \tilde{D} \approx -\frac{e^{i\pi/4}}{(2\pi)^{1/2}}\frac{1}{\sin\tilde{\theta}}.$$

Inserting (9.636) into (9.628) we see that (9.635) and the diffracted field u_D agree in the neighborhood of the shadow boundary since $u_I(x, 0, 0) = 1$.

The field near the *reflection boundary* is considered in the exercises.

EXAMPLE 9.17. SCATTERING BY A CIRCULAR CYLINDER

A *plane wave* $u_I = e^{ikx}$ is incident upon a *circular cylinder* of radius a, whose axis coincides with the z-axis. The total field $u = u_I + u_S$ where u_S is the *scattered wave* satisfies the reduced wave equation (9.418)—with $n = 1$—and vanishes on the surface of the cylinder. This problem can clearly be formulated as a two-dimensional one.

Thus we consider the function $u = u(x, y)$ that satisfies the reduced wave equation

$$(9.637) \qquad u_{xx} + u_{yy} + k^2 u = 0$$

in the exterior of the circle $x^2 + y^2 = a^2$. On the circle we have

$$(9.638) \qquad u(x, y) = 0; \qquad x^2 + y^2 = a^2.$$

With $u = u_I + u_S$ where $u_I = e^{ikx}$, we require that the scattered field u_S satisfy the radiation condition at infinity.

Assuming k is large, we determine an asymptotic solution for u_S in the manner described. The scattered field u_S is expanded as

$$(9.639) \qquad u_S(x, y) = e^{ik\varphi} \sum_{j=0}^{\infty} v_j (ik)^{-j}.$$

On the circle the phase term $\varphi(x, y)$ satisfies the (initial) condition

$$(9.640) \qquad \varphi(x, y) = x; \qquad x^2 + y^2 = a^2$$

as well as an outgoing condition. The amplitude terms $v_j(x, y)$ satisfy the initial conditions

$$(9.641) \qquad v_0(x, y) = -1; \qquad x^2 + y^2 = a^2;$$

$$(9.642) \qquad v_j(x, y) = 0; \qquad x^2 + y^2 = a^2; \qquad j \geqslant 1.$$

To solve for φ we represent the circle in parametric form as

(9.643) $$x = a \cos \tau; \quad y = a \sin \tau; \quad 0 \leqslant \tau \leqslant 2\pi.$$

We note that τ is not an arc length parameter unless $a = 1$. With σ equal to arc length along the rays we must solve the characteristic equations for $x(\sigma, \tau)$, $y(\sigma, \tau)$, $\varphi(\sigma, \tau)$, $p(\sigma, \tau)$, and $q(\sigma, \tau)$ in order to determine $\varphi(x, y)$. The initial conditions for x and y, which are given at $\sigma = 0$, are

(9.644) $$x(0, \tau) = a \cos \tau; \quad y(0, \tau) = a \sin \tau.$$

From (9.640) we have

(9.645) $$\varphi(0, \tau) = x(0, \tau) = a \cos \tau.$$

The initial conditions for p and q must be found from the eiconal equation and strip condition as in Section 2.4. We have from the eiconal equation

(9.646) $$p^2(0, \tau) + q^2(0, \tau) = 1,$$

and the strip condition

(9.647) $$- a \sin \tau = -p(0, \tau) a \sin \tau + q(0, \tau) a \cos \tau.$$

On solving (9.646)–(9.647) we easily conclude that there are two possible sets of solutions. They are

(9.648) $$p(0, \tau) = 1; \quad q(0, \tau) = 0$$

and

(9.649) $$p(0, \tau) = -\cos 2\tau; \quad q(0, \tau) = -\sin 2\tau.$$

The appropriate solutions must be determined from the outgoing condition. This states that if $\mathbf{N}$ is the exterior unit normal to the circle, then $\nabla \varphi \cdot \mathbf{N} = [p, q] \cdot \mathbf{N} > 0$. The normal $\mathbf{N}$ for $x^2 + y^2 = a^2$ is $\mathbf{N} = [\cos \tau, \sin \tau]$. Thus

(9.650) $$\nabla \varphi \cdot \mathbf{N} = [p, q] \cdot \mathbf{N} = \cos \tau$$

if (9.648) is used, and

(9.651) $$\nabla \varphi \cdot \mathbf{N} = [p, q] \cdot \mathbf{N} = -\cos \tau$$

if (9.649) is used. Therefore, since $\cos \tau > 0$ for $0 \leqslant \tau < \pi/2$ and $3\pi/2 < \tau \leqslant 2\pi$ and $\cos \tau < 0$ for $\pi/2 < \tau < 3\pi/2$, we see that $\nabla \varphi \cdot \mathbf{N} > 0$ on the left semicircle (i.e., $\pi/2 < \tau < 3\pi/2$) if (9.649) is chosen, whereas $\nabla \varphi \cdot \mathbf{N} > 0$ on the right semicircle if (9.648) is selected. At the points $(0, a)$ and $(0, -a)$ where

$\tau = \pi/2$ and $\tau = 3\pi/2$, respectively, the rays of the scattered wave are tangent to the circle.

It has been shown that on the right semicircle (i.e., $x^2 + y^2 = a^2$ with $x > 0$) we must set $p(0, \tau) = 1$ and $q(0, \tau) = 0$. Since the characteristic equations are

$$(9.652) \quad \frac{dx}{d\sigma} = p; \qquad \frac{dy}{d\sigma} = q; \qquad \frac{d\varphi}{d\sigma} = 1; \qquad \frac{dp}{d\sigma} = 0; \qquad \frac{dq}{d\sigma} = 0,$$

we see that $p(\sigma, \tau) = 1$ and $q(\sigma, \tau) = 0$. Then (9.644)–(9.645) imply that

$$(9.653) \qquad x(\sigma, \tau) = \sigma + a\cos\tau; \qquad y(\sigma, \tau) = a\sin\tau$$

and

$$(9.654) \qquad \varphi(\sigma, \tau) = \sigma + a\cos\tau = x.$$

Thus the rays of the scattered wave that issue from the right semicircle are identical to those of the incident wave. The phase term φ equals x, the phase of the incident wave. Further, since $\nabla^2\varphi = 0$, we conclude that

$$(9.655) \qquad v_0(x, y) = -1; \qquad v_j(x, y) = 0; \qquad j \geqslant 1$$

from the transport equations (9.472)–(9.473) and the initial conditions (9.641)–(9.642). Therefore, the scattered field u_S is given as

$$(9.656) \quad u_S(x, y) = -e^{ikx}; \qquad x^2 + y^2 \geqslant a^2; \qquad x > 0, |y| < a,$$

in the region directly behind the circle where no incident rays penetrate. The total field $u = u_I + u_S$ in that region is

$$(9.657) \quad u(x, y) = e^{ikx} - e^{ikx} = 0; \qquad x^2 + y^2 \geqslant a^2, x > 0, |y| < a.$$

The *geometrical optics* approach thus predicts a zero total field or, equivalently, a shadow in the region behind the circle not *illuminated* by the incident rays.

On the left semicircle (i.e., $x^2 + y^2 = a^2$ with $x < 0$) we must have $p(0, \tau) = -\cos 2\tau$ and $q(0, \tau) = -\sin 2\tau$, as has been shown. Since p and q are constant along the characteristics they retain their initial values along these curves. Consequently, the equations for x and y are

$$(9.658) \quad \frac{dx}{d\sigma} = p = -\cos 2\tau; \qquad \frac{dy}{d\sigma} = q = -\sin 2\tau,$$

with $\pi/2 \leqslant \tau \leqslant 3\pi/2$. Noting the initial conditions (9.644), we find that

$$(9.659) \quad x(\sigma, \tau) = a\cos\tau - \sigma\cos 2\tau; \qquad y(\sigma, \tau) = a\sin\tau - \sigma\sin 2\tau.$$

Also, from $d\varphi/d\sigma = 1$ and (9.645) we conclude that

$$(9.660) \qquad \varphi(\sigma, \tau) = \sigma + a\cos\tau; \qquad \frac{\pi}{2} \leqslant \tau \leqslant \frac{3\pi}{2}.$$

It is not possible to express σ and τ as functions of x and y in a simple form. Therefore, we leave our result for the phase in parametric form.

The *rays* of the scattered or reflected field u_S are determined from (9.659), and their direction is given by (9.658). The direction of the incident rays is that of the unit vector $\mathbf{i}$. The exterior unit normal vector to the circle is $\mathbf{N} =$ [$\cos\tau, \sin\tau$]. It is a simple matter to show that the *angle of incidence* α_I of the incident ray equals the *angle of reflection* α_R of the reflected ray. These angles are defined in Figure 9.9. In fact, we have $\alpha_I = \alpha_R = |\pi - \tau|$ with $\pi/2 \leqslant \tau \leqslant 3\pi/2$. When $\tau = \pi/2$ and $\tau = 3\pi/2$, the reflected rays coincide with the incident rays. That is, the incident rays are tangent to the circle at the points $(0, a)$ and $(0, -a)$ and their extension beyond these points into the region $x > 0$ coincides with the reflected rays. We note that the equality of the angles of incidence and reflection is a general principle of geometrical optics. The region covered by the reflected and the incident rays is given as $x^2 + y^2 \geqslant a^2$, with $-\infty < y < \infty$ for $x \leqslant 0$ and $|y| \geqslant a$ for $x > 0$. This region is referred to as the illuminated region.

We now proceed to determine the leading order amplitude term v_0 of the reflected wave u_S. To do so we express the transport equation for v_0, which is given as

$$(9.661) \qquad 2\frac{\partial v_0}{\partial \sigma} + v_0\nabla^2\varphi = 0,$$

in the ray coordinates σ and τ determined from (9.659). The Laplacian ∇^2 must be expressed in the ray coordinate system so that $\nabla^2\varphi$ can be evaluated.

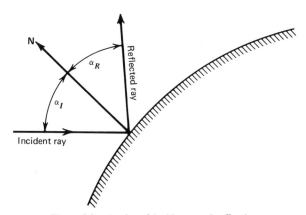

Figure 9.9. **Angles of incidence and reflection.**

It is readily found that

$$(9.662) \qquad \nabla^2 \varphi = \frac{1}{\sigma - (a/2)\cos \tau}; \qquad \frac{\pi}{2} < \tau < \frac{3\pi}{2}.$$

Inserting (9.662) into (9.661) and using the initial condition $v_0(0, \tau) = -1$, we obtain

$$(9.663) \qquad v_0(\sigma, \tau) = -\sqrt{\frac{(a/2)\cos \tau}{(a/2)\cos \tau - \sigma}}; \qquad \frac{\pi}{2} < \tau < \frac{3\pi}{2},$$

as is easily verified. Thus the scattered field is given as

$$(9.664)$$
$$u_S(\sigma, \tau) = -\sqrt{\frac{(a/2)\cos \tau}{(a/2)\cos \tau - \sigma}} \exp[ik(\sigma + a\cos \tau)]; \qquad \frac{\pi}{2} < \tau < \frac{3\pi}{2},$$

to leading order.

The curve

$$(9.665) \qquad \sigma = \frac{a}{2}\cos \tau; \qquad \frac{\pi}{2} \leqslant \tau \leqslant \frac{3\pi}{2}$$

in the envelope of the reflected ray system (9.659), as is easily verified by evaluating the Jacobian of the transformation (9.659) or, equivalently, by using the results in the Appendix of Chapter 2. When $\pi/2 < \tau < 3\pi/2$, $\cos \tau < 0$ so that σ is negative in (9.665). Therefore, the envelope lies within the circle $x^2 + y^2 = a^2$ for those values of τ, since $\sigma \geqslant 0$ corresponds to points on or outside the circle. However, if $\tau = \pi/2$ or $\tau = 3\pi/2$, we have $\cos \pi/2 = \cos 3\pi/2 = 0$, so that $\sigma = 0$ there. Consequently, the envelope or caustic curve (9.665) intersects the circle at the points $(0, a)$ and $(0, -a)$. It is for this reason that u_S is not well defined at $\tau = \pi/2$ and $\tau = 3\pi/2$.

The asymptotic (geometrical optics) result we have obtained is the following. In the *illuminated region* the solution is expressed in the ray coordinates σ and τ as

$$(9.666) \qquad u(\sigma, \tau) \approx \exp[ik(a\cos \tau - \sigma \cos 2\tau)]$$
$$- \sqrt{\frac{(a/2)\cos \tau}{(a/2)\cos \tau - \sigma}} \exp[ik(\sigma + a\cos \tau)],$$

and this is valid for $\pi/2 < \tau < 3\pi/2$ and $\sigma \geqslant 0$. In the *shadow region* we have

$$(9.667) \qquad u(x, y) \approx 0; \qquad x^2 + y^2 \geqslant 0; \qquad x > 0, |y| < a.$$

The lines $y = \pm a$, $x \geqslant 0$ are the *shadow boundaries* that separate the illuminated region from the shadow region. The asymptotic solution is not continuous across these lines.

The field near the shadow boundaries may be studied by the use of the boundary layer theory. The parabolic equation (9.610) obtained in the preceding problem may be derived. However, the solution must be matched not only with the incident and reflected waves in the illuminated region and the zero field in the shadow region. The field in the shadow boundary must also be matched with the field near the *points of diffraction* $(0, a)$ and $(0, -a)$. The field near these points may be analyzed by deriving an equation of the form (9.381) by the use of boundary layer methods. The problem is rather complicated and has undergone much investigation. We note that the geometrical theory of diffraction introduces *surface diffracted rays* which yield an additional nonzero but exponentially small field in the shadow and illuminated regions. However, we do not pursue these matters here any further.

Our discussion has so far been limited to the linear reduced wave equation (9.418) and its asymptotic solutions. However, in the theory of *nonlinear optics* one is led to consider a nonlinear version of (9.418). If a strong *laser beam* propagates in a medium, it may happen as a result of various effects that the index of refraction is altered as the beam traverses the medium. Assuming the medium to be otherwise homogeneous, the *index of refraction n* is found to become a function of the *intensity* of the field. If u represents the (complex) field, we have $n = n(|u|^2)$ and the field satisfies the *nonlinear reduced wave equation*

$$(9.668) \qquad \nabla^2 u + k^2 n^2(|u|^2) u = 0$$

in two or three space dimensions. We again assume that the wave number k is large and look for asymptotic solutions of (9.668). It is immediately possible to construct *plane wave* solutions of (9.668). For example,

$$(9.669) \qquad u = A \exp\left[ikn(A^2)x \right]$$

where the amplitude term A is a real constant satisfies (9.668). We note that in contrast to the linear problem studied in the foregoing, the phase in (9.669) is a function of the amplitude.

Proceeding as in the linear case, since there are solutions of (9.668) in the form (9.669) we look for a solution of the nonlinear reduced wave equation in the form

$$(9.670) \qquad u = A e^{ik\varphi}$$

where the amplitude term A and the phase term φ are now assumed to be real valued (we no longer require that A be a constant). Inserting (9.670) into

(9.668) and equating real and imaginary parts to zero gives

(9.671)
$$\left[(\nabla\varphi)^2 - n^2(A^2)\right]A - \frac{1}{k^2}\nabla^2 A = 0,$$

(9.672)
$$2\nabla\varphi \cdot \nabla A + A\nabla^2\varphi = 0.$$

In the geometrical optics approximation we let $k \to \infty$ in (9.671) and obtain

(9.673)
$$(\nabla\varphi)^2 = n^2(A^2),$$

(9.674)
$$\nabla \cdot (A^2\nabla\varphi) = 0$$

where (9.490) has been used in (9.672). These equations are referred to as the equations of *nonlinear geometrical optics*. In contrast to the situation in linear optics, (9.673)–(9.674) are coupled and must be solved simultaneously. Nevertheless, the *eiconal equation* (9.673) determines phase surfaces φ = constant and rays orthogonal to these surfaces. Also, the *transport equation* (9.674) determines the amplitude in terms of the divergence and convergence of the rays, as was the case in the linear problem. However, we cannot determine the rays without knowing the amplitude and vice versa.

We do not consider how the system (9.673)–(9.674) may be solved in general. Instead, we examine how a *beam* propagates in the nonlinear medium on the basis of the geometrical optics equations. A beam is characterized as follows. Let φ = constant represent a particular phase surface or wave front. The amplitude A is assumed to have a maximum at a point P on the wave front, and it decreases sharply in value as we move away from P in all directions on the wave front. Thus the significant values of the field are concentrated near P on the wave front φ = constant, and these values are propagated along the rays in the form of a beam. We now assume that the index of refraction $n(A^2)$ decreases as the amplitude A decreases. Thus since the amplitude falls off as we move away from P on φ = constant, so does $n(A^2)$. On considering the two neighboring wave fronts φ = constant and $\varphi + d\varphi$ = constant, the distance between them measured on the (orthogonal) rays is given as

(9.675)
$$d\sigma = \frac{1}{n(A^2)}d\varphi,$$

in view of (9.476). Since $d\varphi$ is fixed, we find that as A^2 and, consequently, $n(A^2)$ decrease (on the wave front φ = constant) as we move away from the point P, the distance $d\sigma$ between the two wave fronts increases. That is, φ = constant and $\varphi + d\varphi$ = constant are nearest to each other at P and become increasingly separated away from P. The rays, however, always remain orthogonal to the wave fronts. On applying this argument to a succession of wave fronts, we obtain the situation pictured in Figure 9.10. The wave fronts become increasingly concave and the (orthogonal) rays begin to converge. If the field is symmetric with respect to the point P, the rays may eventually focus at some point. In any case, the rays begin to converge and this property of nonlinear optics is known as the *self-focusing effect*. Even if the wave front φ = constant is plane or initially convex, the dependence of the index of refraction n on the amplitude will bend the rays and force them to converge, possibly even to a focus. A medium for which the index of refraction has this property is called a *focusing medium*.

However, as the rays converge, the amplitude of the beam begins to increase in view of the relation between the amplitude and the ray structure. If the amplitude A becomes extremely large or possibly singular, the geometrical optics approximation that predicts the self-focusing effect no longer remains valid and diffraction effects must be included. That is, we must retain second derivative terms in (9.671). In the linear problems discussed earlier the inclusion of diffraction effects was shown to remove the singularities predicted by the geometrical optics approach. Here, because we have a nonlinear problem, it may happen that a beam that undergoes focusing in the geometrical optics approximation may actually focus and become singular even if diffraction effects are included.

To see that it is possible to obtain beams that do not undergo self-focusing, we construct an exact solution of (9.668) or, equivalently, of (9.671)–(9.672), which represents a beam. We consider the two-dimensional case and assume that $n(|u|^2)$ is given as

$$(9.676) \qquad\qquad n^2(|u|^2) = n_0^2 + n_1|u|^2;$$

that is, we have a quadratic nonlinearity. The constant term n_0 is the linear index of refraction that occurs when nonlinear effects can be neglected. The constant n_1 is positive, so that $n(A^2)$ decreases as A^2 decreases. To solve

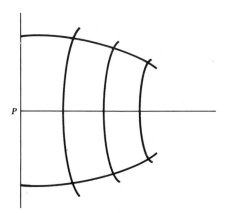

Figure 9.10. The self-focusing effect.

(9.671)–(9.672) we set

(9.677) $$\varphi(x, y) = \lambda x; \qquad A(x, y) = \tilde{A}(y)$$

where λ is a constant. Then (9.672) is identically satisfied and (9.671) becomes

(9.678) $$\tilde{A}''(y) - k^2(\lambda^2 - n_0^2)\tilde{A}(y) + k^2 n_1 \tilde{A}^3(y) = 0.$$

It is easily verified that a solution of (9.678) that vanishes together with its derivatives as $|y| \to \infty$ is

(9.679) $$\tilde{A}(y) = \left[\frac{2(\lambda^2 - n_0^2)}{n_1}\right]^{1/2} \operatorname{sech}\left[k(\lambda^2 - n_0^2)^{1/2} y\right]$$

if we assume that $|\lambda| > n_0 > 0$. The amplitude term $\tilde{A} = |u|$ decays exponentially as $|y| \to \infty$ and is significantly different from zero only in the interval $|y| \lesssim O[k^{-1}(\lambda^2 - n_0^2)^{-1/2}]$. Thus the solution

(9.680) $$u(x, y) = \left[\frac{2(\lambda^2 - n_0^2)}{n_1}\right]^{1/2} \operatorname{sech}\left[k(\lambda^2 - n_0^2)^{1/2} y\right] \exp[ik\lambda x]$$

represents a *beam* that travels without change of amplitude along the x-axis. We have shown in Example 9.14 that a general plane wave solution of the linear reduced wave equation is described by a parabolic equation in the far field. Although the geometric optics result in that problem predicted that the plane wave travels without change of amplitude to leading order, the diffraction effects characterized by the parabolic equation tended to diffuse and spread the field out at large distances. In the nonlinear problem considered here, the diffraction effects again tend to spread out the beam, whereas the nonlinear effects tend to focus the beam. Since these effects appear to be balanced in (9.680) and the beam travels without spreading, the solution (9.680) is referred to as a *self-trapped beam*.

It should be noted that if the index $n(A^2)$ is such that it increases as A decreases, this discussion would show that the rays diverge as a result of the nonlinearity. An initially thin beam would begin to spread out as it moved through the nonlinear medium. Therefore, we refer to such media as *defocusing media*. For example, if $n^2(|u|^2) = n_0^2 + n_1|u|^2$ with $n_1 < 0$, the index n has the defocusing property. Self-trapped solutions of the form (9.680) are not possible for such a medium since the amplitude A is real by assumption, and if $n_1 < 0$, the amplitude term (9.679) is purely imaginary. In fact, since both diffraction and nonlinear effects tend to spread out the beam, we would not expect that a self-trapped beam would occur in a defocusing medium.

To determine whether the self-trapped beam (9.680) can be expressed in some simpler form, we rewrite the hyperbolic secant in terms of exponentials and easily conclude that

$$(9.681) \quad u(x, y) = 2\left[\frac{2(\lambda^2 - n_0^2)}{n_1}\right]^{1/2} \exp\left[ik\lambda x - k(\lambda^2 - n_0^2)^{1/2}|y|\right]$$

$$\times \sum_{j=0}^{\infty} (-1)^j \exp\left[-2jk(\lambda^2 - n_0^2)^{1/2}|y|\right]$$

if $|y| > 0$. Since the amplitude term is exponentially small, we could not generate (9.681) by looking for an asymptotic solution of (9.668) in the form $u = Ae^{ik\varphi}$ where A is expanded in inverse powers of k. However, since λ is arbitrary (apart from the requirement that $\lambda^2 > n_0^2$) we may consider solutions of the form (9.680) in which $\lambda^2 \approx n_0^2 + c^2/k^2$, where c is a constant. Then (9.680) becomes

$$(9.682) \quad u(x, y) \approx \frac{c}{k}\left[\frac{2}{n_1}\right]^{1/2} \text{sech}[cy]\exp\left[i\left(kn_0 x + \frac{c^2 x}{2kn_0}\right)\right].$$

This suggests that if we ask for solutions of (9.668) in the form $|u| = O(k^{-r})$ with $r > 0$, an asymptotic approach similar to that used for the linear problem can be applied to the nonlinear problem. We note that with the above assumption on $|u| = A$, we are essentially considering a weakly nonlinear problem since $k \gg 1$ by assumption.

With $|u| = O(k^{-r})$, $r > 0$, we assume that $n(|u|^2)$ can be expanded as

$$(9.683) \qquad n^2(|u|^2) = n_0^2 + n_1|u|^2 + n_2|u|^4 + \cdots$$

where the coefficients n_j ($j = 1, 2, \ldots$) are all constants and $n_1 > 0$. As we have indicated, with n_1 positive the index of refraction (9.683) corresponds to a focusing medium if $|u|$ is small. We define the amplitude $|u|$ as

$$(9.684) \qquad |u| = A = \frac{1}{k^r}a, \qquad r > 0$$

and with $u = Ae^{ik\varphi}$ as in (9.670) we obtain for φ and a,

$$(9.685) \qquad \left\{\left[(\nabla\varphi)^2 - n_0^2\right]a - \frac{1}{k^{2r}}n_1 a^3 - \frac{1}{k^{4r}}n_2 a^5 - \cdots\right\}$$

$$-\frac{1}{k^2}\nabla^2 a = 0,$$

$$(9.686) \qquad\qquad 2\nabla\varphi \cdot \nabla a + a\nabla^2\varphi = 0.$$

To proceed with the solution of (9.685)–(9.686) we must specify the exponent r and then expand φ and a in inverse powers of k. The most meaningful and interesting choices for r are $r = \frac{1}{2}$ and $r = 1$. With $r = \frac{1}{2}$ we find that φ and a should be expanded in powers of k^{-1}, whereas if $r = 1$, the expansion of φ and a should contain powers of k^{-2}. Now if $r = 1$, we see from (9.685) that the cubic term a^3 and the term $\nabla^2 a$ are of the same order in k. Thus at the second level of approximation the (leading) nonlinear term and the diffraction term $\nabla^2 a$ are balanced. A particular case of this occurs in the result (9.682) where $A = O(1/k)$ and we have a self-trapped wave. However, if $r = \frac{1}{2}$, the nonlinear effect due to a^3 enters at the second level of approxima-tion. The diffraction effect due to $\nabla^2 a$ does not occur until the third level of approximation as we shall see. We consider the case where $r = \frac{1}{2}$. The case where $r = 1$ plays a role in the following example.

With $r = \frac{1}{2}$ in (9.685), we expand φ and a as

$$(9.687) \qquad \varphi = \sum_{j=0}^{\infty} \varphi_j k^{-j},$$

$$(9.688) \qquad a = \sum_{j=0}^{\infty} a_j k^{-j},$$

and insert these expansions in (9.685)–(9.686). Collecting like powers of $1/k$ and equating their coefficients to zero gives

$$(9.689) \qquad (\nabla \varphi_0)^2 = n_0^2,$$

$$(9.690) \qquad 2\nabla \varphi_0 \cdot \nabla a_0 + a_0 \nabla^2 \varphi_0 = 0,$$

$$(9.691) \qquad 2\nabla \varphi_0 \cdot \nabla \varphi_1 = n_1 a_0^2,$$

$$(9.692) \qquad 2\nabla \varphi_0 \cdot \nabla a_1 + a_1 \nabla^2 \varphi_0 = -2\nabla \varphi_1 \cdot \nabla a_0 - a_0 \nabla^2 \varphi_1,$$

(9.693)

$$2\nabla \varphi_0 \cdot \nabla \varphi_2 = n_2 a_0^4 + 3n_1 a_0 a_1 + \frac{1}{a_0} \nabla^2 a_0 - \frac{2a_1}{a_0} \nabla \varphi_0 \cdot \nabla \varphi_1 - (\nabla \varphi_1)^2,$$

and

(9.694)

$$2\nabla \varphi_0 \cdot \nabla a_2 + a_2 \nabla^2 \varphi_0 = -2\nabla \varphi_2 \cdot \nabla a_0 - a_0 \nabla^2 \varphi_2 - 2\nabla \varphi_1 \cdot \nabla a_1 - a_1 \nabla^2 \varphi_1$$

as the leading equations. We note that the nonlinear term $n_1 a_0^2$ enters in the equation for φ_1, whereas the diffraction term $\nabla^2 a_0$ enters in the equation for φ_2. The preceding equations and all further equations for the φ_j and a_j are to be solved recursively. The equations (9.689) and (9.690) for φ_0 and a_0 are just the eiconal and transport equations of linear geometrical optics. Each of these equations reduces to an ordinary differential equation along the rays associ-ated with the phase term φ_0.

Rather than discuss the solution of these equations for a general class of boundary value problems, we consider a specific two-dimensional problem in the following example. This example exhibits some of the basic features of beam propagation in a nonlinear medium.

EXAMPLE 9.18. THE PROPAGATION OF A BEAM IN A NONLINEAR MEDIUM

The x, y-plane is assumed to be divided into a *linear* and *nonlinear region*. The *interface* between the two regions is the line $x = 0$. A *general plane wave* is incident on the nonlinear region $x > 0$ from the linear region $x < 0$. We do not consider the field in the linear region but only in the nonlinear region.

At the interface $x = 0$ the field $u(x, y)$ equals the incident field u_I and this is assumed to be

$$(9.695) \qquad u(0, y) = u_I(0, y) = \frac{1}{\sqrt{k}} E \exp\left[-\frac{y^2}{\alpha^2}\right]$$

where E and α are real constants. The form of $u_I(x, y)$ for $x < 0$ does not concern us; however, it is a plane wave whose phase term vanishes at $x = 0$. We express the nonlinear field $u(x, y)$ in the form

$$(9.696) \qquad u = \frac{1}{\sqrt{k}} a e^{ik\varphi}$$

where $a(x, y)$ and $\varphi(x, y)$ are real valued and are to be expanded in inverse powers of k as in (9.687) and (9.688). The field u in the region $x > 0$ satisfies the nonlinear reduced wave equation (9.668) with n expanded as in (9.683). The terms φ_j and a_j in the expansion of φ and a satisfy (9.689)–(9.694) for $j = 0, 1,$ and 2. The initial values for the φ_j and a_j are found from (9.695) to be

$$(9.697) \qquad \varphi_j(0, y) = 0, \qquad j \geq 0$$

and

$$(9.698) \qquad a_0(0, y) = E \exp\left[-\frac{y^2}{\alpha^2}\right]; \qquad a_j(0, y) = 0; \qquad j \geq 1.$$

We also require that (9.696) be outgoing from the interface $x = 0$ toward $x > 0$, so that $\varphi_0(x, y)$ must satisfy the outgoing condition at $x = 0$.

Now $\varphi_0(x, y)$ satisfies the eiconal equation (9.689), the initial condition $\varphi_0(0, y) = 0$, and the outgoing condition at $x = 0$. We immediately conclude that

$$(9.699) \qquad \varphi_0(x, y) = n_0 x.$$

The equation (9.690) for $a_0(x, y)$ then becomes

(9.700)
$$2n_0 \frac{\partial a_0}{\partial x} = 0.$$

In view of (9.698) we find that

(9.701)
$$a_0(x, y) = E \exp\left[-\frac{y^2}{\alpha^2} \right].$$

To leading order we have

(9.702)
$$u(x, y) \approx \frac{E}{\sqrt{k}} \exp\left[-\frac{y^2}{\alpha^2} \right] e^{ikn_0 x}.$$

This represents a beam propagating in the x-direction. The *beam half-width* may be given as α since for $|y| > \alpha$ the amplitude term of u is exponentially small. At this level of approximation neither the *nonlinear* nor the *diffraction effects* have been accounted for and we consider those effects next.

The equation (9.691) for $\varphi_1(x, y)$ has the form

(9.703)
$$2n_0 \frac{\partial \varphi_1}{\partial x} = n_1 a_0^2 = n_1 E^2 \exp\left[-\frac{2y^2}{\alpha^2} \right].$$

Noting (9.697) gives

(9.704)
$$\varphi_1(x, y) = \frac{1}{2} \frac{n_1}{n_0} E^2 \exp\left[-\frac{2y^2}{\alpha^2} \right] x.$$

Thus the phase term φ is given as

(9.705) $$\varphi(x, y) \approx \varphi_0 + \frac{1}{k}\varphi_1 = n_0 x + \frac{1}{2k} \frac{n_1}{n_0} E^2 \exp\left[-\frac{2y^2}{\alpha^2} \right] x.$$

The rays, which are orthogonal to the wave fronts, have the direction of $\nabla\varphi$, and this is

(9.706)
$$\nabla\varphi \approx \left[n_0 + \frac{n_1}{2n_0 k} a_0^2 \right] \mathbf{i} + \left[-\frac{2n_1 y}{n_0 k \alpha^2} a_0^2 x \right] \mathbf{j}.$$

At $y = 0$, $\nabla\varphi$ has the direction of the vector $\mathbf{i}$. For $y \neq 0$, we see from (9.706) that for fixed y as x increases, the $\mathbf{j}$ component of $\nabla\varphi$ becomes increasingly negative if $y > 0$ and increasingly positive if $y < 0$. At $x = 0$, $\nabla\varphi$ has no $\mathbf{j}$ component. This indicates the presence of a self-focusing effect, in that the rays start bending towards the x-axis as x increases, even though they were originally parallel at $x = 0$. We note that if $n_1 < 0$, the reverse is true and defocusing takes place. We also observe the dependence of the phase on the amplitude in (9.705). This represents a strictly *nonlinear effect*.

The equation (9.692) for a_1 becomes

(9.707) $$2n_0 \frac{\partial a_1}{\partial x} = \left[\frac{2n_1 E^3 x}{n_0 \alpha^2} - \frac{16n_1 E^3 y^2 x}{n_0 \alpha^4} \right] \exp\left[-\frac{3y^2}{\alpha^2} \right]$$

so that

(9.708) $$a_1(x, y) = \left[\frac{n_1 E^3 x^2}{2n_0^2 \alpha^2} - \frac{4n_1 E^3 y^2 x^2}{n_0^2 \alpha^4} \right] \exp\left[-\frac{3y^2}{\alpha^2} \right],$$

since $a_1(0, y) = 0$. This can be written as

(9.709) $$a_1(x, y) = \left[\frac{n_1 E^2 x^2}{2n_0^2 \alpha^2} - \frac{4n_1 E^2 y^2 x^2}{n_0^2 \alpha^4} \right] \exp\left[-\frac{2y^2}{\alpha^2} \right] a_0(x, y)$$

and we obtain

(9.710)

$$a(x, y) \approx a_0 + \frac{1}{k} a_1 = \left\{ 1 + \left(\frac{n_1 E^2 x^2}{2kn_0^2 \alpha^2} - \frac{4n_1 E^2 y^2 x^2}{kn_0^2 \alpha^4} \right) \exp\left[-\frac{2y^2}{\alpha^2} \right] \right\}$$
$$\times E \exp\left[-\frac{y^2}{\alpha^2} \right].$$

The presence of terms of order x^2 in (9.709) and, consequently, in (9.710) shows that $a_1(x, y)$ undergoes *secular growth* as $x \to \infty$. Since the exponential term in (9.709) is small for large $|y|$ and the term of order y^2 is small for small $|y|$, we conclude that the first term in the brackets in (9.709) is the most significant for large x. With x fixed, the term that corresponds to it in (9.710) attains its maximum at $y = 0$. Thus it can be said that when

(9.711) $$\frac{n_1 E^2 x^2}{2kn_0^2 \alpha^2} \approx 1,$$

we have the first value of x at which a_0 and $(1/k)a_1$ are essentially of the same order of magnitude and the asymptotic series for $a(x, y)$ begins to become disordered.

We denote the value of x determined from (9.711) by x_f and we see that

(9.712) $$x_f \approx \frac{\alpha n_0 \sqrt{2k}}{E\sqrt{n_1}}.$$

Apart from a (dimensionless) constant factor, this determines the value of x at which the initially plane wave is expected to focus and we refer to it as a *focal length*. We note, however, that if $n_1 < 0$ we would obtain a similar expression for x with $\sqrt{n_1}$ replaced by $\sqrt{|n_1|}$, at which the secular behavior becomes significant. Yet with $n_1 < 0$ we have a defocusing medium and no focus is expected to appear. Nevertheless, we associate the region where secular growth becomes significant with the onset of the self-focusing region for the nonlinear medium. We now demonstrate that such a relationship does exist.

Since our geometrical optics result for φ and a breaks down when $x = O(\sqrt{k})$, as follows from (9.712), we study that region using a boundary layer approach. It is easily seen from (9.693) that $\varphi_2(x, y)$ contains a term of the order x^3. Thus $(1/k)\varphi_1 + (1/k^2)\varphi_2$ are of the same order of magnitude when $x = O(\sqrt{k})$ since $\varphi_1 = O(x)$. However, the leading term $\varphi_0 = n_0 x$ is never of the same order of magnitude as the other terms in the expansion of φ. Thus secular behavior occurs in the expansions of both φ and a, but φ_0 is not involved in it. As a result we shall modify only the series expansion of φ beginning with the φ_1 term. Since $(1/k)\varphi_1 = O(1/\sqrt{k})$ in the secular region $x = O(\sqrt{k})$, we set

$$\varphi = n_0 x + \frac{1}{\sqrt{k}}\hat{\varphi}$$

in (9.685)–(9.686), where $r = \tfrac{1}{2}$. Then

(9.713)

$$\left\{ \frac{2}{\sqrt{k}} n_0 \hat{\varphi}_x a + \frac{1}{k}(\nabla\hat{\varphi})^2 a - \frac{1}{k}n_1 a^3 - \frac{1}{k^2}n_2 a^5 + \cdots \right\} - \frac{1}{k^2}\nabla^2 a = 0,$$

(9.714)

$$2n_0 a_x + \frac{2}{\sqrt{k}}\nabla\hat{\varphi} \cdot \nabla a + \frac{1}{\sqrt{k}} a\nabla^2\hat{\varphi} = 0.$$

Since $x = O(\sqrt{k})$ in the secular region we set

(9.715)

$$x = \sqrt{k}\,\xi$$

in (9.713)–(9.714). Then $\partial/\partial x = (1/\sqrt{k})\partial/\partial\xi$ so that to leading order

(9.716)

$$\frac{1}{k}\left\{2n_0\hat{\varphi}_\xi + \hat{\varphi}_y^2 - n_1\hat{a}^2\right\}\hat{a} + O\left(\frac{1}{k^2}\right) = 0$$

(9.717)

$$\frac{1}{\sqrt{k}}\left\{2n_0\hat{a}_\xi + 2\hat{\varphi}_y\hat{a}_y + \hat{a}\hat{\varphi}_{yy}\right\} + O\left(\frac{1}{k}\right) = 0,$$

where $\hat{a} = \hat{a}(\xi, y)$ and $\hat{\varphi} = \varphi(\xi, y)$. To solve, we expand $\hat{a}$ and $\hat{\varphi}$ in powers of

$1/\sqrt{k}$, and obtain to leading order for $\hat{a}_0$ and $\hat{\varphi}_0$,

$$(9.718) \qquad 2n_0\frac{\partial\hat{\varphi}_0}{\partial\xi} + \left(\frac{\partial\hat{\varphi}_0}{\partial y}\right)^2 - n_1\hat{a}_0^2 = 0,$$

$$(9.719) \qquad 2n_0\frac{\partial\hat{a}_0}{\partial\xi} + 2\frac{\partial\hat{\varphi}_0}{\partial y}\frac{\partial\hat{a}_0}{\partial y} + \hat{a}_0\frac{\partial^2\hat{\varphi}_0}{\partial y^2} = 0.$$

Although these equations are coupled as was the case for the nonlinear geometrical optics equations (9.673)–(9.674), they are, nevertheless, easier to solve.

A number of solutions of (9.718)–(9.719) have been obtained by Akhmanov et al. and other authors. We shall consider one of these solutions and match it with our geometrical optics results. To carry this out we require that the field be considered only near the center of the beam (i.e., near $y = 0$). Thus

$$(9.720) \qquad a_0^2(x, y) = E^2\exp\left[-\frac{2y^2}{\alpha^2}\right] \approx E^2\left[1 - \frac{2y^2}{\alpha^2}\right]$$

and

$$(9.721) \quad \varphi_1(x, y) = \frac{1}{2}\frac{n_1}{n_0}E^2\exp\left[-\frac{2y^2}{\alpha^2}\right]x \approx \frac{1}{2}\frac{n_1}{n_0}E^2x - \frac{n_1E^2}{n_0\alpha^2}y^2x.$$

An exact solution of (9.718)–(9.719) can be found that has the form

$$(9.722) \qquad \hat{\varphi}_0(\xi, y) = \gamma(\xi) + \tfrac{1}{2}y^2\beta(\xi),$$

$$(9.723) \qquad \hat{a}_0^2(\xi, y) = \left[\frac{E^2}{f(\xi)}\right]\left[1 - \frac{2y^2}{\alpha^2 f^2(\xi)}\right],$$

where γ, β, and f are specified on substituting (9.722)–(9.723) into (9.718)–(9.719). We match $\hat{\varphi}_0$ and $\hat{a}_0^2$ with φ_1 and a_0^2 by noting that $x = \sqrt{k}\,\xi$, so that for small ξ we have moderate or large values of x. On comparing the equations for the two sets of variables, we find $\gamma(\xi)$ and $\beta(\xi)$ must tend to zero as $\xi \to 0$, whereas $f(\xi)$ must tend to unity as $\xi \to 0$. Thus we set

$$(9.724) \qquad \gamma(0) = \beta(0) = 0; \qquad f(0) = 1.$$

Before inserting (9.722)–(9.723) into (9.718)–(9.719) we multiply (9.719) by $\hat{a}_0$ to simplify the calculations. We obtain

$$(9.725) \qquad n_0y^2\beta'(\xi) + 2n_0\gamma'(\xi) + y^2\beta^2(\xi) - \frac{n_1E^2}{f(\xi)}$$

$$+ \frac{2n_1E^2y^2}{\alpha^2f^3(\xi)} = 0$$

from (9.718) and

(9.726)
$$\frac{E^2\beta(\xi)}{f(\xi)} - \frac{2E^2y^2\beta(\xi)}{\alpha^2 f^3(\xi)} - \frac{n_0 E^2 f'(\xi)}{f^2(\xi)}$$
$$+ \frac{6n_0 E^2 y^2 f'(\xi)}{\alpha^2 f^4(\xi)} - \frac{4E^2 y^2 \beta(\xi)}{\alpha^2 f^3(\xi)} = 0$$

from the modified form of (9.719).

Now if we set

(9.727)
$$\gamma'(\xi) = \frac{n_1 E^2}{2n_0 f(\xi)}$$

in (9.725), the two terms in that equation that do not contain a factor y^2 cancel out. We are left with

(9.728)
$$n_0 \beta'(\xi) + \beta^2(\xi) + \frac{2n_1 E^2}{\alpha^2 f^3(\xi)} = 0.$$

Similarly, if we set

(9.729)
$$\beta(\xi) = \frac{n_0 f'(\xi)}{f(\xi)},$$

the two terms in (9.726) that do not contain a factor y^2 drop out. It then turns out that with this choice of $\beta(\xi)$ the remaining terms in (9.726) cancel out, so that this equation is satisfied regardless of the form of $f(\xi)$. Therefore, $f(\xi)$ must be specified from (9.728). Inserting (9.729) into (9.728) yields

(9.730)
$$f''(\xi) = -\frac{2n_1 E^2}{n_0^2 \alpha^2 f^2(\xi)}.$$

A first integral of (9.730) may be obtained by multiplying through by $f'(\xi)$. This gives

(9.731)
$$[f'(\xi)]^2 = \frac{4n_1 E^2}{n_0^2 \alpha^2}\left[\frac{1}{f(\xi)} - 1\right],$$

where we have used (9.729) to conclude that $f'(0) = 0$ since $\beta(0) = 0$ and $f(0) = 1$. Now (9.731) is exactly solvable by means of a special transformation that we do not exhibit. With $f(0) = 1$ we obtain in implicit form

(9.732)
$$f(\xi) = \sin^2\left[\frac{\pi}{2} + \sqrt{f(\xi)[1 - f(\xi)]} + \frac{E\sqrt{2n_1}\,\xi}{\alpha n_0}\right].$$

Then $\gamma(\xi)$ and $\beta(\xi)$ can be specified from (9.727) and (9.729) but we do not carry this out.

Since $f(\xi)$ appears in the denominator in the expression (9.723) for $\hat{a}_0^2$, we see that when $f(\xi) = 0$, the amplitude blows up. This corresponds to a *focal point* for the field. To determine the location of this focal point we set $f(\xi) = 0$ in (9.732) and obtain as the first focal length

$$(9.733) \qquad \hat{x}_f = \sqrt{k}\,\xi_f = \frac{\pi}{2}\,\frac{\alpha n_0\sqrt{k}}{E\sqrt{2n_1}}$$

since $\sin \pi = 0$. There are, in fact, infinitely many focal points predicted by (9.732) but the others are not of interest to us.

We want to match the preceding *boundary layer results* with the previously given *geometrical optics results*. This requires that we consider $\gamma(\xi)$, $\beta(\xi)$, and $f(\xi)$ for small ξ. Since $f(0) = 1$, $f'(0) = 0$, and $f''(0) = -2n_1E^2/n_0^2\alpha^2$, in view of (9.730) we have

$$(9.734) \qquad f(\xi) = f\left(\frac{1}{\sqrt{k}}x\right) \approx 1 - \frac{n_1E^2}{n_0^2\alpha^2 k}x^2.$$

As a result we find that

$$(9.735) \qquad \beta(\xi) = \beta\left(\frac{1}{\sqrt{k}}x\right) \approx -\frac{2n_1E^2x}{n_0\alpha^2\sqrt{k}}$$

and

$$(9.736) \qquad \gamma(\xi) = \gamma\left(\frac{1}{\sqrt{k}}x\right) \approx \frac{n_1E^2x}{2n_0\sqrt{k}}.$$

Inserting these expressions into (9.722) yields

$$(9.737) \qquad \varphi(x, y) \approx \varphi_0(x, y) + \frac{1}{\sqrt{k}}\hat{\varphi}_0\left(\frac{1}{\sqrt{k}}x, y\right) \approx n_0 x$$

$$+ \frac{1}{k}\left[\frac{1}{2}\frac{n_1}{n_0}E^2x - \frac{n_1E^2y^2}{n_0\alpha^2}x\right],$$

which agrees with the geometrical optics result (9.705) when that is evaluated near the axis of the beam as in (9.721).

For $\hat{a}_0(\xi, y)$ we have

$$(9.738) \quad \hat{a}_0(\xi, y) = \hat{a}_0\left(\frac{1}{\sqrt{k}}x, y\right) \approx E\left[1 - \frac{n_1 E^2 x^2}{n_0^2 \alpha^2 k}\right]^{-1/2}\left[1 - \frac{2y^2}{\alpha^2}\right]^{1/2}$$

$$\approx E\left[1 + \frac{n_1 E^2 x^2}{2 n_0^2 \alpha^2 k}\right],$$

for small y. This agrees with the expression (9.710) for $a(x, y)$ near the axis of the beam. Furthermore, if we determine the focal point for $\hat{a}_0$ to the level of approximation used in (9.738), we find that the *focal length* is

$$(9.739) \qquad\qquad\qquad \hat{x}_f \approx \frac{\alpha n_0 \sqrt{k}}{E\sqrt{n_1}}$$

since $f(\xi)$ vanishes at this point if it is expressed as in (9.734). Comparing (9.739) with (9.733) we find that they differ by a factor of $\pi/2\sqrt{2} \approx 1.1$. The difference is not substantial. The result given in (9.712) on the basis of the geometrical optics approach differs from (9.739) by a factor of $\sqrt{2}$. We see from (9.738) that this difference in results is due to the factor $\frac{1}{2}$ which occurs when the binomial expansion is used in (9.738). Nevertheless, all expressions for the focal length contain the same dependence on the parameters α, n_0, n_1, E, and k.

Even though the boundary layer equations (9.718)–(9.719) predict that focusing occurs, as we have seen, it may yet be that focusing is prevented if diffraction effects, which have so far been neglected, are taken into account. That is, we must consider equations for the field that include second derivative terms in the amplitude. The self-trapped solution (9.682) shows that plane wave solutions (in an approximate form) do exist that do not undergo self-focusing. To analyze this problem we now obtain the terms φ_2 and a_2 in the expansion of the geometrical optics solution (9.696).

The phase term φ_2 satisfies (9.693). Using the expressions obtained above for φ_0, φ_1, a_0, and a_1 and the initial condition $\varphi_2(0, y) = 0$, we easily obtain

$$(9.740) \quad \varphi_2(x, y) = \left[-\frac{1}{n_0 \alpha^2} + \frac{2y^2}{n_0 \alpha^4}\right]x$$

$$+ \left[\frac{n_2 x}{2 n_0} + \frac{n_1^2 x^3}{6 n_0^3 \alpha^2} - \frac{n_1^2 x}{8 n_0^3} - \frac{2 n_1^2 y^2 x^3}{n_0^3 \alpha^4}\right]E^4\exp\left[-\frac{4y^2}{\alpha^2}\right].$$

We then insert (9.740) into (9.694) for a_2 and integrate, retaining only the most significant terms near $y = 0$ and for x large. This gives

$$(9.741) \qquad a_2(x, y) \approx -\frac{x^2 E}{n_0^2 \alpha^4} \exp\left[-\frac{y^2}{\alpha^2}\right]$$

$$+ \left[\frac{n_1^2 x^4 E^5}{2 n_0^4 \alpha^4} + \frac{n_2 x^2 E^5}{n_0^2 \alpha^2}\right] \exp\left[-\frac{5 y^2}{\alpha^2}\right].$$

In the absence of nonlinear effects (i.e., $n_1 = n_2 = 0$) the first term on the right in (9.741) is the only contribution to a_2. It represents a *diffraction effect*. Since $a_1 = 0$ if $n_1 = 0$, we have

$$(9.742) \qquad a(x, y) \approx a_0 + \frac{1}{k^2} a_2 \approx \left[1 - \frac{x^2}{k^2 n_0^2 \alpha^4}\right] E \exp\left[-\frac{y^2}{\alpha^2}\right].$$

Thus for a linear medium, secular effects occur when

$$(9.743) \qquad x_d \approx k n_0 \alpha^2$$

where x_d may be termed a *diffraction length* (apart from a possible numerical factor, it agrees with the results in the literature). We have shown in Example 9.14 that when $x = O(k)$, the field resulting from a general plane wave must be described by means of a *Schrödinger equation*. However, since diffraction effects occur when $x = O(k)$ and self-focusing occurs when $x = O(\sqrt{k})$ in view of (9.712), we might expect that diffraction cannot, in general, influence and counteract self-focusing.

With n_1 and n_2 not equal to zero, we obtain near $y = 0$,

$$(9.744) \qquad a(x, y) \approx a_0 + \frac{1}{k} a_1 + \frac{1}{k^2} a_2$$

$$\approx E + \frac{n_1 E^3 x^2}{2 n_0^2 \alpha^2 k} - \frac{E x^2}{n_0^2 \alpha^4 k^2} + \frac{n_2 E^5 x^2}{n_0^2 \alpha^2 k^2}.$$

Now secular effects occur when $(1/k) a_1 \approx a_0$ in the nonlinear problem and when $(1/k^2) a_2 \approx a_0$ in the (strictly) linear problem. The combined effects in (9.744) that result in $a_0 \approx (1/k) a_1 \approx (1/k^2) a_2$ yield

$$(9.745) \qquad \left[\frac{n_1 E^2}{2 n_0^2 \alpha^2 k} + \frac{n_2 E^4}{n_0^2 \alpha^2 k^2} - \frac{1}{n_0^2 \alpha^4 k^2}\right] x^2 \approx 1.$$

Noting (9.712) and (9.743), we have

$$(9.746) \qquad \frac{1}{x^2} \approx \frac{1}{x_f^2} - \frac{1}{x_d^2} + \left(\frac{E^2 \sqrt{n_2}}{n_0 \alpha k}\right)^2,$$

and we assume that $n_2 > 0$. Arguing as above, we could state that (9.746) determines a first value of x at which self-focusing becomes significant if we include diffraction and higher order nonlinear effects. We have established a relationship between the secularity and self-focusing regions previously. Thus (9.746) shows that diffraction need not prevent self-focusing but it does delay its onset. The higher order nonlinear effect enhances the self-focusing property and brings the focal region closer to the origin. This result is not unexpected in view of our general discussion of the self-focusing effect. If k is large and the other terms in (9.745) are of moderate size, the diffraction and higher order nonlinear effects do not appear to be significant.

An interesting situation arises if $x_f = x_d$ so that the focal and diffraction lengths are equal. This implies that

$$(9.747) \qquad\qquad E^2 = \frac{2}{n_1 \alpha^2 k}$$

so that E is small for large k. With this choice for E, the last term in (9.744) is of order $k^{-4} E$ so that

$$(9.748) \qquad a \approx a_0 + \frac{1}{k} a_1 + \frac{1}{k^2} a_2 \approx E + O(k^{-4} E).$$

Also, in (9.745) we may eliminate the n_2 term since it is of order k^{-4} compared to the other two bracketed terms which are of order k^{-2}. Consequently, (9.746) becomes

$$(9.749) \qquad\qquad \frac{1}{x^2} \approx 0,$$

and the focal point is moved off to infinity at this level of approximation. Furthermore, we have

$$(9.750) \qquad \varphi \approx \varphi_0 + \frac{1}{k} \varphi_1 + \frac{1}{k^2} \varphi_2 \approx n_0 x + O(k^{-4})$$

for $y \approx 0$ as is easily seen. Thus with E given as in (9.747) we obtain

$$(9.751) \quad u(x, y) = \frac{1}{\sqrt{k}} a e^{ik\varphi} \approx \left[\frac{2}{n_1 \alpha^2 k^2} \right]^{1/2} \exp\left[-\frac{y^2}{\alpha^2} \right] \exp[ikn_0 x]$$

using the given approximation. We have replaced $a \approx E$ by $a \approx E \exp[-y^2/\alpha^2]$ in (9.748) since the preceding was valid near $y = 0$ and the field, which has the form of a beam, is exponentially small away from $y = 0$.

We have shown that if E is given by (9.747), the resulting field is a beam that travels without change of shape to a high level of approximation for large

k. Since this is effectively a self-trapped beam for which an (exact) balance between the nonlinear self-focusing effect and the defocusing diffraction effect exists, it is of interest to compare (9.751) with the approximate form (9.682) of the exact self-trapped solution found earlier. The parameter c that occurs in (9.682) may be interpreted as an inverse beam half-width for that solution, since when $cy \geq 1$, $\text{sech}(cy)$ is small. Thus if $c = 1/\alpha$, the amplitude term in (9.682) evaluated at the center of the beam is equal to E as defined in (9.747). We recall that α is the half-width of the beam (9.751). Even though the beams (9.682) and (9.751) have somewhat different phase and amplitude representations, they have a common beam half-width α and have the same amplitude E at the center of the beam. In addition, they are both (approximately) general plane waves and are self-trapped.

We now present a boundary layer analysis of the field in the region $x = O(k)$ where diffraction effects become important. The phase in that region should equal $n_0 x$ plus lower order correction terms, and the amplitude should remain bounded or decreasing as k increases. From (9.704) and (9.708) we see that $(1/k)\varphi_1 = O(1)$ and $(1/k)a_1 = O(k)$ when $x = O(k)$. However, $\varphi_0 = O(k)$ and $a_0 = O(1)$ when $x = O(k)$. Since we cannot have $(1/k)a_1$ be of higher order than a_0, it must be assumed that $a_0 = E\exp(-y^2/\alpha^2) = O(k^{-s})$ for some positive s and this means that $E = O(k^{-s})$. Then $(1/k)\varphi_1 = O(k^{-2s})$ and $(1/k)a_1 = O(k^{1-3s})$, so that these correction terms are of reduced orders in k.

To proceed we set

(9.752)
$$\varphi = n_0 x + \frac{1}{k}\tilde{\varphi}$$

in (9.685)–(9.686), since when $x = O(k)$ we have seen that the correction term to $\varphi_0 = n_0 x$ is one order of k lower than φ_0 itself. With $r = \frac{1}{2}$ in (9.685)–(9.686), we have

(9.753)
$$\left\{ \frac{2}{k} n_0 \tilde{\varphi}_x a + \frac{1}{k^2}(\nabla\tilde{\varphi})^2 a - \frac{1}{k} n_1 a^3 + \cdots \right\} - \frac{1}{k^2}\nabla^2 a = 0,$$

(9.754)
$$2n_0 a_x + \frac{2}{k}\nabla\tilde{\varphi}\cdot\nabla a + \frac{1}{k}a\nabla^2\tilde{\varphi} = 0.$$

Then with

(9.755)
$$x = k\xi,$$

we set

(9.756)
$$a = \frac{1}{k^s}\hat{a}; \qquad \tilde{\varphi} = \frac{1}{k^{2s-1}}\hat{\varphi}$$

where $s > 0$ and is to be determined. This implies that the amplitude a is $O(k^{-s})$ and the correction term to $\varphi_0 = n_0 x$ is $O(k^{-2s})$ as was required previously. We obtain

(9.757)

$$\frac{1}{k^{1+3s}}\left\{2n_0\frac{\partial\hat{\varphi}}{\partial\xi}\hat{a} + \left(\frac{\partial\hat{\varphi}}{\partial y}\right)^2\hat{a} - n_1\hat{a}^3\right\} - \frac{1}{k^{2+s}}\frac{\partial^2\hat{a}}{\partial y^2} + \cdots = 0,$$

(9.758)

$$\frac{1}{k^{1+s}}\left[2n_0\frac{\partial\hat{a}}{\partial\xi}\right] + \frac{1}{k^{3s}}\left[2\frac{\partial\hat{\varphi}}{\partial y}\frac{\partial\hat{a}}{\partial y} + \hat{a}\frac{\partial^2\hat{\varphi}}{\partial y^2}\right] + \cdots = 0$$

where we have neglected lower order terms. We must choose $s = \frac{1}{2}$ so that the significant terms in (9.757)–(9.758) include the leading terms in (9.716)–(9.717) valid in the focal region. However, we do have an additional second derivative term in the amplitude to account for diffraction.

The leading terms in the expansions of $\hat{\varphi}$ and $\hat{a}$, which we denote by $\hat{\varphi}_0$ and $\hat{a}_0$, satisfy

(9.759)

$$2n_0\frac{\partial\hat{\varphi}_0}{\partial\xi} + \left(\frac{\partial\hat{\varphi}_0}{\partial y}\right)^2 - n_1 a_0^2 - \left(\frac{1}{\hat{a}_0}\right)\frac{\partial^2\hat{a}_0}{\partial y^2} = 0,$$

(9.760)

$$2n_0\frac{\partial\hat{a}_0}{\partial\xi} + 2\frac{\partial\hat{\varphi}_0}{\partial y}\frac{\partial\hat{a}_0}{\partial y} + a_0\frac{\partial^2\hat{\varphi}_0}{\partial y^2} = 0.$$

Introducing the function

(9.761)

$$\hat{v}(\xi, y) = \hat{a}_0(\xi, y)\exp[i\hat{\varphi}_0(\xi, y)],$$

we easily find that

(9.762)

$$2in_0\hat{v}_\xi + \hat{v}_{yy} + n_1|\hat{v}|^2\hat{v} = 0.$$

This *nonlinear Schrödinger equation* reduces (as it should) to the linear Schrödinger equation (9.552) when $n_1 = 0$. Both equations describe the far field in the region $x = O(k)$ where diffraction effects are important. It terms of $u(x, y)$ our result is now given as

(9.763)

$$u(x, y) \approx \frac{1}{k}\hat{v}\left(\frac{1}{k}x, y\right)\exp[ikn_0 x]$$

where $\hat{v}$ satisfies the equation

(9.764)

$$2ikn_0\hat{v}_x + \hat{v}_{yy} + n_1|\hat{v}|^2\hat{v} = 0.$$

The nonlinear Schrödinger equation (9.762) is one of a class of nonlinear equations for which an exact solution of the initial value problem can be found. The inverse scattering method and related methods that yield these solutions are too complicated to be considered here. A self-trapped solution of (9.762) may be found by setting

(9.765) $$\hat{\varphi}_0 = \hat{\lambda}\xi; \qquad \hat{a}_0 = \hat{a}_0(y)$$

in (9.759)–(9.760). Then (9.760) is identically satisfied and (9.759) reduces to the ordinary differential equation

(9.766) $$2n_0\hat{\lambda}\hat{a}_0(y) - n_1\hat{a}_0^3(y) - \hat{a}_0''(y) = 0.$$

This has the form of (9.678), and a solution that vanishes as $|y| \to \infty$ is

(9.767) $$\hat{a}_0(y) = \left[\frac{4n_0\hat{\lambda}}{n_1}\right]^{1/2} \operatorname{sech}\left[(2n_0\hat{\lambda})^{1/2} y\right]$$

in view of (9.679).

The field $u(x, y)$ is given as

(9.768) $$u(x, y) \approx \left[\frac{4n_0\hat{\lambda}}{n_1 k^2}\right]^{1/2} \operatorname{sech}\left[(2n_0\hat{\lambda})^{1/2} y\right] \exp\left[i\left(kn_0 x + \frac{\hat{\lambda}x}{k}\right)\right],$$

on using (9.761) and (9.763). In we set $\hat{\lambda} = c^2/2n_0$, this expression is identical to (9.682). However, we are interested in obtaining a solution of (9.759)–(9.760) that matches (9.751). To do so we set

(9.769) $$\hat{\varphi}_0 = 0; \qquad \hat{a}_0 = \hat{E}\exp\left[-\frac{y^2}{\alpha^2}\right]$$

where $\hat{E}$ is a constant that is to be determined. Inserting (9.769) into (9.759)–(9.760) yields

(9.770) $$-\frac{2\hat{E}}{\alpha^2}\left[1 - \frac{2y^2}{\alpha^2}\right]\exp\left[-\frac{y^2}{\alpha^2}\right] = -n_1\hat{E}^3\exp\left[-\frac{3y^2}{\alpha^2}\right]$$

from (9.759), since (9.760) is satisfied identically. Near the center of the beam where $y \approx 0$, we may approximate $\exp[-2y^2/\alpha^2]$ by $1 - 2y^2/\alpha^2$. Doing so in (9.770) shows that $\hat{E}$ must be given as

(9.771) $$\hat{E}^2 = \frac{2}{n_1\alpha^2}.$$

Thus the field $u(x, y)$ near $y = 0$ is

$$(9.772) \qquad u(x, y) \approx \left[\frac{2}{n_1 \alpha^2 k^2} \right]^{1/2} \exp\left[-\frac{y^2}{\alpha^2} \right] \exp[ikn_0 x].$$

The expression (9.772) for u agrees with (9.751) which was obtained from the geometrical optics solution under the assumption that the self-focusing and diffraction lengths are equal. The field (9.772) represents a self-trapped beam since the amplitude is independent of x and decays exponentially in $|y|$. If we define $(2n_0 \hat{\lambda})^{-1/2} = \alpha$ to be the beam half-width of (9.768), we obtain for (9.768)

$$(9.773) \quad u(x, y) \approx \left[\frac{2}{n_1 \alpha^2 k^2} \right]^{1/2} \operatorname{sech}\left[\frac{y}{\alpha} \right] \exp\left[i\left(kn_0 x + \frac{x}{2n_0 \alpha^2 k} \right) \right].$$

The fact that (9.772) and (9.773) are both self-trapped beams of half-width α and $|u(x,0)|^2 = 2/n_1 \alpha^2 k^2 = E_0^2$, suggests that E_0 plays a critical role in determining whether beams are self-focused, self-trapped, or effectively propagate as in a linear medium. If a beam $u(x, y)$ has $|u(x,0)| > E_0$, the focusing effect is stronger than the defocusing diffraction effects and self-focusing occurs. If $|u(x,0)| = E_0$, self-trapping occurs and if $|u(x,0)| < E_0$, the nonlinear effects contribute small corrections to linear wave propagation. The concept of a *critical intensity* E_0^2 associated with beam propagation has been discussed in the literature and is somewhat consistent with these results, but is difficult to verify in general.

We conclude our discussion of asymptotic methods for equations with a large parameter by considering an initial value problem for the Klein–Gordon equation. It is shown how the methods presented to deal with elliptic equations can be carried over to hyperbolic equations. Rather than give a general discussion, we concentrate on a specific problem for the Klein–Gordon equation. As we have indicated in Examples 3.7 and 5.13, this equation is representative of dispersive wave motion. It is for dispersive equations that our method is applicable and most useful.

EXAMPLE 9.19. AN ASYMPTOTIC SOLUTION OF THE KLEIN–GORDON EQUATION

With $\hat{u} = \hat{u}(\sigma, \tau)$ we begin by considering the *Klein–Gordon equation*

$$(9.774) \qquad \hat{u}_{\tau\tau} - \gamma^2 \hat{u}_{\sigma\sigma} + c^2 \hat{u} = 0, \qquad -\infty < \sigma < \infty, \tau > 0$$

with the initial conditions

(9.775) $$\hat{u}(\sigma,0) = 0; \quad \hat{u}_\tau(\sigma,0) = \delta(\sigma)$$

where $\delta(\sigma)$ is the Dirac delta function and γ and c are specified constants. To obtain a large parameter in the equation we set

(9.776) $$\sigma = kx; \quad \tau = kt$$

where $k \gg 1$, and this gives

(9.777) $$u_{tt} - \gamma^2 u_{xx} + k^2 c^2 u = 0$$

where $u = u(x, t) = \hat{u}[kx, kt]$. The initial conditions for u are

(9.778) $$u(x,0) = 0; \quad u_t(x,0) = \delta(x)$$

since $\hat{u}_\tau(\sigma, 0) = \dfrac{1}{k} u_t(x, 0) = \delta[kx] = \dfrac{1}{k}\delta(x)$.

The solutions $\hat{u}$ and u of these problems yield the *causal fundamental solutions* for the appropriate Klein–Gordon equations as follows from the results of Chapter 7. The transformation (9.776) indicates that we are to study these fundamental solutions at large times and large distances from the origin. In Example 5.13 the method of stationary phase was applied to the Klein–Gordon equation to examine its solutions for large $|x|$ and t. The asymptotic result given in that example suggests that we look for a solution of (9.777)–(9.778) in the form

(9.779) $$u(x, t) = a(x, t)e^{ik\varphi(x, t)}$$
$$= \left[\sum_{j=0}^{\infty} a_j(x, t)(ik)^{-j}\right]\exp[ik\varphi(x, t)].$$

We note that the assumed expansion has the same form as that considered for the reduced wave equation.

On inserting (9.779) into (9.777) and equating like powers of $1/k$ we obtain

(9.780) $$\varphi_t^2 - \gamma^2\varphi_x^2 - c^2 = 0,$$

(9.781) $$2\frac{\partial\varphi}{\partial t}\frac{\partial a_0}{\partial t} - 2\gamma^2\frac{\partial\varphi}{\partial x}\frac{\partial a_0}{\partial x} + a_0\left[\varphi_{tt} - \gamma^2\varphi_{xx}\right] = 0,$$

(9.782) $$2\frac{\partial\varphi}{\partial t}\frac{\partial a_j}{\partial t} - 2\gamma^2\frac{\partial\varphi}{\partial x}\frac{\partial a_j}{\partial x} + a_j\left[\varphi_{tt} - \gamma^2\varphi_{xx}\right]$$
$$= -\frac{\partial^2 a_{j-1}}{\partial t^2} + \gamma^2\frac{\partial^2 a_{j-1}}{\partial x^2}; \quad j = 1, 2, \ldots.$$

The equation (9.780) for the phase term $\varphi(x, t)$ is known as the *dispersion equation*. If we put

$$(9.783) \qquad\qquad \omega = -\varphi_t; \qquad \lambda = \varphi_x$$

in (9.780) we obtain

$$(9.784) \qquad\qquad \omega^2 = \gamma^2\lambda^2 + c^2,$$

and this leads to the *dispersion relation* $\omega = \omega(\lambda)$ as defined in (3.146). The equations (9.781)–(9.782) are the transport equations for the amplitude terms a_j. We shall solve the dispersion equation using the method of characteristics and solve the leading order transport equation for a_0.

To begin, we solve (9.780) for φ_t and obtain

$$(9.785) \qquad\qquad \varphi_t = \pm\sqrt{\gamma^2\varphi_x^2 + c^2}\,.$$

Each of these equations must be dealt with separately. We study the case with the minus sign in some detail and then state the results for the other case. Thus we consider

$$(9.786) \qquad\qquad \varphi_t + \sqrt{\gamma^2\varphi_x^2 + c^2} = 0.$$

With $p = \varphi_x$ and $q = \varphi_t$ we have

$$(9.787) \qquad\qquad F(p, q) = q + \sqrt{\gamma^2 p^2 + c^2} = 0.$$

Noting the results of Section 2.4 we obtain the characteristic equations for x, t, φ, p, and q as

$$(9.788) \qquad \frac{dx}{ds} = F_p = \frac{\gamma^2 p}{\sqrt{\gamma^2 p^2 + c^2}}; \qquad \frac{dt}{ds} = F_q = 1;$$

$$(9.789) \qquad \frac{d\varphi}{ds} = pF_p + qF_q = -\frac{\gamma^2 p^2}{q} + q = \frac{c^2}{q} = \frac{-c^2}{\sqrt{\gamma^2 p^2 + c^2}};$$

$$(9.790) \qquad \frac{dp}{ds} = 0; \qquad \frac{dq}{ds} = 0.$$

Here s is a parameter along the characteristic curves. The curves $x = x(s)$, $y = y(s)$ are called *rays*.

Since the initial data for $u(x, t)$ are concentrated at the point $x = 0$, the data for the system (9.788)–(9.790) are concentrated at the origin $(x, t) = (0, 0)$.

Therefore, we set

$$(9.791) \qquad x(0) = 0; \qquad t(0) = 0; \qquad \varphi(0) = \varphi_0$$

where φ_0 is a constant that has to be specified. Also,

$$(9.792) \qquad q(0) + \sqrt{\gamma^2 p^2(0) + c^2} = 0.$$

Since the initial curve reduces to a point in this case, there is no *strip condition* that p and q must satisfy. Now (9.790) shows that $p(s)$ and $q(s)$ are constant along the characteristics. Thus (9.792) determines q in terms of the parameter p where $-\infty < p < \infty$.

With the given data we easily find that

$$(9.793) \qquad x = \frac{\gamma^2 p}{\sqrt{\gamma^2 p^2 + c^2}} s,$$

$$(9.794) \qquad t = s,$$

and

$$(9.795) \qquad \varphi = \frac{-c^2}{\sqrt{\gamma^2 p^2 + c^2}} s + \varphi_0.$$

The parameter p identifies the specific characteristic curves and rays. Since $s = t$, we can replace (9.793)–(9.794) by

$$(9.796) \qquad x = \frac{\gamma^2 p}{\sqrt{\gamma^2 p^2 + c^2}} t, \qquad -\infty < p < \infty,$$

and these are the rays for this problem. Also,

$$(9.797) \qquad \varphi = \frac{-c^2}{\sqrt{\gamma^2 p^2 + c^2}} t + \varphi_0.$$

Setting $G(p) = \gamma^2 p / \sqrt{\gamma^2 p^2 + c^2}$ we find that $G'(p) > 0$ for all p, so that $G(p)$ is an increasing function. As $p \to -\infty$, $G(p) \to -\gamma$, and as $p \to \infty$, $G(p) \to \gamma$. Thus

$$(9.798) \qquad -\gamma < \frac{\gamma^2 p}{\sqrt{\gamma^2 p^2 + c^2}} < \gamma, \qquad -\infty < p < \infty.$$

Since $dx/dt = G(p)$ is the ray velocity on a ray $p = $ constant, we see that

$|dx/dt| < \gamma$. Now γ is the maximum speed of disturbances for the Klein–Gordon equation as was seen in Chapter 7. Also, given the dispersion relation $\omega = \omega(\lambda) = \sqrt{\gamma^2\lambda^2 + c^2}$ for the Klein–Gordon equation, the quantity $\omega'(\lambda) = \gamma^2\lambda/\sqrt{\gamma^2\lambda^2 + c^2}$ is known as the group velocity. Identifying p with λ, we find that $G(p) = \omega'(p)$ is the group velocity. Thus the rays (9.796) may be termed group lines, and they are located within the characteristic sector with vertex at the origin and bounded by the characteristics $x = \pm\gamma t$ of the Klein–Gordon equation, as shown in Figure 9.11.

The ray equation (9.796) may be used to express p as a function of x and t. If the result is inserted in (9.797), we obtain

$$(9.799) \qquad \varphi(x, t) = -\frac{c}{\gamma}\sqrt{\gamma^2 t^2 - x^2} + \varphi_0.$$

This shows explicitly that φ is not real valued outside the characteristic sector and that $\varphi = \varphi_0$ on the lines $x = \pm\gamma t$.

Turning to the solution of the transport equations we show that they reduce to ordinary differential equations along the rays. Since $\varphi_x = p$ and $\varphi_t = q$, we have

$$(9.800) \quad 2\frac{\partial\varphi}{\partial t}\frac{\partial a_j}{\partial t} - 2\gamma^2\frac{\partial\varphi}{\partial x}\frac{\partial a_j}{\partial x} = 2q\left[\frac{\partial a_j}{\partial t} - \frac{\gamma^2 p}{q}\frac{\partial a_j}{\partial x}\right] = 2q\frac{da_j}{dt}$$

in view of (9.788). Also, we easily show that

$$(9.801) \qquad \varphi_{tt} - \gamma^2\varphi_{tt} = \frac{q}{t}.$$

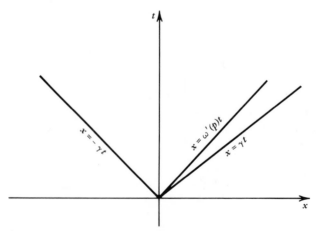

Figure 9.11. The characteristic sector and a ray.

We solve only the equation (9.781) for $a_0(x, t)$, and this becomes

$$(9.802) \qquad q \left[2 \frac{da_0}{dt} + \frac{1}{t} a_0 \right] = 0,$$

along a ray p = constant. Thus the general solution of (9.802) is

$$(9.803) \qquad a_0 = \frac{H(p)}{\sqrt{t}}.$$

Since p depends on x and t in the special form x/t, as follows from (9.796), we find that the general solution of (9.781) is, in fact,

$$(9.804) \qquad a_0(x, t) = \frac{f(x/t)}{\sqrt{t}},$$

where f is an arbitrary function.

To complete the solution of the problem we must add the asymptotic result that corresponds to choosing the plus sign in (9.785). Proceeding as above, we easily show that

$$(9.805) \qquad \varphi(x, t) = \frac{c}{\gamma} \sqrt{\gamma^2 t^2 - x^2} + \hat{\varphi}_0,$$

$$(9.806) \qquad a_0(x, t) = \frac{\hat{f}(x/t)}{\sqrt{t}}$$

in that case, with $\hat{\varphi}_0$ and $\hat{f}$ as yet undetermined. The asymptotic solution of (9.777)–(9.778) is the sum of both the results obtained and to leading order we have

$$(9.807) \qquad u(x, t) \approx \frac{1}{\sqrt{t}} f\left(\frac{x}{t} \right) \exp\left[ik \left(-\frac{c}{\gamma} \sqrt{\gamma^2 t^2 - x^2} + \varphi_0 \right) \right]$$
$$+ \frac{1}{\sqrt{t}} \hat{f}\left(\frac{x}{t} \right) \exp\left[ik \left(\frac{c}{\gamma} \sqrt{\gamma^2 t^2 - x^2} + \hat{\varphi}_0 \right) \right].$$

If we are to specify the unknown functions and constants in (9.807), we must relate $u(x, t)$ to the initial data (9.778). The form of the data has been used only to determine that all rays issue from the origin $(x, t) = (0, 0)$, but no further information about $u(x, 0)$ and $u_t(x, 0)$ has been used. However, since the amplitude terms in (9.807) are singular at $t = 0$, $u(x, t)$ is not valid near the initial line so that no direct matching with the initial data can be carried out. This is consistent with the fact that our results are expected to be valid at large times and at large distances from $x = 0$.

The use of boundary layer methods near the initial line leads to an interesting result. If we simply put $t = (1/k)\tau$ in (9.777) to emphasize the neighborhood of the initial line, it is easily found that the boundary layer result

$$(9.808) \qquad v(x, \tau) = \frac{1}{c}\delta(x)\sin(c\tau)$$

is inadequate for the purpose of matching with (9.807). Since (9.807) is not valid for $|x| > \gamma t$, we see that not only t but also x must be stretched to emphasize the origin $(x, t) = (0, 0)$. The stretching clearly has the form $x = (1/k)\sigma$ and $t = (1/k)\tau$ and is identical with (9.776). Thus the boundary layer problem is just the original problem (9.774)–(9.775), which must now be solved exactly. Although this result appears to raise doubts about the usefulness of our asymptotic approach, it must be emphasized that the asymptotic method also works if the equation has variable coefficients. In that case, the boundary layer problem would reduce to the study of an equation with constant coefficients so that a simplification is achieved.

The exact solution of (9.777)–(9.778) in the region $|x| \leqslant \gamma t$ is found from (7.351) to be

$$(9.809) \qquad u(x, t) = \frac{1}{2\gamma}J_0\left[\frac{ck}{\gamma}\sqrt{\gamma^2 t^2 - x^2}\right].$$

Using the large argument asymptotic form for $J_0(z)$, that is,

$$J_0(z) \approx \sqrt{\frac{2}{\pi z}}\,\sin\left(z + \frac{\pi}{4}\right), \qquad z \to \infty,$$

we find that

$$(9.810)$$

$$u(x, t) \approx \frac{1}{2i}(2\pi\gamma kc)^{-1/2}(\gamma^2 t^2 - x^2)^{-1/4}\exp\left[\frac{ikc}{\gamma}\sqrt{\gamma^2 t^2 - x^2} + i\frac{\pi}{4}\right]$$

$$- \frac{1}{2i}(2\pi\gamma kc)^{-1/2}(\gamma^2 t^2 - x^2)^{-1/4}\exp\left[-\frac{ikc}{\gamma}\sqrt{\gamma^2 t^2 - x^2} - i\frac{\pi}{4}\right].$$

Comparing (9.807) and (9.810) shows that we must set

$$(9.811) \qquad f\left(\frac{x}{t}\right) = \hat{f}\left(\frac{x}{t}\right) = (8\pi\gamma kc)^{-1/2}\left[\gamma^2 - \frac{x^2}{t^2}\right]^{-1/4}$$

$$(9.812) \qquad \varphi_0 = -\hat{\varphi}_0 = \frac{1}{k}\frac{\pi}{4}.$$

It is seen from (9.810) that $u(x, t)$ when expanded asymptotically is not only singular at $(x, t) = (0, 0)$ but also on the characteristics $x = \pm\gamma t$. Thus

(9.807) is valid only in the interior of the characteristic sector. Near $x = \pm \gamma t$, which determine the wave fronts for this problem, the form (9.809) must be used.

9.5. THE PROPAGATION OF DISCONTINUITIES AND SINGULARITIES FOR HYPERBOLIC EQUATIONS

Discontinuities and *singularities* in the data for (linear) *hyperbolic equations* are propagated along *characteristics*, as we have demonstrated on a number of occasions. The behavior of solutions in the neighborhood of characteristics across which discontinuities or other rapid variations of the solution occur can be studied separately. Although the results obtained do not, in general, yield a full description of the solution, they do describe it in a region of interest such as at a wave front. Other approximate methods of the type given in Sections 9.4 and 5.7 may be used to complete the description of the solution in certain other regions of interest.

To begin, we consider the *hyperbolic equation*

$$(9.813) \qquad u_{tt} - \nabla \cdot (p \nabla u) + 2\lambda u_t + qu = 0.$$

This has the general form of the hyperbolic equations considered in Chapter 4 except that we have put $\rho = 1$ in (4.10), as well as $F = 0$, and added a damping term $2\lambda u_t$ where λ is a positive function of the space variables only. The coefficients p and q are also functions of the space variables only and satisfy the conditions given in Section 4.1.

We study the form of the solutions of (9.813) near a *wave front surface* $\varphi(x, y, t) = 0$, and restrict our discussion to two space dimensions. It is assumed that the solution u vanishes on one side of the surface and is a nonzero function that satisfies (9.813) on the other side of the surface. The solution may be represented as

$$(9.814) \qquad u(x, y, t) = v(x, y, t) H[\varphi],$$

where $H[\varphi]$ is the Heaviside function that vanishes when $\varphi < 0$ and equals unity when $\varphi \geqslant 0$. This requires that v be a solution of (9.813) in the region $\varphi > 0$ and that the value of v on $\varphi = 0$ equals the jump in the solution u across the wave front.

To determine what conditions must be placed on φ and v we insert (9.814) into (9.813). This gives, on using $H'[\varphi] = \delta[\varphi]$,

$$(9.815) \quad \delta'[\varphi]\{\varphi_t^2 - p(\nabla\varphi)^2\}v + \delta[\varphi]\{2\varphi_t v_t - 2p\nabla\varphi \cdot \nabla v + 2\lambda\varphi_t v$$
$$- (\nabla\varphi \cdot \nabla p)v + [\varphi_{tt} - p\nabla^2\varphi]v\}$$
$$+ H[\varphi]\{v_{tt} - \nabla \cdot (p\nabla v) + 2\lambda v_t + qv\} = 0.$$

Since v is a solution of (9.813), the coefficient of $H[\varphi]$ in (9.815) must vanish where $H[\varphi] \neq 0$. The delta function $\delta[\varphi]$ and its derivative $\delta'[\varphi]$ both vanish when $\varphi \neq 0$. Since at $\varphi = 0$, $\delta'[\varphi]$ is more singular than $\delta[\varphi]$, we require that their coefficients vanish separately so that (9.815) is satisfied. This implies that $\varphi = 0$ must be a solution of the characteristic equation for (9.813)

$$(9.816) \qquad\qquad \varphi_t^2 - p(\nabla\varphi)^2 = 0.$$

This result is not unexpected since we have already shown that all discontinuities must occur across characteristics.

The vanishing of the coefficient of $\delta[\varphi]$ in (9.815) implies that on the characteristic surface $\varphi(x, y, t) = 0$ the function v must satisfy

$$(9.817) \quad 2\varphi_t v_t - 2p\nabla\varphi \cdot \nabla v + 2\lambda\varphi_t v + \left[\varphi_{tt} - \nabla \cdot (p\nabla\varphi) \right] v = 0.$$

The appropriate solution $\varphi(x, y, t) = 0$ of (9.816) may be determined by the method of characteristics. The relevant equations are given in Exercise 2.4.7. The initial conditions for determining the requisite characteristic function φ must be found from the data for the hyperbolic equation (9.813). Then (9.817) can be solved as we demonstrate. Again the conditions on v in (9.817) come from the data for (9.813).

We now apply the method of characteristics to (9.816). Let $\alpha = \varphi_t$, $\beta = \varphi_x$, and $\gamma = \varphi_y$. Then we have

$$(9.818) \qquad F(x, y, \alpha, \beta, \gamma) = \alpha^2 - p(x, y)(\beta^2 + \gamma^2) = 0.$$

The characteristic curves for (9.816) are called *bicharacteristics* to distinguish them from the characteristics of the hyperbolic equation (9.813). With $x(s)$, $y(s)$, $t(s)$, $\varphi(s)$, $\alpha(s)$, $\beta(s)$, and $\gamma(s)$ representing the bicharacteristics of (9.813) we have the equations (see Exercise 2.4.7)

$$(9.819) \qquad \frac{dx}{ds} = -2p\beta; \qquad \frac{dy}{ds} = -2p\gamma; \qquad \frac{dt}{ds} = 2\alpha;$$

$$(9.820) \qquad \frac{d\varphi}{ds} = 2\left[\alpha^2 - p(\beta^2 + \gamma^2)\right] = 0;$$

$$(9.821) \qquad \frac{d\alpha}{ds} = 0; \qquad \frac{d\beta}{ds} = p_x\beta^2; \qquad \frac{d\gamma}{ds} = p_y\gamma^2$$

where $d\varphi/ds = 0$ since $F = 0$ along the bicharacteristics.

From (9.820) we see that $\varphi = $ constant on the bicharacteristics. Thus if initially $\varphi = 0$, we conclude that $\varphi(x, y, t) = 0$ is the resulting characteristic surface. Also, (9.821) shows that $\alpha = $ constant along the bicharacteristics. Then (9.819) indicates that the parameter s can be replaced by the time parameter t along the bicharacteristics, with $dt = 2\alpha\, ds$. Thus (9.819) is replaced by

$$(9.822) \qquad\qquad \frac{dx}{dt} = -\frac{p\beta}{\alpha}; \qquad \frac{dy}{dt} = -\frac{p\gamma}{\alpha};$$

and a similar transformation can be carried out for (9.820) and (9.821).

The bicharacteristics can be used to reduce (9.817) to an ordinary differential equation. Given the (bicharacteristic) curve $x = x(t)$, $y = y(t)$, $t = t$ on the characteristic surface $\varphi(x, y, t) = 0$, we have

$$(9.823) \quad 2\varphi_t v_t - 2p\varphi_x v_x - 2p\varphi_y v_y + 2\lambda\varphi_t v$$
$$= 2\alpha v_t - 2p\beta v_x - 2p\gamma v_y + 2\lambda\alpha v$$
$$= 2\alpha\left[v_t - \frac{p\beta}{\alpha}v_x - \frac{p\gamma}{\alpha}v_y + \lambda v\right] = 2\alpha\left[\frac{dv}{dt} + \lambda v\right],$$

in view of (9.822), with $v = v[x(t), y(t), t]$. Thus (9.817) reduces to

$$(9.824) \qquad \frac{dv}{dt} + \lambda v + \frac{1}{2\alpha}\left[\varphi_{tt} - \nabla \cdot (p\nabla\varphi)\right]v = 0$$

along the bicharacteristics. We recall that $\alpha = $ constant on the bicharacteristics.

In the special case that $p = $ constant and $\varphi = 0$ is the plane wave front

$$(9.825) \qquad \varphi = pt - x\cos\theta - y\sin\theta - c = 0$$

where θ and c are constants, (9.824) becomes

$$(9.826) \qquad \frac{dv}{dt} + \lambda v = 0.$$

Since $\lambda > 0$, this implies that v—when evaluated on the characteristic—decays (exponentially) with increasing time. As a result, the presence of the damping term $2\lambda u_t$ in (9.813) has the effect of smoothing out the discontinuity (9.814).

Although this result only describes the variation of the solution within the region of discontinuity, it is possible to extend it to obtain an expression valid near the discontinuity region as well. To do so we consider the characteristic surfaces $\varphi = $ constant, which are parallel to the characteristic surface $\varphi = 0$. For small values of φ it is possible to expand $v(x, y, t)$ in a Taylor series of the form

$$(9.827) \qquad v(x, y, t) = \sum_{j=0}^{\infty} v_j \frac{\varphi^j}{j!}$$

where the v_j are evaluated on the characteristic $\varphi = 0$. Effectively, we are introducing two coordinates on the characteristic surface $\varphi = 0$ and using φ itself as the third coordinate. We insert (9.827) into (9.814) and obtain

$$(9.828) \qquad u(x, y, t) = \left(\sum_{j=0}^{\infty} v_j \frac{\varphi^j}{j!}\right) H[\varphi].$$

The generalized functions

(9.829) $$H_j(x) = \left[\frac{x^j}{j!}\right] H(x), \qquad j = 0, 1, 2 \ldots,$$

where $H(x)$ is the Heaviside function, have the property that

(9.830) $$H_j'(x) = H_{j-1}(x), \qquad j \geq 1,$$

as we now demonstrate. Let $\theta(x)$ be a test function (see Section 7.2) and then

(9.831) $$\left(H_j(x), \theta(x)\right) = \int_0^\infty \frac{1}{j!} x^j \theta(x)\, dx.$$

The derivative of $H_j(x)$ is defined as

(9.832) $$\left(H_j'(x), \theta(x)\right) = -\left(H_j(x), \theta'(x)\right) = -\frac{1}{j!} \int_0^\infty x^j \theta'(x)\, dx$$

$$= -\frac{1}{j!} x^j \theta(x)\Big|_0^\infty + \frac{1}{(j-1)!} \int_0^\infty x^{j-1} \theta(x)\, dx$$

$$= \left(H_{j-1}(x), \theta(x)\right); \qquad j \geq 1,$$

since $\theta(x)$ vanishes identically for sufficiently large x. Thus (9.830) is verified for $j \geq 1$. With $j = 0$, $H_0(x) = H(x)$ and $H_0'(x) = \delta(x)$.

Thus we can write (9.828) as

(9.833) $$u(x, y, t) = \sum_{j=0}^\infty v_j H_j[\varphi].$$

Inserting (9.833) into (9.813) and using (9.830), we again find that $\delta'[\varphi]$ is the most singular term and this implies that φ must satisfy the characteristic equation (9.816). The remaining terms are

(9.834)

$$\delta[\varphi]\left\{2\frac{\partial \varphi}{\partial t}\frac{\partial v_0}{\partial t} - 2p\nabla\varphi \cdot \nabla v_0 + 2\lambda\frac{\partial \varphi}{\partial t} v_0 + \left[\varphi_{tt} - \nabla \cdot (p\nabla\varphi)\right] v_0\right\}$$

$$+ \sum_{j=1}^\infty H_{j-1}[\varphi]\left\{2\frac{\partial \varphi}{\partial t}\frac{\partial v_j}{\partial t} - 2p\nabla\varphi \cdot \nabla v_j + 2\lambda\frac{\partial \varphi}{\partial t} v_j\right.$$

$$+ \left[\varphi_{tt} - \varphi \cdot (p\nabla\varphi)\right] v_j + \frac{\partial^2 v_{j-1}}{\partial t^2} - \nabla \cdot (p\nabla v_{j-1})$$

$$\left. + 2\lambda\frac{\partial v_{j-1}}{\partial t} + q v_{j-1}\right\} = 0.$$

Equating the coefficients of $\delta[\varphi]$ and $H_j[\varphi]$ to zero yields the system of *transport equations*

(9.835)
$$2\frac{\partial \varphi}{\partial t} \frac{\partial v_0}{\partial t} - 2p\nabla\varphi \cdot \nabla v_0 + 2\lambda\frac{\partial \varphi}{\partial t}v_0$$
$$+ \left[\varphi_{tt} - \nabla \cdot (p\nabla\varphi)\right]v_0 = 0,$$

(9.836)
$$2\frac{\partial \varphi}{\partial t} \frac{\partial v_j}{\partial t} - 2p\nabla\varphi \cdot \nabla v_j + 2\lambda\frac{\partial \varphi}{\partial t}v_j + \left[\varphi_{tt} - \nabla \cdot (p\nabla\varphi)\right]v_j$$
$$= -\frac{\partial^2 v_{j-1}}{\partial t^2} + \nabla \cdot (p\nabla v_{j-1}) - 2\lambda\frac{\partial v_{j-1}}{\partial t} - qv_{j-1}; \qquad j \geq 1.$$

In view of (9.823), each of the transport equations can be expressed as an ordinary differential equation along the bicharacteristics.

This problem can be simplified somewhat if the surface $\varphi = 0$ is expressed in the form

(9.837)
$$\varphi(x, y, t) = t - \psi(x, y) = 0.$$

Then the curves $\psi(x, y) = $ constant represent the position of the wave fronts at different times t. The functions v_j in (9.827) that are evaluated on the characteristic may be taken to be functions of x and y. Thus (9.833) becomes

(9.838)
$$u(x, y, t) = \sum_{j=0}^{\infty} v_j(x, y)H_j[t - \psi(x, y)].$$

The characteristic equation (9.816) becomes

(9.839)
$$(\nabla\psi)^2 = \frac{1}{p},$$

and this is essentially the *eiconal equation*. The transport equations (9.835) and (9.836) take the form

(9.840)
$$2p\nabla\psi \cdot \nabla v_0 + 2\lambda v_0 + \nabla \cdot (p\nabla\psi)v_0 = 0$$

and

(9.841)
$$2p\nabla\psi \cdot \nabla v_j + 2\lambda v_j + \nabla \cdot (p\nabla\psi)v_j$$
$$= \nabla \cdot (p\nabla v_{j-1}) - qv_{j-1}; \qquad j \geq 1.$$

Apart from the term $2\lambda v_j$ in (9.840) and (9.841), these equations are similar to the transport equations obtained in Section 9.4. If $\lambda = 0$ in (9.840) and we multiply through by v_0 we can write the equation as

$$\nabla \cdot (p\nabla\psi v_0^2) = 0.$$

By integrating along the rays determined from (9.839) as was done in the preceding section, the variation of v_0 along the rays can be found. With $\lambda > 0$, the medium is dissipative and v_0 depends not only on the convergence or divergence of the rays but undergoes exponential decay along each ray.

The expressions (9.833) or (9.838) are to be interpreted as weak solutions of (9.813) since they involve generalized functions. To see how initial conditions are to be assigned for the bicharacteristic or ray equations and for the transport equations for the v_j, we consider the one-dimensional telegrapher's equation in Example 9.20. The results that can be obtained for various problems involving that equation suggest how the general problem ought to be treated. A method for the discussion of the propagation of singularities other than jump discontinuities for (9.813) is also given.

EXAMPLE 9.20.   SINGULAR SOLUTIONS OF THE TELEGRAPHER'S EQUATION

We begin by considering an initial and boundary value problem for the *telegrapher's equation*

$$(9.842) \qquad u_{tt} - \gamma^2 u_{xx} + 2\lambda u_t = 0; \qquad x > -a - ct, t > 0,$$

where γ, λ, a, and c are positive constants with $c < \gamma$. The initial data are

$$(9.843) \qquad u(x,0) = H(x); \qquad u_t(x,0) = \gamma\delta(x); \qquad x > -a$$

and on the *moving boundary* $x = -a - ct$, $u(x, t)$ satisfies

$$(9.844) \qquad\qquad u(-a - ct, t) = 0, \qquad t > 0.$$

We require that $c < \gamma$ so that the boundary line is time-like (see Section 6.5) and, therefore, only one condition need be assigned on it.

In the absence of the damping term $2\lambda u_t$, (i.e., if $\lambda = 0$), the solution of (9.842)–(9.844) is

$$(9.845) \qquad u(x, t) = H(x + \gamma t) - H\left[\frac{c - \gamma}{c + \gamma}(x - \gamma t) - \frac{2a\gamma}{c + \gamma}\right],$$

as is easily verified. The second term in (9.845) is the wave reflected from the boundary line so that the singularity in the initial data gives rise to an *incident* and *reflected wave*.

With $\lambda > 0$ in (9.842), we consider a solution $u(x, t)$ expanded in the form (9.833) and given as

$$(9.846) \qquad\qquad u(x, t) = \sum_{j=0}^{\infty} v_j(x, t) H_j[\varphi(x, t)].$$

Inserting (9.846) into (9.842) and equating coefficients of the $H_j[\varphi]$, $\delta[\varphi]$, and $\delta'[\varphi]$ to zero yields the *characteristic equation*

$$(9.847) \qquad \varphi_t^2 - \gamma^2 \varphi_t^2 = 0,$$

and the *transport equations*

$$(9.848) \qquad 2\varphi_t \frac{\partial v_0}{\partial t} - 2\gamma^2 \varphi_x \frac{\partial v_0}{\partial x} + 2\lambda \varphi_t v_0 + \left[\varphi_{tt} - \gamma^2 \varphi_{xx} \right] v_0 = 0,$$

$$(9.849) \qquad 2\varphi_t \frac{\partial v_j}{\partial t} - 2\gamma^2 \varphi_x \frac{\partial v_j}{\partial x} + 2\lambda \varphi_t v_j + \left[\varphi_{tt} - \gamma^2 \varphi_{xx} \right] v_j$$

$$= -\frac{\partial^2 v_{j-1}}{\partial t^2} + \gamma^2 \frac{\partial^2 v_{j-1}}{\partial x^2} - 2\lambda \frac{\partial v_{j-1}}{\partial t}, \qquad j \geq 1.$$

Since it takes a finite time for the disturbance resulting from the initial data to reach the boundary, we first use (9.846) only to account for the initial conditions. Afterwards, we deal with the boundary conditions. In addition, we note that (9.846) is a formal solution of (9.842) if (9.847)–(9.849) are satisfied. That is, we do not require that $\varphi(x, t) = 0$ but we can consider any characteristic $\varphi(x, t) = $ constant. Thus even though the results are expected to be useful and meaningful with $v_j(x, t)$ evaluated on the characteristic $\varphi = 0$, we apply the initial data (9.843) directly to (9.846) so that we are effectively dealing with the whole family of characteristic lines $\varphi(x, t) = $ constant.

At $t = 0$ we have

$$(9.850) \qquad u(x,0) = H(x) = \sum_{j=0}^{\infty} v_j(x,0) H_j[\varphi(x,0)],$$

$$(9.851) \qquad u_t(x,0) = \gamma\delta(x) = \sum_{j=0}^{\infty} \left\{ v_j(x,0)\varphi_t(x,0) H_{j-1}[\varphi(x,0)] \right.$$

$$\left. + \frac{\partial v_j(x,0)}{\partial t} H_j[\varphi(x,0)] \right\}.$$

On comparing terms in these equations, we find that $\varphi(x,0)$ should be chosen as

$$(9.852) \qquad \varphi(x,0) = x.$$

Since the solutions of (9.847) are $\varphi = x \pm \gamma t = $ constant, we conclude that there are two possible solutions of (9.847) and (9.852):

$$(9.853) \qquad \tilde{\varphi}(x, t) = x - \gamma t; \qquad \hat{\varphi}(x, t) = x + \gamma t.$$

Since there is no basis for selecting one of these characteristics above the other, we must use both of them in our result.

Thus (9.846) must be replaced by

$$(9.854) \qquad u(x, t) = \sum_{j=0}^{\infty} \tilde{v}_j(x, t) H_j[\tilde{\varphi}] + \sum_{j=0}^{\infty} \hat{v}_j(x, t) H_j[\hat{\varphi}].$$

Consequently, we are led to consider two solutions valid near the characteristic lines $\tilde{\varphi}(x, t) = 0$ and $\hat{\varphi}(x, t) = 0$, each of which issues from the singular data point $(x, t) = (0, 0)$.

If (9.854) is inserted into (9.842), we find that $\tilde{v}_j$ and $\hat{v}_j$ each satisfy the transport equations (9.848)–(9.849) with φ replaced by $\tilde{\varphi}$ and $\hat{\varphi}$, respectively. We determine only leading terms $\tilde{v}_0$, $\hat{v}_0$, $\tilde{v}_1$, and $\hat{v}_1$. Inserting (9.854) into the initial data—as in (9.850)–(9.851)—we obtain

$$(9.855)$$
$$H[x] = \left[\tilde{v}_0(x,0) + \hat{v}_0(x,0)\right] H_0(x) + \left[\tilde{v}_1(x,0) + \hat{v}_1(x,0)\right] H_1(x) + \cdots$$

and

$$(9.856) \qquad \gamma\delta[x] = \left[-\gamma\tilde{v}_0(x,0) + \gamma\hat{v}_0(x,0)\right] H_{-1}(x)$$
$$+ \left[-\gamma\tilde{v}_1(x,0) + \gamma\hat{v}_1(x,0) + \frac{\partial\tilde{v}_0(x,0)}{\partial t}\right.$$
$$\left. + \frac{\partial\hat{v}_0(x,0)}{\partial t}\right] H_0(x) + \cdots .$$

Since $H_0(x) = H(x)$ and $H_{-1}(x) = H'(x) = \delta(x)$, equating coefficients in (9.855)–(9.856) gives

$$(9.857) \qquad \tilde{v}_0(x,0) + \hat{v}_0(x,0) = 1; \qquad -\tilde{v}_0(x,0) + \hat{v}_0(x,0) = 1;$$

and this yields

$$(9.858) \qquad\qquad \tilde{v}_0(x,0) = 0; \qquad \hat{v}_0(x,0) = 1.$$

Clearly, the solution of the transport equation for $\tilde{v}_0(x, t)$ is $\tilde{v}_0(x, t) = 0$. The equation for $\hat{v}_0(x, t)$ is

$$(9.859) \qquad\qquad 2\gamma\frac{\partial\hat{v}_0}{\partial t} - 2\gamma^2\frac{\partial\hat{v}_0}{\partial x} + 2\lambda\gamma\hat{v}_0 = 0$$

with the initial condition (9.858). Using the method of characteristics, or otherwise, we obtain

(9.860) $\hat{v}_0(x, t) = e^{-\lambda t}.$

Then the equations for the data $\tilde{v}_1(x, 0)$ and $\hat{v}_1(x, 0)$ are found from (9.855)–(9.856) to be

(9.861) $\tilde{v}_1(x, 0) + \hat{v}_1(x, 0) = 0; \qquad -\tilde{v}_1(x, 0) + \hat{v}_1(x, 0) = \dfrac{\lambda}{\gamma};$

so that

(9.862) $\tilde{v}_1(x, 0) = -\dfrac{\lambda}{2\gamma}; \qquad \hat{v}_1(x, 0) = \dfrac{\lambda}{2\gamma}.$

Since $\tilde{v}_0(x, t) = 0$, the transport equation for $\tilde{v}_1(x, t)$ is

(9.863) $-2\gamma \dfrac{\partial \tilde{v}_1}{\partial t} - 2\gamma^2 \dfrac{\partial \tilde{v}_1}{\partial x} - 2\lambda \gamma \tilde{v}_1 = 0.$

For $\hat{v}_1(x, t)$ we have

(9.864) $2\gamma \dfrac{\partial \hat{v}_1}{\partial t} - 2\gamma^2 \dfrac{\partial \hat{v}_1}{\partial x} + 2\lambda \gamma \hat{v}_1 = \lambda^2 e^{-\lambda t}.$

Using (9.862) we easily find that

(9.865) $\tilde{v}_1(x, t) = -\dfrac{\lambda}{2\gamma} e^{-\lambda t}; \qquad \hat{v}_1(x, t) = \left[\dfrac{\lambda}{2\gamma} + \dfrac{\lambda^2 t}{2\gamma} \right] e^{-\lambda t}.$

We have, therefore, found for small t,

(9.866) $u(x, t) = e^{-\lambda t} H[x + \gamma t] + \left[\dfrac{\lambda}{2\gamma} + \dfrac{\lambda^2 t}{2\gamma} \right] e^{-\lambda t} [x + \gamma t] H[x + \gamma t]$

$$-\frac{\lambda}{2\gamma} e^{-\lambda t} [x - \gamma t] H[x - \gamma t] + \cdots .$$

This shows that the jump in $u(x, t)$ across $\hat{\varphi} = x + \gamma t = 0$ equals $e^{-\lambda t}$, whereas there is no jump in $u(x, t)$ across $\tilde{\varphi} = x - \gamma t = 0$.

We now introduce the boundary condition (9.844) into our discussion. The solution found in (9.866) is evaluated on the boundary line and this gives

$$(9.867) \quad u(-a - ct, t) = e^{-\lambda t} H[-a + (\gamma - c)t]$$

$$+ \left[\frac{\lambda}{2\gamma} + \frac{\lambda^2 t}{2\gamma} \right] e^{-\lambda t} H_1[-a + (\gamma - c)t]$$

$$- \frac{\lambda}{2\gamma} e^{-\lambda t} H_1[-a - (\gamma + c)t] + \cdots.$$

Now on $x = -a - ct$,

$$(9.868) \qquad H_j[\tilde{\varphi}] = H_j[-a - (\gamma + c)t] = 0, \qquad t > 0$$

since $H_j[x]$ vanishes for $x < 0$. Also, on $x = -a - ct$,

$$(9.869) \qquad H_j[\hat{\varphi}] = H_j[-a + (\gamma - c)t] = 0; \qquad t < \frac{a}{\gamma - c}.$$

Therefore, the terms $H_j[\tilde{\varphi}]$ that represent *right traveling waves* or disturbances do not play a role in the boundary condition since these waves do not intersect the boundary line. The terms $H_j[\hat{\varphi}]$ represent *left traveling waves* or disturbances that do not reach the boundary until the time $t = a/(\gamma - c) > 0$. Thus (9.866) remains valid until that time.

If $t > a/(\gamma - c)$, the solution (9.866) must be modified. We express it as

$$(9.870) \qquad u(x, t) = U(x, t) + \sum_{j=0}^{\infty} \tilde{v}_j H_j[\tilde{\varphi}] + \sum_{j=0}^{\infty} \hat{v}_j H_j[\hat{\varphi}]$$

where $U(x, t)$ is to be specified and the two series are as previously given in terms of the initial data. The boundary condition (9.844) implies that

$$(9.871) \quad U(-a - ct, t) = - \sum_{j=0}^{\infty} \hat{v}_j(-a - ct, t) H_j[-a + (\gamma - c)t],$$

in view of (9.868). Since $u(x, t)$ is given by (9.866) in the region $t < a/(\gamma - c)$, we conclude that $U(x, t)$ vanishes identically in that region.

The boundary value (9.871) for $U(x, t)$ suggests that we expand $U(x, t)$ as

$$(9.872) \qquad\qquad U(x, t) = \sum_{j=0}^{\infty} v_j(x, t) H_j[\varphi].$$

Since U must satisfy (9.842), φ must be a solution of the characteristic equation (9.847) and the v_j must satisfy the transport equations (9.848)–(9.849). From

(9.871) we determine that

$$(9.873) \qquad \varphi(-a - ct, t) = -a + (\gamma - c)t$$

and

$$(9.874) \qquad v_j(-a - ct, t) = -\hat{v}_j(-a - ct, t),$$

with the $\hat{v}_j$ having been specified in terms of the initial data.

Of the two solutions of the characteristic equation (9.847) that satisfy (9.873), the appropriate one is

$$(9.875) \qquad \varphi(x, t) = \frac{c - \gamma}{c + \gamma}(x - \gamma t) - \frac{2a\gamma}{c + \gamma}.$$

The second solution contains characteristic lines that intersect the x-axis in the region $x > -a$. Since $U(x, 0) = 0$ in that region, the $v_j(x, t)$ corresponding to that solution must be chosen to vanish.

With $\varphi(x, t)$ given by (9.875), $v_0(x, t)$ satisfies

$$(9.876) \qquad \frac{\partial v_0}{\partial t} + \gamma \frac{\partial v_0}{\partial x} + \lambda v_0 = 0.$$

Since $\hat{v}_0(x, t) = e^{-\lambda t}$, we easily conclude from (9.874) that

$$(9.877) \qquad v_0(x, t) = -e^{-\lambda t}.$$

The equation for $v_1(x, t)$ is

$$(9.878) \qquad \frac{\partial v_1}{\partial t} + \gamma \frac{\partial v_1}{\partial x} + \lambda v_1 = \frac{\lambda^2}{2\gamma}\left(\frac{c + \gamma}{c - \gamma}\right)e^{-\lambda t},$$

and the boundary condition is

$$(9.879) \qquad v_1(-a - ct, t) = -\left[\frac{\lambda}{2\gamma} + \frac{\lambda^2 t}{2\gamma}\right]e^{-\lambda t}.$$

We readily find that

$$(9.880) \qquad v_1(x, t) = \left[\frac{\lambda^2 c}{\gamma(c^2 - \gamma^2)}(x - \gamma t + a) - \frac{\lambda}{2a}\right]e^{-\lambda t}$$

$$+ \frac{\lambda^2}{2\gamma}\left(\frac{c + \gamma}{c - \gamma}\right)te^{-\lambda t}.$$

Combining these results we find the leading terms of our solution to be

(9.881)

$$u(x, t) = e^{-\lambda t} H[x + \gamma t] - e^{-\lambda t} H\left[\frac{c - \gamma}{c + \gamma}(x - \gamma t) - \frac{2a\gamma}{c + \gamma}\right]$$

$$+ \left[\frac{\lambda}{2\gamma} + \frac{\lambda^2 t}{2\gamma}\right] e^{-\lambda t}[x + \gamma t] H[x + \gamma t]$$

$$- \frac{\lambda}{2\gamma} e^{-\lambda t}[x - \gamma t] H[x - \gamma t]$$

$$+ \left[-\frac{\lambda}{2a} + \frac{\lambda^2 c}{\gamma(c^2 - \gamma^2)}(x - \gamma t + a) + \left(\frac{c + \gamma}{c - \gamma}\right)\frac{\lambda^2 t}{2\gamma}\right] e^{-\lambda t}$$

$$\times \left[\frac{c - \gamma}{c + \gamma}(x - \gamma t) - \frac{2a\gamma}{c + \gamma}\right] H\left[\frac{c - \gamma}{c + \gamma}(x - \gamma t) - \frac{2a\gamma}{c + \gamma}\right] + \cdots.$$

We now bring (9.881) into a form where each coefficient of H_0 and H_1 is evaluated on the appropriate characteristic. To do so we express $v_1(x, t)$—see (9.880)—as

$$(9.882) \quad v_1(x, t) = \left[\frac{\lambda^2 ac}{\gamma(c - \gamma)^2} - \frac{\lambda}{2a}\right] e^{-\lambda t} + \frac{\lambda^2}{2\gamma}\left(\frac{c + \gamma}{c - \gamma}\right) t e^{-\lambda t}$$

$$+ \frac{\lambda^2 c}{\gamma(c - \gamma)^2} \varphi(x, t) e^{-\lambda t}$$

where φ is defined in (9.875). Then (9.881) can be written as

$$(9.883) \quad u(x, t) = e^{-\lambda t} H[x + \gamma t] - e^{-\lambda t} H\left[\frac{c - \gamma}{c + \gamma}(x - \gamma t) - \frac{2a\gamma}{c + \gamma}\right]$$

$$+ \left[\frac{\lambda}{2\gamma} + \frac{\lambda^2 t}{2\gamma}\right] e^{-\lambda t} H_1[x + \gamma t] - \frac{\lambda}{2\gamma} e^{-\lambda t} H_1[x - \gamma t]$$

$$+ \left[\frac{\lambda^2 ac}{\gamma(c - \gamma)^2} - \frac{\lambda}{2a} + \frac{\lambda^2 t}{2\gamma}\left(\frac{c + \gamma}{c - \gamma}\right)\right]$$

$$\times e^{-\lambda t} H_1\left[\frac{c - \gamma}{c + \gamma}(x - \gamma t) - \frac{2a\gamma}{c + \gamma}\right] + \cdots.$$

In contrast to (9.845), which gives the full (weak) solution of (9.842)–(9.844) in the case where $\lambda = 0$, the expression (9.883) is only valid near the three characteristics that occur in the expansion of $u(x, t)$. We note that (9.883) reduces to (9.845) when $\lambda = 0$, but with $\lambda > 0$ there are additional waves of the form $H_j(x - \gamma t)$ that have no counterpart in (9.845). Each term in (9.883) is multiplied by an exponential decay factor $e^{-\lambda t}$. The jumps in $u(x, t)$ across $x + \gamma t = 0$ and $[(c - \gamma)/(c + \gamma)](x - \gamma t) - [2a\gamma/(c + \gamma)] = 0$ are $e^{-\lambda t}$ and $-e^{-\lambda t}$, respectively. The jump across the third characteristic curve $x - \gamma t = 0$ is seen to be zero. In contrast to the case $\lambda = 0$ where the jumps are $+1$ or -1, we see that these jumps decay exponentially in the damped case where $\lambda > 0$.

If the characteristics $\tilde{\varphi} = 0$, $\hat{\varphi} = 0$, and $\varphi = 0$ are each expressed in a form $t = \psi(x)$, the points at which $\psi(x) = $ constant are the wave fronts for the problem (9.842)–(9.844). The aforementioned jumps are, in fact, the jumps in the solution across these wave fronts at different times t. The expansion (9.883) not only determines the jump in $u(x, t)$ across the wave fronts but also describes the wake behind these fronts. As indicated in Example 5.14, for instance, what remains in the wake of the wave fronts at large times is a diffusion effect. This conclusion cannot be reached, however, by considering only the leading terms in the given expansion.

Adopting a slightly different point of view we express the solution of (9.842)–(9.844) as

(9.884)

$$u(x, t) = e^{-\lambda t}H[x + \gamma t] - e^{-\lambda t}H\left[\frac{c - \gamma}{c + \gamma}(x - \gamma t) - \frac{2a\gamma}{c + \gamma}\right] + v(x, t)$$

where we have retained the two leading terms in (9.883). If we insert (9.884) into (9.842), we obtain

(9.885)

$$v_{tt} - \gamma^2 v_{xx} + 2\lambda v_t = \lambda^2 e^{-\lambda t}H[x + \gamma t] + \lambda^2 e^{-\lambda t}H\left[\frac{c - \gamma}{c + \gamma}(x - \gamma t) - \frac{2a\gamma}{c + \gamma}\right].$$

At $t = 0$ (9.843) implies

(9.886) $v(x, 0) = 0;$ $v_t(x, 0) = \lambda H[x];$ $x > -a;$

and on the boundary $x = -a - ct$,

(9.887) $v(-a - ct, t) = 0;$ $t > 0.$

The data for $v(x, t)$ are less singular than those for $u(x, t)$. Thus $v(x, t)$ is expected to be a smoother function than $u(x, t)$ since it is the solution of a problem with smoother data. If additional terms from the series (9.883) are included in (9.884), the resulting function $v(x, t)$ becomes even smoother. Thus this method may be used to take into account singularities in the solution and to reduce the problem to one that has smooth data and a smooth solution.

In the preceding problem the data (9.843) were expressed directly in terms of $H(x)$ and $\delta(x)$. If the data are given in terms of a discontinuous function

$F(x)$ defined as

(9.888)
$$F(x) = \begin{cases} f(x); & x < 0 \\ g(x); & x > 0 \end{cases},$$

and $f(x)$ can be defined or extended as a smooth function for all x, we can write

(9.889)
$$F(x) = f(x) + H(x)[g(x) - f(x)].$$

Each jump in the function $F(x)$ can be accounted for by a Heaviside function in this manner. For each Heaviside function we must introduce an expansion of the form (9.846). Also, derivatives of functions with jump discontinuities can be expressed in terms of the Dirac delta function as was shown in Example 7.1.

As a specific example we consider the initial value problem for the *telegrapher's equation*

(9.890) $$u_{tt} - \gamma^2 u_{xx} + 2\lambda u_t = 0, \qquad -\infty < x < \infty, t > 0$$

with the initial data

(9.891) $$u(x, 0) = \begin{cases} -1; & x < -1 \\ \sin x; & -1 < x < 1; \\ 1; & x > 1 \end{cases} \qquad u_t(x, 0) = 0.$$

The function $u(x, 0)$ can be written as

(9.892) $$u(x, 0) = -1 + [1 + \sin x] H[x + 1] + [1 - \sin x] H[x - 1],$$

as is easily checked.

To account for the jumps in the data at $x = 1$ and $x = -1$ we introduce the expansions

(9.893) $$U(x, t) = \sum_{j=0}^{\infty} v_j(x, t) H_j[\varphi(x, t)] + \sum_{j=0}^{\infty} w_j(x, t) H_j[\psi(x, t)].$$

Inserting (9.893) into (9.890), we conclude that both φ and ψ must satisfy the characteristic equation (9.847) and the v_j and w_j satisfy the (appropriate) transport equations (9.848)–(9.849). At $t = 0$,

(9.894) $$U(x, 0) = \sum_{j=0}^{\infty} v_j(x, 0) H_j[\varphi(x, 0)] + \sum_{j=0}^{\infty} w_j(x, 0) H_j[\psi(x, 0)],$$

and this can account only for the singular terms in the data (9.892). This implies that we should set

$$(9.895) \qquad \varphi(x,0) = x + 1; \qquad \psi(x,0) = x - 1.$$

Again we are led to consider both solutions of the characteristic equation (9.847) that satisfy (9.895) and they are

$$(9.896) \qquad \tilde{\varphi}(x,t) = x + 1 - \gamma t; \qquad \hat{\varphi}(x,t) = x + 1 + \gamma t;$$
$$(9.897) \qquad \tilde{\psi}(x,t) = x - 1 - \gamma t; \qquad \hat{\psi}(x,t) = x - 1 + \gamma t.$$

For each characteristic line $\tilde{\varphi} = \hat{\varphi} = \tilde{\psi} = \hat{\psi} = 0$ we introduce a series of the form (9.846), and thus we replace (9.893) by

$$(9.898)$$

$$U(x,t) = \sum_{j=0}^{\infty} \tilde{v}_j(x,t) H(x + 1 - \gamma t) + \sum_{j=0}^{\infty} \hat{v}_j(x,t) H(x + 1 + \gamma t)$$

$$+ \sum_{j=0}^{\infty} \tilde{w}_j(x,t) H(x - 1 - \gamma t) + \sum_{j=0}^{\infty} \hat{w}_j(x,t) H(x - 1 + \gamma t).$$

Considering only the leading terms we have

$$(9.899) \quad U(x,0) = \left[\tilde{v}_0(x,0) + \hat{v}_0(x,0)\right] H[x + 1]$$
$$+ \left[\tilde{w}_0(x,0) + \hat{w}_0(x,0)\right] H[x - 1] + \cdots$$
$$= [1 + \sin x] H[x + 1] + [1 - \sin x] H[x - 1],$$
$$(9.900) \quad U_t(x,0) = \left[-\gamma\tilde{v}_0(x,0) + \gamma\hat{v}_0(x,0)\right] \delta[x + 1]$$
$$+ \left[-\gamma\tilde{w}_0(x,0) + \gamma\hat{w}_0(x,0)\right] \delta[x - 1] + \cdots = 0,$$

since we are only concerned with the singular part of the data. Equating coefficients of like terms in (9.899)–(9.900) and solving the resulting equations yields

$$(9.901) \qquad \tilde{v}_0(x,0) = \hat{v}_0(x,0) = \tfrac{1}{2}[1 + \sin x];$$
$$(9.902) \qquad \tilde{w}_0(x,0) = \hat{w}_0(x,0) = \tfrac{1}{2}[1 - \sin x].$$

The transport equations for $\tilde{v}_0$, $\hat{v}_0$, $\tilde{w}_0$, and $\hat{w}_0$ are easily solved, and we obtain, on using (9.901)–(9.902),

$$(9.903) \quad U(x,t) = \tfrac{1}{2}[1 + \sin(x - \gamma t)] e^{-\lambda t} H[x + 1 - \gamma t]$$
$$+ \tfrac{1}{2}[1 + \sin(x + \gamma t)] e^{-\lambda t} H[x + 1 + \gamma t]$$
$$+ \tfrac{1}{2}[1 - \sin(x - \gamma t)] e^{-\lambda t} H[x - 1 - \gamma t]$$
$$+ \tfrac{1}{2}[1 - \sin(x + \gamma t)] e^{-\lambda t} H[x - 1 + \gamma t] + \cdots.$$

The coefficients of the Heaviside functions in (9.903) are not evaluated on the relevant characteristics $x + 1 \pm \gamma t = 0$ or $x - 1 \pm \gamma t = 0$. Although our original discussion of these expansion forms required that the transport coefficients be evaluated on the characteristics, we showed in the preceding that these expansions are formal solutions even if the coefficients are not evaluated on the characteristics. However, each of the coefficients in (9.903) can be expanded in powers of the relevant characteristics. For instance,

$$(9.904) \quad \tilde{v}_0(x, t) = \tfrac{1}{2} e^{-\lambda t} [1 + \sin(x - \gamma t)]$$
$$= \tfrac{1}{2} e^{-\lambda t} \{[1 - \sin 1] - \tfrac{1}{2} [\cos 1][x + 1 - \gamma t] + \cdots \}$$

and similar expansions can be carried out for $\hat{v}_0$, $\tilde{w}_0$, and $\hat{w}_0$. If this is done, we obtain to leading terms.

$$(9.905) \quad U(x, t) = \tfrac{1}{2} [1 - \sin 1] e^{-\lambda t} \{ H[x + 1 - \gamma t] + H[x + 1 + \gamma t]$$
$$+ H[x - 1 - \gamma t] + H[x - 1 + \gamma t] \} + \cdots$$

and the jump across each characteristic is $\tfrac{1}{2}[1 - \sin 1]e^{-\lambda t}$.

The solution $u(x, t)$ of the given problem (9.890)–(9.891) may be written as

$$(9.906) \qquad\qquad u(x, t) = v(x, t) + U(x, t)$$

with $U(x, t)$ as given earlier. In view of (9.891)–(9.892), the initial data for $v(x, t)$ are

$$(9.907) \qquad v(x, 0) = -1 + [\sin x + \sin 1] H[x + 1]$$
$$- [\sin x - \sin 1] H[x - 1] + \cdots,$$
$$(9.908) \qquad v_t(x, 0) = \lambda [\sin 1 - 1]\{H[x + 1] + H[x - 1]\} + \cdots.$$

In addition, $U(x, t)$ gives rise to an inhomogeneous term that occurs in the equation for $v(x, t)$. Although $u(x, 0)$ is discontinuous at $x = -1$ and $x = 1$, $v(x, 0)$ is continuous for all x. Consequently, $v(x, t)$ is a smoother function than $u(x, t)$, but we do not solve for $v(x, t)$.

A further question that should be dealt with is how to treat problems in which the data do not lead to jump discontinuities but result in stronger or, possibly, weaker singularities in the solution. For instance, the initial value problem for the telegrapher's equation in Section 1.2 has $u(x, 0) = \delta(x)$ so that the solution is expected to have a delta function singularity. Also, the problem considered for $v(x, t)$ in (9.906) leads to a solution with discontinuous derivatives but that is continuous everywhere.

A more general class of problems can be treated by observing that (9.846) can be replaced by the expansion

$$(9.909) \qquad\qquad u(x, t) = \sum_{j=0}^{\infty} v_j(x, t) S_j[\varphi(x, t)]$$

where the (generalized) functions $S_j[x]$ are related to each other by

$$(9.910) \qquad S_j'[x] = S_{j-1}[x],$$

but are otherwise arbitrary. If we set $S_0[x] = H[x]$, the Heaviside function, (9.910) yields $S_j[x] = H_j[x]$ and (9.909) reduces to (9.846). If (9.909) is inserted into the telegrapher's equation (9.842), (9.910) is used, and coefficients of the $S_j[\varphi]$ are equated to zero, we conclude that φ must satisfy the characteristic equation (9.847) and the v_j satisfy the transport equations (9.848)–(9.849).

For a given problem we may use the general series (9.909) and then decide on the basis of the data what form S_0 and the S_j should take. As an example, we consider the problem of Section 1.2 in which $u(x, t)$ satisfies

$$(9.911) \qquad u_{tt} - \gamma^2 u_{xx} + 2\lambda u_t = 0, \qquad -\infty < x < \infty, t > 0$$

with the data

$$(9.912) \qquad u(x, 0) = \delta(x); \qquad u_t(x, 0) = 0.$$

Expanding $u(x, t)$ as in (9.909) we find that

$$(9.913) \qquad u(x, 0) = \sum_{j=0}^{\infty} v_j(x, 0) S_j[\varphi(x, 0)] = \delta(x).$$

Since the S_j are increasingly less singular for increasing j, we find that

$$(9.914) \qquad S_0[x] = \delta(x); \qquad \varphi(x, 0) = x,$$

and consequently

$$(9.915) \qquad S_1[x] = H[x]; \qquad S_j[x] = H_{j-1}[x]; \qquad j \geqslant 1;$$

with the H_j defined as in (9.829). Again, (9.914) implies that we must consider the two characteristics

$$(9.916) \qquad \tilde{\varphi}(x, t) = x - \gamma t = 0; \qquad \hat{\varphi}(x, t) = x + \gamma t = 0.$$

It is easily shown on using the transport equations for $\tilde{v}_0$ and $\hat{v}_0$ that

$$(9.917) \qquad u(x, t) = \tfrac{1}{2} e^{-\lambda t} \delta[x + \gamma t] + \tfrac{1}{2} e^{-\lambda t} \delta[x - \gamma t] + \cdots,$$

to leading terms. It is of interest to note that (9.917) confirms the result given in (1.88).

As another example we consider the initial value problem that yields the *causal fundamental solution* for the telegrapher's equation as follows from Chapter 7. The function $u(x, t)$ satisfies (9.911) and the data are

$$(9.918) \qquad\qquad u(x,0) = 0; \qquad u_t(x,0) = \delta(x).$$

Expanding $u(x, t)$ as in (9.909) we have

$$(9.919) \qquad\qquad u(x,0) = \sum_{j=0}^{\infty} v_j(x,0) S_j[\varphi(x,0)] = 0,$$

$$(9.920)$$
$$u_t(x,0) = v_0(x,0)\varphi_t(x,0) S_{-1}[\varphi(x,0)]$$
$$+ \sum_{j=0}^{\infty} \left\{ v_{j+1}(x,0)\varphi_t(x,0) + \frac{\partial v_j(x,0)}{\partial t} \right\} S_j[\varphi(x,0)] = \delta(x).$$

Since $S_{-1}[x] = S_0'[x]$ is the most singular term in (9.920), we conclude that

$$(9.921) \qquad\qquad S_0'[x] = \delta(x); \qquad \varphi(x,0) = x;$$

so that

$$(9.922) \qquad\qquad S_0[x] = H[x].$$

Consequently, the series (9.909) for this problem is identical with (9.846) since $S_j[x] = H_j[x]$.

Adapting the results (6.226)–(6.227) or (7.353) it is easy to show that the causal fundamental solution of the telegrapher's equation is

$$(9.923) \quad u(x,t) = \frac{1}{2\gamma} e^{-\lambda t} I_0 \left[\frac{\lambda}{\gamma} \sqrt{\gamma^2 t^2 - x^2} \right] H(\gamma t - x) H(\gamma t + x)$$

where $I_0(z)$ is the modified Bessel function of zero order. On using the Taylor series for $I_0(z)$ and considering (9.923) near each characteristic $x - \gamma t = 0$ and $x + \gamma t = 0$ for $t > 0$, it can be shown that the results of this expansion method agree completely with the expression obtained from the exact solution (9.923). To accomplish this, however, the terms in the expansion of (9.918) must be evaluated on the appropriate characteristics $x - \gamma t = 0$ or $x + \gamma t = 0$. This is considered in the exercises.

An important conclusion that can be drawn from the preceding example is that we need not consider only solutions of (9.813) having the expansion form (9.833). Instead, we can expand u in the more general form

$$(9.924) \qquad\qquad u = \sum_{j=0}^{\infty} v_j S_j[\varphi]$$

where the $S_j[x]$ satisfy the relations

(9.925)
$$S_j'[x] = S_{j-1}[x].$$

This represents a formal solution of (9.813) if φ satisfies the *characteristic equation*

(9.926)
$$\varphi_t^2 - p(\nabla\varphi)^2 = 0$$

and the v_j satisfy the *transport equations*

(9.927) $\quad 2\varphi_t\dfrac{\partial v_j}{\partial t} - 2p\nabla\varphi \cdot \nabla v_j + 2\lambda\varphi_t v_j + \left[\varphi_{tt} - \nabla \cdot (p\nabla\varphi)\right]v_j$

$$= -\frac{\partial^2 v_{j-1}}{\partial t^2} + \nabla \cdot (p\nabla v_{j-1}) - 2\lambda\frac{\partial v_{j-1}}{\partial t} - qv_{j-1}, \quad j \geqslant 0$$

where it is assumed that $v_{-1} = 0$, so that v_0 satisfies an homogeneous equation.

The expansion (9.924) no longer need be restricted to describing the *propagation of singularities*. For example, if we set

(9.928)
$$S_0[\varphi] = e^{i\omega\varphi}; \quad S_j[\varphi] = \frac{1}{(i\omega)^j}e^{i\omega\varphi}$$

where ω is a large (real) parameter, (9.924) may be regarded as an asymptotic solution of (9.813). It would occur if the problem contains *rapidly oscillating data*. For instance, if initially we have

(9.929)
$$u(x, y, 0) = f(x, y)e^{i\omega g(x, y)}; \quad u_t(x, y, 0) = 0$$

in the two-dimensional case where ω is large, we look for a solution of (9.813) in the form (9.924) with the S_j given by (9.928). From the initial data (9.929) we conclude that $\varphi(x, y, 0) = g(x, y)$. There are two solutions $\check{\varphi}(x, y, t)$ and $\hat{\varphi}(x, y, t)$ of (9.926) that satisfy this initial condition. For each of those functions we construct a series (9.924) with coefficients $\check{v}_j$ and $\hat{v}_j$, respectively. On inserting these series into the data (9.929) we easily determine initial conditions for $\check{v}_j$ and $\hat{v}_j$, each of which satisfies the transport equations (9.927).

With $\lambda = q = 0$ and $p = \gamma^2 = \text{constant}$ in (9.813) the choice

(9.930)
$$\varphi = t - \psi(x, y); \quad v_j = v_j(x, y)$$

with the S_j defined as in (9.928) leads to the *geometrical optics expansion* for the *reduced wave equation* as is immediately seen. In fact, the series

(9.931)
$$ve^{-i\omega\psi} = \sum_{j=0}^{\infty} v_j(x, y)\frac{1}{(i\omega)^j}e^{-i\omega\psi(x, y)}$$

is a solution of the reduced wave equations as follows on setting

$$(9.932) \qquad u(x, y, t) = e^{i\omega t} v(x, y) e^{i\omega \psi(x, y)}.$$

With $\omega = -\hat{\omega}$, (9.932) reduces to the form considered in Section 9.4. In view of the fact that solutions of (9.813) undergo rapid variations across (characteristic) singularity regions, it is not surprising that solutions of the type (9.924) and (9.928) with rapidly oscillating terms are similar in form to solutions that describe the propagation of singularities.

In the present section, however, we are only interested in discussing the propagation of singularities for the solutions of (9.813). Before proceeding to a discussion of some more complicated problems associated with (9.813) in two space dimensions, we consider two simple problems for the wave and the Klein–Gordon equations.

Let $u(x, y, t)$ satisfy the *wave equation*

$$(9.933) \qquad u_{tt} - \gamma^2(u_{xx} + u_{yy}) = 0, \qquad -\infty < x, y < \infty, t > 0$$

where γ is a constant. The initial data for u are

$$(9.934) \quad u(x, y, 0) = \begin{cases} \alpha; & x^2 + y^2 < a^2 \\ \beta; & x^2 + y^2 > a^2 \end{cases}; \qquad u_t(x, y, 0) = 0.$$

With $\rho^2 = x^2 + y^2$, (9.934) can be written as

$$(9.935) \quad u(x, y, 0) = \alpha + (\beta - \alpha) H[\rho - a]; \qquad u_t(x, y, 0) = 0$$

where $H[x]$ is the Heaviside function and α and β are constants. We may expand u as

$$(9.936) \qquad u(x, y, t) = \alpha + \sum_{j=0}^{\infty} v_j(x, y, t) S_j[\varphi(x, y, t)].$$

At $t = 0$ we have

$$(9.937)$$

$$u(x, y, 0) = \alpha + \sum_{j=0}^{\infty} v_j(x, y, 0) S_j[\varphi(x, y, 0)] = \alpha + (\beta - \alpha) H[\rho - a].$$

This suggests that we set

$$(9.938) \qquad S_0[\varphi] = H[\varphi]; \qquad S_j[\varphi] = H_j[\varphi]$$

with the H_j defined as in (9.829) and

(9.939) $$\varphi(x, y, 0) = \rho - a.$$

Since φ must also satisfy the characteristic equation $\varphi_t^2 - \gamma^2(\varphi_x^2 + \varphi_y^2) = 0$, we obtain two possible solutions:

(9.940) $$\tilde{\varphi}(x, y, t) = \rho - a - \gamma t; \qquad \hat{\varphi}(x, y, t) = \rho - a + \gamma t.$$

To account for $\tilde{\varphi}$ and $\hat{\varphi}$ we replace (9.936) by

(9.941) $$u(x, y, t) = \alpha + \sum_{j=0}^{\infty} \tilde{v}_j H_j[\tilde{\varphi}] + \sum_{j=0}^{\infty} \hat{v}_j H_j[\hat{\varphi}].$$

Inserting (9.941) into the initial conditions (9.935), we easily conclude that

(9.942) $$\tilde{v}_0(x, y, 0) = \hat{v}_0(x, y, 0) = \tfrac{1}{2}(\beta - \alpha).$$

The transport equations for $\tilde{v}_0$ and $\hat{v}_0$ are

(9.943) $$\frac{\partial \tilde{v}_0}{\partial t} + \gamma \frac{\partial \tilde{v}_0}{\partial \rho} + \frac{\gamma}{2\rho} \tilde{v}_0 = 0,$$

(9.944) $$\frac{\partial \hat{v}_0}{\partial t} - \gamma \frac{\partial \hat{v}_0}{\partial \rho} - \frac{\gamma}{2\rho} \hat{v}_0 = 0,$$

and the solutions of (9.942)–(9.944) are

(9.945) $$\tilde{v}_0(x, y, t) = \tfrac{1}{2}(\beta - \alpha) \left[\frac{\rho - \gamma t}{\rho} \right]^{1/2};$$

$$\hat{v}_0(x, y, t) = \tfrac{1}{2}(\beta - \alpha) \left[\frac{\rho + \gamma t}{\rho} \right]^{1/2}.$$

These expressions may be expanded in powers of $\tilde{\varphi}$ and $\hat{\varphi}$. Then

(9.946) $$\tilde{v}_0(x, y, t) = \frac{1}{2}(\beta - \alpha) \left[\frac{a}{a + \gamma t} \right]^{1/2} + O[\tilde{\varphi}],$$

(9.947) $$\hat{v}_0(x, y, t) = \frac{1}{2}(\beta - \alpha) \left[\frac{a}{a - \gamma t} \right]^{1/2} + O[\hat{\varphi}].$$

Thus the leading terms in the expansion of $u(x, y, t)$ are

(9.948) $$u(x, y, t) = \alpha + \frac{1}{2}(\beta - \alpha) \left[\frac{a}{a + \gamma t} \right]^{1/2} H[\rho - a - \gamma t]$$

$$+ \frac{1}{2}(\beta - \alpha) \left[\frac{a}{a - \gamma t} \right]^{1/2} H[\rho - a + \gamma t] + \cdots .$$

The characteristic surface $\hat{\varphi} = \rho - a - \gamma t = 0$ represents *cylindrical wave fronts* that travel outward from the circle $\rho = a$ with speed $d\rho/dt = \gamma$. The jump in u across the wave fronts decays like $t^{-1/2}$ for increasing t. The characteristic surface $\hat{\varphi} = \rho - a + \gamma t = 0$ yields cylindrical wave fronts that travel inward from $\rho = a$ with speed γ. At the time $t = a/\gamma$ the wave fronts collapse on the point $\rho = 0$, and (9.948) is singular there since the coefficient of $H[\rho - a + \gamma t]$ blows up. The singularity results because the origin is a focal point for the bicharacteristics. We do not consider how (9.948) should be modified to deal with this singularity.

Turning now to a three-dimensional example, we consider the causal fundamental solution for the *Klein–Gordon equation*

(9.949)
$$u_{tt} - \gamma^2[u_{xx} + u_{yy} + u_{zz}] + c^2 u = 0, \qquad -\infty < x, y, z < \infty, t > 0.$$

It follows from our discussion in Chapter 7 that it can be determined as a solution of (9.949) with the initial data

(9.950) $u(x, y, z, 0) = 0; \qquad u_t(x, y, z, 0) = \delta(x)\delta(y)\delta(z).$

It is given as

(9.951) $u(x, y, z, t) = \dfrac{1}{4\pi\gamma r}\delta[\gamma t - r] - \dfrac{cJ_1\left[\dfrac{c}{\gamma}\sqrt{\gamma^2 t^2 - r^2}\right]}{4\pi\gamma^2\sqrt{\gamma^2 t^2 - r^2}} H[\gamma t - r],$

for $t > 0$, where $r^2 = x^2 + y^2 + z^2$ and J_1 is the Bessel function of order one. We do not derive (9.951) which may be obtained by using Fourier transforms or otherwise.

Using the Taylor series for $J_1(z)$ we have

(9.952)
$$u(x, y, z, t) = \frac{1}{4\pi\gamma r}\delta[\gamma t - r]$$

$$- \frac{c}{4\pi\gamma^2}\sum_{j=0}^{\infty}\frac{(-1)^j(c/2\gamma)^{2j+1}}{j!(j+1)!}[\gamma^2 t^2 - r^2]^j H[\gamma t - r]$$

$$= \frac{1}{4\pi\gamma r}\delta[\gamma t - r]$$

$$- \frac{c}{4\pi\gamma^2}\sum_{j=0}^{\infty}\frac{(-1)^j(c/2\gamma)^{2j+1}}{j!(j+1)!}(\gamma t + r)^j[\gamma t - r]^j H[\gamma t - r].$$

Since $[\gamma t + r]^j = [2r + \gamma t - r]^j$ and can be expanded in powers of $\gamma t - r$, we see that (9.952) has the form

$$(9.953) \qquad u(x, y, z, t) = \sum_{j=0}^{\infty} v_j(x, y, z, t) S_j[\gamma t - r]$$

where

$$(9.954) \qquad S_0[x] = \delta[x]; \qquad S_j[x] = H_j[x]; \qquad j \geqslant 1;$$

with $H_j[x]$ defined as in (9.892). The transport coefficients v_j are evaluated on the characteristic surface $\varphi = \gamma t - r = 0$. The leading terms are

$$(9.955) \qquad v_0 = \frac{1}{4\pi\gamma r}; \qquad v_1 = -\frac{c^2}{8\pi\gamma^3}.$$

We now attempt to reproduce these results by using the direct expansion method and set

$$(9.956) \qquad u = \sum_{j=0}^{\infty} v_j S_j[\varphi]$$

and insert this into (9.949). Then φ must satisfy the characteristic equation

$$(9.957) \qquad \varphi_t^2 - \gamma^2 (\nabla\varphi)^2 = 0,$$

and the v_j satisfy the transport equations

$$(9.958) \quad 2\varphi_t \frac{\partial v_j}{\partial t} - 2\gamma^2 \nabla\varphi \cdot \nabla v_j + \left[\varphi_{tt} - \gamma^2 \nabla^2\varphi\right] v_j$$

$$= -\frac{\partial^2 v_{j-1}}{\partial t^2} + \gamma^2 \nabla^2 v_{j-1} - c^2 v_{j-1}, \qquad j \geqslant 0$$

where $v_{-1} = 0$. Since the data for this problem are concentrated at $(x, y, z) = (0, 0, 0)$, we must choose $\varphi = 0$ to be the forward characteristic cone

$$(9.959) \qquad \varphi = \gamma t - r = 0.$$

The equation for v_0 becomes

$$(9.960) \qquad \frac{\partial v_0}{\partial t} + \gamma \frac{\partial v_0}{\partial r} + \frac{\gamma}{r} v_0 = 0,$$

and this has a solution

(9.961)
$$v_0 = \frac{V_0}{r}$$

where V_0 may be an arbitrary function of $\gamma t - r$. For v_1 we have

(9.962)
$$\frac{\partial v_1}{\partial t} + \gamma \frac{\partial v_1}{\partial r} + \frac{\gamma}{r} v_1 = -\frac{c^2}{2\gamma} \frac{V_0}{r},$$

and this has the solution

(9.963)
$$v_1 = \frac{V_1}{r} - \frac{c^2 V_0}{2\gamma^2},$$

with V_1 being an arbitrary function of $\gamma t - r$.

As has been done in the examples considered, it is necessary to specify the S_j and the initial values for the v_j in terms of the data (9.950) for $u(x, y, z, t)$. To do so we first express the delta function in (9.950) in a form appropriate for spherical coordinates. This can be shown to be

(9.964)
$$\delta(x)\delta(y)\delta(z) = \frac{1}{4\pi r^2}\delta(r).$$

If $f(x, y, z)$ is multiplied by the right side of (9.964) and is integrated over a neighborhood of the origin using spherical coordinates, we obtain $f(0, 0, 0)$, as is easily seen. Thus

(9.965) $\quad u_t(x, y, z, 0) = \gamma v_0(x, y, z, 0) S_0'[-r]$

$$+ \sum_{j=0}^{\infty} \left\{ \frac{\partial v_j(x, y, z, 0)}{\partial t} + \gamma v_{j+1}(x, y, z, 0) \right\} S_j[-r]$$

$$= \frac{1}{4\pi r^2}\delta[r].$$

Since $S_0'[-r]$ is the most singular term in this series we set

(9.966)
$$S_0'[-r] = \frac{1}{r}\delta[r],$$

as we have already determined that $v_0 = V_0/r$. Noting that $r\delta'(r) + \delta(r) = 0$ and $\delta(-r) = \delta(r)$, we have

(9.967)
$$S_0'[r] = -\frac{1}{r}\delta[-r] = -\frac{1}{r}\delta[r] = \delta'[r],$$

so that $S_0 = \delta$ and $S_j = H_j$ as given in (9.954). Having specified S_0, we find from (9.965) that

$$(9.968) \qquad v_0(x, y, z, 0) = \frac{1}{4\pi\gamma r},$$

which implies that

$$(9.969) \qquad V_0 = \frac{1}{4\pi\gamma}$$

in (9.961).

It should be noted that the second initial condition for u yields

$$(9.970) \qquad u(x, y, z, 0) = \sum_{j=0}^{\infty} v_j(x, y, z, 0) S_j[-r] = 0.$$

This seems to imply that the v_j must vanish initially, in contradiction to (9.968), for instance. However, considering only the leading term we have

$$(9.971) \quad v_0 S_0[-r] = \frac{1}{4\pi\gamma r}\delta[r] = \frac{r}{\gamma}\frac{1}{4\pi r^2}\delta[r] = \frac{r}{\gamma}\delta(x)\delta(y)\delta(z) = 0$$

where the fact that $f(x, y, z)\delta(x)\delta(y)\delta(z) = f(0,0,0)\delta(x)\delta(y)\delta(z)$ and (9.964) have been used. The difficulties arising for this problem result from the fact that $t = 0, r = 0$ is a *focal point* for the bicharacteristics. We do not determine any further transport coefficients but note that with $V_1 = 0$ and $V_0 = 1/4\pi\gamma$, v_1 as given in (9.963) agrees with (9.955) which is determined from the exact solution. In the following example, with which we conclude this section, it is shown how *boundary layer theory* can be used to determine initial data for the transport coefficients v_j.

EXAMPLE 9.21. POINT SOURCE PROBLEMS FOR HYPERBOLIC EQUATIONS IN TWO DIMENSIONS

The solution of the initial value problem for the *wave equation*

$$(9.972) \qquad u_{tt} - \gamma^2(u_{xx} + u_{yy}) = 0, \qquad -\infty < x, y < \infty, t > 0$$

with

$$(9.973) \qquad u(x, y, 0) = 0; \qquad u_t(x, y, 0) = \delta(x)\delta(y)$$

was given in (7.325) as

$$(9.974) \quad u(x, y, t) = \frac{1}{2\pi\gamma}[\gamma^2 t^2 - \rho^2]^{-1/2} H[\gamma t - \rho], \qquad t > 0$$

where $H(x)$ is the Heaviside function and $\rho^2 = x^2 + y^2$. This represents a *causal fundamental solution* for the wave equation in two dimensions, as well as the solution of an instantaneous point source problem. The source point is at the origin and the source acts at the time $t = 0$.

We would like to express (9.974) in the form of the series (9.924). The function $\varphi = \gamma t - \rho$ is a solution of the characteristic equation $\varphi_t^2 - \gamma^2(\nabla\varphi)^2 = 0$ for (9.972). The expression $[\gamma^2 t^2 - \rho^2]^{-1/2}$ can be expanded in the form of a binomial series as

(9.975)

$$\left[\gamma^2 t^2 - \rho^2\right]^{-1/2} = \left[\gamma t - \rho\right]^{-1/2}\left[\gamma t + \rho\right]^{-1/2}$$

$$= \varphi^{-1/2}\left[2\rho + \varphi\right]^{-1/2} = \left[2\rho\varphi\right]^{-1/2}\left\{1 - \frac{1}{4\rho}\varphi + \cdots\right\}$$

where $\varphi = \gamma t - \rho$, and the further terms contain increasing integral powers of φ. Inserting (9.975) into (9.974) suggests that we define $S_0[\varphi]$ as

(9.976)
$$S_0[\varphi] = \frac{1}{\sqrt{\varphi}}H[\varphi].$$

With $S_j'[x] = S_{j-1}[x]$ it is easily shown that

(9.977)
$$S_j[\varphi] = \frac{2^j \varphi^{(2j-1)/2}}{1 \cdot 3 \cdot 5 \cdots (2j-1)}H[\varphi], \qquad j \geq 1.$$

For example, $S_1[x] = 2\sqrt{x}\,H[x]$, and generalized differentiation gives

(9.978)
$$S_1'[x] = \frac{1}{\sqrt{x}}H[x] + 2\sqrt{x}\,\delta[x] = \frac{1}{\sqrt{x}}H[x] = S_0[x]$$

since $2\sqrt{x}\,\delta(x) = 0\delta[x] = 0$. Thus u can be expanded as

(9.979)
$$u(x, y, t) = \sum_{j=0}^{\infty} v_j(x, y, t)S_j[\gamma t - \rho] = \frac{\rho^{-1/2}}{2\pi\gamma\sqrt{2}}S_0[\gamma t - \rho]$$

$$-\frac{\rho^{-3/2}}{16\pi\gamma\sqrt{2}}S_1[\gamma t - \rho] + \cdots,$$

with the S_j defined as in the foregoing.

These results are now used to obtain the solution of the point source problem for the equation

(9.980)
$$u_{tt} - \gamma^2(u_{xx} + u_{yy}) + 2\lambda u_t + c^2 u = 0, \qquad -\infty < x, y < \infty, t > 0$$

where $\lambda > 0$ and all the coefficients are constants. As for the wave equation, it can be determined from the solution of an initial value problem with the data (9.973). We expand u as in (9.924) and obtain the *characteristic equation*.

$$(9.981) \qquad \varphi_t^2 - \gamma^2 \left(\varphi_x^2 + \varphi_y^2 \right) = 0$$

and the *transport equations*

$$(9.982) \quad 2\varphi_t \frac{\partial v_j}{\partial t} - 2\gamma^2 \nabla\varphi \cdot \nabla v_j + 2\lambda\varphi_t v_j + \left[\varphi_{tt} - \gamma^2 \nabla^2\varphi \right] v_j$$

$$= - \frac{\partial^2 v_{j-1}}{\partial t^2} + \gamma^2 \nabla^2 v_{j-1} - 2\lambda \frac{\partial v_{j-1}}{\partial t} - c^2 v_{j-1}, \qquad j \geq 0$$

where $v_{-1} = 0$. Since the data are concentrated at the origin for $t = 0$ we set

$$(9.983) \qquad \varphi = \gamma t - \rho,$$

so that $\varphi = 0$ yields the *forward characteristic cone*.

The transport equation for v_0 is

$$(9.984) \qquad \frac{\partial v_0}{\partial t} + \gamma \frac{\partial v_0}{\partial \rho} + \lambda v_0 + \frac{\gamma}{2\rho} v_0 = 0.$$

This has the general solution

$$(9.985) \qquad v_0 = V_0 \rho^{-1/2} e^{-\lambda t}$$

where V_0 may be a function of $\gamma t - \rho$. For v_1 we obtain

$$(9.986)$$

$$\frac{\partial v_1}{\partial t} + \gamma \frac{\partial v_1}{\partial \rho} + \lambda v_1 + \frac{\gamma}{2\rho} v_1 = \left[\left(\frac{\lambda^2 - c^2}{2\gamma} \right) \rho^{-1/2} + \frac{\gamma}{8} \rho^{-5/2} \right] V_0 e^{-\lambda t}.$$

With

$$(9.987) \qquad v_1 = \rho^{-1/2} e^{-\lambda t} \hat{v}_1$$

we have

$$(9.988) \qquad \frac{\partial \hat{v}_1}{\partial t} + \gamma \frac{\partial \hat{v}_1}{\partial \rho} = V_0 \left[\frac{\lambda^2 - c^2}{2\gamma} + \frac{\gamma}{8\rho^2} \right],$$

and this has the general solution

$$(9.989) \qquad \hat{v}_1 = V_1 + \left(\frac{\lambda^2 - c^2}{2\gamma} \right) t V_0 - \frac{1}{8\rho} V_0$$

where V_1 may be a function of $\gamma t - \rho$. Thus

$$(9.990) \quad v_1 = V_1 \rho^{-1/2} e^{-\lambda t} + \left[\frac{(\lambda^2 - c^2)t}{2\gamma} - \frac{1}{8\rho} \right] V_0 \rho^{-1/2} e^{-\lambda t}.$$

Collecting these results we obtain for the solution of (9.980) with the data (9.973),

$$(9.991) \quad u = V_0 \rho^{-1/2} e^{-\lambda t} S_0[\gamma t - \rho]$$

$$+ \left\{ V_1 + \left(\frac{\lambda^2 - c^2}{2\gamma} t - \frac{1}{8\rho} \right) V_0 \right\} \rho^{-1/2} e^{-\lambda t} S_1[\gamma t - \rho] + \cdots$$

where V_0, V_1, and the S_j are as yet unspecified. One way to determine these unknowns is the *method of canonical problems* developed by J. B. Keller in his *geometrical theory of diffraction* (see Section 9.4). Roughly speaking, it states that in view of the similarity of the initial value problems for (9.980) and the wave equation (9.972) near the source point, where the second derivative terms in (9.980) are most significant, we may compare and identify terms in the expansions (9.991) and (9.979) of their solutions when evaluated near the source point. If we do so, we find that S_0 and the S_j in (9.991) should be defined as (9.976) and (9.977), respectively. Also, we should set $V_0 = 1/2\pi\gamma\sqrt{2}$. Then with $\lambda = c = 0$ and $V_1 = 0$ in (9.991) we find that it agrees completely with (9.979). We cannot specify V_1 by using this method. This is typical of the results given by this approach where the leading and, generally, the most significant term is the only one that can be specified. Nevertheless, we conclude that

$$(9.992)$$

$$u = \frac{\rho^{-1/2} e^{-\lambda t}}{2\pi\gamma\sqrt{2}} S_0[\gamma t - \rho]$$

$$+ \left\{ V_1 + \frac{1}{2\pi\gamma\sqrt{2}} \left(\frac{\lambda^2 - c^2}{2\gamma} t - \frac{1}{8\rho} \right) \right\} \rho^{-1/2} e^{-\lambda t} S_1[\gamma t - \rho] + \cdots$$

with S_0 and S_1 defined by (9.976)–(9.977). The term V_1 in (9.992) is not specified. We do see from this result that the *damping term* $2\lambda u_t$ in (9.980) gives rise to the exponential decay term $e^{-\lambda t}$, whereas the effect of the *dispersive term* $c^2 u$ is not felt until the second term in the expansion.

Alternatively, we may use boundary layer theory to specify the unknown terms in (9.991). Since the difficulties arise at the source point when $t = 0$, we set (with $0 < \epsilon \ll 1$)

$$(9.993) \quad t = \epsilon\tau; \qquad x = \epsilon\sigma; \qquad y = \epsilon\eta; \qquad \rho = \epsilon\hat{\rho} = \epsilon\sqrt{\sigma^2 + \eta^2};$$

to emphasize the neighborhood of $(x, y, t) = (0, 0, 0)$. With

(9.994) $$w(\sigma, \eta, \tau) = u(\epsilon\sigma, \epsilon\eta, \epsilon\tau),$$

we have instead of (9.980):

(9.995) $$w_{\tau\tau} - \gamma^2[w_{\sigma\sigma} + w_{\eta\eta}] + 2\lambda\epsilon w_\tau + \epsilon^2 c^2 w = 0.$$

It is apparent that the stretching transformation (9.993) could have contained any positive but identical power of ϵ for each variable so long as the leading equation resulting from the transformation was the wave equation. Recalling that $\delta(\epsilon z) = (1/\epsilon)\delta(z)$ and noting that $u_t = (1/\epsilon)u_\tau$ we have

(9.996) $$w(\sigma, \eta, 0) = 0; \qquad w_\tau(\sigma, \eta, 0) = \frac{1}{\epsilon}\delta(\sigma)\delta(\eta)$$

on using the data (9.973).

In view of (9.996) we introduce the boundary layer expansion

(9.997) $$w = \frac{1}{\epsilon}w_0 + w_1 + \epsilon w_2 + \cdots$$

into (9.995) and (9.996). Then w_0 satisfies the wave equation

(9.998) $$\frac{\partial^2 w_0}{\partial\tau^2} - \gamma^2\left[\frac{\partial^2 w_0}{\partial\sigma^2} + \frac{\partial^2 w_0}{\partial\eta^2}\right] = 0$$

and the initial conditions

(9.999) $$w_0(\sigma, \eta, 0) = 0; \qquad \frac{\partial w_0(\sigma, \eta, 0)}{\partial\tau} = \delta(\sigma)\delta(\eta).$$

The solution of this problem is given in (9.974) as

(9.1000) $$w_0(\sigma, \eta, \tau) = \frac{1}{2\pi\gamma}[\gamma^2\tau^2 - \rho^2]^{-1/2} H[\gamma\tau - \rho].$$

The equation for w_1 is

(9.1001) $$\frac{\partial^2 w_1}{\partial\tau^2} - \gamma^2\nabla^2 w_1 = -2\lambda\frac{\partial w_0}{\partial\tau},$$

and w_1 has homogeneous initial data

(9.1002) $$w_1(\sigma, \eta, 0) = \frac{\partial w_1(\sigma, \eta, 0)}{\partial\tau} = 0.$$

Since w_0 is a solution of the wave equation, it is easy to show that a particular solution of (9.1001) is

$$(9.1003) \qquad w_1(\sigma, \eta, \tau) = -\lambda \tau w_0(\sigma, \eta, \tau).$$

Also, since w_0 vanishes at $\tau = 0$ we see that w_1 satisfies the initial conditions (9.1002), so that (9.1003) is the solution of (9.1001)–(9.1002).

Collecting results, we have

$$(9.1004) \quad w(\sigma, \eta, \tau] = \frac{1}{2\pi\gamma\epsilon}[1 - \lambda\epsilon\tau][\gamma^2\tau^2 - \hat{\rho}^2]^{-1/2}H[\gamma\tau - \hat{\rho}]$$

up to order ϵ^2. This expression must be matched with the expansion (9.991) for $u(x, y, t)$. To do so, $u(x, y, t)$ must be expressed in terms of σ, η and τ to give

$$(9.1005)$$
$$u(\epsilon\sigma, \epsilon\eta, \epsilon\tau) = V_0(\epsilon\hat{\rho})^{-1/2}e^{-\lambda\epsilon\tau}S_0[\epsilon(\gamma\tau - \hat{\rho})]$$
$$+ \{V_1(\epsilon\hat{\rho})^{-1/2} - \tfrac{1}{8}V_0(\epsilon\hat{\rho})^{-3/2}\}e^{-\lambda\epsilon\tau}S_1[\epsilon(\gamma\tau - \hat{\rho})] + \cdots$$

to leading terms. The exponential $e^{-\lambda\epsilon\tau}$ must be expanded in powers of ϵ. Also, $[\gamma^2\tau^2 - \hat{\rho}^2]^{-1/2}$ must be expanded out as in (9.975). If this is carried out in (9.1004) and (9.1005), we easily conclude that

$$(9.1006) \qquad V_0 = \frac{1}{2\pi\gamma\sqrt{2}}; \qquad V_1 = 0; \qquad S_0[x] = \frac{1}{\sqrt{x}}H[x];$$

so that the $S_j[x]$ are given as in (9.977). We note that since $\epsilon > 0$, $H[\epsilon(\gamma\tau - \hat{\rho})]$ $= H[\gamma\tau - \hat{\rho}]$. This result agrees with (9.992) which was obtained by using the method of canonical problems, except that we have now determined the value of V_1 also.

To conclude our discussion of this example we consider the *causal fundamental solution* for a problem with variable coefficients. The equation is

$$(9.1007) \quad u_{tt} - \gamma^2\nabla \cdot [(1 + a^2\rho^2)\nabla u] = 0; \qquad -\infty < x, y < \infty, t > 0$$

where a and γ are positive constants and $\rho^2 = x^2 + y^2$. The initial data are again (9.973). We expand u as

$$(9.1008) \qquad u(x, y, t) = \sum_{j=0}^{\infty} v_j(x, y, t)S_j[\varphi(x, y, t)].$$

Then φ satisfies the characteristic equation

$$(9.1009) \qquad \varphi_t^2 - \gamma^2(1 + a^2\rho^2)(\nabla\varphi)^2 = 0$$

and v_0 satisfies the transport equation

$$(9.1010) \qquad 2\varphi_t \frac{\partial v_0}{\partial t} - 2\gamma^2 (1 + a^2\rho^2) \nabla\varphi \cdot \nabla v_0$$
$$+ \left\{ \varphi_{tt} - \gamma^2 \nabla \cdot \left[(1 + a^2\rho^2) \nabla\varphi \right] \right\} v_0 = 0.$$

We do not consider the other transport equations.

Since the data are concentrated at a point, we expect that $\varphi = 0$ is the *forward characteristic conoid*. It is easy to show that it is given by

$$(9.1011) \qquad \varphi(x, y, t) = \gamma t - \frac{1}{a} \sin^{-1}(a\rho) = 0.$$

Since the problem has no angular dependence, we assume that $v_0 = v_0(\rho, t)$. On the conoid (9.1011) we have

$$(9.1012) \qquad \frac{dv_0}{d\rho} = \frac{1}{\gamma\sqrt{1 + a^2\rho^2}} \frac{\partial v_0}{\partial t} + \frac{\partial v_0}{\partial\rho},$$

and (9.1010) can be written as

$$(9.1013) \qquad \frac{dv_0}{d\rho} + \left(\frac{1 + 2a^2\rho^2}{2\rho(1 + a^2\rho^2)} \right) v_0 = 0.$$

Using a partial fraction expansion, (9.1013) can be solved as

$$(9.1014) \qquad v_0 = V_0 \rho^{-1/2} (1 + a^2\rho^2)^{-1/4}$$

where V_0 is to be specified and is constant on the conoid (9.1011).

We have found that

$$(9.1015) \quad u(x, y, t) = V_0 \rho^{-1/2} (1 + a^2\rho^2)^{-1/4} S_0 \left[\gamma t - \frac{1}{a} \sin^{-1}(a\rho) \right] + \cdots,$$

but V_0 and S_0 have not been specified. Near the source point $(\rho, t) = (0, 0)$,

$$(9.1016) \qquad \gamma t - \frac{1}{a} \sin^{-1}(a\rho) \approx \gamma t - \rho, \qquad \rho \approx 0,$$

$$(9.1017) \qquad \rho^{-1/2} (1 + a^2\rho^2)^{-1/4} \approx \rho^{-1/2}, \qquad \rho \approx 0.$$

Since (9.1007) has the form of the wave equation (9.972) for small ρ and (9.1015) reduces to the form (9.979), we conclude on the basis of the *method of canonical problems* that S_0 is given as in (9.976) and V_0 as in (9.1006). Thus

(9.1015) becomes

$$(9.1018) \quad u(x, y, t) = \frac{1}{2\pi\gamma\sqrt{2}} \rho^{-1/2} (1 + a^2\rho^2)^{-1/4} \left[\gamma t - \frac{1}{a} \sin^{-1}(a\rho) \right]^{-1/2}$$

$$\times H \left[\gamma t - \frac{1}{a} \sin^{-1}(a\rho) \right] + \cdots$$

to leading order.

We have shown in our general discussion and in the examples how the behavior of solutions of hyperbolic equations near singularity regions can be discussed without having to find the full solution. Further examples are given in the exercises.

9.6. ASYMPTOTIC SIMPLIFICATION OF EQUATIONS

In the preceding sections of this chapter we discussed techniques for obtaining approximate solutions of initial and boundary value problems for partial differential equations. In this, the concluding section of the text, we consider methods for replacing the equations themselves by simpler, more easily solvable, equations. Although the use of perturbation and asymptotic methods also results in the solution of simplified equations, the approach we use differs from that employed in the aforementioned methods. We show how to construct the simplified equations but do not, in general, indicate how they may be used to solve specific initial and boundary value problems. As shown, general information about the behavior of the solutions of the given full equations can be inferred from their simplified forms.

To fix ideas we discuss the *dissipative wave equation* (5.274), that is, with constant α and c,

$$(9.1019) \qquad\qquad u_{tt} - c^2 u_{xx} + u_t - \alpha u_x = 0.$$

Assuming that the point (x_0, t_0) is a discontinuity or singular point for the data for (9.1019), we study its effect on the solutions of (9.1019) by setting $x - x_0 = \epsilon\sigma$ and $t - t_0 = \epsilon\tau$. This yields

$$(9.1020) \qquad\qquad u_{\tau\tau} - c^2 u_{\sigma\sigma} + \epsilon[u_\tau - \alpha u_\sigma] = 0.$$

Although we have, in effect, introduced a boundary layer stretching to obtain (9.1020) we do not proceed as in boundary layer theory by expanding u in powers of ϵ. Instead, we introduce an approximate factorization of (9.1020). If $\epsilon = 0$, then (9.1020) can be factored exactly as was done in (2.1), but if $\epsilon \neq 0$ we cannot factor (9.1020) in a simple fashion, yet since ϵ is small it can be factored approximately. Letting $\partial_\sigma = \partial/\partial\sigma$ and $\partial_\tau = \partial/\partial\tau$ in (9.1020) we have

(9.1021) $\left[\partial_\tau^2 - c^2\partial_\sigma^2 + \epsilon\partial_\tau - \epsilon\alpha\partial_\sigma\right]u$

$$= \left[\partial_\tau - c\partial_\sigma + \epsilon a + O(\epsilon^2)\right]\left[\partial_\tau + c\partial_\sigma + \epsilon b + O(\epsilon^2)\right]u = 0$$

where a and b are to be specified. (We may assume that a and b are constants.) Multiplying out the two first order operators in (9.1021) and comparing the results with the second order operator in (9.1021), we find that

(9.1022) $$a = \frac{c - \alpha}{2c}; \qquad b = \frac{c + \alpha}{2c}.$$

Since the order of the operators in (9.1021) can be interchanged, we see that to order ϵ^2, (9.1020) is equivalent to the two first order equations

(9.1023) $$\begin{cases} u_\tau - cu_\sigma + \epsilon\left(\dfrac{c - \alpha}{2c}\right)u = O(\epsilon^2) \\[2mm] u_\tau + cu_\sigma + \epsilon\left(\dfrac{c + \alpha}{2c}\right)u = O(\epsilon^2) \end{cases}.$$

Neglecting the $O(\epsilon^2)$ terms in (9.1023), we obtain the solutions

(9.1024) $$u(\sigma, \tau) = \begin{cases} f(\sigma + c\tau)\exp\left[-\left(\dfrac{c - \alpha}{2c}\right)\epsilon\tau\right] \\[2mm] g(\sigma - c\tau)\exp\left[-\left(\dfrac{c + \alpha}{2c}\right)\epsilon\tau\right] \end{cases}$$

where f and g are arbitrary functions. If $|\alpha| < c$, these solutions decay exponentially as $t = \epsilon\tau \to \infty$, and it is easy to show that the initial value problem for (9.1019) or (9.1020) is unstable if $|\alpha| \geqslant c$. Since discontinuities in the solutions are carried along the characteristics $\sigma \pm c\tau = $ constant, we see that (9.1024) is useful in describing the propagation of discontinuities in the solutions of (9.1019). In fact, the neglected $O(\epsilon^2)$ terms in (9.1023) are antiderivative or integral operators that have the effect of smoothing out singularities in the solutions. The expressions for u in (9.1024) have the general form of the solutions given in the preceding section.

To study the solutions of (9.1019) at large times and in the far field we set $x = \sigma/\epsilon$ and $t = \tau/\epsilon$ and obtain

(9.1025) $$\epsilon\left[u_{\tau\tau} - c^2u_{\sigma\sigma}\right] + u_\tau - \alpha u_\sigma = 0.$$

Then using (9.1025),

(9.1026) $u_{\tau\tau} = \alpha u_{\sigma\tau} - \epsilon\left[u_{\tau\tau\tau} - c^2u_{\sigma\sigma\tau}\right]$

$$= \frac{\partial}{\partial\sigma}\left[\alpha u_\tau\right] + O(\epsilon) = \alpha^2 u_{\sigma\sigma} + O(\epsilon),$$

and (9.1025) can be replaced by

$$(9.1027) \qquad u_\tau - \alpha u_\sigma + \epsilon(\alpha^2 - c^2)u_{\sigma\sigma} = O(\epsilon^2).$$

If $|\alpha| < c$, so that the initial value problem for (9.1019) is stable, we see that on neglecting the $O(\epsilon^2)$ terms in (9.1027) we have a diffusion equation. In the original variables this equation has the form

$$(9.1028) \qquad u_t - \alpha u_x + \epsilon(\alpha^2 - c^2)u_{xx} = 0$$

and this shows that for large x and t the dissipative wave equation yields a *wave motion* along the line $x + \alpha t = 0$ modified by a diffusion effect represented by the second derivative term. That is, if $\epsilon = 0$ in (9.1028), we have undirectional wave motion along $x + \alpha t = 0$. The $O(\epsilon^2)$ terms neglected in (9.1027) are higher derivatives of u. Since the diffusion effect smooths out the solutions, these terms can be disregarded if they are multiplied by higher powers of ϵ. Our conclusions are consistent with those obtained in Example 5.14 and the discussion in Section 1.2.

There are two types of wave motion associated with (9.1019). The *principal part* of (9.1019)—that is, $u_{tt} - c^2 u_{xx}$— yields waves moving to the right or left with speed c. The *reduced part* $u_t - \alpha u_x$ gives rise to a traveling wave moving with speed $|\alpha|$. We have shown that wave motions with both speeds play a role in the solution of (9.1019) in different regions. In (9.1023) we obtained approximate equations corresponding to waves moving with speed c and in (9.1027)–(9.1028) we obtained an equation characterizing wave motion with the speed $|\alpha|$. In this section we show how to construct such simplified approximating equations that retain certain aspects of the given equations. Since our method is most easily presented in terms of matrix theory, we restrict our discussion to systems of equations.

With $v = u_x$ and $w = u_t$, the dissipative equation (9.1019) can be written as a first order system

$$(9.1029) \qquad \mathbf{u}_t + A\mathbf{u}_x + B\mathbf{u} = \mathbf{0},$$

where

$$(9.1030) \qquad \mathbf{u} = \begin{bmatrix} v \\ w \end{bmatrix}; \qquad A = \begin{bmatrix} 0 & -1 \\ -c^2 & 0 \end{bmatrix}; \qquad B = \begin{bmatrix} 0 & 0 \\ -\alpha & 1 \end{bmatrix}.$$

Then (9.1020) is equivalent to

$$(9.1031) \qquad \mathbf{u}_\tau + A\mathbf{u}_\sigma + \epsilon B\mathbf{u} = \mathbf{0}$$

and (9.1025) corresponds to

$$(9.1032) \qquad \epsilon[\mathbf{u}_\tau + A\mathbf{u}_\sigma] + B\mathbf{u} = \mathbf{0}.$$

We introduce decompositions of (9.1031) and (9.1032) that correspond to the reduced equations obtained on setting $\epsilon = 0$ in each of these systems of equations. To carry out these decompositions we require a result from matrix theory.

Let M be an n by n matrix and $C(\lambda) = |M - \lambda I|$ its *characteristic polynomial*. We factor $C(\lambda)$ as $C(\lambda) = C_1(\lambda)C_2(\lambda)$ where the factors $C_1(\lambda)$ and $C_2(\lambda)$ contain no common eigenvalues. Introducing the partial fraction decomposition

$$(9.1033) \qquad \frac{1}{C(\lambda)} = \frac{c_1(\lambda)}{C_1(\lambda)} + \frac{c_2(\lambda)}{C_2(\lambda)},$$

we have

$$(9.1034) \qquad 1 = c_1(\lambda)C_2(\lambda) + c_2(\lambda)C_1(\lambda).$$

Then the matrices

$$(9.1035) \qquad P_1 = c_1(M)C_2(M); \qquad P_2 = c_2(M)C_1(M)$$

are *projection matrices* that project vectors into the *eigenspaces* spanned by the eigenvectors corresponding to the factors $C_1(\lambda)$ and $C_2(\lambda)$, respectively. They have the following additional properties:

$$(9.1036) \qquad P_1 + P_2 = I; \qquad P_1P_2 = P_2P_1 = 0; \qquad P_1M = MP_1;$$
$$P_2M = MP_2; \qquad P_1^2 = P_1; \qquad P_2^2 = P_2.$$

We do not require any additional facts regarding these matrices for our discussion beyond what was stated. We assume in our discussion that the matrix M has n linearly independent eigenvectors.

As an example, we construct projection operators for the matrices A and B defined in (9.1030). The characteristic polynomial for A is $C(\lambda) = |A - \lambda I| = \lambda^2 - c^2 = (\lambda + c)(\lambda - c)$. With $C_1(\lambda) = \lambda + c$ and $C_2(\lambda) = \lambda - c$, we have

(9.1037)
$$\frac{1}{\lambda^2 - c^2} = \frac{(-1/2c)}{\lambda + c} + \frac{(1/2c)}{\lambda - c} = \frac{(-1/2c)(\lambda - c) + (1/2c)(\lambda + c)}{\lambda^2 - c^2}.$$

Thus $c_1(\lambda) = -c_2(\lambda) = 1/2c$, $c_1(\lambda)C_2(\lambda) = (-1/2c)(\lambda - c)$, and $c_2(\lambda)C_1(\lambda) = (1/2c)(\lambda + c)$. Finally, the projection operators are given as

$$(9.1038) \qquad P_1 = -\frac{1}{2c}[A - cI] = \frac{1}{2c}\begin{bmatrix} c & 1 \\ c^2 & c \end{bmatrix},$$

$$(9.1039) \qquad P_2 = \frac{1}{2c}[A + cI] = \frac{1}{2c}\begin{bmatrix} c & -1 \\ -c^2 & c \end{bmatrix}.$$

The matrix A has two eigenvalues $\lambda_1 = -c$ and $\lambda_2 = c$. The corresponding eigenvectors are

$$(9.1040) \quad \mathbf{r}_1 = \begin{bmatrix} 1 \\ c \end{bmatrix}; \quad A\mathbf{r}_1 = -c\mathbf{r}_1; \quad \mathbf{r}_2 = \begin{bmatrix} 1 \\ -c \end{bmatrix}; \quad A\mathbf{r}_2 = c\mathbf{r}_2;$$

and they are linearly independent. We have $P_1\mathbf{r}_1 = \mathbf{r}_1$, $P_1\mathbf{r}_2 = 0$, $P_2\mathbf{r}_1 = 0$, and $P_2\mathbf{r}_2 = \mathbf{r}_2$. Since any two-component vector can be expressed as a linear combination of $\mathbf{r}_1$ and $\mathbf{r}_2$, this shows that P_1 and P_2 project vectors into the required eigenspaces (these are one-dimensional in this case). Also, it is easily seen that P_1 and P_2 have the properties given in (9.1036).

For the matrix B in (9.1030) we have $C(\lambda) = |B - \lambda I| = \lambda(\lambda - 1)$. With $C_1(\lambda) = \lambda$ and $C_2(\lambda) = \lambda - 1$ we have

$$(9.1041) \qquad \frac{1}{\lambda(\lambda - 1)} = \frac{-1}{\lambda} + \frac{1}{\lambda - 1} = \frac{-(\lambda - 1) + \lambda}{\lambda(\lambda - 1)}$$

so that

$$(9.1042) \quad P_1 = -[B - I] = \begin{bmatrix} 1 & 0 \\ \alpha & 0 \end{bmatrix}; \quad P_2 = B = \begin{bmatrix} 0 & 0 \\ -\alpha & 1 \end{bmatrix}.$$

The eigenvectors $\mathbf{r}_1$ and $\mathbf{r}_2$ corresponding to the eigenvalues $\lambda_1 = 0$ and $\lambda_2 = 1$, respectively, of B are

$$(9.1043) \quad \mathbf{r}_1 = \begin{bmatrix} 1 \\ \alpha \end{bmatrix}; \quad B\mathbf{r}_1 = 0; \quad \mathbf{r}_2 = \begin{bmatrix} 0 \\ 1 \end{bmatrix}; \quad B\mathbf{r}_2 = \mathbf{r}_2.$$

Again, $\mathbf{r}_1$ and $\mathbf{r}_2$ are linearly independent, and P_1 and P_2 have all the properties required of projection operators.

We now use these projection operators to construct *asymptotic decompositions* of (9.1031) and (9.1032). Beginning with (9.1031) we set

$$(9.1044) \qquad \mathbf{u}(\sigma, \tau) = V(\sigma, \tau)\mathbf{r}_1 + \epsilon W(\sigma, \tau)\mathbf{r}_2$$

where V and W are scalar functions that may depend on ϵ and $\mathbf{r}_1$ and $\mathbf{r}_2$ are the eigenvectors of A given in (9.1040). Inserting (9.1044) into (9.1031) gives

$$(9.1045) \qquad [V_\tau - cV_\sigma]\mathbf{r}_1 + \epsilon\left\{ (W_\tau + cW_\sigma)\mathbf{r}_2 + V\begin{bmatrix} 0 \\ c - \alpha \end{bmatrix} \right\}$$
$$+ \epsilon^2\left\{ -W\begin{bmatrix} 0 \\ c + \alpha \end{bmatrix} \right\} = 0.$$

Applying the projection operator P_1 to (9.1045)—that is, multiplying across on the left by P_1—gives

$$(9.1046) \qquad \left\{ V_\tau - cV_\sigma + \epsilon\left(\frac{c - \alpha}{2c}\right)V - \epsilon^2\left(\frac{c + \alpha}{2c}\right)W \right\}\mathbf{r}_1 = 0,$$

since $P_1\mathbf{r}_2 = \mathbf{0}$. Applying P_2 to (9.1045) yields

$$(9.1047) \quad \epsilon\left\{W_\tau + cW_\sigma - \left(\frac{c-\alpha}{2c}\right)V\right\}\mathbf{r}_2 + \epsilon^2\left(\frac{c+\alpha}{2c}\right)W\mathbf{r}_2 = \mathbf{0}.$$

Retaining terms up to order ϵ in (9.1046) as was done in our discussion of the scalar example, we find that

$$(9.1048) \qquad V_\tau - cV_\sigma + \epsilon\left(\frac{c-\alpha}{2c}\right)V = 0,$$

and this agrees with (9.1023). The leading term of (9.1047) shows that W can be expressed in terms of V apart from an arbitrary function of $\sigma - c\tau$. If this arbitrary function is disregarded, we can replace W in (9.1046) by a function of V given in integral form. Higher approximations may be obtained by expanding W in a power series in ϵ and solving (9.1047) for the coefficients as functions of V, but we do not pursue this matter here. We do note that if $\mathbf{r}_1$ and $\mathbf{r}_2$ are interchanged in (9.1044), we obtain the second form of the equations for u given in (9.1023) as the equation satisfied by V.

Next we consider the system (9.1032) and again represent $\mathbf{u}$ as in (9.1044), with $\mathbf{r}_1$ and $\mathbf{r}_2$ as defined in (9.1043). Inserting (9.1044) into (9.1032) yields

$$(9.1049) \quad \epsilon\left\{V_\tau\mathbf{r}_1 + V_\sigma\begin{bmatrix} -\alpha \\ -c^2 \end{bmatrix} + W\mathbf{r}_2\right\} + \epsilon^2\left\{W_\tau\mathbf{r}_2 + W_\sigma\begin{bmatrix} -1 \\ 0 \end{bmatrix}\right\} = \mathbf{0},$$

since $B\mathbf{r}_1 = \mathbf{0}$. Applying the projection operator P_1, as given in (9.1042), to (9.1049) gives

$$(9.1050) \qquad \{V_\tau - \alpha V_\sigma - \epsilon W_\sigma\}\mathbf{r}_1 = \mathbf{0},$$

as is easily seen. Applying $P_2 = B$ to (9.1049) results in

$$(9.1051) \qquad \{(\alpha^2 - c^2)V_\sigma + W\}\mathbf{r}_2 + \epsilon\{W_\tau + \alpha W_\sigma\}\mathbf{r}_2 = \mathbf{0}.$$

We solve (9.1051) by expanding W as

$$(9.1052) \qquad W = \sum_{j=0}^{\infty} W_j\epsilon^j.$$

Then W_0 has the form

$$(9.1053) \qquad W_0(\sigma, \tau) = (c^2 - \alpha^2)V_\sigma.$$

The expansion (9.1052) must also be inserted into (9.1050), and, retaining only terms up to order ϵ, we have

(9.1054) $V_\tau - \alpha V_\sigma - \epsilon(c^2 - \alpha^2)V_{\sigma\sigma} = 0.$

This result has the form of (9.1027). Clearly, the W_j in (9.1052) are all expressed in terms of increasingly higher order derivatives of V, as is seen from the form of (9.1051). Thus in neglecting $O(\epsilon^2)$ terms to obtain (9.1054) we are disregarding presumably smoother higher derivative terms in V, in agreement with the discussion given for the scalar version of this example.

For (9.1032), we do not consider the effect of interchanging $\mathbf{r}_1$ and $\mathbf{r}_2$ in the assumed solution form (9.1044) because on setting $\epsilon = 0$ in (9.1032) we obtain $B\mathbf{u} = \mathbf{0}$, and this has the solution $\mathbf{u} = V\mathbf{r}_1$ where V is arbitrary. It is this expression that was perturbed around in (9.1044). Similarly, with $\epsilon = 0$ in (9.1031) we have $\mathbf{u}_\tau + A\mathbf{u}_\sigma = \mathbf{0}$, and this has the solutions $\mathbf{u} = \tilde{V}(\sigma + c\tau)\mathbf{r}_1$ and $\mathbf{u} = \hat{V}(\sigma - c\tau)\mathbf{r}_2$ with arbitrary $\tilde{V}$ and $\hat{V}$. Therefore, we considered perturbations corresponding to each of the eigenvectors $\mathbf{r}_1$ and $\mathbf{r}_2$ in our discussion of (9.1031).

The system obtained on setting $\epsilon = 0$ in (9.1031) or (9.1032) is known as the *reduced system*. We constructed asymptotic simplifications of the full systems corresponding to solutions for the reduced system. The solutions of the reduced system are related to the properties of a relevant coefficient matrix. Based on a decomposition of that matrix according to its eigenvectors (even though this was not done explicitly), a decomposition of the full system was carried out. We now show how this technique may be applied to a general class of systems of equations.

We consider systems of equations of the form

(9.1055) $\mathbf{u}_t + A\mathbf{u}_x + \epsilon N[\mathbf{u}] = \mathbf{0}$

where $\mathbf{u}$ is an n-component vector, A is a constant n by n matrix, and N is a linear or nonlinear differential operator. The matrix A is assumed to have n linearly independent eigenvectors. With $C(\lambda) = |A - \lambda I|$ as the characteristic polynomial of A, we assume it can be factored as $C(\lambda) = C_1(\lambda)C_2(\lambda)$ where $C_1(\lambda)$ and $C_2(\lambda)$ have no common eigenvalues as roots. Further, it is assumed that $0 < m < n$ independent eigenvectors are associated with the eigenvalues of $C_1(\lambda)$ and the remaining $n - m$ independent eigenvectors are associated with $C_2(\lambda)$. The eigenspaces spanned by these two sets of eigenvectors have the property that any n-component vector can be expressed as a sum of two vectors, each of which is one of the eigenspaces.

Proceeding as before, but replacing M by A in the equations, we can construct the projection operators P_1 and P_2 associated with the eigenspaces corresponding to $C_1(\lambda)$ and $C_2(\lambda)$, respectively. If $\mathbf{v}$ lies in the eigenspace related to $C_1(\lambda)$ and $\mathbf{w}$ lies in the eigenspace related to $C_2(\lambda)$, we have $P_1\mathbf{v} = \mathbf{v}$, $P_1\mathbf{w} = \mathbf{0}$, $P_2\mathbf{w} = \mathbf{w}$, and $P_2\mathbf{v} = \mathbf{0}$ as easily follows from the properties of P_1 and P_2. We recall that the projection operators were defined in (9.1035).

With $\mathbf{v}$ and $\mathbf{w}$ lying in the two eigenspaces defined previously, we look for a solution of (9.1055) in the form

$$(9.1056) \qquad \mathbf{u} = \mathbf{v} + \epsilon\mathbf{w}.$$

In doing so it is assumed that we are interested in perturbing $\mathbf{u}$ around a solution of the reduced system $\mathbf{u}_t + A\mathbf{u}_x = \mathbf{0}$ which lies in the eigenspace associated with $\mathbf{v}$.

Inserting (9.1056) into (9.1055) yields

$$(9.1057) \qquad \mathbf{v}_t + A\mathbf{v}_x + \epsilon\{\mathbf{w}_t + A\mathbf{w}_x + N[\mathbf{v} + \epsilon\mathbf{w}]\} = \mathbf{0}.$$

We assume that $N[\mathbf{v} + \epsilon\mathbf{w}]$ can be expanded in powers of ϵ with $N[\mathbf{v} + \epsilon\mathbf{w}] = N_0[\mathbf{v}] + O(\epsilon)$. Multiplying (9.1057) by the projection operator P_1 gives

$$(9.1058) \qquad \mathbf{v}_t + A\mathbf{v}_x + \epsilon P_1 N[\mathbf{v} + \epsilon\mathbf{w}] = \mathbf{0}.$$

To obtain (9.1058) we have used the facts that $P_1 A = AP_1$ and that P_1 is a constant matrix. Thus, for example, $P_1 A\mathbf{v}_x = AP_1\mathbf{v}_x = A(\partial/\partial x)[P_1\mathbf{v}] = A\mathbf{v}_x$ and $P_1\mathbf{w}_t = (\partial/\partial t)[P_1\mathbf{w}] = \mathbf{0}$. Similarly, we obtain on multiplying (9.1057) by P_2,

$$(9.1059) \qquad \epsilon\{\mathbf{w}_t + A\mathbf{w}_x + P_2 N[\mathbf{v} + \epsilon\mathbf{w}]\} = \mathbf{0}.$$

Next, we expand $\mathbf{w}$ as

$$(9.1060) \qquad \mathbf{w} = \sum_{j=0}^{\infty} \mathbf{w}_j \epsilon^j$$

and insert this series into (9.1058) and (9.1059). Using the expansion of $N[\mathbf{v} + \epsilon\mathbf{w}]$ in powers of ϵ, we equate like powers of ϵ in (9.1059). This yields

$$(9.1061) \qquad \frac{\partial \mathbf{w}_0}{\partial t} + A\frac{\partial \mathbf{w}_0}{\partial x} + P_2 N_0[\mathbf{v}] = \mathbf{0}$$

for $\mathbf{w}_0$, and similar equations for the remaining $\mathbf{w}_j$. The equations for $\mathbf{w}_j$ are to be solved only in terms of $\mathbf{v}$ and any solutions of the homogeneous versions of these equations (which do not depend on $\mathbf{v}$) are to be discarded. The resulting expressions for the $\mathbf{w}_j$ are then inserted into (9.1058). This yields the *simplified system* for $\mathbf{v}$ since $\mathbf{v}$ belongs to a lower dimensional space than $\mathbf{u}$. Thus (9.1058) effectively contains fewer equations in fewer unknowns than (9.1055).

The equation (9.1058) for $\mathbf{v}$ may have a very complicated form. However, we do not expand $\mathbf{v}$ in powers of ϵ since this—in combination with (9.1060)—would yield the conventional perturbation solution of (9.1055). That solution may contain secular behavior that we are trying to avoid. Instead, we truncate (9.1058) at some order of ϵ. This new equation might then be used as a basis for a perturbation result. For example, retaining only $O(\epsilon)$ terms in (9.1058) gives

$$(9.1062) \qquad\qquad \mathbf{v}_t + A\mathbf{v}_x + \epsilon P_1 N_0[\mathbf{v}] = \mathbf{0}.$$

This equation corresponds to (9.1048) in the example considered earlier. Combined with (9.1061), this yields a leading order approximation to a solution of (9.1055). The leading order perturbation result can always be retrieved by expanding $\mathbf{v}$ in a series in powers of ϵ. Thus (9.1058) represents a more general result and it can be constructed for each eigenspace associated with the matrix A.

For example, if $\lambda = \lambda_1$ is a simple eigenvalue of A and $\mathbf{r}_1$ is the corresponding eigenvector, we may consider $C_1(\lambda) = \lambda - \lambda_1$. The eigenspace for $C_1(\lambda)$ is then one-dimensional and is spanned by $\mathbf{r}_1$. Then we can set $\mathbf{v} = V\mathbf{r}_1$, where V is a scalar function, in (9.1056). Since P_1 projects vectors into the eigenspace spanned by $\mathbf{r}_1$, we have $P_1 N[\mathbf{v} + \epsilon\mathbf{w}] = \alpha(V, \mathbf{w})\mathbf{r}_1$ where α is a scalar function of V and $\mathbf{w}$. Thus (9.1058) takes the form

$$(9.1063) \qquad\qquad [V_t + \lambda_1 V_x + \epsilon\alpha(V, \mathbf{w})]\mathbf{r}_1 = \mathbf{0},$$

and this is a scalar equation for V once the $\mathbf{w}_j$ are specified in terms of V.

In the two examples considered we show how equations of the form (9.1055) arise in fluid dynamics and the theory of water waves. The aforementioned theory is then used to obtain simplified forms for these equations. We note that the simplification method presented can also be applied if the matrix A in (9.1055) is a slowly varying function of x and t. Also, a related simplification method can be used even if the given equation is not in the form (9.1055) as we have shown for the dissipative wave equation. However, we restrict our discussion only to the preceding constant coefficient case. An extension of this method to higher space dimensions is not obvious since the reduced system would generally contain more than one coefficient matrix to be dealt with in such manner. If one of the spatial variables can be distinguished from the others for some reason, this procedure can be carried out but this is not considered here.

EXAMPLE 9.22. THE NAVIER–STOKES AND BURGERS' EQUATIONS

The *Navier–Stokes equations* that describe *viscous fluid flow* in one dimension were given in Example 8.10 as [see (8.378)–(8.381)]

$$(9.1064) \qquad\qquad \rho_t + \rho u_x + u\rho_x = 0,$$

$$(9.1065) \qquad\qquad \rho u_t + \rho u u_x + p_x = \tfrac{4}{3}\mu u_{xx},$$

$$(9.1066) \qquad\qquad \rho c_v[T_t + uT_x] + pu_x = kT_{xx} + \tfrac{4}{3}\mu u_x^2,$$

$$(9.1067) \qquad\qquad p = R\rho T.$$

Here ρ is the density, u is the velocity, p is the pressure, and T is the temperature. The parameter μ is the viscosity coefficient, k is the coefficient of heat conduction, c_p and c_v are the specific heats at constant pressure and volume, respectively, and $R = c_p - c_v$ is the gas constant.

To study these equations we *perturb* ρ and T around a constant density and temperature and u around a zero velocity. Thus we set

$$(9.1068) \qquad u = \epsilon \hat{u}; \qquad \rho = \rho_0 + \epsilon \hat{\rho}; \quad T = T_0 + \epsilon \hat{T}$$

where $\epsilon > 0$ is a small parameter. Also, we consider the solution in the far field and at large times and let

$$(9.1069) \qquad x = \frac{\sigma}{\epsilon}; \qquad t = \frac{\tau}{\epsilon}.$$

Inserting (9.1068)–(9.1069) into (9.1064)–(9.1067) and using (9.1067) to express p as a function of ρ and T, we readily obtain

$$(9.1070) \qquad \hat{\rho}_\tau + \rho_0 \hat{u}_\sigma + \epsilon [\hat{\rho} \hat{u}_\sigma + \hat{u} \hat{\rho}_\sigma] = 0,$$

$$(9.1071) \qquad \hat{u}_\tau + \left(\frac{RT_0}{\rho_0} \right) \hat{\rho}_\sigma + R \hat{T}_\sigma + \epsilon \left[\left(\frac{R}{\rho_0} \right) \hat{T} \hat{\rho}_\sigma + \left(\frac{1}{\rho_0} \right) \hat{\rho} \hat{u}_\tau + \hat{u} \hat{u}_\sigma \right.$$

$$\left. + \left(\frac{R}{\rho_0} \right) \hat{\rho} \hat{T}_\sigma - \left(\frac{4\mu}{3\rho_0} \right) \hat{u}_{\sigma\sigma} \right] = 0,$$

$$(9.1072) \qquad \hat{T}_\tau + \left(\frac{RT_0}{c_v} \right) \hat{u}_\sigma + \epsilon \left[\hat{u} \hat{T}_\sigma + \left(\frac{1}{\rho_0} \right) \hat{\rho} \hat{T}_\tau + \left(\frac{RT_0}{c_v \rho_0} \right) \hat{\rho} \hat{u}_\sigma \right.$$

$$\left. + \left(\frac{R}{c_v} \right) \hat{T} \hat{u}_\sigma - \left(\frac{k}{\rho_0 c_v} \right) \hat{T}_{\sigma\sigma} \right] = 0.$$

In these equations we have only retained terms up to order ϵ. If we set $\gamma = c_p/c_v$, the ratio of the specific heats, we find that with $c_p = 1/(\gamma - 1)$, we have $c_v = 1/\gamma(\gamma - 1)$, $R = 1/\gamma$, and $R/c_v = \gamma - 1$. By rescaling the variables and parameters in (9.1070)–(9.1072), it is possible to eliminate the constants ρ_0 and T_0. Thus we may set $\rho_0 = T_0 = 1$. Also, for simplicity of notation we drop the carets on the variables $\hat{\rho}$, $\hat{u}$, and $\hat{T}$ and replace σ and τ by x and t.

As a result, the system (9.1070)–(9.1072) can be written in the form

$$(9.1073) \qquad \mathbf{u}_t + A\mathbf{u}_x + \epsilon [B\mathbf{u}_x + C\mathbf{u}_t + D\mathbf{u}_{xx}] = 0,$$

where

(9.1074)

$$
\mathbf{u} = \begin{bmatrix} \rho \\ u \\ T \end{bmatrix}; \quad
A = \begin{bmatrix} 0 & 1 & 0 \\ \dfrac{1}{\gamma} & 0 & \dfrac{1}{\gamma} \\ 0 & \gamma - 1 & 0 \end{bmatrix}; \quad
D = \begin{bmatrix} 0 & 0 & 0 \\ 0 & -\dfrac{4\mu}{3} & 0 \\ 0 & 0 & k(1 - \gamma)\gamma \end{bmatrix},
$$

(9.1075) $\quad B = \begin{bmatrix} u & \rho & 0 \\ \dfrac{T}{\gamma} & u & \dfrac{\rho}{\gamma} \\ 0 & (\gamma - 1)(T + \rho) & u \end{bmatrix}; \quad C = \begin{bmatrix} 0 & 0 & 0 \\ 0 & \rho & 0 \\ 0 & 0 & \rho \end{bmatrix}.$

This system is of the form (9.1055), and we note that the operator multiplying ϵ is nonlinear. The linear system $\mathbf{u}_t + A\mathbf{u}_x = \mathbf{0}$ obtained on setting $\epsilon = 0$ in (9.1073) yields the *equations of linear acoustics*.

The characteristic polynomial for A is $C(\lambda) = |A - \lambda I| = \lambda(1 - \lambda)(1 + \lambda)$ so that the eigenvalues of A are $\lambda_0 = 0$, $\lambda_1 = 1$, and $\lambda_2 = -1$. The corresponding linearly independent eigenvectors are

(9.1076) $\quad \mathbf{r}_0 = \begin{bmatrix} 1 \\ 0 \\ -1 \end{bmatrix}; \quad \mathbf{r}_1 = \begin{bmatrix} 1 \\ 1 \\ \gamma - 1 \end{bmatrix}; \quad \mathbf{r}_2 = \begin{bmatrix} 1 \\ -1 \\ \gamma - 1 \end{bmatrix}.$

Using these results we easily find that the reduced system

(9.1077) $$\mathbf{u}_t + A\mathbf{u}_x = \mathbf{0}$$

has the solutions

(9.1078) $\quad \mathbf{u} = \alpha(x)\mathbf{r}_0; \quad \mathbf{u} = \beta(x - t)\mathbf{r}_1; \quad \mathbf{u} = \delta(x + t)\mathbf{r}_2$

where α, β, and δ are arbitrary functions. The general solution of (9.1077) is a linear combination of these three solutions and it shows that two waves traveling to the right and to the left with unit speed occur in the solution of the acoustic equations.

Using the (conventional) *perturbation method* to solve (9.1073) we set

(9.1079) $$\mathbf{u} = \mathbf{u}_0 + \epsilon\mathbf{u}_1 + \cdots .$$

Then $\mathbf{u}_0$ satisfies the reduced equation (9.1077). If we choose $\mathbf{u}_0 = \beta(x - t)\mathbf{r}_1$, we readily find that $\mathbf{u}_1$ satisfies

(9.1080) $\quad \dfrac{\partial \mathbf{u}_1}{\partial t} + A\dfrac{\partial \mathbf{u}_1}{\partial x} + \beta\beta'\hat{B}\mathbf{r}_1 - \beta\beta'\hat{C}\mathbf{r}_1 + \beta''D\mathbf{r}_1 = \mathbf{0}$

where

$$(9.1081) \quad \hat{B} = \begin{bmatrix} 1 & 1 & 0 \\ \dfrac{\gamma - 1}{\gamma} & 1 & \dfrac{1}{\gamma} \\ 0 & \gamma(\gamma - 1) & 1 \end{bmatrix}; \quad \hat{C} = \begin{bmatrix} 0 & 0 & 0 \\ 0 & 1 & 0 \\ 0 & 0 & 1 \end{bmatrix}.$$

Either by expressing $\mathbf{u}_1$ and all the vectors in (9.1080) as linear combinations of the eigenvectors (9.1076) or by using the projection operators associated with the eigenspaces for A, the system (9.1080) can be solved. It is then seen that *secular terms* result in the expression for $\mathbf{u}_1$ and the expansion (9.1079) becomes invalid for large t, specifically when $t = O(1/\epsilon)$. We do not carry out the solution of (9.1080). Although we have shown for several problems in this chapter how secular effects can be removed, we do not apply those methods here. Instead, we show how the asymptotic simplification method presented enables us to avoid any problems with secular terms.

The given perturbation result suggests that as we follow the wave $\mathbf{u}_0 = \beta(x - t)\mathbf{r}_1$ that satisfies the reduced system for a long time, the nonlinear and higher derivative terms in (9.1073) become significant. To determine their effect, we look for a solution of (9.1073) in the form

$$(9.1082) \qquad \mathbf{u} = \beta \mathbf{r}_1 + \epsilon[\alpha \mathbf{r}_0 + \delta \mathbf{r}_2].$$

This corresponds to (9.1056) with $\mathbf{v} = \beta \mathbf{r}_1$ and $\mathbf{w} = \alpha \mathbf{r}_0 + \delta \mathbf{r}_2$. The functions $\beta(x, t)$, $\alpha(x, t)$, and $\delta(x, t)$ are to be specified, but $\beta \mathbf{r}_1$ lies in the eigenspace spanned by $\mathbf{r}_1$ whereas $\alpha \mathbf{r}_0 + \delta \mathbf{r}_2$ is a typical vector in the eigenspace spanned by $\mathbf{r}_0$ and $\mathbf{r}_2$. To obtain the projection operators for these eigenspaces we set $C(\lambda) = |A - \lambda I| = C_1(\lambda)C_2(\lambda)$ where $C_1(\lambda) = 1 - \lambda$ and $C_2(\lambda) = \lambda(1 + \lambda)$. The eigenvalue of A corresponding to $C_1(\lambda)$ is $\lambda_1 = 1$, and the remaining eigenvalues $\lambda_0 = 0$ and $\lambda_2 = -1$ correspond to the factor $C_2(\lambda)$. Since

$$(9.1083) \qquad \frac{1}{\lambda(1 - \lambda)(1 + \lambda)} = \frac{1/2}{1 - \lambda} + \frac{1 + \lambda/2}{\lambda(1 + \lambda)},$$

we find that the projection operator P_1 which projects vectors into the eigenspace spanned by $\mathbf{r}_1$ is

$$(9.1084) \quad P_1 = \frac{1}{2}A[I + A] = \frac{1}{2\gamma}\begin{bmatrix} 1 & \gamma & 1 \\ 1 & \gamma & 1 \\ \gamma - 1 & \gamma(\gamma - 1) & \gamma - 1 \end{bmatrix}.$$

The projection operator P_2 corresponding to the eigenspace of $\mathbf{r}_0$ and $\mathbf{r}_2$ is

$$(9.1085) \qquad P_2 = I - P_1 = \frac{1}{2\gamma} \begin{bmatrix} 2\gamma - 1 & -\gamma & -1 \\ -1 & \gamma & -1 \\ 1 - \gamma & \gamma(1 - \gamma) & 1 + \gamma \end{bmatrix}$$

Recalling the definition of $\mathbf{u}$ given in (9.1074), we see that (9.1082) implies

$$(9.1086) \quad \rho = \beta + O(\epsilon); \quad u = \beta + O(\epsilon); \quad T = (\gamma - 1)\beta + O(\epsilon).$$

If these expressions for ρ, u, and T are inserted into the matrices B and C as given in (9.1075), we obtain

$$(9.1087) \qquad\qquad B = \beta\hat{B} + O(\epsilon); \quad C = \beta\hat{C} + O(\epsilon)$$

with $\hat{B}$ and $\hat{C}$ given in (9.1081). Thus on inserting (9.1082) into (9.1073) we have

$$(9.1088) \qquad [\beta_t + \beta_x]\mathbf{r}_1 + \epsilon\{[\delta_t - \delta_x]\mathbf{r}_2 + \alpha_t\mathbf{r}_0$$
$$+ \beta\beta_x\hat{B}\mathbf{r}_1 + \beta\beta_t\hat{C}\mathbf{r}_1 + \beta_{xx}D\mathbf{r}_1\} = O(\epsilon^2).$$

On multiplying (9.1088) with the projection operator P_1 we obtain

$$(9.1089) \quad \beta_t + \beta_x + \epsilon\left\{\left(\frac{\gamma^2 + 3\gamma - 1}{2\gamma}\right)\beta\beta_x + \left(\frac{2\gamma - 1}{2\gamma}\right)\beta\beta_t \right.$$
$$\left. - \left[\frac{2}{3}\mu + \frac{1}{2}k(\gamma - 1)^2\right]\beta_{xx}\right\} = O(\epsilon^2),$$

since $P_1\mathbf{r}_0 = P_1\mathbf{r}_2 = \mathbf{0}$. Now (9.1089) implies that $\beta_t = -\beta_x + O(\epsilon)$, thus we can replace (9.1089) by

$$(9.1090) \quad \beta_t + \beta_x + \epsilon\left\{\left(\frac{\gamma + 1}{2}\right)\beta\beta_x - \left[\frac{2}{3}\mu + \frac{1}{2}k(\gamma - 1)^2\right]\beta_{xx}\right\} = O(\epsilon^2).$$

Neglecting the $O(\epsilon^2)$ terms in (9.1090) we obtain *Burgers' equation* for *right traveling waves*. It contains both a nonlinear term $\beta\beta_x$ and a second derivative term β_{xx}. The combination of these terms in the equation leads to some interesting solutions, as we show when we discuss Burgers' equation.

Next we multiply (9.1088) by P_2 and this yields

$$(9.1091) \qquad [\delta_t - \delta_x]\mathbf{r}_2 + \alpha_t\mathbf{r}_0 + \beta\beta_x P_2\hat{B}\mathbf{r}_1$$
$$+ \beta\beta_t P_2\hat{C}\mathbf{r}_1 + \beta_{xx}P_2 D\mathbf{r}_1 = O(\epsilon).$$

since $P_2 r_1 = 0$. This equation is to be solved by expanding α and δ as

$$(9.1092) \qquad \alpha = \sum_{j=1}^{\infty} \alpha_j \epsilon^j; \qquad \delta = \sum_{j=1}^{\infty} \delta_j \epsilon^j.$$

It is easiest to solve (9.1091) by constructing projection operators $\hat{P}_0$ and $\hat{P}_2$ that project vectors into the eigenspaces spanned by r_0 and r_2, respectively. Using this method and the fact that $P_1 + \hat{P}_0 + \hat{P}_2 = I$, we easily establish that

(9.1093)

$$\hat{P}_0 = \frac{1}{\gamma} \begin{bmatrix} \gamma - 1 & 0 & -1 \\ 0 & 0 & 0 \\ 1 - \gamma & 0 & 1 \end{bmatrix}; \qquad \hat{P}_2 = \frac{1}{2\gamma} \begin{bmatrix} 1 & -\gamma & 1 \\ -1 & \gamma & -1 \\ \gamma - 1 & \gamma(1 - \gamma) & \gamma - 1 \end{bmatrix}.$$

Then $\hat{P}_0 r_0 = r_0$, $\hat{P}_0 r_2 = 0$, $\hat{P}_2 r_2 = r_2$, and $\hat{P}_2 r_0 = 0$. Since $\beta_t = -\beta_x + O(\epsilon)$, we may express (9.1091) as

$$(9.1094) \quad [\delta_t - \delta_x]r_2 + \alpha_t r_0 + \beta \beta_x [P_2 \hat{B} - P_2 \hat{C}]r_1 + \beta_{xx} P_2 D r_1 = O(\epsilon).$$

Inserting (9.1092) into (9.1094) and multiplying by $\hat{P}_0$ yields for α_0,

$$(9.1095) \qquad \frac{\partial \alpha_0}{\partial t} + (1 - \gamma)\beta \beta_x + k(\gamma - 1)^2 \beta_{xx} = 0.$$

Multiplying (9.1094) by $\hat{P}_2$, we obtain for δ_0,

$$(9.1096) \quad \frac{\partial \delta_0}{\partial t} - \frac{\partial \delta_0}{\partial x} + \left(\frac{\gamma^2 - 3\gamma + 4}{2\gamma} \right)\beta \beta_x + \left[\frac{2}{3}\mu - \frac{1}{2}k(\gamma - 1)^2 \right]\beta_{xx} = 0.$$

We do not consider the equations for α_j and δ_j with $j \geq 1$.

If we set

$$(9.1097) \qquad \mathbf{u} = \delta r_2 + \epsilon \{\alpha r_0 + \beta r_1\},$$

it may be shown by following this procedure that $\delta(x, t)$ satisfies a *Burgers' equation* for *left traveling waves*. The derivation is carried out in the exercises. However, if we set

$$(9.1098) \qquad \mathbf{u} = \alpha r_0 + \epsilon \{\beta r_1 + \delta r_2\},$$

we find that α satisfies a *heat conduction equation* to the previous level of approximation as we now show.

With $\mathbf{u}$ given by (9.1098) we have

$$(9.1099) \quad \rho = \alpha + O(\epsilon); \qquad u = O(\epsilon); \qquad T = (\gamma - 1)\alpha + O(\epsilon),$$

and the matrices B and C become

$$(9.1100) \quad B = \alpha \begin{bmatrix} 0 & 1 & 0 \\ \dfrac{\gamma - 1}{\gamma} & 0 & \dfrac{1}{\gamma} \\ 0 & \gamma(\gamma - 1) & 0 \end{bmatrix} + O(\epsilon) \equiv \alpha\tilde{B} + O(\epsilon),$$

and $C = \alpha\hat{C} + O(\epsilon)$ with $\hat{C}$ defined as in (9.1081). Thus inserting (9.1098) into (9.1073) yields

$$(9.1101) \quad \alpha_t \mathbf{r}_0 + \epsilon\{(\beta_t + \beta_x)\mathbf{r}_1 + (\delta_t - \delta_x)\mathbf{r}_2$$
$$+ \alpha\alpha_x \tilde{B}\mathbf{r}_0 + \alpha\alpha_t \hat{C}\mathbf{r}_0 + \alpha_{xx} D\mathbf{r}_0\} = O(\epsilon^2).$$

Multiplying (9.1101) by the projection operator $\hat{P}_0$ gives

$$(9.1102) \quad \alpha_t + \epsilon\left\{\frac{1}{\gamma}\alpha\alpha_t - k(\gamma - 1)\alpha_{xx}\right\} = O(\epsilon^2).$$

Since $\alpha_t = O(\epsilon)$, we may replace (9.1102) by

$$(9.1103) \quad \alpha_t - \epsilon k(\gamma - 1)\alpha_{xx} = O(\epsilon^2).$$

Since $\gamma > 1$, this has the form of the heat conduction equation if the $O(\epsilon^2)$ terms are neglected. On using the projection operators P_1 and $\hat{P}_2$ in (9.1101) we may express β and δ in terms of α, but this is not carried out here.

Truncating (9.1090) at the $O(\epsilon^2)$ level, we find that $\beta(x, t)$ satisfies the *Burgers' equation*

$$(9.1104) \quad \beta_t + \beta_x + \epsilon\left\{\left(\frac{\gamma + 1}{2}\right)\beta\beta_x - \left[\frac{2}{3}\mu + \frac{1}{2}k(\gamma - 1)^2\right]\beta_{xx}\right\} = 0.$$

The leading terms α_0 and δ_0 in the expansions of α and δ are expressed in terms of β by means of (9.1095) and (9.1096). The perturbed density, velocity, and temperature as defined in (9.1068) are thus given as

$$(9.1105) \quad \hat{\rho} = \beta + \epsilon\alpha_0 + \epsilon\delta_0; \qquad \hat{u} = \beta - \epsilon\delta_0;$$
$$\hat{T} = (\gamma - 1)\beta - \epsilon\alpha_0 + \epsilon(\gamma - 1)\delta_0.$$

In order to compare our present results with those obtained in Example 8.10, we discuss the preceding approximations as they relate to the velocity $u = \epsilon\hat{u}$—see (9.1068). We have

$$(9.1106) \quad u = \epsilon\hat{u} = \epsilon\beta - \epsilon^2\delta_0,$$

and to leading order we set $u = \epsilon\beta$. Multiplying (9.1104) by ϵ gives

$$(9.1107) \quad u_t + \left[1 + \left(\frac{\gamma + 1}{2}\right)u\right]u_x - \epsilon\left[\frac{2}{3}\mu + \frac{1}{2}k(\gamma - 1)^2\right]u_{xx} = 0.$$

The diffusion term u_{xx} in (9.1107) disappears if $\epsilon = 0$ or if $\mu = k = 0$. Then (9.1107) takes the form of the quasilinear one-dimensional wave equation (8.373) except that $c_0 = 1$ in our case. The equation (8.373) was obtained on the basis of Euler's equations of motion where viscosity and conductivity are neglected (i.e., we have $\mu = k = 0$). It was also assumed that the gas is polytropic; that is, (8.356) is valid. Consequently, (9.1107) is seen to represent a modification of the quasilinear equation (8.373) whose solutions were called *simple waves* in Example 8.10. Since both (8.373) and (9.1107) can be solved exactly, we can compare their solutions for identical initial values and determine the role of the diffusion term in smoothing out the solutions.

With

$$(9.1108) \qquad\qquad v = \tfrac{2}{3}\mu + \tfrac{1}{2}k(\gamma - 1)^2,$$

(9.1107) becomes

$$(9.1109) \qquad\qquad u_t + \left[1 + \left(\frac{\gamma + 1}{2}\right)u\right]u_x - \epsilon v u_{xx} = 0.$$

Putting ϵ or v equal to zero in (9.1109), the solution of the initial value problem for the resulting first order equation with the initial condition

$$(9.1110) \qquad\qquad u(x,0) = h(x)$$

is given implicitly as

$$(9.1111) \qquad\qquad u = h\left[x - \left(1 + \frac{\gamma + 1}{2}u\right)t\right].$$

It was observed by Cole and Hopf that if we set

$$(9.1112) \qquad\qquad 1 + \left(\frac{\gamma + 1}{2}\right)u = -2\epsilon v\frac{v_x}{v}$$

in (9.1109), then v satisfies the *heat conduction equation*

$$(9.1113) \qquad\qquad v_t = \epsilon v v_{xx},$$

as is readily verified. The initial condition (9.1110) for $u(x,0)$ takes the form

$$(9.1114) \quad v(x,0) = \exp\left\{-\frac{1}{2\epsilon v}\int_0^x\left[1 + \frac{\gamma + 1}{2}h(s)\right]ds\right\} \equiv H(x),$$

and the solution of the initial value problem (9.1113)–(9.1114) is given as

$$(9.1115) \qquad v(x, t) = \frac{1}{\sqrt{4\pi\epsilon v t}} \int_{-\infty}^{\infty} H(\xi) \exp\left[-\frac{(x - \xi)^2}{4\epsilon v t}\right] d\xi.$$

The solution $u(x, t)$ of (9.1109) and (9.1110) can then be obtained from (9.1112) as

(9.1116)

$$u(x, t) = \frac{[2/(\gamma + 1)] \int_{-\infty}^{\infty} [(x - \xi)/t] \exp[-g/2\epsilon v] \, d\xi}{\int_{-\infty}^{\infty} \exp[-g/2\epsilon v] \, d\xi} - \left(\frac{2}{\gamma + 1}\right)$$

where

$$(9.1117) \qquad g = \int_0^{\xi} \left[1 + \frac{\gamma + 1}{2} h(s)\right] ds + \frac{(x - \xi)^2}{2t}.$$

We study (9.1116) for the special *initial condition*

$$(9.1118) \qquad u(x, 0) = h(x) = \begin{cases} h_1, & x > 0 \\ h_2, & x < 0 \end{cases}$$

where $0 < h_1 < h_2$ and h_1, h_2 are constants. Then g can be expressed in terms of the functions g_i defined as

$$(9.1119) \qquad g_i = a_i \xi + \frac{(x - \xi)^2}{2t} = \frac{[\xi - x + a_i t]^2}{2t}$$

$$+ a_i\left[x - \frac{a_1 + a_2}{2} t\right] + \frac{a_1 a_2}{2} t; \qquad i = 1, 2$$

where $a_i = 1 + [(\gamma + 1)/2]h_i$ $(i = 1, 2)$, in the form

$$(9.1120) \qquad g = \begin{cases} g_1, & \xi > 0 \\ g_2, & \xi < 0 \end{cases}$$

In terms of the *fundamental solution* $G(x, t)$ of the heat equation

$$(9.1121) \qquad G(x, t) = \frac{1}{\sqrt{4\pi t}} \exp\left[-\frac{x^2}{4t}\right],$$

we can express (9.1116) as

(9.1122)

$$1 + \frac{\gamma + 1}{2} u(x, t)$$

$$= \frac{\exp[\hat{\lambda}(x - \tilde{\lambda}t)/\epsilon v] \int_{-\infty}^{0} [(x - \xi)/t] G_2(\xi) \, d\xi + \int_{0}^{\infty} [(x - \xi)/t] G_1(\xi) \, d\xi}{\exp[\hat{\lambda}(x - \tilde{\lambda}t)/\epsilon v] \int_{-\infty}^{0} G_2(\xi) \, d\xi + \int_{0}^{\infty} G_1(\xi) \, d\xi}$$

where $\tilde{\lambda} = (a_1 + a_2)/2$, $\hat{\lambda} = (a_1 - a_2)/2$ and $G_i(\xi) \equiv G(x - a_i t - \xi, \epsilon v t)$, $(i = 1, 2)$, as is easily verified.

Now as $\epsilon v \to 0$ we have

(9.1123) $\quad \lim_{\epsilon v \to 0} G_i(\xi) = \lim_{\epsilon v \to 0} G(x - a_i t - \xi, \epsilon v t) = \delta(x - a_i t - \xi),$

as follows from the well known property of the fundamental solution $G(x, t)$. Also, since $a_1 < a_2$, the exponentials in (9.1122) tend to infinity as $\epsilon v \to 0$ if $x - \tilde{\lambda}t < 0$, and to zero if $x - \tilde{\lambda}t > 0$. Noting these results we readily find that

(9.1124) $\quad \lim_{\epsilon v \to 0} u(x, t) = \begin{cases} h_1; & x - \left[1 + \dfrac{\gamma + 1}{4}(h_1 + h_2)\right]t > 0 \\[2mm] h_2; & x - \left[1 + \dfrac{\gamma + 1}{4}(h_1 + h_2)\right]t < 0, \end{cases}$

since

(9.1125) $\qquad\qquad \tilde{\lambda} = \dfrac{a_1 + a_2}{2} = 1 + \dfrac{\gamma + 1}{4}(h_1 + h_2).$

It is expected that (9.1124) is a solution of the reduced equation for (9.1109), that is,

(9.1126) $\qquad\qquad u_t + \left[1 + \left(\dfrac{\gamma + 1}{2}\right)u\right]u_x = 0$

with the initial condition (9.1118). However, the solution (9.1111) of (9.1126) shows that u is a constant on the lines $x - \{1 + [(\gamma + 1)/2]u\}t = \xi = \text{constant}$. And at $t = 0$, $u = h_2$ if $x < 0$, and $u = h_1$ if $x > 0$, so that

(9.1127) $\qquad\qquad u(x, t) = \begin{cases} h_1; & x - a_1 t > 0 \\ h_2; & x - a_2 t < 0 \end{cases}.$

Since $a_1 < a_2$, these regions in the (x, t)-plane are given as shown in Figure 9.12. In the sector between $x = a_1 t$ and $x = a_2 t$ the solution (9.1127) is

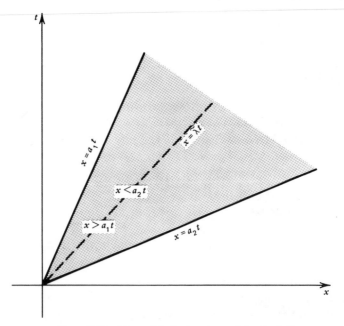

Figure 9.12. The multivalued region and the shock line.

multivalued and since this multivaluedness occurs as soon as t exceeds zero, this solution is not valid for any time $t > 0$.

However, the result (9.1124) yields a (discontinuous) solution of (9.1126) that satisfies the initial condition (9.1118) and has a jump across the line $x = \tilde{\lambda}t$ that lies between $x = a_1 t$ and $x = a_2 t$ as pictured in Figure 9.12. The jump in $u(x, t)$ travels to the right at the speed $\tilde{\lambda}$ and this does not correspond to either of the two characteristic speeds a_1 and a_2 for this problem. The traveling wave solution (9.1124) of (9.1126) is called a *shock wave*. Thus by considering the limit as $\epsilon v \to 0$ of the Burgers' equation (9.1109), we have shown how to obtain a single-valued solution of (9.1126) with the data (9.1118).

We have already indicated in our discussion of quasilinear first order equations in Chapter 2 that the introduction of shock waves permits us to extend solutions into regions where the method of characteristics predicts that they are multivalued. Although there are methods for determining the appropriate shock waves for (9.1126) and other first order quasilinear equations without considering the limit of a second order equation as we have done, our approach is the most satisfactory. This is especially so if the appropriate second order equation is available, as is the case in the example we have considered.

An alternative approach to determining the shock wave solution (9.1124) of (9.1126), which is also useful for more general (quasilinear) equations that do not have an exact solution as in the case of Burgers' equations, is to look for a *traveling wave* solution

(9.1128) $$u(x, t) = v(x - \lambda t)$$

in (9.1109) with the function v and the constant λ to be specified. We also require that

(9.1129) $$v(x - \lambda t) \rightarrow \begin{cases} h_1; & x - \lambda t \rightarrow \infty \\ h_2; & x - \lambda t \rightarrow -\infty \end{cases}$$

with h_1 and h_2 defined as in (9.1118).
  Inserting (9.1128) into (9.1109) gives

(9.1130) $$(1 - \lambda)v' + \left(\frac{\gamma + 1}{2}\right)vv' = \epsilon vv''.$$

Integrating once in (9.1130) and noting that $v' \rightarrow 0$ as $|x - \lambda t| \rightarrow +\infty$, in view of (9.1129), we have

(9.1131) $$(1 - \lambda)v + \left(\frac{\gamma + 1}{4}\right)v^2 + v_0 = \epsilon vv',$$

where v_0 is a constant of integration. As $x - \lambda t \rightarrow \infty$,

(9.1132) $$(1 - \lambda)h_1 + \left(\frac{\gamma + 1}{4}\right)h_1^2 + v_0 = 0$$

and as $x - \lambda t \rightarrow -\infty$,

(9.1133) $$(1 - \lambda)h_2 + \left(\frac{\gamma + 1}{4}\right)h_2^2 + v_0 = 0.$$

  Solving (9.1132)–(9.1133) for λ and v_0 yields

(9.1134) $$\lambda = 1 + \left(\frac{\gamma + 1}{4}\right)(h_1 + h_2) = \tilde{\lambda}$$

as in (9.1125) and

(9.1135) $$v_0 = \left(\frac{\gamma + 1}{4}\right)h_1 h_2.$$

Inserting these values into (9.1131), we note that this equation is separable. Using partial fractions we easily find that

(9.1136)
$$u(x, t) = v(x - \tilde{\lambda}t) = h_1 + \frac{h_2 - h_1}{1 + \exp\{[(\gamma + 1)/4\epsilon v](h_2 - h_1)(x - \tilde{\lambda}t)\}}$$

and this satisfies (9.1129). Furthermore, as $\epsilon v \to 0$, (9.1136) tends to (9.1124) since $h_2 > h_1$, as is readily checked. Thus for small but nonzero ϵv, this solution which is easier to evaluate than (9.1122) may be taken to describe how the initial discontinuity (9.1118) propagates as a smooth step whose transition from the value h_2 to the value h_1 moves at the *shock speed* $\bar{\lambda}$.

More general initial data for (9.1109) and the behavior of the corresponding solution (9.1110) as $\epsilon v \to 0$ may be discussed by evaluating the integrals in (9.1116) by the method of steepest descents or the saddle point method. We do not pursue this matter, however.

EXAMPLE 9.23. THE BOUSSINESQ AND KORTEWEG–DEVRIES EQUATIONS

It can be shown that the *theory of water waves* for the case of *shallow water* and waves of *small amplitude* can be (approximately) described by the *Boussinesq equations*

$$(9.1137) \qquad v_t + w_x + \epsilon[vw_x + wv_x] = 0,$$

$$(9.1138) \qquad w_t + v_x + \epsilon[ww_x - \tfrac{1}{3}w_{xxt}] = 0.$$

Here $v(x, t)$ is related to the equation of the water's surface and $w(x, t)$ is related to the water's velocity. The small parameter ϵ results from the fact that we are considering small wave amplitudes and shallow water, or, equivalently, long waves.

With

$$(9.1139)$$

$$\mathbf{u} = \begin{bmatrix} v \\ w \end{bmatrix}; \qquad A = \begin{bmatrix} 0 & 1 \\ 1 & 0 \end{bmatrix}; \qquad B = \begin{bmatrix} w & v \\ 0 & w \end{bmatrix}; \qquad C = \begin{bmatrix} 0 & 0 \\ 0 & -\tfrac{1}{3} \end{bmatrix}$$

the system (9.1137)–(9.1138) can be written as

$$(9.1140) \qquad \mathbf{u}_t + A\mathbf{u}_x + \epsilon[B\mathbf{u}_x + C\mathbf{u}_{xxt}] = \mathbf{0}.$$

The system (9.1140) is of the form (9.1055). If we put $\epsilon = 0$ in (9.1140), we obtain the linear reduced system $\mathbf{u}_t + A\mathbf{u}_x = \mathbf{0}$, which is equivalent to the one-dimensional *wave equation* with unit wave speed. In fact, the characteristic polynomial for A is $C(\lambda) = |A - \lambda| = (\lambda - 1)(\lambda + 1)$. With $\lambda_1 = 1$ and $\lambda_2 = -1$ as the eigenvalues of A, the corresponding eigenvectors are

$$(9.1141) \qquad \mathbf{r}_1 = \begin{bmatrix} 1 \\ 1 \end{bmatrix}; \qquad \mathbf{r}_2 = \begin{bmatrix} 1 \\ -1 \end{bmatrix}.$$

With $\mathbf{u} = \alpha\mathbf{r}_1$ we have $\mathbf{u}_t + A\mathbf{u}_x = (\alpha_t + \alpha_x)\mathbf{r}_1 = \mathbf{0}$, and when $\mathbf{u} = \beta\mathbf{r}_2$, we have $\mathbf{u}_t + A\mathbf{u}_x = (\beta_t - \beta_x)\mathbf{r}_2 = \mathbf{0}$. Thus $\alpha = \alpha(x - t)$ and $\beta = \beta(x + t)$.

Since A has only two (orthogonal) eigenvectors we do not require the full machinery developed in our general discussion. To study the effect of the nonlinear and higher derivative terms on the right traveling wave obtained from the reduced equation, we set

$$(9.1142) \qquad \mathbf{u} = \alpha\mathbf{r}_1 + \epsilon\beta\mathbf{r}_2,$$

where $\alpha(x, t)$ and $\beta(x, t)$ are to specified. We insert (9.1142) into (9.1140) and obtain

$$(9.1143) \quad [\alpha_t + \alpha_x]\mathbf{r}_1 + \epsilon\left\{[\beta_t - \beta_x]\mathbf{r}_2 + \alpha\alpha_x\begin{bmatrix}2\\1\end{bmatrix} - \frac{1}{3}\alpha_{xxt}\begin{bmatrix}0\\1\end{bmatrix}\right\} = O(\epsilon^2),$$

since $v = \alpha + O(\epsilon)$ and $w = \alpha + O(\epsilon)$. Since $\mathbf{r}_1$ and $\mathbf{r}_2$ are orthogonal, we need not construct projection matrices corresponding to their eigenspaces. Instead, we simply take dot products with $\mathbf{r}_1$ and then with $\mathbf{r}_2$ in (9.1143).

Dotting with $\mathbf{r}_1$ in (9.1143) gives

$$(9.1144) \qquad \alpha_t + \alpha_x + \tfrac{3}{2}\epsilon\alpha\alpha_x - \tfrac{1}{6}\epsilon\alpha_{xxt} = O(\epsilon^2).$$

Dotting with $\mathbf{r}_2$ gives

$$(9.1145) \qquad \beta_t - \beta_x + \tfrac{1}{2}\alpha\alpha_x + \tfrac{1}{6}\alpha_{xxt} = O(\epsilon).$$

We solve (9.1145) by expanding β in a series in powers of ϵ with $\beta = \beta_0 + O(\epsilon)$. Thus

$$(9.1146) \qquad \frac{\partial\beta_0}{\partial t} - \frac{\partial\beta_0}{\partial x} + \frac{1}{2}\alpha\alpha_x + \frac{1}{6}\alpha_{xxt} = 0.$$

Since $\alpha_t = -\alpha_x + O(\epsilon)$ in view of (9.1144), we may replace that equation by

$$(9.1147) \qquad \alpha_t + \alpha_x + \tfrac{3}{2}\epsilon\alpha\alpha_x + \tfrac{1}{6}\epsilon\alpha_{xxx} = 0,$$

on dropping the $O(\epsilon^2)$ terms. This is the *Korteweg–deVries equation*. As we have indicated in Example 9.7, this is a *nonlinear dispersive equation*. Thus it differs from the Burgers' equation which is of dissipative type. It turns out that the Korteweg–deVries equation is exactly solvable; however, the method for doing so is substantially more complicated than that given in the preceding example for Burgers' equation and it is not presented here.

We note that if (9.1147) is solved by expanding α in powers of ϵ (i.e., by using the perturbation method), secular terms result in that expansion. Thus we leave (9.1147) as it stands, and in terms of its solution we obtain for v and w,

$$(9.1148) \qquad v = \alpha + \epsilon\beta_0; \qquad w = \alpha - \epsilon\beta_0$$

in view of (9.1142) and (9.1146) where β_0 is solved for in terms of α. If we interchange αr_1 and βr_2 in (9.1142) we again find on proceeding as earlier, that β satisfies a Korteweg–deVries equation for left traveling waves as shown in the exercises.

We have already discussed approximate periodic solutions of the Korteweg–deVries equation in Example 9.7. We now obtain a special traveling wave solution of (9.1147) that represents what is known as a *solitary wave*. This wave consists of a single hump of constant shape that moves at constant speed.

Let

$$(9.1149) \qquad\qquad \alpha(x, t) = af(x - \lambda t)$$

where the constants a and λ and the function f are to be specified. We assume that $f(x - \lambda t)$ vanishes together with its derivatives as $|x - \lambda t| \to \infty$. Inserting (9.1149) into (9.1147) gives

$$(9.1150) \qquad\qquad (1 - \lambda)f' + \frac{3}{2}\epsilon aff' + \frac{\epsilon}{6}f''' = 0.$$

Integrating once in (9.1150) gives

$$(9.1151) \qquad\qquad (1 - \lambda)f + \frac{3}{4}\epsilon af^2 + \frac{\epsilon}{6}f'' = 0,$$

with no constant of integration since f and f'' vanish at infinity. Next we multiply (9.1151) by f' and integrate once more to obtain

$$(9.1152) \qquad\qquad \left(\frac{1 - \lambda}{2}\right)f^2 + \frac{\epsilon}{4}af^3 + \frac{\epsilon}{12}f'^2 = 0.$$

Again, there is no constant of integration. Then (9.1152) can be written as

$$(9.1153) \qquad\qquad \frac{1}{3a}f'^2 = f^2\left[\frac{2(\lambda - 1)}{\epsilon a} - f\right].$$

The solution of (9.1153) that vanishes at infinity is readily verified to be

$$(9.1154) \quad f(x - \lambda t) = \frac{2(\lambda - 1)}{\epsilon a}\operatorname{sech}^2\left\{\left[\frac{3(\lambda - 1)}{2\epsilon}\right]^{1/2}(x - \lambda t)\right\}.$$

If we set

$$(9.1155) \qquad\qquad \hat{a} = \frac{2(\lambda - 1)}{\epsilon},$$

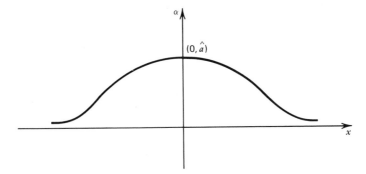

Figure 9.13. The initial waveform.

the solution $\alpha = af(x - \lambda t)$ can be written as

$$(9.1156) \qquad \alpha(x, t) = \hat{a}\,\mathrm{sech}^2\left\{\left(\frac{3}{4}\hat{a}\right)^{1/2}\left[x - \left(1 + \frac{\epsilon\hat{a}}{2}\right)t\right]\right\},$$

with the solitary wave speed given as

$$(9.1157) \qquad \qquad \qquad \lambda = 1 + \frac{\epsilon\hat{a}}{2}.$$

At the time $t = 0$, $\alpha(x, t)$ has the form given in Figure 9.13. The maximum value of $\alpha(x, 0)$ occurs at $x = 0$ and equals $\hat{a}$. The waveform $\alpha(x, 0)$ is symmetric with respect to $x = 0$. This form travels to the right without change in shape at the speed $\lambda = 1 + \epsilon\hat{a}/2$. Thus as $\hat{a}$ increases, so does the wave speed, and higher solitary waves move more rapidly than lower ones. This leads to some interesting problems regarding the interaction of two or more solitary waves moving at different speeds. Also, by modifying the foregoing discussion it is possible to obtain exact periodic traveling wave solutions of the Korteweg–deVries equation. We do not discuss these questions here.

The Burgers', Korteweg–deVries, and nonlinear Schrödinger equations that were discussed in this and the preceding section occur as the relevant approximate equations for a number of significant problems of physical interest. Each of these equations can be solved exactly, and the study of the solution of these and related nonlinear partial differential equations has been the subject of intense investigation in recent years.

EXERCISES FOR CHAPTER 9

Section 9.2

9.2.1. Use the perturbation method to solve the following boundary value problem.

$$u_{xx} + u_{yy} + \epsilon^2 u = 1; \qquad x^2 + y^2 < 1;$$
$$u(x, y) = 0; \qquad x^2 + y^2 = 1.$$

Obtain the first two terms in the expansion.

9.2.2. Obtain the exact solution of the following boundary value problem for the Helmholtz equation in the unit sphere.

$$\nabla^2 u + \epsilon^2 u = 0; \qquad x^2 + y^2 + z^2 < 1;$$
$$u(x, y, z) = 1; \qquad x^2 + y^2 + z^2 = 1.$$

Also, solve the problem using perturbation theory and compare the exact with the approximate solution up to terms of $O(\epsilon^4)$.

9.2.3. Use the perturbation method to solve:

$$\left[1 + \epsilon(x^2 + y^2)\right]u_{xx} + u_{yy} = 1; \qquad x^2 + y^2 < 1;$$
$$u(x, y) = 0; \qquad x^2 + y^2 = 1.$$

9.2.4. Apply the perturbation method to obtain an approximate solution of the following problem.

$$u_{xx} + u_{yy} - \epsilon^2 u^2 = 0; \qquad x^2 + y^2 < 1;$$
$$u(x, y) = 1; \qquad x^2 + y^2 = 1.$$

9.2.5. Let $u(x, y)$ satisfy

$$u_{xx} + (1 + \epsilon y)u_{yy} = 0; \qquad 0 < x < \pi, \quad 0 < y < \pi,$$

and the boundary conditions

$$u(0, y) = u(\pi, y) = u(x, \pi) = 0,$$
$$u(x, 0) = 1; \qquad 0 < x < \pi.$$

Use the perturbation method to obtain an approximate solution.

9.2.6. Let $K(x, y, z; \xi, \eta, \zeta)$ represent the free space Green's function for the modified Helmholtz equation

$$\nabla^2 K - \epsilon^2 K = -\delta(x - \xi)\,\delta(y - \eta)\,\delta(z - \zeta)$$

with $-\infty < x, y, z, \xi, \eta, \zeta < \infty$. Assuming ϵ is small, apply the perturbation method to obtain K in terms of the Green's function for Laplace's equation.

Using the results of Example 6.12, discuss the behavior of the exact and the approximate Green's functions at infinity. Explain the differences in their behavior at infinity as a result of the occurrence of secular terms in the perturbation series.

9.2.7. Consider the boundary value problem of the third kind for the elliptic equation

$$- \nabla \cdot (p\nabla u) + qu = \rho F$$

in a region G with the boundary condition

$$\alpha u + \beta \frac{\partial u}{\partial n}\bigg|_{\partial G} = B$$

with $\alpha > 0$ and $\beta > 0$. Assuming either α or β are uniformly small on ∂G, construct a perturbation method whereby the solution of the given third boundary value problem can be approximated by solutions of Dirichlet or Neumann problems. Comment on the possibility that the data for the reduced Neumann problem are incompatible and explain how to resolve that difficulty.

9.2.8. Use the method of Exercise 9.2.7 to solve Laplace's equation $\nabla^2 u = 0$ in the circle $r^2 = x^2 + y^2 < 1$ with the boundary conditions:

(a) $\epsilon \dfrac{\partial u}{\partial r} + u = 1;$ $r = 1.$

(b) $\dfrac{\partial u}{\partial r} + \epsilon u = 1;$ $r = 1.$

Show that the problem of part (b) cannot be solved by using the perturbation method and obtain the exact solution of the problem to show why the regular perturbation theory fails.

9.2.9. Use the perturbation method to solve the Cauchy problem for the parabolic equation

$$u_t + \epsilon u_x = u_{xx}, \qquad -\infty < x < \infty, \quad t > 0.$$

Determine that the perturbation series exhibits secular behavior and use the method of multiple scales to deal with this problem. Also, convert the given problem to an initial value problem for the heat equation by introducing the change of variable $\tau = t$, $\sigma = x - \epsilon t$. Obtain the solution of the transformed problem and compare it with the perturbation result.

9.2.10. Solve the Cauchy problem for the weakly damped wave equation

$$u_{tt} + \epsilon u_t - c^2 u_{xx} = 0, \qquad -\infty < x < \infty, \quad t > 0$$

with the initial data

$$u(x,0) = f(x); \qquad u_t(x,0) = 0$$

using perturbation theory. Show that secular terms arise in the perturbation expansion and use the method of multiple scales to eliminate them.

9.2.11. Construct a solution of the Cauchy problem (9.73)–(9.74) in the form.

$$u(x, t) = \sum_{n=0}^{\infty} \epsilon^{2n} v_n(t) \cos[(2n + 1)kx].$$

Note that this should lead to the expansion (9.89) but show that $v_0(t)$ can be specified so that the $O(\epsilon^2)$ secular term is eliminated. Indicate how $v_1(t)$ must be chosen to eliminate secular behavior at the $O(\epsilon^4)$ level.

9.2.12. Show that if $c(\epsilon x)$ in the wave equation (9.93) is given as in (9.120), the equation can be transformed into an equation with constant coefficients in the region $x > 0$. Let $\sigma = (1/\epsilon)\log[c_0 + \epsilon x]$ and $\tau = t$ in (9.93) and obtain the equation

$$v_{\tau\tau} - v_{\sigma\sigma} + \epsilon v_\sigma = 0.$$

If the data for the wave equation (9.93) are $u(0, t) = 0$, $u(x,0) = f(x)$, and $u_t(x,0) = 0$, so that the problem is given in the region $x > 0$, $t > 0$, determine the appropriate region and data for the transformed problem for $v(\sigma, \tau)$. Discuss the approximate solution of the initial and boundary value problem for $v(\sigma, \tau)$ in the region $\sigma - \tau > 0$ and compare with the result (9.124).

9.2.13. Consider the initial and boundary value problem for the parabolic equation.

$$u_t - (c_0 + \epsilon x)^2 u_{xx} = 0, \qquad x > 0, \quad t > 0$$

where $c_0 > 0$ and $0 < \epsilon \ll 1$, with the data

$$u(0, t) = 0; \qquad u(x,0) = f(x).$$

Solve by using the perturbation method and show that secular terms occur in the perturbation series. Attempt to eliminate this difficulty by using the method of multiple scales. Use the transformation given in Exercise 9.2.12 to transform the problem into one with constant coefficients that can be solved exactly. Compare the exact result with the perturbation result.

9.2.14. Using the boundary perturbation method, solve the Neumann problem for Laplace's equation

$$u_{xx} + u_{yy} = 0, \qquad 0 < x < l, \quad \epsilon x < y < L$$

with the data

$$\frac{\partial u(0, y)}{\partial x} = \frac{\partial u(l, y)}{\partial x} = \frac{\partial u(x, L)}{\partial y} = 0;$$

$$\frac{\partial u(x, \epsilon x)}{\partial n} = \cos\left[\frac{\pi x}{l}\right]; \qquad 0 < x < l.$$

9.2.15. Solve the following Dirichlet problem using the boundary perturbation approach.

$$u_{xx} + u_{yy} = 0, \qquad\qquad \epsilon \sin y < x < l, \quad y > 0;$$
$$u(\epsilon \sin y, y) = u(l, y) = 0, \qquad y > 0;$$
$$u(x,0) = \sin\left[\frac{\pi x}{l}\right], \qquad\qquad 0 < x < l;$$

$$u(x, y) \text{ bounded as } y \to \infty.$$

9.2.16. Apply the boundary perturbation method to solve the following problem.

$$u_{tt} - c^2 u_{xx} = 0, \qquad\qquad \epsilon t < x < \infty, \quad t > 0;$$
$$u(x,0) = u_t(x,0) = 0, \qquad x > 0;$$
$$u(\epsilon t, t) = A \cos \omega t, \qquad t > 0.$$

Show that secular terms result in the perturbation series. Compare the perturbation result with the exact solution given in (6.129) when $c_0 = \epsilon$.

9.2.17. Solve the Dirichlet problem for Laplace's equation $\nabla^2 u = 0$ in an ellipse $x = (1 + a\epsilon)\cos \sigma; y = (1 + b\epsilon)\sin \sigma$ (i.e., a slightly perturbed unit circle) using the boundary perturbation method. The boundary condition on the ellipse is $u(x, y) = x^2$.

9.2.18. Use the eigenvalue perturbation method to determine up to $O(\epsilon)$ terms the leading eigenvalue and eigenfunction for the following problem.

$$M''(x) + \lambda(1 - \epsilon \sin^2 x) M = 0, \qquad 0 < x < \pi;$$
$$M(0) = M(\pi) = 0.$$

9.2.19. Approximate the leading eigenvalue for the following problem.

$$-\nabla^2 M + \epsilon xy M = \lambda M, \qquad 0 < x < \pi, \quad 0 < y < \pi;$$
$$M(0, y) = M(\pi, y) = \frac{\partial M(x,0)}{\partial y} = \frac{\partial M(x, \pi)}{\partial y} = 0.$$

9.2.20. Use the results of Example 8.3 to approximate the leading eigenvalue for the following problem in a sphere $r^2 = x^2 + y^2 + z^2 = 1$.

$$\nabla^2 M + (\lambda - \epsilon r^2) M = 0, \qquad r < 1;$$
$$M(x, y, z) = 0, \qquad r = 1.$$

9.2.21. Develop a perturbation method to obtain the eigenvalues for the following Sturm–Liouville problems:

$$-\frac{d}{dx}\left(p\frac{dM}{dx}\right) + qM = \lambda M, \qquad 0 < x < l$$

with the boundary conditions

(a) $\epsilon M'(0) - M(0) = 0$; $M(l) = 0$.
(b) $M'(0) - \epsilon M(0) = 0$; $M(l) = 0$.

9.2.22. Apply the method of Exercise 9.2.21 to obtain the leading eigenvalue for each of the following problems:

$$M''(x) + \lambda M(x) = 0, \qquad 0 < x < \pi$$

with the boundary conditions

(a) $\epsilon M'(0) - M(0) = 0$; $M(\pi) = 0$.
(b) $M'(0) - \epsilon M(0) = 0$; $M(\pi) = 0$.

9.2.23. Reproduce the results (9.175)–(9.178) by looking for a solution of the nonlinear Klein–Gordon equation (9.173) in the form

$$w(x, t) = \sum_{n=1}^{\infty} w_n(\Theta)\epsilon^n$$

where $\Theta = kx - \hat{\omega}t$ and

$$\hat{\omega} = \sum_{n=0}^{\infty} \omega_n(k)\epsilon^n.$$

Proceed as in Example 9.7 and choose the ω_n to eliminate secular terms in the expansion of w.

9.2.24. Obtain traveling wave solutions of the nonlinear hyperbolic equation

$$w_{tt} - \gamma^2 w_{xx} - c^2 w - \sigma w^3 = 0$$

by using the expansion forms given in Exercise 9.2.23. (Note that the linearized version of this equation does not have a real dispersion relation for all values of k. Consider the problem only for those k that yield traveling waves in the linearized case.)

9.2.25. The nonlinear system

$$E_{tt} + P_{tt} = \gamma^2 E_{xx},$$

$$P_{tt} + P - \sigma P^3 = c^2 E$$

EXAMPLE 4.4.   FOURIER SINE SERIES

In this example we again consider the eigenvalue equation (4.85), but we simplify the boundary conditions (4.57) and assume that

$$(4.90) \qquad \alpha_1 = \alpha_2 = 1; \qquad \beta_1 = \beta_2 = 0.$$

Then the eigenvalue equation (4.88) reduces to

$$(4.91) \qquad \sin(\sqrt{\lambda}\, l) = 0,$$

and this yields the eigenvalues

$$(4.92) \qquad \lambda_k = \left(\frac{\pi k}{l}\right)^2; \qquad k = 1, 2, \ldots.$$

The normalized eigenfunctions are given as

$$(4.93) \qquad v_k(x) = \sqrt{\frac{2}{l}} \, \sin\left(\frac{\pi k}{l} x\right); \qquad k = 1, 2, \ldots,$$

as is easily verified.

On the basis of the direct determination of the eigenvalues and eigenfunctions for the foregoing Sturm–Liouville problem, the first four properties for the general eigenvalue problem listed above can be verified directly. The expansion of a function in a series of eigenfunctions (4.93) is known as a *Fourier sine series*.

EXAMPLE 4.5.   FOURIER COSINE SERIES

We again consider (4.85) but with boundary conditions (4.57) for which

$$(4.94) \qquad \alpha_1 = \alpha_2 = 0; \qquad \beta_1 = \beta_2 = 1.$$

The eigenvalue equation becomes

$$(4.95) \qquad \sin(\sqrt{\lambda}\, l) = 0$$

and this yields the eigenvalues

$$(4.96) \qquad \lambda_k = \left(\frac{\pi k}{l}\right)^2; \qquad k = 0, 1, 2, \ldots.$$

The corresponding normalized eigenfunctions are found to be

$$(4.97) \quad \begin{cases} v_0(x) = \dfrac{1}{\sqrt{l}} \\ v_k(x) = \sqrt{\dfrac{2}{l}} \, \cos\!\left(\dfrac{\pi k}{l} x\right); \quad k = 1, 2, \ldots \end{cases}$$

In contrast to the preceding example, $\lambda_0 = 0$ is an eigenvalue. Again, the first four properties for the general Sturm–Liouville problem can be verified. The series expansion of a function in terms of the eigenfunctions (4.97) is known as a *Fourier cosine series*.

EXAMPLE 4.6. COMPLETE FOURIER SERIES

In this example we again consider the equation (4.85) but extend the interval of definition of $v(x)$ to $-l \leqslant x \leqslant l$. In addition, the boundary conditions (4.57) are replaced by what are known as periodic boundary conditions. Specifically, we study the eigenvalue problem

$$(4.98) \qquad -v''(x) = \lambda v(x); \qquad -l < x < l,$$

with the (periodic) boundary conditions

$$(4.99) \qquad \begin{cases} v(-l) = v(l) \\ v'(-l) = v'(l) \end{cases}.$$

The eigenvalue problem (4.98)–(4.99) is not of the Sturm–Liouville type. Nevertheless, most of the properties listed for the Sturm–Liouville problem remain valid in this case as we shall see.

Using the general solution

$$(4.100) \qquad v(x; \lambda) = a\cos(\sqrt{\lambda}\, x) + b\sin(\sqrt{\lambda}\, x)$$

of (4.98) (where we have tacitly assumed $\lambda > 0$), we easily conclude that the eigenvalues are given as

$$(4.101) \qquad \lambda_k = \left(\dfrac{\pi k}{l}\right)^2; \qquad k = 0, 1, 2, \ldots$$

on applying (4.99). By considering the solution of (4.98)–(4.99) for $\lambda < 0$ it can be shown that $v(x) = 0$ is the only possible result so that there are no negative eigenvalues. However, $\lambda_0 = 0$ is an eigenvalue in this case.

Since for each λ_k except $\lambda_0 = 0$, $\cos(\sqrt{\lambda_k}\, x)$ and $\sin(\sqrt{\lambda_k}\, x)$ are independent eigenfunctions, we conclude that each $\lambda_k > 0$ is a double eigenvalue. Thus property 3 given for the Sturm–Liouville problem is not valid in this case.

The full set of normalized eigenfunctions in terms of the (new) inner product (defined over $-l < x < l$)

$$(4.102) \qquad (\varphi, \psi) = \int_{-l}^{l} \varphi(x)\psi(x)\, dx,$$

and the induced norm

$$(4.103) \qquad \|\varphi\| = \sqrt{\int_{-l}^{l} \varphi^2(x)\, dx}$$

is given as

$$(4.104) \quad \hat{v}_0(x) = \frac{1}{\sqrt{2l}}; \qquad \hat{v}_k(x) = \left(\frac{1}{\sqrt{l}}\right)\cos\left(\frac{\pi k}{l}x\right); \qquad k = 1, 2, \ldots,$$

$$(4.105) \quad v_k(x) = \left(\frac{1}{\sqrt{l}}\right)\sin\left(\frac{\pi k}{l}x\right); \qquad k = 1, 2, \ldots .$$

For $k \geq 1$, $\hat{v}_k(x)$ and $v_k(x)$ are linearly independent normalized eigenfunctions that correspond to the eigenvalues λ_k. It can be verified directly that

$$(4.106) \qquad (\hat{v}_k, \hat{v}_l) = (v_k, v_l) = 0, \qquad k \neq l;$$

$$(4.107) \qquad (\hat{v}_k, v_l) = 0, \qquad k, l = 0, 1, 2, \ldots .$$

Thus the set of eigenfunctions $\{\hat{v}_k(x), v_l(x)\}$, $k = 0, 1, 2, \ldots, l = 1, 2, \ldots$ is an orthonormal set with respect to the inner product (4.102). Further, it can be shown that the eigenfunctions form a complete set with respect to square integrable functions $v(x)$ over the interval $-l < x < l$. Such functions have a complete Fourier series or, more simply, a *Fourier series* expansion

$$(4.108) \quad v(x) = (v, \hat{v}_0)\hat{v}_0 + \sum_{k=1}^{\infty} \left[(v, \hat{v}_k)\hat{v}_k(x) + (v, v_k)v_k(x)\right]$$

that converges to $v(x)$ in the mean.

If $v(x)$ is continuous and, in addition, has a piecewise continuous first derivative in the interval $-l \leq x \leq l$ and $v(x)$ satisfies the boundary conditions (4.99), the series (4.108) converges absolutely and uniformly to $v(x)$. It may be noted that all essential properties of the Sturm–Liouville eigenvalues and eigenfunctions carry over to the present case.

The trigonometric Fourier series considered in the Examples 4.4–4.6 are of sufficient interest to be studied without relating them to any eigenvalue problem. Their basic common feature is that they all involve expansions in the functions $\{\cos[(\pi k/l)x]\}$, $k = 0, 1, 2, \ldots$ and $\{\sin[(\pi k/l)x]\}$, $k = 1, 2, \ldots$. These functions are all periodic of period $2l$, the cosine functions are all even functions of x, and the sine functions are odd functions of x. [We recall that a function is periodic of period P if $\varphi(x + P) = \varphi(x)$ for all x. It is an even function if $\varphi(-x) = \varphi(x)$ for all x and it is odd if $\varphi(-x) = -\varphi(x)$ for all x.] Thus although the functions $v(x)$ that are expanded in these trigonometric series are defined over the interval $0 \leqslant x \leqslant l$ or $-l \leqslant x \leqslant l$, the Fourier series themselves are defined for all x.

Given that the trigonometric (Fourier) series converge, they may be considered to provide an extension of the definition of the given function $v(x)$ from its given interval to the entire real line. Thus in the case of the (complete) Fourier series, the given function $v(x)$ is extended as the function $V(x)$ as follows

$$(4.109) \qquad \begin{cases} V(x) = v(x); & -l \leqslant x \leqslant l \\ V(x + 2l) = V(x); & -\infty < x < \infty \end{cases}.$$

In the case of the Fourier sine series $v(x)$, which is defined over the interval $0 \leqslant x \leqslant l$, is extended to $V_o(x)$ which is given as

$$(4.110) \qquad \begin{cases} V_o(x) = v(x), & 0 \leqslant x \leqslant l \\ V_o(x) = -v(-x), & -l < x < 0 \\ V_o(x + 2l) = V_o(x), & -\infty < x < \infty \end{cases}.$$

Thus $V_o(x)$ first extends $v(x)$ as an odd function into the interval $-l < x < 0$ and then as an (odd) periodic function of period $2l$ over the entire x-axis. The odd extension is relevant here since all the sine functions are odd.

For the Fourier cosine series we extend $v(x)$, which is defined over the interval $0 \leqslant x \leqslant l$, to the entire x-axis via the function $V_e(x)$ as follows,

$$(4.111) \qquad \begin{cases} V_e(x) = v(x), & 0 \leqslant x \leqslant l \\ V_e(x) = v(-x), & -l < x < 0 \\ V_e(x + 2l) = V_e(x), & -\infty < x < \infty \end{cases}.$$

Thus $V_e(x)$ is an even periodic extension of period $2l$ of the function $v(x)$.

The significance of the above discussion lies in the fact that the Fourier sine series, cosine series, and the (complete) Fourier series obtained in Examples 4.4–4.6 not only represent the given function $v(x)$ over the appropriate interval (i.e., either $0 \leqslant x \leqslant l$ or $-l \leqslant x \leqslant l$) but also the functions $V_o(x)$, $V_e(x)$, and $V(x)$, respectively, over the entire real line. Consequently, if $V(x)$ is an odd periodic function, the (complete) Fourier series must reduce to the form of a sine series. Similarly, if $V(x)$ is periodic and even, the (complete) Fourier series must reduce to a cosine series. Furthermore, the graphs of the

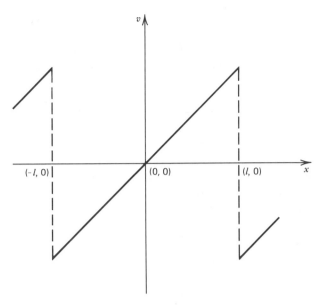

Figure 4.4.   Odd extension of $v(x) = x$.

extended functions $V(x)$, $V_o(x)$, and $V_e(x)$ provide some insight into the pointwise convergence properties of the appropriate Fourier series.

As an example, we consider the function $v(x) = x$ over the interval $0 \leqslant x \leqslant l$. The odd and even periodic extensions of $v(x)$ are shown in Figures 4.4 and 4.5. Now, if $v(x) = x$ in $-l \leqslant x \leqslant l$, the periodic extension of $v(x)$ to $V(x)$ with $V(x + 2l) = V(x)$ is identical to the function $V_o(x)$ graphed in Figure 4.4. Thus in either case the extended function $V_o(x) = V(x)$ has jump discontinuities at the points $x = \pm(2n + 1)l$, $(n = 0, 1, 2, \dots)$. Because of

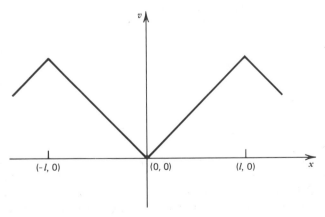

Figure 4.5.   Even extension of $v(x) = x$.

these discontinuities, the pointwise convergence of the Fourier series in these cases is slow away from these points. At these points the Fourier series converges to zero. Note that in the case of the sine series $v(x) = x$ does not satisfy the appropriate boundary condition (4.57) at $x = l$, whereas in the case of the complete Fourier series the function $v(x) = x$ is not periodic over the interval $-l \leqslant x \leqslant l$ since $v(-l) \neq v(l)$. For the even periodic extension, the function $V_e(x)$ is continuous for all x, but is not differentiable at the points $x = nl$ ($n = 0, \pm 1, \pm 2, \dots$). Pointwise convergence is more rapid in this case than for the odd or periodic extension considered previously. However, it is still fairly slow because the curve is not smooth. In fact, with $v(x) = x$ we have $v'(0) = v'(l) = 1$, so that the boundary condition (4.57) appropriate for the cosine eigenfunctions is not satisfied.

If the boundary conditions for the relevant eigenvalue problem are satisfied by the given function $v(x)$, the smoother the function $v(x)$ is, the better the convergence. This is evidenced by the smoothness of the relevant extended functions $V_o(x)$, $V_e(x)$, and $V(x)$. Generally speaking, the greater the number of continuous derivatives the extended functions have, the more rapid the pointwise convergence of the Fourier series. This matter is further examined in the exercises.

EXAMPLE 4.7.   BESSEL FUNCTION EXPANSIONS

We consider a singular Sturm–Liouville problem that arises from the eigenvalue problem for *Bessel's equation* of order n, where n is an integer or zero. The equation we consider is

$$(4.112) \qquad -\frac{d}{dx}\left(x\frac{dv}{dx}\right) + \left(\frac{n^2}{x}\right)v = \lambda x v,$$

on the interval $0 < x < l$. [On replacing x by $\sqrt{\lambda}\, x$, (4.112) becomes Bessel's equation of order n.] Since $p(x) = \rho(x) = x$ and $q(x) = n^2/x$, we see that these functions are positive for $0 < x < l$ but p and ρ vanish at $x = 0$ and $q(x)$ is singular there. Nevertheless, since $x = 0$ is a regular singular point for Bessel's equation, bounded solutions of (4.112) do exist. To obtain these solutions we impose the boundary conditions

$$(4.113) \qquad \begin{cases} v(x) \text{ bounded at } x = 0 \\ v(l) = 0 \end{cases}.$$

The condition at $x = l$ could be replaced by the more general form given in (4.57). In general, the boundedness condition (4.113) at the singular point is supplemented by the condition $\lim_{x \to 0} p(x)v'(x) = 0$. This guarantees that not only are the eigenfunctions bounded, but that eigenfunctions corresponding to different eigenvalues are orthogonal. To see this, we recall from

(4.71)–(4.72) that with $\lambda_i \neq \lambda_j$,

$$(4.114) \qquad (\lambda_i - \lambda_j)(v_i, v_j) = \left[pv_j v_i' - pv_i v_j' \right]\Big|_{x=0}^{x=l}.$$

For the right side to vanish so that orthogonality results, we must not only have v_i and v_j bounded at $x = 0$ and $x = l$ but also $\lim p(x)v_i'(x) = \lim pv_j'(x) = 0$ as $x \to 0$ and $x \to l$ if a singularity exists at $x = 0$ and $x = l$.

For our eigenvalue problem (4.112)–(4.113), the singular solution of Bessel's equation is rejected in view of (4.113) and the eigenfunctions must be determined from the Bessel functions of order n, that is, $J_n(\sqrt{\lambda} x)$. Since $J_0 = 1 + O(x^2)$ and $J_n(\sqrt{\lambda} x) = O(x^n)$ as $x \to 0$, we find that not only are these functions bounded at $x = 0$ but also $xJ_0'(\sqrt{\lambda} x) = O(x^2)$ and $xJ_n'(\sqrt{\lambda} x) = O(x^n)$ as $x \to 0$. Since $p(x) = x$, in this case, the added condition $\lim_{x \to 0} p(x)v'(x) = 0$ is satisfied.

The eigenvalues are specified by the boundary condition $v(l) = 0$. We denote the positive zeros of the Bessel function $J_n(x)$ by α_{kn}, $k = 1, 2, \ldots$, so that $J_n(\alpha_{kn}) = 0$. [There are infinitely many real positive zeros of the Bessel functions $J_n(x)$ with a limit point at infinity and they are tabulated.] Then $v(l) = 0$ implies that $J_n(\sqrt{\lambda} l) = 0$ and the eigenvalues for each n are given by

$$(4.115) \qquad \lambda_{kn} = \left(\frac{\alpha_{kn}}{l} \right)^2; \qquad k = 1, 2, \ldots .$$

Using the properties of the Bessel function, it isn't hard to show that the square of the norm of the eigenfunctions $v_k(x) = J_n(\sqrt{\lambda_{kn}} x)$ is

$$(4.116) \qquad \left\| J_n\left(\sqrt{\lambda_{kn}} x\right) \right\|^2 = \int_0^l x J_n^2\left(\sqrt{\lambda_{kn}} x\right) dx = \left(\frac{l^2}{2}\right) J_{n+1}^2\left(\sqrt{\lambda_{kn}} l\right).$$

In view of the orthogonality property demonstrated above, the set

$$(4.117) \qquad v_{k,n}(x) = \left(\frac{\sqrt{2}}{l}\right) \frac{J_n\left(\sqrt{\lambda_{kn}} x\right)}{\left| J_{n+1}\left(\sqrt{\lambda_{kn}} l\right) \right|}; \qquad k = 1, 2, \ldots,$$

is an orthonormal set of eigenfunctions for the Sturm–Liouville problem (4.112)–(4.113) for each n. These eigenfunctions form a complete set, and any smooth function $v(x)$ defined over the interval $0 \leqslant x \leqslant l$ that satisfies the boundary condition $v(l) = 0$ can be expanded in a convergent series

$$(4.118) \qquad v(x) = \sum_{k=1}^{\infty} (v(x), v_{k,n}(x)) v_{k,n}(x).$$

EXAMPLE 4.8.   LEGENDRE POLYNOMIAL EXPANSIONS

The eigenvalue problem associated with the self-adjoint form of the *Legendre equation* is

$$(4.119) \qquad -\frac{d}{dx}\left[(1 - x^2)\frac{dv}{dx}\right] = \lambda v,$$

where $v(x)$ is defined over the interval $-1 < x < 1$ and satisfies the boundary conditions

$$(4.120) \qquad v(x) \text{ bounded at } x = \pm 1.$$

Since $p(x) = 1 - x^2$ is positive for $-1 < x < 1$ and vanishes at $x = \pm 1$, we have a singular Sturm–Liouville problem in (4.119)–(4.120). The points $x = -1$ and $x = 1$ are regular singular points for Legendre's equation, and bounded solutions for (4.119) exist.

The Frobenius theory of power series solutions for (4.119) shows that the only bounded solutions of the Legendre equation (4.119) over the interval $-1 \leqslant x \leqslant 1$ are polynomial solutions and these occur only if λ assumes the eigenvalues

$$(4.121) \qquad \lambda_k = k(k + 1); \qquad k = 0, 1, 2, \ldots .$$

The eigenfunctions are usually normalized to yield the *Legendre polynomials*. These are polynomials of degree k and are defined as

$$(4.122) \quad P_k(x) = \left(\frac{1}{2^k k!}\right)\frac{d^k}{dx^k}(x^2 - 1)^k = \sum_{n=0}^{k}\frac{(n + k)!}{(k - n)!(n!)^2 2^n}(x - 1)^n$$

with the normalization chosen such that $P_k(1) = 1$ for all k. Since $\lim_{x \to \pm 1}(1 - x^2)P_k'(x) = 0$, the supplementary condition that $p(x)v'(x)$ be finite at the singular points is satisfied [see the discussion leading to (4.114)]. Consequently, the Legendre polynomials form an orthogonal set over the interval $-1 < x < 1$. Further, it can be shown that

$$(4.123) \quad \|P_k(x)\|^2 = \int_{-1}^{1} P_k^2(x)\, dx = \frac{2}{2k + 1}; \qquad k = 0, 1, 2, \ldots .$$

Thus the orthonormal set of eigenfunctions is given as

$$(4.124) \qquad v_k(x) = \sqrt{\frac{2k + 1}{2}}\, P_k(x); \qquad k = 0, 1, 2, \ldots .$$

The set $\{v_k(x)\}$ is complete and any smooth function $v(x)$ defined over the interval $-1 \leqslant x \leqslant 1$ can be expanded in a convergent series of eigenfunctions

$$(4.125) \quad v(x) = \sum_{k=0}^{\infty} \left(v(x), \sqrt{\frac{2k+1}{2}} \, P_k(x) \right) \sqrt{\frac{2k+1}{2}} \, P_k(x).$$

4.4. SERIES SOLUTIONS OF INITIAL AND BOUNDARY VALUE PROBLEMS

In this section the method of separation of variables is applied to specific initial and boundary value problems for partial differential equations. Three examples dealing with equations of hyperbolic, parabolic, and elliptic types are presented, and the solution of each of the problems is discussed in some detail.

EXAMPLE 4.9.   VIBRATIONS OF A FIXED STRING

Let $u(x, t)$ represent the transverse displacement of a tightly stretched string of length l. Under various simplifying assumptions it can be shown that $u(x, t)$ satisfies the *wave equation*

$$(4.126) \qquad u_{tt} - c^2 u_{xx} = 0, \qquad 0 < x < l, t > 0,$$

with $c = \sqrt{T/\rho}$, where T is the tension and ρ is the density of the string, both of which are assumed to be constant.

Now $u = u(x, t)$ is the graph or shape of the string at the time t and $u_x(x, t)$ is the slope at a point x on the string. In the derivation of (4.126) it is assumed the $|u_x| \ll 1$ and this implies that the curvature of the string approximately equals $u_{xx}(x, t)$. Then (4.126) asserts that the vertical acceleration $u_{tt}(x, t)$ of a point x on the string is proportional to the curvature $u_{xx}(x, t)$. In view of Newton's law of motion, this implies that the only forces acting on the string are due to the string's tension. The greater the curvature, the greater the tension forces whose effect is to straighten the string out.

The ends of the string are assumed to be fixed for all time and this yields the boundary conditions

$$(4.127) \qquad u(0, t) = 0; \qquad u(l, t) = 0; \qquad t > 0.$$

The given displacements and velocities at $t = 0$ yield the initial conditions

$$(4.128) \qquad u(x, 0) = f(x); \qquad u_t(x, 0) = g(x); \qquad 0 < x < l.$$

To solve the initial and boundary value problem (4.126)–(4.128) we apply the separation of variables method of Section 4.2 and set $u(x, t) = M(x)N(t)$. Then $M(x)$ satisfies the following eigenvalue problem,

$$(4.129) \qquad -c^2 M''(x) = \lambda M(x); \qquad M(0) = M(l) = 0.$$

An equivalent Sturm–Liouville problem was considered in Example 4.4 and yielded the following results. The eigenvalues for (4.129) are

$$(4.130) \qquad \lambda_k = \left(\frac{\pi k c}{l}\right)^2; \qquad k = 1, 2, 3, \dots,$$

and the eigenfunctions are

$$(4.131) \qquad M_k(x) = \sqrt{\frac{2}{l}} \sin\left(\frac{\pi k}{l} x\right); \qquad k = 1, 2, 3, \dots.$$

From (4.43) we obtain

$$(4.132) \quad N_k(t) = a_k \cos\left(\frac{\pi k c}{l} t\right) + b_k \sin\left(\frac{\pi k c}{l} t\right); \qquad k = 1, 2, 3, \dots,$$

and the corresponding solutions $u_k(x, t)$, which are known as *normal modes*,

$$(4.133)$$
$$u_k(x, t) = \left[a_k \cos\left(\frac{\pi k c}{l} t\right) + b_k \sin\left(\frac{\pi k c}{l} t\right)\right]\sqrt{\frac{2}{l}} \sin\left[\frac{\pi k}{l} x,\right]; \qquad k = 1, 2, \dots.$$

Each of the u_k satisfies the wave equation (4.126) and the boundary conditions (4.127).

To satisfy the initial conditions (4.128) we form the sum (4.45), and set

$$(4.134) \qquad\qquad u(x, t) = \sum_{k=1}^{\infty} u_k(x, t),$$

and apply the initial conditions to $u(x, t)$. This gives

$$(4.135) \quad u(x, 0) = \sum_{k=1}^{\infty} u_k(x, 0) = \sum_{k=1}^{\infty} a_k \sqrt{\frac{2}{l}} \sin\left(\frac{\pi k}{l} x\right) = f(x),$$

$$(4.136) \quad u_t(x, 0) = \sum_{k=1}^{\infty} \frac{\partial u_k(x, 0)}{\partial t} = \sum_{k=1}^{\infty} \left(\frac{\pi k c}{l}\right) b_k \sqrt{\frac{2}{l}} \sin\left(\frac{\pi k}{l} x\right) = g(x).$$

These are Fourier sine series with normalized eigenfunctions $M_k = \sqrt{(2/l)} \sin[(\pi k/l)x]$ and (4.48)–(4.49) imply that

$$(4.137)$$
$$a_k = (f(x), M_k(x)) = \sqrt{\frac{2}{l}} \int_0^l f(x) \sin\left(\frac{\pi k}{l} x\right) dx; \qquad k = 1, 2, \dots,$$

and

(4.138)

$$b_k = \left(\frac{1}{\sqrt{\lambda_k}} g(x), M_k(x) \right) = \frac{\sqrt{2l}}{\pi k c} \int_0^l g(x) \sin\left(\frac{\pi k}{l} x \right) dx; \qquad k = 1, 2, \ldots .$$

This completes the formal determination of the solution.

To assure that the formal series solution (4.134) represents an actual solution of the wave equation and satisfies the initial conditions we must place conditions on the data $f(x)$ and $g(x)$. The function $f(x)$ is required to have two continuous derivatives and a third piecewise continuous derivative and, in addition, we must have $f(0) = f(l) = 0$ and $f''(0) = f''(l) = 0$. The function $g(x)$ must be continuously differentiable and have a piecewise continuous second derivative as well as satisfying the conditions $g(0) = g(l) = 0$. Under these conditions the series (4.134) can be differentiated term by term twice and the resulting series converge uniformly. Thus $u(x, t)$ is a solution of the wave equation since each $u_k(x, t)$ satisfies the equation. The conditions on $f(x)$ and $g(x)$ guarantee that the sine series as well as the series of its derivatives converges, since it is required that $f(x)$ and $g(x)$ satisfy the boundary conditions for the appropriate Sturm–Liouville problems to assure the convergence of the Fourier series as was indicated in the preceding section (see Exercises 4.3.9 and 4.3.10).

We now give a brief interpretation of the solution (4.134). Each of the normal modes $u_k(x, t)$ [i.e., (4.133)] can be expressed as

$$(4.139) \quad u_k(x, t) = \alpha_k \cos\left[\frac{\pi k c}{l} (t + \delta_k) \right] \sin\left(\frac{\pi k}{l} x \right); \qquad k = 1, 2, \ldots,$$

$$(4.140) \quad \alpha_k = \sqrt{\frac{2}{l}} \sqrt{a_k^2 + b_k^2}; \qquad \frac{\pi k c}{l} \delta_k = -\tan^{-1}\left(\frac{b_k}{a_k} \right); \qquad k = 1, 2, \ldots .$$

For each solution $u_k(x, t)$, a fixed point x_0 executes *harmonic vibrations* with amplitude

$$(4.141) \qquad\qquad A_k = \alpha_k \left| \sin\left(\frac{\pi k}{l} x_0 \right) \right|.$$

Solutions of the wave equation are called waves; we refer to $u_k(x, t)$ as a *standing wave*, since each point x on the string oscillates in place as t varies. The points $x_m = ml/k$ $(m = 1, 2, \ldots, k - 1)$ at which $\sin[(\pi k/l)x_m] = 0$ remain fixed in the vibration process and are called the *nodes* of the standing wave u_k. The points $x_n = (2n + 1)l/2k$ $(n = 0, 1, 2, \ldots, k - 1)$ at which $\sin[(\pi k/l)x_n] = \pm 1$, are those where the standing wave has maximum amplitudes and these are called *antinodes*. We also note that

$$(4.142) \qquad \omega_k = \frac{\pi k c}{l} = \sqrt{\lambda_k} \, ; \qquad k = 1, 2, 3, \ldots$$

equals the *frequency* of vibration of the kth *harmonic* $u_k(x, t)$ (the standing wave u_k is also known as the kth harmonic).

The *energy* of the vibrating string at the time t is given by the integral

$$(4.143) \qquad E(t) = \frac{1}{2} \int_0^l \left\{ \rho \left(\frac{\partial u}{\partial t} \right)^2 + T \left(\frac{\partial u}{\partial x} \right)^2 \right\} dx,$$

which represents a sum of the kinetic energy of motion and the potential energy due to the tension. The total energy of the string is a sum of the energies distributed among the harmonics $u_k(x, t)$. It is readily seen that the energy of the kth harmonic $u_k(x, t)$ is

$$(4.144) \quad E_k(t) = \frac{1}{2} \int_0^l \left\{ \rho \left(\frac{\partial u_k}{\partial t} \right)^2 + T \left(\frac{\partial u_k}{\partial x} \right)^2 \right\} dx = \frac{\omega_k^2 m [a_k^2 + b_k^2]}{2l}$$

where $m = \rho l$ is the mass of the string. We observe that $E_k(t)$ is independent of t and thus the total energy

$$(4.145) \qquad E(t) = \sum_{k=1}^{\infty} E_k(t) = \frac{m}{2l} \sum_{k=1}^{\infty} \omega_k^2 (a_k^2 + b_k^2)$$

is also constant in time. That is, $E(t) = E(0)$, so that the initial energy is conserved as was to be expected, since no dissipative effects are assumed to be present. This is consistent with our characterization of the wave equation in Example 3.7 as being of conservative type.

We have seen in Example 2.3 that the solution of the initial value problem for the wave equation can be written as the sum of two *propagating* or *traveling waves*; that is, it has the form

$$(4.146) \qquad u(x, t) = \hat{F}(x - ct) + \hat{G}(x + ct)$$

where $\hat{F}$ and $\hat{G}$ are arbitrary functions. The normal modes $u_k(x, t)$ can also be expressed as the sum of propagating waves as

$$(4.147) \quad u_k(x, t) = \frac{1}{\sqrt{2l}} \left\{ a_k \sin \left[\frac{\pi k}{l} (x + ct) \right] + a_k \sin \left[\frac{\pi k}{l} (x - ct) \right] \right.$$

$$\left. + b_k \cos \left[\frac{\pi k}{l} (x - ct) \right] - b_k \cos \left[\frac{\pi k}{l} (x + ct) \right] \right\}.$$

Although there are four propagating waves in (4.147), through interference they combine to form a standing wave. In this representation, the solution $u(x, t)$ has the form

$$(4.148) \quad u(x, t) = \sum_{k=1}^{\infty} u_k(x, t) = \frac{1}{2} \sum_{k=1}^{\infty} a_k \sqrt{\frac{2}{l}} \sin\left[\frac{\pi k}{l}(x + ct)\right]$$

$$+ \frac{1}{2} \sum_{k=1}^{\infty} a_k \sqrt{\frac{2}{l}} \sin\left[\frac{\pi k}{l}(x - ct)\right]$$

where we have assumed for simplicity that $g(x) = 0$, so that $b_k = 0$ for all k.

Apart from the factor $\frac{1}{2}$, these series are identical in form to the Fourier sine series (4.135), if the argument x is replaced by $x \pm ct$. We note that $f(x)$ is not defined outside the interval $[0, l]$. Proceeding as in Section 4.3, we define the function $F(x)$ as

$$(4.149) \quad \begin{cases} F(x) = f(x); & 0 < x < l \\ F(x) = -f(-x); & -l < x < 0 \\ F(x + 2l) = F(x); & -\infty < x < \infty \end{cases},$$

so that $F(x)$ is an odd periodic extension of $f(x)$ of period $2l$ over the entire x-axis. The Fourier series expansion of $F(x)$ is identical to the sine series (4.135) for $f(x)$, since $F(x)$ is an odd periodic function. Now $F(x)$ is defined for all x so that the Fourier series for the periodic function $F(x \pm ct)$ can be constructed. Its form is precisely that of (4.135) with x replaced by $x \pm ct$. Consequently, the solution (4.148) can be expressed as

$$(4.150) \quad u(x, t) = \tfrac{1}{2}F(x + ct) + \tfrac{1}{2}F(x - ct); \quad 0 < x < l, t > 0.$$

The conditions under which (4.150) represents a solution of the initial and boundary value problem for the wave equation [with $g(x) = 0$] can be shown to be slightly weaker than those given for the standing wave representation of the solution.

The propagating wave representation (4.150) of the solution of the wave equation can be interpreted in terms of waves being continually reflected off the ends of the string. The interference of these waves with one another explains why only standing wave motion seems to occur when the vibration of a string is observed over a length of time.

The Fourier series representation of $F(x)$ retains the discontinuities of $F(x)$ and does not smooth them out as was indicated previously. Thus unless $F(x)$ is a smooth function, the solution $u(x, t)$ may have discontinuities or discontinuous derivatives along the curves $x \pm ct = $ constant, in view of (4.150). This shows that the discontinuities in solutions or derivatives of solution of the wave equation can occur only along the characteristic curves, consistent with the results of Section 3.2. It is shown later in the text how the concept of solution may be generalized to include functions that do not have the required number of derivatives.

EXAMPLE 4.10. HEAT CONDUCTION IN A FINITE ROD

Let $u(x, t)$ represent the temperature in a homogeneous, laterally insulated rod of length l at the point x $(0 \leqslant x \leqslant l)$ and at the time t. Since no heat can escape through the sides of the rod, the problem is effectively one-dimensional and in the absence of heat sources $u(x, t)$ satisfies the *heat equation*

$$(4.151) \qquad u_t = c^2 u_{xx}, \qquad 0 < x < l, t > 0,$$

where c^2 is the coefficient of heat conduction. The ends of the rod at $x = 0$ and $x = l$ are assumed to be kept at zero temperature so that

$$(4.152) \qquad u(0, t) = u(l, t) = 0, \qquad t > 0,$$

whereas initially there is a given temperature distribution

$$(4.153) \qquad u(x, 0) = f(x), \qquad 0 < x < l.$$

We apply the separation of variables technique to solve this problem and proceed as in Example 4.9. With $u_k = M_k(x)N_k(t)$, the eigenvalues λ_k and the eigenfunctions M_k are given as in (4.130)–(4.131) and N_k is found to be [see (4.43)]

$$N_k(t) = a_k \exp\left[-\left(\frac{\pi k c}{l}\right)^2 t \right]; \qquad k = 1, 2, \dots .$$

The solution $u(x, t)$ becomes

(4.154)

$$u(x, t) = \sum_{k=1}^{\infty} u_k(x, t) = \sum_{k=1}^{\infty} M_k(x) N_k(t) = \sqrt{\frac{2}{l}} \sum_{k=1}^{\infty} a_k e^{-(\pi k c/l)^2 t} \sin\left(\frac{\pi k}{l} x\right).$$

Initially, we have

$$(4.155) \qquad u(x, 0) = \sum_{k=1}^{\infty} a_k \sqrt{\frac{2}{l}} \, \sin\left(\frac{\pi k}{l} x\right) = f(x)$$

so that the a_k are defined as in (4.137). Thus the formal solution of (4.151)–(4.153) is completely specified.

To examine the validity of the expansion (4.154), we note for $t > 0$ the terms in the series decay exponentially. If we assume that the initial temperature $f(x)$ is bounded over the interval $[0, l]$, the Fourier coefficients a_k are bounded. As a result, the series for $u(x, t)$ with $t > 0$ can be differentiated term by term as often as required. Since each term $u_k(x, t)$ satisfies the heat equation, so does $u(x, t)$ for $t > 0$ and we also have $u(0, t) = u(l, t) = 0$. To assure that the initial condition is satisfied, we must have the Fourier series (4.155) converge uniformly. To achieve this we require that $f(x)$ be continuous and piecewise continuously differentiable in $0 < x < l$ and that $f(0) = f(l) = 0$. The uniform convergence of the Fourier sine series (4.155) implies that the series (4.154) for $u(x, t)$ converges uniformly for $t \geqslant 0$ and that the solution is continuous for $0 \leqslant x \leqslant l$ and $t \geqslant 0$.

In the vibrating string problem of the previous example (which involved a hyperbolic equation) the solution as given in (4.148) essentially has the form of a Fourier series. This has the effect that any discontinuities in the data are preserved and transmitted along the characteristic lines $x \pm ct = $ constant in (x, t)-space. For the heat equation, however, even if the initial temperature $f(x)$ is discontinuous but bounded, the solution $u(x, t)$ is continuously differentiable as often as is required. Now the more derivatives a function has, the smoother it is, so that heat conduction is seen to be a smoothing process. Even if the heat is initially concentrated near one point on the interval $0 < x < l$, it is distributed instantaneously according to our solution in an even and smooth fashion throughout the interval. Further, as $t \to \infty$, the temperature $u(x, t) \to 0$ since there are no heat sources in this problem and the rod is continually being cooled at its end points.

We have already indicated in our discussion of the diffusion equation in Chapter 1 that disturbances move at infinite speeds for the heat equation, and this fact is borne out by the properties of the solution obtained. It may also be noted that the characteristics of the heat equation are the lines $t = $ constant. Since characteristics are carriers of discontinuities or rapid variations of the solution, there is no mechanism whereby these effects can be transmitted into the region $t > 0$ from the initial line $t = 0$ which is itself a characteristic curve.

EXAMPLE 4.11.   LAPLACE'S EQUATION IN A RECTANGLE

We consider the boundary value problem for *Laplace's equation*

$$(4.156) \qquad\qquad u_{xx} + u_{yy} = 0$$

in the rectangle $0 < x < l$ and $0 < y < \hat{l}$, with the boundary conditions

$$(4.157) \qquad\qquad \begin{cases} u(0, y) = u(l, y) = 0 \\ u(x, 0) = f(x); \qquad u(x, \hat{l}) = g(x). \end{cases}$$

The function $u(x, y)$ may represent the steady state (or equilibrium) displacement of a stretched membrane whose rectangular boundaries are fixed according to (4.157). [Proceeding in a similar fashion, we can solve another problem for (4.156) with nonzero boundary conditions on $x = 0$ and $x = l$ and zero boundary conditions on $y = 0$ and $y = \hat{l}$. On adding the solutions of both problems we can thereby solve Laplace's equation with arbitrary boundary conditions on a rectangle.]

Applying the separation of variables method of Section 4.2 and proceeding as in Example 4.9, we obtain the eigenfunctions $M_k(x)$ as in (4.131) and the eigenvalues $\lambda_k = (\pi k/l)^2$. With $u_k(x, y) = M_k(x)N_k(y)$, the $N_k(y)$ are given as [see (4.43)]

$$(4.158) \quad N_k(y) = \hat{a}_k \exp\left[\frac{\pi k}{l}y\right] + \hat{b}_k \exp\left[-\frac{\pi k}{l}y\right]; \qquad k = 1, 2, \ldots .$$

The functions $N_k(y)$ can be expressed as linear combinations of hyperbolic sine functions and we write $u_k(x, y)$ as

$$(4.159) \quad u_k(x, y) = \left\{a_k \sinh\left[\frac{\pi k}{l}y\right] + b_k \sinh\left[\frac{\pi k}{l}(y - \hat{l})\right]\right\} \sqrt{\frac{2}{l}} \sin\left(\frac{\pi k}{l}x\right)$$

with the a_k and b_k as yet arbitrary. A formal solution of Laplace's equation obtained by superposition is

$$(4.160) \qquad\qquad u(x, y) = \sum_{k=1}^{\infty} u_k(x, y).$$

The boundary conditions at $y = 0$ and $y = \hat{l}$ imply that

$$(4.161) \quad u(x, 0) = \sum_{k=1}^{\infty} b_k \sinh\left[-\frac{\pi k}{l}\hat{l}\right] \sqrt{\frac{2}{l}} \sin\left(\frac{\pi k}{l}x\right) = f(x),$$

$$(4.162) \quad u(x, \hat{l}) = \sum_{k=1}^{\infty} a_k \sinh\left[\frac{\pi k}{l}\hat{l}\right] \sqrt{\frac{2}{l}} \sin\left(\frac{\pi k}{l}x\right) = g(x).$$

The Fourier coefficients of this sine series are determined to give a_k and b_k as [see (4.137)–(4.138)]

$$(4.163) \quad \begin{cases} b_k = \dfrac{(f(x), M_k(x))}{\sinh\left(-\dfrac{\pi k}{l}\hat{l}\right)}; & k = 1, 2, \ldots, \\[4ex] a_k = \dfrac{(g(x), M_k(x))}{\sinh\left(\dfrac{\pi k}{l}\hat{l}\right)}; & k = 1, 2, \ldots . \end{cases}$$

This completes the formal determination of the solution of the boundary value problem.

If we assume that the integrals $\int_0^l |f|\, dx$ and $\int_0^l |g|\, dx$ are bounded by the number m, and note that $\sinh x = (e^x - e^{-x})/2 \approx (1/2)e^x$ as $x \to \infty$, we easily find that for large values of k,

$$(4.164) \quad \left| b_k \sinh\left[\frac{\pi k}{l}(y - \hat{l}) \right] \right| \leqslant \frac{\sqrt{2/l}\, m \exp\left[(\pi k/l)(\hat{l} - y) \right]}{\exp\left[\pi k \hat{l}/l \right]\left[1 - \exp(-2\pi k \hat{l}/l) \right]}$$

$$\leqslant \frac{\sqrt{2/l}\, m \exp\left[-\pi k y/l \right]}{1 - \exp\left[-2\pi \hat{l}/l \right]}$$

with a similar bound valid for $|a_k \sinh[(\pi k/l)y]|$. Thus the terms in the series $u(x, y) = \sum_{k=1}^{\infty} u_k(x, y)$ are bounded by exponentially decaying terms for large k in the open interval $0 < y < \hat{l}$. (The exponential bound for the b_k and a_k terms breaks down at $y = 0$ and $y = \hat{l}$, respectively.) Consequently, the series can be differentiated term by term as often as desired in the interval $0 < y < \hat{l}$ and $u(x, y)$ satisfies Laplace's equation and the boundary conditions at $x = 0$ and $x = l$, since each of the u_k does so. To assure that $u(x, y)$ is continuous up to $y = 0$ and $y = \hat{l}$ and assumes the boundary values there, we must assure that the Fourier series (4.161)–(4.162) are uniformly convergent in $0 \leqslant x \leqslant l$. This is the case if $f(0) = f(l) = g(0) = g(l) = 0$, if $f(x)$ and $g(x)$ are continuous and $f'(x)$ and $g'(x)$ are piecewise continuous in that interval.

For $u(x, y)$ to satisfy Laplace's equation inside the rectangle we merely require that $f(x)$ and $g(x)$ be bounded or that $\int_0^l |f(x)|\, dx$ and $\int_0^l |g(x)|\, dx$ exist, (i.e., f and g may even be singular). Then $u(x, y)$ is defined by the series to be an infinitely differentiable function within the rectangle. However, the boundary values at $y = 0$ and $y = l$ need not be assumed continuously at all points unless the additional conditions on f and g given previously are met. Again, this contrasts with the results obtained for the wave equation, where discontinuities are seen to spread into interior regions from the boundary. Since Laplace's equation has no real characteristics, boundary data discontinuities must be confined to the boundary and the interior solution is smooth. Physically, since Laplace's equation characterizes steady-state or equilibrium situations, we may expect that effects due to discontinuities in the data have smoothed themselves out.

This concludes our discussion of the separation of variables method for homogeneous problems. Further examples involving other eigenvalue problems are considered in the exercises.

4.5. INHOMOGENEOUS EQUATIONS: DUHAMEL'S PRINCIPLE

The method of separation of variables was applied in the foregoing to obtain solutions of initial and boundary value problems for homogeneous partial differential equations. In this section we construct solutions of inhomogeneous equations by using a technique known as *Duhamel's principle*, which effectively relates the problem to one involving a homogeneous equation. The method is valid for Cauchy (i.e., initial value) problems and initial and boundary value problems for hyperbolic and parabolic equations.

We consider equations of the form

$$(4.165) \qquad \begin{cases} \rho(x)u_{tt} + L[u] = g(x,t), & \text{hyperbolic case} \\ \rho(x)u_t + L[u] = g(x,t), & \text{parabolic case} \end{cases}$$

with $L[u]$ defined as in (4.6) or (4.7), $\rho(x) > 0$ and $g(x, t)$ a given forcing term (note that x may be vector or scalar variable). If we consider the Cauchy problem for (4.165), we assume $u(x, t)$ satisfies homogeneous initial conditions at $t = 0$. For the initial and boundary value problem for (4.165) in a bounded region G, we again assume homogeneous initial conditions for $u(x, t)$ at $t = 0$ in addition to the boundary conditions (4.14) or (4.15). Thus

$$(4.166) \qquad \begin{cases} u(x,0) = u_t(x,0) = 0, & \text{hyperbolic case} \\ u(x,0) = 0, & \text{parabolic case} \end{cases}$$

Duhamel's principle proceeds as follows. Consider a homogeneous version of (4.165) that is,

$$(4.167) \qquad \begin{cases} \rho(x)v_{tt} + L[v] = 0, & t > \tau \\ \rho(x)v_t + L[v] = 0, & t > \tau \end{cases}$$

for the function $v(x, t)$, which is assumed to satisfy the same boundary conditions (if any are given) as $u(x, t)$. Let $v(x, t)$ satisfy the following initial conditions given at $t = \tau$ where $\tau \geqslant 0$:

$$(4.168) \qquad \begin{cases} v(x,\tau) = 0, \ v_t(x,\tau) = \dfrac{g(x,\tau)}{\rho(x)}, & \text{hyperbolic case} \\ v(x,\tau) = \dfrac{g(x,\tau)}{\rho(x)}, & \text{parabolic case} \end{cases}$$

where ρ and g are given as above.

It is assumed that the above problem for v can be solved either by separation of variables for the initial and boundary value problem or by other means for the Cauchy problem. Since the solution depends on the parameter τ (i.e., the initial time), we write it as $v = v(x, t; \tau)$. Then Duhamel's principle states that the solution $u(x, t)$ of the given inhomogeneous problem is

$$(4.169) \qquad u(x,t) = \int_0^t v(x,t;\tau)\, d\tau.$$

[A motivation for the method is obtained by noting that the effect of the forcing term $g(x, t)$ can be characterized as resulting from a superposition of impulses at times $t = \tau$ over the time span $0 \leqslant \tau \leqslant t$].

To verify that $u(x, t)$ as given in (4.169) is a solution of the problem, we note that

$$u_t(x, t) = v(x, t; t) + \int_0^t v_t(x, t; \tau) \, d\tau,$$

$$u_{tt}(x, t) = \frac{\partial}{\partial t} [v(x, t; t)] + v_t(x, t; t) + \int_0^t v_{tt}(x, t; \tau) \, d\tau,$$

and

$$L[u] = \int_0^t L[v] \, d\tau.$$

In view of (4.168) we have in the parabolic case

(4.170) $$v(x, t; t) = \frac{g(x, t)}{\rho(x)}$$

and in the hyperbolic case

(4.171) $$v(x, t; t) = 0; \qquad v_t(x, t; t) = \frac{g(x, t)}{\rho(x)}.$$

Therefore,

(4.172) $$\rho u_t + L[u] = g(x, t) + \int_0^t [\rho v_t + L(v)] \, d\tau = g(x, t)$$

and

(4.173) $$\rho u_{tt} + L[u] = g(x, t) + \int_0^t [\rho v_{tt} + L(v)] \, d\tau = g(x, t).$$

Also, $u(x, 0) = 0$ and $u_t(x, 0) = v(x, 0; 0) = 0$ so that $u(x, t)$ as defined by (4.169) satisfies all the conditions of the problem.

Next we consider two examples where we apply Duhamel's principle to a Cauchy problem for the inhomogeneous wave equation and an initial and boundary value problem for the inhomogeneous heat or diffusion equation.

EXAMPLE 4.12.   THE INHOMOGENEOUS WAVE EQUATION

We consider the wave equation with a given forcing term $g(x, t)$,

(4.174) $$u_{tt} - c^2 u_{xx} = g(x, t), \qquad -\infty < x < \infty, t > 0$$

and the homogeneous initial conditions

$$(4.175) \qquad u(x,0) = u_t(x,0) = 0, \qquad -\infty < x < \infty.$$

[Note since we can add to $u(x, t)$ a solution of the homogeneous wave equation with arbitrary data, we are effectively able to solve the general initial value problem for (4.174)].

Applying Duhamel's principle, we consider the function $v(x, t; \tau)$ that satisfies the equation

$$(4.176) \qquad v_{tt} - c^2 v_{tt} = 0, \qquad t > \tau$$

and the initial conditions at $t = \tau$,

$$(4.177) \qquad v(x, \tau; \tau) = 0; \qquad v_t(x, \tau; \tau) = g(x, \tau).$$

From d'Alembert's solution (see Example 2.3) we easily obtain

$$(4.178) \qquad v(x, t; \tau) = \frac{1}{2c} \int_{x-c(t-\tau)}^{x+c(t-\tau)} g(\sigma, \tau) \, d\sigma$$

and from (4.169), $u(x, t)$ takes the form

$$(4.179) \qquad u(x, t) = \frac{1}{2c} \int_0^t \int_{x-c(t-\tau)}^{x+c(t-\tau)} g(\sigma, \tau) \, d\sigma \, d\tau.$$

If we replace the initial data (4.175) by the arbitrary data

$$(4.180) \qquad u(x,0) = F(x); \qquad u_t(x,0) = G(x),$$

the solution of the initial value problem (4.174) and (4.180) is immediately obtained as

$$(4.181) \quad u(x, t) = \frac{1}{2} [F(x + ct) + F(x - ct)] + \frac{1}{2c} \int_{x-ct}^{x+ct} G(\sigma) \, d\sigma$$

$$+ \frac{1}{2c} \int_0^t \int_{x-c(t-\tau)}^{x+c(t-\tau)} g(\sigma, \tau) \, d\sigma \, d\tau$$

At the arbitrary point (x_0, t_0) with $t_0 > 0$, the solution $u(x_0, t_0)$ depends only on values x and t within the *characteristic triangle* pictured in Figure 4.6. An inspection of the arguments of $F(x_0 \pm ct_0)$ and the domains of integration for $G(\sigma)$ and $g(\sigma, \tau)$ in (4.181) shows this to be the case. As a result, the characteristic triangle is called the *domain of dependence* of the solution at the point x_0 and at the time t_0. For a similar reason, the sector pictured in Figure 4.7 that is bounded by the characteristic lines issuing from the point $(x_0, 0)$ is

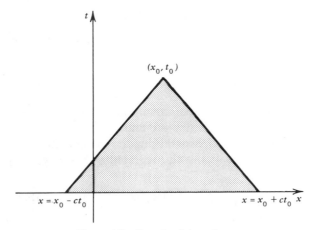

Figure 4.6. Domain of dependence.

called the *domain* or *region of influence* of the initial point $(x_0, 0)$. All points (x, t) that lie in this sector have the point $(x_0, 0)$ within their domain of dependence. In particular, if $g(x, t) = 0$ and $F(x)$ and $G(x)$ are concentrated at the point x_0, say, they have Dirac delta function behavior, we see that the solution $u(x, t)$ vanishes identically outside the domain of influence of the point x_0. Within the sector, $u(x, t)$ may or may not be zero. The two characteristic lines $x \pm ct = x_0 \pm ct_0$ represent wave fronts for the solution, since $u \equiv 0$ ahead of the wavefronts (which move to the right and to the left) and $u \neq 0$, in general, at points (x, t) that lie behind the wave fronts, that is, within the sector.

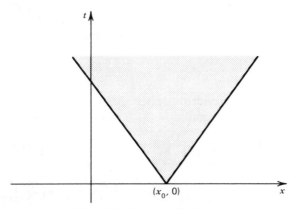

Figure 4.7. Domain of influence.

EXAMPLE 4.13. THE INHOMOGENEOUS HEAT EQUATION

The heat equation in a finite interval $0 < x < l$, with a heat source, has the form

(4.182) $$u_t - c^2 u_{xx} = g(x, t), \qquad 0 < x < l, t > 0.$$

We assume a homogeneous initial condition $u(x, 0) = 0$ and the boundary conditions $u(0, t) = u(l, t) = 0$ at $x = 0$ and $x = l$.

Applying Duhamel's principle and using the solution obtained in (4.154) for the homogeneous heat equation with t replaced by $t - \tau$, we obtain

(4.183) $$v(x, t; \tau) = \sqrt{\frac{2}{l}} \sum_{k=1}^{\infty} a_k(\tau) \exp\left[- \left(\frac{\pi k c}{l} \right)^2 (t - \tau) \right] \sin\left(\frac{\pi k}{l} x \right),$$

with the $a_k(\tau)$ determined from the initial condition

(4.184) $$v(x, \tau; \tau) = g(x, \tau) = \sqrt{\frac{2}{l}} \sum_{k=1}^{\infty} a_k(\tau) \sin\left(\frac{\pi k}{l} x \right).$$

Note that the Fourier coefficients a_k are functions of the parameter τ [i.e., for each value of τ we obtain a Fourier sine series of $g(x, \tau)$]. Then the solution of the initial and boundary value problem for (4.182) is

(4.185)

$$u(x, t) = \int_0^t v(x, t; \tau) \, d\tau$$

$$= \sqrt{\frac{2}{l}} \sum_{k=1}^{\infty} \left\{ \int_0^t a_k(\tau) \exp\left[- \left(\frac{\pi k c}{l} \right)^2 (t - \tau) \right] d\tau \right\} \sin\left(\frac{\pi k}{l} x \right)$$

where summation and integration have been interchanged. This is a valid procedure since the series converges uniformly, say, if we assume $g(x, t)$ is a bounded function. If $u(x, 0) = f(x)$ and is nonzero, we may add the solution (4.154) to (4.185) to obtain the solution of the new problem.

To consider a specific problem we suppose that

(4.186) $$g(x, t) = \sin\left(\frac{\pi x}{l} \right)$$

(i.e., we assume g is independent of time). Then it can be seen by inspection that we have $a_k = 0$, $k > 1$, and $a_1 = \sqrt{l/2}$, so that the solution is

(4.187) $$u(x, t) = \left\{ \int_0^t \exp\left[- \left(\frac{\pi c}{l} \right)^2 (t - \tau) \right] d\tau \right\} \sin\left(\frac{\pi x}{l} \right)$$

$$= \left(\frac{l}{\pi c} \right)^2 \left\{ 1 - \exp\left[- \left(\frac{\pi c}{l} \right)^2 t \right] \right\} \sin\left(\frac{\pi x}{l} \right).$$

[Note that we again have $u(x, 0) = 0$.]

As $t \to \infty$ we see that

(4.188)
$$u(x, t) \underset{t \to \infty}{\to} \left(\frac{l}{\pi c} \right)^2 \sin\left(\frac{\pi x}{l} \right) \equiv v(x).$$

The limiting function $v(x)$ in (4.188) is a solution of the *steady-state problem* for the heat equation with $g = \sin(\pi x/l)$ which is given as

(4.189)
$$- c^2 v_{xx} = \sin\left(\frac{\pi x}{l} \right); \qquad v(0) = v(l) = 0.$$

This problem results when we assume $u(x, t) = v(x)$ (i.e., u is independent of t) and the initial condition is dropped.

More generally, if the inhomogeneous term $g(x, t)$ is independent of time, that is, $g(x, t) = G(x)$, with $G(x)$ arbitrary, we obtain from (4.185)

(4.190)
$$u(x, t) = \sqrt{\frac{2}{l}} \sum_{k=1}^{\infty} \left\{ \left(\frac{l}{\pi k c} \right)^2 a_k \left\{ 1 - \exp\left[- \left(\frac{\pi k c}{l} \right)^2 t \right] \right\} \sin\left(\frac{\pi k}{l} x \right) \right\},$$

where we have used the fact that a_k is independent of τ to carry out the τ integration. As $t \to \infty$ in (4.190) we have

(4.191)
$$\lim_{t \to \infty} u(x, t) = \sqrt{\frac{2}{l}} \sum_{k=1}^{\infty} \left(\frac{l}{\pi k c} \right)^2 a_k \sin\left(\frac{\pi k}{l} x \right) \equiv v(x).$$

It can be verified directly that $v(x)$ satisfies the steady-state problem for the heat equation

(4.192)
$$- c^2 v_{xx} = G(x); \qquad v(0) = v(l) = 0.$$

Now even if $u(x, 0) = f(x) \neq 0$, we have already seen in Example 4.10 that the effect of the nonzero initial heat distribution dies out as $t \to \infty$, so that when the heat source term $g(x, t) = G(x)$ is independent of time, the temperature distribution (when the ends of the rod are fixed at zero temperature) is well approximated by the solution of (4.192).

This completes our discussion of Duhamel's principle. The method of finite Fourier transforms and other techniques for handling inhomogeneous problems are discussed in the following section.

4.6. EIGENFUNCTION EXPANSIONS: FINITE FOURIER TRANSFORMS

The method of separation of variables presented in Section 4.2 led to the representation of the solution of the given problem in a series of eigenfunctions of the differential operator L defined in (4.6)–(4.7). To generate the appropriate eigenvalue problem it was necessary that both the equation and the boundary conditions be homogeneous. For the self-adjoint operators L under consideration, the eigenfunctions have the property of completeness, which means that under certain conditions a function $f(x)$ can be expanded in a (convergent) series of eigenfunctions. The orthogonality of the eigenfunctions renders the determination of the coefficients in the series to be quite simple.

In this section we reverse the above approach and begin directly by expressing the solution of the problem as a series of eigenfunctions. The coefficients in the series are arbitrary and must be determined from the differential equation and the data for the problem. Rather than substituting the series directly into the equation, we convert the partial differential equation into a hierarchy of ordinary differential equations for the determination of the unknown coefficients, termed Fourier coefficients. Thereby, the equation is transformed into a collection of equations for these Fourier coefficients, and this technique is often called the method of *finite Fourier transforms*. The reason for this becomes clearer when we discuss (infinite) Fourier transforms in Chapter 5. This method applies to inhomogeneous equations with inhomogeneous initial and boundary data. Technically, therefore, this method enables us to solve the most general problems formulated in Section 4.1.

We consider equations of the form

$$(4.193) \qquad\qquad \rho K u + L u = \rho F$$

where L is defined as in (4.6) or (4.7) and $\rho(x) > 0$ and F are given functions. The differential operator K can be given as $K = \partial/\partial t$, $K = \partial^2/\partial t^2$, or $K = -\partial^2/\partial y^2$, so that (4.193) has the form of (4.8), (4.10) or (4.19), respectively. Or $K = 0$ and (4.193) reduces to (4.9). Thus equations of parabolic, hyperbolic, or elliptic type can be treated by this method. The boundary conditions for the problem are of inhomogeneous type as given in (4.11) or (4.12) and initial conditions are assigned in the hyperbolic and parabolic cases as in (4.16) or (4.18). Conditions of the form (4.20) for the elliptic case can also be given.

The solution u of each of the foregoing problems is expanded in a series of eigenfunctions $M_k(x)$ determined from the eigenvalue problem

$$(4.194) \qquad\qquad LM_k = \lambda_k \rho M_k, \qquad k = 1, 2, \ldots$$

with homogeneous boundary conditions of the form (4.14) or (4.15). It is assumed that the M_k form a complete orthonormal set for the purposes of this discussion. Thus $(M_k, M_j) = 0$ for $k \neq j$ and $(M_k, M_k) = 1$ for all k. We expand u as

$$(4.195) \qquad\qquad u = \sum_{k=1}^{\infty} N_k M_k(x),$$

where the *Fourier coefficients* N_k (which may depend on t or on y) are to be specified. In terms of the inner product (4.33), the N_k are formally given as

$$(4.196) \qquad\qquad N_k = (u, M_k),$$

since the M_k are orthonormalized, but the solution u is not as yet known. To specify the N_k we do not substitute the series (4.195) directly into the equation (4.193) since the series may not be differentiable term by term. Instead, we transform (4.193) into a system of equations for the Fourier coefficients N_k.

To do so, we multiply (4.193) by $M_k(x)$ and integrate over the underlying region G to obtain (on using notation appropriate for the two or three-dimensional case)

$$(4.197) \qquad \iint_G \rho M_k Ku \, dv = K \iint_G \rho u M_k \, dv = K(u, M_k)$$

$$= -\iint_G M_k Lu \, dv + \iint_G \rho F M_k \, dv,$$

where we have pulled the differential operator K out of the integral. Now from (4.31) we have

$$(4.198) \quad -\iint_G M_k Lu \, dv = -\iint_G u L M_k \, dv + \int_{\partial G} p\left(M_k \frac{\partial u}{\partial n} - u \frac{\partial M_k}{\partial n}\right) ds.$$

Then using (4.194) and the boundary condition satisfied by u and M_k in (4.198) we readily obtain

$$(4.199) \qquad -\iint_G M_k Lu \, dv = -\lambda_k(u, M_k) + \int_{\partial G}\left(\frac{p}{\beta}\right) M_k B \, ds.$$

We have assumed that $\beta \neq 0$ in the boundary conditions. The appropriate expression for the case $\beta = 0$ is easily derived. It will also be assumed in our discussion that zero is not an eigenvalue for any of the problems we consider.

Combining (4.199) and (4.197), we obtain the hierarchy of ordinary differential equations for N_k,

$$(4.200) \quad KN_k + \lambda_k N_k = F_k + \int_{\partial G}\left(\frac{p}{\beta}\right) M_k B \, ds, \qquad k = 1, 2, \ldots,$$

where F_k is the Fourier coefficient of F [i.e., $F_k = (F, M_k)$]. In the elliptic case that corresponds to (4.9) and for which $K = 0$ in (4.193), the equations (4.200) are algebraic and since $\lambda_k > 0$ for all k by assumption, the N_k are uniquely determined. For the elliptic case corresponding to (4.19), (4.200) is a second order equation and the N_k must satisfy the boundary conditions obtained by setting $u = f$ and $u = g$ in (4.196) at $y = 0$ and $y = l$, respectively, in view of (4.20).

For the parabolic case, $K = \partial/\partial t$ and we have for $N_k(t)$,

$$(4.201) \qquad \frac{dN_k(t)}{dt} + \lambda_k N_k(t) = F_k(t) + \int_{\partial G}\left(\frac{p}{\beta}\right)M_k B(x, t)\, ds,$$

with the initial condition

$$(4.202) \qquad N_k(0) = (f, M_k)$$

which is obtained from the eigenfunction expansion of $u(x, 0) = f(x)$. If we set $B_k(t) = \int_{\partial G}(p/\beta)M_k B(x, t)\, ds$, the solution of this problem is

(4.203)

$$N_k(t) = (f, M_k)\exp[-\lambda_k t] + \int_0^t [F_k(\tau) + B_k(\tau)]\exp[-\lambda_k(t - \tau)]\, d\tau.$$

In the hyperbolic case, $K = \partial^2/\partial t^2$ and we obtain the initial conditions from

$$(4.204) \qquad \begin{cases} u(x, 0) = f(x) = \displaystyle\sum_{k=1}^{\infty} N_k(0) M_k(x) \\[2mm] u_t(x, 0) = g(x) = \displaystyle\sum_{k=1}^{\infty} N_k'(0) M_k(x) \end{cases}$$

so that at $t = 0$ we have for $N_k(t)$,

$$(4.205) \qquad N_k(0) = (f, M_k); \qquad N_k'(0) = (g, M_k).$$

The solution of the equation

$$(4.206) \qquad \frac{d^2 N_k(t)}{dt^2} + \lambda_k N_k(t) = F_k(t) + B_k(t)$$

satisfying (4.205) is

$$(4.207) \quad N_k(t) = (f, M_k)\cos\left(\sqrt{\lambda_k}\, t\right) + \frac{1}{\sqrt{\lambda_k}}(g, M_k)\sin\left(\sqrt{\lambda_k}\, t\right)$$

$$+ \frac{1}{\sqrt{\lambda_k}}\int_0^t [F_k(\tau) + B_k(\tau)]\sin\left[\sqrt{\lambda_k}(t - \tau)\right]\, d\tau.$$

The foregoing results for the parabolic and hyperbolic cases reduce to those obtained in Section 4.2 by means of separation of variables when F_k and B_k are equated to zero.

We note that the solutions u obtained as a series of eigenfunctions cannot, in general, be expected to be *classical* solutions that satisfy the differential equation (4.193) and assume the initial and boundary values pointwise. Thus we observe that each of the eigenfunctions $M_k(x)$ satisfies a homogeneous boundary condition, whereas the solution u given as a sum of these eigenfunctions may satisfy an inhomogeneous boundary condition. Nevertheless, these boundary conditions are certainly accounted for in the equations (4.200) for the N_k. Thus convergence up to the boundary may have to be interpreted in a *generalized* sense, perhaps using mean square convergence. Additionally, although there may be pointwise convergence in interior regions, the lack of convergence up to the boundary generally slows the rate of convergence for the series everywhere. Consequently, we present a method (following Example 4.15) that often enables us to circumvent the foregoing difficulty by converting the problem to one with homogeneous boundary conditions.

EXAMPLE 4.14.   HYPERBOLIC EQUATIONS: RESONANCE

We consider the inhomogeneous hyperbolic equation

$$(4.208) \qquad \rho(x)u_{tt} + Lu = \rho(x)M_i(x)\sin \omega t,$$

where $M_i(x)$ is one of the eigenfunctions determined from (4.194) and $F(x, t) = M_i(x)\sin \omega t$ is a periodic forcing term in t with frequency of vibration $\omega = $ constant. The initial conditions as well as the boundary conditions are assumed to be homogeneous; that is, $f(x) = g(x) = B(x, t) = 0$.
    Now

$$(4.209) \qquad F_k(t) = (F, M_k) = \sin \omega t(M_i, M_k) = \delta_{ik}\sin \omega t$$

where $\delta_{ik} = 0$ for $i \neq k$ and $\delta_{ii} = 1$ for all i. Since $(f, M_k) = (g, M_k) = B_k = 0$ for all k in this problem, we have from (4.207)

$$(4.210) \qquad N_i(t) = \frac{1}{\sqrt{\lambda_i}} \int_0^t \sin(\omega \tau)\sin\left[\sqrt{\lambda_i}\,(t - \tau)\right] d\tau$$

$$= \frac{1}{\omega^2 - \lambda_i}\left[\frac{\omega}{\sqrt{\lambda_i}}\sin\left(\sqrt{\lambda_i}\,t\right) - \sin(\omega t)\right]$$

and $N_k(t) = 0$ for $k \neq i$. Thus the series (4.195) reduces to the single term

$$(4.211) \qquad u(x, t) = \frac{1}{\omega^2 - \lambda_i}\left[\frac{\omega}{\sqrt{\lambda_i}}\sin\left(\sqrt{\lambda_i}\,t\right) - \sin(\omega t)\right]M_i(x),$$

which is, in fact, the solution of the problem. Clearly, (4.211) is valid only if

$\omega \neq \sqrt{\lambda_i}$. If $\omega = \sqrt{\lambda_i}$, the above solution is not valid, but we may obtain the solution in that case by going to the limit as $\omega \to \sqrt{\lambda_i}$ in (4.211). Using l'Hospital's rule we obtain

$$(4.212) \qquad u(x, t) = \frac{1}{2\sqrt{\lambda_i}} \left[\frac{\sin\left(\sqrt{\lambda_i}\, t\right)}{\sqrt{\lambda_i}} - t\cos\left(\sqrt{\lambda_i}\, t\right) \right] M_i(x).$$

To interpret these results, we observe that the numbers $\omega_k = \sqrt{\lambda_k}$ ($k = 1, 2, \ldots$) represent the natural frequencies of vibration of the solution of the homogeneous initial and boundary value problem for (4.208) (i.e., with $F = 0$) in view of (4.43). Thus when $\omega \neq \omega_i$ (one of the natural frequencies), the solution (4.211) oscillates with the imposed external frequency ω, as well as the natural frequency $\omega_i = \sqrt{\lambda_i}$. The external energy fed into the system is distributed between these two frequencies. However, if $\omega = \omega_i = \sqrt{\lambda_i}$, we find that the amplitude of the oscillatory solution increases unboundedly with t as $t \to \infty$. This effect is known as *resonance*. It results because the entire external energy is concentrated within a single natural frequency of vibration and the continuous input builds up its amplitude.

EXAMPLE 4.15. POISSON'S EQUATION IN A CIRCLE

The inhomogeneous Laplace's equation

$$(4.213) \qquad \nabla^2 u = -F,$$

is often referred to as *Poisson's equation*. We consider the boundary value problem for (4.213) in a circle of radius R with center at the origin and the homogeneous boundary condition $u = 0$ on the circle. Expressing the problem in polar coordinates, we obtain for $u(r, \theta)$ the equation

$$(4.214) \qquad \frac{\partial^2 u}{\partial r^2} + \frac{1}{r}\frac{\partial u}{\partial r} + \frac{1}{r^2}\frac{\partial^2 u}{\partial \theta^2} = -F(r, \theta),$$

within the circle (i.e., for $r < R$) and the boundary condition

$$(4.215) \qquad u(R, \theta) = 0.$$

In view of our discussion following equation (4.193), this problem can be solved in terms of the eigenfunctions M_k determined from the problem $\nabla^2 M_k = -\lambda_k M_k$ for $r < R$ with $M_k = 0$ on $r = R$. Since this higher dimensional eigenvalue problem is not considered until Chapter 8, we present an alternative approach to this problem that involves an expansion in terms of one-dimensional eigenfunctions.

The solution of (4.214)–(4.215) is expected to be single valued at any point within the circle, so we must have

$$(4.216) \qquad u(r, \theta + 2\pi) = u(r, \theta),$$

with a similar result for $F(r, \theta)$. That is, $u(r, \theta)$ and $F(r, \theta)$ are periodic of period 2π in θ. This suggests that the eigenvalue problem associated with $\hat{L}u = -\partial^2 u / \partial \theta^2$ with periodic boundary conditions of period 2π may be appropriate here. [That is, (4.214) can be written as $\hat{L}u - r^2 u_{rr} - ru_r = r^2 F$ and the solution can be expanded in a series of eigenfunctions for the operator $\hat{L}$.] This eigenvalue problem was discussed in Example 4.6, where with $l = \pi$ we obtained the eigenfunctions $\{\cos k\theta\}$, $k = 0, 1, 2, \ldots$ and $\{\sin k\theta\}$, $k = 1, 2, \ldots$, and the eigenvalues $\lambda_k = k^2$, $k = 0, 1, 2, \ldots$. Using the orthonormal set of eigenfunctions as defined in (4.104)–(4.105) we obtain the eigenfunction expansion

$$(4.217) \quad u(r, \theta) = \frac{1}{\sqrt{2\pi}} a_0(r) + \sum_{k=1}^{\infty} \frac{1}{\sqrt{\pi}} \{a_k(r)\cos k\theta + b_k(r)\sin k\theta\}$$

which corresponds to (4.195). Even though this eigenvalue problem is not of Sturm–Liouville type, the finite Fourier transform method can be applied.

To begin, we note that the Fourier coefficients a_k and b_k are given in terms of u as

$$(4.218)$$

$$a_0(r) = \left(u, \frac{1}{\sqrt{2\pi}}\right) = \frac{1}{\sqrt{2\pi}} \int_{-\pi}^{\pi} u(r, \theta) \, d\theta,$$

$$(4.219)$$

$$a_k(r) = \left(u, \frac{1}{\sqrt{\pi}}\cos k\theta\right) = \frac{1}{\sqrt{\pi}} \int_{-\pi}^{\pi} u(r, \theta)\cos(k\theta) \, d\theta; \qquad k = 1, 2, \ldots,$$

$$(4.220)$$

$$b_k(r) = \left(u, \frac{1}{\sqrt{\pi}}\sin k\theta\right) = \frac{1}{\sqrt{\pi}} \int_{-\pi}^{\pi} u(r, \theta)\sin(k\theta) \, d\theta; \qquad k = 1, 2, \ldots .$$

The equations for the Fourier coefficients a_k and b_k are obtained from (4.214) on multiplying across by the eigenfunctions and integrating from $-\pi$ to π, as

$$(4.221) \qquad \frac{d^2 a_k(r)}{dr^2} + \frac{1}{r}\frac{da_k(r)}{dr} - \frac{k^2}{r^2} a_k(r) = -A_k(r); \qquad k = 0, 1, 2,$$

and

$$(4.222) \qquad \frac{d^2 b_k(r)}{dr^2} + \frac{1}{r}\frac{db_k(r)}{dr} - \frac{k^2}{r^2} b_k(r) = -B_k(r); \qquad k = 1, 2, \ldots,$$

where A_k and B_k are the Fourier coefficients of $F(r, \theta)$. Since $r = 0$ is a singular point in the equations for $a_k(r)$ and $b_k(r)$ and we require the solution $u(r, \theta)$ to be bounded at $r = 0$, we obtain the boundary condition

$$(4.223) \qquad a_k(r), \, b_k(r) \text{ bounded at } r = 0; \qquad k = 0, 1, 2, \ldots.$$

The boundary condition (4.215) implies

$$(4.224) \qquad a_k(R) = b_k(R) = 0; \qquad k = 0, 1, 2, \ldots.$$

Noting that (4.221)–(4.222) are both inhomogeneous forms of Euler's equation

$$(4.225) \qquad C_k''(r) + \frac{1}{r} C_k'(r) - \frac{k^2}{r^2} C_k(r) = 0; \qquad k = 0, 1, 2, \ldots$$

for which a fundamental set of solutions is

$$(4.226) \qquad C_0(r) = \begin{cases} \text{constant} \\ \log r \end{cases}; \qquad C_k(r) = \begin{cases} r^k \\ r^{-k} \end{cases}; \qquad k \geqslant 1;$$

we obtain on using variation of parameters, the following solutions of the boundary value problems (4.221)–(4.224):

$$(4.227) \quad a_0(r) = \int_0^r \log\left(\frac{R}{r}\right) A_0(t) t \, dt + \int_r^R \log\left(\frac{R}{t}\right) A_0(t) t \, dt,$$

$$(4.228) \quad a_k(r) = \frac{1}{2k} \left\{ \int_0^r \left[\left(\frac{R}{r}\right)^k - \left(\frac{r}{R}\right)^k \right] \left(\frac{t}{R}\right)^k A_k(t) t \, dt \right.$$
$$\left. + \int_r^R \left[\left(\frac{R}{t}\right)^k - \left(\frac{t}{R}\right)^k \right] \left(\frac{r}{R}\right)^k A_k(t) t \, dt \right\}; \qquad k \geqslant 1$$

and

$$(4.229) \quad b_k(r) = \frac{1}{2k} \left\{ \int_0^r \left[\left(\frac{R}{r}\right)^k - \left(\frac{r}{R}\right)^k \right] \left(\frac{t}{R}\right)^k B_k(t) t \, dt \right.$$
$$\left. + \int_r^R \left[\left(\frac{R}{t}\right)^k - \left(\frac{t}{R}\right)^k \right] \left(\frac{r}{R}\right)^k B_k(t) t \, dt \right\}; \qquad k \geqslant 1.$$

Thus the Fourier coefficients in the series (4.217) for $u(r, \theta)$ are completely specified. Under suitable conditions on $F(r, \theta)$, say, $\int_0^R \int_{-\pi}^{\pi} F^2(r, \theta) r \, dr \, d\theta < \infty$, the Fourier series (4.217) can be shown to converge uniformly. Then $u(r, \theta)$ is continuous and $u(R, \theta) = 0$.

An interesting special case for the parabolic and hyperbolic problems discussed arises when the functions $F(x, t)$ and $B(x, t)$ are independent of t. Since these terms cause the problems to be inhomogeneous, we say that inhomogeneities are *stationary* (i.e., time independent). The effect of this assumption is that the terms F_k and B_k are time independent constants. As a result, the integrals in (4.203) and (4.207) can be evaluated, and we obtain for $N_k(t)$,

$$(4.230) \quad N_k(t) = \left[(f, M_k) - \frac{1}{\lambda_k}(F_k + B_k) \right] e^{-\lambda_k t} + \frac{1}{\lambda_k}[F_k + B_k],$$

and

(4.231)

$$N_k(t) = \left[(f, M_k) - \frac{1}{\lambda_k}(F_k + B_k) \right] \cos\left(\sqrt{\lambda_k}\, t \right) + \frac{1}{\sqrt{\lambda_k}}(g, M_k) \sin\left(\sqrt{\lambda_k}\, t \right)$$

$$+ \frac{1}{\lambda_k}[F_k + B_k],$$

in the parabolic and hyperbolic cases, respectively.

In both cases we can introduce the decomposition of $N_k(t)$ in the form

$$(4.232) \qquad\qquad N_k(t) = \tilde{N}_k(t) + \hat{N}_k$$

where $\hat{N}_k = (1/\lambda_k)[F_k + G_k]$ is independent of t. Correspondingly, the solution $u(x, t)$ can be formally expressed as

$$(4.233) \qquad u(x, t) = \sum_{k=1}^{\infty} \tilde{N}_k(t) M_k(x) + \sum_{k=1}^{\infty} \hat{N}_k M_k(x).$$

If we set $F(x, t) = \hat{F}(x)$ and $B(x, t) = \hat{B}(x)$ since F and B are independent of t and define

$$(4.234) \qquad\qquad v(x) = \sum_{k=1}^{\infty} \hat{N}_k M_k(x),$$

it is readily seen that $v(x)$ is the solution of the stationary version of (4.193); that is,

$$(4.235) \qquad\qquad Lv = \rho(x)\hat{F}(x),$$

and the boundary condition (in two or three dimensions)

$$(4.236) \qquad\qquad \alpha(x)v(x) + \beta(x) \left. \frac{\partial v}{\partial n} \right|_{\partial G} = \hat{B}(x).$$

[On solving (4.235)–(4.236) by the finite transform method, we obtain (4.234).]

Then if we put

(4.237) $$u(x, t) = w(x, t) + v(x)$$

with $v(x)$ defined as above, $w(x, t)$ satisfies the homogeneous equation

(4.238) $$\rho Kw + Lw = 0$$

with the boundary condition

(4.239) $$\alpha w + \beta \left. \frac{\partial w}{\partial n} \right|_{\partial G} = 0,$$

where $K = \partial/\partial t$ and $K = \partial^2/\partial t^2$ in the parabolic and hyperbolic cases, respectively. The respective initial conditions are

(4.240) $$w(x, 0) = f(x) - v(x)$$

and

(4.241) $$w(x, 0) = f(x) - v(x); \qquad w_t(x, 0) = g(x).$$

Once the possibility of the decomposition (4.237) has been recognized, it may be introduced directly when solving initial and boundary value problems with stationary inhomogeneities. Since the equation for $w(x, t)$ and the boundary conditions are homogeneous, the rate of convergence of the formal series solution of the problem for $w(x, t)$ is accelerated. Additionally, if the problem involves only one space dimension, we obtain an ordinary differential equation for $v(x)$. If the solution for $v(x)$ can be obtained without the use of eigenfunction expansions, we expect that the solution of the given problem $u = w + v$ takes a form better suited for numerical evaluation.

In a similar vein, recognizing that the nonhomogeneity of the boundary conditions weakens the convergence rate of the eigenfunction expansion (4.195), the following method is occasionally useful. Given the boundary condition (4.11) or (4.12), we seek a function $V(x, t)$ such that

(4.242) $$\alpha(x)V + \beta(x) \left. \frac{\partial V}{\partial n} \right|_{\partial G} = B(x, t)$$

or

(4.243) $$\begin{cases} \alpha_1 V(0, t) - \beta_1 V_x(0, t) = g_1(t) \\ \alpha_2 V(l, t) + \beta_2 V_x(l, t) = g_2(t) \end{cases}.$$

The function $V(x, t)$ must be differentiable as often as required in the given differential equation. If such a function can be constructed we set

$$(4.244) \qquad u(x, t) = W(x, t) + V(x, t)$$

and find that $W(x, t)$ satisfies the equation

$$(4.245) \qquad \rho KW + LW = \rho F - \rho KV - LV,$$

with initial conditions of the form (4.240) or (4.241) (where v is replaced by V and a term $-V_t$ is added to the second equation in (4.241)), depending on whether the equation is parabolic or hyperbolic. However, the boundary conditions for W are homogeneous since, for example,

$$\alpha W + \beta \left. \frac{\partial W}{\partial n} \right|_{\partial G} = \alpha(u - V) + \beta \left. \frac{\partial}{\partial n}(u - V) \right|_{\partial G} = B - B = 0,$$

in the three-dimensional case. Duhamel's principle or the finite transform method can now be applied to solve for $W(x, t)$.

EXAMPLE 4.16.   THE INHOMOGENEOUS HEAT EQUATION

We consider the fully inhomogeneous initial and boundary value problem for the one-dimensional heat equation in a finite interval. We have

$$(4.246) \qquad u_t - c^2 u_{xx} = g(x, t), \qquad 0 < x < l, t > 0$$

$$(4.247) \qquad u(x, 0) = f(x), \qquad 0 < x < l$$

$$(4.248) \qquad u(0, t) = g_1(t), \qquad u(l, t) = g_2(t)$$

where the functions $g(x, t)$, $f(x)$, $g_1(t)$, and $g_2(t)$ are prescribed.

Using linear interpolation we readily construct the function

$$(4.249) \qquad V(x, t) = \frac{1}{l}\left[xg_2(t) + (l - x)g_1(t) \right],$$

which satisfies the boundary conditions (4.248). Then with $u = W + V$, we find that W satisfies the equation

$$(4.250) \qquad W_t - c^2 W_{xx} = g(x, t) - \frac{1}{l}\left[xg_2'(t) + (l - x)g_1'(t) \right]$$

for $0 < x < l$ and $t > 0$ with the initial condition

$$(4.251) \qquad W(x, 0) = f(x) - \frac{1}{l}\left[xg_2(0) + (l - x)g_1(0) \right]$$

and the homogeneous boundary conditions

$$(4.252) \qquad\qquad W(0, t) = W(l, t) = 0.$$

It may be noted that if $g(x, t) \equiv 0$ and the functions $g_1(t)$ and $g_2(t)$ are constants independent of time so that the inhomogeneities are stationary, we have

$$(4.253) \qquad V(x, t) \equiv V(x) = \frac{1}{l}\left[xg_2 + (l - x)g_1 \right].$$

The function $V(x)$ is a solution of the stationary form of the heat equation $u_{xx} = 0$ with the boundary conditions $u(0, t) = g_1$ and $u(l, t) = g_2$. However, this does not mean that there is an equivalence between the two decomposition methods given above in all cases in which the given equation is homogeneous and the boundary conditions are stationary. For example, another possible choice for $V(x, t)$ for the above problem is

$$(4.254) \qquad V(x, t) = \frac{1}{\sinh l}\{(\sinh x)g_2(t) + [\sinh(l - x)]g_1(t)\},$$

and this does not reduce to a solution of the stationary case if g_1 and g_2 are constant.

4.7. NONLINEAR STABILITY THEORY: EIGENFUNCTION EXPANSIONS

In this section we consider a *nonlinear heat conduction equation* in one dimension and study the *stability* of the equilibrium or zero temperature distribution. That is, we wish to determine if small perturbations at the initial time around the zero temperature distribution decay to zero as t increases or grow, perhaps, into a new stationary solution with increasing t. We are concerned with a problem in a finite interval with the ends of the (insulated) rod being kept at zero temperature for all time. However, to compare results of the present stability analysis with that carried out in Section 3.5, we briefly consider the stability problem for a rod of infinite extent.

Let $u(x, t)$ represent the temperature distribution of an insulated rod and assume that there is a nonlinear heat source of strength $-\hat{\lambda}u(1 - u^2)$. Then the (nonlinear) heat equation to be studied has the form

$$(4.255) \qquad\qquad u_t - u_{xx} = \hat{\lambda}u(1 - u^2),$$

where $\hat{\lambda}$ is a parameter that depends on the properties of the rod. Since we are interested in the development in time of small initial perturbations around the equilibrium solution $u_0(x, t) = 0$ of (4.255), we introduce the initial condition

$$(4.256) \qquad\qquad u(x,0) = \epsilon h(x),$$

where $h(x)$ is uniformly bounded and $0 < \epsilon \ll 1$.

We are interested in two problems. The basic problem is that in which the rod is of finite extent (it is assumed to have length π), and the boundary conditions at the end points $x = 0$ and $x = \pi$ are

$$(4.257) \qquad\qquad u(0, t) = 0; \qquad u(\pi, t) = 0.$$

In this case the equations (4.255)–(4.256) are valid in the interval $0 < x < \pi$. The other problem deals with a rod of infinite extent so that (4.255)–(4.256) are given over the interval $-\infty < x < \infty$. Since the data for either problem are small (of order of magnitude ϵ), we look for a solution of each problem in the form

$$(4.258) \qquad\qquad u(x, t) = \epsilon w(x, t)$$

and trace its evolution in time. If this *perturbation* around the solution $u_0 = 0$ remains small as t increases, the zero solution is *stable*. However, if $w(x, t)$ grows without bound as time increases, the equilibrium solution is *unstable*. Since we are unable to solve either of the above nonlinear problems exactly, we use approximate methods for studying the stability of the equilibrium solutions of the above problems.

Inserting (4.258) into (4.255) and dividing by ϵ gives

$$(4.259) \qquad\qquad w_t - w_{xx} = \hat{\lambda} w - \epsilon^2 \hat{\lambda} w^3.$$

We first carry out a *linear stability analysis*. Assuming the solution $u = \epsilon w$ is stable so that it does not grow without bound and noting that $\epsilon^2 \ll 1$, we *linearize* the equation (4.259) by dropping the term proportional to w^3. The resulting linear equation is

$$(4.260) \qquad\qquad w_t - w_{xx} = \hat{\lambda} w,$$

and we perform a stability analysis of (4.260) with the initial and boundary conditions carried over from the nonlinear problem. If the solutions of the linear problem are stable, our assumption leading to the neglect of the nonlinear term in (4.259) is valid for all time. We then say that the nonlinear problem is stable and is well approximated by the linearized version of the problem.

However, if the linearized problem exhibits instability, the solution $w(x, t)$ grows in time so that, eventually, the term $\hat{\lambda} \epsilon^2 w^3$ may attain an order of magnitude equal to that of the term w. Consequently, the nonlinear term in (4.259) cannot be neglected and the full nonlinear equation must be used.

Although the linear stability analysis predicts that the equilibrium solution $u_0 = 0$ is unstable and that the perturbation grows without bound, a nonlinear analysis may show that, in fact, the perturbation grows only until it reaches another equilibrium solution. This is demonstrated for the initial and boundary value problem for (4.255).

We begin by considering a *linear stability analysis* of the *Cauchy problem* for (4.255). Thus we study the normal mode solutions

$$(4.261) \qquad w(x, t) = a(k)\exp[ikx + \lambda(k)t]$$

of the linearized equation (4.260). Inserting (4.261) into (4.260) gives

$$(4.262) \qquad \lambda(k) = \hat{\lambda} - k^2.$$

Since $\hat{\lambda}$ is a real constant, the stability index Ω defined as in Section 3.5 is given by $\Omega = \hat{\lambda}$. Thus (4.260) is strictly stable if $\hat{\lambda} < 0$ and is unstable if $\hat{\lambda} > 0$. For $\hat{\lambda} > 0$, the linear stability analysis predicts unbounded growth for the perturbation ϵw and indicates that the equilibrium solution $u_0 = 0$ is unstable.

To see that the perturbation ϵw need not necessarily grow without bound but may reach another (stable) equilibrium solution, we proceed as follows. We first note that $u_1 = 1$ is a solution of (4.255). To check for stability of this (additional) equilibrium solution we set

$$(4.263) \qquad u(x, t) = 1 + \epsilon v(x, t),$$

with $0 < \epsilon \ll 1$ and analyze the linearized equation for $v(x, t)$; that is,

$$(4.264) \qquad v_t - v_{xx} = -2\hat{\lambda}v.$$

For the normal mode solutions (4.261) (with w replaced by v) we now obtain

$$(4.265) \qquad \lambda(k) = -2\hat{\lambda} - k^2$$

so that the stability index $\Omega = -2\hat{\lambda}$. Thus for $\hat{\lambda} > 0$, in which case the linear stability analysis showed the zero (equilibrium) solution $u_0 = 0$ to be unstable, we now have stability for the equilibrium solution $u_1 = 1$. This suggests that perturbations of the zero solution do not grow unboundedly but grow only until they reach the stable equilibrium $u_1 = 1$. Although we do not prove this is the case in general, we exhibit an explicit solution of a Cauchy problem for (4.255) where this is exactly what happens.

Let $h(x) = 1$ in (4.256) and look for a solution of (4.255) that is independent of x; that is, $u = U(t)$. Then $U(t)$ satisfies the ordinary differential equation

$$(4.266) \qquad U'(t) = \hat{\lambda} U(t)\left[1 - U(t)^2\right],$$

with the initial condition

$$(4.267) \qquad U(0) = \epsilon.$$

The equation (4.266) can be solved by separation of variables and the solution of (4.266)–(4.267) is

$$(4.268) \qquad u(x, t) \equiv U(t) = \frac{\epsilon e^{\hat{\lambda} t}}{\sqrt{1 + \epsilon^2\left(e^{2\hat{\lambda} t} - 1\right)}}.$$

For small values of t, we may expand this solution in powers of ϵ and find that $u \approx \epsilon e^{\hat{\lambda} t} = \epsilon w(x, t)$, where $w(x, t)$ satisfies (4.260) and $w(x, 0) = 1$. As $t \to \infty$, $w(x, t) \to \infty$ if $\hat{\lambda} > 0$, but the solution (4.268) does not tend to infinity but approaches the equilibrium solution $u_1 = 1$ as is easily seen. If $\hat{\lambda} < 0$, the term $e^{2\hat{\lambda} t}$ in the denominator of (4.268) tends to zero as $t \to \infty$. Thus since $e^{2\hat{\lambda} t} - 1 \approx 0$ for small t and $\epsilon^2 \ll 1$, we conclude that the solution $u(x, t)$ given in (4.268) is well approximated by $\epsilon w(x, t) = \epsilon e^{\hat{\lambda} t}$ for all time. That is, the linearization procedure is valid for all time if the linearized problem is stable.

Turning now to the consideration of the *boundary value problem* (4.255)–(4.257) in the interval $0 < x < \pi$, we begin by applying *linear stability theory*. We look for normal mode solutions of the linearized equation (4.260) that satisfy the homogeneous boundary conditions (4.257). Although (4.261) yields solutions of (4.260) for all real k, to find solutions that vanish at $x = 0$ and $x = \pi$ we must take linear combinations of the normal modes and restrict the values of k. We obtain

$$(4.269) \qquad w_n(x, t) = h_n \exp\left[(\hat{\lambda} - n^2)t\right]\sin nx, \qquad n = 1, 2, \ldots,$$

where h_n is a constant and we have set $k = n$ since for these values of k $w(x, t)$ vanishes at $x = 0$ and $x = \pi$. [The solutions (4.269) are precisely what results if separation of variables is applied to (4.260) with the boundary conditions (4.257)].

The solution of the initial and boundary value problem for the equation (4.260) with the conditions (4.256)–(4.257) is a linear combination of the (normal mode) solutions (4.269). Thus the stability properties are determined from the discrete set of solutions $w_n(x, t)$ given in (4.269), rather than from the continuous set (4.261) with $-\infty < k < \infty$, which is relevant for the Cauchy problem. We see that if $\hat{\lambda} < 1$, the $w_n(x, t)$ in (4.269) all decay to zero as $t \to \infty$, whereas if $\hat{\lambda} > 1$, at least one of the $w_n(x, t)$ grows exponentially as

$t \to \infty$. Consequently, the linear stability analysis predicts that the equilibrium solution $u_0 = 0$ is unstable to small perturbations if $\hat{\lambda} > 1$, whereas if $\hat{\lambda} < 1$, it is stable and all perturbations eventually die out as t increases. Again, if $\hat{\lambda} > 1$, the linear analysis that predicts instability becomes invalid after a finite time since the nonlinear term in (4.259) becomes equally important with the linear term in view of the growth of $w(x, t)$. Therefore, a *nonlinear analysis* of the growth of the perturbation term $u = \epsilon w$ must be carried out.

The values $\hat{\lambda} = 0$ and $\hat{\lambda} = 1$ given for the Cauchy and the initial and boundary value problems for the linearized equation (4.260) determine the *threshold of instability* (in the linear theory) for the given problems. These values of $\hat{\lambda}$ are denoted by $\hat{\lambda}_c$, the *critical value* of $\hat{\lambda}$. That is, if $\hat{\lambda} > \hat{\lambda}_c$, we have instability and if $\hat{\lambda} < \hat{\lambda}_c$, we have stability. It should be noted that the critical value is determined not only from the given (linearized) equation but also by the boundary conditions if any are given. As shown above, the critical values $\hat{\lambda}_c$ differ in the case where no boundary conditions are assigned and in the case where boundary conditions are given.

The initial and boundary value problem (4.255)–(4.257) cannot be solved exactly. There are several approximate methods for analyzing this nonlinear problem, but we consider only one approach which is based on eigenfunction expansions or equivalently finite Fourier transforms.

The operator $L = -\partial^2/\partial x^2$ in (4.255), together with the boundary conditions (4.257), is associated with the set of eigenfunctions

$$(4.270) \qquad M_k(x) = \sqrt{\frac{2}{\pi}} \sin kx; \qquad k = 1, 2, \ldots$$

[i.e., $LM_k = \lambda_k M_k$ with $\lambda_k = k^2$ and $M_k(0) = M_k(\pi) = 0$]. The normalized set $\{M_k(x)\}$ is complete and we represent the function $w(x, t)$ (recall that $u = \epsilon w$) in the form

$$(4.271) \qquad w(x, t) = \sum_{k=1}^{\infty} N_k(t) M_k(x)$$

where

$$(4.272) \qquad N_k(t) = (w, M_k).$$

Using the initial condition (4.256), we have

$$(4.273) \qquad N_k(0) = (h, M_k).$$

[The inner product $(f, g) = \int_0^\pi fg \, dx$ in this problem.]

We obtain an equation for $N_k(t)$ by taking a finite sine transform in (4.259); that is, we multiply by $M_k(x)$ and integrate between 0 and π. This gives

$$(4.274) \quad N_k'(t) + k^2 N_k(t) = \hat{\lambda} N_k(t) - \hat{\lambda}\epsilon^2 \int_0^\pi w^3 M_k(x)\, dx; \qquad k = 1, 2, \ldots.$$

The integral term yields

$$(4.275) \qquad \int_0^\pi w^3 M_k\, dx = \int_0^\pi \left[\sum_{i=1}^\infty N_i(t) M_i(x)\right]^3 M_k(x)\, dx$$

$$= \sum_{i,\,j,\,l=1}^\infty \left[a_{ijl}^{(k)} N_i(t) N_j(t) N_l(t)\right],$$

where the coefficients $a_{ijl}^{(k)}$ are obtained by cubing the series and integrating term by term. Thus we obtain the infinite system of coupled equations for $N_k(t)$,

$$(4.276) \quad N_k'(t) + (k^2 - \hat{\lambda}) N_k(t) = -\hat{\lambda}\epsilon^2 \sum_{i,\,j,\,l=1}^\infty a_{ijl}^{(k)} N_i(t) N_j(t) N_l(t)$$

with the initial conditions (4.273).

Neglecting the terms of order ϵ^2 in the equations (4.276) is equivalent to linearizing the problem and has the effect of uncoupling the equations. Then the terms $N_k(t) M_k(x)$ have the form (4.269) where k is replaced by n. We again observe that for $\hat{\lambda} < \hat{\lambda}_c = 1$ all the $N_k(t)$ in the linearized case tend to zero as $t \to \infty$, whereas for $\hat{\lambda} > \hat{\lambda}_c = 1$, at least one of the $N_k(t)$ grows exponentially in t.

We wish to examine the behavior of the solution if $\hat{\lambda}$ is slightly larger than the critical value $\hat{\lambda}_c$; that is, $\hat{\lambda} \approx 1$ but $\hat{\lambda} > 1$. Then $k^2 - \hat{\lambda} > 0$ for $k \geqslant 2$ but $1 - \hat{\lambda} < 0$. It is of interest to study the behavior of the solution in the region where according to the linear theory a transition from stability to instability takes place.

Since for $k \geqslant 2$, $k^2 - \hat{\lambda} > 0$, we take as a first approximation

$$(4.277) \qquad N_k(t) = (h, M_k)\exp\left[(\hat{\lambda} - k^2)t\right]; \qquad k \geqslant 2;$$

that is, we neglect the terms of order ϵ^2 in (4.274) with $k \geqslant 2$. A similar approximation in the equation for $N_1(t)$ leads to an exponentially growing term of the form (4.277) with $k = 1$. Inserting these expressions for $N_k(t)$ ($k = 1, 2, \ldots$) into the series on the right side of (4.276) shows that all the terms in the sum decay exponentially except for the terms $a_{111}^{(k)} N_1^3(t)$.

The linearized form of the equation for $N_1(t)$ implies it is exponentially growing, whereas all other $N_k(t)$ decay exponentially, so we retain the term

$\hat{\lambda}\epsilon^2 a_{111}^{(1)} N_1^3(t)$ in the equation for $N_1(t)$ and obtain

$$(4.278) \qquad N_1'(t) + (1 - \hat{\lambda}) N_1(t) + \hat{\lambda}\epsilon^2 a_{111}^{(1)} N_1^3(t) = 0.$$

For the coefficient $a_{111}^{(1)}$ we have from (4.275)

$$(4.279) \qquad a_{111}^{(1)} = \int_0^\pi [M_1(x)]^4 \, dx = \frac{4}{\pi^2} \int_0^\pi \sin^4 x \, dx = \frac{3}{2\pi}.$$

Thus $N_1(t)$ satisfies

$$(4.280) \qquad N_1'(t) + (1 - \hat{\lambda}) N_1(t) + \frac{3\hat{\lambda}}{2\pi} \epsilon^2 N_1^3(t) = 0.$$

To solve (4.280) we multiply across by $N_1(t)$ and obtain

$$(4.281) \qquad \frac{1}{2} \frac{d[N_1^2(t)]}{dt} + (1 - \hat{\lambda}) N_1^2(t) + \frac{3\hat{\lambda}\epsilon^2}{2\pi} N_1^4(t) = 0,$$

which is a *generalized Riccati equation* for N_1^2. The solution of (4.281) satisfying $N_1(0) = (h, M_1)$ is found to be

$$(4.282) \quad N_1(t) = \frac{(h, M_1) e^{(\hat{\lambda} - 1)t}}{\left[1 + \left(3\hat{\lambda}\epsilon^2 (h, M_1)^2 / 2\pi(\hat{\lambda} - 1)\right)\left(e^{2(\hat{\lambda} - 1)t} - 1\right)\right]^{1/2}}$$

For small values of t this expression reduces to

$$(4.283) \qquad N_1(t) \approx (h, M_1) e^{(\hat{\lambda} - 1)t},$$

which is the solution of the linearized equation, as was expected. However, as $t \to \infty$, since $\hat{\lambda} - 1 > 0$ we have

$$(4.284) \qquad N_1(t) \to \frac{(h, M_1)}{|(h, M_1)|} \left(\sqrt{\frac{2\pi}{3}}\right) \left[\frac{\hat{\lambda} - 1}{\hat{\lambda}\epsilon^2}\right]^{1/2}$$

so that $N_1(t)$ tends to a finite (stationary) value, rather than growing without bound as predicted by the linear theory.

We remark that the above results are self-consistent since we now find that the terms $a_{111}^{(k)} N_1^3(t)$ in the equations (4.276) for $k \geqslant 2$ are uniformly bounded for all t and it is correct to approximate $N_k(t)$ as in (4.277) since the right side of (4.276) is of order ϵ^2.

Thus we obtain from (4.271)

$$(4.285)$$

$$u(x, t) = \epsilon w(x, t) = \epsilon \sum_{k=1}^\infty N_k(t) M_k(x) \approx \frac{(h, M_1)}{|(h, M_1)|} \left(\frac{2}{\sqrt{3}}\right) \left[\frac{\hat{\lambda} - 1}{\hat{\lambda}}\right]^{1/2} \sin x$$

as $t \to \infty$ in view of (4.277) and (4.284). We assume that $(h, M_1) \neq 0$ and (4.285) shows that the solution at large time depends only on the sign (i.e., $(h, M_1)/|(h, M_1)|$) of the leading term of the Fourier series of $u(x, 0)$ and not on its magnitude. The large time behavior of the solution is independent of t, so that the solution $u(x, t)$ with $\hat{\lambda} \geq \hat{\lambda}_c$ does not grow unboundedly as predicted by the linear theory. Instead, it approaches a *steady state* whose approximate description, (correct to order ϵ) is given in (4.285). We note that (4.285) is an approximate solution of the stationary form of (4.255) (i.e., with $u_t = 0$) when $\hat{\lambda} \approx \hat{\lambda}_c = 1$.

The foregoing approximation method may be extended to deal with nonlinear heat sources of the form $\hat{\lambda} f(u)$, where $f(0) = 0$ and $f(u)$ has a Taylor expansion around $u = 0$. Also, hyperbolic and elliptic equations with similar types of nonlinearities can be treated by this method. Some examples are presented in the exercises.

EXERCISES FOR CHAPTER 4

Section 4.1

4.1.1. Derive the heat or diffusion equation (4.5) from a "balance law" appropriate to the one-dimensional case.

4.1.2. Show that if $\hat{\rho} = c\rho$, where c is a constant, the equation

$$\rho u_{tt} + \hat{\rho} u_t + Lu = \rho F$$

can be brought into the general form (4.10) if we set $u = \exp[\alpha t]v$ and choose α appropriately.

4.1.3. Carry out the transformation of Exercise 4.1.2 for the telegrapher's equation

$$u_{tt} - \gamma^2 u_{xx} + 2\hat{\lambda} u_t = 0,$$

where $\hat{\lambda}$ is a positive constant.

Section 4.2

4.2.1. Verify the result (4.28) in the two-dimensional case if $L = -\nabla^2$, $u = 1 - r^2$, $w = (1 - r)^2$, where $r^2 = x^2 + y^2$, if the region G is the interior of the circle $r = 1$. Note that both u and w vanish on the boundary $r = 1$.

4.2.2. Let $u = (1 - r)^2$ and $w = \cos \pi r$, so that $\partial u/\partial n$ and $\partial w/\partial n$ both vanish on the circle $r = 1$. With $L = -\nabla^2$ and G given as the region $r < 1$, verify that the integral over the region G in (4.28) vanishes.

4.2.3. With $R^2 = x^2 + y^2 + z^2$, let $u = \exp[1 - R]$ and $w = (R - 1)^2$, so that $\partial u / \partial n + u$ and $\partial w / \partial n + w$ both vanish on the sphere $R = 1$. If G is the interior of the unit sphere (i.e., $R < 1$) and $L = -\nabla^2 + 10$, show that (4.28) is valid for the above functions.

4.2.4. If G is the unit square $0 < x < 1$, $0 < y < 1$, and $u(x, y) = xy(1 - x)(1 - y)$ so that $u(x, y)$ vanishes on ∂G, show that (4.38) is satisfied if $L = -\nabla^2$.

4.2.5. Let $q = 0$ in (4.6) and $\alpha = 0$ in (4.14). Determine that $M(x) = c = $ constant is an eigenfunction for (4.26) that corresponds to the eigenvalue $\lambda = 0$.

4.2.6. Show that if $\lambda_0 = 0$ occurs as an eigenvalue when the elliptic problem (4.19)–(4.20) is solved by separation of variables, the system (4.54) has no solution unless f and g satisfy compatibility conditions. (Use the results of Exercise 4.2.5 to show this.) Conclude that the solution is not unique if the compatibility conditions are met and verify that these results are consistent with those obtained in Exercise 3.4.6.

4.2.7. Show that $M_j(x, y) = 1$ and $M_k(x, y) = \cos \pi x \cos \pi y$ are both eigenfunctions for the problem

$$- (M_{xx} + M_{yy}) = \lambda M, \qquad 0 < x < 1, 0 < y < 1,$$

with the boundary condition $\partial M / \partial n = 0$ on the unit square. Determine the corresponding eigenvalues and verify directly that the orthogonality condition $(M_j, M_k) = 0$ is satisfied.

4.2.8. Show that $M = 1$ is an eigenfunction for the problem

$$- \nabla \cdot (p \nabla M) - M = \lambda M$$

in the region G with $\partial M / \partial n = 0$ on ∂G. Determine the corresponding eigenvalue and explain why the occurrence of a negative eigenvalue does not contradict (4.42).

4.2.9. Assuming the eigenfunctions $M_k(x)$ are uniformly bounded, interpret the fact that $u(x, t)$ as given in (4.52) vanishes as $t \to \infty$ if the λ_k are all positive in terms of the physical significance of the heat conduction problem and its boundary conditions. Consider also the special case referred to in Exercise 4.2.5 which yields $\lambda = 0$ as an eigenvalue.

4.2.10. Consider the two-dimensional eigenvalue problem in polar coordinates r and θ,

$$LM = -\nabla^2 M = \lambda M$$

in the region $r < l$ with the boundary condition $M(l, \theta) = 0$. By expressing the Laplacian in polar coordinates and assuming that M is independent of θ, show that $M(r)$ satisfies Bessel's equation of order zero. In the general case, with

$M = M(r, \theta)$, set $M = M_1(r)M_2(\theta)$ and use separation of variables to show that if $M(r, \theta + 2\pi) = M(r, \theta)$, the function $M_1(r)$ must satisfy a Bessel equation of integral order. In both of these eigenvalue problems we require that the solutions be bounded at $r = 0$ (see Example 4.7 for a further discussion).

4.2.11. Express Laplace's equation $\nabla^2 u = 0$ in spherical coordinates (r, θ, φ) with $r \geq 0$, $0 \leq \theta \leq 2\pi$, and $0 \leq \varphi \leq \pi$. Assume $u = u(r, \varphi)$ (i.e., u is independent of θ) and set $u = N(r)M(\varphi)$. Use separation of variables to show that $M(\varphi)$ satisfies the equation

$$ -\frac{d}{d\varphi}\left[\sin \varphi M'(\varphi)\right] = \lambda \sin \varphi M(\varphi), $$

with λ equal to the separation constant. Let $x = \cos \varphi$ and conclude that $\hat{M}(x) = M(\varphi)$ satisfies the Legendre equation. If we require that $\hat{M}(x)$ be bounded at $x = -1$ and $x = +1$, we obtain the eigenvalue problem discussed in Example 4.8.

Section 4.3

4.3.1. Show that (with $\rho(x) = 1$), the function $\varphi(x) = 1/\sqrt{x}$ is integrable (as an improper integral) over the interval $0 < x < 1$, but that it is not square integrable.

4.3.2. Determine that the set of square integrable functions $\hat{\varphi}_k(x) = \sin(\pi k \log x/\log 2)$, $k \geq 1$, is orthogonal over the interval $1 < x < 2$ with respect to the inner product with weight function $\rho(x) = 1/x$. Obtain the norms of these functions and construct an orthonormal set.

4.3.3. Show that the set of complex valued functions $\hat{\varphi}_k(x) = \exp[ikx]$, $k = 0, \pm 1, \pm 2, \ldots$, is orthogonal with respect to the Hermitian inner product (4.62) over the interval $-\pi < x < \pi$, if the weight function $\rho(x) = 1$. Determine the norms of the $\hat{\varphi}_k(x)$ and construct an orthonormal set.

4.3.4. Obtain the Fourier coefficients of the function $\varphi(x) = x$ with respect to the orthonormal set of functions obtained in Exercise 4.3.2.

4.3.5. Adapt the argument given following (4.32) to show that the boundary terms in (4.71) vanish.

4.3.6. Conclude from (4.77) that if $\lambda = 0$ is an eigenvalue of the (regular) Sturm–Liouville problem, the corresponding eigenfunction $v(x)$ must satisfy

$$ \int_0^l \left[pv'^2 + qv^2\right] dx - pvv'\Big|_0^l = 0, $$

and we must have $qv^2(x) = p[v'(x)]^2 = 0$, since the boundary contributions are nonnegative. Noting that $v(x)$ cannot vanish identically, show that $q(x) = 0$ and $v(x) = $ constant. This must, in turn, imply that $\alpha_1 = \alpha_2 = 0$, since the boundary conditions could not otherwise be satisfied.

4.3.7. Suppose there are two linearly independent eigenfunctions $v(x)$ and $w(x)$ that correspond to the eigenvalue λ. Since they both satisfy the same equation and boundary conditions, show that we must have

$$\alpha_1 v(0) - \beta_1 v'(0) = 0; \qquad \alpha_1 w(0) - \beta_1 w'(0) = 0.$$

Noting that α_1 and β_1 cannot both vanish, show that the determinant of the above system, which equals the Wronskian of v and w evaluated at zero, must vanish. Thereby, conclude that $v(x)$ and $w(x)$ are linearly dependent, so that each eigenvalue must be simple.

4.3.8. Consider the transcendental equation (4.88) for the determination of the eigenvalues in Example 4.3. Let $\lambda = \rho^2$ and show that the equation can be written as

$$\tan(\rho l) = \frac{\rho(\alpha_1\beta_2 + \alpha_2\beta_1)}{\rho^2\beta_1\beta_2 - \alpha_1\alpha_2}.$$

Let $x = \tan(\rho l)$ and $x = \rho(\alpha_1\beta_2 + \alpha_2\beta_1)/(\rho^2\beta_1\beta_2 - \alpha_1\alpha_2)$. Graph both functions in the (ρ, x)-plane and show thereby that there are an infinite number of eigenvalues λ_k and that $\lambda_k \to \infty$ as $k \to \infty$. Show that for large k, $\lambda_k \approx (\pi k/l)^2$.

4.3.9. Use (4.56) and integration by parts to show that if $v(x)$ satisfies the same boundary conditions as the eigenfunctions $v_k(x)$ and is sufficiently smooth, the Fourier coefficients of v have the form

$$(v, v_k) = \frac{1}{\lambda_k}\left(v_k, \frac{1}{\rho}Lv\right); \qquad \lambda_k \neq 0.$$

It can be shown that $\lambda_k \approx c^2 k^2$ as $k \to \infty$ for any (regular) Sturm–Liouville problem. (The constant c may be different for each problem.) Conclude thereby that if $|v_k(x)|$ and $|(v_k, (1/\rho)Lv)|$ are uniformly bounded, the Fourier series (4.79) converges uniformly and absolutely. (Hint: Use the Weierstrass M-test.)

4.3.10. Verify the results of Exercise 4.3.9 for the case of the Fourier sine series assuming $v(x)$ has two continuous derivatives and $v(0) = v(l) = 0$.

4.3.11. Determine the eigenvalues and eigenfunctions of the problem

$$v'' + \lambda v = 0; \qquad v'(0) = 0; \quad v(l) = 0.$$

Show that the properties 1–4 in the text are valid for this problem and normalize the set of eigenfunctions.

4.3.12. Determine the eigenvalues (approximately) and the eigenfunctions of the problem

$$v'' + \lambda v = 0; \qquad v(0) = 0; \qquad v'(l) + \beta v(l) = 0,$$

where $\beta > 0$. Show that the properties 1–4 in the text are satisfied and normalize the set of eigenfunctions.

4.3.13. Obtain the eigenvalues and eigenfunctions of the problem

$$\left[(1 + x)^2 v'\right]' + \lambda v = 0; \qquad v(0) = v(l) = 0.$$

Verify the properties 1–4 in the text and normalize the eigenfunctions. [Hint: Let $t = 1 + x$ to transform the given equation into a Cauchy (homogeneous) equation.]

4.3.14. Obtain the Fourier sine and cosine series of the following functions given over the interval $0 < x < l$.

    (a) $v(x) = x$.

    (b) $v(x) = x(x - l)$.

    (c) $v(x) = x^2(x - l)^2$.

    (d) $v(x) = \cos\left(\dfrac{5\pi x}{l}\right)$.

Construct the even and odd extensions of each of the above functions. Compare the rates of convergence for each of the Fourier series based on the smoothness properties of the even and odd extensions.

4.3.15. Obtain the complete Fourier series expansion of the function $v(x) = \sin^2(\pi x/l)(x - l)^2$, defined in the interval $-l < x < l$. Discuss the convergence properties of the series.

4.3.16. With $\lambda = \lambda_{kn}$, $v = J_n[\sqrt{\lambda_{kn}}\,x]$ and $t = \sqrt{\lambda_{kn}}\,x$ in (4.112), multiply across by tJ_n' and integrate from 0 to $\sqrt{\lambda_{kn}}\,l$ to obtain

$$-\int_0^{\sqrt{\lambda_{kn}}\,l} \frac{d}{dt}\left[(tJ_n')^2\right] dt + n^2 \int_0^{\sqrt{\lambda_{kn}}\,l}(J_n^2)' \, dt = \int_0^{\sqrt{\lambda_{kn}}\,l} t^2 (J_n^2)' \, dt.$$

Noting that $J_n' = -J_{n+1} + (n/x)J_n(x)$, integrating by parts in the third integral above, and using the properties of the Bessel function J_n given in the text, obtain the normalization constants (4.116).

4.3.17. Given the Taylor expansion

$$J_0(x) = \sum_{j=0}^{\infty} \frac{(-1)^j}{(j!)^2}\left(\frac{x}{2}\right)^{2j}$$

show that the first three (positive) zeros of $J_0(x)$ are (approximately) $\alpha_{10} = 2.41$, $\alpha_{20} = 5.52$, and $\alpha_{30} = 8.65$. [Note that the series for $J_0(x)$ is alternating.]

4.3.18. Use the asymptotic formula

$$J_n(\sqrt{\lambda}\,x) \approx \sqrt{\frac{2}{\pi\sqrt{\lambda}\,x}} \, \cos\left(\sqrt{\lambda}\,x - \frac{\pi}{2}n - \frac{\pi}{4}\right),$$

which is valid as $\sqrt{\lambda}\, x \to \infty$, to show that the large eigenvalues for Example 4.7 are approximately given as $\lambda_{kn} \approx (\pi k/l)^2$.

4.3.19. Expand $v(x) = l^2 - x^2$ in a series of eigenfunctions $v_{k,0}(x)$ as given in (4.117)–(4.118). Evaluate the Fourier coefficients by using integration by parts and the formulas $[x^n J_n(x)]' = x^n J_{n-1}(x)$ and $J_{-n}(x) = (-1)^n J_n(x)$.

4.3.20. Obtain the first five Legendre polynomials $P_0(x), \ldots, P_4(x)$ and show that they are mutually orthogonal.

4.3.21. Use the derivative definition of the Legendre polynomials $P_k(x)$ given in (4.122) (this is known as *Rodrigues' formula*) to obtain the normalization constants (4.123). [Hint: Integrate by parts as often as necessary.]

4.3.22. Expand the following functions in a series of Legendre polynomials.

(a) $v(x) = x$.

(b) $v(x) = 5 - 4x + 10x^3$.

(c) $v(x) = x^4$.

Section 4.4

4.4.1. Determine the motion of a *plucked string*. That is, find a solution $u(x, t)$ of the wave equation (4.126) with $u(0, t) = u(l, t) = 0$ and the initial conditions

$$f(x) = \begin{cases} \dfrac{h}{a}x, & 0 < x < a \\[2mm] \dfrac{h(l - x)}{l - a}, & a \leqslant x < l \end{cases} \quad ; \qquad g(x) = 0.$$

4.4.2. Solve the initial and boundary value problem for the wave equation (4.126) with $u(0, t) = u(l, t) = 0$ and the initial conditions

$$f(x) = 0; \qquad g(x) = \delta(x - a); \qquad 0 < a < l,$$

where $\delta(x)$ is the Dirac delta function. This yields the motion of a string due to a point impulse at $x = a$ administered at the time $t = 0$.

4.4.3. Solve (4.126)–(4.128) if

$$f(x) = 0; \qquad g(x) = x(x - l).$$

4.4.4. Solve (4.126)–(4.128) if

$$f(x) = \sin^2\left(\frac{\pi x}{l}\right); \qquad g(x) = 0.$$

4.4.5. Express the solution of Exercise 4.4.1 in the form (4.150) and use this result to describe the displacement $u(x, t)$ of the string at various times $t > 0$.

4.4.6. Use separation of variables to obtain the general form of the (series) solution of the problem

$$u_{tt} - c^2 u_{xx} = 0; \qquad 0 < x < l, t > 0$$

with the initial conditions

$$u(x,0) = f(x); \qquad u_t(x,0) = g(x)$$

and the boundary conditions

$$u_x(0, t) = u_x(l, t) = 0.$$

4.4.7. Obtain the solution of the problem in Exercise 4.4.6 if $f(x) = 0$ and $g(x) = 1$. Interpret the result.

4.4.8. Solve the problem of Exercise 4.4.6 if

$$f(x) = x^2(x - l)^2; \qquad g(x) = 0.$$

4.4.9. Using separation of variables solve the following initial and boundary value problem:

$$u_{tt} - c^2 u_{tt} = 0, \qquad 0 < x < l, t > 0;$$
$$u(x,0) = f(x); \qquad u_t(x,0) = g(x);$$
$$u(0, t) = 0; \qquad u_x(l, t) = 0.$$

4.4.10. Solve the problem in Exercise 4.4.9 if

$$f(x) = x(x - l)^2, \qquad g(x) = 0.$$

4.4.11. Apply the method of separation of variables to solve the wave equation (4.126) with the initial data (4.128) and the boundary data

$$u(0, t) = 0; \qquad u_x(l, t) + \beta u(l, t) = 0; \qquad \beta > 0.$$

4.4.12. Obtain the solution of the problem in Exercise 4.4.11 if the initial data are

$$f(x) = \sin^2\left(\frac{\pi x}{l}\right); \qquad g(x) = 0.$$

4.4.13. Multiply the wave equation (4.126) by u_t and use the identity $u_{xx}u_t = (u_t u_x)_x - \frac{1}{2}(u_x^2)_t$ to obtain

$$\frac{1}{2}\frac{\partial}{\partial t}\left[u_t^2 + c^2 u_x^2\right] - c^2\frac{\partial}{\partial x}\left[u_t u_x\right] = 0.$$

Integrate over the interval $0 < x < l$ and show that if either u or u_x vanishes at $x = 0$ and $x = l$, the energy integral (4.143) is a constant (recall that $T = c^2\rho$). Obtain an appropriate energy integral if $u_x(0, t) - \beta_1 u(0, t) = 0$ and $u_x(l, t) + \beta_2 u(l, t) = 0$ with $\beta_1, \beta_2 > 0$.

4.4.14. Verify the result (4.145).

4.4.15. Use separation of variables to solve the following problem for the telegrapher's equation:

$$v_{tt} - \gamma^2 v_{xx} + 2\hat{\lambda}v_t = 0, \qquad 0 < x < l, t > 0,$$

where $\hat{\lambda} > 0$, with

$$v(x,0) = f(x); \qquad v_t(x,0) = g(x)$$
$$v(0, t) = v(l, t) = 0.$$

Show that the solution $v(x, t)$ tends to zero as $t \to \infty$.

4.4.16. Solve the initial and boundary value problem for the Klein–Gordon equation:

$$u_{tt} - \gamma^2 u_{xx} + c^2 u = 0; \qquad 0 < x < l, t > 0;$$
$$u(x,0) = f(x); \qquad u_t(x,0) = g(x),$$
$$u(0, t) = u(l, t) = 0.$$

Use separation of variables.

4.4.17. Solve the heat equation (4.151) with the boundary data (4.152) and the initial data

$$u(x,0) = f(x) = x(x - l).$$

4.4.18. Solve the problem (4.151)–(4.153) if

$$f(x) = \delta(x - a), \qquad 0 < a < l,$$

where $\delta(x)$ is the Dirac delta function.

4.4.19. Apply separation of variables to solve,

$$u_t - c^2 u_{xx} = 0, \qquad 0 < x < l, t > 0;$$
$$u(x,0) = f(x);$$
$$u_x(0, t) = u_x(l, t) = 0.$$

The boundary conditions imply that no heat escapes through the ends of the rod. Show that as $t \to \infty$ we have

$$\lim_{t \to \infty} u(x, t) = \frac{1}{l} \int_0^l f(x) \, dx.$$

This represents the average of the initial temperature distribution.

4.4.20. Solve the problem in Exercise 4.4.19 if

(a) $f(x) = x$.

(b) $f(x) = \sin^2(\pi x / l)$.

4.4.21. By integrating the heat equation (4.151) over the interval $0 < x < l$, show that if $u_x(0, t) = u_x(l, t) = 0$, we have $\int_0^l u(x, t) = $ constant. Explain why this is consistent with the result in Exercise 4.4.19 regarding the limit of $u(x, t)$ as $t \to \infty$.

4.4.22. Use separation of variables to solve the following problem for the heat equation:

$$u_t - c^2 u_{xx} = 0, \qquad 0 < x < l, t > 0;$$
$$u(x, 0) = f(x);$$
$$u(0, t) = 0; \qquad u_x(l, t) + \beta u(l, t) = 0; \qquad \beta > 0.$$

The boundary condition of the third kind at $x = l$ results from *Newton's law of cooling* if there is convective heat exchange between the rod and a medium adjacent to the rod at $x = l$ that is kept at zero temperature.

4.4.23. Solve the problem in Exercise 4.4.22 if

$$f(x) = x.$$

4.4.24. If the lateral portion of a rod undergoes convective heat exchange with a medium kept at zero temperature, the temperature $u(x, t)$ in the rod satisfies the equation

$$u_t - c^2 u_{xx} + a^2 u = 0, \qquad 0 < x < l, t > 0.$$

Assume that $u(0, t) = u(l, t) = 0$, $u(x, 0) = f(x)$, and that $a = $ constant. Find the temperature $u(x, t)$ by the method of separation of variables. Obtain the limit as $t \to \infty$ of the temperature $u(x, t)$.

4.4.25. Use separation of variables to solve Laplaces equation $\nabla^2 u = 0$ in the rectangle $0 < x < l, 0 < y < \hat{l}$, with the boundary conditions of the second kind

$$u_x(0, y) = u_x(l, y) = 0, \qquad 0 < y < \hat{l};$$
$$u_y(x, 0) = f(x); \qquad u_y(x, \hat{l}) = g(x); \qquad 0 < x < l.$$

Determine conditions on $f(x)$ and $g(x)$ for a solution to exist. Discuss the rate of convergence of the solution.

4.4.26. Solve the following boundary value problem for Laplace's equation:

$$u_{xx} + u_{yy} = 0, \qquad 0 < x < l, 0 < y < \hat{l};$$

$$u(0, y) = u(l, y) = 0, \qquad 0 < y < \hat{l};$$

$$u_y(x, 0) = f(x); \qquad u_y(x, \hat{l}) = g(x); \qquad 0 < x < l.$$

4.4.27. Solve Laplace's equation $\nabla^2 u = 0$ in the rectangle $0 < x < l$, $0 < y < \hat{l}$ with the boundary conditions

$$u(0, y) = 0; \qquad u_x(l, y) + \beta u(l, y) = 0; \qquad \beta > 0;$$

$$u(x, 0) = f(x); \qquad u(x, \hat{l}) = 0.$$

4.4.28. Let $f(x) = x(x - l)$ and $g(x) = 0$. Solve the following boundary value problems:

    (a) Example 4.11.

    (b) Exercise 4.4.25.

    (c) Exercise 4.4.26.

4.4.29. Consider Laplace's equation $\nabla^2 u = 0$ in the unbounded region $0 < x < l$, $y > 0$. Let

$$u(0, y) = u(l, y) = 0, \qquad y > 0,$$

and

$$u(x, 0) = f(x), \qquad 0 < x < l,$$

as well as requiring that $u(x, y)$ be (uniformly) bounded as $y \to \infty$ for $0 < x < l$. Show that the solution of this problem can be obtained by separation of variables. Find the limit of $u(x, y)$ as $y \to \infty$.

4.4.30. Apply separation of variables to solve the boundary value problem for the elliptic equation

$$u_{xx} + u_{yy} - c^2 u = 0, \qquad 0 < x < l, 0 < y < \hat{l}$$

in the given rectangle, with the boundary conditions

$$u(0, y) = u(l, y) = 0;$$

$$u(x, 0) = f(x); \qquad u(x, \hat{l}) = g(x).$$

4.4.31. Use separation of variables to show that the reduced wave equation

$$u_{xx} + u_{yy} + k^2 u = 0, \qquad 0 < x < l, 0 < y < \hat{l},$$

with the (homogeneous) boundary condition $u(x, y) = 0$ on $x = 0$, $x = l$, $y = 0$, $y = \hat{l}$, can have nonzero solutions for certain values of k. Determine these values of $k = k_n$. (They correspond to the eigenvalues for Laplace's equation in a rectangle.) By applying Green's theorem [see (4.28)] conclude that if $k = k_n$, the Dirichlet problem for the reduced wave equation has no solution unless the boundary values satisfy a compatibility or orthogonality condition. [Let $w = w_n$ be a solution of $\nabla^2 w_n + k_n^2 w_n = 0$, $w_n = 0$ on the boundary of the rectangle, in the application of (4.28)]. Show that if a solution does exist for $k = k_n$, it is not unique.

4.4.32. Solve the initial and boundary value problem for the two-dimensional wave equation in a disk,

$$u_{tt} - c^2[u_{xx} + u_{yy}] = 0; \qquad x^2 + y^2 < l^2; \qquad t > 0,$$

with the initial conditions

$$u(x, y, 0) = f(x, y); \qquad u_t(x, y, 0) = g(x, y)$$

and the boundary condition

$$u(x, y, t) = 0; \qquad x^2 + y^2 = l^2; \qquad t > 0.$$

Use separation of variables and the results of Exercise 4.2.10 and Example 4.7. (The solution of this problem describes the *vibration of a circular membrane with a fixed edge*.)

4.4.33. Solve the initial and boundary value problem for the heat equation in a disk,

$$u_t - c^2[u_{xx} + u_{yy}] = 0; \qquad x^2 + y^2 < l^2, t > 0,$$

with the initial condition

$$u(x, y, 0) = f(x, y)$$

and the boundary condition

$$u(x, y, t) = 0; \qquad x^2 + y^2 = l^2; \qquad t > 0.$$

Use separation of variables and see Exercise 4.4.32.

4.4.34. Express $\nabla^2 u = 0$ in spherical coordinates (r, θ, φ) with $r \geq 0$, $0 \leq \theta \leq 2\pi$, and $0 \leq \varphi \leq \pi$. Consider the boundary value problem for $\nabla^2 u = 0$ within the sphere $r < l$ for the function $u = u(r, \theta, \varphi)$ with the boundary condition

$$u(l, \theta, \varphi) = f(\varphi), \qquad 0 \leq \varphi \leq \pi,$$

(i.e., the boundary values are independent of θ). Look for a solution in the form $u = U(r, \varphi)$ and use separation of variables to construct an eigenfunction expansion of the solution in terms of Legendre polynomials (see Exercise 4.2.11 and Example 4.8).

Section 4.5

4.5.1. Solve the initial value problem

$$u_{tt} - c^2 u_{xx} = g(x, t), \qquad -\infty < x < \infty, t > 0$$

with $u(x, 0) = u_t(x, 0) = 0$ if $g(x, t)$ has the following form:
   (a) $g(x, t) = (\sin x)e^{-t}$.
   (b) $g(x, t) = t$.
   (c) $g(x, t) = \delta(x)\cos t$.
   (d) $g(x, t) = \delta(x - a)\delta(t - b)$, $\qquad a, b = $ constant, $b > 0$.
In (c) and (d), $\delta(z)$ is the Dirac delta function.

4.5.2. Use Duhamel's principle to solve,

$$\begin{aligned}
u_{tt} - c^2 u_{xx} &= g(x, t), \qquad 0 < x < l, t > 0; \\
u(x, 0) &= u_t(x, 0) = 0; \\
u(0, t) &= u(l, t) = 0.
\end{aligned}$$

4.5.3. Solve the problem in Exercise 4.5.2 if $g(x, t)$ assumes the following forms:
   (a) $g(x, t) = x(x - l)$.

   (b) $g(x, t) = \sin(\omega x)\sin(\omega c t)$; $\qquad \omega \neq \dfrac{\pi k}{l}$.

   (c) $g(x, t) = \sin\left(\dfrac{\pi}{l} x\right)\sin\left(\dfrac{\pi c}{l} t\right)$.

4.5.4. Apply Duhamel's principle to obtain the solution of the following problem:

$$\begin{aligned}
u_t - c^2 u_{xx} &= g(x, t), \qquad 0 < x < l, t > 0; \\
u(x, 0) &= 0; \\
u_x(0, t) &= u_x(l, t) = 0.
\end{aligned}$$

4.5.5. Consider the special case in which $g = g(x)$ (i.e., it is independent of t) in Exercise 4.5.4 and examine the limit of the solution $u(x, t)$ as $t \to \infty$. What condition must $g(x)$ satisfy so that this limit exists?

4.5.6. Let $g(x, t) = \cos(\pi x/l)$ in Exercise 4.5.4 and obtain the solution of the initial and boundary value problem in that case.

4.5.7. Use Duhamel's principle to solve the telegrapher's equation

$$u_{tt} - c^2 u_{xx} + 2\hat{\lambda} u_t = g(x), \qquad 0 < x < l, t > 0$$

where $\hat{\lambda} > 0$, if $u(x, 0) = u_t(x, 0) = 0$ and $u(0, t) = u(l, t) = 0$. Find the limit of the solution as $t \to \infty$ and show that it satisfies an appropriate steady-state problem. Verify this result if $g(x) = \sin(\pi x/l)$.

4.5.8. Use Duhamel's principle to obtain the solution of the inhomogeneous Klein-Gordon equation

$$u_{tt} - u_{xx} + c^2 u = g(x, t), \qquad 0 < x < l, t > 0,$$

with $u(0, t) = u(x, 0) = u(l, t) = u_t(x, 0) = 0$. Use the results of exercise 4.4.16.

Section 4.6

4.6.1. Obtain the coefficients N_k in the eigenfunction expansion (4.195) for the one-dimensional hyperbolic and parabolic problems.

4.6.2. It was assumed in the text that the eigenvalues λ_k are all positive. Discuss the necessary modifications if $\lambda_0 = 0$ is an eigenvalue and the term $N_0 M_0(x)$ occurs in the expansion (4.195).

4.6.3. Use the finite sine transform to solve the problem:

$$u_{tt} - c^2 u_{xx} = xe^{-t}, \qquad 0 < x < l, t > 0,$$
$$u(0, t) = \sin t; \qquad u(l, t) = 1,$$
$$u(x, 0) = u_t(x, 0) = 0.$$

4.6.4. Apply the finite cosine transform to solve the problem:

$$u_{tt} - c^2 u_{xx} = 0, \qquad 0 < x < l, t > 0;$$
$$u_x(0, t) = t; \qquad u_x(l, t) = 0;$$
$$u(x, 0) = u_t(x, 0) = 0.$$

4.6.5. Use the finite sine transform to solve the problem:

$$u_{tt} - c^2 u_{xx} = F(x)\sin \omega t, \qquad 0 < x < l, t > 0,$$

where $\omega \neq \pi kc/l$ and the initial and boundary data are

$$u(x,0) = u_t(x,0) = u(0,t) = u(l,t) = 0.$$

Consider the limit of the solution $u(x, t)$ as $\omega \to \pi kc/l$ and obtain a resonance effect.

4.6.6. Consider the two-dimensional Laplace's equation $\nabla^2 u = 0$ in the disk $r < R$ with the boundary condition $u(R, \theta) = f(\theta)$. Solve this problem by using finite Fourier transforms.

4.6.7. Use the finite sine transform to solve the first boundary value or *Dirichlet problem* for Laplace's equation in a rectangle:

$$u_{xx} + u_{yy} = 0, \qquad 0 < x < l, 0 < y < \hat{l};$$

$$u(x,0) = f(x); \qquad u(x, \hat{l}) = g(x);$$

$$u(0, y) = h(y); \qquad u(l, y) = r(y).$$

4.6.8. Apply the finite cosine transform to solve the secondary boundary value or *Neumann problem* for Poisson's equation in a rectangle:

$$u_{xx} + u_{yy} = -F(x, y), \qquad 0 < x < l, 0 < y < \hat{l};$$

$$u_x(0, y) = h(y); \qquad u_x(l, y) = r(y);$$

$$u_y(x,0) = f(x); \qquad u_y(x, \hat{l}) = g(x).$$

Determine conditions on $F(x, y)$ and the boundary values of $u(x, y)$ so that a solution exists.

4.6.9. Solve the *Neumann problem* for Laplace's equation $\nabla^2 u = 0$ in the disk $r < R$ if

$$\frac{\partial u(R, \theta)}{\partial r} = \sin^3 \theta.$$

4.6.10. Use the finite sine transform to solve the following problem for the heat equation:

$$u_t - c^2 u_{xx} = e^{-t}, \qquad 0 < x < l, t > 0;$$

$$u(x,0) = 0;$$

$$u(0, t) = \alpha, u(l, t) = \beta,$$

where α and β are constants. Obtain the behavior of the solution as t gets large.

4.6.11. Use the finite cosine transform to solve the following problem:

$$u_t - c^2 u_{xx} = 0, \qquad 0 < x < l, t > 0;$$
$$u(x, 0) = 0;$$
$$u_x(0, t) = 0; \qquad u_x(l, t) = e^{-t}.$$

4.6.12. Consider the problem:

$$u_t - c^2 u_{xx} = 0, \qquad 0 < x < l, t > 0;$$
$$u(x, 0) = 0;$$
$$u(0, t) = 0; \qquad u(l, t) = 1.$$

(a) Solve the problem by using (4.249) to eliminate the inhomogeneous boundary terms.

(b) Solve the problem by using the finite sine transform.

(c) Compare the rates of convergence of the series solutions obtained in parts (a) and (b).

4.6.13. Given the following problem:

$$u_t - c^2 u_{xx} = 0, \qquad 0 < x < l, t > 0;$$
$$u(x, 0) = 0;$$
$$u(0, t) = 0; \qquad u(l, t) = e^{-t}.$$

Determine conditions on c and l such that there is a solution of the homogeneous heat equation in the form $u(x, t) = v(x)e^{-t}$ that satisfies the above boundary conditions. Use this solution to solve the given initial and boundary value problem.

4.6.14. Obtain a solution of the heat equation

$$u_t - c^2 u_{xx} = 0, \qquad 0 < x < l, t > 0$$

with the boundary conditions

$$u(0, t) = 0; \qquad u(l, t) = t, \qquad t > 0$$

in the form of a polynomial in x and t. Show how this solution can be used to solve the initial and boundary value problem for the heat equation with the above boundary data and $u(x, 0) = f(x)$, $0 < x < l$. Discuss the behavior of the solution of the latter problem for large t.

4.6.15. Solve the problem,

$$u_{tt} - c^2 u_{xx} = 0, \qquad 0 < x < l, t > 0,$$
$$u(x, 0) = u_t(x, 0) = 0;$$
$$u(0, t) = t; \qquad u(l, t) = 1,$$

by introducing a change of the dependent variable that renders the boundary conditions homogeneous.

Section 4.7

4.7.1. Show that (4.284) is a solution of the equation (4.280).

4.7.2. Obtain the solution (4.282) of the equation (4.280).

4.7.3. Consider the nonlinear heat equation

$$u_t - u_{xx} = \hat{\lambda} u(1 - u), \qquad -\infty < x < \infty, t > 0.$$

Carry out a linear stability analysis around the two solutions $u_0(x, t) = 0$ and $u_1(x, t) = 1$.

4.7.4. Solve the Cauchy problem for

$$u_t - u_{xx} = \hat{\lambda} u(1 - u), \qquad -\infty < x < \infty, t > 0$$

with the initial condition

$$u(x, 0) = \epsilon, \qquad 0 < \epsilon \ll 1$$

by looking for a solution independent of x. Expand the solution for small ϵ and compare the result with that obtained by solving a linearized version of the problem. Obtain the limit of the solution of the nonlinear problem as $t \to \infty$.

4.7.5. Apply the method of eigenfunction expansions to solve (approximately) the nonlinear heat equation.

$$u_t - u_{xx} = \hat{\lambda} u(1 - u), \qquad 0 < x < \pi, t > 0$$

if

$$u(x, 0) = \epsilon h(x), \qquad 0 < \epsilon \ll 1,$$
$$u(0, t) = u(\pi, t) = 0.$$

Determine the critical value $\hat{\lambda}_c$ of $\hat{\lambda}$ and solve the problem if $\hat{\lambda}$ is slightly larger than $\hat{\lambda}_c$.

4.7.6. Consider the nonlinear elliptic equation

$$u_{xx} + u_{yy} = -\hat{\lambda} u(1 - u^2), \qquad 0 < x < \pi, y > 0$$

with the boundary conditions

$$u(0, y) = u(\pi, y) = 0; \qquad u(x, 0) = \epsilon h(x).$$

Requiring that $u(x, y) \to 0$ as $y \to \infty$, apply a linear stability analysis to obtain a critical value $\hat{\lambda}_c$ for $\hat{\lambda}$. Expand the solution in the form

$$u(x, y) = \epsilon \sum_{k=1}^{\infty} N_k(y) \left(\sqrt{\frac{2}{\pi}} \sin kx \right)$$

and obtain an (approximate) nonlinear equation for $N_1(y)$ and linear equations for the $N_k(y)$ with $k \geqslant 2$. Discuss the behavior of the nonlinear equation for $N_1(y)$ as far as possible without necessarily solving it and consider the behavior of the solution $u(x, y)$ as $y \to \infty$.

4.7.7. Consider the nonlinear hyperbolic equation

$$u_{tt} - u_{xx} = \hat{\lambda} u(1 - u^2), \qquad -\infty < x < \infty, t > 0.$$

(a) Show that if the equation is linearized for small u and $\hat{\lambda} > 0$, there exist exponentially growing solutions.

(b) Set $u = 1 + \epsilon v$ and obtain a linearized equation for v in the form of the Klein–Gordon equation if $\hat{\lambda} > 0$, and show that it is neutrally stable.

4.7.8. Consider the initial and boundary value problem for the nonlinear hyperbolic equation

$$u_{tt} - u_{xx} = \hat{\lambda} u(1 - u^2), \qquad 0 < x < \pi, t > 0;$$
$$u(x, 0) = \epsilon h(x), u_t(x, 0) = 0;$$
$$u(0, t) = u(\pi, t) = 0.$$

Use a linear stability analysis around the solution $u_0 = 0$ to determine a critical value $\hat{\lambda}_c$ such that exponentially growing normal mode solutions exist if $\hat{\lambda} > \hat{\lambda}_c$. Using eigenfunction expansions in the manner presented in the text, attempt a discussion of this nonlinear initial and boundary value problem if $\hat{\lambda}$ is slightly greater than $\hat{\lambda}_c$.

5

INTEGRAL TRANSFORMS

5.1. INTRODUCTION

The preceding chapter dealt with initial and boundary value problems for partial differential equations given over bounded spatial regions. The method of separation of variables and the closely related finite Fourier transform method were used to obtain solutions of these problems. The present chapter deals for the most part with partial differential equations defined over unbounded spatial regions. The tools we shall use for solving Cauchy and initial and boundary value problems are integral transforms. Specifically, we shall consider the Fourier transform, the Fourier sine and cosine transforms, the Hankel transform, and the Laplace transform.

Instead of proceeding directly to a discussion of each transform to be considered, we begin by showing how the separation of variables method motivates some of the results and suggests which transforms are appropriate for given problems.

We again consider the second order equations (4.13), (4.17), and (4.19), of hyperbolic, parabolic, and elliptic types, respectively. Applying separation of variables we set $u(x, t) = M(x)N(t)$ (with t replaced by y in the elliptic case) and obtain as before

$$(5.1) \qquad LM(x) = \lambda^2 \rho(x) M(x),$$

where we have replaced the separation constant λ by λ^2 for convenience in our further discussion. For N we have the equations

$$(5.2) \qquad \begin{cases} N''(t) + \lambda^2 N(t) = 0, & \text{hyperbolic case} \\ N'(t) + \lambda^2 N(t) = 0, & \text{parabolic case} \\ N''(y) - \lambda^2 N(y) = 0, & \text{elliptic case} \end{cases}$$

Since we are dealing with unbounded spatial regions, we shall generally require that the solutions $M(x)$ of (5.1) be bounded or vanish at infinity. If the region is a semi-infinite interval or a half-plane, one of the boundary conditions given in Chapter 4 will be applied on the finite portion of the boundary. [It may also happen that conditions of this type are applied to (5.2) rather than (5.1).] All boundary conditions are assumed to be of homogeneous type, so that $M(x) = 0$ satisfies (5.1) and the boundary conditions. To obtain nonzero $M(x)$, restrictions must be placed on the parameter λ^2. The values of λ for which nonzero $M(x)$ can be found are the eigenvalues and the corresponding $M(x)$ are the eigenfunctions for the eigenvalue problem.

The fundamental distinction between the eigenvalue problem for (5.1) over an unbounded region and that for a bounded region considered in Chapter 4 is that, in general (and this is the case for the problems considered here), the *spectrum* (i.e., the set of eigenvalues) is *continuous* in the unbounded case, whereas in the bounded case the spectrum is *discrete* as we have seen in Chapter 4. Consequently, if λ_k ($k = 1, 2, \ldots$) is the set of eigenvalues in the discrete case and $u_k = M_k N_k$ is the corresponding set of solutions of the given differential equation, the general solution is obtained by superposition of the u_k as

$$(5.3) \qquad u = \sum_{k=1}^{\infty} u_k = \sum_{k=1}^{\infty} M_k N_k.$$

In the case of a continuous spectrum, the eigenvalues λ range over the set D, which may be the interval $-\infty < \lambda < \infty$, $0 \leqslant \lambda < \infty$, or some other uncountable set. Let $u(\lambda) = M(\lambda)N(\lambda)$ be the separated solution corresponding to λ of the given partial differential equation. We then obtain a general solution by the formal superposition

$$(5.4) \qquad u = \int_D u(\lambda) \, d\lambda = \int_D M(\lambda)N(\lambda) \, d\lambda.$$

Since the spectrum is continuous and an orthogonality property for the eigenfunctions $M(\lambda)$ does not occur in a simple and natural way (as was the case for the discrete spectrum), the specification of the $N(\lambda)$ in (5.4) in terms of initial and boundary data cannot be carried out in a general way as was done in Chapter 4.

Therefore, we shall concentrate on a number of specific eigenvalue problems for (5.1) in this chapter which lead to the consideration of Fourier, Hankel, and Laplace transforms. (In this respect our discussion parallels that given in Section 4.3 for the Sturm–Liouville problem.) Only the most basic properties of each of the transforms will be presented and their use will be demonstrated in a number of examples. Once it is determined that a particular transform is relevant for the solution of a given problem, the solution is obtained by transforming the given equation and solving for the transform function. (This approach is identical to that used in the finite Fourier transform method of Chapter 4.) Since a simple and useful expression for the solution does not often result once its transform is known, we present approximate methods for evaluating Laplace and Fourier transforms and integrals in the concluding sections of this chapter.

5.2. ONE-DIMENSIONAL FOURIER TRANSFORMS

We consider the one-dimensional form of (5.1) with $\rho = p = 1$ and $q = 0$, defined over the infinite interval $-\infty < x < \infty$; that is,

$$(5.5) \qquad M''(x) + \lambda^2 M(x) = 0; \qquad -\infty < x < \infty,$$

with the auxiliary condition that $M(x)$ remain bounded as $|x| \to \infty$. The general solution of (5.5) in complex form is

$$(5.6) \qquad M(x; \lambda) = \alpha(\lambda)e^{i\lambda x} + \beta(\lambda)e^{-i\lambda x},$$

where α and β are constants. The only restriction placed on the eigenvalue parameter λ by the boundedness condition is that λ be real valued. Thus the eigenvalues λ lie in the interval

$$(5.7) \qquad -\infty < \lambda < \infty,$$

so that the spectrum is continuous.

Corresponding to the eigenfunction expansions given in Chapter 4, we would like to be able to represent an arbitrary function $f(x)$ (under suitable conditions) in terms of the eigenfunctions $M(x; \lambda)$ in the form

$$(5.8) \quad f(x) = \int_{-\infty}^{\infty} \left[\alpha(\lambda)e^{i\lambda x} + \beta(\lambda)e^{-i\lambda x} \right] d\lambda = \int_{-\infty}^{\infty} \gamma(\lambda)e^{-i\lambda x} \, d\lambda,$$

where $\gamma(\lambda) = \beta(\lambda) + \alpha(-\lambda)$ [the integral in $\gamma(\lambda)$ is obtained from the first integral via a simple change of variables]. Unfortunately, since the eigenfunctions for this eigenvalue problem are not orthogonal (in an elementary sense), it is not immediately apparent how to determine $\gamma(\lambda)$ in terms of $f(x)$ in (5.8). [An "orthogonality" property for the eigenfunctions $M(x; \lambda)$ is defined when Dirac delta functions are discussed in Chapter 7.]

To see how $\gamma(\lambda)$ in (5.8) is determined, we may examine the expansion (4.108) for the complete Fourier series of the function $v(x)$ given in Example 4.6. Expressing the trigonometric functions in complex form and letting $l \to \infty$, it can be shown (see the exercises) that a plausible result is the *Fourier integral formula*

$$(5.9) \qquad f(x) = \frac{1}{2\pi} \int_{-\infty}^{\infty} \int_{-\infty}^{\infty} e^{-i\lambda(x-t)} f(t) \, dt \, d\lambda,$$

where we have used $f(x)$ in place of $v(x)$. If we define the function $F(\lambda)$ in terms of the inner integral in (5.9) to be

$$(5.10) \qquad F(\lambda) = \frac{1}{\sqrt{2\pi}} \int_{-\infty}^{\infty} e^{i\lambda x} f(x) \, dx$$

(where we have changed the variable of integration from t to x), we obtain

$$(5.11) \qquad f(x) = \frac{1}{\sqrt{2\pi}} \int_{-\infty}^{\infty} e^{-i\lambda x} F(\lambda) \, d\lambda.$$

The coefficients in (5.10)–(5.11) have been chosen to be $1/\sqrt{2\pi}$ for the purposes of symmetry.

The function $F(\lambda)$ in (5.10) is called the *Fourier transform* of $f(x)$ and (5.11) is the *inversion formula* that gives the function $f(x)$ in terms of its transform $F(\lambda)$. The function $f(x)$ is referred to as the inverse transform of $F(\lambda)$. By comparing (5.11) with (5.8) we conclude that $\gamma(\lambda) = (1/\sqrt{2\pi})F(\lambda)$.

Once the Fourier integral formula has been obtained, sufficient conditions for its validity can be verified directly. Thus if $f(x)$ is piecewise differentiable on each finite interval and $\int_{-\infty}^{\infty} |f(x)| \, dx < \infty$, it can be shown that the integral in (5.9) converges pointwise to the function $f(x)$ at points of continuity of $f(x)$. At a point x_0 where $f(x)$ has a jump discontinuity, the integral converges to $\frac{1}{2}f(x_0 -) + \frac{1}{2}f(x_0 +)$. If we weaken the conditions on $f(x)$ and merely require that it be square integrable over $-\infty < x < \infty$, it can be shown that the integral converges to $f(x)$ in the mean square sense.

In applications of Fourier transforms it is often necessary to find the inverse transform of $H(\lambda) = F(\lambda)G(\lambda)$ where $F(\lambda)$ and $G(\lambda)$ are transforms of the known functions $f(x)$ and $g(x)$, respectively. The inverse transform of $H(\lambda)$ is given as

$$(5.12) \qquad \frac{1}{\sqrt{2\pi}} \int_{-\infty}^{\infty} e^{-i\lambda x} F(\lambda)G(\lambda) \, d\lambda$$

$$= \frac{1}{2\pi} \int_{-\infty}^{\infty} \int_{-\infty}^{\infty} e^{-i\lambda(x-t)} f(t) G(\lambda) \, dt \, d\lambda$$

$$= \frac{1}{2\pi} \int_{-\infty}^{\infty} f(t) \int_{-\infty}^{\infty} e^{-i\lambda(x-t)} G(\lambda) \, d\lambda \, dt$$

$$= \frac{1}{\sqrt{2\pi}} \int_{-\infty}^{\infty} f(t) g(x-t) \, dt,$$

where we have assumed the interchange of orders of integration is valid. This result is known as the *convolution integral theorem* for Fourier transforms.

Putting $G(\lambda) = \overline{F(\lambda)}$, the complex conjugate of $F(\lambda)$ in (5.12), it is easily shown that we obtain the *Parseval equation* for Fourier transforms

$$(5.13) \qquad \int_{-\infty}^{\infty} |F(\lambda)|^2 \, d\lambda = \int_{-\infty}^{\infty} |f(x)|^2 \, dx.$$

This result should be compared with the Parseval equation (4.69) for Fourier series.

An additional result that relates the Fourier transform of derivatives of functions to the transform of the functions themselves is important for applications of transforms to differential equations. For the transform of $f'(x)$ we have

$$(5.14) \quad \frac{1}{\sqrt{2\pi}} \int_{-\infty}^{\infty} e^{i\lambda x} f'(x)\, dx = \frac{-i\lambda}{\sqrt{2\pi}} \int_{-\infty}^{\infty} e^{i\lambda x} f(x)\, dx = -i\lambda F(\lambda),$$

where $F(\lambda)$ is the Fourier transform of $f(x)$, assuming that $f(x)$ is a smooth function that vanishes at infinity. More generally, if $f^{(n)}(x)$ is the nth derivative of $f(x)$, and $f(x)$ and its first $n - 1$ derivatives are smooth functions that vanish at infinity, we have

$$(5.15) \quad \frac{1}{\sqrt{2\pi}} \int_{-\infty}^{\infty} e^{i\lambda x} f^{(n)}(x)\, dx = (-i\lambda)^n F(\lambda); \qquad n = 1, 2, 3, \dots.$$

The formulas (5.14)–(5.15) are obtained on integrating by parts.

Further useful properties of Fourier transforms and transforms of specific functions may be found in tables of Fourier transforms. We now consider several examples whose solutions involve Fourier transforms. In each example one of the independent variables has the infinite interval $(-\infty, +\infty)$ as its domain of definition.

EXAMPLE 5.1. AN ORDINARY DIFFERENTIAL EQUATION

A simple example that exhibits the essential features of the Fourier transform method for solving differential equations is given by the following boundary value problem for the ordinary differential equation

$$(5.16) \qquad y''(x) - k^2 y(x) = -f(x), \qquad -\infty < x < \infty,$$

where k is a constant and $f(x)$ is specified. We require that

$$(5.17) \qquad\qquad y(x) \to 0 \text{ as } |x| \to \infty.$$

Assuming $F(\lambda)$ is the Fourier transform of $f(x)$ we set

$$(5.18) \qquad\qquad Y(\lambda) = \frac{1}{\sqrt{2\pi}} \int_{-\infty}^{\infty} e^{i\lambda x} y(x)\, dx,$$

so that $Y(\lambda)$ is the Fourier transform of the solution $y(x)$.

To solve this boundary value problem, we (Fourier) transform the equation (5.16) by multiplying across by $(1/\sqrt{2\pi})e^{i\lambda x}$ and integrating both sides with respect to x from $-\infty$ to $+\infty$. By our assumption on $y(x)$ at infinity we have, on using (5.15),

$$(5.19) \qquad \frac{1}{\sqrt{2\pi}} \int_{-\infty}^{\infty} e^{i\lambda x} y''(x)\, dx = (-i\lambda)^2 Y(\lambda) = -\lambda^2 Y(\lambda).$$

The transformed equation is, therefore, algebraic and is given as

$$(5.20) \qquad -(\lambda^2 + k^2) Y(\lambda) = -F(\lambda),$$

so that

$$(5.21) \qquad Y(\lambda) = \frac{F(\lambda)}{\lambda^2 + k^2} \equiv F(\lambda)G(\lambda)$$

with $G(\lambda) = 1/(\lambda^2 + k^2)$.

To find $y(x)$ we must invert $Y(\lambda)$. This can be done using the convolution theorem (5.12) if the inverse transform of $G(\lambda)$ is known. Using a table of transforms or by direct verification based on complex integration theory it can be shown that

$$(5.22) \qquad \frac{1}{\sqrt{2\pi}} \int_{-\infty}^{\infty} e^{-i\lambda x} G(\lambda)\, d\lambda = \frac{1}{\sqrt{2\pi}} \int_{-\infty}^{\infty} \frac{e^{-i\lambda x}}{\lambda^2 + k^2}\, d\lambda = \frac{\sqrt{2\pi}}{2k} e^{-k|x|}$$

with $k > 0$. [Note that once the inverse transform of $G(\lambda)$ is given, as in (5.22), the Fourier transform of that function is easy to evaluate and thereby shown to equal $G(\lambda)$.] Using (5.22) in the convolution theorem we obtain the solution of (5.16)–(5.17) in the form

$$(5.23) \quad y(x) = \frac{1}{\sqrt{2\pi}} \int_{-\infty}^{\infty} e^{-i\lambda x} Y(\lambda)\, d\lambda = \frac{1}{\sqrt{2\pi}} \int_{-\infty}^{\infty} e^{-i\lambda x} F(\lambda)G(\lambda)\, d\lambda$$

$$= \frac{1}{2k} \int_{-\infty}^{\infty} e^{-k|x-t|} f(t)\, dt.$$

It is of interest to apply the solution formula (5.23) for two special choices of $f(x)$. First we set $f(x) = 1$ and then we set $f(x) = \delta(x - \xi)$, where $\delta(x - \xi)$ is the Dirac delta function. With $f(x) = 1$ we have

$$(5.24) \qquad y(x) = \frac{1}{2k} \int_{-\infty}^{\infty} e^{-k|x-t|}\, dt$$

$$= \frac{1}{2k} \int_{-\infty}^{x} e^{-k(x-t)}\, dt + \frac{1}{2k} \int_{x}^{\infty} e^{k(x-t)}\, dt = \frac{1}{k^2}.$$

Putting $f(x) = \delta(x - \xi)$ and recalling the properties of the delta function given in Example 1.2 gives

$$(5.25) \qquad y(x) = \frac{1}{2k} \int_{-\infty}^{\infty} e^{-k|x-t|} \delta(t - \xi) \, dt = \frac{1}{2k} e^{-k|x-\xi|}.$$

However, each of these "solutions" has a shortcoming. The function $y(x) = 1/k^2$ satisfies (5.16) but fails to vanish at infinity, as required by the boundary condition (5.17). The function $y(x) = (1/2k)e^{-k|x-\xi|}$ satisfies (5.16) at all $x \neq \xi$ and vanishes at infinity, but is not differentiable at $x = \xi$.

We do not expect the above "solutions" to satisfy all the conditions of the boundary value problem (5.16)–(5.17) since in both cases the Fourier transforms of the functions $f(x)$ are not defined in a conventional sense. The function $f(x) = 1$ has a Fourier transform equal to $\sqrt{2\pi} \, \delta(\lambda)$. This can be seen by putting $F(\lambda) = \sqrt{2\pi} \, \delta(\lambda)$ in the inversion formula (5.11). Similarly, the Fourier transform of $f(x) = \delta(x - \xi)$ is $F(\lambda) = (1/\sqrt{2\pi})e^{i\lambda\xi}$. In either case we are required to deal with *generalized functions* and without extending the theory of Fourier transforms to include such functions and an appropriate formulation of the corresponding boundary value problems, it is not clear that the solution formula can be applied in such cases. Nevertheless, the solution $y(x) = 1/k^2$ in the case where $f(x) = 1$ is certainly plausible (even if it does not vanish at infinity), and the solution $y(x) = (1/2k)e^{-k|x-\xi|}$ when $f(x) = \delta(x - \xi)$ is known as Green's function for the boundary value problem (5.16)–(5.17). (Green's functions play an important role in the theory of boundary value problems for both ordinary and partial differential equations.) Although a direct approach can be developed to deal with each of the above cases which may or may not depend on Fourier transform theory, we now show how each of the foregoing solutions can be obtained as limits of sequences of solutions of the boundary value problem (5.16)–(5.17), each of which satisfies the conditions of the problem. As such, the two solutions obtained previously are called *generalized solutions*.

We define the sequence of functions

$$(5.26) \qquad f_N(x) = \begin{cases} \alpha(N); & |x| \leqslant N \\ 0; & |x| > N \end{cases}$$

where $N > 0$ and consider two cases. First, we set $\alpha = 1$ for all N and find that $\lim_{N \to \infty} f_N(x) = 1$ for all x. This corresponds to the case $f(x) = 1$ given above. Second, we set $\alpha(N) = 1/2N$ and obtain in accordance with our discussion in Example 1.2, $\lim_{N \to 0} f_N(x) = \delta(x)$. This corresponds to the case $f(x) = \delta(x)$ where we have put $\xi = 0$ for simplicity.

Clearly, each of the $f_N(x)$ has a Fourier transform, and the problem (5.16)–(5.17) may be solved in the manner given above with $f = f_N$. The solution $y_N(x)$ corresponding to $f_N(x)$ is found to be

$$(5.27) \qquad y_N(x) = \frac{\alpha(N)}{2k} \int_{-N}^{N} e^{-k|x-t|} \, dt$$

$$= \begin{cases} \dfrac{[1 - e^{-kN}\cosh kx]\alpha(N)}{k^2}; & |x| \leqslant N \\[2ex] \dfrac{[e^{-k|x|}\sinh kN]\alpha(N)}{k^2}; & |x| > N \end{cases}.$$

For each $y_N(x)$ we have $\lim_{|x| \to \infty} y_N(x) = 0$ so that (5.17) is satisfied.

First we put $\alpha(N) = 1$ in (5.27). Then as $N \to \infty$, the pointwise limit of $y_N(x)$ is clearly $y_N(x) \to 1/k^2$ for all finite x. Next we set $\alpha(N) = 1/2N$ and the limit as $N \to 0$ is found to be, on using l'Hospital's rule, $y_N(x) \to (1/2k)e^{-k|x|}$. (This limit equals the Green's function with $\xi = 0$.) Since the limits are nonuniform over the infinite interval, the smoothness properties and the vanishing at infinity are not preserved in both cases, yet we have obtained these solutions as limits of strict solutions of the boundary value problem. It may be stated as a general principle that solutions of a given problem obtained by formal and plausible means can be characterized as limits obtained in an appropriate fashion of a sequence of strict solutions of the problem. We shall have further occasion to construct solutions of problems by formal means, but we will not always indicate how they may be obtained as a limit of strict solutions. It should be remarked that in Chapter 4, the eigenfunction expansions did not always constitute strict solutions of the given problem and had to be interpreted as generalized solutions of the given problem even though this was not always explicitly indicated.

EXAMPLE 5.2. THE CAUCHY PROBLEM FOR THE HEAT EQUATION

We consider the heat (or diffusion) equation

$$(5.28) \qquad u_t - c^2 u_{xx} = 0, \qquad -\infty < x < \infty, \, t > 0,$$

where c^2 is a constant, with the initial condition

$$(5.29) \qquad u(x,0) = f(x), \qquad -\infty < x < \infty,$$

and solve this problem using a Fourier transform in the x-variable. We assume $f(x)$ has a Fourier transform and that u and u_x vanish at infinity so that (5.15) is applicable. It will be found, however, that the solution formula obtained is valid under weaker conditions than are necessary for the Fourier transform method to be applicable.

Denoting the Fourier transform of $u(x, t)$ by

$$(5.30) \qquad U(\lambda, t) = \frac{1}{\sqrt{2\pi}} \int_{-\infty}^{\infty} e^{i\lambda x} u(x, t) \, dx,$$

we multiply through by $(1/\sqrt{2\pi})e^{i\lambda x}$ in (5.28) and integrate with respect to x from $-\infty$ to $+\infty$. Using (5.15) we obtain an ordinary differential equation for $U(\lambda, t)$,

$$(5.31) \qquad \frac{1}{\sqrt{2\pi}} \int_{-\infty}^{\infty} u_t e^{i\lambda x} \, dx + (c\lambda)^2 U(\lambda, t)$$

$$= \frac{\partial}{\partial t} U(\lambda, t) + (c\lambda)^2 U(\lambda, t) = 0.$$

Fourier transforming the initial condition (5.29) yields

$$(5.32) \qquad U(\lambda, 0) = F(\lambda) = \frac{1}{\sqrt{2\pi}} \int_{-\infty}^{\infty} e^{i\lambda x} f(x) \, dx.$$

The solution of the initial value problem (5.31)–(5.32) is

$$(5.33) \qquad U(\lambda, t) = F(\lambda) \exp\left[- (c\lambda)^2 t \right].$$

To find $u(x, t)$ we must invert the Fourier transform $U(\lambda, t)$. We have

$$(5.34) \qquad u(x, t) = \frac{1}{\sqrt{2\pi}} \int_{-\infty}^{\infty} e^{-i\lambda x - \lambda^2 c^2 t} F(\lambda) \, d\lambda$$

$$= \frac{1}{2\pi} \int_{-\infty}^{\infty} \int_{-\infty}^{\infty} e^{-i\lambda(x-s) - \lambda^2 c^2 t} f(s) \, d\lambda \, ds,$$

where we have used (5.32) and interchanged the order of integration. To evaluate the inner integral we first transform it as

$$(5.35) \qquad \int_{-\infty}^{\infty} e^{-i\lambda(x-s) - \lambda^2 c^2 t} \, d\lambda = \int_{-\infty}^{0} e^{-i\lambda(x-s) - \lambda^2 c^2 t} \, d\lambda$$

$$+ \int_{0}^{\infty} e^{-i\lambda(x-s) - \lambda^2 c^2 t} \, d\lambda = 2\int_{0}^{\infty} e^{-\lambda^2 c^2 t} \cos[\lambda(x-s)] \, d\lambda.$$

The last integral in (5.35) can be evaluated explicitly. Let the integral $I(\alpha)$ be defined as

$$(5.36) \qquad I(\alpha) = 2\int_{0}^{\infty} e^{-\lambda^2 c^2 t} \cos[\alpha\lambda] \, d\lambda.$$

In view of the exponential decay of the integrand we can obtain $dI/d\alpha$ by differentiating under the integral sign. We have

$$(5.37) \quad \frac{dI(\alpha)}{d\alpha} = -2\int_0^\infty \lambda e^{-\lambda^2 c^2 t}\sin[\alpha\lambda]\, d\lambda = \frac{1}{c^2 t}\int_0^\infty \sin[\alpha\lambda]\, d\left(e^{-\lambda^2 c^2 t}\right)$$

$$= -\frac{\alpha}{2c^2 t}I(\alpha).$$

Now $I(0)$ is given as

$$(5.38) \qquad\qquad I(0) = 2\int_0^\infty e^{-\lambda^2 c^2 t}\, d\lambda = \sqrt{\frac{\pi}{c^2 t}}.$$

The solution of the initial value problem (5.37)–(5.38) is easily found to be

$$(5.39) \qquad I(\alpha)\big|_{\alpha = x - s} = 2\int_0^\infty e^{-\lambda^2 c^2 t}\cos[\lambda(x - s)]\, d\lambda$$

$$= \sqrt{\frac{\pi}{c^2 t}}\,\exp\left[-\frac{(x - s)^2}{4c^2 t}\right].$$

Introducing (5.39) into (5.34) yields the solution of the Cauchy problem (5.28)–(5.29) as

$$(5.40) \qquad u(x, t) = \frac{1}{\sqrt{4\pi c^2 t}}\int_{-\infty}^\infty \exp\left[-\frac{(x - s)^2}{4c^2 t}\right]f(s)\, ds.$$

The term

$$(5.41) \qquad\qquad G(x - \xi, t) = \frac{1}{\sqrt{4\pi c^2 t}}\exp\left[-\frac{(x - \xi)^2}{4c^2 t}\right]$$

is known as the *fundamental solution* of the heat or diffusion equation. With the formal substitution $f(s) = \delta(s - \xi)$ in (5.40) we obtain the fundamental solution (5.41). Thus (5.41) is a solution of the heat equation corresponding to a point source of heat at the initial time $t = 0$ located at the point $x = \xi$. The solution (5.40) can be thought of as a superposition of solutions due to point sources distributed along the x-axis with density $f(x)$.

We have already encountered a solution of the form (5.41) in Section 1.1. In fact, given the diffusion equation for $v(x, t)$,

$$(5.42) \qquad\qquad v_t + cv_x = \frac{1}{2}Dv_{xx},$$

considered in (1.21), if we set

$$(5.43) \qquad\qquad \hat{x} = x - ct, \qquad \hat{t} = t,$$

and $u(\hat{x}, \hat{t}) = v(x, t)$, then u satisfies (5.28) with c^2 replaced by $\frac{1}{2}D$ and $x, t \to \hat{x}, \hat{t}$. Further, $u(\hat{x}, 0) = u(x, 0) = v(x, 0)$. Thus with the initial condition (1.24) for $v(x, t)$ [i.e., $v(x, 0) = \delta(x)$], we obtain the solution (1.28) for $v(x, t)$ on comparing with the fundamental solution (5.41).

Either from the solution (5.40) with $f(x) = 0$ outside a finite interval or from the fundamental solution (5.41), one sees that $u(x, t)$ is instantaneously greater than zero at each x, for any time $t > 0$. [If $u(x, t)$ represents temperature, it is generally assumed that $u(x, 0) = f(x) \geqslant 0$.] Thus heat propagates at infinite speed according to the heat equation (5.28). A similar observation was made and discussed in connection with the diffusion equation of Section 1.1.

For the nonhomogeneous heat equation

$$(5.44) \qquad u_t - c^2 u_{xx} = F(x, t), \qquad -\infty < x < \infty, t > 0,$$

with homogeneous initial data

$$(5.45) \qquad\qquad u(x, 0) = 0,$$

we easily obtain, on using Duhamel's principle,

$$(5.46) \qquad u(x, t) = \int_0^t \int_{-\infty}^{\infty} G(x - s, t - \tau) F(s, \tau) \, ds \, d\tau,$$

where $G(x - s, t - \tau)$ is defined as in (5.41). The result (5.46) can also be obtained by the use of Fourier transforms. The fundamental solution (5.41) is thus seen to play an important role in the solution of the Cauchy problem for the heat equation. As we shall see, it also plays a role in the solution of problems over semi-infinite intervals and even problems over a finite interval.

The function $G(x - \xi, t)$ is also of interest for the reason that as $t \to 0$, $G(x - \xi, t) \to \delta(x - \xi)$. Thus we have in $G(x - \xi, t)$ a smooth function which in the limit tends to the Dirac delta function. This is in contrast to the functions $\delta_\epsilon(x)$ defined in Example 1.2, which tend to $\delta(x)$ as $\epsilon \to 0$ but have jump discontinuities at $x = \pm\epsilon$ and are not smooth functions.

EXAMPLE 5.3.   THE CAUCHY PROBLEM FOR THE WAVE EQUATION

The Cauchy problem for the wave equation asks for a function $u(x, t)$ that satisfies

$$(5.47) \qquad u_{tt} - c^2 u_{xx} = 0, \qquad -\infty < x < \infty, t > 0$$

and has the initial values

$$(5.48) \qquad u(x,0) = f(x), \qquad -\infty < x < \infty;$$

$$(5.49) \qquad u_t(x,0) = g(x), \qquad -\infty < x < \infty.$$

The Fourier transformed problem is

$$(5.50) \qquad \frac{\partial^2 U(\lambda, t)}{\partial t^2} + (c\lambda)^2 U(\lambda, t) = 0; \qquad t > 0$$

with

$$(5.51) \qquad U(\lambda, 0) = F(\lambda),$$

$$(5.52) \qquad \frac{\partial U(\lambda, 0)}{\partial t} = G(\lambda),$$

where U, F, and G are Fourier transforms of u, f, and g, respectively. It is assumed that the operations necessary to obtain (5.50)–(5.52) [e.g., the use of (5.15)] are valid.

The solution of the initial value problem (5.50)–(5.52) for $U(\lambda, t)$ is easily found to be

$$(5.53) \qquad U(\lambda, t) = \left[\frac{1}{2} F(\lambda) + \frac{1}{2i\lambda c} G(\lambda) \right] e^{i\lambda ct}$$

$$+ \left[\frac{1}{2} F(\lambda) - \frac{1}{2i\lambda c} G(\lambda) \right] e^{-i\lambda ct}.$$

Inverting this transform gives the solution

$$(5.54)$$

$$u(x, t) = \frac{1}{\sqrt{2\pi}} \int_{-\infty}^{\infty} e^{-i\lambda x} U(\lambda, t) \, d\lambda$$

$$= \frac{1}{2} \left[\frac{1}{\sqrt{2\pi}} \int_{-\infty}^{\infty} e^{-i\lambda(x-ct)} F(\lambda) \, d\lambda + \frac{1}{\sqrt{2\pi}} \int_{-\infty}^{\infty} e^{-i\lambda(x+ct)} F(\lambda) \, d\lambda \right.$$

$$\left. + \frac{1}{c\sqrt{2\pi}} \int_{-\infty}^{\infty} e^{-i\lambda(x-ct)} \frac{G(\lambda)}{i\lambda} \, d\lambda - \frac{1}{c\sqrt{2\pi}} \int_{-\infty}^{\infty} e^{-i\lambda(x+ct)} \frac{G(\lambda)}{i\lambda} \, d\lambda \right].$$

The integrals involving $G(\lambda)$ in (5.54) are to be interpreted in the *Cauchy principal value* sense. [The Cauchy principal value of the integral $\int_{-\infty}^{\infty} k(\lambda) \, d\lambda$ is defined as $\lim_{N \to \infty} \int_{-N}^{N} k(\lambda) \, d\lambda$.] If we express the function $g(x)$ as

$$(5.55) \qquad g(x) = \frac{1}{\sqrt{2\pi}} \int_{-\infty}^{\infty} e^{-i\lambda x} G(\lambda) \, d\lambda,$$

the indefinite integral of $g(x)$ is given as

$$(5.56) \qquad \int^x g(s)\, ds = -\frac{1}{\sqrt{2\pi}} \int_{-\infty}^{\infty} e^{-i\lambda x} \frac{G(\lambda)}{i\lambda}\, d\lambda,$$

with the integral in $G(\lambda)$ interpreted in the Cauchy principal value sense. Then (5.54) can be expressed as

$$(5.57) \qquad u(x,t) = \frac{1}{2}[f(x-ct)+f(x+ct)]$$
$$-\frac{1}{2c}\int^{x-ct} g(s)\, ds + \frac{1}{2c}\int^{x+ct} g(s)\, ds$$
$$= \frac{1}{2}[f(x-ct)+f(x+ct)] + \frac{1}{2c}\int_{x-ct}^{x+ct} g(s)\, ds,$$

on using (5.11) and (5.56). The expression (5.57) is just d'Alembert's solution which was obtained in Example 2.2 by another method.

EXAMPLE 5.4.   LAPLACE'S EQUATION IN A HALF - PLANE

We begin by considering the first boundary value problem, which is also referred to as *Dirichlet's problem*, for Laplace's equation in a half-plane. That is, we require that $u(x, y)$ satisfy

$$(5.58) \qquad u_{xx} + u_{yy} = 0, \qquad -\infty < x < \infty, \; y > 0$$

with the boundary condition

$$(5.59) \qquad u(x,0) = f(x), \qquad -\infty < x < \infty.$$

and the additional condition

$$(5.60) \qquad u(x,y) \to 0 \text{ as } |x| \to \infty, \; y \to \infty.$$

To solve (5.58)–(5.60) we introduce the Fourier transform in x of the function $u(x, y)$ defined as

$$(5.61) \qquad U(\lambda, y) = \frac{1}{\sqrt{2\pi}} \int_{-\infty}^{\infty} e^{i\lambda x} u(x, y)\, dx.$$

Then the Fourier transformed problem (5.58)–(5.60) becomes

$$(5.62) \qquad \frac{\partial^2 U}{\partial y^2} - \lambda^2 U = 0, \qquad 0 < y < \infty$$

with the conditions

(5.63) $U(\lambda, 0) = F(\lambda), U(\lambda, y) \to 0$ as $y \to \infty$.

Here $F(\lambda)$ is the transform of $f(x)$ and $U(\lambda, y) \to 0$ as $y \to \infty$ is a consequence of the assumption (5.60). The solution of (5.62)–(5.63) is

(5.64) $U(\lambda, y) = F(\lambda)e^{-|\lambda|y}$

where the $|\lambda|$ occurs, since we require that $U \to 0$ as $y \to \infty$ and λ ranges from $-\infty$ to $+\infty$. Inverting the transform $U(\lambda, y)$ gives

(5.65) $u(x, y) = \dfrac{1}{\sqrt{2\pi}} \displaystyle\int_{-\infty}^{\infty} e^{-i\lambda x - |\lambda|y} F(\lambda) \, d\lambda$

$$= \frac{1}{2\pi} \int_{-\infty}^{\infty} \int_{-\infty}^{\infty} e^{-i\lambda(x - t) - |\lambda|y} f(t) \, d\lambda \, dt,$$

on using the inverse transform for $F(\lambda)$. The inner integral can be evaluated explicitly as

(5.66)

$$\int_{-\infty}^{\infty} e^{-i\lambda(x - t) - |\lambda|y} \, d\lambda = \int_{-\infty}^{0} e^{\lambda[-i(x - t) + y]} \, d\lambda + \int_{0}^{\infty} e^{\lambda[-i(x - t) - y]} \, d\lambda$$

$$= \frac{e^{\lambda[-i(x - t) + y]}}{-i(x - t) + y} \bigg|_{\lambda = -\infty}^{\lambda = 0} + \frac{e^{\lambda[-i(x - t) - y]}}{-i(x - t) - y} \bigg|_{\lambda = 0}^{\lambda = \infty} = \frac{2y}{(x - t)^2 + y^2}.$$

Thus we have

(5.67) $u(x, y) = \dfrac{y}{\pi} \displaystyle\int_{-\infty}^{\infty} \dfrac{f(t) \, dt}{(x - t)^2 + y^2}.$

It may be verified directly that for $y > 0$, (5.67) satisfies Laplace's equation. However, a careful examination of the integral is required to show that the boundary condition (5.59) is satisfied. Although the coefficient y of the integral vanishes at $y = 0$, the integral is singular at the point $t = x$ when $y = 0$. The combination of these two effects leads to the satisfaction of the boundary condition. We do not carry out this verification.

The second boundary value problem for (5.58) is also referred to as the *Neumann problem* for Laplace's equation. The boundary condition (5.59) is replaced by

(5.68) $\dfrac{\partial u(x, 0)}{\partial y} = g(x), \qquad -\infty < x < \infty,$

and we require that $u(x, y)$ be bounded in the upper half-plane. Solving this problem directly leads to certain difficulties with the convergence of the relevant Fourier transforms that we prefer to avoid. Instead, we solve this problem by using a general principle known as *Stokes' Rule*. For our problem it takes the following form.

Let $u(x, y)$ satisfy

$$(5.69) \qquad u_{xx} + u_{yy} = 0, \qquad -\infty < x < \infty, \, y > 0$$

and the boundary condition

$$(5.70) \qquad u_y(x, 0) = g(x).$$

Put

$$(5.71) \qquad v(x, y) = \frac{\partial u(x, y)}{\partial y}.$$

Then

$$(5.72) \qquad \nabla^2 v = \nabla^2 u_y = \frac{\partial}{\partial y} \nabla^2 u = 0$$

and

$$(5.73) \qquad v(x, 0) = u_y(x, 0) = g(x).$$

Thus if we can solve the Dirichlet problem (5.72)–(5.73) for $v(x, y)$, the solution of the Neumann problem (5.69)–(5.70) is given as

$$(5.74) \qquad u(x, y) = \int^y v(x, s) \, ds,$$

and is determined up to an arbitrary constant. Using the solution (5.67) of the Dirichlet problem we have

$$(5.75) \qquad u(x, y) = \frac{1}{\pi} \int^y \int_{-\infty}^\infty \frac{s g(t)}{(x - t)^2 + s^2} \, dt \, ds$$

$$= \frac{1}{\pi} \int_{-\infty}^\infty \left(\int^y \frac{s \, ds}{(x - t)^2 + s^2} \right) g(t) \, dt$$

$$= \frac{1}{2\pi} \int_{-\infty}^\infty \log\left[(x - t)^2 + y^2 \right] g(t) \, dt.$$

It can be verified that (5.75) satisfies (5.69)–(5.70).

We have considered the application of Fourier transforms to some basic but simple problems. In each case it was possible to obtain the solution in a fairly straightforward manner by inverting relevant Fourier transforms. Equations of parabolic, hyperbolic, and elliptic types were chosen to show the applicability of the Fourier transform technique for equations of each of the three basic types. Further discussion of the solutions obtained in the examples is given later in the text and in the exercises.

In general applications of transform methods it is not expected that the inverse transforms can be simplified as far as was done in the above examples. However, approximation methods are available that yield results in various regions of interest in the domain of the dependent variable. Certain methods of this type are presented in Section 5.7.

5.3. FOURIER SINE AND COSINE TRANSFORMS

On considering the one-dimensional form of (5.1) over the semi-infinite interval $0 < x < \infty$ with $\rho = p = 1$ and $q = 0$, we obtain the equation

$$(5.76) \qquad M''(x) + \lambda^2 M(x) = 0, \qquad 0 < x < \infty.$$

Two sets of boundary conditions are considered for the function $u(x, t)$ [or $u(x, y)$] which gives rise to the eigenvalue problem under consideration when the variables are separated as in Section 5.1. First, we assume that u vanishes at $x = 0$ and is bounded as $x \to \infty$. This implies the (homogeneous) boundary conditions

$$(5.77) \qquad M(0) = 0, \qquad M(x) \text{ bounded as } x \to \infty,$$

for the solution $M(x)$ of (5.76). Second, we assume that u_x vanishes at $x = 0$ and that u is bounded as $x \to \infty$. This yields the (homogeneous) boundary conditions

$$(5.78) \qquad M'(0) = 0, \qquad M(x) \text{ bounded as } x \to \infty,$$

for (5.76).

For the eigenvalue problem (5.76) and (5.77) we immediately obtain the eigenfunctions

$$(5.79) \qquad M(x; \lambda) = \sin \lambda x$$

with the continuous spectrum

$$(5.80) \qquad 0 < \lambda < \infty.$$

Negative values of λ need not be considered since $\sin(-\lambda x) = -\sin(\lambda x)$.

For the eigenvalue problem (5.76) and (5.78) we obtain the eigenfunctions

$$(5.81) \qquad\qquad M(x; \lambda) = \cos \lambda x$$

with the continuous spectrum

$$(5.82) \qquad\qquad 0 \leqslant \lambda < \infty.$$

Again, negative values of λ are not included since $\cos(-\lambda x) = \cos(\lambda x)$.

Let the function $f(x)$ defined for $x > 0$ satisfy conditions equivalent to those given in Section 5.2 for Fourier transformable functions. Then corresponding to the eigenvalue problem (5.76) and (5.77) we have the representation

$$(5.83) \qquad\qquad F_s(\lambda) = \sqrt{\frac{2}{\pi}} \int_0^\infty \sin(\lambda x) f(x)\, dx,$$

with $F_s(\lambda)$ defined as the *Fourier sine transform* of $f(x)$. The *inversion formula* giving $f(x)$ in terms of its transform is

$$(5.84) \qquad\qquad f(x) = \sqrt{\frac{2}{\pi}} \int_0^\infty \sin(\lambda x) F_s(\lambda)\, d\lambda.$$

The formula (5.84) yields an odd extension of $f(x)$ to the entire x-axis [i.e., $f(-x) = -f(x)$]. By considering the Fourier transform (5.10) and its inversion formula (5.11) as applied to the odd extension of $f(x)$, we obtain the formulas (5.83)–(5.84), as is shown in the exercises. Thus (5.84) converges to $f(x)$ at points of continuity of $f(x)$ and to its mean value at points of jump discontinuity.

Corresponding to the eigenvalue problem (5.76) and (5.78) we have the representation

$$(5.85) \qquad\qquad F_c(\lambda) = \sqrt{\frac{2}{\pi}} \int_0^\infty \cos(\lambda x) f(x)\, dx,$$

with $F_c(\lambda)$ defined as the *Fourier cosine transform* of $f(x)$. The *inversion formula* for the cosine transform is

$$(5.86) \qquad\qquad f(x) = \sqrt{\frac{2}{\pi}} \int_0^\infty \cos(\lambda x) F_c(\lambda)\, d\lambda.$$

These formulas result from the Fourier integral formulas (5.10)–(5.11) when applied to an even function $f(x)$ [i.e., $f(-x) = f(x)$], as is shown in the exercises. It is seen that (5.86) yields an even extension of $f(x)$ to the entire axis and again, the convergence properties for (5.86) are equivalent to those for the Fourier transform.

In applications of the sine and cosine transforms to differential equations, it is again necessary to express the transform of derivatives of $f(x)$ in terms of the transform of $f(x)$. For the sine transform we have

(5.87)
$$\sqrt{\frac{2}{\pi}} \int_0^\infty \sin(\lambda x) f'(x)\, dx$$

$$= \sqrt{\frac{2}{\pi}}\, f(x)\sin(\lambda x)\Big|_{x=0}^{x=\infty} - \lambda \sqrt{\frac{2}{\pi}} \int_0^\infty \cos(\lambda x) f(x)\, dx,$$

(5.88)
$$\sqrt{\frac{2}{\pi}} \int_0^\infty f''(x)\sin(\lambda x)\, dx$$

$$= \sqrt{\frac{2}{\pi}}\, f'(x)\sin(\lambda x)\Big|_{x=0}^{x=\infty} - \lambda \sqrt{\frac{2}{\pi}} \int_0^\infty f'(x)\cos(\lambda x)\, dx$$

$$= \sqrt{\frac{2}{\pi}}\, f'(x)\sin(\lambda x)\Big|_{x=0}^{x=\infty} - \lambda \sqrt{\frac{2}{\pi}}\, f(x)\cos(\lambda x)\Big|_{x=0}^{x=\infty}$$

$$- \lambda^2 \sqrt{\frac{2}{\pi}} \int_0^\infty \sin(\lambda x) f(x)\, dx.$$

Thus if we assume that $f(x)$ and $f'(x)$ vanish at infinity, we obtain on using the definitions (5.83) and (5.85),

(5.89)
$$\sqrt{\frac{2}{\pi}} \int_0^\infty \sin(\lambda x) f'(x)\, dx = -\lambda F_c(\lambda),$$

(5.90)
$$\sqrt{\frac{2}{\pi}} \int_0^\infty \sin(\lambda x) f''(x)\, dx = \lambda \sqrt{\frac{2}{\pi}}\, f(0) - \lambda^2 F_s(\lambda).$$

We note that the Fourier sine transform of a first derivative of a function is given in terms of the Fourier cosine transform of the function itself. The Fourier sine transform of a second derivative is given in terms of the sine transform of the function. There is an additional boundary term $\lambda\sqrt{2/\pi}\, f(0)$ that vanishes if $f(0) = 0$; that is, if $f(x)$ satisfies the boundary condition of the relevant eigenvalue problem. The formulas (5.89) and (5.90) show that the use of the sine transform is not expected to be effective unless the differential equation contains only derivatives of even orders in the transformed independent variable.

A similar result is true for the Fourier cosine transform. We easily find on using (5.83) and (5.85) that

$$(5.91) \qquad \sqrt{\frac{2}{\pi}} \int_0^\infty f'(x)\cos(\lambda x)\, dx = \sqrt{\frac{2}{\pi}}\, f(0) + \lambda F_s(\lambda),$$

$$(5.92) \qquad \sqrt{\frac{2}{\pi}} \int_0^\infty f''(x)\cos(\lambda x)\, dx = -\sqrt{\frac{2}{\pi}}\, f'(0) - \lambda^2 F_c(\lambda),$$

if we assume that $f(x)$ and $f'(x)$ vanish at infinity. Again, if $f'(0) = 0$ so that $f(x)$ satisfies the boundary condition at $x = 0$ of the associated eigenvalue problem, the cosine transform of $f''(x)$ is given strictly in terms of the cosine transform of the function $f(x)$ as shown in (5.92).

We now consider several applications that involve the use of Fourier cosine and sine transforms. It should be noted that some problems that can be solved by use of Fourier transforms in one of the variables, such as Example 5.4, can also be solved by the use of cosine or sine transforms in the other variable(s). Cosine or sine transforms can be used when the transformed variable is restricted to a semi-infinite interval, and the choice of the cosine or sine transform is dictated by the boundary conditions, as indicated. However, in applications to partial differential equations, since more than one variable occurs, it can happen that one variable is unrestricted whereas the other variable is bounded on one side. Thus either the Fourier transform or the cosine or sine transform may be used to solve the given problem.

Example 5.5. The Heat Equation in a Semi - Infinite Interval

We consider the heat (or diffusion) equation

$$(5.93) \qquad u_t - c^2 u_{xx} = 0; \qquad 0 < x < \infty, t > 0$$

where c^2 is a constant, over the semi-infinite interval $0 < x < \infty$, with the initial condition

$$(5.94) \qquad u(x,0) = f(x), \qquad 0 < x < \infty$$

and either the boundary condition of the first kind

$$(5.95) \qquad u(0, t) = g(t), \qquad t > 0$$

or the boundary condition of the second kind

$$(5.96) \qquad u_x(0, t) = h(t), \qquad t > 0.$$

For the first boundary value problem [i.e., where $u(0, t) = g(t)$] we apply the Fourier sine transform in x, since the term $u(0, t)$ corresponds to the $f(0)$

term that occurs in (5.90). For the second boundary value problem [i.e., where $u_x(0, t) = h(t)$] we use the Fourier cosine transform, in view of the relation between $u_x(0, t)$ and $f'(0)$ in (5.92).

Applying the sine transform to (5.93), we multiply through in (5.93) by $\sqrt{2/\pi}\,\sin(\lambda x)$ and integrate from 0 to ∞ to obtain

$$(5.97) \qquad \frac{\partial U_s(\lambda, t)}{\partial t} + (\lambda c)^2 U_s(\lambda, t) = \lambda\sqrt{\frac{2}{\pi}}\,g(t)$$

on using (5.90) and (5.95). The function $U_s(\lambda, t)$ is the sine transform of $u(x, t)$; that is,

$$(5.98) \qquad U_s(\lambda, t) = \sqrt{\frac{2}{\pi}} \int_0^\infty \sin(\lambda x) u(x, t)\, dx.$$

From the sine transform of the initial condition (5.94) we have

$$(5.99) \qquad U_s(\lambda, 0) = F_s(\lambda).$$

The solution of the initial value problem (5.97) and (5.99) is

$$(5.100) \quad U_s(\lambda, t) = F_s(\lambda)e^{-\lambda^2 c^2 t} + \lambda\sqrt{\frac{2}{\pi}} \int_0^t e^{-\lambda^2 c^2(t-\tau)}g(\tau)\, d\tau.$$

The inverse transform of $U_s(\lambda, t)$ [see (5.84)] yields the solution $u(x, t)$ as

(5.101)

$$\begin{aligned}
u(x, t) &= \sqrt{\frac{2}{\pi}} \int_0^\infty U_s(\lambda, t)\sin(\lambda x)\, d\lambda \\
&= \sqrt{\frac{2}{\pi}} \int_0^\infty F_s(\lambda)\sin(\lambda x)e^{-\lambda^2 c^2 t}\, d\lambda \\
&\quad + \frac{2}{\pi} \int_0^\infty \int_0^t \lambda e^{-\lambda^2 c^2(t-\tau)}g(\tau)\sin(\lambda x)\, d\tau\, d\lambda \\
&= \frac{2}{\pi} \int_0^\infty \int_0^\infty \left[e^{-\lambda^2 c^2 t}\sin(\lambda s)\sin(\lambda x) \right] f(s)\, d\lambda\, ds \\
&\quad + \frac{2}{\pi} \int_0^t \int_0^\infty \lambda e^{-\lambda^2 c^2(t-\tau)}\sin(\lambda x)g(\tau)\, d\lambda\, d\tau \\
&= \frac{1}{\pi} \int_0^\infty \int_0^\infty e^{-\lambda^2 c^2 t}\{\cos[\lambda(x-s)] - \cos[\lambda(x+s)]\}f(s)\, d\lambda\, ds \\
&\quad + \frac{x}{\sqrt{4\pi c^2}} \int_0^t \exp\left(-\frac{x^2}{4c^2(t-\tau)} \right) \frac{g(\tau)}{(t-\tau)^{3/2}}\, d\tau \\
&= \int_0^\infty [G(x-s, t) - G(x+s, t)]f(s)\, ds \\
&\quad - 2c^2 \int_0^t \frac{\partial G(x, t-\tau)}{\partial x}g(\tau)\, d\tau,
\end{aligned}$$

where we have used the notation $G(x, t)$ for the fundamental solution of the heat equation defined in (5.41). The foregoing result was obtained by interchanging the order of integration in both integrals and evaluating the inner integrals following the approach presented in (5.35)–(5.39). We have also used the identity $2 \sin(\lambda x)\sin(\lambda s) = \cos[\lambda(x - s)] - \cos[\lambda(x + s)]$.

Using Duhamel's principle and (5.101) it is easy to see that the solution of the inhomogeneous heat equation

$$(5.102) \qquad v_t - c^2 v_{xx} = F(x, t); \qquad x > 0, t > 0$$

with data $v(x, 0) = f(x)$ and $v(0, t) = g(t)$ is

$$(5.103)$$

$$v(x, t) = u(x, t) + \int_0^t \int_0^\infty [G(x - s, t - \tau) - G(x + s, t - \tau)] F(s, \tau) \, ds \, d\tau,$$

where $u(x, t)$ is given by (5.101).

In the case of uniform initial temperature, $u(x, 0) = f(x) = u_0 = $ constant, we readily obtain the following result on changing the variable of integration,

$$(5.104) \quad \int_0^\infty [G(x - s, t) - G(x + s, t)] f(s) \, ds$$

$$= \frac{u_0}{\sqrt{\pi}} \left\{ \int_{-x/2c\sqrt{t}}^\infty e^{-r^2} \, dr - \int_{x/2c\sqrt{t}}^\infty e^{-r^2} \, dr \right\}$$

$$= \frac{2u_0}{\sqrt{\pi}} \int_0^{x/2c\sqrt{t}} e^{-r^2} \, dr = u_0 \Phi\left(\frac{x}{2c\sqrt{t}} \right),$$

where $\Phi(z)$ is the (tabulated) *error function* integral defined as

$$(5.105) \qquad \Phi(z) \equiv \text{erf}(z) = \frac{2}{\sqrt{\pi}} \int_0^z e^{-r^2} \, dr,$$

and normalized such that $\Phi(\infty) = 1$ since $\int_0^\infty e^{-r^2} dr = \sqrt{\pi}/2$. Thus the solution of the initial and boundary value problem (5.93)–(5.95) with $u(x, 0) = u_0$ and $u(0, t) = 0$ is given as

$$(5.106) \qquad u(x, t) = u_0 \Phi\left(\frac{x}{2c\sqrt{t}} \right).$$

We note that $u(x, 0) = u_0 \Phi(\infty) = u_0$, $u(0, t) = u_0 \Phi(0) = 0$ and $u(x, t) \to 0$ as $t \to \infty$. Even though $f(x) = u_0$ does not have a sine transform, the solution (5.106) satisfies all conditions of the problem. Again it can be shown that (5.106) is the limit of a sequence of problems with Fourier transformable initial data but we do not demonstrate this.

We observe that if the initial temperature $f(x)$ is not constant but $|f(x)| < M < \infty$ for all $x > 0$ (as we indeed assume), we have

$$(5.107) \quad \left| \int_0^\infty [G(x - s, t) - G(x + s, t)] f(s)\, ds \right|$$

$$\leqslant M \int_0^\infty [G(x - s, t) - G(x + s, t)]\, ds$$

$$= M\Phi\left(\frac{x}{2c\sqrt{t}}\right) \to 0 \text{ as } t \to \infty$$

since $G(x - s, t) \geqslant G(x + s, t)$ for $x, s \geqslant 0$ and $\Phi(z) \to 0$ as $z \to 0$. [We have used the extended mean value theorem for integrals applied to the finite interval $[0, N]$ and then let $N \to \infty$ to obtain the inequality in (5.107).] Thus as $t \to \infty$, the solution $u(x, t)$ [i.e., (5.101)] of the initial and boundary value problem (5.93)–(5.95) tends to the *steady state*

$$(5.108) \quad u(x, t) \approx -2c^2 \int_0^t \frac{\partial G(x, t - \tau)}{\partial x} g(\tau)\, d\tau, \qquad t \gg 1,$$

in the sense that the effect of the initial temperature distribution $u(x, 0) = f(x)$ is dissipated (the term steady state does not signify time independence for this problem). In fact, (5.108) is the solution of the steady-state problem for (5.93) over the interval $0 < x < \infty$, where no initial condition is prescribed and the boundary condition is given for all $t > -\infty$ as

$$(5.109) \quad u(0, t) = \begin{cases} g(t), & t \geqslant 0 \\ 0, & t < 0 \end{cases}.$$

If the initial temperature $u(x, 0) = f(x) = 0$ and the boundary temperature $u(0, t) = g(t) = u_1 = $ constant, it is readily verified (as shown in the exercises) that the solution (5.101) can be expressed as

$$(5.110) \quad u(x, t) = u_1\left[1 - \Phi\left(\frac{x}{2c\sqrt{t}}\right)\right] \equiv u_1 \operatorname{erfc}\left(\frac{x}{2c\sqrt{t}}\right)$$

The bracketed term in (5.110) is often referred to as the *complementary error function* and it tends to unity as $t \to \infty$. Thus from (5.107) we conclude that the temperature $u(x, t) \to 0$ as $t \to \infty$, if the boundary temperature $g(t) = 0$ and the initial temperature $u(x, 0) = f(x)$ is uniformly bounded. If the boundary temperature $g(t) = u_1 = $ constant, and the initial temperature is uniformly bounded, (5.110) shows that the temperature $u(x, t)$ tends to the constant state u_1 as $t \to \infty$.

Finally, we remark that the function

$$(5.111) \quad w(x, t) = -2c^2 \frac{\partial G(x, t)}{\partial x},$$

which occurs in the last integral in the solution (5.101), is readily verified to be a solution of the homogeneous heat equation (5.93) that satisfies the conditions

$$(5.112) \qquad\qquad \lim_{t\to 0+} w(x,t) = 0, \qquad x > 0,$$

and

$$(5.113) \qquad\qquad \lim_{x\to 0+} w(x,t) = 0, \qquad t > 0.$$

However, as we approach the origin $(x,t) = (0,0)$ along the curve $x = 2c\sqrt{t}$ we have

$$(5.114) \quad \lim_{t\to 0+} -2c^2 \left.\frac{\partial G(x,t)}{\partial x}\right|_{\substack{x=2c\sqrt{t}}} = \lim_{\substack{t\to 0+\\ x=2c\sqrt{t}}} \left[\frac{x}{2\sqrt{\pi}\,ct^{3/2}} \exp\left(-\frac{x^2}{4c^2 t} \right) \right]$$

$$= \lim_{t\to 0+} \frac{e^{-1}}{\sqrt{\pi}\,t} \to \infty.$$

Thus $w(x,t)$ is unbounded at the origin. Although $w(x,t)$ is a solution of the heat equation that satisfies homogeneous initial and boundary conditions, in view of (5.112)–(5.113), it is unbounded at the origin. To be certain that solutions of the initial and boundary value problem (5.93)–(5.95) are uniquely determined, we must require the solution $u(x,t)$ to be bounded. Otherwise, we could add the function $w(x,t)$ multiplied by an arbitrary constant to any solution, without altering the initial and boundary data for the problem.

Using the Fourier cosine transform and proceeding as above, we readily obtain the solution of the second boundary value problem (5.93), (5.94), and (5.96) in the form

$$(5.115)$$
$$u(x,t) = \int_0^\infty [G(x-s,t) + G(x+s,t)] f(s)\,ds$$
$$- 2c^2 \int_0^t G(x, t-\tau) h(\tau)\,d\tau$$
$$+ \int_0^t \int_0^\infty [G(x-s,t-\tau) + G(x+s,t-\tau)] F(s,\tau)\,ds\,d\tau,$$

where the last term occurs if the inhomogeneous problem for the heat equations [i.e., (5.102)] is considered.

With constant initial temperature $u(x,0) = f(x) = u_0 =$ constant and $h = F = 0$ in (5.115) we have, after some manipulation of the integrals,

$$(5.116) \qquad u(x,t) = \int_0^\infty [G(x-s,t) + G(x+s,t)] u_0\,ds$$
$$= u_0 \left[\frac{2}{\sqrt{\pi}} \int_0^\infty e^{-r^2}\,dr \right] = u_0.$$

This is to be expected since $\partial u(0, t)/\partial x$ determines the amount of heat passing through the boundary $x = 0$ and $\partial u(0, t)/\partial x = 0$. Also, $F(x, t) = 0$ so that there are no heat sources. Thus no heat escapes through the boundary $x = 0$ and no heat is generated. Consequently, the temperature remains fixed at $u(x, t) = u_0$.

It should also be noted that the function

$$(5.117) \qquad \hat{w}(x, t) = -2c^2 G(x, t)$$

that occurs in the second integral in (5.115) has the property that

$$(5.118) \qquad \lim_{t \to 0+} \hat{w}(x, t) = 0, \qquad x > 0$$

and

$$(5.119) \qquad \lim_{x \to 0+} \frac{\partial \hat{w}(x, t)}{\partial x} = 0, \qquad t > 0,$$

as is readily seen. Also, $\hat{w}(x, t)$ is a solution of the heat equation (5.93). However, $\hat{w}(x, t)$ is unbounded at $(x, t) = (0, 0)$ if we approach the origin along the curve given in (5.114), so we must require the solution of the second boundary value problem to be bounded in order to obtain a unique solution.

The solution of the third boundary value problem for the heat equation in the semi-infinite interval with the boundary condition

$$(5.120) \qquad u_x(0, t) - hu(0, t) = r(t),$$

where $h > 0$ is a constant and $r(t)$ is given, is considered in the exercises.

Some of the above results are rederived in Chapter 7 using Green's function methods and have been expressed in terms of the fundamental solution $G(x, t)$ for the purposes of comparison with later results.

EXAMPLE 5.6.   LAPLACE'S EQUATION IN A STRIP AND A QUARTER - PLANE

We consider Laplace's equation in a semi-infinite strip

$$(5.121) \qquad u_{xx} + u_{yy} = 0, \qquad 0 < x < \infty, 0 < y < \alpha,$$

with the boundary data

$$(5.122) \quad u(0, y) = 0; \qquad u(x, y) \to 0 \text{ as } x \to \infty, \text{ uniformly in } y;$$

$$(5.123) \quad u(x, \alpha) = 0;$$

$$(5.124) \quad u(x, 0) = f(x).$$

Since $u(x, y)$ is given on the boundary $x = 0$, the Fourier sine transform is appropriate for this problem. Let the sine transform of $u(x, y)$ be defined as

$$(5.125) \qquad U_s(\lambda, y) = \sqrt{\frac{2}{\pi}} \int_0^\infty u(x, y)\sin(\lambda x) \, dx,$$

and multiply through in (5.121) by $\sqrt{2/\pi}\,\sin(\lambda x)$ and integrate from 0 to ∞. Using (5.90) and noting that $u(0, y) = 0$, we obtain

$$(5.126) \qquad \frac{\partial^2 U_s(\lambda, y)}{\partial y^2} - \lambda^2 U_s(\lambda, y) = 0.$$

With

$$(5.127) \qquad U_s(\lambda, 0) = F_s(\lambda),$$

where $F_s(\lambda)$ is the sine transform of $f(x)$ and

$$(5.128) \qquad U_s(\lambda, \alpha) = 0,$$

in view of (5.123), we obtain for $U_s(\lambda, y)$,

$$(5.129) \qquad U_s(\lambda y) = F_s(\lambda)\frac{\sinh[\lambda(\alpha - y)]}{\sinh[\lambda\alpha]}.$$

The inverse sine transform yields

$$(5.130) \quad u(x, y) = \sqrt{\frac{2}{\pi}} \int_0^\infty U_s(\lambda, y)\sin(\lambda x) \, d\lambda$$

$$= \frac{2}{\pi}\int_0^\infty \int_0^\infty f(s)\sin(\lambda s)\sin(\lambda x)\frac{\sinh[\lambda(\alpha - y)]}{\sinh[\alpha\lambda]} \, ds \, d\lambda.$$

Note that if $\alpha\lambda \gg 1$, we have

$$(5.131) \qquad \frac{\sinh[\lambda(\alpha - y)]}{\sinh[\alpha\lambda]} \approx e^{-\lambda y}; \qquad \alpha\lambda \gg 1.$$

Thus the λ-integral in (5.130) converges exponentially. The s-integral also converges well if $f(x) \to 0$ as $x \to \infty$ as is, in fact, required by the boundary condition (5.122). We remark that if there are inhomogeneous boundary conditions at $y = \alpha$ and $x = 0$, the resulting problem can be solved in a similar fashion. Also, if we replace the Dirichlet boundary condition in (5.122) with the Neumann condition $u_x(0, y) = g(y)$ at $x = 0$ the problem can be solved with the use of the Fourier cosine transform.

In the limit as $\alpha \to \infty$ in the above problem we obtain the quarter-plane problem for Laplace's equation, that is,

$$(5.132) \qquad u_{xx} + u_{yy} = 0, \qquad x > 0, y > 0$$

with the boundary conditions

$$(5.133) \quad u(0, y) = 0; \qquad u(x, y) \to 0 \text{ as } x \to \infty, \text{ uniformly in } y;$$
$$(5.134) \quad u(x, 0) = f(x); \qquad x > 0.$$

This problem may be solved by a direct application of the sine transform or by going to the limit as $\alpha \to \infty$ in the solution (5.130). Using the latter approach and noting (5.131), we obtain as the solution of (5.132)–(5.134),

$$(5.135) \qquad u(x, y) = \frac{2}{\pi} \int_0^\infty \int_0^\infty f(s) \sin(\lambda s) \sin(\lambda x) e^{-\lambda y} \, ds \, d\lambda.$$

Interchanging the order of integration, we have for the inner integral,

$$(5.136) \quad 2 \int_0^\infty \sin(\lambda s) \sin(\lambda x) e^{-\lambda y} \, d\lambda$$
$$= \int_0^\infty e^{-\lambda y} \{\cos[\lambda(x - s)] - \cos[\lambda(x + s)]\} \, d\lambda$$
$$= \frac{y}{y^2 + (x - s)^2} - \frac{y}{y^2 + (x + s)^2},$$

on adapting the results (5.36)–(5.39). Then the solution $u(x, y)$ takes the form

$$(5.137) \quad u(x, y) = \frac{y}{\pi} \int_0^\infty \left\{ \frac{1}{(x - s)^2 + y^2} - \frac{1}{(x + s)^2 + y^2} \right\} f(s) \, ds.$$

The formula (5.137) may be compared with that given in (5.67) for the solution of the half-plane problem for Laplace's equation. If we let $f(t)$ in (5.67) be defined as an odd function of t [i.e., $f(-t) = -f(t)$], the integral (5.67) can easily be transformed into (5.137). This is equivalent to extending the above quarter-plane problem to a half-plane problem by extending the solution $u(x, y)$ to the full infinite interval as an odd function [i.e., $u(-x, y) = -u(x, y)$] with $u(0, y) = 0$. Then if $f(x)$ is also extended as an odd function, we would expect the solutions (5.67) and (5.137) to agree, as they indeed do. A further connection between half and quarter-plane problems is considered when we discuss Green's functions in Chapter 7.

5.4. HIGHER DIMENSIONAL FOURIER TRANSFORMS

Higher dimensional Fourier transforms may be characterized (as in the preceding sections) in terms of higher dimensional eigenvalue problems for (5.1) with

$\rho = p = 1$ and $q = 0$ over the entire space. Alternatively, they may be obtained by a repeated application of one-dimensional Fourier transforms in each of the variables. The conditions of validity for the transforms are then readily carried over from the one-dimensional case. Instead of discussing the properties of the higher dimensional transforms, we merely define the transforms and the inversion formulas and cite the necessary properties as they are required in the following examples to be considered.

Let $\mathbf{x} = [x_1, \ldots, x_n]$ and $\boldsymbol{\lambda} = [\lambda_1, \ldots, \lambda_n]$ be n-component vectors. Under hypotheses on $f(\mathbf{x})$ analogous to those given in the one-dimensional case we have for the *Fourier transform* $F(\boldsymbol{\lambda})$ of $f(\mathbf{x})$ the representation

$$(5.138) \qquad F(\boldsymbol{\lambda}) = \frac{1}{(\sqrt{2\pi})^n} \int_{-\infty}^{\infty} \cdots \int_{-\infty}^{\infty} e^{i\boldsymbol{\lambda}\cdot\mathbf{x}} f(\mathbf{x})\, d\mathbf{x}$$

and the *inversion formula*

$$(5.139) \qquad f(\mathbf{x}) = \frac{1}{(\sqrt{2\pi})^n} \int_{-\infty}^{\infty} \cdots \int_{-\infty}^{\infty} e^{-i\boldsymbol{\lambda}\cdot\mathbf{x}} F(\boldsymbol{\lambda})\, d\boldsymbol{\lambda}.$$

The integrals (5.138)–(5.139) are both n-dimensional, $d\mathbf{x}$ and $d\boldsymbol{\lambda}$ are n-dimensional volume elements, and $\boldsymbol{\lambda}\cdot\mathbf{x} = \sum_{i=1}^{n}\lambda_i x_i$ is the scalar or dot product of the vectors $\boldsymbol{\lambda}$ and $\mathbf{x}$. The formulas (5.14)–(5.15) relating the transforms of derivatives to the transforms of given functions are valid in the higher dimensional case as well. Here they must be applied to transforms of partial derivatives of $f(\mathbf{x})$ and we require that f and its partial derivatives vanish at infinity.

It is also possible to consider transforms equivalent to the sine and cosine transforms given above if the problem is given over a semi-infinite space. We remark that although Fourier transforms lead to a formally simple approach to the solution of initial and boundary value problems for partial differential equations, the evaluation and simplification of the resulting integral representation of the solution is generally not a simple task. This was already observed in the examples we have considered. We restrict our discussion only to two and three-dimensional Fourier transforms in the examples we consider.

EXAMPLE 5.7. THE CAUCHY PROBLEM FOR THE THREE AND TWO - DIMENSIONAL WAVE EQUATIONS: HADAMARD'S METHOD OF DESCENT

The Cauchy or initial value problem for the wave equation in three dimensions is formulated as

$$(5.140) \quad u_{tt} - c^2[u_{xx} + u_{yy} + u_{zz}] = 0; \qquad t > 0,\ -\infty < x, y, z < \infty,$$

where c^2 is a constant, with the initial conditions

(5.141) $$u(x, y, z, 0) = 0$$

(5.142) $$u_t(x, y, z, 0) = f(x, y, z).$$

We consider only the case where $u(x, y, z, 0) = 0$, since according to *Stokes' rule* (see Example 5.4), the solution of the wave equation (5.140) with data $u(x, y, z, 0) = f(x, y, z)$ and $u_t(x, y, z, 0) = 0$ is given as $\partial u / \partial t$ where $u(x, y, z, t)$ satisfies (5.140)–(5.142). This is easily verified. Then the solution of the general initial value problem for (5.140) is just the sum of these two solutions.

To solve (5.140)–(5.142) by the Fourier transform method, we multiply (5.140) by $1/(\sqrt{2\pi})^3 \exp[i(\lambda_1 x + \lambda_2 y + \lambda_3 z)]$ and integrate with respect to x, y, and z from $-\infty$ to ∞. Let $U(\lambda, t)$ denote the Fourier transform of $u(x, y, z, t)$. Using the analogues of the formulas (5.14)–(5.15), we obtain the transformed equation

(5.143) $$\frac{\partial^2 U(\lambda, t)}{\partial t^2} + c^2(\lambda_1^2 + \lambda_2^2 + \lambda_3^2)U(\lambda, t) = 0,$$

on assuming that u and its first partial derivatives vanish at infinity. The initial conditions for $U(\lambda, t)$ are

(5.144) $$U(\lambda, 0) = 0; \qquad U_t(\lambda, 0) = F(\lambda),$$

where $F(\lambda)$ is the Fourier transform of $f(x, y, z)$.

The solution of (5.143)–(5.144) is

(5.145) $$U(\lambda, t) = F(\lambda)\frac{\sin[|\lambda|ct]}{|\lambda|c},$$

where $|\lambda| = \sqrt{\lambda_1^2 + \lambda_2^2 + \lambda_3^2}$. We determine $u(x, y, z, t)$ by inverting the Fourier transform and have

(5.146) $$u(x, y, z, t) = \frac{1}{(\sqrt{2\pi})^3} \iiint_{-\infty}^{\infty} \frac{F(\lambda)}{|\lambda|c} \sin[|\lambda|ct] e^{-i\lambda \cdot x} \, d\lambda$$

where $\lambda = [\lambda_1, \lambda_2, \lambda_3]$, $x = [x, y, z]$ and $d\lambda = d\lambda_1 d\lambda_2 d\lambda_3$.

By writing $\sin[|\lambda|ct] = (1/2i)(e^{i|\lambda|ct} - e^{-i|\lambda|ct})$, we can express $u(x, y, z, t)$ as

(5.147)

$$u(x, y, z, t) = \frac{1}{2i(\sqrt{2\pi})^3} \iiint_{-\infty}^{\infty} \frac{F(\lambda)}{|\lambda|c}$$

$$\times \left\{ \exp\left[i|\lambda|\left(ct - \frac{\lambda \cdot x}{|\lambda|}\right)\right] - \exp\left[-i|\lambda|\left(ct + \frac{\lambda \cdot x}{|\lambda|}\right)\right] \right\} d\lambda$$

The terms $\exp[\pm i|\lambda|(ct \mp \lambda \cdot x/|\lambda|)]$ in (5.147) represent plane wave solutions of the wave equation (5.140). That is, the solutions remain constant on planes $\lambda \cdot x = $ constant that move parallel to themselves at the speed c. Thus (5.147) represents a superposition of plane wave solutions traveling in all possible directions. (It should also be noted that these plane waves represent normal mode solutions of the wave equation.)

The integral (5.147) can be simplified by expressing $F(\lambda)$ in terms of $f(x, y, z)$. It is then possible to integrate out four of the resulting integrals. Instead of carrying out this process, we merely quote the final result and indicate in the exercises an alternative, simpler method for deriving the final expression.

In terms of spherical coordinates with their origin at the point (x, y, z) defined as

(5.148) $\xi = x + r\sin\varphi\cos\theta;$ $\eta = y + r\sin\varphi\sin\theta;$ $\zeta = z + r\cos\varphi,$

the solution is

(5.149) $u(x, y, z, t) =$

$$\frac{t}{4\pi} \int_0^\pi \int_0^{2\pi} f(x + ct\sin\varphi\cos\theta, \; y + ct\sin\varphi\sin\theta, \; z + ct\cos\varphi)\sin\varphi \, d\theta \, d\varphi,$$

Given a sphere of radius a with center at (x, y, z), we define the average value of the function $f(\xi, \eta, \zeta)$ over the sphere to be given as

(5.150) $$M_a[f] = \frac{1}{4\pi a^2} \iint_S f \, ds,$$

where $4\pi a^2$ is the area of the sphere S and ds is the area element on the sphere. In the spherical coordinates (5.148) the area element is given as $ds = a^2 \sin\varphi \, d\theta \, d\varphi$. Then it is easy to show that the solution (5.149) can be written as

(5.151) $$u(x, y, z, t) = tM_{ct}[f],$$

that is, it is the average value of f over the sphere of radius ct with center at the point (x, y, z) multiplied by t.

Recalling Stokes' rule and using superposition of the two solutions we obtain as the solution of the wave equation (5.140) with initial data

(5.152) $u(x, y, z, 0) = g(x, y, z);$ $u_t(x, y, z, 0) = f(x, y, z),$

the following result;

(5.153) $$u(x, y, z, t) = tM_{ct}[f] + \frac{\partial}{\partial t}\{tM_{ct}[g]\}.$$

The expression (5.153) is often called *Kirchhoff's formula* for the solution of the wave equation.

Using Duhamel's principle (see Section 4.5), we readily obtain the solution of the inhomogeneous wave equation

$$(5.154) \qquad u_{tt} - c^2[u_{xx} + u_{yy} + u_{xx}] = G(x, y, z, t)$$

with initial data (5.152), as

$$(5.155) \quad u(x, y, z, t) = tM_{ct}[f] + \frac{\partial}{\partial t}\{tM_{ct}[g]\}$$

$$+ \int_0^t (t - \tau) M_{c(t-\tau)}[G(x, y, z, \tau)] \, d\tau.$$

The integral term on the right in (5.155) can be expressed as a triple integral over the sphere $r = ct$ with center at (x, y, z) and radius ct, where $r = \sqrt{(\xi - x)^2 + (\eta - y)^2 + (\zeta - z)^2}$. In rectangular coordinates, the integral takes a form referred to as a *retarded potential*:

$$(5.156)$$

$$\int_0^t (t - \tau) M_{c(t-\tau)}[G] \, d\tau = \frac{1}{4\pi c^2} \iiint_{r \leqslant ct} \frac{1}{r} G\left(\xi, \eta, \zeta, t - \frac{r}{c}\right) d\xi \, d\eta \, d\zeta.$$

In the retarded potential the contributions to the solution at the point (x, y, z) and at time t come from the points on a sphere of radius ct with center at (x, y, z). Since the speed of wave propagation is c, ct is the time it takes for the effect of points on the sphere to reach the center of the sphere.

Before discussing the solution (5.153), we obtain from it by means of *Hadamard's method of descent* the solution of the initial value problem for the wave equation in two dimensions. In this method we look for a solution of the wave equation (5.140) independent of z with initial data

$$(5.157) \quad u(x, y, t)\bigg|_{t=0} = g(x, y); \qquad \frac{\partial u(x, y, t)}{\partial t}\bigg|_{t=0} = f(x, y),$$

so that $u(x, y, t)$ satisfies the wave equation in two dimensions

$$(5.158) \qquad u_{tt} - c^2[u_{xx} + u_{yy}] = 0; \qquad t > 0, -\infty < x, y < \infty.$$

We introduce $f(x, y)$ and $g(x, y)$ into the solution (5.153) and because the data are independent of z we may take the center of the sphere $r = ct$ to be at $(x, y, 0)$. Introducing spherical coordinates on the sphere $r = \sqrt{(\xi - x)^2 + (\eta - y)^2 + \zeta^2} = ct$ with the surface element ds, we can write

(5.153) as

(5.159) $$u(x, y, t) = \frac{1}{4\pi c^2 t} \iint_{r=ct} f(\xi, \eta) \, ds$$

$$+ \frac{1}{4\pi c^2} \frac{\partial}{\partial t} \left\{ \frac{1}{t} \iint_{r=ct} g(\xi, \eta) \, ds \right\}.$$

The sphere $r = ct$ has its center at $(x, y, 0)$ and the (ξ, η)-plane cuts it in the circle $\sqrt{(\xi - x)^2 + (\eta - y)^2} = ct$. Since the integrands in (5.159) are independent of ζ we can integrate out the ζ dependence by effectively projecting the upper and lower hemispheres onto the circle in the (ξ, η)-plane. Using the relation

(5.160) $$ds = \sec \gamma \, d\xi \, d\eta,$$

between the surface element ds on the sphere and the area element $d\xi \, d\eta$ in the plane, where

(5.161) $$\sec \gamma = \frac{ct}{\sqrt{(ct)^2 - (\xi - x)^2 - (\eta - y)^2}}$$

and noting that we get the same contribution from each hemisphere, we obtain

(5.162) $$u(x, y, t) = \frac{1}{2\pi c} \iint_{r \leqslant ct} \frac{f(\xi, \eta)}{\sqrt{(ct)^2 - r^2}} \, d\xi \, d\eta$$

$$+ \frac{1}{2\pi c} \frac{\partial}{\partial t} \iint_{r \leqslant ct} \frac{g(\xi, \eta)}{\sqrt{(ct)^2 - r^2}} \, d\xi \, d\eta,$$

where $r = \sqrt{(\xi - x)^2 + (\eta - y)^2}$. The integration in (5.162) is carried out over the interior of the circle of radius ct centered at (x, y). For the nonhomogeneous problem

(5.163) $$u_{tt} - c^2[u_{xx} + u_{yy}] = G(x, y, t), \qquad t > 0, -\infty < x, y < \infty,$$

with homogeneous initial data, Duhamel's principle yields the solution

(5.164) $$u(x, y, t) = \frac{1}{2\pi c} \int_0^t \iint_{r \leqslant c(t-\tau)} \left\{ \frac{G(\xi, \eta, \tau)}{\sqrt{c^2(t-\tau)^2 - r^2}} \right\} d\xi \, d\eta \, d\tau,$$

where $r = \sqrt{(\xi - x)^2 + (\eta - y)^2}$.

This method is called the method of descent since we descend from three to two space dimensions to construct a solution for the lower dimensional problem in terms of the solution of the higher dimensional one. It is also possible to descend from two to one dimension and construct d'Alembert's solution from the solution (5.162).

We now discuss the solutions (5.153) and (5.162) of the three and two-dimensional wave equations, respectively. The basic distinction between the two solutions is that (5.153) represents integration over the surface of a sphere whereas (5.162) represents integration over the interior of a circle. This points up a sharp difference between wave propagation in three and two dimensions.

We suppose that the initial data f and g are concentrated in a neighborhood of a point P_0 in the two or the three-dimensional case. That is, we assume that they vanish outside an arbitrarily small neighborhood of P_0. In the three-dimensional case, (5.153) states that the solution $u(x, y, z, t)$ of the wave equation, at the time t and at the point (x, y, z), depends on the data f and g only at points intersected by the sphere of radius ct with center at (x, y, z). If this sphere does not intersect the given neighborhood of P_0, the solution $u(x, y, z, t) = 0$. As t increases, the sphere must eventually intersect a neighborhood of P_0 (no matter where (x, y, z) is located). The first time a disturbance is felt at (x, y, z) represents the time t it takes the wave front, originating at the point of the neighborhood of P_0 nearest to (x, y, z), to reach the point (x, y, z). Since the distance from the nearest point in the neighborhood of P_0 to (x, y, z) is ct, the speed of the propagation of the wave front is c. Because the data f and g are concentrated near P_0, as t increases the sphere will ultimately no longer intersect the given neighborhood of P_0 and the solution $u(x, y, z, t)$ will again be equal to zero.

The preceding discussion shows that sharp signals or disturbances can occur in three-dimensional wave motion. Not only does it take a finite time for disturbances to travel from one point to another, but after the disturbance has passed, the solution returns to zero and its effect is no longer felt. This phenomenon is known as *Huygens' principle*. Huygens used this principle to construct solutions of wave propagation problems. By considering a disturbance surface (or a sharp wave front) at the time t_0, one constructs a sphere of radius ct around each point on the surface. Then the envelope of these spheres gives the location of the disturbance surface at the time $t_0 + t$. Although this construction only gives the front of the disturbance in general, in the case of concentrated sources (i.e., initial data concentrated at a point P_0) the construction gives the full disturbance.

For the two-dimensional case, if the initial data f and g are concentrated in a neighborhood of the point P_0, the solution (5.162) shows that $u(x, y, t)$ will be zero until the circle with center at (x, y) and radius ct intersects the neighborhood of P_0. Once the circle begins to intersect the neighborhood of P_0, the solution at the point (x, y) will never vanish as t increases. This results because the integration in (5.162) is carried out over the interior of the circle, not just its boundary. Thus although sharp wave fronts do exist in two-dimensional wave propagation, since it takes a finite time for the effect of the data near P_0 to reach the point (x, y), the effects of the disturbance linger and do not disappear sharply as in the three-dimensional case. Instead they diffuse slowly to zero at $t \to \infty$. As a result, Huygens' principle in the above form is not valid in the two-dimensional case. However, Huygens' construction as given can be used to give the location of the wave front.

The solutions given for the two and three-dimensional wave equations can be used to determine domains of dependence and influence as done earlier in the one-dimensional case, but this is not carried out here. Also, as seen in the exercises, Hadamard's method can be applied to other equations.

EXAMPLE 5.8.   THE MODIFIED HELMHOLTZ EQUATION

We consider the (inhomogeneous) elliptic equation

$$(5.165) \qquad u_{xx} + u_{yy} - k^2 u = f(x, y), \qquad -\infty < x, y < \infty,$$

where the constant $k > 0$, $f(x, y)$ is specified, and $u(x, y)$ satisfies the additional condition.

$$(5.166) \qquad u(x, y) \to 0 \text{ as } |x|, |y| \to \infty.$$

We assume that $f(x, y)$ has a Fourier transform. The equation (5.165) is sometimes referred to as the *modified Helmholtz equation*, in contrast to the case where $-k^2$ is replaced by $+k^2$ which is called the *Helmholtz equation*. The equation (5.165) can arise as a stationary version of two-dimensional diffusion equations.

To solve (5.165)–(5.166) we apply the two-dimensional Fourier transform to (5.165) and easily obtain

$$(5.167) \qquad U(\lambda) = -\frac{F(\lambda)}{\lambda_1^2 + \lambda_2^2 + k^2}$$

where $\lambda = [\lambda_1, \lambda_2]$, and $U(\lambda)$ and $F(\lambda)$ are the Fourier transforms of $u(x, y)$ and $f(x, y)$, respectively. [We have assumed in obtaining (5.167) that not only u but also its first partial derivatives vanish at infinity.]

Inverting the transform gives the solution

$$(5.168) \quad u(x, y) = -\frac{1}{2\pi} \int_{-\infty}^{\infty} \int_{-\infty}^{\infty} e^{-i(\lambda_1 x + \lambda_2 y)} \frac{F(\lambda)}{\lambda_1^2 + \lambda_2^2 + k^2} d\lambda_1 \, d\lambda_2.$$

Since $F(\lambda)$ has the form

$$(5.169) \qquad F(\lambda) = \frac{1}{2\pi} \int_{-\infty}^{\infty} \int_{-\infty}^{\infty} e^{i(\lambda_1 \xi + \lambda_2 \eta)} f(\xi, \eta) \, d\xi \, d\eta,$$

we obtain, on inserting (5.169) into (5.168),

$$(5.170) \quad u(x, y) = -\frac{1}{4\pi^2} \iiiint\limits_{-\infty}^{\infty} \exp\{-i[\lambda_1(x - \xi) + \lambda_2(y - \eta)]\}$$

$$\times \frac{f(\xi, \eta)}{\lambda_1^2 + \lambda_2^2 + k^2} d\lambda_1 d\lambda_2 d\xi d\eta$$

on interchanging the order of integration. Introducing polar coordinates in the inner integral in (5.170), that is,

$$(5.171) \qquad \lambda_1 = \rho \cos \varphi; \qquad \lambda_2 = \rho \sin \varphi$$

and also expressing $x - \xi$ and $y - \eta$ in polar form

$$(5.172) \qquad x - \xi = r \cos \theta; \qquad y - \eta = r \sin \theta,$$

we readily obtain

$$(5.173) \qquad \lambda_1(x - \xi) + \lambda_2(y - \eta) = \rho r \cos(\varphi - \theta).$$

Then transforming to polar coordinates gives

$$(5.174) \quad \int_{-\infty}^{\infty} \int_{-\infty}^{\infty} \frac{\exp\{-i[\lambda_1(x - \xi) + \lambda_2(y - \eta)]\}}{\lambda_1^2 + \lambda_2^2 + k^2} d\lambda_1 d\lambda_2$$

$$= \int_0^{\infty} \int_0^{2\pi} \frac{e^{-i\rho r \cos(\varphi - \theta)}}{\rho^2 + k^2} \rho \, d\varphi \, d\rho$$

Using a well-known integral representation of the Bessel function of order zero $J_0(z)$,

$$(5.175) \qquad J_0(z) = \frac{1}{2\pi} \int_0^{2\pi} e^{iz \cos(\varphi - \theta)} d\varphi; \qquad \theta = \text{constant},$$

we obtain, since $J_0(z) = J_0(-z)$,

$$(5.176) \qquad \int_0^{\infty} \int_0^{2\pi} \frac{e^{-i\rho r \cos(\varphi - \theta)}}{\rho^2 + k^2} \rho \, d\varphi \, d\rho = 2\pi \int_0^{\infty} \frac{\rho J_0(\rho r) \, d\rho}{\rho^2 + k^2}.$$

A further result from the theory of Bessel functions yields

$$(5.177) \qquad 2\pi \int_0^{\infty} \frac{\rho J_0(\rho r)}{\rho^2 + k^2} d\rho = -2\pi K_0(kr),$$

where K_0 is the modified Bessel function of the second kind, some of whose properties are given in Chapter 6. When (5.177) is introduced into (5.170) we obtain

$$(5.178) \qquad u(x, y) = \frac{1}{2\pi} \iint_{-\infty}^{\infty} K_0(kr) f(\xi, \eta) \, d\xi \, d\eta,$$

where $r = \sqrt{(x - \xi)^2 + (y - \eta)^2}$.

With $f(x, y) = \delta(x - x_0)\delta(y - y_0)$, the two-dimensional Dirac delta function, (5.178) reduces to

$$(5.179) \quad u(x, y) = \frac{1}{2\pi} K_0(kr); \qquad r = \sqrt{(x - x_0)^2 + (y - y_0)^2}.$$

This is known as the *free-space Green's function* for the modified Helmholtz equation. As shown in the exercises, (5.165) becomes a modified Bessel equation when expressed in a polar coordinate form independent of the angular variable. We return to this problem in later chapters.

5.5. HANKEL TRANSFORMS

We have seen in Chapter 4 and the exercises for that chapter that certain problems lead naturally to the representation of the solutions in terms of Bessel functions. In bounded regions we are led to the Bessel function expansions considered in Chapter 4. For unbounded regions, in carrying out separation of variables in two or three space dimensions in cases where the Laplacian operator is expressed in polar or cylindrical coordinates, we are often led to consider the following eigenvalue problem for the equation involving the radial variable:

$$(5.180) \qquad -(rv'(r))' + \frac{n^2}{r} v(r) = \lambda^2 r v(r); \qquad 0 < r < \infty,$$

where λ^2 is the eigenvalue parameter and the equation is written in the self-adjoint form (4.56). (This is Bessel's equation of order n in the variable λr.) Here n can be any positive integer or zero. We require that the eigenfunctions $v(r)$ are bounded for all r. This implies that the eigenfunctions are

$$(5.181) \qquad v(r) = J_n(\lambda r)$$

where $J_n(z)$ is the Bessel function of order n and the eigenvalues λ can be any nonnegative real number (i.e., $0 \leqslant \lambda < \infty$) so that the spectrum is continuous. Since $J_n(-z) = (-1)^n J_n(z)$, we need not consider negative values of λ.

A function $f(r)$ defined for $r \geq 0$ and satisfying the conditions given in Section 5.3 has the representation in terms of the eigenfunctions (5.181),

(5.182)
$$f(r) = \int_0^\infty \lambda J_n(\lambda r) F(\lambda) \, d\lambda; \qquad n \geq 0$$

where the *Hankel transform* $F(\lambda)$ is given by

(5.183)
$$F(\lambda) = \int_0^\infty r J_n(\lambda r) f(r) \, dr; \qquad n \geq 0.$$

[We note that an alternative form of these formulas given in some texts, with the terms λ and r in (5.189)–(5.183) both replaced by $\sqrt{\lambda r}$, thereby giving the formulas a more symmetric form.] The formulas (5.182)–(5.183) can be obtained from the two-dimensional Fourier transform formulas, and this derivation is considered in the exercises.

In applying the Hankel transform to a given equation we are often led to consider an expression of the form

(5.184)
$$\int_0^\infty r J_n(\lambda r) \left[f_{rr} + \frac{1}{r} f_r - \frac{n^2}{r^2} f \right] dr = -\lambda^2 F(\lambda)$$
$$+ \left[r f_r J_n(\lambda r) - \lambda r f J_n'(\lambda r) \right]_{r=0}^{r=\infty}.$$

This result is obtained on integrating by parts and is closely related to (4.114). In general, the bracketed term on the right is required to vanish. This occurs if f and f_r are bounded at the origin and $\sqrt{r} f$ and $\sqrt{r} f_r$ vanish at infinity. Note that for $n \geq 0$ the Bessel function $J_n(\lambda r)$ is bounded at zero and vanishes like $1/\sqrt{r}$ at infinity. In certain problems, we require that f be singular at $r = 0$ but in such a fashion that the $\lim_{r \to 0} r f_r J_n$ and $\lim_{r \to 0} r f J_n'$ be finite.

Although comprehensive tables of Hankel transforms are available, we list several results for use in the examples. The equation

(5.185)
$$\int_0^\infty e^{-\lambda z} J_0(\lambda r) \, d\lambda = \frac{1}{\sqrt{z^2 + r^2}}; \qquad z, r > 0$$

shows that the Hankel transform ($n = 0$) of $f(r, z) = 1/\sqrt{r^2 + z^2}$ is $F(\lambda, z) = (1/\lambda) e^{-\lambda z}$. Using the transform formula (5.183) we obtain

(5.186)
$$\int_0^\infty \frac{r J_0(\lambda r)}{\sqrt{r^2 + z^2}} \, dr = \frac{1}{\lambda} e^{-\lambda z}.$$

The result (5.185) can be verified by using the integral representation of the

Bessel function $J_0(\lambda r)$ and interchanging orders of integration. A second result that can be obtained by using the series expansion of the Bessel function $J_n(\lambda r)$ and integrating term by term is

$$(5.187) \qquad \int_0^\infty \lambda J_n(\lambda r) e^{-t\lambda^2} \lambda^n \, d\lambda = \frac{1}{2t} \left(\frac{r}{2t}\right)^n \exp\left(-\frac{r^2}{4t}\right).$$

We cite additional results as they are required and now consider several examples in which Hankel transforms are used to solve problems for equations of elliptic, hyperbolic, and parabolic types.

EXAMPLE 5.9. LAPLACE'S EQUATION: AN AXIAL SOURCE

We consider Laplace's equation in three dimensions with a source concentrated on the z-axis. Introducing cylindrical coordinates (r, θ, z) and noting the axial symmetry we ask for a solution of

$$(5.188) \qquad \nabla^2 u = u_{rr} + \frac{1}{r} u_r + u_{zz} = 0; \qquad r > 0, -\infty < z < \infty$$

with the conditions on the z-axis given as

(5.189)

$$\lim_{r \to 0} r^2 u(r, z) = 0; \qquad \lim_{r \to 0} r \frac{\partial u}{\partial r}(r, z) = -f(z); \qquad -\infty < z < \infty,$$

so that $f(z)$ is a measure of the strength of the source. [Note that in (5.188) the θ-dependence of the Laplacian is dropped because of the assumed axial symmetry. [A discussion of concentrated source problems is given in Section 6.7.]

Applying the zero-order Hankel transform to (5.188) [in view of (5.184)] we multiply across by $rJ_0(\lambda r)$ and integrate from 0 to ∞. We use $J_0(\lambda r)$ in this case because the term $(n^2/r^2)u$ does not occur in equation (5.188). Defining the Hankel transform $U(\lambda, z)$ of $u(r, z)$ as

$$(5.190) \qquad\qquad U(\lambda, z) = \int_0^\infty rJ_0(\lambda r) u(r, z) \, dr$$

and using (5.184) we obtain

$$(5.191) \quad 0 = \int_0^\infty rJ_0(\lambda r) \left[u_{rr} + \frac{1}{r} u_r + u_{zz}\right] dr = -\lambda^2 U(\lambda, z) + \frac{\partial^2 U(\lambda, z)}{\partial z^2}$$
$$+ \left[ru_r J_0(\lambda r) - \lambda ru J_0'(\lambda r)\right]_{r=0}^{r=\infty}.$$

Now $J_0(0) = 1$, $J_0' = -J_1$ and $J_1(\lambda r) \approx \lambda r/2$ as $r \to 0$ as is well-known. Assuming that the contributions from the limit at infinity vanish in (5.191) [$u(r, z)$ is assumed to be of that form] and noting that $ruJ_0'(\lambda r) \approx -(\lambda r^2/2)u$ as $r \to 0$, we obtain on using (5.189),

$$(5.192) \qquad U_{zz} - \lambda^2 U = -f(z), \qquad -\infty < z < \infty.$$

Requiring that $U(\lambda, z) \to 0$ as $|z| \to \infty$, we have precisely the problem considered in Example 5.1 at the beginning of this chapter. The solution is given as

$$(5.193) \qquad U(\lambda, z) = \frac{1}{2\lambda} \int_{-\infty}^{\infty} e^{-\lambda|z-s|} f(s) \, ds,$$

on comparing with (5.23). Inverting the transform $U(\lambda, z)$ gives

$$(5.194)$$

$$u(r, z) = \int_0^{\infty} \lambda J_0(\lambda r) U(\lambda, z) \, d\lambda = \frac{1}{2} \int_{-\infty}^{\infty} \int_0^{\infty} e^{-\lambda|z-s|} J_0(\lambda r) f(s) \, d\lambda \, ds$$

$$= \frac{1}{2} \int_{-\infty}^{\infty} \frac{f(s) \, ds}{\sqrt{r^2 + (z-s)^2}}$$

on interchanging the order of integration and using (5.185).

EXAMPLE 5.10.   THE WAVE EQUATION IN TWO DIMENSIONS

We consider the initial value problem for the two-dimensional wave equation over the entire space with radially symmetric initial data. We introduce polar coordinates (r, θ) and look for a solution $u(r, t)$ independent of θ. Thus we consider

$$(5.195) \qquad u_{tt} - c^2 \left(u_{rr} + \frac{1}{r} u_r \right) = 0, \qquad r > 0, t > 0$$

with initial data

$$(5.196) \qquad u(r, 0) = f(r); \qquad u_t(r, 0) = g(r).$$

Applying the (zero-order) Hankel transform to (5.195)–(5.196) gives

$$(5.197) \qquad U_{tt} + c^2 \lambda^2 U = 0, \qquad t > 0$$

$$(5.198) \qquad U(\lambda, 0) = F(\lambda); \qquad U_t(\lambda, 0) = G(\lambda)$$

where U, F, and G are the (zero-order) Hankel transforms of u, f, and g, respectively, and (5.184) with $n = 0$ has been used. [It is assumed that u is such that the boundary terms in (5.184) vanish.] Solving (5.197)–(5.198) gives

(5.199) $$U(\lambda, t) = F(\lambda)\cos(\lambda ct) + \frac{G(\lambda)}{\lambda c}\sin(\lambda ct).$$

Inverting the transform yields

(5.200)
$$u(r, t) = \int_0^\infty \lambda F(\lambda)\cos(\lambda ct)J_0(\lambda r)\, d\lambda + \frac{1}{c}\int_0^\infty G(\lambda)\sin(\lambda ct)J_0(\lambda r)\, d\lambda.$$

For specific choices of $f(r)$ and $g(r)$ we can occasionally evaluate the integrals in closed form. Thus if

(5.201) $\quad f(r) = \dfrac{A}{\sqrt{1 + (r^2/\alpha^2)}}\,; \qquad g(r) = 0, \qquad A, \alpha = \text{constant}$

we have on using (5.186)

(5.202) $$F(\lambda) = \frac{A\alpha}{\lambda}e^{-\lambda\alpha}; \qquad \alpha > 0;$$

(5.203) $$G(\lambda) = 0.$$

This gives

(5.204) $$u(r, t) = A\alpha \int_0^\infty e^{-\lambda\alpha}\cos(\lambda ct)J_0(\lambda r)\, d\lambda.$$

The integral may be evaluated by expressing it as

(5.205) $\quad \displaystyle\int_0^\infty e^{-\lambda\alpha}\cos(\lambda ct)J_0(\lambda r)\, d\lambda = \text{Re}\left[\int_0^\infty e^{-\lambda(\alpha + ict)}J_0(\lambda r)\, d\lambda\right]$

where Re[ ] = the real part of the bracketed quantity. Using (5.185) we have

(5.206) $$u(r, t) = \text{Re}\left[\frac{A\alpha}{\sqrt{(\alpha + ict)^2 + r^2}}\right]$$

By requiring the complex square root in (5.206) to be positive when $t = 0$ (i.e. when it is real valued), the initial conditions (5.201) are satisfied by this solution. [On equating the bracketed term in (5.206) to the complex number $a + ib$, squaring both terms and solving the resulting algebraic equations, the real value of (5.206) may be determined, but we do not carry this out].

EXAMPLE 5.11. THE DIFFUSION EQUATION WITH AXIALLY SYMMETRIC DATA

We consider the three-dimensional diffusion equation in cylindrical coordinates where the concentration $u(x, y, z, t)$ has an initial distribution that depends only on $r = \sqrt{x^2 + y^2}$. We introduce cylindrical coordinates (r, θ, z) and look for a solution $u(r, t)$ of the equation

$$(5.207) \qquad u_t = D\left(u_{rr} + \frac{1}{r}u_r\right), \qquad r > 0, t > 0$$

where $D > 0$ is the diffusion constant, with the initial condition

$$(5.208) \qquad u(r, 0) = f(r), \qquad r \geqslant 0.$$

Applying the (zero-order) Hankel transform to (5.207) and (5.208) we obtain on using the notation of Example 5.10,

$$(5.209) \qquad \frac{\partial U}{\partial t} + D\lambda^2 U = 0, \qquad t > 0$$

$$(5.210) \qquad U(\lambda, 0) = F(\lambda).$$

Thus

$$(5.211) \qquad U(\lambda, t) = F(\lambda)e^{-D\lambda^2 t}$$

and

$$(5.212) \quad u(r, t) = \int_0^\infty \lambda J_0(\lambda r) U(\lambda, t)\, d\lambda = \int_0^\infty \lambda e^{-D\lambda^2 t} J_0(\lambda r) F(\lambda)\, d\lambda$$

$$= \int_0^\infty \int_0^\infty \lambda s J_0(\lambda r) J_0(\lambda s) e^{-D\lambda^2 t} f(s)\, d\lambda\, ds$$

where we have interchanged the order of integration in the last integral. We cite a known result to evaluate the inner integral. That is,

$$(5.213) \quad \int_0^\infty \lambda e^{-D\lambda^2 t} J_0(\lambda r) J_0(\lambda s)\, d\lambda = \frac{1}{2Dt}\exp\left[-\frac{(r^2 + s^2)}{4Dt}\right]I_0\left(\frac{rs}{2Dt}\right)$$

where $I_0(z)$ is the modified zero order Bessel function discussed in Chapter 6. We note the property of $I_0(z)$ that $I_0(0) = 1$, so that for $s = 0$ in (5.213) [since $J_0(0) = 1$] we find that (5.213) reduces to (5.187) with $n = 0$ and $D = 1$. Inserting (5.213) into (5.212) gives

$$(5.214) \quad u(r, t) = \frac{1}{2Dt}\int_0^\infty \exp\left[-\frac{(r^2 + s^2)}{4Dt}\right]I_0\left(\frac{rs}{2Dt}\right)f(s)s\, ds.$$

It is of interest to show how this solution reduces to the fundamental solution for the two-dimensional heat or diffusion equation if we allow $f(r)$ to represent a concentrated source at $r = 0$. We require that with $f(r) \geqslant 0$

$$(5.215) \qquad \lim_{\epsilon \to 0} 2\pi \int_0^\epsilon f(r) r \, dr = 1,$$

where we assume the source is concentrated in a circle of radius ϵ and let the radius tend to zero but keep the source strength fixed at unity. It may be assumed that $f(r)$ vanishes for $r > \epsilon$. Using the generalized mean value theorem for integrals we have [since $I_0(z) > 0$ for all z]

$$(5.216) \qquad u(r, t) = \frac{1}{2Dt} \exp\left[-\frac{(r^2 + \hat{s}^2)}{4Dt} \right] I_0\left(\frac{r\hat{s}}{2Dt} \right) \int_0^\epsilon f(s) s \, ds$$

where $\hat{s} = \hat{s}(\epsilon) \to 0$ as $\epsilon \to 0$. Then as $\epsilon \to 0$ we obtain

$$(5.217) \qquad u(r, t) = \frac{1}{4\pi Dt} \exp\left[-\frac{r^2}{4Dt} \right] = \frac{1}{4\pi Dt} \exp\left[-\frac{x^2 + y^2}{4Dt} \right]$$

on using the result (5.215) and $I_0(0) = 1$. This is the fundamental solution of the diffusion equation in the two-dimensional case. It should be compared with the one-dimensional fundamental solution $G(x, t)$ defined in (5.41).

5.6. LAPLACE TRANSFORMS

The Laplace transform together with its inversion formula may be derived from the Fourier transform (5.10)–(5.11) by admitting complex transform parameters and placing certain restrictions on the transformed functions. As a result, it has many properties in common with the Fourier transform. However, we do not demonstrate how the theory of Laplace transforms follows from that for Fourier transforms, but merely state the basic results and properties of Laplace transforms.

Given a function $f(x)$ defined for $x \geqslant 0$ and integrable over any finite (positive) interval (say, it is piecewise continuous), and such that $e^{-kx}|f(x)|$ is integrable over the interval $0 < x < \infty$ for some real k, we define $F(\lambda)$ as

$$(5.218) \qquad F(\lambda) = \int_0^\infty e^{-\lambda x} f(x) \, dx,$$

where we require that $Re(\lambda) > k$ so that (5.218) converges. $F(\lambda)$ is called the (one-sided) *Laplace transform* of $f(x)$. The parameter λ is not required to be real valued. [We note that one occasionally considers bilateral Laplace transforms where the integral in (5.218) extends from $-\infty$ to $+\infty$. We consider only one-sided Laplace transforms as these play the most important role in the solution of initial and boundary value problems for partial differential equations.]

The complex *inversion formula* for the Laplace transform is

$$(5.219) \qquad f(x) = \frac{1}{2\pi i} \int_L e^{\lambda x} F(\lambda)\, d\lambda,$$

where $x > 0$ and L is a line in the complex λ-plane, generally taken to be parallel to the imaginary axis such that $Re(\lambda) > k$, having the direction indicated in Figure 5.1. (The λ-plane is given as $\lambda = \lambda_1 + i\lambda_2$, with λ_1 and λ_2 as the real and imaginary axes, respectively.) Given the transform $F(\lambda)$, the function $f(x)$ can often be determined from (5.219) by using the residue theory of complex variables. Otherwise, approximation methods are available to deal with (5.219). As we do not presuppose a knowledge of complex integration theory, the formula (5.219) does not play a (significant) role in our discussion. However, Laplace transforms are extensively tabulated, and we can often invert Laplace transforms by reducing $F(\lambda)$ to a form for which the inverse transform is known. In this respect the linearity of the Laplace transform, that is,

$$(5.220) \qquad \int_0^\infty e^{-\lambda x}\left[c_1 f_1(x) + c_2 f_2(x)\right] dx = c_1 \int_0^\infty e^{-\lambda x} f_1(x)\, dx$$
$$+ c_2 \int_0^\infty e^{-\lambda x} f_2(x)\, dx,$$

and the convolution theorem (see Exercise 5.6.3) play an important role. Thus if $F(\lambda)$ can be expressed as a finite (or even an infinite) sum each of whose terms has a known inverse transform, we can invert $F(\lambda)$ given that the inverse

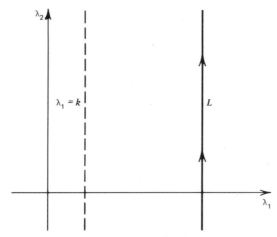

Figure 5.1.   The line L.

of the sum is the sum of the inverses. It is not always a straightforward matter to carry out this procedure, as we shall see.

In the context of our approach of characterizing transforms in terms of eigenvalue problems, it may be noted that the Laplace transform can be related to the eigenvalue problem associated with (5.2) for the hyperbolic and parabolic cases. We do not pursue this point, however.

An important result for Laplace transforms relates the transform of derivatives to the transform of the given function. If $f^{(n)}(x) = d^n f(x)/dx^n$ we have

(5.221)
$$\int_0^\infty e^{-\lambda x} f^{(n)}(x)\, dx = \lambda^n \int_0^\infty e^{-\lambda x} f(x)\, dx - \lambda^{n-1} f'(0) - \cdots - f^{(n-1)}(0),$$

on using integration by parts. This result shows why Laplace transforms are especially useful in dealing with initial and initial boundary value problems for hyperbolic and parabolic equations. Additional general properties of Laplace transforms are found in tables of transforms. We now consider an example in some detail.

EXAMPLE 5.12.   THE HEAT EQUATION IN A FINITE INTERVAL

We again consider the heat equation in a finite interval $0 < x < l$,

(5.222) $u_t - c^2 u_{xx} = 0,$ $0 < x < l, t > 0$

with the homogeneous boundary conditions

(5.223) $u(0, t) = 0;$ $u(l, t) = 0;$ $t > 0$

and the initial condition

(5.224) $u(x, 0) = f(x).$

This problem was solved by separation of variables in Example 4.10. However, we now obtain the solution in a different form and compare both results.

The problem (5.222)–(5.224) is solved by applying the Laplace transform in the t-variable. We multiply in (5.222) by $e^{-\lambda t}$ and integrate with respect to t from 0 to ∞. Defining the transform of $u(x, t)$ as

(5.225) $U(\lambda, x) = \int_0^\infty e^{-\lambda t} u(x, t)\, dt$

and using (5.221) we have

(5.226) $\lambda U(\lambda, x) - f(x) - c^2 \dfrac{\partial^2 U(\lambda, x)}{\partial x^2} = 0;$ $0 < x < l.$

From the transform of the boundary conditions (5.223) we obtain

(5.227) $U(\lambda, 0) = U(\lambda, l) = 0.$

The solution of the boundary value problem (5.226)–(5.227) is found to be

(5.228)

$$U(\lambda, x) = \frac{1}{c\sqrt{\lambda}\,\sinh(\sqrt{\lambda}\,l/c)} \left\{ \int_0^x \sinh\left(\frac{\sqrt{\lambda}\,s}{c}\right) \sinh\left[\frac{\sqrt{\lambda}}{c}(l - x)\right] f(s)\, ds \right.$$

$$\left. + \int_x^l \sinh\left(\frac{\sqrt{\lambda}\,x}{c}\right) \sinh\left[\frac{\sqrt{\lambda}}{c}(l - s)\right] f(s)\, ds \right\}.$$

This result may be obtained by using the Green's function or variation of parameters method for ordinary differential equation.

The inversion of the transform $U(\lambda, x)$ is not straightforward, even if the inversion formula (5.219) is used, unless the transform is simplified first. In view of the discussion following (5.219) we may assume that λ (or $|\lambda|$) is not small. Consequently, we have the following result:

(5.229)

$$\frac{1}{\sinh(\sqrt{\lambda}\,l/c)} = \frac{2}{\exp(\sqrt{\lambda}\,l/c) - \exp(-\sqrt{\lambda}\,l/c)} = \frac{2\exp(-\sqrt{\lambda}\,l/c)}{1 - \exp(-2\sqrt{\lambda}\,l/c)}$$

$$= 2\exp\left[\frac{-\sqrt{\lambda}\,l}{c}\right] \sum_{k=0}^{\infty} \exp\left[\frac{-\sqrt{\lambda}\,2kl}{c}\right]$$

on using the series $1/(1-x) = \sum_{k=0}^{\infty} x^k$ which is valid for $|x|<1$. Then it may be assumed that λ is large enough so that $\exp[-2\sqrt{\lambda}\,l/c]<<1$. Further, the hyperbolic functions in the integrals (5.228) can also be expressed in terms of exponentials. We combine the exponentials and assume that in the inversion process, summation and integration, as well as orders of integration, can be reversed (this can be shown to be valid for this problem). Further, we find from a table of Laplace transforms that

(5.230) $$\int_0^{\infty} e^{-\lambda t} \left[\frac{1}{\sqrt{\pi t}} \exp\left(\frac{-\alpha^2}{4t}\right) \right] dt = \frac{1}{\sqrt{\lambda}} e^{-\alpha\sqrt{\lambda}}; \qquad \alpha > 0.$$

Using (5.230), the solution $u(x, t)$ can then be written as

(5.231)

$$u(x, t) = \int_0^l \sum_{k=-\infty}^{\infty} \{G(x - s - 2kl, t) - G(x + s - 2kl, t)\} f(s)\, ds,$$

where $G(x - s, t)$ is the fundamental solution of the heat equation defined in (5.41).

The solution (5.231) has a form quite different from that given in terms of Fourier sine series in Example 4.10. By the uniqueness theorem for the initial and boundary value problem for the heat equation demonstrated in Chapter 8, both solutions must be identical. The equivalence of the both sums is demonstrated in the exercises at the end of Chapter 7 based on the *Poisson summation formula*. We remark that both solution forms are useful in evaluating the solution $u(x, t)$ for different ranges of values of t.

Because of the rapid exponential decay of the terms in the Fourier series (4.153), the series converges very rapidly for large values of t. However, the series of fundamental solutions converges most rapidly for small values of t since all the terms in the series except $G(x - s, t)$, which has delta function behavior in the interval $0 < x < l$, vanish at $t = 0$. Thus for the purposes of computation each of the series is useful.

As we have seen in the foregoing example, it is not always a simple matter to invert a Laplace transform resulting from the solution of a partial differential equation. Consequently, it is of interest to examine methods that yield approximate expressions for inverse Laplace transforms. The *Abelian* and *Tauberian asymptotic theories* for Laplace transforms that relate the values of the transform $F(\lambda)$ as $\lambda \to \infty$ and $\lambda \to 0$ to the values of the function $f(t)$ as $t \to 0$ and $t \to \infty$, respectively, are useful in that regard. [We replace x by t in (5.218)–(5.219) in the present discussion.] Given the presence of the exponential term in the formula for $F(\lambda)$ and assuming $f(t)$ is reasonably behaved, we expect that the main contributions to the integral for $F(\lambda)$ come from small values of λt. Thus if λ is large, t must be small and vice versa.

A useful Abelian result states that if as $t \to 0$ we have

$$(5.232) \qquad f(t) \approx \sum_{n=0}^{\infty} \alpha_n t^n, \qquad t \to 0,$$

then for the transform $F(\lambda)$ we have, as $\lambda \to \infty$,

$$(5.233) \qquad F(\lambda) \approx \sum_{n=0}^{\infty} \alpha_n \frac{n!}{\lambda^{n+1}}, \qquad \lambda \to \infty.$$

The converse is also true, so that (5.233) implies (5.232). The result (5.233) can be obtained by formally substituting (5.232) in the integral $F(\lambda)$ and integrating term by term. The series (5.232) and (5.233) need not converge, but may have only asymptotic validity. This means that for small t and large λ, $f(t)$ and $F(\lambda)$ are well approximated by the leading terms in the given series even if the series diverge. (See Section 5.7 and Chapter 9 for a more complete discussion of the meaning of asymptotic equalities and series.)

For small values of λ, a useful Tauberian result which requires a knowledge of complex variables theory for its application is as follows. Let $F(\lambda)$ be analytic for $\text{Re}(\lambda) > k$. Assume further that $F(\lambda)$ has isolated poles in the finite λ-plane located at the set of points $\{\lambda_n\}$ with $\lambda_1 > \lambda_2 > \lambda_3 > \dots$. With $f(t)$ assumed to be real, suppose that $F(\lambda)$ has real residues $\{\beta_n\}$ at the poles $\{\lambda_n\}$. Then we have

$$(5.234) \qquad f(t) \approx \sum_{n=1}^{\infty} \beta_n e^{\lambda_n t},$$

which is valid as $t \to \infty$. This result may be obtained by a formal application of residue theory to the inversion formula (5.219) for $f(t)$. [Again, the series (5.234) may have only asymptotic validity.] We do not prove the above results but instead refer to the literature.

In the preceding example we expanded the transform function $U(\lambda, x)$ for large λ and obtained the series (5.231) which was found to be useful for small t. This represents an Abelian result, but the series (5.231) converges and is not just asymptotically valid.

To obtain a Tauberian result for the preceding example we note that the transform function $U(\lambda, x)$ is apparently singular at the zeros of $\sinh(\sqrt{\lambda}\, l/c)$. These are located at the points $\lambda_n = -(\pi c/l)^2 n^2$ $(n = 0, 1, 2, \dots)$. It is easily seen, however, that $\lambda_0 = 0$ is not a singular point of $U(\lambda, x)$. Since the zeros $\{\lambda_n\}$ $(n \geq 1)$ are all simple, $U(\lambda, x)$ has simple poles at these points. The λ_n satisfy the condition $\lambda_1 > \lambda_2 > \dots$ and it can be verified that the residues $\{\beta_n(x)\}$ of $U(\lambda, x)$ at $\{\lambda_n\}$ are all real so that the conditions for the aforementioned Tauberian theorem are met. The Tauberian result $u(x, t) \approx \sum_{n=1}^{\infty} \beta_n(x)\exp[\lambda_n t]$ can be shown to be identical with the Fourier sine series (4.154). Again, that series is convergent and not merely asymptotic and is useful for large t. Further examples are given in the exercises.

5.7. ASYMPTOTIC APPROXIMATION METHODS FOR FOURIER INTEGRALS

In this section we discuss two useful approximation methods for evaluating integrals that result from the application of Fourier transform techniques. We give only a brief discussion of these methods and apply them to two problems relating to dispersive and dissipative wave motion (these concepts were defined in Section 3.5). We refer to the literature for additional details regarding the validity and applicability of these and other asymptotic methods for evaluating integrals. We say that the function $f(x)$ is asymptotic to the function $g(x)$ as $x \to \alpha$ and denote the asymptotic equality by

$$f(x) \approx g(x); \qquad x \to \alpha$$

if the ratio $f(x)/g(x)$ tends to unity as $x \to \alpha$.

We first consider the *method of stationary phase* which we apply to a solution of the Cauchy problem for the Klein–Gordon equation representative of dispersive wave motion. Then we consider a method developed by Sirovich (which is related to the given Tauberian theory for Laplace transforms) and apply it to a solution of the Cauchy problem for the telegrapher's equation representative of dissipative wave motion. (Some properties of both these equations were given in Example 3.7.) Neither of these methods requires complex variables theory for its application.

It was seen in our discussion of transform methods in this chapter that one cannot, in general, expect to evaluate integrals resulting from transform methods in closed form. Consequently, the methods discussed are essential in that they yield simple and useful representations of the solutions.

EXAMPLE 5.13. DISPERSIVE WAVE MOTION: THE METHOD OF STATIONARY PHASE

The *method of stationary phase* is an asymptotic approximation method for evaluating integrals of the form

$$(5.235) \qquad I(k) = \int_{-\infty}^{\infty} \exp[ik\varphi(t)] f(t)\, dt,$$

where k is a large real parameter, $\varphi(t)$ and $f(t)$ are real valued functions, with $\varphi(t)$ called the *phase term*. (We do not consider the most general version of this method.) Since k is large and $\varphi(t)$ real valued, the integrand in (5.235) oscillates rapidly and cancellation occurs over most of the domain of integration. However, near the points t where $\varphi'(t) = 0$, that is, the *stationary points* of the phase, no such cancellation occurs because of the diminished oscillation of the integrand. Thus the major contribution to the integral comes from the neighborhood of these points.

We assume that $\varphi(t)$ and $f(t)$ are smooth functions and proceed as follows. Let t_0 be a stationary point and assume that $\varphi''(t_0) \neq 0$ and $f(t_0) \neq 0$. Expand $\varphi(t)$ and $f(t)$ in a Taylor series around $t = t_0$ and obtain [since $\varphi'(t_0) = 0$]

$$(5.236) \quad \varphi(t) = \varphi(t_0) + \tfrac{1}{2}\varphi''(t_0)(t - t_0)^2 + \ldots; \qquad f(t) = f(t_0) + \ldots.$$

Only the terms given in (5.236) are retained in the (leading order) stationary phase approximation. Let $\epsilon > 0$ be a small number. Then the method of stationary phase asserts that the main contribution to the integral $I(k)$ for large k is given as

$$
\begin{aligned}
(5.237) \quad I(k) &\approx \int_{t_0-\epsilon}^{t_0+\epsilon} \exp\left[ik\varphi(t_0) + \frac{ik}{2}\varphi''(t_0)(t - t_0)^2 \right] f(t_0)\, dt \\
&= \exp[ik\varphi(t_0)] f(t_0) \int_{t_0-\epsilon}^{t_0+\epsilon} \exp\left[\frac{ik}{2}\varphi''(t_0)(t - t_0)^2 \right] dt \\
&\approx \exp[ik\varphi(t_0)] f(t_0) \int_{-\infty}^{\infty} \exp\left[\frac{ik}{2}\varphi''(t_0)(t - t_0)^2 \right] dt,
\end{aligned}
$$

where we have replaced the finite limits $t_0 \pm \epsilon$ by $\pm \infty$ since the main contribution to the infinite integral according to the discussion comes from the neighborhood of t_0. The last integral in (5.237) can be evaluated and we obtain the asymptotic result

$$(5.238) \quad I(k) \approx e^{ik\varphi(t_0)} f(t_0) \exp\left[\frac{i\pi}{4} \frac{\varphi''(t_0)}{|\varphi''(t_0)|}\right]\left[\frac{2\pi}{k|\varphi''(t_0)|}\right]^{1/2}; \quad k \to \infty.$$

The formula (5.238) represents the dominant contribution to the integral $I(k)$ for $k \gg 1$ if $t = t_0$ is the only stationary point. If there is more than one stationary point, there is a contribution of the form (5.238) from each such point and we sum all the contributions. If $\varphi''(t_0) = 0$ so that there is a higher order stationary point, the above method fails and a different discussion which requires the expansion of $\varphi(t)$ up to cubic or higher powers is necessary to yield the contribution from that stationary point. If the phase $\varphi(t)$ has no stationary points, we conclude that $I(k) \approx 0$ for large k. [We do not present here the precise conditions for the validity of (5.238) nor do we discuss the case of a higher order stationary point.]

We now apply the method of stationary phase to the solution of the Cauchy problem for the *Klein–Gordon equation*

$$(5.239) \quad u_{tt} - \gamma^2 u_{xx} + c^2 u = 0, \quad t > 0, \ -\infty < x < \infty,$$

with initial data $u(x,0)$ and $u_t(x,0)$ specified at $t = 0$. Let $U(\lambda, t)$ be the Fourier transform of $u(x, t)$, that is,

$$(5.240) \quad U(\lambda, t) = \frac{1}{\sqrt{2\pi}} \int_{-\infty}^{\infty} e^{i\lambda x} u(x, t) \, dx.$$

The Fourier transformed equation (5.239) is given as

$$(5.241) \quad \frac{\partial^2 U(\lambda, t)}{\partial t^2} + (\gamma^2 \lambda^2 + c^2) U(\lambda, t) = 0.$$

Introducing the Fourier transforms of the initial data, we easily determine that $U(\lambda, t)$ can be written as

$$(5.242) \quad U = U_+(\lambda, t) + U_-(\lambda, t) \equiv F_+(\lambda) \exp\left[i\sqrt{\gamma^2 \lambda^2 + c^2}\, t\right]$$
$$+ F_-(\lambda) \exp\left[-i\sqrt{\gamma^2 \lambda^2 + c^2}\, t\right].$$

The terms $F_\pm(\lambda)$ can be specified in terms of the transforms of the initial data for $u(x, t)$. Since we are only interested in a general characterization of dispersive wave motion, we are not concerned with the specific form of the functions $F_{\pm}(\lambda)$. We do require, however, that $F_{\pm}(\lambda)$ be smooth functions. This is the case if the initial data vanish sufficiently rapidly at infinity, for example.

Inverting the Fourier transform $U(\lambda, t)$ gives

$$(5.243) \quad u(x, t) = \frac{1}{\sqrt{2\pi}} \int_{-\infty}^{\infty} F_{+}(\lambda)\exp(i[\omega_{+}(\lambda)t - \lambda x]) \, d\lambda$$

$$+ \frac{1}{\sqrt{2\pi}} \int_{-\infty}^{\infty} F_{-}(\lambda)\exp(i[\omega_{-}(\lambda)t - \lambda x]) \, d\lambda$$

where

$$(5.244) \qquad\qquad \omega_{\pm}(\lambda) = \pm\sqrt{\gamma^2\lambda^2 + c^2} = \pm\omega(\lambda).$$

The equation $\omega = \omega(\lambda)$ is the dispersion relation for the Klein–Gordon equation and the terms $\exp\{i[\pm\omega(\lambda)t - \lambda x]\}$ are the normal mode solutions. (In comparing with the results in Example 3.7, the parameter k used there is replaced here by $-\lambda$.) The solution (5.243) is thus a superposition of normal mode solutions with the phase term

$$(5.245) \qquad\qquad \varphi_{\pm}(x, t, \lambda) = \pm\omega(\lambda)t - \lambda x.$$

The *phase velocity*, that is, the velocity dx/dt for which $\varphi_{\pm}(x, t, \lambda)$ remains constant, is given as

$$(5.246) \qquad\qquad \frac{dx}{dt} = \frac{x}{t} = \pm\frac{\omega(\lambda)}{\lambda} = \pm\frac{\sqrt{\gamma^2\lambda^2 + c^2}}{\lambda}.$$

We see that for each real value of λ we have $|dx/dt| > \gamma$, so that the phase speed of the normal modes exceeds the characteristic or wave front speed γ of the Klein–Gordon equation. The *characteristic speed* represents the maximum speed of propagation of disturbances for hyperbolic equations. This property has already been demonstrated for the wave equation, and is shown to be valid for more general hyperbolic equations such as (5.239) later in the text. Thus the physical significance of the phase velocity must be examined more closely.

Noting that the general solution (5.243) of the Klein–Gordon equation is a superposition of the normal modes with phase terms (5.245), each of which has a different phase speed, it appears that the appropriate object to consider is not a single normal mode but a group of normal modes or a *wave packet*. A wave packet is obtained by superposing a collection of normal modes with wave numbers λ ranging over some interval. Since each mode has a different phase velocity, the group of normal modes disperses as it propagates and fails to retain its shape. We now show by using the method of stationary phase, that although individual modes in a group travel at different velocities, the wave packet as a whole retains its group character and travels with a velocity appropriately known as the *group velocity*. (It can be shown that an *energy* associated with the wave packet travels with group velocity, but we do not demonstrate this.) We apply the stationary phase method to each of the integrals in the solution (5.243), choosing x and t, which are parameters in the integrals, to be large. A single stationary point λ is found for each choice of x and t. Since the method determines the major contribution to each integral as coming from values of λ near the stationary point, we thereby obtain a wave packet whose properties we examine.

Both integrals in (5.243) are of the same form, apart from the sign in front of $\omega(\lambda)$, so it is sufficient to restrict our attention to the integral with the plus sign and we assume for the purposes of our discussion that only the first integral in (5.243) occurs in the solution $u(x, t)$. (A simple relationship that exists between both integrals is given in the exercises.) Let

$$(5.247) \quad I(x, t) = \frac{1}{\sqrt{2\pi}} \int_{-\infty}^{\infty} F_+(\lambda) \exp\left\{ i\left[\sqrt{\gamma^2\lambda^2 + c^2}\, t - \lambda x \right] \right\} d\lambda,$$

and assume that both x and t are large. If we were to set $x = \alpha t$ where α is constant, then either x or t would play the role of the large parameter k in the integral (5.235) so that the stationary phase method could be applied to (5.247) and we do so without bringing the integral into the exact form (5.235).

The stationary points of the phase $\varphi_+(x, t, \lambda)$ are determined from

$$(5.248) \qquad \frac{d\varphi_+}{d\lambda} = \frac{d}{d\lambda}\left[\sqrt{\gamma^2\lambda^2 + c^2}\, t - \lambda x \right]$$

$$= \frac{d}{d\lambda}\left[\omega(\lambda)t - \lambda x \right] = \omega'(\lambda)t - x$$

$$= \frac{\gamma^2\lambda t}{\sqrt{\gamma^2\lambda^2 + c^2}} - x = 0.$$

This implies that

$$(5.249) \qquad \frac{x}{t} = \omega'(\lambda) = \frac{\gamma^2\lambda}{\sqrt{\gamma^2\lambda^2 + c^2}}$$

and the stationary value λ is given as

$$(5.250) \qquad \lambda = \frac{c}{\gamma}\left(\frac{x}{t}\right)\left[\gamma^2 - \left(\frac{x}{t}\right)^2\right]^{-1/2}$$

The expression $\omega'(\lambda) = \gamma^2\lambda / \sqrt{\gamma^2\lambda^2 + c^2}$ is known as the *group velocity*. As is seen from (5.249), there is a common stationary value λ for all (x, t) that satisfy the equation $x/t = \omega'(\lambda)$ [that is, different values of x and t do not lead to different stationary values λ if the x and t are related by $x = \omega'(\lambda)t$]. It follows from the stationary phase method that the solution retains a fixed character or form at a point x moving with velocity $\omega'(\lambda)$, if the wave packet with wave numbers λ near the stationary value is observed. We note that the group velocity $\omega'(\lambda)$ is less than γ (the characteristic speed) for all values of λ. In fact, there are (real) stationary values λ only for x and t such that $\gamma^2 t^2 - x^2 > 0$ as is seen from (5.250). If x and t are such that $\gamma^2 t^2 - x^2 < 0$, the stationary phase result is $I(x, t) \approx 0$ since there are no stationary points.

Now the lines $x = \pm \gamma t$ represent the boundaries of the domain of influence of the point $(0, 0)$ for the Klein–Gordon equation (this has already been shown for the wave equation and is shown to be valid for the Klein–Gordon equation in Chapter 7). Thus in the context of the stationary phase result we have $I(x, t) \approx 0$ and, equivalently, the solution $u(x, t) \approx 0$ outside the domain of influence corresponding to a concentrated or point initial source at $x = 0$. This suggests that for large x and t, the solution $u(x, t)$ (under the foregoing assumptions on the initial data) appears to correspond to the solution of a problem with initial data concentrated at the origin. The group velocity lines $x = \omega'(\lambda)t$ lie within the domain of influence, as shown in Figure 5.2. (This contrasts with the phase velocity lines all of which lie outside the domain of influence.) As the values x and t approach the characteristic lines $x = \pm \gamma t$, (5.250) shows that the stationary value λ tends to infinity. Then the stationary phase method, as presented, fails and a different discussion, which may be found in the literature, is necessary to discuss the solution near the characteristic lines. We observe that as $|\lambda| \to \infty$ (5.249) implies that $|\omega'(\lambda)| \to \gamma$, so that the group velocity tends to the characteristic velocity.

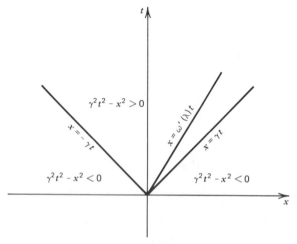

Figure 5.2.   The group line.

To complete the determination of the asymptotic approximation of $I(x, t)$ we note that $\omega''(\lambda) = \gamma^2 c^2 (\gamma^2 \lambda^2 + c^2)^{-3/2} > 0$ and $\varphi_+''(\lambda) = \omega''(\lambda) t \geqslant 0$ so that $\varphi_+''(\lambda)/|\varphi_+''(\lambda)| = 1$. Consequently, we obtain from (5.238)

$$(5.251) \quad I(x, t) \approx \frac{1}{[t\omega''(\lambda)]^{1/2}} F_+(\lambda) \exp\left\{ i\left[\omega(\lambda)t - \lambda x + \frac{\pi}{4} \right] \right\}$$

$$= \frac{1}{\sqrt{t}} H_+\left(\frac{x}{t} \right) \exp\left[i\frac{c}{\gamma} \sqrt{\gamma^2 t^2 - x^2} + \frac{i\pi}{4} \right],$$

where (5.250) was used to express λ as a function of x/t and $H_+ = F_+[\omega'']^{-1/2}$. This asymptotic formula determines the amplitude and the phase of a wave packet moving with the group velocity $dx/dt = \omega'(\lambda)$ $(-\infty < \lambda < \infty)$ and is valid for $|x| < \gamma t$. The amplitude of the wave packet decays like $1/\sqrt{t}$ as $t \to \infty$. As $x \to \pm \gamma t$ or $|\lambda| \to \infty$, we see that $\omega''(\lambda) \to 0$, so that (5.251) becomes invalid near the characteristic lines as was indicated. In the region $|x| \geqslant \gamma t$, that is, outside the characteristic sector indicated in Figure 5.2, we have $I(x, t) \approx 0$ since the integral has no stationary points.

We conclude our discussion by observing that the initial value problem for the Klein–Gordon can be solved exactly in terms of Bessel functions (this is demonstrated in Chapter 7). However, the asymptotic characterization of dispersive wave propagation previously given and the significance of the group velocity and wave packets are not apparent from the exact solution without further analysis. In fact, in a discussion of asymptotic methods for partial differential equations in Chapter 9 we shall use the general form of (5.251) to construct approximate solutions of the Klein–Gordon equation directly. Finally, we emphasize that (5.251) is only valid for large times t as our construction shows.

EXAMPLE 5.14.   DISSIPATIVE WAVE MOTION

We consider the Cauchy problem for the dissipative wave equation (or the telegrapher's equation),

$$(5.252) \qquad u_{tt} - u_{xx} + u_t = 0, \qquad t > 0, \ -\infty < x < \infty,$$

with smooth initial data $u(x, 0)$ and $u_t(x, 0)$ at $t = 0$. Using an asymptotic approximation method for evaluating the integrals that result when the Cauchy problem for (5.252) is solved by means of Fourier transforms, we determine the behavior of $u(x, t)$ as t tends to infinity. An exact solution of the Cauchy problem can be given in terms of modified Bessel functions as shown in Chapter 7. However, the large time behavior of the solution obtained by our approach is not apparent from the Bessel function representation unless further analyses are carried out. Because of the dissipative character of the equation, the method of stationary phase is not suitable for this problem and another approach is used.

We define the Fourier transform $U(\lambda, t)$ of $u(x, t)$ as in (5.240) and obtain the transformed equation of (5.252) as

$$(5.253) \qquad \frac{\partial^2 U}{\partial t^2} + \frac{\partial U}{\partial t} + \lambda^2 U = 0, \qquad t > 0.$$

The initial values $U(\lambda, 0)$ and $\partial U(\lambda, 0)/\partial t$ of $U(\lambda, t)$ are given in terms of the transforms of $u(x, 0)$ and $u_t(x, 0)$. We are not concerned with the specific form of the data since we want to characterize the solution in general terms. Solving (5.253) we have

$$(5.254) \qquad U(\lambda, t) = F_+(\lambda)\exp\left[\left(-\tfrac{1}{2} + \tfrac{1}{2}\sqrt{1 - 4\lambda^2}\right)t\right]$$
$$+ F_-(\lambda)\exp\left[\left(-\tfrac{1}{2} - \tfrac{1}{2}\sqrt{1 - 4\lambda^2}\right)t\right],$$

and the inverse transform yields the solution as

$$(5.255)$$
$$u(x, t) = \frac{1}{\sqrt{2\pi}} \int_{-\infty}^{\infty} F_+(\lambda)\exp\left[\left(-\frac{1}{2} + \frac{1}{2}\sqrt{1 - 4\lambda^2}\right)t - i\lambda x\right] d\lambda$$
$$+ \frac{1}{\sqrt{2\pi}} \int_{-\infty}^{\infty} F_-(\lambda)\exp\left[\left(-\frac{1}{2} - \frac{1}{2}\sqrt{1 - 4\lambda^2}\right)t - i\lambda x\right] d\lambda,$$

where $F_+(\lambda)$ and $F_-(\lambda)$ are specified in terms of the initial data. We assume that

$$(5.256) \qquad \int_{-\infty}^{\infty} |F_\pm(\lambda)| \, d\lambda < M < \infty.$$

To discuss the behavior of the solution for large values of t, we first notice that for $\lambda^2 > \tfrac{1}{4}$ we have $\sqrt{1 - 4\lambda^2} = i\sqrt{4\lambda^2 - 1}$, since $4\lambda^2 - 1 > 0$. Thus

$$(5.257)$$
$$\exp\left[\left(-\frac{1}{2} \pm \frac{1}{2}\sqrt{1 - 4\lambda^2}\right)t\right] = e^{-(t/2)}\exp\left[\pm \frac{i}{2}(4\lambda^2 - 1)^{1/2}t\right]; \qquad \lambda^2 > \frac{1}{4}$$

and, consequently, we obtain

$$(5.258) \qquad \left| \frac{1}{\sqrt{2\pi}} \int_{\lambda^2 > \frac{1}{4}} F_\pm(\lambda)\exp\left[\pm \frac{i}{2}(4\lambda^2 - 1)^{1/2}t - i\lambda x\right] e^{-(t/2)} \, d\lambda \right|$$

$$\leqslant \frac{e^{-(t/2)}}{\sqrt{2\pi}} \int_{\infty}^{\infty} |F_\pm(\lambda)| d\lambda \leqslant \frac{1}{\sqrt{2\pi}} M e^{-t/2}$$

in view of (5.256). Therefore, the contributions to both integrals in (5.255) for values of λ such that $\lambda^2 > \frac{1}{4}$, are exponentially small for large t. For $\lambda^2 \leqslant \frac{1}{4}$ we have

$$(5.259) \qquad \exp\left[\left(-\tfrac{1}{2} - \tfrac{1}{2}\sqrt{1 - 4\lambda^2}\right)t\right] \leqslant e^{-t/2}; \qquad \lambda^2 \leqslant \tfrac{1}{4},$$

$$(5.260) \qquad \exp\left[\left(-\tfrac{1}{2} + \tfrac{1}{2}\sqrt{1 - 4\lambda^2}\right)t\right] \leqslant e^0 = 1; \qquad \lambda^2 \leqslant \tfrac{1}{4},$$

as is easily seen. Thus the exponential in (5.259) is maximal at $\lambda = \pm \frac{1}{2}$, whereas that in (5.260) has its maximum at $\lambda = 0$. Combining the result (5.259) with (5.258) we obtain an estimate for the $F_-(\lambda)$ integral in (5.255) over the full interval of integration. We have

$$(5.261) \qquad \left| \frac{1}{\sqrt{2\pi}} \int_{-\infty}^{\infty} F_-(\lambda) \exp\left[\left(-\frac{1}{2} - \frac{1}{2}\sqrt{1 - 4\lambda^2}\right)t - i\lambda x\right] d\lambda \right|$$

$$\leqslant \frac{e^{-t/2}}{\sqrt{2\pi}} \int_{-\infty}^{\infty} \left| F_-(\lambda) \right| d\lambda.$$

Therefore, the entire $F_-(\lambda)$ integral is bounded by $(M/\sqrt{2\pi})e^{-t/2}$ and we have shown it to decay exponentially uniformly in x as $t \to \infty$. The $F_+(\lambda)$ integral does not exhibit uniform decay as $t \to \infty$, and we expect that the major contribution to the solution $u(x, t)$ as $t \to \infty$ must come from this integral. Furthermore, since the exponential (5.260) that occurs in this integral decays for all $\lambda \neq 0$, we expect that the major contribution to this integral for large t must come from the neighborhood of $\lambda = 0$.

To evaluate the first integral in (5.255) asymptotically, we introduce a method developed by Sirovich for approximately evaluating Fourier integrals of the type considered in this example and which generally arise in problems of dissipative type. This method differs from the method of stationary phase in that the integrals considered decay exponentially as the relevant parameter gets large. Thus the major contribution comes from the neighborhood of the point in the interval of integration where the integral has its maximum value. In the stationary phase method, the integral oscillates rapidly as the relevant parameter gets large and the main contributions come from the points where the oscillation is minimal. Sirovich's method is closely related to approaches used in approximating Laplace transform integrals, since for those integrals we also have exponential decay as the transform parameter gets large.

We now present *Sirovich's method* and then apply it to the $F_+(\lambda)$ integral in (5.255). We consider the integral

$$(5.262) \qquad I(x, t) = \int_{-\infty}^{\infty} F(\lambda) \exp\left[-g(\lambda)t - i\lambda x\right] d\lambda$$

for which the following conditions are satisfied:

(5.263)

(i)
$$\int_{-\infty}^{\infty} |F(\lambda)|\,d\lambda < M < \infty,$$

(ii)
$$\max_{\text{all } \lambda} |F(\lambda)| < M,$$

(iii)
$$\text{Re}[g(\lambda)] \geq 0,$$

(iv)
$$g(\lambda) = 0 \text{ if and only if } \lambda = 0,$$

(v)
$$g(\lambda) = i\alpha\lambda + \beta\lambda^2 + O(|\lambda|^3); \quad \alpha, \beta \text{ real, } \beta > 0, \lambda \to 0,$$

(vi)
$$g(\lambda) \text{ continuous in } \lambda.$$

Then as $t \to \infty$ we have

$$(5.264) \quad I(x, t) = \int_{-\infty}^{\infty} \exp\left[-\beta\lambda^2 t - i\alpha\lambda t - i\lambda x\right] F(\lambda)\,d\lambda + O\left(\frac{1}{t^{1-\delta}}\right),$$

where δ is a small positive constant. The order term $O[\cdots]$ is understood to mean that if $F(x) = O[G(x)]$, then $|F(x)/G(x)| \to A$ for some constant A as $x \to \alpha$. Thus $O(1/t^{1-\delta})$ means that as $t \to \infty$ the error term in (5.264) decays like $A/t^{1-\delta}$. To simplify the integral in (5.264) we introduce a further approximation and expand $F(\lambda)$ in a Taylor series around $\lambda = 0$ retaining only the leading term $F(0)$. We then obtain

$$(5.265) \quad I(x, t) \approx \int_{-\infty}^{\infty} \exp\left[-\beta\lambda^2 t - i\alpha\lambda t - i\lambda x\right] F(0)\,d\lambda.$$

This integral can be evaluated exactly [see (5.35)–(5.39)] and we have

$$(5.266) \quad I(x, t) \approx \sqrt{\frac{\pi}{\beta t}}\, F(0) \exp\left[-\frac{(x + \alpha t)^2}{4\beta t}\right]; \quad t \to \infty.$$

For a discussion of the errors involved in replacing $F(\lambda)$ by $F(0)$ in (5.264) and for proofs, further developments, and refinements of this method, we refer to Sirovich's work.

We are now in a position to apply the preceding method to the $F_+(\lambda)$ integral in (5.255). To do so we identify $F_+(\lambda)/\sqrt{2\pi}$ with $F(\lambda)$ and the term $\frac{1}{2} - \frac{1}{2}\sqrt{1 - 4\lambda^2}$ with $g(\lambda)$ in (5.262). We assume that $F_+(\lambda)$ satisfies the first two conditions in (5.263) so that, in addition to being absolutely integrable, $F_+(\lambda)$ must be uniformly bounded. Further, we have

(5.267)
$$\text{Re}[g(\lambda)] = \text{Re}\left[\frac{1}{2} - \frac{1}{2}\sqrt{1 - 4\lambda^2}\right] \geq 0; \quad -\infty < \lambda < \infty,$$

(5.268)
$$g(\lambda) = \frac{1}{2} - \frac{1}{2}\sqrt{1 - 4\lambda^2} = 0 \quad \text{only at} \quad \lambda = 0,$$

(5.269)
$$g(\lambda) = \frac{1}{2} - \frac{1}{2}\sqrt{1 - 4\lambda^2} = \frac{1}{2} - \frac{1}{2}\left[1 - 2\lambda^2 + O(|\lambda|^3)\right]$$
$$= \lambda^2 + O(|\lambda|^3) \quad \text{as} \quad \lambda \to 0,$$

on using the binomial expansion. Also, $g(\lambda)$ is continuous for all λ. Thus all the conditions on $g(\lambda)$ are met, and we have $\alpha = 0$ and $\beta = 1$ in condition (v) in (5.263). Then (5.266) yields

(5.270)

$$\frac{1}{\sqrt{2\pi}} \int_{-\infty}^{\infty} F_+(\lambda) \exp\left[\left(-\tfrac{1}{2} + \tfrac{1}{2}\sqrt{1 - 4\lambda^2}\right)t - i\lambda x\right] d\lambda \approx \frac{F_+(0)}{\sqrt{2t}} e^{-(x^2/4t)}$$

which is valid as $t \to \infty$.

It may be remarked that to apply Sirovich's method to the integral in (5.262), we simply expand $g(\lambda)$ in a Taylor series around $\lambda = 0$ retaining terms up to quadratic order only. This yields the exponent in (5.264). The conditions given on g and F insure that over the full range of integration the maximum contribution for large t comes from the neighborhood of $\lambda = 0$.

Combining the results (5.261) and (5.270) we obtain

(5.271) $$u(x, t) \approx \frac{F_+(0)}{\sqrt{2t}} \exp\left(-\frac{x^2}{4t}\right); \qquad t \to \infty.$$

Recalling the form of the fundamental solution of the heat equation [see (5.41) with $\xi = 0$ and $c = 1$], that is,

(5.272) $$G(x, t) = \frac{1}{\sqrt{4\pi t}} \exp\left(-\frac{x^2}{4t}\right),$$

we find that

(5.273) $$u(x, t) \approx \sqrt{2\pi}\, F_+(0) G(x, t), \qquad t \to \infty.$$

Since the exponential in (5.271) essentially differs from zero only in the parabolic region $x^2/4t = O(1)$, the large time behavior of the solution $u(x, t)$ is as shown in Figure 5.3.

The solution (5.273) has the form of a constant multiple of the fundamental solution of the heat or diffusion equation with a source point at the origin. (Note that at large times the solution appears to represent the effect of a source at the origin just as it was the case for the large time solution of the Klein–Gordon equation.) The *diffusion effect* (5.273) occurs in the wake of the wave fronts or characteristics $x = \pm t$ after a long time, and yields the major contribution to the solution $u(x, t)$ since the *dissipative effect* has damped out everything else. Although (5.273) also dies out as $t \to \infty$, it decays only algebraically like $t^{-1/2}$ for $x^2 \le 4t$ rather than exponentially as occurs for all other values of x.

For the more general *dissipative wave equation* or telegrapher's equation,

$$(5.274) \qquad u_{tt} - c^2 u_{xx} + u_t - \alpha u_x = 0, \qquad t > 0, \ -\infty < x < \infty,$$

where α and c are positive constants and appropriate initial data $u(x,0)$ and $u_t(x,0)$ are assigned, the Fourier transform method readily yields

(5.275)

$$u(x,t) = \frac{1}{\sqrt{2\pi}} \int_{-\infty}^{\infty} F_+(\lambda)\exp\left[\left(-\frac{1}{2} + \frac{1}{2}\sqrt{1 - 4c^2\lambda^2 - 4i\lambda\alpha}\,\right)t - i\lambda x\right] d\lambda$$

$$+ \frac{1}{\sqrt{2\pi}} \int_{-\infty}^{\infty} F_-(\lambda)\exp\left[\left(-\frac{1}{2} - \frac{1}{2}\sqrt{1 - 4c^2\lambda^2 - 4i\lambda\alpha}\,\right)t - i\lambda x\right] d\lambda,$$

with $F_\pm(\lambda)$ specified in terms of the initial data. The coefficient of t in the exponential of the first integral has the following expansion around $\lambda = 0$:

(5.276)

$$-\tfrac{1}{2} + \tfrac{1}{2}\sqrt{1 - 4c^2\lambda^2 - 4i\lambda\alpha} = -\tfrac{1}{2} + \tfrac{1}{2}\left[1 - \tfrac{1}{2}(4c^2\lambda^2 + 4i\lambda\alpha)\right.$$

$$\left. - \tfrac{1}{8}(4c^2\lambda^2 + 4i\lambda\alpha)^2 + \cdots\right]$$

$$= -\tfrac{1}{2} + \tfrac{1}{2} - i\lambda\alpha - c^2\lambda^2 + \alpha^2\lambda^2 + O(\lambda^3)$$

$$= -i\alpha\lambda + (\alpha^2 - c^2)\lambda^2 + O(\lambda^3).$$

Thus unless $\alpha^2 < c^2$, the coefficient of λ^2 is positive, and the exponential grows as t increases if λ is small. Then for $c^2 < \alpha^2$ the Cauchy problem for (5.274) is unstable since there are normal modes for small λ that grow unboundedly as t increases.

We assume, therefore, that $\alpha^2 < c^2$ in (5.274). It can then be shown that the second integral in (5.275) decays exponentially as $t \to \infty$ and that Sirovich's

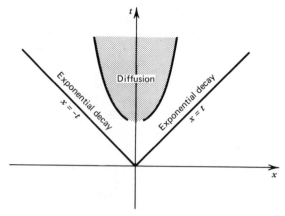

Figure 5.3. The large time behavior.

method can be applied to the first integral. We obtain

$$(5.277) \quad u(x, t) \approx \frac{F_+(0)}{\sqrt{2(c^2 - \alpha^2)t}} \exp\left(-\frac{(x + \alpha t)^2}{4(c^2 - \alpha^2)t}\right); \quad t \to \infty.$$

As was shown in Example 5.2, the expression (5.277) is—apart from a multiplicative constant—the fundamental solution of the diffusion equation (5.42) if we set $c = -\alpha$ and $D/2 = c^2 - \alpha^2$ in that equation. In contrast to the result obtained for (5.252), the diffusion effect is now concentrated along the line $x = -\alpha t$ rather than along the t-axis. To observe the diffusion as $t \to \infty$, we must let $x \to -\infty$ such that $x + \alpha t$ remains small. Since $\alpha < c$, the line of diffusion $x + \alpha t = 0$ lies within the characteristic lines $x = \pm ct$. As the *diffusion effect* may be thought to be traveling along the line $x = -\alpha t$ with velocity $-\alpha$, we may consider (5.274) to possess two speeds of wave propagation. For small times t, disturbances travel at the characteristic speed c to the right and/or to the left. At large times t, the *dissipative effects* have damped out most of the initial disturbances and what remains in the wake of the wave fronts is a diffusion effect characterized by (5.277) that propagates with speed α to the left.

The relationship between dissipative wave propagation and diffusion has been discussed in the models presented in Chapter 1. The foregoing asymptotic results yield a further demonstration of the close relationship between both effects. We have further occasion to touch upon this question later in the text.

To conclude our discussion of asymptotic methods, we remark that even when exact solutions of problems are available, asymptotic approximations yield useful and often significant insights into the behavior of solutions in different regions. In addition, by suggesting general forms for solutions in different regions, it is often possible to construct approximate solutions of equations (say, when they have variable coefficients) for which no exact solution is available by assuming that these solutions have the same general asymptotic form. This idea is exploited in Chapter 9 when we discuss direct asymptotic methods for partial differential equations.

EXERCISES FOR CHAPTER 5

Section 5.2

5.2.1. Show that the complete Fourier series of Example 4.6 may be expressed in complex form as

$$f(x) = \sum_{k=-\infty}^{\infty} c_k \exp\left(-i\frac{\pi k}{l}x\right)$$

with the Fourier coefficients c_k given as

$$c_k = \frac{1}{2l} \int_{-l}^{l} f(t) \exp\left(i\frac{\pi k}{l} t\right) dt.$$

This result may be obtained directly by using the orthogonality property of the functions $\exp[i(\pi k/l)x]$ over the interval $-l < x < l$, in the manner of Exercise 4.3.3.

5.2.2. Show that the results in Exercise 5.2.1 may be expressed as

$$f(x) = \frac{1}{2l} \sum_{k=-\infty}^{\infty} \int_{-l}^{l} f(t) \exp\left[-i\frac{\pi k}{l}(x - t)\right] dt.$$

Assume that the λ-axis $(-\infty < \lambda < \infty)$ is subdivided into intervals $[\lambda_{k-1}, \lambda_k]$, with $\lambda_k = \pi k/l$, whose length is $\Delta\lambda_k = \pi/l$. Rewrite the expression as

$$f(x) = \frac{1}{2\pi} \sum_{k=-\infty}^{\infty} \int_{-l}^{l} f(t) \exp[-i\lambda_k(x - t)] dt \, \Delta\lambda_k,$$

and show that as $l \to \infty$ this (formal) Riemann sum tends to the Fourier integral formula (5.9) if the limit exists.

5.2.3. Show that the Fourier transform of the function $f_N(x)$ defined in (5.26) is given as

$$F_N(\lambda) = \sqrt{\frac{2}{\pi}} \, \alpha(N) \frac{\sin \lambda N}{\lambda}.$$

5.2.4. Let $\alpha(N) = 1/2N$ in Exercise 5.2.3 and show that the limit of $F_N(\lambda)$ as $N \to 0$ is $1/\sqrt{2\pi}$. Conclude that the Fourier transform of $\delta(x)$ should be $1/\sqrt{2\pi}$.

5.2.5. Obtain the Fourier transform of $f(x) = \exp(-|x|)$ and verify the result in (5.22).

5.2.6. Adapt the results of Example 5.2 to obtain the Fourier transform of $f(x) = \exp[-c^2 x^2]$.

5.2.7. Use the formula (5.14) and the result of Exercise 5.2.6 to obtain the Fourier transform of the error integral

$$\Phi(x) = \frac{2}{\sqrt{\pi}} \int_0^x \exp[-s^2] \, ds.$$

5.2.8. Obtain the (closed form) solution of the problem (5.16)–(5.17) if $f(x)$ is defined as

$$f(x) = \begin{cases} 1, & -a < x < a \\ 0, & |x| > a \end{cases}.$$

5.2.9. Obtain the result (5.46) by using

(a) Duhamel's principle.

(b) Fourier transforms.

5.2.10. Show that the fundamental solution $G(x - \xi, t)$ of the heat equation as defined in (5.41) has the property that as $t \to 0$,

$$\lim_{t \to 0} G(x - \xi, t) = 0; \qquad x \neq \xi.$$

Also demonstrate that for $t > 0$ we have

$$\int_{-\infty}^{\infty} G(x - \xi, t) \, dx = 1.$$

Conclude from these results that $G(x - \xi, t)$ tends to the Dirac delta function $\delta(x - \xi)$ as $t \to 0$.

5.2.11. Assuming u and its derivatives vanish as $|x| \to \infty$, integrate the heat equation (5.28) over the x-axis and obtain the result

$$\int_{-\infty}^{\infty} u(x, t) \, dx = \int_{-\infty}^{\infty} f(x) \, dx,$$

where $u(x, 0) = f(x)$. Using Exercise 5.2.10, verify this result for the solution given in (5.40).

5.2.12. Let the initial temperature $f(x)$ in (5.40) be given as

$$f(x) = \begin{cases} u_0; & |x| < a \\ 0; & |x| > a \end{cases},$$

where $u_0 = $ constant. Express (5.40) in terms of the error integral $\Phi(x)$ defined in Exercise 5.2.7.

5.2.13. Use Fourier transforms to solve

$$u_{tt} - c^2 u_{xx} = F(x, t), \qquad -\infty < x < \infty, t > 0,$$
$$u(x, 0) = u_t(x, 0) = 0.$$

5.2.14. Solve the initial value problem for the hyperbolic equation

$$u_{tt} - c^2 u_{xx} - a^2 u = 0, \qquad -\infty < x < \infty, t > 0$$

with the initial data

$$u(x, 0) = f(x); \qquad u_t(x, 0) = 0$$

(you may leave the answer in terms of the transform of the initial data).

5.2.15. Use Fourier transforms to solve

$$u_{tt} + u_{xxxx} = 0, \qquad -\infty < x < \infty, t > 0,$$
$$u(x,0) = f(x); \qquad u_t(x,0) = g(x).$$

5.2.16. Let $f(t) = 1$ in (5.67) and evaluate the integral in terms of $\tan^{-1}z$ to show that the solution becomes $u(x, y) = 1$. Note that $u(x, y)$ does not tend to zero as $y \to \infty$ in this case and that $f(x) = 1$ does not have a Fourier transform.

5.2.17. Noting the result in Exercise 5.2.16 and the fact that

$$\lim_{y \to 0} K(x - t, y) = 0, \qquad x \neq t$$

where $K(x - t, y) = (y/\pi)/[(x - t)^2 + y^2]$, show that $K(x - t, y)$ tends to $\delta(x - t)$ as $y \to 0$.

5.2.18. Use Fourier transforms to solve the boundary value problem

$$u_{xx} + u_{yy} - c^2 u = 0, \qquad -\infty < x < \infty, y > 0$$

with the boundary conditions

$$u(x,0) = f(x); \qquad u(x, y) \text{ bounded as } y \to \infty.$$

5.2.19. Solve Laplace's equation in a strip using Fourier transforms:

$$u_{xx} + u_{yy} = 0, \qquad -\infty < x < \infty, 0 < y < L,$$
$$u(x,0) = e^{-|x|}; \qquad u(x, L) = 0,$$
$$u(x, y) = 0 \text{ as } |x| \to \infty.$$

Section 5.3

5.3.1. Show that if $f(x)$ is an odd function of x [i.e., $f(-x) = -f(x)$], the formulas (5.10)–(5.11) for the Fourier transform yield the formulas (5.83)–(5.84) for the Fourier sine transform.

5.3.2. Show that if $f(x)$ is an even function of x [i.e., $f(-x) = f(x)$], the formulas (5.10)–(5.11) yield the formulas (5.85)–(5.86) for the Fourier cosine transform.

5.3.3. Determine the Fourier sine and cosine transforms of the following functions:

(a) $f(x) = e^{-x}$.

(b) $f(x) = \begin{cases} 1, & 0 \leqslant x \leqslant a \\ 0, & x > a \end{cases}$.

(c) $f(x) = xe^{-x}$.

(d) $f(x) = e^{-x}\cos x$.

5.3.4. Use the Fourier sine transform to solve the boundary value problem for the ordinary differential equation

$$y''(x) - k^2 y(x) = e^{-x}, \qquad 0 < x < \infty, \quad k \neq 1;$$
$$y(0) = 1; \qquad y(x) \to 0 \qquad \text{as} \qquad x \to \infty.$$

5.3.5. Use the Fourier cosine transform to solve the problem in Exercise 5.3.4 if we replace the condition $y(0) = 1$ by the boundary condition $y'(0) = 1$.

5.3.6. Use Duhamel's principle to obtain the result (5.103).

5.3.7. Show that the solution (5.101) takes the form (5.110) if $f(x) = 0$ and $g(t) = u_1 = $ constant.

5.3.8. Suppose that $u(x, t)$ satisfies the heat equation (5.93), the initial condition (5.94), and the boundary condition (5.120). Let

$$v(x, t) = u_x(x, t) - hu(x, t).$$

Show that $v(x, t)$ is also a solution of the heat equation; that is,

$$v_t - c^2 v_{xx} = 0, \qquad 0 < x < \infty, t > 0$$

with the initial condition

$$v(x, 0) = f'(x) - hf(x), \qquad x > 0$$

and the boundary condition

$$v(0, t) = r(t), \qquad t > 0.$$

Assuming that $u(x, t)$ is bounded as $x \to \infty$, show that the solution of the given problem for u is given in terms of $v(x, t)$ as

$$u(x, t) = e^{hx} \int_{\infty}^{x} e^{-hs} v(s, t) \, ds.$$

5.3.9. Use the Fourier sine transform to solve the following initial and boundary value problem for the wave equation:

$$u_{tt} - c^2 u_{xx} = 0, \qquad 0 < x < \infty, t > 0;$$
$$u(x, 0) = u_t(x, 0) = 0;$$
$$u(0, t) = g(t).$$

5.3.10. Solve the following problem for the wave equation using the Fourier cosine transform:

$$u_{tt} - c^2 u_{xx} = 0, \qquad 0 < x < \infty, t > 0;$$
$$u_x(0, t) = h(t);$$
$$u(x, 0) = u_t(x, 0) = 0.$$

5.3.11. Solve the Dirichlet and Neumann problems for Laplace's equation in a half-plane given in Example 5.4 by the use of Fourier sine and cosine transforms, respectively.

5.3.12. Solve the boundary value problem (5.132)–(5.134) for Laplace's equation using the Fourier sine transform.

5.3.13. Use the Fourier sine transform to solve the problem:

$$u_{xx} + u_{yy} = 0, \qquad 0 < x < \infty, 0 < y < \alpha;$$

$$u(0, y) = 1; \qquad u(x, y) \to 0 \quad \text{as} \quad x \to \infty;$$

$$u(x, \alpha) = 0; \qquad u(x, 0) = e^{-x}.$$

5.3.14. Apply the Fourier cosine transform to solve the problem:

$$u_{xx} + u_{yy} = 0, \qquad 0 < x < \infty, 0 < y < \infty;$$

$$u_x(0, y) = 0, \qquad y > 0;$$

$$u(x, 0) = \begin{cases} 1, & 0 < x < 1 \\ 0, & x > 1 \end{cases}.$$

5.3.15. Use Stokes' rule in the manner indicated in Example 5.4 to construct a solution of the problem:

$$u_{xx} + u_{yy} = 0, \qquad 0 < x < \infty, 0 < y < \infty,$$

$$u(0, y) = 0; \qquad u_y(x, 0) = g(x)$$

from the solution of the problem (5.132)–(5.134) given in (5.137).

Section 5.4

5.4.1. Express the three-dimensional wave equation in spherical (spatial) coordinates and show that if we look for a solution $u = u(r, t)$ where r is the radial variable, the function $v(r, t) = ru(r, t)$ is a solution of the one-dimensional wave equation

$$v_{tt} - c^2 v_{rr} = 0.$$

Obtain the general solution

$$u = \frac{1}{r} F(r - ct) + \frac{1}{r} G(r + ct),$$

where F and G are arbitrary functions. These solutions represent spherical (propagating) waves.

5.4.2. Consider the function $\delta_\epsilon(x)$ defined in (1.25) and the integral

$$I(x, y, z, t; \epsilon) = \frac{1}{4\pi c} \iiint\limits_{-\infty}^{\infty} f(\xi, \eta, \zeta) \frac{1}{r} \delta_\epsilon(r - ct) \, d\xi \, d\eta \, d\zeta,$$

where $r^2 = (x - \xi)^2 + (y - \eta)^2 + (z - \zeta)^2$. Since $\delta_\epsilon(r - ct)$ vanishes when $|r - ct| > \epsilon$, the integral extends only over a finite region. Conclude, in view of the fact that the integral represents a superposition of spherical waves, that $I(x, y, z, t; \epsilon)$ is a solution of the three-dimensional wave equation. Introduce the spherical coordinates defined in (5.148) and show that the integral takes the form

$$I = \frac{1}{4\pi c} \int_0^\pi \int_0^{2\pi} \int_0^\infty f(x + r\sin\varphi\cos\theta, \, y + r\sin\varphi\sin\theta, \, z + r\cos\varphi)$$

$$\times \delta_\epsilon(r - ct) r\sin\varphi \, dr \, d\theta \, d\varphi.$$

Carry out the limit as $\epsilon \to 0$, use the property (1.26) of δ_ϵ and the fact that $\lim_{\epsilon \to 0} \delta_\epsilon(r - ct)$ vanishes when $r \neq ct$, and conclude that $I(x, y, z, t; \epsilon)$ tends to (5.149) as $\epsilon \to 0$.

5.4.3. Verify by direct differentiation that (5.149) is a solution of (5.140)–(5.142).

5.4.4. Solve the Cauchy problem for the wave equation

$$u_{tt} - c^2[u_{xx} + u_{yy} + u_{zz}] = 0, \qquad -\infty < x, y, z < \infty, t > 0$$

if the initial data are

$$u(x, y, z, 0) = 0; \qquad u_t(x, y, z, 0) = \begin{cases} 1, & r < a \\ 0, & r > a \end{cases}$$

where $r^2 = x^2 + y^2 + z^2$.

5.4.5. Solve the following problem for the wave equation using the spherical wave solutions obtained in Exercise 5.4.1.

$$u_{tt} - c^2 \nabla^2 u = 0, \qquad -\infty < x, y, z, < \infty, t > 0;$$
$$u(x, y, z, 0) = 1; \qquad u_t(x, y, z, 0) = r^2$$

where $r^2 = x^2 + y^2 + z^2$.

5.4.6. Obtain the solution of the Cauchy problem for the two-dimensional wave equation

$$u_{tt} - c^2(u_{xx} + u_{yy}) = 0, \qquad -\infty < x, y < \infty, t > 0;$$
$$u(x, y, 0) = 0; \qquad u_t(x, y, 0) = x^2 + y^2.$$

Determine the value of $u(0, 0, t)$.

5.4.7. Solve the following Cauchy problem.

$$u_{tt} - c^2(u_{xx} + u_{yy}) = 0, \qquad -\infty < x, y < \infty, t > 0;$$
$$u(x, y, 0) = x; \qquad u_t(x, y, 0) = 0.$$

5.4.8. Apply the method of descent to solve the Cauchy problem for the Klein–Gordon equation:

$$v_{tt} - c^2 v_{xx} + a^2 v = 0, \qquad -\infty < x < \infty, t > 0;$$
$$v(x, 0) = 0; \qquad v_t(x, 0) = f(x).$$

Let $u(x, y, t) = \cos(ay/c)v(x, t)$ and show that $u(x, y, t)$ satisfies the wave equation in two dimensions. Then descend from two dimensions to one.

5.4.9. Use the method of descent to solve the Cauchy problem for the following hyperbolic equation:

$$v_{tt} - c^2 v_{xx} - a^2 v = 0, \qquad -\infty < x < \infty, t > 0;$$
$$v(x, 0) = 0; \qquad v_t(x, 0) = f(x).$$

Let $u(x, y, t) = \exp(ay/c)v(x, t)$, show that $u(x, y, t)$ satisfies the two-dimensional wave equation, and descend to one dimension.

5.4.10. Express the Laplacian in polar coordinates and show that (5.179) is a solution of the homogeneous form of (5.165) when $r > 0$.

5.4.11. Use the two-dimensional Fourier transform to solve the following Cauchy problem for the heat equation.

$$u_t - c^2(u_{xx} + u_{yy}) = 0, \qquad -\infty < x, y < \infty, t > 0,$$
$$u(x, y, 0) = f(x, y).$$

5.4.12. Consider the following Cauchy problem.

$$u_{tt} - c^2(u_{xx} + u_{yy} + u_{zz}) = F(x, y, z)e^{-i\omega t}, \qquad -\infty < x, y, z < \infty, t > 0;$$
$$u(x, y, z, 0) = u_t(x, y, z, 0) = 0$$

where $\omega =$ constant. Assume that $F(x, y, z)$ vanishes outside the sphere $r = R$. Use the retarded potential (5.156) to show that for large t the solution of this problem has the form

$$u(x, y, z, t) = \frac{e^{-i\omega t}}{4\pi c^2} \iiint\limits_{r < R} F(\xi, \eta, \zeta) \frac{\exp[ikr]}{r} d\xi \, d\eta \, d\zeta; \qquad t \to \infty$$

where $k = \omega/c$. Show that $v = (1/r)\exp[ikr]$ is a solution of the reduced wave equation

$$\nabla^2 v + k^2 v = 0,$$

when $r \neq 0$.

Section 5.5

5.5.1. Consider the Fourier transform formulas (5.138)–(5.139) in the case $n = 2$. Let

$$x_1 = r \cos \theta; \qquad x_2 = r \sin \theta; \qquad \lambda_1 = \lambda \cos \varphi; \qquad \lambda_2 = \lambda \sin \varphi;$$

put $f(x_1, x_2) = \exp[-in\theta] f(r)$. Show that the transform $F(\lambda_1, \lambda_2)$ [i.e., (5.138)] takes the form

$$F(\lambda_1, \lambda_2) = \frac{1}{2\pi} \int_0^\infty \int_{-\pi}^\pi \exp[i\lambda r \sin(\theta + \varphi) - in\theta] f(r) r \, d\theta \, dr.$$

From the integral representation

$$J_n(\lambda r) = \frac{1}{2\pi} \int_{-\pi}^\pi \exp[i\lambda r \sin \theta - in\theta] \, d\theta$$

for the Bessel function $J_n(\lambda r)$ of integral order, conclude that

$$e^{-in\varphi} F(\lambda_1, \lambda_2) = \int_0^\infty r J_n(\lambda r) f(r) \, dr = F(\lambda).$$

Insert $e^{in\varphi} F(\lambda)$ and $e^{-in\theta} f(r)$ into the inversion formula (5.139) to obtain

$$f(r) = \frac{1}{2\pi} \int_0^\infty \int_{-\pi}^\pi \exp[in(\theta + \varphi) - i\lambda r \sin(\theta + \varphi)] F(\lambda) \lambda \, d\varphi \, d\lambda.$$

Use the integral representation of $J_n(\lambda r)$ given previously to derive (5.182).

5.5.2. Verify the result given in (5.184).

5.5.3. Integrate by parts in (5.185) to obtain the result

$$\int_0^\infty e^{-\lambda z} J_1(\lambda r) \, d\lambda = \frac{1}{r} \left(1 - \frac{z}{\sqrt{z^2 + r^2}} \right).$$

(Hint: $J_0' = -J_1$.)

5.5.4. Evaluate the integral (5.194) if $f(z) = 1$.

5.5.5. Use the Hankel transform to solve the problem

$$u_{rr} + \frac{1}{r}u_r + u_{zz} - k^2 u = 0, \qquad r > 0, \, -\infty < z < \infty$$

with the boundary conditions (5.189) given on the z-axis.

5.5.6. Use the Hankel transform to construct a (formal) solution of the following problem

$$u_t = D\left(u_{rr} + \frac{1}{r}u_r\right) + F(r, t), \qquad r > 0, \, t > 0$$

where $D > 0$ and $u(r, 0) = f(r)$.

5.5.7. Solve the Cauchy problem for the damped wave equation

$$u_{tt} + a^2 u_t - c^2\left(u_{rr} + \frac{1}{r}u_r\right) = 0, \qquad r > 0, \, t > 0$$

with the initial conditions

$$u(r, 0) = f(r); \qquad u_t(r, 0) = g(r); \qquad r > 0$$

by using the Hankel transform.

Section 5.6

5.6.1. Determine the Laplace transforms of the following functions:
(a) $f(x) = 1$.
(b) $f(x) = e^{ax}$.
(c) $f(x) = \cos \omega x$.
(d) $f(x) = \sin \omega x$.
(e) $f(x) = (1/n!)x^n$.
(f) $f(x) = J_0(ax)$.
(g) $f(x) = \delta(x - a); \qquad a > 0$.

5.6.2. Show that if $F(\lambda)$ is the Laplace transform of $f(x)$, then the Laplace transform of $e^{-ax}f(x)$ is $F(\lambda + a)$.

5.6.3. Let $f(x)$ and $g(x)$ have Laplace transforms $F(\lambda)$ and $G(\lambda)$, respectively. Show that the Laplace transform of $\int_0^x f(s)g(x - s)\, ds$ is given as $F(\lambda)G(\lambda)$. (This is the *convolution theorem* for the Laplace transform.)

5.6.4. Solve, using the Laplace transform:

$$u_t - c^2 u_{xx} = 0, \qquad 0 < x < \infty, \, t > 0;$$
$$u(x, 0) = f(x); \qquad u(0, t) = 0.$$

5.6.5. Use the Laplace transform to solve the following problem:

$$u_t - c^2 u_{xx} = 0, \qquad 0 < x < l, t > 0;$$
$$u(0, t) = e^{-t}; \qquad u_x(l, t) = 0; \qquad t > 0;$$
$$u(x, 0) = 0, \qquad 0 < x < l.$$

5.6.6. Solve, using the Laplace transform:

$$u_{tt} - c^2 u_{xx} = 0, \qquad 0 < x < \infty, t > 0;$$
$$u(x, 0) = u_t(x, 0) = 0, \qquad u(0, t) = f(t).$$

5.6.7. Solve the Cauchy problem for the telegrapher's equation using the Laplace transform:

$$u_{tt} - \gamma^2 u_{xx} + 2\hat{\lambda} u_t = 0, \qquad -\infty < x < \infty, t > 0;$$
$$u(x, 0) = f(x); \qquad u_t(x, 0) = g(x).$$

5.6.8. Use a Laplace transform in the t-variable and a Fourier sine transform in the x-variable to obtain a solution of the following problem.

$$u_{tt} - c^2[u_{xx} + u_{yy}] = 0, \qquad -\infty < y < \infty, x > 0, t > 0;$$
$$u(x, y, 0) = u_t(x, y, 0) = 0, \qquad -\infty < y < \infty, x > 0;$$
$$u(0, y, t) = e^{-t}, \qquad t > 0, \qquad -\infty < y < \infty.$$

5.6.9. Verify the relationship (5.232)–(5.233) by formally integrating term by term in the formula (5.218).

5.6.10. Use the Laplace transform to obtain the exact solution of the following problem:

$$u_t - c^2 u_{xx} = 0, \qquad 0 < x < l, t > 0;$$
$$u(0, t) = u(l, t) = 0;$$
$$u(x, 0) = x + \sin\left(\frac{3\pi}{l} x\right).$$

Use the Abelian result (5.232)–(5.233) to discuss the solution of this problem for small t on the basis of the behavior of the Laplace transform. Compare your results with those obtained from the exact solution.

5.6.11. Show that if the Laplace transform has only a finite number of poles and satisfies the conditions stated in connection with the Tauberian result given in the text, then the (finite) sum (5.234) results from a partial fraction expansion of the transform function $F(\lambda)$.

5.6.12. Demonstrate that the Laplace transform of the solution of the following problem satisfies the conditions required for the validity of the Tauberian result given in the text.

$$u_t - c^2 u_{xx} = 0, \qquad 0 < x < l, t > 0;$$
$$u(0, t) = u(l, t) = 1, \qquad 0 < t;$$
$$u(x, 0) = 0.$$

Use the Tauberian result (5.234) to obtain an expansion of the solution $u(x, t)$ for large t and show that it is, in fact, an exact solution of the given problem.

5.6.13. Solve, using the Laplace transform,

$$u_{tt} - c^2 u_{xx} = e^{-t}, \qquad 0 < x < \infty, t > 0;$$
$$u(x, 0) = u_t(x, 0) = 1, \qquad 0 < x < \infty;$$
$$u(0, t) = \sin t, \qquad t > 0.$$

Section 5.7

5.7.1. The Bessel function of integral order $J_n(x)$ has the integral representation

$$J_n(x) = \frac{1}{2\pi} \int_{-\pi}^{\pi} \exp[ix \sin \theta - in\theta] \, d\theta.$$

Apply the method of stationary phase to obtain the asymptotic formula

$$J_n(x) \approx \sqrt{\frac{2}{\pi x}} \cos\left[x - \frac{\pi n}{2} - \frac{\pi}{4}\right],$$

valid as $x \to \infty$. (Hint: There are two stationary points.)

5.7.2. The Airy function $Ai(x)$ has the integral representation

$$Ai[x] = \frac{1}{\pi} \int_0^{\infty} \cos\left[\frac{1}{3}\tau^3 + x\tau\right] d\tau.$$

Let $x < 0$ and $\tau = \sqrt{-x}\, t$ and use the method of stationary phase to show that the Airy function has the asymptotic expansion

$$Ai[x] \approx \frac{1}{\sqrt{\pi}} |x|^{-1/4} \sin\left[\frac{2}{3}|x|^{3/2} + \frac{\pi}{4}\right],$$

as $x \to -\infty$. (Hint: There are two stationary points.)

5.7.3. Express the functions $F_+(\lambda)$ and $F_-(\lambda)$ in (5.243) in terms of the transforms of the initial data $u(x,0) = f(x)$ and $u_t(x,0) = g(x)$. Show that $F_+(-\lambda) = \overline{F_-(\lambda)}$ (the complex conjugate) and that the solution $u(x,t)$ in (5.243) is real. Obtain the asymptotic form of $u(x,t)$ by considering the contributions from the $F_+(\lambda)$ and $F_-(\lambda)$ integrals.

5.7.4. Show that

$$u(x,t) = J_0\left[\frac{c}{\gamma}\sqrt{\gamma^2 t^2 - x^2}\right]$$

is a solution of the Klein–Gordon equation (5.239). Let $x = \alpha t$ in the Bessel function with $|\alpha| < \gamma$, and expand J_0 asymptotically for large x and t. Show that the result agrees with that obtained in Exercise 5.7.3 for an appropriate choice of F_+ and F_-.

5.7.5. Use the Fourier transform to obtain the solution of the Cauchy problem for the linearized Korteweg–deVries equation (see Example 9.23)

$$u_t + a^2 u_x + b^2 u_{xxx} = 0, \qquad -\infty < x < \infty, t > 0$$

with the initial condition

$$u(x,0) = f(x), \qquad -\infty < x < \infty.$$

Apply the method of stationary phase to discuss the solution $u(x,t)$ for large x and t.

5.7.6. Use the Fourier transform to solve the initial value problem for the dissipative wave equation

$$u_{tt} - u_{xx} - a^2 u_{xxt} = 0, \qquad -\infty < x < \infty, t > 0$$

with the initial data

$$u(x,0) = f(x); \qquad u_t(x,0) = a^2 f''(x); \qquad -\infty < x < \infty.$$

Apply Sirovich's method to obtain an asymptotic expression for the solution as t gets large.

5.7.7. Apply the Fourier transform to solve the boundary value problem for the elliptic equation

$$u_{xx} + u_{tt} - u_t = f(x,t), \qquad -\infty < x, t < \infty,$$

assuming $u(x,t)$ and $f(x,t)$ are suitably behaved at infinity. Apply Sirovich's method to discuss the asymptotic behavior of the solution $u(x,t)$ for large t.

5.7.8. Verify that the solution (5.275) of the initial value problem for (5.274) has the behavior given in (5.277) for large t.

6

INTEGRAL RELATIONS

The problems considered in the preceding chapters dealt mostly with partial differential equations whose coefficients, inhomogeneous terms, and initial and/or boundary data were smooth functions. Consequently, the solutions were expected to be smooth functions as well.

It is often the case, however, that the medium for which the differential equation models some physical property is heterogeneous and some of its characteristics, such as density or conductivity, change discontinuously across some region. For example, this situation arises when we consider the vibration of a composite string composed of two strings of different constant densities joined at some point. Then the coefficients in the equation for the vibrating string may be singular at the point where the strings are attached.

In addition, it is of interest to consider problems where the inhomogeneous term in the equation, which may represent a forcing term or a source (or sink), is concentrated over some lower dimensional region such as a curve or a point. Also, the data for the problem may have discontinuities or singularities. As a result, the solutions of these problems are no longer expected to be smooth functions in general. Thus it becomes necessary to attach a meaning to "solutions" of differential equations that are not differentiable as often as required by the equation. In some cases these solutions are not even continuous everywhere.

This chapter deals with the foregoing questions by showing how the given differential equations can be replaced by equivalent *integral relations*. These relations will be derived directly from the partial differential equations and may involve derivatives of the unknown function. In any case, fewer derivatives than are required in the solution of the partial differential equation are needed for the solution of the equivalent integral relation. Thereby, we weaken for some problems the concept of solution of a differential equation and the results obtained are called *generalized* or *weak solutions*.

The wave equation will be discussed in some detail and an equivalent integral wave equation will be derived. We have seen in Chapter 3 that hyperbolic equations, in contrast to elliptic and parabolic equations, do not smooth out discontinuities or singularities in the data. Since the wave equation is a protototype equation of hyperbolic type, we concentrate our discussion on it.

Finally, we introduce the concept of energy integrals for partial differential equations of each of the three types. We show how they can be used to prove uniqueness and continuous dependence of the solutions on the data.

Our discussion will be restricted to second order partial differential equations of the form considered in Chapter 4. However, the ideas and methods introduced can be carried over to other equations.

6.1. INTEGRAL RELATIONS

In Section 4.1 it was shown how to derive the parabolic equation (4.4) by applying a *balance law* over a certain region. By assuming all terms occurring in the balance law are continuous (or more generally smooth), it is possible to derive the parabolic equation (4.4). If the smoothness requirement is not met throughout the region, for example, there is some subregion across which the properties of the medium undergo a sharp discontinuous change, the arguments leading to (4.4) may fail. In that case, the balance law can be used to derive *matching conditions* for the solutions of the partial differential equation (4.4) that remains valid on both sides of the region of discontinuity. These conditions show how to connect solutions across the region of discontinuity. A similar approach may be applied if the inhomogeneous term of the equation is singular in some region.

Generally speaking, if the partial differential equation is derived from some physical principle applied over an arbitrary region as in Section 4.1, it is possible to take into account the foregoing special situations and incorporate the appropriate results into the derivation. We have derived various equations in Chapter 1 from a different point of view, one from which it is not obvious how to take into account discontinuities or other singularities in our derivation. Therefore, we adopt the following approach. We assume the partial differential equation is given and convert it into an equivalent integral relation (thereby, we reverse the commonly used method of derivation of the differential equation from the integral relation). A detailed discussion is given for the hyperbolic equation (4.10) and the appropriate results are given without derivation for the parabolic and elliptic equations (4.8) and (4.9), respectively. No other equations are discussed.

We consider the hyperbolic equation (4.10) in two or three dimensions; that is,

$$(6.1) \qquad \rho \frac{\partial^2 u}{\partial t^2} - \nabla \cdot (p \nabla u) = \rho F - qu,$$

where ρ, p, and q are functions of x only and have the properties given in

Chapter 4. Let R be a closed bounded region in (x, t)-space (x is a two or three-dimensional variable) with ∂R as its boundary. We integrate (6.1) over R to obtain

$$(6.2) \qquad \int_R \int (\rho u_{tt} - \nabla \cdot (p \nabla u)) \, dv = \int_R \int (\rho F - qu) \, dv.$$

Let the gradient operator in (x, t)-space be defined as

$$(6.3) \qquad \tilde{\nabla} = \left[\nabla, \frac{\partial}{\partial t} \right],$$

so that (6.2) may be written as [since $\rho = \rho(x)$]

$$(6.4) \qquad \int_R \int \tilde{\nabla} \cdot [p \nabla u, -\rho u_t] \, dv = \int_R \int [qu - \rho F] \, dv,$$

where $[p \nabla u, -\rho u_t]$ is a three or four-component vector. Applying the divergence theorem to the first integral in (6.4) gives

$$(6.5) \qquad \int_{\partial R} [p \nabla u, -\rho u_t] \cdot \mathbf{n} \, ds = \int_R \int (qu - \rho F) \, dv,$$

where $\mathbf{n}$ is the exterior unit normal vector to ∂R. Equation (6.5) is the equivalent integral relation (or, more precisely, the integro-differential equation) we are seeking.

We now choose a special form for the region R that commonly occurs when the partial differential equations are derived using a balance law or a similar physical principle. Fixing t, we choose a closed and bounded region R_x in x-space, and extend it parallel to itself between $t = t_0$ and $t = t_1$ in (x, t)-space to form the region R shown in Figure 6.1. Let ∂R_x be the boundary of R_x and let $\partial \tilde{R}_x$ be the lateral boundary of R. Also, denote the upper and lower caps of R by ∂R_1 and ∂R_0, respectively (note that $\partial R_0 \equiv R_x$). The exterior unit normal vector $\mathbf{n}$ on these surfaces assumes the following form. With $\mathbf{n}_x$ equal to the exterior unit normal vector to ∂R_x, we find that on $\partial \tilde{R}_x$, $\mathbf{n} = [\mathbf{n}_x, 0]$, on ∂R_1, $\mathbf{n} = [0, 1]$, whereas on ∂R_0, $\mathbf{n} = [0, -1]$. Then (6.5) takes the form

$$(6.6) \quad \int_{\partial R_0} \rho u_t \, ds - \int_{\partial R_1} \rho u_t \, ds + \int_{\partial \tilde{R}_x} (p \nabla u) \cdot \mathbf{n}_x \, ds = \int_R \int (qu - \rho F) \, dv.$$

The integration over ∂R_0 and ∂R_1 in (6.6) is effectively taken over the region R_x. On the lateral surface $\partial \tilde{R}_x$ we have $ds = ds_x \, dt$, where ds_x is the surface differential over ∂R_x, and in the region R we have $dv = dx \, dt$ (here dx is an area or volume element in two or three dimensions, respectively.) The equation